Crop Improvement

Khalid Rehman Hakeem • Parvaiz Ahmad
Munir Ozturk
Editors

Crop Improvement

New Approaches and Modern Techniques

Springer

Editors
Khalid Rehman Hakeem
Universiti Putra Malaysia
Serdang, Selangor
Malaysia

Munir Ozturk
Universiti Putra Malaysia
Serdang, Selangor
Malaysia

Dr. Parvaiz Ahmad
Government Anantnag
College, Srinagar, India

ISBN 978-1-4614-7027-4
ISBN 978-1-4614-7028-1 (eBook)
DOI 10.1007/978-1-4614-7028-1
Springer New York Dordrecht Heidelberg London

Library of Congress Control Number: 2013939183

Printed on acid-free paper

Springer is part of Springer Science+Business Media (www.springer.com)

Foreword

Crop improvement is now one of the most significant subject matters in agriculture, which includes the genetic alteration of plants to satisfy ever increasing human need. A large number of genetic techniques were developed and refined in the twentieth-century. It has been suggested that many of the limitations of conventional breeding can be overcome with advances in molecular biology. The aim in crop development is to support innovative and excellent research to underpin the development of improved crop varieties that deliver increased productivity and consistent, high quality end products. The limitations of the new breeding methods include technical problems, such as the difficulty of transformation, problems of gene expression, or the lack of knowledge concerning suitable genes to transfer.

Biotechnology is generally accepted as the use of living systems and organisms to develop or make useful products. Increases in crop yield is one of the most obvious applications of modern biotechnology in agriculture, it is also recognized the most difficult one. Many of the genetic characteristics associated with yield (e.g., enhanced growth) are controlled by a large number of genes, each of which has a minimal effect on the overall yield. Most of the current commercial applications of modern biotechnology in agriculture are related to reduce the dependence of farmers on agrochemicals. There is a need much scientific work to be done in this area.

The book contains state-of-the-art new research results in crop improvement and related disciplines in crop development. It provides up-to-date information for researchers, educators, graduate students and industry. It consists of 17 Chapters. The *first Chapter* provides the reader with extensive information on *A. rhizogene, s* which is responsible for the development of hairy root disease in a wide range of dicotyledonous plants and its T-DNA system components. *Second Chapter* talks about recent advances in bioinformatic tools, together with advance molecular technology under clear biological categories. *Chapter 3* deals with tissue culture, which is employed for large-scale propagation of disease free clones and gene pool conservation. It covers *in vitro* propagation and role of biotechnology in crop improvement. *Chapter 4* describes about mutagenesis, which is a crucial step in crop improvement program.

Chapter 5 explains the importance of biofertilizers in sustainable ecosystem. Biofertilizers are now gaining ground as they are used to maintain the soil health, curtail the environmental pollution and cut down on the use of chemicals in agriculture. *Chapter 6* indicates the importance of Arbuscular mycorrhizal fungi (AMF) for soil quality and tolerance of plants to biotic and abiotic stresses. Biotic stress is the subject matter of *Chap. 7* is the subject matter of how wheat genetic variability is obtained. New and useful genetic variations exist in the wild wheat progenitor species that can be utilized for the enhancement of the existing wheat breeding pools and improve yield stability. This was followed by *Chapter 8* dealing specifically with Variability in *Fusarium* Species causing wilt disease in crops. Abiotic stresses including salinity are a major threat to agricultural productivity and hence global food security are described in *Chapter 9*. Crop plants have adopted specialized strategies to reduce the impact of stress.

Chapter 10 is devoted to wheat grain quality advances in the genomics of grain quality are considered crucial for defining genes and their networks underpinning functional flour qualities. *Chapter 11* talks about N use efficiency (NUE) in agriculture and future development. The use of N in agriculture and its significance in the sustaining human society is addressed, especially in the developing countries. *Chapters 12 and 13* covers the issue of heavy metals toxicity in soils, uptake by plants. They throw light on the arsenic toxicity in plants and their tolerance mechanism in plants.

Chapter 14 describes the in vitro production of secondary metabolites using elicitor *in Catharanthus roseus.* Elicitation has been carried out in a large number of medicinal plants, this article deals with the *Catharanthus roseus*, as it is an important source of anticancer compounds Vinblastine (VLB) and Vincristine (VCR). Handling soybeans under stress is the topic of *Chapter 15.* Soybean is among the most important leguminous plants with the ability to establish symbiotic association with the N-fixing bacteria, *Bradyrhizobium japonicum*. One of the most important processes, affecting the performance of soybean under stress is the inhibited exchange of the signal molecules, specifically genistein, between the host legume and *B. japonicum* during the initiation of symbiosis. *Chapter 16* is a review on the *genus Atriplex*. This review is a contribution to the knowledge on the ecological and socio-economical potential of some plant genus *Atriplex*. The last *Chapter 17* deals with' the role of polyamines in stress responses'. Genetic manipulation of crop plants for altered regulation of PA biosynthesis/catabolism may lead to improved stress tolerance potential.

This book will be a new contribution on crop improvement and be useful for scientists and graduate students in the area.

Prof. Dr. Ahmet Ruhi Mermut
President, Federation of the European Soil Scientists, Turkey & Prof. of Soil Sciences, University of Saskatchewan, Canada.

Preface

The improvement of crop species has been a basic pursuit since cultivation began thousands of years ago. To feed an ever increasing world population will require a great increase in food production. Wheat, corn, rice, potato and few others are expected to lead as the most important crops in the world. Enormous efforts are made all over the world to document as well as use these resources. Everybody knows that the introgression of genes in wheat provided the foundation for the "Green Revolution". Later also demonstrated the great impact that genetic resources have on production. Several factors are contributing to high plant performance under different environmental conditions, therefore an effective and complementary use of all available technological tools and resources is needed to meet the challenge.

The developments in biotechnology, genomic research, and molecular marker applications has brought to the forefront an interdisciplinary science that is revolutionizing 21st century crop improvement. Many new genomics technologies like next generation sequencing, omics technologies have emerged as powerful tools for understanding genome variation in crop species at different molecular levels.

The era of genomics seems to be upon us and new techniques will probably enable us to access the genetic basis of metabolomics associated traits much more rapidly. The information and developments related to the metabolomics, transcriptomics analysis and extensive phenotyping of genetically diverse populations together with bioinformatics is going to prove of great help in the field of crop biotechnology. These technologies will unveil the metabolic pathways for under-resourced crop species.

In this book attempt has been made to bring together chapters from different authors and highlight the current status of crop productivity in the light of developments in crop biotechnology, and at the same time provide information on some recent genomic tools and novel genetic and breeding approaches with a final aim of crop improvement. Emphasis has been laid on the topics related to advances in crop biotechnology, the key principles influencing the current practice in crop improvement programs and elucidate the nature of new approaches as well as modern techniques in crop improvement and how molecular plant breeding opens new avenues for research and is contributing to discoveries in this field.

We hope that a new generation of researchers will benefit much from this book and share the respect for the crop plants we all live by and concern for the maintenance of diversity.

The final objective of this book is to refresh and emphasize the fact that we are compelled to save our biodiversity, otherwise plant breeding possibilities will decrease to the extent that it will cost us much.

Dr. Khalid Rehman Hakeem
Dr. Parvaiz Ahmad
Prof. Munir Ozturk

Contents

About the Editors

Dr. Khalid Rehman Hakeem, PhD is Post Doctorate Research Fellow at Universiti Putra Malaysia (UPM), Serdang, Selangor, Malaysia. He has completed his Post Graduation (Env. Botany) as well as Ph.D. (Botany) from Jamia Hamdard, New Delhi, India in 2006 and 2011 respectively. Dr. Hakeem has more than 6 years of teaching/research experience in Plant Eco-Physiology, Biotechnology, as well as Environmental sciences. Recipient of several fellowships at both national and international levels, Dr. Hakeem has so far edited/authored more than seven books with International publishers. He has also to his credit more than 20 research publications in peer reviewed journals and 10 book chapters. Dr. Hakeem is currently engaged in studying the plant processes at ecophysiological as well as proteomic levels.

Dr. Parvaiz Ahmad, PhD is Assistant professor in Botany at A.S. College, Srinagar, Jammu and Kashmir, India. He has completed his post-graduation in Botany in 2000 from Jamia Hamdard New Delhi India. After receiving Doctorate degree from Indian Institute of Technology (IIT) Delhi, India he joined International Centre for Genetic Engineering and Biotechnology, New Delhi in 2007. His main research area is stress physiology and molecular biology. He has published more than 30 research papers in peer reviewed journals and 16 book chapters. He is also editor of 6 volumes (5 with Springer and 1 with Studium Press-New Delhi, India). He is recipient of Junior Research Fellowship and Senior Research Fellowship by CSIR, India, and has been awarded Young Scientist Award under Fast Track scheme in 2007, by the Department of Science and Technology, Govt. of India. Dr. Parvaiz is actively engaged in studying the molecular and physio-biochemical responses of different agricultural and horticultural plants under environmental stress.

Prof. Munir Ozturk, PhD is a Consultant Fellow at Faculty of Forestry, Universiti Putra Malaysia. He has completed his Post-Graduation in Botany in 1964 Jammu & Kashmir University, India, Ph.D. and D.Sc. in 1970, 1975 in the field of Eco-Physiology from Ege University, Izmir, Turkey. Dr. Ozturk was appointed as full Professor in 1980 at the Ege University and served as Chairman of Botany from 1983–1986, and served as Director of Centre for Environmental Studies between

1990–1998. His main research areas are Eco-Physiology, Phytoremediation, Sabkha Ecosystems and Medicinal-Aromatic Plant Diversity. He has published more than 400 papers in national and international journals, more than 50 book chapters and has edited more than 10 books with Springer and Cambridge Scholars. He is fellow of the Islamic World Academy of Sciences.

Contributors

Bilal Ahmad Padder Division of Plant Pathology, SKAUST-K, Srinagar 191121, India
e-mail: bapadder@rediffmail.com

Chedly Abdelly Laboratoire des Plantes Extrêmophiles, Centre de Biotechnologie de Borj-Cedria (CBBC), BP 901, 2050 Hammam-Lif, Tunisia
e-mail: chedly.abdelly@cbbc.rnrt.tn

Junaid Aslam Department of Biotechnology, Hamdard University, New Delhi 110062, India
e-mail: Junaidg1@gmail.com

Mohamed Mahgoub Azooz Department of Biological Sciences, Faculty of Science, King Faisal University, Hofuf, Saudi Arabia
e-mail: azzozm@yahoo.com

Department of Botany, Faculty of Science, South Valley University, 83523 Qena, Egypt

Maali Benzarti Laboratoire des Plantes Extrêmophiles, Centre de Biotechnologie de Borj-Cedria (CBBC), BP 901, 2050 Hammam-Lif, Tunisia
e-mail: maali23@yahoo.fr

Mehraj ul Din Shah Division of Plant Pathology, SKAUST-K, Srinagar 191121, India
e-mail: mehraj547@rediffmail.com

Ruby Chandna Department of Botany, Jamia Hamdard New Delhi, New Delhi 110062, India
e-mail: rubychandna@gmail.com

Majeed–ul-Hassan Chesti Regional Agricultural Research Station, SKUAST-J, Tandwal, Rajouri 185131, Jammu and Kashmir, India
e-mail: chestimh2008@yahoo.co.in

Lou F. De Filippis Centre for Environmental Sustainability (CENS), Department of Environmental Sciences, University of Technology, Sydney NSW 2007, P O Box 123, Sydney, Australia
e-mail: lou.defilippis@uts.edu.au

Ahmed Debez Laboratoire des Plantes Extrêmophiles, Centre de Biotechnologie de Borj-Cedria (CBBC), BP 901, 2050 Hammam-Lif, Tunisia
e-mail: ahmed_debez@yahoo.com

Ilhan Dogan Department of Molecular Biology and Genetics, Faculty of Science, Izmir Institute of Technology, 35430 Urla, Izmir, Turkey
e-mail: ilhandogan@iyte.edu.tr

Muriel da Silva Folli-Pereira Departamento de Microbiologia, Universidade Federal de Viçosa, 36570-000 Viçosa, Minas Gerais, Brasil
e-mail: mfolli@gmail.com

Asiya Hamid Department of Botany, GDC Budgam, Srinagar, Jammu and Kashmir, India
e-mail: hameed901@gmail.com

Cristina Maria Nobre Sobral de Vilhena da Cruz Houghton Faculdade de Ciências, Universidade de Lisboa, 1749-016, Lisboa, Portugal
e-mail: cristinacruzhoughton@gmail.com

Muhammad Iqbal Department of Botany, Faculty of Science, Jamia Hamdard, New Delhi 110062, India
e-mail: miqbal@jamiahamdard.ac.in

Sumiya Jamsheed Department of Botany, Faculty of Science, Jamia Hamdard, New Delhi 110062, India
e-mail: sumiyajkhan@gmail.com

Maria Catarina Megumi Kasuya Departamento de Microbiologia, Universidade Federal de Viçosa, 36570-000 Viçosa, Minas Gerais, Brasil
e-mail: mkasuya@ufv.br

Alvina Gul Kazi Atta-ur-Rahman School of Applied Biosciences, National University of Sciences and Technology (NUST), Islamabad, Pakistan
e-mail: alvina_gul@yahoo.com

Shivani Koul Department of Botany, Faculty of Science, Jamia Hamdard, New Delhi 110062, India
e-mail: shivanee_kaul@rediffmail.com

Tariq Mahmood Department of Plant Sciences, Quaid-i-Azam University, Islamabad, Pakistan
e-mail: tmahmood@qau.edu.pk

Mahmooduzzafar Department of Botany, Hamdard University, New Delhi 110062, India
e-mail: mahmooduzzafar@jamiahamdard.ac.in

Mohammad Miransari AbtinBerkeh Limited Co., Imam Blvd., Shariati Blvd., #107, 3973173831, Tehran, Iran
e-mail: miransari1@gmail.com

Mehrabad Rudehen, Imam Ali Blvd., Mahtab Alley, 55, 3978147395, Tehran, Iran

Lydice Sant'Anna Meira-Haddad Centro de Ciências Agrárias, Ambientais e Biológicas, Universidade Federal do Recôncavo da Bahia, Cruz das Almas, BA, Brasil
e-mail: lydicemeira@yahoo.com.br

Abdul Mujib Department of Botany, Hamdard University, New Delhi 110062, India
e-mail: amujib@jamiahamdard.ac.in

Abdul Mujeeb-Kazi National Institute of Biotechnology and Genetic Engineering (NIBGE), Faisalabad, Pakistan
e-mail: m.kazi@cgiar.org

Ibrahim Ilker Ozyigit Department of Biology, Faculty of Science & Arts, Marmara University, Goztepe, 34722 Istanbul, Turkey
e-mail: ilkozyigit@marmara.edu.tr

Talat Parween Department of Bioscience, Jamia Millia Islamia, New Delhi 110025, India
e-mail: talat.jmi@gmail.com

Javed Qadri Regional Agricultural Research Station, SKUAST-J, Tandwal, Rajouri 185131, Jammu and Kashmir, India
e-mail: qadrijavid2008@gmai.com

Tabasum N. Qadri Department of Botany, GDC Budgam, Srinagar, Jammu and Kashmir, India
e-mail: tabasumqadri@yahoo.co.in

Awais Rasheed Department of Plant Sciences, Quaid-i-Azam University, Islamabad, Pakistan
e-mail: awaispbg@gmail.com

Saiema Rasool Department of Botany, Faculty of Science, Jamia Hamdard, New Delhi 110062, India
e-mail: saimu083@yahoo.co.in

Muneeb U. Rehman Department of Medical Elementology and Toxicology, Faculty of Science, Jamia Hamdard, New Delhi 110062, India
e-mail: reiazrehman@yahoo.co.in

Reiaz ul Rehman P.G Programme in Bioresources, Department of Botany, University of Kashmir, Srinagar 190006, India
e-mail: reiazrehman@yahoo.co.in

Khalid Rehman Hakeem P.G Programme in Bioresources, Department of Botany, University of Kashmir, Srinagar 190006, India

Faculty of Forestry, Universiti Putra Malaysia, 43400 Serdang, Selangor, Malaysia
e-mail: hakeem.bot@gmail.com

Kilani Ben Rejeb Laboratoire des Plantes Extrêmophiles, Centre de Biotechnologie de Borj-Cedria (CBBC), BP 901, 2050 Hammam-Lif, Tunisia
e-mail: kilanib@yahoo.com

Physiologie Cellulaire et Moléculaire des Plantes, UR5, EAC 7180 CNRS, Université Pierre et Marie Curie (UPMC), Case 156, 4 place Jussieu, 75252 Paris cedex 05, France

Rajib Roychowdhury Department of Biotechnology, Visva-Bharati, Santiniketan 731235, West Bengal, India
e-mail: rajibroychowdhurybu@gmail.com

Günther F. E. Scherer Institute of Floriculture and Wood Science, Section of Molecular Developmental Physiology, Leibniz University of Hannover, Herrenhäuser Str. 2, 30419, Hannover, Germany
e-mail: scherer@zier.uni-hannover.de

Iti Sharma Department of Bioscience and Biotechnology, Banasthali University, Banasthali 304022, India
e-mail: itisharma@hotmail.com

Tariq Omar Siddiqi Department of Botany, Faculty of Science, Jamia Hamdard, New Delhi 110062, India
e-mail: tosiddiqui@jamiahamdard.ac.in

Zahid Hameed Siddiqui Department of Botany, Zakir Husain Delhi College, University of Delhi, Jawahar Lal Nehru Marg, New Delhi 110002, India
e-mail: siddiquizh84@gmail.com

Penna Suprasanna Nuclear Agriculture and Biotechnology Division, Bhabha Atomic Research Centre, Trombay, Mumbai 400085, India
e-mail: prasanna@barc.gov.in

Jagatpati Tah Botany Department (UGC-CAS), The University of Burdwan, Burdwan 713104, West Bengal, India
e-mail: jagatpati_tah@yahoo.com

Inayatullah Tahir Department of Botany, University of Kashmir, Srinagar 190006, India
e-mail: inayatullahtahir@gmail.com

Ebru Artam Tarhan Department of Biology, Faculty of Science & Arts, Marmara University, Goztepe, 34722 Istanbul, Turkey
e-mail: eartamtarhan@gmail.com

Mushtaq Ahmad Teli Division of Plant Pathology, SKAUST-K, Srinagar 191121, India
e-mail: ahmadm9969@gmail.com

Prasad Shekhar Variyar Food Technology Division, Bhabha Atomic Research Centre, Trombay, Mumbai 400085, India
e-mail: pvariyar@barc.gov.in

Parvaiz Ahmad Department of Botany, GDC Anantnag, Jammu and Kashmir 192101, India
e-mail: parvaizbot@yahoo.com

Rinukshi Wimalasekara Institute of Floriculture and Wood Science, Section of Molecular Developmental Physiology, Leibniz University of Hannover, Herrenhäuser Str. 2, 30419, Hannover, Germany
e-mail: rinukshi@yahoo.co.in

Chapter 1
Agrobacterium rhizogenes-Mediated Transformation and Its Biotechnological Applications in Crops

Ibrahim Ilker Ozyigit, Ilhan Dogan and Ebru Artam Tarhan

Abstract The history of *Agrobacterium*-related plant biotechnology goes back for more than three decades with the discovery of molecular mechanisms of crown gall disease in plants. After 1980s, gene technologies began developing rapidly and today, related with the improved gene transfer methods, plant biotechnology has become one of the most important branches in science. Till now, the most important genes related with agricultural affairs have been utilized for cloning of plants with the deployment of different techniques used in genetic engineering. Especially, *Agrobacterium tumefaciens* was used extensively for transferring desired genetic materials to plants rapidly and effectively by the researchers to create transgenic plants. Recognition of the biology of *Agrobacterium* species and newly developed applications of their T-DNA systems has been a great step in plant biotechnology. This chapter provides the reader with extensive information on *A. rhizogenes* which is responsible for the development of hairy root disease in a wide range of dicotyledonous plants and its T-DNA system. This knowledge will be useful in improving utilization of crops and the formulation of new and up-graded transgenic based food products.

Introduction

The increase in demand for food is dramatic with an expanding population growth in the world. According to latest projections, continued increase at the current rate of the population is expected to reach between 7.5 and 10.5 billion by

I. I. Ozyigit (✉) · E. Artam Tarhan
Department of Biology, Faculty of Science & Arts, Marmara University,
Goztepe, 34722 Istanbul, Turkey
e-mail: ilkozyigit@marmara.edu.tr

I. Dogan
Department of Molecular Biology and Genetics, Faculty of Science,
Izmir Institute of Technology, 35430 Urla, Izmir, Turkey
e-mail: ilhandogan@iyte.edu.tr

E. Artam Tarhan
e-mail: eartamtarhan@gmail.com

K. R. Hakeem et al. (eds.), *Crop Improvement*, DOI 10.1007/978-1-4614-7028-1_1,

2050 (Census 2012). Climate changes in terms of shifting weather patterns will result in decreased water availability and in conjunction with this, providing food for this inevitable future population size will be a very hard task without adding new arable lands (Milly et al. 2005). To deal with this challenge one of the major solutions is plant breeding, which has been used since ancient times in order to create desired genotypes and phenotypes for specific objectives. The main goals of conventional plant breeding are improvement of crop yield and quality, agricultural convenience and resistance to the parasites. While the conventional plant breeding efforts used in the past were sufficient, nowadays with the increasing demand additional and supplementary technology necessities emerged (Gepts 2002). As a result of industrial revolution and its reflection to the biological and agricultural sciences, plant biotechnology reached spectacular success with understanding of how genes operate and function in plant. The first genetically modified crops were obtained in the early 1980s by using *Agrobacterium tumefaciens* following the plant regeneration systems, production of novel chimeric genes and transformation vectors. Multidisciplinary studies of academic institutions and agricultural seed companies took the leadership on genetic engineering and biotechnological progresses of crop plants (Özcan et al. 2004). Although, many political, regulatory, ethical and religious obstacles are still present, the adoption rate of crop biotechnology in the area of agriculture is high at global level. Crop biotechnology involves a different set of technologies such as industrial use of recombinant DNA, cell fusion and tissue engineering. *Agrobacterium*-mediated transformation has always been the most commonly used method for novel transgenic technologies. Till now, a number of commercially valuable crops like tomato, potato, rice, wheat, maize, cotton, soybean, alfalfa, barley, carrot, sugarcane, pepper and broccoli were obtained using *Agrobacterium*-mediated transformation (Ozyigit 2012).

Characteristics of *Agrobacterium rhizogenes*

Certain bacterial species are capable of transferring some of their genes to higher plants ending up with insertion and permanent integration in the nuclear genome (Broothaerts et al. 2005; Kumar et al. 2006). Members of genus *Agrobacterium* are widely known for their ability of forming a wide variety of different neoplastic diseases, including crown gall (*A. tumefaciens* and *A. vitis*), hairy root (*A. rhizogenes*) and cane gall (*A. rubi*) (Gelvin 2009; Ozyigit 2012). Among them, the first identified one was *A. rhizogenes* (formerly *Phytomonas rhizogenes*) in 1930s belonging to the family Rhizobiaceae in the alpha-2 subclass of Proteobacteria (Riker et al. 1930; Hildebrand 1934; Conn 1942; White 1972; Kersters and De Ley 1984; Woese et al. 1984; Willems and Collins 1993).

A. rhizogenes is a rod-shaped Gram-negative, non-spore forming (0.6–1 μm by 1.5–3.0 μm in size) soil bacterium that occurs singly or in pairs and is motile by means of one to six peritrichous flagella (Conn 1942; Meyer et al. 2000; Tzfira and

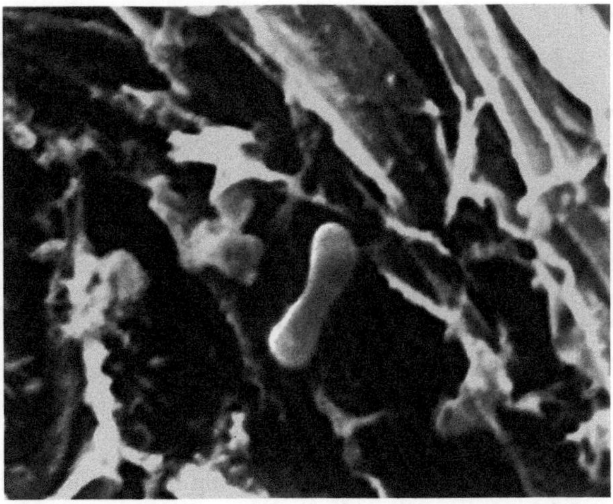

Fig. 1.1 Scanning electron micrograph of attachment of *Agrobacterium rhizogenes* strain R1000 to sunflower (*Helianthus annuus* L.) cotyledonary node cell

Citovsky 2000; Giri and Giri 2007; Murugesan et al. 2010) (Fig. 1.1). It is a close relative of the better known *A. tumefaciens*, which is the best-characterized species among the genus *Agrobacterium* (Rao 2009; Ozyigit 2012) (Fig. 1.1).

All *A. rhizogenes* strains are characterized by the presence of a large root inducing (Ri) plasmid containing a highly conserved "core" DNA region required for hairy root formation (Filetici et al. 1987; Gelvin 2003; Veena and Taylor 2007). Like the crown gall disease, which is caused by *A. tumefaciens* (Ream 2002; McCullen and Binns 2006; Ozyigit 2012) *A. rhizogenes* causes hairy root (root-mat) disease in infected plants through genetic transformation (Weller and Stead 2002; Weller et al. 2005).

Hairy Root Disease

The "hairy root" is the term first used in 1900 by Stewart et al. (as quoted by Hildebrandt 1934). The distinctive symptom of hairy root disease is the formation of a mass of roots. Following the *A. rhizogenes* infection, hairy root formation occurs as a result of protruding large numbers of small roots as fine hairs directly from the infection site (Chandra 2012) (Fig. 1.2). Besides the plagiotropic root growth, hairy-root disease is characterized as short internodes, a high degree of lateral branching, wrinkled leaves, reduced apical dominance, reduced fertility, profusion of root hairs, abnormal flower production, advanced flowering, increased number of flowers, enhanced growth rates and changed secondary metabolite accumulation (Ackermann 1977; Tepfer 1983; Balandrin et al. 1985; Charlwood and Charlwood 1991; Pellegrineschi et al. 1994; Flores et al. 1999; Lee et al. 2001; Keil 2002; Casanova et al. 2004; Veena and Taylor 2007) (Fig. 1.2).

Fig. 1.2 Hairy root formation induced by *A. rhizogenes* strain 8196 in potato (*Solanum tuberosum* L.) callus cultures (**a**), regenerated tobacco (*Nicotiana tabacum* L.) plantlets (**b**). (From Arican)

In nature, when plants are suffering from wounds, phenolic compounds are released from wounded sides and that cause attraction for *A. rhizogenes*. The bacterium moves toward the wounded sites by chemotaxis and infect plant cells. Subsequent infection at wound site followed by transfer of a particular DNA segment (T-DNA) from the root-inducing (Ri) plasmid (pRi) of the bacteria (Kumar et al. 2006). *A. rhizogenes*-induced roots have the unique property of being able to grow *in vitro* without exogenous plant growth regulators (Lee et al. 2001; Rao and Ravishankar 2002). With this unique ability, by the utilization of *A. rhizogenes* strains in *in vitro* plant organ cultures, broad range difficulties were eliminated and as a result, fast growing organs with the capable of producing extensive branching and main metabolites even higher than the mother plant or new metabolites undetected in the mother plant or in other kinds of *in vitro* cultures were generated (Doran 2002; Nader et al. 2006; Bensaddek et al. 2008).

Over the three decades, hairy roots have been applied in a wide range of fundamental studies of plant biochemistry, molecular biology, and physiology, as well as for agricultural, horticultural, and large-scale tissue culture purposes (Doran 2002). In general, hairy root cultures have been used extensively in root nodule research (Diaz et al. 1989; Quandt et al. 1993; Diouf et al. 1995; Hu and Du 2006; Hirotaka and Hiroshi 2003; Aarrouf et al. 2012), production of artificial seeds (Uozumi and Kobayashi 1997), plant secondary metabolites and proteins (Aarrouf et al. 2012), plant breeding and plant improvement, experimental systems to study responses to chemicals (Downs et al. 1994; Mugnier 1997), plant morphology and development (Bandyopadhyay et al. 2007; Turgut-Kara and Ari 2008; Hasancebi et al. 2011; Aarrouf et al. 2012), detoxifing environmental pollutants (Rugh 2001), validate and analyze the functions of genes conferring resistance to root specific pathogens (Remeeus et al. 1998; Hwang et al. 2000; Alpizar et al. 2006; Aarrouf et al. 2012) and study interactions with other organisms such as nematodes (Kifle et al. 1999), mycorrhizal fungi and root pathogens (Mugnier 1997; Christey 2001). Besides these sights, enhanced rooting in plants helps establishment or surviving transplant shocks or abiotic stress like drought, salinity and heavy metal stress (Bulgakov, 2008; Li et al. 2011).

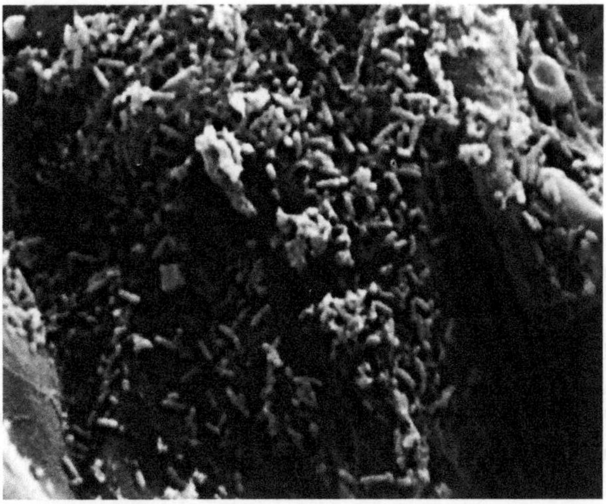

Fig. 1.3 Scanning Electron Micrograph of *A. rhizogenes* strain 8196 colonizing sunflower (*H. annuus* L.) cotyledonary node cell wall

The Mechanism of Hairy Root Formation

The overall process of hairy roots disease by *A. rhizogenes* wild strains is defined by the following four steps. Chimiotactism is the first step leading to induced movement of *Agrobacterium* towards to the plant cells. The following step is binding of *Agrobacterium* to the surface components of the cell wall (Fig. 1.3). After binding, transfer and integration of the transfer-DNA (T-DNA) into the plant genome is completed. The last step is subsequent induction of root formation and growth (Zupan et al. 1996). The information gained in the first three steps is better understood because of the similarities in biological processes and existing models of pathogenesis provided by extensive studies of *A. tumefaciens* stain C58 (Tomilov et al. 2007; Abarca-Grau et al. 2011). The compositions as well as structures are broadly similar for Ri and the Ti plasmids from *A. rhizogenes* and *A. tumefaciens*, respectively (Gelvin 2003; Ozyigit 2012) (Fig. 1.3).

Comparative studies showed a high degree of homology between Ri and Ti plasmids indicating that there are conserved regions between the two types of plasmids. This shows general mechanisms such as activation, processing, and movement of the T-DNA from the bacteria to the plant cell are highly sustained. A segment in both Ri and Ti plasmids called T-DNA consists of highly homologous 24-bp direct repeats known as border sequences (Yadav et al. 1982; Filichkin and Gelvin 1993; Ziemienowicz 2001; Veena and Taylor 2007; Chandra 2012). During infection with *Agrobacterium*, T-DNA is transferred from the bacterium to the plant cell (Rao et al. 2009). The wild-type T-DNA encodes oncogenes and opine catabolism genes, which cause neoplastic growth of tissues and the production of opines (Guyon et al. 1980, 1993; Costantino et al. 1994; Gaudin et al. 1994; Weising and Kahl 1996; Hong et al. 1997; Lee et al. 2001; Rao and Ravishankar 2002; Veena and Taylor 2007). Also, another segment known as the virulence (vir) region in the Ti-plasmid is involved in transferring of DNA into the plant genome (Bulgakov et al. 2004). Hairy roots are capable of growing in the absence of exogenous plant hormones on the plant cells due to the presence of T-DNA. *Agrobacterium* species are highly adapted for sophisticated parasitic relationship with host plants and thus found to establish a unique ecological niche by genetically engineering (Vilkar et al. 1987).

Gall Proteins

One of the similarities of Ri and Ti plasmid is that bearing nearly identical organization of the vir operons (Zhu et al. 2000). Only noticeable difference can be seen is neither genomes nor Ri plasmids of *A. rhizogenes* contains *vir*E1 and *vir*E2 genes (Moriguchi et al. 2001; Hodges et al. 2004). As known from studies about *A. tumafaciens* VirE2 is a single-stranded DNA binding protein and VirE1 acts as a chaperone of VirE2. The VirE2 covers single-stranded T-DNA (T-strands) from nuclease attack (Rossi et al. 1996; Ozyigit 2012) and involves nuclear import of T-DNA to the plant cells (Yusibov et al. 1994; Rossi et al. 1996; Zupan et al. 1996; Gelvin 1998). *vir*E genes play critical roles in pathogenesis of *A. tumefaciens* (Christie et al. 1988; Citovsky et al. 1992; Ward and Zambryski 2001; Duckely and Hohn 2003; Ozyigit 2012). However, the absence of *vir*E genes or no other homolog genes in the *A. rhizogenes* genome clearly shows that *vir*E genes are not necessary in the mechanism of hairy root induction (Moriguchi et al. 2001). Recent studies imply that despite sharing no homology, the *GALLS* gene located on the Ri plasmid can substitute VirE2 function in *A. tumefaciens* (Hodges et al. 2004). GALLS protein differs from VirE2 with ATP-binding and helicase motifs resembling to those in TraA protein involved in conjugation. Both GALLS and VirE2 contain nuclear localization sequences and a C-terminal type IV secretion signal. Mutations in these domains lead to loss of GALLS ability to substitute for VirE2 (Sinkar et al. 1988; Hodges et al. 2006). However, mechanism of GALLS protein in *A. rhizogenes* is still not fully known. All these facts reveal that in spite of differences in their virulence systems, the Ti and Ri plasmids are share a common ancestor. However, the way of T-DNA transfer and those other variations in T-DNA processing also show signs of independent evolution from each other. Current understanding of the molecular bases of the differences between hairy root and gall formation will be accelerated by further studies on genome sequencing and comparison of various *Agrobacterium* strains (Hodges et al. 2006).

Ri Plasmid

Ri plasmid in all *A. rhizogenes* strains has a region known as T-DNA which carries genes (*rol*-genes) involved in root initiation and development and genes essential for opine biosynthesis (Slightom et al. 1986; Hansen et al. 1994a). *Agrobacterium*

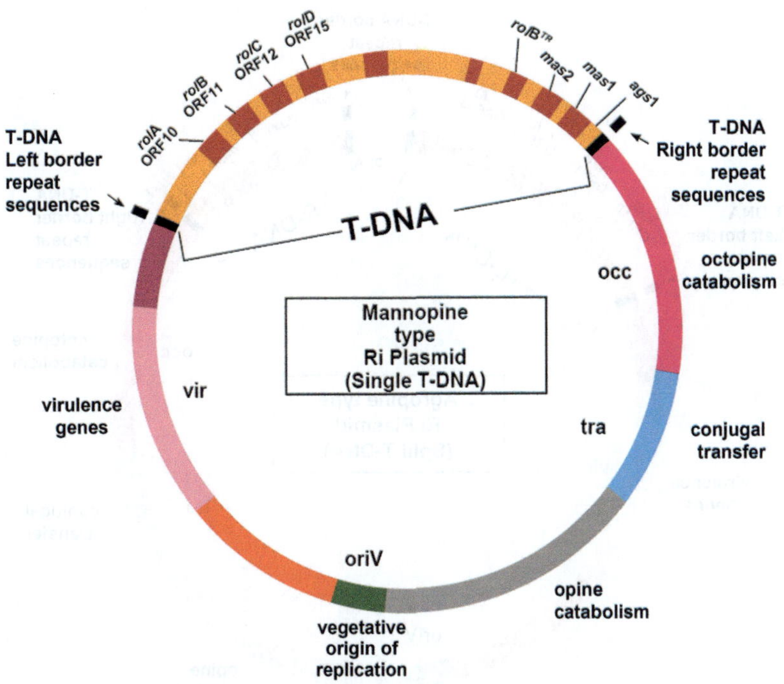

Fig. 1.4 Schematic representation of Mannopine type Ri plasmid of *A. rhizogenes*

T-DNA makes up a small region (approximately 200 kb) of Ti/Ri plasmids which are involved in functions not only for Ti/Ri plasmid conjugation, opine synthesis and catabolism, but also initiation, transfer and integration of the T-DNA (Ozyigit 2012). Although T-DNA contains genes with bacterial origin, these genes have eukaryotic regulatory sequences enabling their expression in infected plant cells (Giri and Narasu 2000). After integration of T-DNA into genomic DNA of the plant cell, T-DNA expresses enzymes that direct the synthesis of unusual amino acid sugar derivatives known as opines, which used by the *Agrobacterium* as nutrient source (Petit et al. 1983; Dessaux et al. 1992; Gartland 1995; Moyano et al. 1999; Navarrete et al. 2006; Bensaddek et al. 2008; Ozyigit 2012).

There are at least two classes of opines produced by *A. rhizogenes* strains. One such class is represented by opines of agropine group, and the other class being the agrocinopine group. Most of the *A. rhizogenes* strains are capable of producing agrocinopine type opines and all or a few strains of producing agropine type opines. The agropine-type opines including agropine, mannopine, agropinic acid and mannopinic acid are produced by the strains known as the agropine-type whereas all agropine-type opines excluding agropine are produced by the strains known as the mannopine-type (Figs. 1.4, 1.5) (White et al. 1982; Petit et al. 1983; Tempe et al. 1984; Savka et al. 1990; Gartland, 1995; Navarrete et al. 2006).

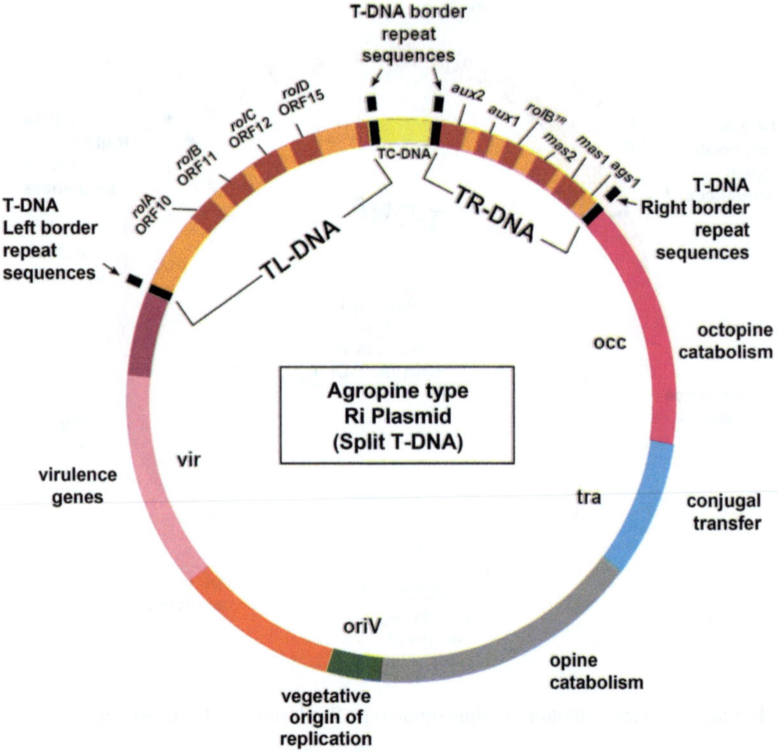

Fig. 1.5 Schematic representation of Agropine type Ri plasmid of *A. rhizogenes*

The most common *A. rhizogenes* strains which represented by Ri plasmids are agropine-type: pRiA4, pRi1855, pRiHRI, pRi15834, and pRiLBA9402, mannopine-type: pRi8196, cucumopine type: pRi2659 and mikimopine-type pRi1724. Although mikimopine and cucumopine are stereo-isomers, there is no homology between opine biosynthetic genes on the nucleotide level (Filetici et al. 1987; Davioud et al. 1988; Gartland 1995; Ouartsi et al. 2004; Veena and Taylor 2007) (Fig. 1.4).

Among the different known strains of *A. rhizogenes*, K47, K599 and HRI are hyper-virulent types known to be capable of infecting a broad range of plant hosts. More research on the virulence factors of these strains needs to be done for understanding of whether they are located on the chromosome(s), plasmid(s) or both (Petit et al. 1983; Isogai et al. 1988; Porter 1991; Suzuki et al. 2001). Also, there are differences between *A. rhizogenes* strains in terms of polarity of infection of the plant tissue. For example, root growth can be induced by some strains of *A. rhizogenes* only on the apical surfaces of carrot root discs and yield no detectable outgrowth on the basal surfaces, whereas root proliferation can be induced by others both inoculation of apical and basal surfaces (Cardarelli et al. 1985; Ryder et al. 1985; Capone et al. 1989; Limami et al. 1998). Based on these findings, various *A. rhizogenes* strains were further classified as polar and non-polar types. Agropine type strains are non-polar whereas all other strains are polar. Agropine type strains give rise to the formation of the hairy roots regardless of the orientation of the disc and the strains other than agropine type form hairy roots when the disc is placed inverted orientation. The presence of second T-DNA encoding genes responsible for auxin production possibly causes observed variation in the polarity of infection in the plant cells transformed by the agropine-type Ri plasmid (Meyer et al. 2000; Veena and Taylor 2007) (Fig. 1.5).

Ri T-DNA Genes

Independent transformations of both left T-DNA (T_L-DNA) (about 15–20 kb) and right T-DNA (T_R-DNA) (about 8–20 kb) to the plant genome termed as "split" T-DNA are carried out by Agropine strains pRi, whereas mannopine strains only transfer a single T-DNA (T_L-DNA). T_L-DNA of pRi contains the four *rol* genes, designated as *rol*A, *rol*B, *rol*C and *rol*D (Schmulling et al. 1988; Petersen et al. 1989; Gelvin 2003; Bensaddek et al. 2008). In Ri plasmid, T_L-DNA and T_R-DNA are separated from each other by at least 15 kb of non-integrated DNA, which is represented by T-Central DNA (TC-DNA) as seen in Fig. 1.5.

The phenotype of hairy root is related with the genes whose products act as the determinants located on T_L-DNA (Tepfer 1984; Taylor et al. 1985; Jouanin et al. 1987b; Nakamura et al. 1988; Schmulling et al. 1988; Sinkar et al. 1988) whereas the genes on the T_R-DNA would only play a role in root induction (Cardarelli et al. 1985; Ryder et al. 1985; Cardarelli et al. 1987a; Smulders et al. 1991). Two fragments, defined as T_L-DNA and T_R-DNA, can be transferred and integrated independently into the plant genome during the infection process. However, the integration capacity of T_L-DNA was much higher than T_R-DNA (Chilton et al. 1982; Costantion et al. 1984; David et al. 1984; Grant et al. 1991; Phelep et al. 1991; Nilsson and Olsson 1997; Holefors et al. 1998; Sevon and Oksman-Caldentey 2002; Kumar et al. 2006; Navarrete et al. 2006; Bensaddek et al. 2008). Furthermore, the present findings imply that a higher number of Ri-T-DNA copies integrated into the plant genome increase the phenotypic effect in the Ri-line (Christensen et al 2008).

T_R-DNA

It was found that the right T-DNA (T_R-DNA) contains genes homologous to T-DNA of *A. tumefaciens* Ti plasmid (Huffman et al. 1984; Jouanin 1984; Vilaine and Casse-Delbart 1987; Hansen et al. 1991; Chandra 2012). Among them, the most important genes are those homologous to the *tms*1 and *tms*2 of the Ti-plasmid. *tms*1 and *tms*2 genes play important roles in auxin biosynthesis in *A. tumefaciens* (Inze et al. 1984; Schröder et al. 1984; Thomashow et al. 1984, 1986; Vilaine and Casse-Delbart 1987). Homology, mutagenesis and complementation experiments show that the two

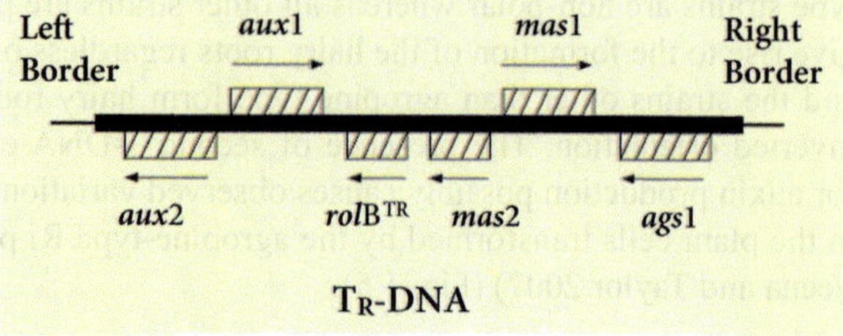

Fig. 1.6 Schematic representation of gene locations on T_R-DNA

morphogenic loci located on the T_R-DNA are counterpart of the tms loci located on the Ti plasmids and involve in hairy root tumorigenesis (White et al. 1985). In *A. rhizogenes* infected *Nicotiana glauca* tissue, the transcripts of the tms loci of Ri plasmids are found to be similar in size to those transcripts found in the tms region of Ti-plasmids (Willmitzer et al. 1983; Taylor et al. 1985; Vilaine and Casse-Delbart 1987). Similar transcripts were also found in carrot plants regenerated from tissues infected with *A. rhizogenes* (De Paolis et al. 1985; Vilaine and Casse-Delbart 1987). The root induction is probably due to auxin biosynthesis carried out by the aux loci located on T_R-DNA. The aux loci are found to be homologous to the tms loci of *A. tumefaciens* T-DNA (Vilaine and Casse-Delbart 1987).

*aux*1, *aux*2, *rol*B^{TR}, *mas*1, *mas*2, and *ags* genes located on the T_R-DNA are responsible for the biosynthesis of agropine and auxin, which cause differences in hairy root growth and morphology when compared to non-transformed roots (Fig. 1.6). It was also reported that the presence of these genes on transformed plant cells caused increase auxin sensitivity (Grant et al. 1991; Lambert and Tepfer 1992; van der Salm et al. 1997; Hansen et al. 1997; Meyer et al. 2000; Alpizar et al. 2006; Nemoto et al. 2009).

Sequence analysis revealed two open reading frames corresponding to proteins of 749 amino acids as *aux*1 gene protein and 466 amino acids *aux*2 gene protein (De Paolis et al. 1985; Camilleri and Jouanin 1991; Gaudin and Jouanin 1995; Christensen et al. 2008; Chandra 2012). Auxin biosynthetic pathway comprises two steps. The *t2m* (tryptophan 2- monooxygenase) gene product encoded by the *aux*1 catalyzes the conversion of tryptophan to indole-3-acetamide (IAM) (Comai and Kosuge 1982; Van Onckelen et al. 1986; Camilleri and Jouanin 1991). Then, IAM is converted to indole-3-acetic acid (IAA) by IAM hydrolase, the product of the *aux*2 (Jouanin 1984; Schröder et al. 1984; Thomashow et al. 1984). The T-DNA of mannopine, cucumopine and mikimopine type strains in Ri plasmids do not carry *aux* genes. Since these strains are still capable to induce a "hairy-root" phenotype, it can be said that the presence of the *aux* genes on T_R-DNA is not necessary to generate hairy root phenotype. It has been demonstrated that the *aux* genes are required to support the "hairy root" phenotype and to extend the host range of the bacterium (White et al. 1985; Cardarelli et al. 1987b; Hansen et al. 1991; Sevon and Oksman-Caldentey 2002).

Hybridization experiments also revealed that the genes encoding agropine biosynthesis (*ags*) are also located on the T_R-DNA region (Willmitzer et al. 1982; Huffman et al. 1984; Lahners et al. 1984; Vilaine and Casse-Delbart 1987; Giri and

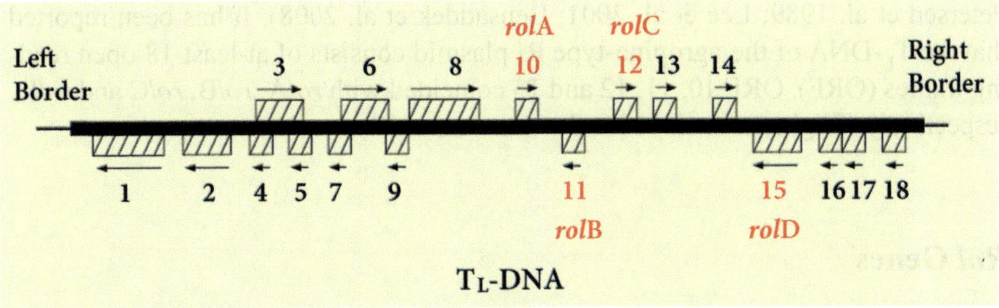

Fig. 1.7 Schematic representation of gene locations on T_L-DNA

Narasu 2000; Christey 2001). Deletion of the right border of nopaline-type or octopine-type T-DNA in Ri plasmids appears to affect virulence. Also, mutations created within this region have the same effect as removing the tms loci of Ti plasmid resulted with being avirulent on plants. The deletion of T_L-DNA in Ri plasmids is being less susceptible to oncogenic transformation than the T_R-DNA deletion (Vilaine and Casse-Delbart 1987). Expression of the T_R-DNA alone can induce root formation in some plants, but the resulting phenotype is not as strong as when both T_L- and T_R-DNA are introduced together (Vilaine and Casse-Delbart 1987).

T_L-DNA

The size of T_L-DNA of agropine type Ri-plasmid is about 19–20 kb in length but, unlike the T_R-DNA, it does not appear to be closely related to any other characterized loci of Ti-plasmids (Huffman et al. 1984; Vilaine and Casse-Delbart 1987; Aoki and Syono 1999; Chandra 2012). In many species, T_L-DNA size seems almost constant, except in *Nicotiana tabacum* consisting shorter T_L-DNA (Jouanin et al. 1987b). The mannopine/cucumopine type T-DNAs and the agropine type T_L-DNA contain two strongly conserved regions which flank an only partially homologous central region (Filetici et al. 1987; Brevet and Tempe 1988; Aoki and Syono 1999; Chandra 2012). A substance carrying out stimulation of hairy root differentiation under the influence of endogenous auxin is synthesized by genes of T_L-DNA (Ooms et al. 1986; Shen et al. 1988; Giri and Narasu 2000; Mishra and Ranjan 2008).

As a result of mutagenesis in T_L-DNA of Ri plasmid, the loss or attenuation of virulence is shown (White et al. 1985). The T_L-DNA of Ri plasmids carrying several loci is identified to be essential for hairy root induction (so-called *rol* genes for root oncogenic loci) (Fig. 1.7). Transposon mutagenesis in the T_L-DNA has identified at least four genes (*rol*A, *rol*B, *rol*C and *rol*D) involved in tumorigenesis as affecting some plants (White et al. 1985; Estramareix et al. 1986; Slightom et al. 1986; Vilaine and Casse-Delbart 1987; Meyer et al. 2000; Christensen et al. 2008). All *rol* genes have been shown to carry out formation of hairy root phenotype (White et al. 1985; Cardarelli 1987a; Jouanin 1987a; Vilaine et al. 1987a; Schmulling et al. 1988;

Petersen et al. 1989; Lee et al. 2001; Bensaddek et al. 2008). It has been reported that the T_L-DNA of the agropine-type Ri plasmid consists of at least 18 open reading frames (ORF). ORF 10, 11, 12 and 15 coincided with *rol*A, *rol*B, *rol*C and *rol*D, respectively (Slightom et al. 1986; Scorza et al. 1994).

Rol Genes

The T-DNAs have many other genes other than those opine and hormone synthesis genes. Although their functions are not well characterized, they are known to have very strong effects on growth. At least four genetic loci (rolA, B, C and D) were identified in the T-DNA regions of pRiA4 by a series of deletions and transposon insertions studies and shown to play important roles of root-inducing properties of *A. rhizogenes* on the T_L-DNA (Table 1.1) (White et al. 1985). The *rol* genes located on the T_L-DNA of Ri plasmid modify auxin and cytokinin biosynthesis and/or endogenous hormone levels and their expressions stimulate the formation of roots in transformed tissues (Nilsson et al. 1993a; Maurel et al. 1994; Moritz and Schmülling 1998; Shen et al. 1990; Bonhomme et al. 2000; Ishizaki et al. 2002; Hong et al. 2006; Bensaddek et al. 2008). Studies have focused on characterizing the three *rol* genes named as *rol*A, *rol*B, and *rol*C because they are considered essential for the hairy root initiation based on transposon "loss-of-function" analysis (White et al. 1985). Induced adventitious root formation by *rol*A, *rol*B and *rol*C genes is shown on tobacco, kalanchoe and tomato leaves (Cardarelli et al. 1987a; Spena et al. 1987; Vilaine et al. 1987; Spano et al. 1988; van Altvorst et al. 1992; Kiyokawa et al. 1994) and plants carrying these genes are morphologically equivalent to those carrying the whole T_L-DNA (Spano et al. 1988). Inactivation or overexpression of various *rol* genes in stable transgenic lines or hairy-root cultures exhibits different variations in plant phenotypes and root morphology (Schmulling et al. 1988; Martin-Tanguy et al. 1996; Casanova et al. 2004).

*rol*A

The *rol*A gene is found on all Ri plasmids and encodes a small protein with a molecular mass of approximately 11 kDa (Nilsson and Olsson 1997). The *rol*A gene sequence length differs in various *A. rhizogenes* strains ranges from 279 to 423 bp (Meyer et al. 2000). Analysis of amino acid sequences showed that *rol*A encodes a protein with basic isoelectric point (PI 11.2). It also contains a frequent sequence motif common in DNA-binding proteins (Suzuki 1989) and proposed to function as a regulatory transcription factor (Levesque et al. 1988; Veena and Taylor 2007).

A dramatic reduction in several classes of hormones, including auxin, cytokinin, gibberellic acid (GA) and abscisic acid triggered by the expression of *rol*A gene is

Table 1.1 Oncogenes of *A. rhizogenes*, their encoded proteins, functions and phenotypic changes in host plants

Gene	Protein	Function	Phenotype
*rol*A	Sequence motif common in DNA-binding proteins Regulatory transcription factor	Inhibits cell elongation via diffusible factor Decreases hormone concentrations Increase sensitivity to auxin Modulating hormone physiology of GA Interfere polyamine metabolism Correlate with plasma membrane H^+ ATPase activity	Stunted growth, dark green wrinkled leaves with an altered length to width ratio, condensed inflorescences, retarded onset of flowering, compact reduced number of flowers
*rol*B	Localizes to plasma membrane	Alterations in the reception/transduction of the auxin signal Stimulates new meristem formation Induce secondary metabolism	Fast growth, root meristem neoformation, high branching and plagiotropism
*rol*C	Phloem-specific expression in the root, low expression in the leaf, and no expression in the shoot tip	Reduces cell size Reduces abscisic acid (ABA), polyamine, and ethylene levels Formation of shoot meristems Regulate sugar metabolism and transport Stimulate the production of high levels of secondary metabolites	Increased branching, dwarfed plants with short internodes, reduced epidermal cell size in internodes, lanceolate leaves, early flowering, reduced flower size and reduced pollen production
*rol*D	Only expresses in Agropine type strains Cytosolic protein Exhibits poor tissue- or organ-specific expression	Incapable of inducing root formation on its own Provide defense response as a result of environmental stress	Increased flowering, reduced rooting, elongating and expanding tissues of each organ but not on apical meristem, callus growth giving rise to initiation of tumor resemble formation
rolB^{TR}	CX5R motif is absent N-terminal part contain 14 amino acids	*rol*B homolog on TR-DNA in the agropine type Ri plasmid	Wrinkled leaves bent strongly downward, formed shoots at the base of the stem and retarded growth
ORF3n	Modification of phenolic enzymes and involve secondary metabolism and/or the transport of hormones	Negative regulator to the dedifferentiation of tissues	Retarded flowering, less dense inflorescences, altered internode elongation and leaf morphology and necrotic tips of upper leaves, sepals and bracts no sign of necrosis on the basal leaves

Table 1.1 (continued)

Gene	Protein	Function	Phenotype
ORF8	Fusion protein consisting of N-terminal domain (NORF8) and C-terminal part (CORF8) Tryptophan monooxygenase activity	Modifies sucrose transport N-terminal domain causes sugar/starch accumulation C-terminal domain reduces sugar/starch accumulation	Growth retardation, chlorotic and necrotic leaves and accumulation of high levels of sugars (glucose, fructose and sucrose) and starch
ORF13	Contains a conservative retinoblastoma (RB)-binding motif LxCxE	Hormone homeostasis and regulation of the cell cycle Increases number of mitoses in shoot apical meristem Induces dedifferentiation (prerequisite to competence) Graft transmissible	Induce cell proliferation such as dense green and rapidly proliferating callus, including irregular formation of leaves, severe leaf nervure, shortened and variable internode length, abnormal and asymmetric flowers, agravitropic root growth and a reduced cell number and cell size in the root
ORF13a	Tissue specific manner in plants, primarily in leaf vascular tissues May interact directly with DNA SPXX repeat motif	Necessary for root induction Regulatory function of itself	Not yield a visible phenotype
ORF14	Auxin like effect	Act together with ORF13 to induce root induction	No morphological change

observed in *N. tabacum*. The reduction ratio depends on tissue type and growth stage of the plant (Dehio et al. 1993). It was demonstrated that despite low level of auxin concentration, auxin sensitivity is enhanced in transgenic plants (Maurel et al. 1991; Vansuyt et al. 1992). Additionally, the effects of *rol*A can be attenuated, probably through methylation (Martin-Tanguy et al. 1996; Lee et al. 2001). Inactivation of *rol*A leads to the formation of long, straight roots giving a less compact appearance on *Kalanchoe daigremontiana* leaves (Vilaine and Casse-Delbart 1987). Transgenic *N. tabacum* plants are also show stunted growth, dark green wrinkled leaves with an altered length to width ratio, condensed inflorescences, retarded onset of flowering, a reduced number of flowers and compact styles (Dehio et al. 1993).

A. rhizogenes infected plant tissues are 100 times more sensitive to auxin than normal phenotype exhibiting plant tissues. This suggests that the increased sensitivity of transformed plants should not be due to a particular insertion position of the *rol*A gene in the transgenic plant genome, but rather reflects the effect(s) of the *rol*A gene product (Vansuyt et al. 1992). It was found that *N. tabacum* leaves of *rol*A transgenic clones show 40–60% reduction of GA content compared to wild-type leaves. The reduction of GA content is indirectly cause stem elongation and planar leaf blade growth (Dehio et al. 1993). When the wild-types of *N. tabacum* treated by gibberellin biosynthesis inhibitors, *rol*A expressing plants and wild types show similar phenotypes. On the other hand, when *rol*A transgenic plants treated with GA, the phenotype of transgenic plant not completely restored (Dehio and Schell 1993; Dehio et al. 1993). All these shows that the *rol*A gene has been considered in playing an important role in modulating hormone physiology of GA and polyamine metabolism (Sun et al. 1991; Dehio and Schell 1993; Dehio et al. 1993; Prinsen et al. 1994; Martin-Tanguy et al. 1996; Veena and Taylor 2007). It was thought that the sensitivity of auxin response might correlate with plasma membrane H^+ ATPase activity observed in *rol*A expressing transgenic plants (Maurel et al. 1991; Vansuyt et al. 1992).

There is data suggesting that there is an antagonism between *rol*A and *rol*B genes in general. An observation of additional transcripts ranging from 2.1 to 2.8 kb in size explains this antagonism (Durand-Tardif et al. 1985). Size of transcription of *rol*A would be more than 2 kb. This would span the whole *rol*B sequence, leading to the generation of an antisense message for *rol*B. Its occurrence could be the major cause of antagonism between *rol*A and *rol*B in the transformed plant cells. Probably, existence of a mechanism prevents co-expression of *rol*A and *rol*B (Capone et al. 1989; van Altvorst et al. 1992; Veena and Taylor 2007).

*rol*B

The *rol*B gene size ranging 765 (strain 8196) to 840 bp (strain 2659) length depending on the strain and encodes 254–279 amino acid protein which has molecular weight of 30 kDa localized in the plasma membrane (Filippini et al. 1996; Meyer et al. 2000; Veena and Taylor 2007). *rol*B gene is present in all Ri plasmids with approximately 60% identity between strains (Meyer et al. 2000). RolB proteins encoded by pRi1724 and pRi2659 have a 17 amino acid longer N-terminal stretch than the RolB proteins encoded by pRi1855 (pRiA4) (Meyer et al. 2000). The physical presence of the *rol*B gene in T_L-DNA segment of Ri plasmid of the infecting *Agrobacterium* in leaf tissues of plants regenerated from selected rhizoclones was demonstrated by a positive PCR amplification (Pal et al. 2012).

The reports have shown that the RolB may have a critical role in early steps of hairy-root induction (Bellincampi et al. 1996). The root induction is totally alleviated when *rol*B gene is inactivated in the pRiA4 on kalanchoe leaves (White et al. 1985). *rol*B also has capacity nearly as much as the wild type *A. rhizogenes* T-DNA for enhancing rooting and hairy root formation on wounded *N. tabacum* stems (Cardarelli et al. 1987b; Bellincampi et al. 1996; Altamura and Tomassi 1998; Binns and Costantino 1998) and leaves (Spena et al. 1987).

Phenotypical abnormalities such as root meristem neoformation on leaf discs and fast growth of *rol*B-transgenic plants and growth pattern of *rol*B-induced roots are characterized by fast growth, high branching, and plagiotropism. As a result of these observations firstly suggested that there is a similarity between the auxin-mediated effects and morphogenic effects of *rol*B. However, further studies demonstrated that an auxin-induced hyperpolarization at the plasma membrane is exhibited by *rol*B-transformed plants. The morphogenic effects of *rol*B involve changes in either the responsiveness to auxin or in auxin content (Cardarelli et al. 1987b; Shen et al. 1988; Capone et al. 1989; Maurel et al. 1991). Activation of auxin-induced hyperpolarization through H^+ ATPase protein pump at the plasma membrane appears to be related to the proton excretion (Ephritikhine et al. 1987; Keller and Van Volkenburgh 1998). *rol*B gene causes transformed plant cells to bind more auxin than wild type and the additional auxin-binding activity is completely abolished by using anti-RolB antibodies (Filippini et al. 1994; Shoja 2010).

Estruch et al. (1991) reported that RolB protein exhibits a β-glucosidase activity able to hydrolyze biologically active indole-3-glucosidese. It can be explained by the increased auxin perception and sensitivity with releasing the hormone from β-glucoside conjugates. As a result of increase concentration of auxin cause the phenotypic alterations observed in *rol*B transgenic tissues (Shen et al. 1988, 1990; Maurel et al. 1991, 1994; Meyer et al. 2000). However, later studies showed that neither the intracellular concentration nor the metabolism of auxin was changed by *rol*B expression in plant cells. Rather, the increased auxin sensitivity of *rol*B-transformed cells results from alterations in the reception/transduction of the auxin signal (Nilsson et al. 1993b; Schmülling et al. 1993; Delbarre et al. 1994; Bellincampi et al. 1996; Veena and Taylor 2007).

Overexpressing *rol*B gene under a constitutive promoter in transgenic plants suppresses adventitious root induction (Spena et al. 1987) and necrosis in callus tissues and leaves of young plants (Schmulling et al. 1988). Both callus and root formations at wound sites are cancelled if mutations occur in *rol*B gene (Vilaine and Casse-Delbart 1987). Normal growth of these organs depends upon the expression level of *rol*B gene necessary for active growth of hairy roots. A high or low level of expression correlates with impaired growth of these organs (Tanaka et al. 2001; Veena and Taylor 2007).

A. rhizogenes rol genes enhance the biosynthesis of certain groups of secondary metabolites in transformed plant cells. It was shown that *rol*B is apparently the most powerful inducer of secondary metabolism and at the same time, the most important inhibitor of callus growth (Palazon et al. 1998; Bonhomme et al. 2000; Bulgakov et al. 2002a; Shkryl et al. 2008; Shoja 2010). *rol*B gene mediated stimulatory effect on resveratrol and anthraquinone production suppresses with the tyrosine phosphatase inhibitors proven that RolB also has tyrosine phosphatase activity (Filippini et al. 1996; Kiselev et al. 2007).

*rol*C

The *rol*C gene sequences vary in different strains but their sizes are similar and ranging between 537 bp (strain 8196) to 543 bp (strain 2659, 1724 and A4). *rol*C gene encodes 178–180 amino acid protein (approximately 20 kDa) that share more than 65 % identity with each other (Meyer et al. 2000).

*rol*C transformed plants exhibited reduced apical dominance leading to increased branching, dwarfed plants with short internodes, lanceolate leaves, early flowering, reduced flower size and reduced pollen production (Schmulling et al. 1988). Dwarfing was caused by reduced epidermal cell size in internodes (Oono et al. 1990). Regulation of expression of *rol*C is complex, and varies depending upon the existence of the complete T-DNA sequences. In addition, root production was increased compared to untransformed plants, but decreased compared to plants transformed with the complete set of *rol* genes (Palazòn et al. 1998). Expressing *rol*C shows phloem-specific expression in the root, low expression in the leaf, and no expression in the shoot tip (Schmulling et al. 1988; Estruch et al. 1991). However, *rol*C is highly expressed in leaves when the whole T-DNA is present (Durand-Tardif et al. 1985; Leach and Aoyagi 1991). More recently, *rol*C gene has been shown to play a role in formation of shoot meristems, hence suggesting its important role in the formation of pluripotent stem cells (Gorpenchenko et al. 2006).

The *rol*C promoter is utilized extensively for phloem-specific gene expression making it a useful tool in some biotechnological programs on pathogen resistance. Replication of many plant viruses, including luteoviruses, reoviruses and most geminiviruses transmitted by hemipteran vectors occur exclusively in phloem-associated tissues. Therefore, by introducing an insecticidal gene that is toxic to hemipteran vectors under the control of phloem-specific *rol*C is a promising way for the control of such viruses through its expression in transgenic plants (Graham et al. 1997). Similarly, a plant lectin with insecticidal activity is encoded by *ASAL* (*Allium sativum* leaf agglutinin) gene and under control of the *rol*C promoter, it confers resistance against various hemipteran pests in transgenic rice, tobacco and chickpea plants (Saha et al. 2007).

*rol*C is known to stimulate rooting by an auxin-like effect of the gene (Schmulling et al. 1988; Zuker et al. 2001; Casanova et al. 2003). An increase in auxin sensitivity may lead to occurrence of the auxin-like effect. In fact, in comparison between *rol*C transgenic *N. tabacum* protoplasts and their wild-type counterparts showed that more sensitivity was recorded in transgenic *N. tabacum* in the measurement of transmembrane hyperpolarization in response to auxin (Maurel et al. 1991; Shoja 2010).

Also, abscisic acid (ABA), polyamine, and ethylene levels are extensively reduced due to *rol*C expression. The promoter of *rol*C activated by sucrose was found to be very high (Yokoyama et al. 1994; Faiss et al. 1996), implying that *rol*C may be influencing the source-sink relationship of a plant by regulating sugar metabolism and transport (Nilsson et al. 1996a, b; Martin-Tanguy 2001).

Alike *rol*B, the *rol*C gene is able to stimulate the production of high levels of secondary metabolites such as tropane alkaloids (Bonhomme et al. 2000), pyridine alkaloids, indole alkaloids (Palazon et al. 1998), ginsenosides (Bulgakov et al. 1998) and anthraquinone phytoalexins (Bulgakov et al. 2002b; Shkryl et al. 2008; Shoja 2010) in transgenic plants.

*rol*D

The *rol*D gene is found only in T_L-DNA of agropine type Ri plasmids. It is also the only *rol* gene that is incapable of inducing root formation on its own (Mauro et al. 1996). The *rol*D gene size 1,032 bp and encodes a protein of 344 amino acids (Meyer et al. 2000; Christey 2001). This is a cytosolic protein with a sequence similar to ornithine cyclodeaminase (OCD) that catalyzes the conversion of ornithine to proline. Proline is an osmoprotectant and its accumulation is considered to be a defense response as a result of environmental stress in many plant species (Mauro et al. 1996; Trovato et al. 2001; Bettini et al. 2003). High levels of proline accumulation are in flowers suggesting a role in flowering (Trovato et al. 2001). The pleiotropic effects induced by expression of *rol*D gene in transgenic plants are increased flowering and reduced rooting (Mauro et al. 1996; Trovato et al. 2001). Although flower yield is accelerated, the flowers show heteromorphic incompability, which prevents self-fertilization. Production of viable seeds is achieved through manually-selfed plants (Mauro et al. 1996). However, it should be noted that these experiments were conducted using the *rol*D sequence from pRi1855. It has been reported that the induction of flowering is not performed by *rol*D from pRiHRI (Lemcke and Schmulling 1998). *rol*D exhibits poor tissue- or organ-specific expression in comparison with other *rol* genes but is shown to have a predominantly developmental expression pattern (Vilaine and Casse-Delbart 1987). Activity is seen in the elongating and expanding tissues of each organ in adult plants, but never in apical meristems. As the plants age, expression decreases and ceases at senescence. The mutations in *rol*D appear to accentuate callus growth giving rise to initiation of tumor formation resembling the Ti-plasmid infection (Trovato et al. 1997).

*rol*B^{TR} *(rol*B *Homologue in T_R-DNA)*

A *rol*B homolog on T_R-DNA in the agropine type Ri plasmid was discovered and named as *rol*B^{TR}. Excluding the 5′ or 3′ flanking sequences, there is a 53 % nucleotide similarity between *rol*B^{TR} and *rol*B in their sequences (Bouchez and Camilleri 1990). The expression of *rol*B^{TR} in *N. tabacum* is shown to cause phenotypical alterations such as wrinkled leaves bent strongly downward, formed shoots at the base of the stem and retarded growth is observed which are different than *rol*B phenotype. Two big differences were noted by the alignment of protein sequences of *rol*B and *rol*B^{TR}. First, a CX5R motif is absent in *rol*B^{TR} and second, N-terminal part of RolB^{TR} contains 14 amino acids and mutations in the corresponding sequence in *rol*B^{TR} gene cause abolishment of the altered phenotype (Lemcke and Schmulling 1998).

ORF Genes

Besides *rol* (root locus) genes, there are several ORFs (Open Reading Frames) located on the T_L-DNA (Slightom et al. 1986). Many of 18 open reading frames (ORFs) nucleotide sequences identified on T_L-DNA region contain 5′ and 3′ regulatory elements similar to those found in eukaryotic genes. They have at least 255 nucleotides and start with the initiation codon ATG (Slightom et al. 1986; Holefors et al. 1998). In many cases, CCAAT and TATA elements were situated upstream of putative transcriptional initiation codons and poly(A) addition (AATAAA) elements were present in presumed 3′-noncoding regions (Slightom et al. 1986). The sequence length of coding regions of ORFs differ in ranging from 255 bp (ORF 9) up to 2280 bp (ORF8) and encode protein products ranging in size from 9,600 to 85,000 daltons, respectively. The results from analysis of insertion mutants within the T-DNA region (White et al. 1985) and transformation experiments with individual or combinations of the ORFs have showed that the open reading frames ORF10, 11 and 12, corresponding to the genes *rol*A, *rol*B and *rol*C, were able to promote the formation of hairy root syndrome (Table 1.1) (Jouanin et al. 1987b; Vilaine et al. 1987; Spena et al. 1987; Spano et al. 1988; Schmulling et al. 1988). Besides this, it has been showed that ORF3n, ORF8 and ORF13 DNA sequences are highly conserved among all known Ri plasmids, indicating that they alter plant morphogenesis or response of transgenic tissues to plant hormones (Lemcke and Schmulling 1998; Veena and Taylor 2007). The sensitivity to auxin and cytokinin in combination or auxin alone can be lowered by expressions of both ORF3n and ORF8 (Lemcke and Schmulling 1998).

ORF3n

Expression of ORF3n in transgenic *N. tabacum* caused retarded flowering, less dense inflorescences, altered internode elongation and leaf morphology and necrotic tips of upper leaves, sepals and bracts (Lemcke and Schmulling 1998). Appearance of localized necrosis was noticed on the tips of apical narrow leaves whereas there was no sign of necrosis on the basal leaves. Additionally, senescence was not altered in these leaves, and bracts became necrotic as a whole. On sepals, the necrosis emerged on the tips just when the corolla was visible through the calyx (Koltunow et al. 2001; Lemcke and Schmulling 1998). The ORF3n protein (48.7 kDa) resembles phenolic-modifying enzymes and may be involved in secondary metabolism and/or the transport of hormones (Binns et al. 1987; Jacobs and Rubery 1988; Lemcke and Schmulling 1998). A cessation was observed in the shoot formation from ORF3n callus in response to auxin and cytokinin. Also, plantlets transferred to the medium containing auxin and cytokinin showed decreased sensitivity leading to small and fewer calli than controls. Thus, it has been proposed that ORF3n may act to negative regulator to the dedifferentiation of tissues as a reaction to auxin and cytokinin, which may favor the formation of *rol* gene-induced roots from such cells during pathogenesis (Britton et al. 2008; Dodueva 2007).

ORF8

The ORF8 gene has the longest sequence of T_L-DNA and coding for a protein containing 780 amino acids (Slightom et al. 1986). The ORF8 protein has one of the most conserved amino acid sequences (81 % similarity) between different strains like pRiA4 and pRi2659 (Ouartsi et al. 2004).

The protein encoded by the ORF8 gene is a natural fusion protein consisting of N-terminal domain (NORF8) of 213 amino acids homologous to RolB protein of the *A. rhizogenes* strain A4 T-DNA and the C-terminal part (CORF8) of approximately 506–524 amino acids shows homology to the IaaM proteins of various other bacteria (Yamada et al. 1985; Slightom et al. 1986; Levesque et al. 1988; Dodueva 2007; Shoja 2010). *iaa*M genes that homologues to the coding sequence of CORF8 codes for a tryptophan monooxygenase which catalyzes the formation of indole-3-acetamide (IAM) from tryptophan (Lemcke et al. 2000).

Furthermore, ORF8 possesses a 200 amino acid stretch at its N-terminus that shows homology with the *rol*B gene product (33.5 % amino acid identity) (Levesque et al. 1988). The N-terminal part (NORF8) of this protein functions in carbohydrate metabolism such that when only NORF8 was expressed, transformed plant showed growth retardation, chlorotic and necrotic leaves and accumulation of high levels of sugars (glucose, fructose and sucrose) and starch (Otten and Helfer 2001).

However, some studies show that the auxin content can be elevated by the genes found in the T_L-DNA region on the T-DNA in some hosts, independent of the presence of the T_R-DNA (Lemcke et al. 2000). Presumably this occurs because of conversion of IAM to IAA in cells expressing only t2m protein (Klee et al. 1987; Prinsen et al. 1990). Besides this, as a characteristic functional motif of the t2m proteins that catalyzes decarboxylation of tryptophan to indole-3-acetamide exhibits 23-aminoacid- long a flavin adenine dinucleotide (FAD) binding site was identified by Levesque et al. (1988). The experimental data obtained from plants and bacteria suggest that the gene product of ORF8 of *A. rhizogenes* T_L-DNA has t2m activity responsible for the increased IAM content in transgenic tissues (Lemcke et al. 2000). Moreover, there is a physical connection between N- and C-regions of ORF8 protein required for the emergence of a specific phenotype in transgenic plants consisting ORF8 gene. This suggests a distinct specific function for the whole protein (Umber et al. 2005; Dodueva 2007).

ORF13 and ORF14

The ORF13 and ORF14 genes are found to be highly conserved among *A. rhizogenes* strains (Stieger et al. 2004). It has been demonstrated that alone A4-*rol*ABC genes carried by an *Agrobacterium* strain are showed to be incapable of inducing rooting on carrot disc and *aux* genes located on the T_R-DNA or ORF13 and ORF14 located on T_L-DNA are also required for rooting (Cardarelli et al. 1987b; Capone et al. 1989). In *N. tabacum* leaf discs harboring *rol*B and ORF13 genes had capacity to induce rooting almost as well as the full length of T_L-DNA (Aoki and Syono 1999). The results obtained via co-inoculation of leaf discs achieved using the *rol*A, *rol*B and *rol*C with either ORF13 or ORF14 showed a limited root induction on carrot disks (Capone et al. 1989). A comparison from the studies showed that there is no homology between ORF13/ORF14 and auxin biosynthetic genes. Furthermore, unlike the genes controlling biosynthesis of auxin (Camilleri and Jouanin 1991), ORF13 and ORF14 have no activity for the induction of roots on *N. tabacum* leaf discs (Cardarelli et al. 1987b). A highly divergent gene family known as plast gene family is constituted by *rol*B, *rol*C, ORF13 and ORF14. They have similar functions and are thought to be evolutionary related (Levesque et al. 1988).

The ORF13 gene is approximately 600 bp in size, encoding a 197–200 amino acid protein, whose expression leads to higher levels in leaves and roots (Durand-Tardif et al. 1985; Veena and Taylor 2007). ORF13 gene leads to the formation of induce cell proliferation such as dense green and rapidly proliferating callus on transformed carrot root and tobacco leaf discs (Capone et al. 1989; Frundt et al. 1998; Dodueva 2007). Wound-inducible and organ-specific expression of ORF13 in transgenic plants lead to a variety of characteristic modifications including irregular formation of leaves, severe leaf nervure, shortened and variable internode length, abnormal and asymmetric flowers, agravitropic root growth and a reduced cell number and cell size in the root (Hansen et al. 1993, 1997; Lemcke and Schmulling 1998; Veena and Taylor 2007). Accelerated expression level in ORF13 gene triggered a more severe reduction of growth in stem and roots through TC-dependent overproduction of the ORF13 gene product, affecting both cell number and cell size in the root. Interestingly, growth and gravitropism was normal in the ORF13 high expressers (Lemcke and Schmulling 1998).

Expression of ORF13 provokes specific phenotype similar to cytokinin-treated plants however free or bound cytokinin content of the transformed tissues shows no difference from wild-type (Medford et al. 1989; Hansen et al. 1993; Lemcke and Schmulling 1998). Furthermore, the shoot part of the ORF13 transformed plant does not resemble cytokinin-overproducing plants, indeed the growth reduction results from the inhibition of cell division in the apical meristems and development of leaves (Lemcke and Schmülling 1998). Some of the phenotypic alterations in transgenic plants are thought to arise from interaction of ORF13 with hormone signaling pathways. ORF13 may play roles in hormone homeostasis and regulation of the cell cycle in infected cells (Veena and Taylor 2007). The observations and grafting of transgenic shoots onto wild type plants revealed that ORF13 may cause the production of a diffusible factor with cytokinin-like activity (Hansen et al. 1993; Dodueva 2007).

Since the only T-DNA gene that induces cell proliferation is ORF13, when inoculated with both carrot discs and tobacco leaf discs produce green callus (Hansen et al. 1993; Frundt et al. 1998). Application of exogenous cytokinin increases the number of roots produced from ORF13 tobacco leaf discs, but does not change root induction on untransformed, even though there was no difference in endogenous cytokinin levels (Specq et al. 1994; Lemcke and Schmulling 1998, Britton et al. 2008).

Furthermore, endoreduplication was reduced in ORF13 plants (Meyer et al. 2000), indicating an interaction of ORF13 with cell cycle control. Stieger et al. (2004) claimed that a proliferative effect of ORF13 expression in the shoot apical meristem (SAM) caused increased number of mitoses and showed no influence on meristem structure. In consequence, the reductions of cell and meristem sizes and the retardation in the formation of leaf primordia were observed. Smaller leaf sizes can be explained by an earlier cessation of leaf growth, but not explained with a reduced size of leaf cells, since the number of epidermal leaf cells per square millimeter was remain unaltered. Enhanced number of cell divisions in the shoot apical meristems and accelerated production of leaf primordia were seen in plant expressing ORF13. ORF13 is involved in the inference of the cell cycle regulation leading to an earlier stop in organ growth in the developing leaves. Furthermore, earlier flowering of plants expressing ORF13 may arrest leaf initiation and leaf expansion, explaining the fewer leaves formed in ORF13 plants (Stieger et al. 2004).

It has been also revealed that ORF13 protein contains a conservative retinoblastoma (RB)-binding motif LxCxE (Meyer et al. 2000). This motif was found in all members of the ORF13 family, including agropine-, mannopine-, cucumopine-, and mikimopine-type Ri plasmids (Stieger et al. 2004). When mutations are introduced into the Rb motif, normal leaf size is restored, but plants still show stunting and reduced apical dominance. It was also observed that ORF13 expression leads to the formation of spur between minor veins on leaves and petals *N. tabacum* (Meyer et al. 2000). Similar structures are formed on leaves, when *KNOX* (KNOTTED1-like homeobox) genes are overexpressed (Sinha et al. 1993; Chuck et al. 1996; Sentoku et al. 2000; Stieger et al. 2004). It was explained that cytokinin-like phenotype such as the formation of spikes, stunted growth, loss of apical dominance, fusion of organs, and stem fasciations observed as consequences of ectopic expression of *KNOX* genes which are induced by ORF1 and cell cycle regulations (Stieger et al. 2004).

Among the additional ORFs in the T_L-DNA, there are two genes, which may also contribute to the hairy root phenotype, ORF13a and ORF14. ORF13a is located between ORF13 and ORF14 on the opposite strand. Expression of this gene is taken place in a tissue specific manner in plants, primarily in leaf vascular tissues (Hansen et al. 1994b). ORF13a is necessary for root induction (Capone et al. 1989). ORF13a containing motifs common to phorphorylated gene regulatory proteins codes for a protein that may interact directly with DNA (Hansen et al. 1994b). Despite a higher expression rate of ORF13a was found in roots compared to leaves, its expression did not yield a visible phenotype (Lemcke and Schmulling 1998; Veena and Taylor 2007). The putative protein encoded by ORF13a has a SPXX repeat motif and is considered to have a regulatory function for this gene (Hansen et al. 1994b). ORF14 is in the same gene family as *rol*B, *rol*C, ORF8 and ORF13 (Levesque et al. 1988). Although overexpression of ORF14 in transgenic carrot and tobacco produced no morphological changes (Lemcke and Schmulling 1998), it has been shown that the *rol* genes and ORF13 act together to induce root induction (Capone et al. 1989; Aoki and Syono 1999) (Table 1.1).

A. rhizogenes and Crop Biotechnology

Genes can be transferred between species and in conjunction with this fact; plant improvements for many decades have been relied heavily upon gene transfer. Either by natural selection or through the efforts of plant breeders, development of plants has always depended upon creating, evaluating and selecting of right combination of alleles. Transgenic plants possessing useful features such as resistance to diseases, insects and pests have been developed by transferring such traits to crop varieties from different species.

Since 1970, rapid progress being made in developing tools for recombinant DNA technology has led to the creation of genetically modified plants. Genetically modified crops have been developed for improving various agricultural, nutritional and food processing traits and used commercially all over the world (Miflin 2000; Kuiper et al. 2001; James 2006; Olempska-Beer et al. 2006). Establishment of plant tissue culture techniques are the most important and preliminary steps for many direct (electroporation, biolistic, microinjection, etc.) and indirect (virus- or bacteria-mediated) gene transfer methods in biotechnology and these methods are used successfully by a lot of laboratories around the world (Ozyigit 2012). The particle bombardment and electroporation transformation methods were favored DNA delivery systems because they do not show any plant host range problems and very effective with high DNA delivery rate (Hauptmann et al. 1987; Birch 1997; Taylor and Fauquet 2002; Turgut-Kara and Ari 2010). However with these methods, gene silencing/co-suppression can be occurred as a result of high copy number of DNA inserted in host cells (Block 1993; Yasuda et al. 2005). On the other hand, *Agrobacterium*-based plant transformation is very effective method of creating plants at low cost, simple to use and with low copy number inserted. Limited number of host range is the only disadvantage (Lessard et al. 2002; Chandra 2012). For achieving transformation of plants, *Agrobacterium* based technology has been used since the mid-1990s increasingly (Hiei et al. 1994). *Agrobacterium*-mediated transformation in generating transgenic plants has been employed as a major DNA delivery system for novel transgenic technologies starting with the transformations of dicotyledonous (Zambryski et al. 1983) and monocotyledonous (Hiei et al. 1994) species in the 1980–1990s. Increasing understanding of *Agrobacterium*-plant relationship (Gelvin 2003) and the mechanisms of transgene integration and genetic recombination in plants (Vain 2007) will lead to achieve further advances in these areas. Conducting efficient and controlled research on targeted gene replacement/alteration, overexpression and mis-expression could provide valuable resource to define gene regulation/function and traits in further in crops. Achievements on *Agrobacterium*-based transformation technologies enable large-scale transgenic studies in a range of important plant and crop species (such as indica rice, wheat, barley, etc.) (Vain 2007) and also bring opportunity to define and select plant cultivars, which could not be obtained by conventional breeding methods (Christou 1997).

For many crops, aim of breeding program is altering plant forms. Establishment of plants with reduced size is favorable in many crops ranging from fruit trees to annual bedding plants (Mayo 1987). Breeding strategies empowered by genetic engineering will lead to the development of more useful and productive crops for plant breeders. While transferring genes to plants for being resistant against diseases and insects, they might have been affected in other ways having altered properties (Oono et al. 1987; Spena et al. 1987; Schmulling et al. 1988; Fladung 1990; Smigocki and Hammerschlag 1991; Scorza et al. 1994). Legumes are not only providing a main source of protein and oil for human and animal nutrition but also contributing to the biological fixation of nitrogen. Moreover, a better understanding of plant-microbe interactions such as symbiotic nitrogen fixation, mycorrhizal associations, and legume-pathogen interactions can be possible with legume studies (Chilton et al. 1982; Christey 2001). Studies on aspects of hairy roots in legumes showed that proliferous root growth and abundant lateral branching are important for improving nitrogen fixation (Cheng et al. 1992).

Most plant structures, such as the hypocotyl, leaf, stem, stalk, petiole, shoot tip, cotyledon, protoplast, storage root, and tuber, have shown capacity to be infected and genetically transformed by *A. rhizogenes* resulting in stimulation of hairy root formation (Mugnier 1988; Han et al. 1993; Bajrovic et al. 1995; Arican et al. 1998; Drewes and Staden 1995; Giri et al. 2001; Krolicka et al. 2001; Azlan et al. 2002; Veena and Taylor 2007). Applications of plant biotechnology favor hairy-root cultures because of their special properties such as fast growth, short doubling time, ease of maintenance, and ability to synthesize a range of chemical compounds and proteins. Hairy root cultures are usually able to produce the same compounds found in wild-type roots of the parent plant, without the loss of concentration (Kim et al. 2002; Veena and Taylor 2007). Above all, hairy roots have an ability to regenerate stable transgenic plants either by a process of somatic embryogenesis or adventitious bud formation, so that genetically modified generations can be achieved (Spano and Costantino 1982; Tepfer 1984; Han et al. 1993; Cho and Wildholm 2002).

It is also known that modification of the cell hormonal balances occurring in response to infection causes root formation at the infected site (Gaudin et al. 1994; Aarrouf et al. 2012). However, the response varies depending upon the strain and its interaction with the plant. One of the most important advantages is that hairy root formation can be used as a verification of transformation. The use of antibiotic resistance markers in the development of transgenic plants is given rise to substantial public attention because of their unknown effects (Christey 2001).

Hairy roots have been used for infection of bacteria, fungi and nematodes and shown to successfully complete their life cycles (Cho et al. 1998; Collier et al. 2005). The resistance genes of nematode have been studied through using hairy roots (Cai et al. 1995; Remeeus et al. 1998; Kifle et al. 1999; Hwang et al. 2000). Development of plants using hairy roots have become of interest because of great potential for building up tolerance to biotic stresses and abiotic stresses (Porter 1991). Hairy root cultures provide an advantage related with making possible the analysis of the changes in enzyme activities and their isoenzyme patterns (Messner and Boll 1993; Kärkönen et al. 2002; Talano et al. 2006).

A variety of dicotyledonous plants are susceptible to *A. rhizogenes*. As a result of stable transformation, root cultures have been established from a range of species of plants (Tepfer 1990). In 1997, Christey reported plant species that had been genetically modified produced from hairy roots of 60 different taxa, representing 51 species from 41 genera and 23 families including Pinaceae Fabaceae, Brassicaceae and Solanaceae Araliaceae, Caricaceae and Rutaceae. In 2001, it was reported that, transgenic plants have been derived via transgenesis using in 89 different taxa, representing 79 species from 55 genera and 27 families (Christey 2001). Because lack of susceptibility, monocotyledonous plants are not a host for *A. rhizogenes* for and still there is no example for transgenic monocotyledonous plant except onion (Dommisse et al. 1990) and asparagus (Hernalsteens et al. 1993; Christey 2001). According to Web of Science, currently there are more than 500 studies conducted on *A. rhizogenes*. Table 1.2 summarizes the studies conducted, the plants and the genes transferred via *A. rhizogenes* in chronological order.

Conclusion and Future Perespective

This chapter deals with current research on *A. rhizogenes*-mediated transformation and its applications in crops. *A. rhizogenes* is responsible for the development of hairy root disease in a wide range of dicotyledonous plants and characterized by a proliferation of excessively branching roots. Containing case studies demonstrating the result of *A. rhizogenes*-mediated transformation includes biosynthesis pathways in plants created a valuable platform in the last years. Furthermore, the plants transformed with *A. rhizogenes* are become increasingly popular for offering approaches to create cost-effective options in mass-producing desired plant metabolites and expressing foreign proteins. The data from numerous proof-of-concept studies including improved the nutritional quality, agronomical characteristics, production of plant-derived products encourages for the realization of scaling up *Agrobacterium* based practices. Recently, transgenic plants produced by *Agrobacterium*-mediated transformation have also been shown to have immense potential for applications in phytoremediation. This chapter highlights recent progresses in the field of *A. rhizogenes*-mediated transformation and outlines future perspectives for the exploitation of it.

Acknowledgement Authors are grateful to Professor Nermin Gözükırmızı, Professor Şule Ari, Associate Professor Ercan Arican and Dr. Neslihan Turgut-Kara at Istanbul University, Department of Molecular Biology and Genetics for providing hairy root pictures of their previous studies and *Agrobacterium rhizogenes* strains (8196 and R1000) which had been given by Associate Professor Kemal Melik Taşkın (Çanakkale 18 Mart University, Biology Department) to Istanbul University Data Collection. Then there were those people at Marmara University, School of Medicine, Department of Histology and Embryology who helped with techniques for obtaining SEM micrographs. We are grateful to all of them, in particular to: Professor Feriha Ercan, Research Assistant Özlem T. Çilingir and Yücel Öztürk. We like to acknowledge Designer Recep Cenk Tarhan and Biologist-Designer İlke Ertem who spent hours of their time helping with the figures and diagrams, Research Assistants Sezen İğdelioğlu and Onur Zorluer for assistance with compiling the references.

Table 1.2 Summary of the studies conducted, the plants and the genes transferred via *A.rhizogenes* in chronological order

Daucus carota	Carrot	*rol*	David et al. 1984
Kalanchoe daigremontiana	Devil's backbone	*rol*	White et al. 1985
Arabidopsis thaliana	Mouse ear cress	*rol*	Pavingerova and Ondrej 1986
Cucumis sativus	Cucumber	NPTII	Trulson et al. 1986
Lycopersicon esculentum	Tomato	NPTII	Shahin et al. 1986
Petunia hybrida	Petunia	*rol*	Ondrej and Biskova 1986
Armoracia lapathifolia	Horseradish	*rol*	Noda et al. 1987
Lycopersicon peruvianum	–	NPTII	Morgan et al. 1987
Nicotiana debneyi	Debney's tobacco	NPTII	Davey et al. 1987
Nicotiana plumbaginifolia	–	NPTII	Davey et al. 1987
Solanum nigrum	Black nightshade	NPTII	Davey et al. 1987
Anagallis arvensis	Pimpernel	*rol*	Mugnier 1988
Convolvulus arvensis	Morning glory	*rol*	Mugnier 1988
Foeniculum vulgare	Fennel	*rol*	Mugnier 1988
Linum usitatissimum	Flax	*rol*	Zhan et al. 1988
Nicotiana glauca	Tree tobacco	*rol*	Sinkar et al. 1988
Nicotiana hesperis	–	*rol*	Walton and Belshaw 1988
Brassica oleracea var. *acephala*	Ornamental kale	*rol*	Hosoki et al. 1989
Catharanthus roseus	Periwinkle	*rol*	Brillanceau et al. 1989
Glycine argyrea	Wild soybean	NPTII	Rech et al. 1989
Glycine canescens	Wild soybean	NPTII	Rech et al. 1989
Lotus corniculatus	Bird's-Foot trefoil	GUS	Forde et al. 1989
Solanum tuberosum	Potato	NPT II, GUS	Visser et al. 1989
Stylosanthes humilis	Townsville stylo	NPT II	Manners and way 1989
Trifolium repens	White clover	*rol*	Diaz et al. 1989
Brassica napus	Rapeseed	NPTII	Boulter et al. 1990
Nicotiana rustica	Mapacho	ODS	Hamill et al. 1990
Nicotiana tabacum	Tobacco	NPTII	Hatamoto et al. 1990
Vicia faba	Fava bean	NPTII	Ramsay and Kumar 1990
Actinidia deliciosa	Kiwifruit	*rol*	Rugini et al. 1991
Allocasuarina verticillata	Drooping she-oak	*rol*	Phelep et al. 1991
Cichorium intybus	Chicory	*rol*	Sun et al. 1991
Hyoscyamus muticus	Egyptian henbane	*rol*	Oksman-Caldentey et al. 1991
Medicago arborea	Tree medick	HPT	Damiani and Aricioni 1991
Medicago sativa	Alfalfa/lucerne	*rol*	Golds et al. 1991
Olea europaea	Olive	*rol*	Rugini et al. 1996
Onobrychis viciifolia	Sainfoin	*rol*	Golds et al. 1991
Pistacia vera	Pistachio	*rol*	Rugini and Mariotti 1991
Malus domestica	Apple	*rol*B	Rugini and Mariotti 1991
Solanum dulcamara	Nightshade	NPTII, *rol*	McInnes et al. 1991
Anthyllis vulneraria	Kidney vetch	NPTII, *ipt*	Stiller et al. 1992
Atropa belladonna	Deadly nightshade	*bar*	Saito et al. 1992
Brassica campestris	Turnip	NPT II	Christey and Sinclair 1992

Table 1.2 (continued)

Brassica campestris var. *rapifera*	Turnip	GUS, NPTII, ALS	Christey and Sinclair 1992
Brassica oleracea var. *acephala*	Forage kale	GUS, NPTII, ALS	Christey and Sinclair 1992
Malus pumila	Apple	*rol*	Lambert and Tepfer 1992
Medicago truncatula	Barrel clover	NPTII	Thomas et al. 1992
Papaver somniferum	Opium poppy	*rol*	Yoshimatsu and Shimomura 1992
Coffea arabica	Coffea	*rol*	Spiral et al. 1993
Eucalyptus sp.	Eucalyptus	*rol*	MacRae and van Staden 1993
Glycine max	Soybean	GUS	Olhoft et al. 2007
Ipomoea batatas	Sweet potato	NPTII, GUS	Otani et al. 1993
Populus trichocarpa × *P. deltoides*	Cottonwood	NPTII	Han et al. 1993
Robinia pseudoacacia	Black locust	NPTII	Han et al. 1993
Vicia hirsuta	Hairy vetch	*rol*	Quandt et al. 1993
Vigna aconitifolia	Moth bean	SbPRP1	Suzuki et al. 1993; Lee et al. 1993
Diospyros kaki	Japanese persimmon	*rol*	Tao et al. 1994
Larix decidua	European larch	NPTII, *aro*A, BT	Shin et al. 1994
Pelargonium graveolens	Lemon geranium	*rol*	Pellegrineschi et al. 1994
Rosa hybrida	Hybrid tea rose	NPTII, GUS	Firoozabady et al. 1994
Rubia peregrina	Wild madder	ICS	Downs et al. 1994
Vinca minor	Lesser periwinkle	NPTII, GUS	Tanaka et al. 1994
Vitis vinifera	Grapevine	NPTII, GUS	Nakano et al. 1994
Casuarina glauca	Swamp she-oak	GUS	Diouf et al. 1995
Gentiana scabra	Japanese gentian	*rol*	Suginuma and Akihama 1995
Solanum tuberosum L.	Potato	*rol*	Bajrovic et al. 1995
Rudbeckia hirta	Black-Eyed susan	*rol*	Daimon and Mii 1995
Verticordia grandis	Scarlet featherflower	NPTII, GUS	Stummer et al. 1995
Citrus sinensis	Sweet orange	*rol*	Li et al. 1996
Ajuga reptans	Blue bugle	GUS	Uozumi et al. 1996
Begonia tuberhybrida	Begonia	*rol*	Kiyokawa et al. 1996
Brassica campestris	Turnip	GUS	Christey et al. 1997
Brassica oleracea	Wild cabbage	GUS	Christey et al. 1997
Carica papaya	Papaya	NPTII, GUS	Cabrera-Ponce et al. 1996
Eustoma grandiflorum	Lisianthus	NPTII, GUS	Handa 1992
Ipomoea trichocarpa	Blue morning glory	NPTII, GUS	Otani et al. 1993
Juglans regia	Walnut	*rol*B	Caboni et al. 1996
Lotus angustissimus	Slender bird's-foot trefoil	NPTII, GUS	Nenz et al. 1996
Pelargonium fragrans	Nutmeg geranium	*rol*	Pellegrineschi and Davolio-Mariani 1996

Table 1.2 (continued)

Pelargonium odoratissimum	Apple geranium	*rol*	Pellegrineschi and Davolio-Mariani 1996
Pelargonium quercifolium	Oak-Leaved geranium	*rol*	Pellegrineschi and Davolio-Mariani 1996
Pinus contorta	Lodgepole pine	*rol*	Yibrah et al. 1996
Pinus halepensis	Aleppo pine	*rol*	Tzfira et al. 1996
Pinus nigra	Austrian pine	*rol*	Mihaljevic et al. 1996
Populus tremula	Aspen	NPTII, GUS	Tzfira et al. 1996
Rosa sp.	Rose	*rol*	Van der Salm et al. 1997
Scoparia dulcis	Licorice weed	*rol*	Yamazaki et al. 1996
Aconitum heterophyllum	Indian atees	*rol*	Giri et al. 1997
Artemisia annua	Sweet wormwood	*rol*	Banerjee et al. 1997
Brassica napus	Oilseed rape	GUS, NPTII, ALS	Christey et al. 1997
Brassica oleracea	Wild cabbage	GUS, NPTII	Christey et al. 1997
Datura arborea	Angel's trumpets	*rol*	Giovannini et al. 1997
Datura sanguinea	Red Angel's trumpets	*rol*	Giovannini et al. 1997
Digitalis lanata	Grecian foxglove	*rol*	Pradel et al. 1997
Gentiana cruciata	Gentian	GUS	Momčilović et al. 1997
Gentiana purpurea	Purple gentian	*rol*	Momčilović et al. 1997
Gentiana triflora × *G. scabra*	–	*rol*	Hosokawa et al. 1997
Lotus japonicus	Lotus japonicus	*rol*	Stiller et al. 1997
Nierembergia scoparia	Tall cupflower	*rol*	Godo et al. 1997
Peganum harmala	Harmal	TDS	Berlin et al. 1993
Antirrhinum majus	Snapdragon	bar, NPTII	Hoshino and Mii 1998
Arachis hypogaea L.	Groundnut	*rol*	Akasaka et al. 1998
Astragalus sinicus	Chinese milk vetch	GUS	Cho et al. 1998
Citrus aurantifolia	Mexican lime	NPTII, GUS	Pérez-Molphe-Balch and Ochoa-Alejo 1998
Nicotiana spp.	–	*rol*	Palazon et al. 1998
Panax ginseng	Ginseng	*rol*	Yang and Choi 2000
Prunus avium	Sweet cherry	*rol*	Gutierrez-Pesce et al. 1998
Brassica campestris var. *pekinensis*	Chinese cabbage	NPTII, EAS	Christey et al. 1999
Brassica oleracea L. var. *italica*	Broccoli	*rol*	Henzi et al. 1999
Brassica oleracea var. *botrytis*	Cauliflower	NPTII, GUS	Christey et al. 1999
Brassica oleracea var. *capitata*	Cabbage	NPTII, GUS	Christey et al. 1999
Brassica oleracea var. *gemmifera*	Brussels sprouts	NPTII	Christey et al. 1999
Brassica oleracea var. *italica*	Broccoli	NPTII, EAS	Christey et al. 1999
Gentiana punctata	Spotted gentian	GUS	Vinterhalter et al. 1999
Pimpinella anisum	Anise	*rol*	Andarwulan and Shetty 1999
Pyrus communis	Pear	*rol*C	Bell et al. 1999

Table 1.2 (continued)

Rubia tinctorum	Common madder	*rol*	Ercan et al. 1999
Ulmus spp.	Elm	*rol*	Rinallo et al. 1999
Ziziphus jujuba	Jujube	*rol*	Hatta et al. 1996
Crotalaria juncea	Sunn hemp	*rol*	Ohara et al. 2000
Trifolium pratense	Red clover	*rol*	Díaz et al. 2000
Brassica napus var. *rapifera*	Swede (Rutabaga)	bar	Christey and Braun 2001
Oryza sativa var. *japonica*	Japanese Rice	*rol*A, NPTII	Lee et al. 2001
Spinacia oleracea	Spinach	*rol*	Ishizaki et al. 2002
Citrus aurantium	Bergamot orange	*rol*	Chavez-Vela et al. 2003
Ginkgo biloba	Ginkgo	*rol*	Ayadi and Tremouillaux-Guiller 2003
Rauvolfia micrantha	–	*rol*	Sudha et al. 2003
Sesbania rostrata	Pea	*rol*	Van de Velde et al. 2003
Aesculus hippocastanum	Horse-chestnut	GUS	Zdravkovic-Korac et al. 2004
Alstroemeria sp.	Peruvian lily	NPTII, GUS, *rol*	Akutsu et al. 2004
Camptotheca acuminata	Happy tree	*rol*	Lorence et al. 2004
Genista tinctoria	Greenweed	*rol*	Luczkiewicz and Kokotkiewicz 2005
Typha latifolia	Common bulrush	*rol*	Nandakumar et al. 2005
Brassica oleracea var. *sabauda*	Savoy cabbage	GUS	Sretenovıc-Rajicic et al. 2006
Brassica oleracea var. *sabauda*	Savoy cabbage	*rol*	Sretenovic-Rajicic et al. 2006
Eustoma grandiflorum	Lisianthus	*rol*	Popa et al. 2006
Echinacea purpurea	Purple coneflower	*rol*B	Wang et al. 2006
Phaseolus vulgaris	Common bean	GFP, GUS	Estrada-Navarrete et al. 2006
Tylophora indica	Indian ipecac	*rol*	Chaudhuri et al. 2006
Asimina triloba	Pawpaw	*rol*B, C	Ayala-Silva et al. 2007
Pueraria candollei	–	*rol*B	Medina-Bolivar et al. 2007
Beta vulgaris	Red beet	NPTII	Thimmaraju et al. 2008
Glycyrrhiza glabra	Licorice	*rol*	Mehrotra et al. 2008
Musa sp.	Banana	*rol*	Matsumoto et al. 2009
Plumbago rosea	Plumbago	*rol*	Satheeshkumar et al. 2009
Podophyllum hexandrum	Himalayan mayapple	*rol*	Lin et al. 2003
Psoralea corylifolia	Babchi	*rol*	Shinde et al. 2009
Drosera burmannii	Tropical sundew	*rol*	Putalun et al. 2010
Echium rauwolfii	Echium rauwolfii	*rol*	Abd El-Mawla 2010
Fagopyrum esculentum	Buckwheat	GUS	Kim et al. 2010
Mangifera indica	Mango	*rol*	Chavarri et al. 2010
Przewalskia tangutica	–	*rol*	Lan and Quan 2010
*Corchorus capsularis*t	Jute	GUS	Chattopadhyay et al. 2011
Nasturtium officinale	Watercresses	*rol*	Park et al. 2011
Prunus sp.	–	Egfp, NPTII	Bosselut et al. 2011
Amaranthus spinosus	Spiny amaranth	*rol*B	Pal et al. 2012
Capsicum annuum	Pepper	GFP	Aarrouf et al. 2012
Clitoria ternatea	Butterfly pea	*rol*	Swain et al. 2012

References

Aarrouf J, Castro-Quezada P, Mallard S, Caromel B, Lizzi Y, Lefebvre V (2012) *Agrobacterium rhizogenes*-dependent production of transformed roots from foliar explants of pepper (*Capsicum annuum*): a new and efficient tool for functional analysis of genes. Plant Cell Rep 31:391–401

Abarca-Grau AM, Penyalver R, Lopez MM, Marco-Noales E (2011) Pathogenic and non-pathogenic *Agrobacterium tumefaciens*, *A. rhizogenes* and *A. vitis* strains form biofilms on abiotic as well as on root surfaces. Plant Pathol 60:416–425

Abd El-Mawla AMA (2010) Effect of certain elicitors on production of pyrrolizidine alkaloids in hairy root cultures of *Echium rauwolfii*. Pharmazie 65:224–226

Ackermann C (1977) Pflanzen aus *Agrobacterium rhizogenes* tumoren and *Nicotiana tabacum*. Plant Sci Lett 8:23–30

Akasaka Y, Mii M, Daimon H (1998) Morphological alterations and root nodule formation in *Agrobacterium rihzogenes*-mediated transgenic hairy roots of peanut (*Arachis hypogaea* L.). Ann Bot 81:355–362

Akutsu M, Ishizaki T, Sato H (2004) Transformation of the monocot *Alstroemeria* by *Agrobacterium rhizogenes*. Mol Breeding 13:69–78

Alpizar E, Dechamp E, Espeout S, Royer M, Lecouls AC, Nicole M, Bertrand B, Lashermes P, Etienne H (2006) Efficient production of *Agrobacterium rhizogenes*-transformed roots and composite plants for studying gene expression in coffee roots. Plant Cell Rep 25:959–967

Altamura MM, Tomassi M (1998) Auxin, photoperiod and putrescine affect flower eoformation in normal and *rol*B-transformed tobacco thin cell layers. Plant Physiol Biochem 36:441–448

Andarwulan N, Shetty K (1999) Phenolic synthesis in differentiated tissue cultures of untransformed and *Agrobacterium-transformed* roots of anise (*Pimpinella anisum* L.). J Agric Food Chem 47:1776–1780

Aoki S, Syono K (1999) Short communication synergistic function of *rol*B, *rol*C, ORF13 and ORF14 of T_L-DNA of *Agrobacterium rhizogenes* in hairy root induction in *Nicotiana tabacum*. Plant Cell Physiol 40(2):252–256

Arican E, Bajrovic K, Gozukirmizi N (1998) Effects of naphthalene acetic acid on transformation frequency of potato and tobacco via *Agrobacterium rhizogenes*. Biotechnol Biotec Eq 12(1):29–33

Ayadi R, Tremouillaux-Guiller J (2003) Root formation from transgenic calli of *Ginkgo biloba*. Tree Physiol 23:713–718

Ayala-Silva T, Bey CA, Dortch G (2007) *Agrobacterium rhizogenes* mediated transformation of *Asimina triloba* L. cuttings. Pak J Biol Sci 10:132–136

Azlan GJ, Marziah M, Radzali M, Johari (2002) Establishment of *Physalis minima* hairy roots culture for the production of physalins. Plant Cell Tiss Org 69:271–278

Bajrovic K, Arı Ş, Arıcan E, Kazan K, Gözükırmızı N (1995) Biotechnol Biotec Eq 1:29–32

Balandrin MF, Klocke JA, Wurtele ES, Bollinger WH (1985) Natural plant chemicals: sources of industrial and medicinal materials. Science 228:1154–1160

Bandyopadhyay M, Jha S, Tepfer D (2007) Changes in morphological phenotypes and withanolide composition of Ri-transformed roots of *Withania somnifera*. Plant Cell Rep 26:599–609

Banerjee S, Zehra M, Gupta MM, Kumar S (1997) *Agrobacterium rhizogenes* mediated transformation of *Artemisia annua* -production of transgenic plants. Planta Med 63:467–469

Bellincampi D, Cardarelli M, Zaghi D, Serino G, Salvi G, Gatz C, Cervone F, Altamura MM, Constantino P, De-Lorenzo G (1996) Oligogalacturonides prevent rhizogenesis in *rol*B transformed tobacco explants by inhibiting auxin-induced expression of the *rol*B gene. Plant Cell 8:477–487

Bell RL, Scorza R, Srinivasan C, Webb K (1999) Transformation of 'Beurre Bosc' pear with the *rolC* gene. J Amer Soc Hort Sci 124:570–574

Bensaddek L, Villarreal ML, Fliniaux MA (2008) Induction and growth of hairy roots for the production of medicinal compounds. Electron J Integr Biosci 3(1):2–9

Berlin J, Ruegenhagen C, Dietze P, Fecker LF, Goddijn OJM, Hoge JHC (1993) Increased production of seratonin by suspension and root cultures of *Peganum harmala* transformed with a tryptophan decarboxylase cDNA clone from *Cathranthus roseus.* Transgenic Res 2:336–344

Bettini P, Michelotti S, Bindi D, Giannini R, Capuana M, Buiatti M (2003) Pleiotropic effect of the insertion of the *Agrobacterium rhizogenes rol*D gene in tomato (*Lycopersicon esculentum* Mill.). Theor Appl Genet 107:831–836

Binns AN, Costantino P (1998) The *Agrobacterium* oncogenes. In: Spaink HP, Kondorosi A, Hooykaas PJ (eds) The *Rhizobiaceae*: molecular biology of model plant-associated bacteria. Kluwer Academic Publishers, Dordrecht, pp. 251–266

Binns AN, Chen RH, Wood HN, Lynn DG (1987) Cell division promoting activity of naturally occurring dehydrodiconiferyl glucosides: do cell wall components control cell division? Proc Natl Acad Sci USA 84:980–984

Birch RG (1997) Plant transformation: problems and strategies for practical application. Annu Rev Plant Physiol Plant Mol Biol 48:297–326

Block M (1993) The cell biology of plant transformation: current state, problems, prospects and the implications for the plant breeding. Euphytica 71(1–2):1–14

Bonhomme V, Laurain-Mattar D, Lacoux J, Fliniaux MA, Jacquin-Dubreuil A (2000) Tropane alkaloid production by hairy roots of *Atropa belladonna* obtained after transformation with *Agrobacterium rhizogenes* 15834 and *Agrobacterium tumefaciens* containing *rol*A, B, C genes only. J Biotech 81(2–3):151–158

Bosselut N, Van Ghelder C, Claverie M, Voisin R, Onesto JP, Rosso MN, Esmenjaud D (2011) *Agrobacterium rhizogenes*-mediated transformation of *Prunus* as an alternative for gene functional analysis in hairy-roots and composite plants. Plant Cell Rep 30(7):1313–1326

Bouchez D, Camilleri C (1990) Identification of a putative *rol*B gene on the TR-DNA of the *Agrobacterium rhizogenes* A4 Ri plasmid. Plant Mol Biol 14:617–619

Boulter ME, Croy E, Simpson P, Shields R, Croy RRD, Shirsat AH (1990) Transformation of *Brassica napus* L. (oilseed rape) using *Agrobacterium tumefaciens* and *Agrobacterium rhizogenes*- a comparison. Plant Sci 70:91–99

Brevet J, Tempe J (1988) Homology mapping of T-DNA regions on three *Agrobacterium rhizogenes* Ri plasmids by electron microscope heteroduplexstudies. Plasmid 19:75–83

Brillanceau MH, David C, TempeÂ J (1989) Genetic transformation of *Catharanthus roseus* G. Don by *Agrobacterium rhizogenes*. Plant Cell Rep 8:63–66

Britton MT, Escobar MA, Dandekar M (2008) The oncogenes of *Agrobacterium tumefaciens* and *Agrobacterium rhizogenes*. In: Tzfira T, Citovsky V (eds) *Agrobacterium*: from biology to biotechnology. Springer, Heidelberg, pp. 525–563

Broothaerts W, Mitchell HJ, Weir B, Kaines S, Smith LMA, Yang WM, Jorge E, Roa-rodriguez CJ, Richard A (2005) Gene transfer to plants by diverse species of bacteria. Nature 433:629–633

Bulgakov VP (2008) Functions of *rol* genes in plant secondary metabolism. Biotechnol Adv 26:318–324

Bulgakov VP, Khodakovskaya MV, Labetskaya NV, Tchernoded GK, Zhuravlev YN (1998) The impact of plant *rol*C oncogene on ginsenoside production by ginseng hairy root cultures. Phytochemistry 49:1929–1934

Bulgakov VP, Tchernoded GK, Mischenko NP, Khodakovskaya MV, Glazunov VP, Zvereva EV, Fedoreyev SA, Zhuravlev YN (2002a) Effect of salicylic acid, methyl jasmonate, ethephon and cantharidin on anthraquinone production by *Rubia cordifolia* callus cultures transformed with the *rol*B and *rol*C genes. J Biotechnol 97:213–221

Bulgakov VP, Kusaykin M, Tchernoded GK, Zvyagintseva TN, Zhuravlev YN (2002b) Carbohydrase activities of the *rol*C-gene transformed and non-transformed ginseng cultures. Fitoterapia 73:638–643

Bulgakov VP, Tchernoded GK, Mischenko NP, Shkryl YN, Fedoreyev SA, Zhuravlev YN (2004) The *rol*B and *rol*C genes activate synthesis of anthraquinones in *Rubia cordifolia* cells by mechanism independent of octadecanoid signaling pathway. Plant Sci 166:1069–1075

Caboni E, Lauri P, Tonelli M, Falasca G, Damiano C (1996) Root induction by *Agrobacterium rhizogenes* in walnut. Plant Sci 118:203–208

Cabrera-Ponce JL, Vegas-Garcia A, Herrera-Estrella L (1996) Regeneration of transgenic papaya plants via somatic embryogenesis induced by *Agrobacterium rhizogenes*. In Vitro Cell Dev Biol Plant 32:86–90

Cai G, Li G, Ye H (1995) Hairy root culture of *Artemisia annua* L. by Ri plasmid transformation and biosynthesis of artemisinin. Chinese J Biotechnol 11(4):227–235

Camilleri C, Jouanin L (1991) The TR-DNA region carrying the auxin synthesis genes of the *Agrobacterium rhizogenes* agropine type plasmid pRiA4: nucleotide sequence analysis and introduction into tobacco plants. Mol Plant Microbe Interact 4:155–162

Capone IL, Spano L, Cardarelli M, Bellincampi D, Petit A, Constantino P (1989) Induction and growth properties of carrot roots with different complements of *Agrobacterium rhizogenes* T-DNA genes. Plant Mol Biol 13:43–52

Cardarelli M, Spanò L, De Paolis A, Mauro ML, Vitali G, Costantino P (1985) Identification of the genetic locus responsible for non-polar root induction by *Agrobacterium rhizogenes* 1855. Plant Mol Biol 5:385–391

Cardarelli M, Mariotti D, Pomponi M, Spano L, Capone I, Costantino P (1987a) *Agrobacterium rhizogenes* T-DNA genes capable of inducing hairy root phenotype. Mol Gen Genet 209(3):475–480

Cardarelli M, Spano L, Mariotti D, Mauro ML, Van Sluys MA, Costantino P (1987b) The role of auxin in hairy root induction. Mol Gen Genet 208:457–463

Casanova E, Zuker A, Trillas MI, Moysset L, Vainstein A (2003) The *rol*C gene in carnation exhibits cytokinin- and auxin-like activities. Sci Hortic 97:321–331

Casanova E, Valdes AE, Zuker A, Fernandez B, Vainstein A, Trillas MI, Moysset L (2004) *rol*C-transgenic carnation plants: adventitious organogenesis and levels of endogenous auxin and cytokinins. Plant Sci 167(3):551–560

Census (2012) The official website of U.S. Department of Commerce, U.S. Census Bureau-World POPClock Projection. http://www.census.gov. Accessed: 25. May 2011

Chandra S (2012) Natural plant genetic engineer *Agrobacterium rhizogenes*: role of T-DNA in plant secondary metabolism. Biotechnol Lett 34(3):407–415

Charlwood BV, Charlwood KA (1991) Terpenoid production in plant cell culture. In: Harborne JB, Tomas-Barberan FE (eds) Ecological chemistry and biochemistry of plant terpenoids. Clarendon Press, Oxford, pp 95–132

Chattopadhyay T, Roy S, Mitra A, Maiti MK (2011) Development of a transgenic hairy root system in jute (*Corchorus capsularis* L.) with GUSA reporter gene through *Agrobacterium rhizogenes* mediated co-transformation. Plant Cell Rep 30(4):485–493

Chaudhuri KN, Ghosh B, Tepfer D, Jha S (2006) Spontaneous plant regeneration in transformed roots and calli from *Tylophora indica*: changes in morphological phenotype and tylophorine accumulation associated with transformation by *Agrobacterium rhizogenes*. Plant Cell Rep 25(10):1059–1066

Chavarri M, García AV, Zambrano AY, Gutiérrez Z, Demey JR (2010) Insertion of *Agrobacterium rhizogenes rolB* gene in Mango. Interciencia 35(7):521–525

Chavez-Vela NA, Chavez-Ortiz LI, Perez-Molphe Balch E (2003) Genetic transformation of sour orange using *Agrobacterium rhizogenes*. Agrociencia 37:629–639

Cheng M, His DCH, Philips GC (1992) In vitro regeneration of Valencia type peanut (*Arachis hypogaea* L.) from cultured petioles, epicotyl, sections and other seedling explants. Peanut Sci 19:82–87

Chilton MD, Tepfer D, Petit A, David C, Delbart C-F, Tempt J (1982) *Agrobacterium rhizogenes* inserts T-DNA into the genomes of the host plant root cells. Nature 295:432–434

Cho HJ, Wildholm JM (2002) Improved shoot regeneration protocol for hairy roots of the legume *Astragalus sinicus*. Plant Cell Tiss Org 69:259–269

Cho H-J, Widholm JM, Tanaka N, Nakanishi Y, Murooka Y (1998) *Agrobacterium rhizogenes*-mediated transformation and regeneration of the legume *Astragalus sinicus* (Chinese milk vetch). Plant Sci 138:53–65

Christensen B, Sriskandarajah S, Serek M, Müller R (2008) Transformation of *Kalanchoe blossfeldiana* with *rol*-genes is useful in molecular breeding towards compact growth. Plant Cell Rep 27:1485–1495
Christey MC (2001) Use of Ri-mediated transformation for production of transgenic plants. In Vitro Cell Dev Biol Plant 37:687–700
Christey MC, Braun RH (2001) Transgenic vegetable and forage Brassica species: rape, kale, turnip and rutabaga (Swede). In: Bajaj YPS (ed) Biotechnology in agriculture and forestry, Transgenic crops II 47:87–101
Christey MC, Braun RH, Reader JK (1999) Field performance of transgenic vegetable brassicas (*Brassica oleracea* and *B. rapa*) transformed with *Agrobacterium rhizogenes*. Sabrao J Breed Genet 31:93–108
Christey MC, Sinclair BK (1992) Regeneration of trasgenic kale (*Brassica oleracea* var. *acephal*), rap (*B. napus*) and turnip (*B. campestris* var *rapifera*) plants via *Agrobacterium rhizogenes* mediated transformation. Plant Sci 87:161–169
Christey MC, Sinclair BK, Braun RH, Wyke L (1997) Regeneration of transgenic vegetable brassicas (*Brassica oleracea* and *B. campestris*) via Ri-mediated transformation. Plant Cell Rep 16:587–593
Christie PJ, Ward JE, Winans SC, Nester EW (1988) The *Agrobacterium tumefaciens vir*E2 gene product is a single-stranded- DNA-binding protein that associates with T-DNA. J Bacteriol 170:2659–2667
Christou P (1997) Biotechnology applied to grain legumes. Field Crop Res 53:83–97
Chuck G, Lincoln C, Hake S (1996) KNAT1 induces lobed leaves with ectopic meritems when overexpressed in *Arabidopsis*. Plant Cell 8:1277–1289
Citovsky V, Zupan J, Warnick D, Zambryski P (1992) Nuclear localization of *Agrobacterium* VirE2 protein in plant cells. Science 256:1802–1805
Collier R, Fuchs B, Walter N, Kevin LW, Taylor CG (2005) Ex vitro composite plants: an inexpensive, rapid method for root biology. Plant J 43:449–457
Comai L, Kosuge T (1982) Cloning and characterization of *iaa*M, a virulence determinant of *Pseudomonas savastanoi*. J Bacteriol 149:40–46
Conn HJ (1942) Validity of the genus *Alcaligenes*. J Bacteriol 44:353–360
Costantion P, Spano L, Pomponi M, Benevuto E, Ancora G (1984) The T-DNA of *Agrobacterium rhizogenes* is transmitted through meiosis to the progeny of hairy root plants. J Mol Appl Genet 2(5):465–470
Costantino P, Capone I, Cardarelli M, De-Paolis A, Mauro ML, Trovato M (1994) Bacterial plant oncogenes: the *rol* genes' saga. Genetica 94:203–211
Daimon H, Mii M (1995) Plant regeneration and thiophene production in hairy root cultures of *Rudbeckia hirta* L. used as an antagonistic plant to nematodes. Jpn J Crop Sci 64:650–655
Damiani F, Aricioni S (1991) Transformation of *Medicago arborea* L. with *Agrobacterium rhizogenes* binary vector carrying the hygromycin resistance genes. Plant Cell Rep 10:300–303
Davey MR, Mulligan BJ, Gartland KMA, Peel E, Sargent AW, Morgan AJ (1987) Transformation of *Solanum* and *Nicotiana* species using an Ri plasmid vector. J Exp Bot 38:1507–1516
David C, Chilton MD, Tempe J (1984) Conservation of T-DNA in plants regenerated from hairy root cultures. Biotech 2:73–76
Davioud E, Petit A, Tate ME, Ryder MH, Tempe J (1988) Cucumopine-a new T-DNA-encoded opine in hairy root and crown gall. Phytochemistry 27(8):2429–2433
De Paolis A, Mauro ML, Pompon M, Cardarelli M, Spano L, Costantino P (1985) Localization of agropine synthesizing functions in the TR region of the root inducing plasmid of *Agrobacterium rhizogenes* 1855. Plasmid 13:1–7
Dehio C, Grossmann K, Schell J, Schmülling T (1993) Phenotype and hormonal status of transgenic tobacco plants overexpressing the *rol*A gene of *Agrobacterium rhizogenes* T-DNA. Plant Mol Biol 23(6):1199–1210
Dehio C, Schell J (1993) Stable expression of a single-copy *rol*A gene in transgenic *Arabidopsis thaliana* plants allows an exhaustive mutagenic analysis of the transgene-associated phenotype. Mol Gen Genet 241:359–366

Delbarre A, Muller P, Imhoff V, Barbier-Brygoo H, Maurel C, Leblanc N, Perrot-Rechenmann C, Guern J (1994) The *rol*B Gene of *Agrobacterium rhizogenes* does not increase the auxin sensitivity of tobacco protoplasts by modifying the intracellular auxin concentration. Plant Physiol 105:563–569

Dessaux Y, Petit A, Tempe J (1992) Opines in *Agrobacterium* biology. In: Verma DPS (ed) Molecular signals in plant-microbe communications. CRC Press, Boca Raton, pp 109–136

Diaz CL, Melchers LS, Hooykaas PJJ, Lugtenberg BJJ, Kijne JW (1989) Root lectin as a determinant of host-plant specificity in the Rhizobium-legume symbiosis. Nature 338:579–581

Díaz Cl, Spaink HP, Kijne JW (2000) Heterologous rhizobial lipchitin oligosaccharides and chitin oligomers induce cortical cell divisions in red clover roots, transformed with the pea lectin gene. Mol Plant Microbe Interact 13:268–276

Diouf D, Gherbi H, Prin Y, Franche C, Duhoux E, Bogusz D (1995) Hairy root nodulation of *Casuarina glauca*: a system for the study of symbiotic gene expression in an actinorhizal tree. Mol Plant Microbe Interact 8:532–537

Dodueva IE (2007) A Study of expression of the genes involved in systemic control of cell division and differentiation in higher plants on the model of spontaneous tumorigenesis in inbred radish lines (*Raphanus sativus* var. *radicula* Pers.). Cand Sci (Biol) Dissertation, St. Petersburg: Gos. Univ

Dommisse EM, Leung DWM, Shaw ML, Conner AJ (1990) Onion is a monocotyledonous host for *Agrobacterium*. Plant Sci 69:249–257

Doran PM (2002) Properties and applications of hairy root cultures. In: Marja K, Caldentey KM, Barz W (eds) Plant biotechnology and transgenic plants. Marcel Dekker Inc, New York, pp 1–20

Downs CG, Christey MC, Davies KM, King GA, Seelye JF, Sinclair BK, Stevenson DG (1994) Hairy roots of *Brassica napus*: II glutamine synthase over expression alters ammonia assimilation and the response to phosphinothricin. Plant Cell Rep 14:41–46

Drewes FE, Staden JV (1995) Initiation of and solasodine production in hairy root cultures of *Solanum mauritianum*. Scop Plant Growth Regul 17:27–31

Duckely M, Hohn B (2003) The VirE2 protein of *Agrobacterium tumefaciens*: the Yin and Yang of T-DNA transfer. FEMS Microbiol Lett 223:1–6

Durand-Tardif M, Broglie R, Slightom J, Tepfer D (1985) Structure and expression of Ri T-DNA from *Agrobacterium rhizogenes* in *Nicotiana tabacum*. J Mol Biol 186:557–564

Ephritikhine G, Barbier-Brygoo H, Muller JF, Guern J (1987) Auxin effect on the transmembrane potential difference of wild-type and mutant tobacco protoplasts exhibiting a differential sensitivity to auxin. Plant Physiol 83:801–804

Ercan AG, Taski KM, Turgut K, Yuce S (1999) *Agrobacterium rhizogenes*-mediated hairy root formation in some *Rubia tinctorum* L populations grown in Turkey. Turk J Bot 23:373–378

Estrada-Navarrete G, Alvarado-Affantranger X, Olivares JE, Diaz-Camino C, Santana O, Murillo E, Guillen G, Sanchez-Guevara N, Acosta J, Quinto C, Li DX, Gresshoff PM, Sanchez F (2006) *Agrobacterium rhizogenes* transformation of the Phaseolus spp: a tool for functional genomics. Mol Plant Microb Interact 19:1385–1393

Estramareix C, Ratet P, Boulanger F, Richaud F (1986) Multiple mutations in the transferred regions of the *Agrobacterium rhizogenes* root-inducing plasmids. Plasmid 15:245–247

Estruch JJ, Chriqui D, Grossmann K, Schell J, Spena A (1991) The plant oncogene *rol*C is responsible for the release of cytokinins from glucoside conjugates. EMBO J 10:2889–2895

Faiss M, Strnad M, Redig P, Dolzak K, Hanus J, Van Onckelen H, Schmuelling T (1996) Chemically induced expression of the *rol* cencoded β-glucuronidase in transgenic tobacco plants and analysis of cytokinin metabolism: RolC does not hydrolyze endogenous cytokinin glucosides in planta. Plant J 10:33–46

Filetici P, Spano L, Costantino P (1987) Conserved regions in the T-DNA of different *Agrobacterium rhizogenes* root inducing plasmid. Plant Mol Biol 9:19–26

Filichkin SA, Gelvin SB (1993) Formation of a putative relaxation intermediate during T-DNA processing directed by *Agrobacterium tumefaciens* VirD1/D2 endonuclease. Mol Microbiol 8:915–926

Filippini F, Lo Schiavo F, Terzi M, Costantino P, Trovato M (1994) The plant oncogene *rol*B alters binding of auxin to plant cell membranes. Plant Cell Physiol 35:767–771
Filippini F, Rossi V, Marin O, Trovato M, Costantino P, Downey PM, Lo Schiavo F, Terzi M (1996) A plant oncogene as a phosphatase. Nature 379:499–500
Firoozabady E, Moy Y, Courtney-Gutterson N, Robinson K (1994) Regeneration of transgenic rose (*Rosa hybrida*) plants from embryogenic tissue. Bio/Technology 12:609–613
Fladung M (1990) Transformation of diploid and tetraploid potato clones with the *rol*C gene of *Agrobacterium rhizogenes* and the characterization of transgenic plants. Plant Breeding 104:295–304
Flores HE, Vivanco JM, Loyola-Vargas VM (1999) Radicle biochemistry: the biology of root-specific metabolism. Trends Plant Sci 4:220–226
Forde BG, Day HM, Turton JF, Shen WJ, Cullimore V, Oliver JE (1989) Two glutamine synthase genes from *Phaseolus vulgaris* L. display contrasting developmental and spatial patterns of expression in transgenic *Lotus corniculatus* plants. Plant Cell 1:391–401
Frundt C, Meyer AD, Ichikawa T, Meins FJ (1998) Evidence for the ancient transfer of Ri plasmid T-DNA genes between bacteria and plants. In: Syvanen M, Kado CI (eds) Horizontal gene transfer. Chapman and Hall, London, pp 94–106
Gartland JS (1995) *Agrobacterium* virulence. In: Gartland KMA, Davey MR (eds) Methods in molecular biology 44 *Agrobacterium* protocols. Humana Press, New Jersey
Gaudin V, Jouanin L (1995) Expression of *Agrobacterium rhizogenes* auxin biosynthesis genes in transgenic tobacco plants. Plant Mol Biol 28:123–36
Gaudin V, Vrain T, Jouanin L (1994) Bacterial genes modifying hormonal balances in plants. Plant Physiol Biochem 32:11–29
Gelvin SB (1998) *Agrobacterium* VirE2 proteins can form a complex with T strands in the plant cytoplasm. J Bacteriol 180:4300–4302
Gelvin SB (2003) Improving plant genetic engineering by manipulating the host. Trends Biotechnol 21:95–98
Gelvin SB (2009) *Agrobacterium* in the genomics age. Plant Physiol 150:1665–1676
Gepts P (2002) A Comparison between crop domestication, classical plant breeding, and genetic engineering. Crop Sci 42:1780–1790
Giovannini A, Pecchioni N, Rabaglio M, Allavena A (1997) Characterization of ornamental datura plants transformed by *Agrobacterium rhizogenes*. In Vitro Cell Dev Biol Plant 33:101–106
Giri A, Banerjee S, Ahuja PS, Giri CC (1997) Production of hairy roots in *Aconitum heterophyllum* wall using *Agerobacterium rhizogenes*. In Vitro Cell Dev Biol Plant 33:280–284
Giri A, Giri CC, Dhingra V, Narasu ML (2001) Enhanced podophyllotoxin production from *Agrobacterium rhizogenes* transformed cultures of *Podophyllum hexandrum*. Nat Prod Lett 15:229–235
Giri A, Narasu ML (2000) Research review paper transgenic hairy roots: recent trends and applications. Biotechnol Adv 18:1–22
Giri CC, Giri A (2007) Plant biotechnology. Practical Manual I International Publishing House Pvt. Ltd., New Delhi, pp 69–76
Golds TJ, Lee JY, Husnain T, Ghose TK, Davey MR (1991) *Agrobacterium rhizogenes* mediated transformation of the forage legumes *Medicago sativa* and *Onobrychis viciifolia*. J Exp Bot 42:1147–1157
Gorpenchenko TY, Kiselev KV, Bulgakov VP, Tchernoded GK, Bragina EA, Khodakovskaya MV, Koren OG, Batygina TB, Zhuravlev YN (2006) The *Agrobacterium rhizogenes rol*C-gene induced somatic embryogenesis and shoot organogenesis in *Panax ginseng* transformed calluses. Planta 22:3457–3467
Graham LA, Liou YC, Walker VK, Davies PL (Aug 1997) Hyperactive antifreeze protein from beetles. Nature 388(6644):727–728
Grant JE, Dommisse EM, Conner AJ (1991) Gene transfer to plants using *Agrobacterium*. In: Murray DR (ed) Advanced methods in plant breeding and biotechnology. CAB International, Wallingford, pp 50–73

Gutierrez-Pesce P, Taylor K, Muleo R, Rugini E (1998) Somatic embryogenesis and shoot regeneration from transgenic roots of the cherry root stock colt (*Prunus avium*, *P. pseudocerasus*) mediated by pRi 1855 T-DNA of *Agrobacterium rhizogenes*. Plant Cell Rep 17:574–580

Guyon P, Chilton M-D, Petit A, Tempe J (1980) Agropine in "null-type" crown gall tumors: evidence for generality of the opine concept. Proc Natl Acad Sci 77:2693–2697

Guyon P, Petit A, Tempe J, Dessau Y (1993) Transformed plants producing opines specifically promote growth of opine-degrading agrobacteria. Mol Plant Microb Interact 6:92–98

Hamill JD, Robins RJ, Parr AJ, Evans PM, Furze JD, Rhodes MJC (1990) Over expressing a yeast ornithine decarboxylase gene in transgenic roots of *Nicotiana rustica* can lead to enhanced nicotine accumulation. Plant Mol Biol 15:27–38

Han KH, Keathley DE, Davis JM, Gordon MP (1993) Regeneration of a transgenic woody legume *Robinia pseudoacacia* L, (Black locust) and morphological alterations induced by *Agrobacterium rhizogenes* mediated transformation. Plant Sci 88:149–57

Handa T (1992) Regneration and charaterization of prairie gentian (*Eustoma grandiflorum*) plants transformed by *Agrobacterium rhizogenes*. Plant Tiss Cult Lett 9:10–14.

Hansen G, Larribe M, Vaubert D, Tempe J, Biermann BJ, Montoya AL, Chilton MD, Brevet J (1991) *Agrobacterium rhizogenes* pRi8196 T-DNA: mapping and DNA sequence of functions involved in mannopine synthesis and hairy root differentiation (Ri plasmid). Proc Natl Acad Sci 88:7763–7767

Hansen G, Vaubert D, Heron JN, Clerot D, Tempe J, Brevet J (1993) Phenotypic effects of overexpression of *Agrobacterium rhizogenes* T-DNA ORF13 in transgenic tobacco plants are mediated by diffusible factor(s). Plant J 4:581–585

Hansen G, Das A, Chilton MD (1994a) Constitutive expression of the virulence genes improves the efficiency of plant transformation by *Agrobacterium*. Proc Nat Acad Sci 91:7603–7607

Hansen G, Vaubert D, Clerot D, Tempe J, Brevet J (1994b) A new open reading frame, encoding a putative regulatory protein, in *Agrobacterium rhizogenes* T-DNA. C R Acad Sci III 317:49–53

Hansen G, Vaubert D, Clerot D, Brevet J (1997) Wound-inducible and organ-specific expression of ORF13 from *Agrobacterium rhizogenes* 8196 T-DNA in transgenic tobacco plants. Mol Gen Genet 254(3):337–343.

Hasancebi S, Turgut Kara N, Cakir O, Ari S (2011) Micropropagation and root culture of Turkish endemic *Astragalus chrysochlorus* (Leguminosae). Turk J Bot 35:203–210

Hatamoto H, Boulter ME, Shirsat AH, Croy EJ, Ellis JR (1990) Recovery of morphologically normal transgenic tobacco from hairy roots co-transformed with *Agrobacterium rhizogenes* and a binary vector plasmid. Plant Cell Rep 9:88–92

Hatta M, Beyl CA, Garton S, Diner AM (1996) Induction of roots on jujube softwood cuttings using *Agrobacterium rhizogenes*. J Hortic Sci 71(6):881–886

Hauptmann RM, Ozias-Akins P, Vasil V, Tabaeizadeh Z, Rogers SG, Horsch RB, Vasil IK, Fraley RT (1987) Transient expression of electroporated DNA in monocotyledonous and dicotyledonous species. Plant Cell Rep 6(4):265–270

Henzi MX, Christey MC, McNeil DL, Davies KM (1999) *Agrobacterium rhizogenes*-mediated transformation of broccoli (*Brasica oleracea* L. var *italica*) with an antisense 1-aminocyclopropane-1-carboxylic acid oxidase gene. Plant Sci 143:55–62

Hernalsteens JP, Bytebier B, Van Montagu M (1993) Transgenic asparagus. In: Kung SD, Wu R (eds) Transgenic plants, present status and social and economic impacts, vol 2. San Diego, Academic Press pp 35–46

Hiei Y, Ohta S, Komari T, Kumashiro T (1994) Efficient transformation of rice (*Oryza sativa* L) mediated by *Agrobacterium* and sequence analysis of the boundaries of the T-DNA. Plant J 6:271–282

Hildebrand E (1934) Life history of the hairy-root organism in relation to its pathogenesis on nursery apple trees. J Agric Res 48:857–885

Hirotaka K, Hiroshi K (2003) Gene silencing by expression of hairpin RNA in *Lotus japonicus* roots and root nodules. Mol Plant Microbe Interact 16:663–668

Hodges LD, Cuperus J, Ream W (2004) *Agrobacterium rhizogenes* GALLS protein substitutes for *Agrobacterium tumefaciens* single- stranded DNA-binding protein VirE2. J Bacteriol 186:3065–3007

Hodges LD, Vergunst AC, Neal-McKinney J, den Dulk-Ras A, Moyer DM, Hooykaas PJ, Ream W (2006) *Agrobacterium rhizogenes* GALLS protein contains domains for ATP binding, nuclear localization, and type IV secretion. J Bacteriol 188:8222–8230

Holefors A, Xue ZT, Welander M (1998) Transformation of the apple rootstock M26 with the *rol*A gene and its influence on growth. Plant Sci 136:69–78

Hong S-B, Hwang I, Dessaux Y, Guyon P, Kim K-S, Farrand SK (1997) A T-DNA gene required for agropine biosynthesis by transformed plants is functionally and evolutionarily related to a Ti plasmid gene required for catabolism of agropine by *Agrobacterium* strains. J Bacteriol 179:4831–4840

Hong SB, Peebles CA, Shanks JV, San KY, Gibson SI (2006) Terpenoid indole alkaloid production by Catharanthus roseus hairy roots induced by *Agrobacterium tumefaciens* harboring *rol*ABC genes. Biotechnol Bioeng 93:386–390

Hoshino Y, Mii M (1998) Bialaphos stimulates shoot regeneration from hairy roots of snapdragon (*Antirrhinum majus* L.) transformed by *Agrobacterium rhizogenes*. Plant Cell Rep 17:256–261

Hosokawa K, Matsuki R, Oikawa Y, Yamamura S (1997) Genetic transformation of gentian using wild-type *Agrobacterium rhizogenes*. Plant Cell Tiss Org Cult 51:137–140

Hosoki T, Shiraishi K, Kigo T, Ando M (1989) Transformation and regeneration of ornamental kale (*Brassica oleracea* var. *Acephala* DC) mediated by *Agrobacterium rhizogenes*. Sci Hort 40:259–266

Hu ZB, Du M (2006) Hairy root and its application in plant genetic engineering. J Int Plant Biol 48:121–127

Huffman GA, White FF, Gordon MP, Nester EW (1984) Hairy-root-inducing plasmid: physical map and homology to tumor-inducing plasmids. J Bacteriol 157:269–276

Hwang CF, Bhakta AV, Truesdell GM, Pudlo WM, Williamson VM (2000) Evidence for a role of the N terminus and leucine-rich repeat region of the *Mi* gene product in regulation of localized cell death. Plant Cell 12:1319–1329

Inze D, Follin A, Van Lijsebettens M, Simoens C, Genetello C, Van Montagu M, Schell J (1984) Genetic analysis of the individual T-DNA genes of *Agrobacterium tumefaciens*; further evidence that two genes are involved in indole-3-acetic acid synthesis. Mol Gen Genet 194:265–274

Ishizaki T, Hoshino Y, Masuda K, Oosawa K (2002) Explants of Ri-transformed hairy roots of spinach can develop embryogenic calli in the absence of gibberellic acid, an essential growth regulator for induction of embryogenesis from nontransformed roots. Plant Sci 163:223–231

Isogai A, Fukuchi N, Hayashi M, Kamada H, Harada H, Suzuki A (1988) Structure of a new opine, mikimopine, in hairy root induced by *Agrobacterium rhizogenes*. Agric Bio and Chem 52:3235–3237

Jacobs M, Rubery PH (1988) Naturally occurring auxin transport regulators. Science 241:346–349

James C (2006) Global Status of Commercialized Biotech/GM Crops: 2006. ISAAA Briefs No. 35. ISAAA (International Service for the Acquisition of Agri-Biotech Applications). Ithaca, New York

Jouanin L (1984) Restriction map of an agropine-type Ri plasmid and its homologies to Ti plasmids. Plasmid 12:91–102

Jouanin L, Guerche P, Pamboukdjian N, Tourneur C, Casse Delbart F, Tourneur J (1987a) Structure of T-DNA in plants regenerated from roots transformed by *Agrobacterium rhizogenes* strain A4. Mol Gen Genet 206(3):387–392

Jouanin L, Vilaine F, Tourneur J, Tourneur C, Pautot V, Muller JF, Caboche M (1987b) Transfer of a 4.3-kb fragment of the T_L-DNA of *Agrobacterium rhizogenes* strain A4 confers the pRi-transformed phenotype to regenerated tobacco plants. Plant Sci 53:53–63

Kärkönen A, Koutaniemi S, Mustonen M, Syrjänen K, Brunow G, Kilpeläinen I, Teeri TH, Simola LK (2002) Lignification related enzymes in *Picea abies* suspension cultures. Physiol Plant 114:343–353

Keil M (2002) Fine chemicals from plants. In: Marja K, Caldentey KM, Barz W (eds) Plant biotechnology and transgenic plants. Marcel Dekker Inc, New York, pp 1–20

Keller CP, Van Volkenburgh E (1998) Evidence that auxin-induced growth of tobacco leaf tissues does not involve cell wall acidification. Plant Physiol 118:557–564

Kersters K, De Ley J (1984) Genus III *Agrobacterium* Conn 1942 In Bergey's Manual of Systematic Bacteriology, vol 1. In: Krieg NR, Holt JG (eds) Baltimore: Williams & Wilkins, pp 244–254

Kifle S, Shao M, Jung C, Cai D (1999) An improved transformation protocol for studying gene expression in hairy roots of sugar beet (*Beta vulgaris* L). Plant Cell Rep 18:514–519

Kim YJ, Weathers PJ, Wyslouzil BE (2002) The growth of *Artemisia annua* hairy roots in liquid and gas phase reactors. Biotechnol Bioeng 80:454–464

Kim YK, Hui X, Park WT, Park NI, Young LS, Park SU (2010) Genetic transformation of buckwheat (*Fagopyrum esculentum* M.) with *Agrobacterium rhizogenes* and production of rutin in transformed root cultures. Aust J Crop Sci 4(7):485–490s

Kiselev KV, Dubrovina AS, Veselova MV, Bulgakov VP, Fedoreyev SA, Zhuravlev YN (2007) The *rol*B gene-induced overproduction of resveratrol in *Vitis amurensis* transformed cells. J Biotechnol 128:681–692

Kiyokawa S, Kobayashi K, Kikuchi Y, Kamada H, Harada H (1994) Root-inducing of mikimopine type Ri plasmid pRi1724. Plant Physiol 104:801–802

Kiyokawa S, Kikuchi Y, Kamada H, Harada H (1996) Genetic transformation of *Begonia tuberhybrida* by Ri *rol* genes. Plant Cell Rep 15:606–609

Klee HJ, Horsch RB, Hinchee MA, Hein MB, Hoffmann NL (1987) The effect of over production of two *Agrobacterium tumefaciens* T-DNA auxin biosynthetic gene products in transgenic Petunia plants. Genes Dev 1:86–89

Koltunow AM, Johnson SD, Lynch M, Yoshihara T, Costantino P (2001) Expression of *rol*B in apomictic *Hieracium piloselloides* Vill. causes ectopic meristems in planta and changes in ovule formation, where apomixis initiates at higher frequency. Planta 214:196–205

Krolicka A, Staniszewska II, Bielawski K, Malinski E, Szafranek J, Lojkowska E (2001) Establishment of hairy root cultures of *Ammi majus*. Plant Sci 160:259–264

Kuiper HA, Kleter GA, Noteborn HPJM, Kok EJ (2001) Assessment of the food safety issues related to genetically modified foods. The Plant J 27(6):503–528

Kumar V, Sharma A, Prasad BCN, Gururaj HB, Ravishankar GA (2006) *Agrobacterium rhizogenes* mediated genetic transformation resulting in hairy root formation is enhanced by ultrasonication and acetosyringone treatment. Electron J Biotech 9(4):349–357

Lahners K, Byrne MC, Chilton MD (1984) T-DNA fragments of hairy root plasmid pRi8196 are distantly related to octopine and nopaline Ti plasmid T-DNA. Plasmid 11:130–140

Lambert C, Tepfer D (1992) Use of *Agrobacterium rhizogenes* to create transgenic apple trees having an altered organogenic response to hormones. Theor Appl Genet 85:105–109

Lan XZ, Quan H (2010) Hairy root culture of *Przewalskia tangutica* for enhanced production of pharmaceutical tropane alkaloids. J Med Plants Res 4:1477–1481

Leach F, Aoyagi K (1991) Promoter analysis of the highly expressed *rol*C and *rol*D root-inducing genes of *Agrobacterium rhizogenes*: enhancer and tissue-specific DNA determinants are dissociated. Plant Sci 79:69–76

Lee NG, Stein B, Suzuki H, Verma DPS (1993) Expression of antisense nodulin-35 RNA in *Vigna aconitifolia* transgenic root nodules retards peroxisome development and affects nitrogen availability to the plant. Plant J 3:599–606

Lee S, Blackhall NW, Power JB, Cocking EC, Tepfer D, Davey MR (2001) Genetic and morphological transformation of rice with the *rol*A gene from the Ri T_L-DNA of *Agrobacterium rhizogenes*. Plant Sci 161:917–925

Lemcke K, Schmulling T (1998) Gain of function assays identify non-rol genes from *Agrobacterium rhizogenes* T_L-DNA that alter plant morphogenesis or hormone sensitivity. Plant J 15(3):423–433

Lemcke K, Prinsen E, Van Onckelen H, Schmülling T (2000) The ORF8 gene product of *Agrobacterium rhizogenes* T_L-DNA has tryptophan 2-monooxygenase activity. Mol Plant Microbe Interact 13:787–790

Lessard PA, Kulaveerasingam H, York GM, Strong A, Sinskey AJ (2002) Manipulating gene expression for the metabolic engineering of plants. Metab Eng 4:67–79

Levesque H, Delepelaire P, Rouze P, Slightom J, Tepfer D (1988) Common evolutionary origin of the central portions of the Ri T_L-DNA of *Agrobacterium rhizogenes* and the Ti T-DNAs of *Agrobacterium tumefaciens*. Plant Mol Biol 11:731–744

Li N, Huxtable S, Yang SF, Kung SD (1996) Effects of N-terminal deletions on 1-minocyclopropane-1-carboxylate synthase activity. FEBS Lett 378:286–290

Li D, Zhang Y, Hu X, Shen X, Ma L, Su Z, Wang T, Dong J (2011) Transcriptional profiling of *Medicago truncatula* under salt stress identified a novel CBF transcription factor MtCBF4 that plays an important role in abiotic stress responses. BMC Plant Biology 11:109–138

Limami A, Sun LY, Douat C, Helgeson J, Tepfer D (1998) Natural genetic transformation by *Agrobacterium rhizogenes*. Plant Physiol 118:543–550

Lin HW, Kwok KH, Doran PM (2003) Production of podophyllotoxin using cross-species coculture of *Linum flavum* hairy roots and *Podophyllum hexandrum* cell suspensions. Biotechnol Bioeng 19:1417–1426

Lorence A, Medina-Bolivar F, Nessler CL (2004) Camptothecin and 10-hydroxycamptothecin from *Camptotheca acuminata* hairy roots. Plant Cell Rep 22:437–441

Luczkiewicz M, Kokotkiewicz A (2005) *Genista tinctoria* hairy root cultures for selective production of isoliquiritigenin. Z Naturforsch 60c:867–875

MacRae S, Van Staden J (1993) *Agrobacterium rhizogenes*-mediated transformation to improve rooting ability of eucalypts. Tree Physiol 12:411–418

Manners JM, Way H (1989) Efficient transformation with regeneration of the tropical pasture legume *Stylosanthes humilis* using *Agrobacterium rhizogenes* and a Ti plasmid-binary vector system. Plant Cell Rep 8:341–345

Martin-Tanguy J (2001) Metabolism and function of polyamines in plants: recent development (new approaches). Plant Growth Regul 34:135–148

Martin-Tanguy J, Sun LY, Burtin D, Vernoy R, Rossin N, Tepfer D (1996) Attenuation of the phenotype caused by the root-inducing, left-hand, transferred DNA and its *rol*A gene. Plant Physiol 111:259–267

Matsumoto K, Cabral GB, Teixeira JB, Monte DC (2009) *Agrobacterium*-mediated transient expression system in banana immature fruits. Afr J Biotechnol 8(17):4039–4042

Maurel C, Barbier-Brygoo H, Spena A, Tempe J, Guern J (1991) Single *rol* genes from the *Agrobacterium rhizogenes* T_L-DNA alter some of the cellular responses to auxin in *Nicotiana tabacum*. Plant Physiol 97(1):212–216

Maurel C, Leblanc N, Barbier-Brygoo H, Perrot-Rechenmann C, Bouvier-Durand M, Guern J (1994) Alterations of auxin perception in *rol*B-transformed tobacco protoplasts (time course of *rol*B mRNA expression and increase in auxin sensitivity reveal multiple control by auxin). Plant Physiol 105:1209–1215

Mauro ML, Trovato M, Paolis AD, Gallelli A, Costantino P, Altamura MM (1996) The plant oncogene *rol*D stimulates flowering in transgenic tobacco plants. Dev Biol 180:693–700

Mayo O (1987) The theory of plant breeding, 2nd edn. Clarendon Press, Oxford

McCullen CA, Binns AN (2006) *Agrobacterium tumefaciens* and plant cell interactions and activities required for interkingdom macromolecular transfer. Annu Rev Cell Dev Biol 22:101–127

McInnes E, Morgan AJ, Mulligan BJ, Davey MR (1991) Phenotypic effects of isolated pRiA4 TL-DNA *rol* genes in the presence of intact TR-DNA in transgenic plants of *Solanum dulcamara* L. J Exp Bot 42(10):1279–1286

Medford J, Horgan R, El-Sawi Z, Klee HJ (1989) Alterations of endogenous cytokinins in transgenic plants using a chimeric isopentenyl transferase gene. Plant Cell 1:403–413

Medina-Bolivar F, Condori J, Rimando AM, Hubstenberger J, Shelton K, O'Keefe SF, Bennett S, Dolan MC (2007) Production and secretion of resveratrol in hairy root cultures of peanut. Phytochemistry 68:1992–2003

Mehrotra S, Kukreja AK, Kumar A, Khanuja SPS, Mishra BN (2008) Genetic transformation studies and scale up of hairy root culture of *Glycyrrhiza glabra* in bioreactor. Electron J Biotech 11(2):15

Messner B, Boll M (1993) Elicitor-mediated induction of enzymes of ligninbiosynthesis and formation of lignin-like material in a cell suspension culture of spruce (*Picea abies*). Plant Cell Tiss Org 34:261–269

Meyer A, Tempe J, Costantino P (2000) Hairy root: a molecular overview functional analysis of *Agrobacterium rhizogenes* T-DNA genes. In: Stacey G, Keen N (eds) Plant-microbe interactions, vol 5. APS Press, Minnesota, pp 93–139

Miflin B (2000) Crop improvement in the 21st century. J Exp Biol 51(342):1–8

Mihaljevic S, Stipkovic S, Jelaska S (1996) Increase of root induction in *Pinus nigra* explants using Agrobacteria. Plant Cell Rep 15:610–614

Milly PCD, Dunne KA, Vecchia AV (2005) Global patterns of trends in streamflow and water availability in a changing climate. Nature 438:347–350

Mishra BN, Ranjan R (2008) Growth of hairy-root cultures in various bioreactors for the production of secondary metabolites. Biotechnol Appl Biochem 49:1–10

Morgan AJ, Cox PN, Turner DA, Peel E, Davey MR, Gartland KMA, Mulligan BJ (1987) Transformation of tomato using an Ri plasmid vector. Plant Sci 49:37–49

Moriguchi K, Maeda Y, Satou M, Hardayani NS, Kataoka M, Tanaka N, Yoshida K (2001) The complete nucleotide sequence of a plant root-inducing (Ri) plasmid indicates its chimeric structure and evolutionary relationship between tumor-inducing (Ti) and symbiotic (Sym) plasmids in *Rhizobiaceae*. J Mol Biol 307:771–784

Moritz T, Schmülling T (1998) The gibberellin content of *rol*A transgenic tobacco plants is specifically altered. J Plant Physiol 153:774–776

Moyano E, Fornalé S, Palazón J, Cusidó RM, Bonfill M, Morales C, Piñol MT (1999) Effect of *Agrobacterium rhizogenes* T-DNA on alkaloid production in *Solanaceae* plants. Phytochemistry 52(7):1287–1292

Momčilović I, Grubišić D, Kojić M, Nešković M (1997) Agrobacterium rhizogenes -mediated transformation and plant regeneration of four Gentiana species. Plant Cell Tiss Org Organ Cult 50(1):1–6

Mugnier AJ (1988) Establishment of new axenic hairy root lines by inoculation with *Agrobacterium rhizogenes*. Plant Cell Rep 7:9–12

Mugnier J (1997) Mycorrhizal interactions and the effects of fungicides, nematicides and herbicides on hairy root cultures. In: Doran PM (ed) Hairy roots: culture and applications. Harwood Academic Publishers, Amsterdam, pp 123–132

Murugesan S, Manoharan C, Vijayakumar R, Panneerselvam A (2010) Isolation and characterization of *Agrobacterium rhizogenes* from the root nodules of some leguminous. Intl J Microbiol Res 1(3):92–96

Nader BL, Taketa AT, Pereda-Miranda R, Villarreal ML (2006) Production of triterpenoids in liquid-cultivated hairy roots of *Galphimia glauca*. Planta Med 72:842–844

Nakamura T, Handa T, Oono Y, Kanaya K, Michikawa M, Uchimiya H (1988) Organ-specific mRNA in transgenic tobacco plants possessing T-DNA of Ri plasmids. Plant Sci 56:213–218

Nakano M, Hoshino Y, Mii M (1994) Regeneration of transgenic plants of grape vine (*Vitis vinifera* L.) via *Agrobacterium rhizogenes* mediated transformation of embryogenic calli. J Exp Bot 45(274):649–656

Nandakumar R, Suzanne LC, Rogers MD (2005) *Agrobacterium*-mediated transformation of the wetland monocot *Typha latifolia* L (Broadleaf cattail). Plant Cell Rep 23:744–750

Navarrete GE, Affantranger XAl, Olivares JE, Camino CD, Santana O, Murillo E, Guillen G, Guevara NS, Acosta J, Quinto C, Li D, Gresshoff PM, Sanchez F (2006) *Agrobacterium rhizogenes* transformation of the phaseolus spp.: a tool for functional genomics. Mol Plant Microbe Interact 19(12):1385–1393

Nemoto K, Hara M, Suzuki M, Seki H, Oka A, Muranaka T, Mano Y (2009) Function of the *aux* and *rol* genes of the Ri plasmid in plant cell division in vitro. Plant Signal Behav 4(12):1145–1147

Nenz E, Pupilli F, Paolocci F, Damiani F, Cenci CA, Arcioni S (1996) Plant regeneration and genetic transformation of *Lotus angustissimus*. Plant Cell Tiss Organ Cult 45:145–152

Nilsson O, Crozier A, Schmülling T, Sandberg G, Olsson O (1993a) Indole-3-acetic acid homeostasis in transgenic tobacco plants expressing the *Agrobacterium rhizogenes rol*B gene. Plant J 3:681–689

Nilsson O, Moritz T, Imbault N, Sandberg G, Olsson O (1993b) Hormonal characterization of transgenic tobacco plants expressing the *rol*C gene of *Agrobacterium rhizogenes* T_L-DNA. Plant Physiol 102:363–371

Nilsson O, Little CH, Sandberg G, Olsson O (1996a) Expression of two heterologous promoters, *Agrobacterium rhizogenes rol*C and cauliflower mosaic virus 35S, in the stem of transgenic hybrid aspen plants during the annual cycle of growth and dormancy. Plant Mol Biol 31:887–895

Nilsson O, Moritz T, Sundberg B, Sandberg G, Olsson O (1996b) Expression of the *Agrobacterium rhizogenes rol*C gene in a deciduous forest tree alters growth and development and leads to stem fasciation. Plant Physiol 112:493–502

Nilsson O, Olsson O (1997) Getting to the root: the role of the *Agrobacterium rhizogenes rol* genes in the formation of hairy roots. Physiol Plant 100:463–473

Noda T, Tanaka N, Mano Y, Nabeshima S, Ohkawa H, Matsui C (1987) Regeneration of horseradish hairy roots incited by *Agrobacterium rhizogenes* infection. Plant Cell Rep 6:283–286

Ohara A, Akasaka Y, Daimon H, Mii M (2000) Plant regeneration from hairy roots induced by infection with *Agrobacterium rhizogenes* in *Crotalaria juncea* L. Plant Cell Rep 19:563–568

Oksman-Caldentey KM, KivelaÈ O, Hiltunen R (1991) Spontaneous shoot organogenesis and plant regeneration from hairy root cultures of *Hyoscyamus muticus*. Plant Sci 78:129–136

Olempska-Beer ZWS, Merker RI, Ditto MD, DiNovi MJ (2006) Food-processing enzymes from recombinant microorganisms-a review. Regul Toxicol Pharmacol 45:144–158

Olhoft PM, Bernal LM, Grist LB, Hill DS, Mankin SL, Shen Y, Kalogerakis M, Wiley H, Toren E, Song H-S, Hillebrand H, Jones T (2007) A novel *Agrobacterium rhizogenes*-mediated transformation method of soybean [*Glycine max* (L.) Merrill] using primary-node explants from seedlings. In Vitro Cell Dev Biol Plant 43:536–549

Ondrej M, Biskova R (1986) Differentiation of Petunia *hybrida tissues* transformed by *Agrobacterium rhizogenes* and *Agrobacterium tumefaciens*. Biol Plant 28:152–155

Ooms G, Twell D, Bossen ME, Hoge JHC, Burrell MM (1986) Developmental regulation of RI T_L-DNA gene expression in roots, shoots and tubers of transformed potato (*Solanum tuberosum* cv. Desiree). Plant Mol Biol 6:321–330

Oono Y, Kanaya K, Uchimiya H (1990) Early flowering in transgenic tobacco plants possessing the rolC gene of *Agrobacterium rhizogenes* Ri plasmid. Jpn J Genet 68:7–16

Oono Y, Handa T, Kanaya K, Uchimiya H (1987) The T_L-DNA gene of Ri plasmids responsible for dwarfness of tobacco plants. Jpn J Genet 62:501–505

Otani M, Mu M, Handa T, Kamada H, Shimada T (1993) Transformation of sweet potato (*Ipomoea batatus* (L.) Lam.) plants by *Agrobacterium rhizogenes*. Plant Sci 94:151–159

Otten L, Helfer A (2001) Biological activity of the *rol*B-like 5′ end of the A4-ORF8 gene from the *Agrobacterium rhizogenes* T_L -DNA. Mol Plant Microbe Interact 14:405–411

Ouartsi A, Clerot D, Meyer A, Dessaux Y, Brevet J, Bonfill M (2004) The T-DNA ORF8 of the cucumopine-type *Agrobacterium rhizogenes* Ri plasmid is involved in auxin response in transgenic tobacco. Plant Sci 166:557–567

Ozyigit II (2012) *Agrobacterium tumefaciens* and its use in plant biotechnology. In: Ashraf M, Ozturk M, Ahmad MSA, Aksoy A (eds) Crop production for agricultural improvement. Springer, The Netherlands, pp 317–361

Özcan S, Uranbey S, Sancak C, Parmaksiz İ, Gürel E, Babaoğlu M (2004) *Agrobacterium* aracılığıyla gen aktarımı. In: Özcan S, Gürel E, Babaoğlu M (eds) Bitki Biyoteknolojisi II (Plant biotechnology, II), Genetik Mühendisliği ve Uygulamaları (Genetic engineering and its applications), Cilt II, 2nd edn. SÜ Vakfi Yayınlari, Turkey, pp 112–159

Pal A., Swain SS, Mukherjee AK, Chand PK (2012) *Agrobacterium* pRi T_L-DNA *rol*B and T_R-DNA opine genes transferred to the spiny Amaranth (*Amaranthus spinosus* L.)—A nutraceutical crop. Food Technol Biotech (In press)

Palazon J, Cusido RM, Roig C, Pinol MT (1998) Expression of the *rol* gene and nicotine production in transgenic hairy roots and their regenerated plants. Plant Cell Rep 17:384–90

Park Nl, JK Kim, WT Park, JW Cho, YP Lim, SU Park (2011) An efficient protocol for genetic transformation of watercress (*Nasturtium officinale*) using *Agrobacterium rhizogenes*. Mol Biol Rep 38:4947–4953

Pavingerova D, Ondrej M (1986) Comparison of hairy root and crown gall tumors of *Arabidopsis thaliana*. Biol Plant 28:149–151

Pellegrineschi A, Davolio-Mariani O (1996) *Agrobacterium rhizogenes*-mediated transformation of scented geranium. Plant Cell Tiss Organ Cult 47:79–86

Pellegrineschi A, Damon JP, Valtorta N, Paillard N, Tepfer D (1994) Improvement of ornamental characters and fragrance production in lemon-scented geranium through genetic transformation by *Agrobacterium rhizogenes*. Nat Biotechnol 12:64–68

Pérez-Molphe-Balch E, Ochoa-Alejo N (1998) Regeneration of transgenic plants of Mexican lime from *Agrobacterium rhizogenes*-transformed tissues. Plant Cell Rep 17:591–596

Petersen SG, Stummann BM, Olesen P, Henningsen KW (1989) Structure and function of root-inducing (Ri) plasmids and their relation to tumor-inducing (Ti) plasmids. Physiol Plantarum 77:427–435

Petit A, David C, Dahl G, Ellis JG, Guyon P, Casse-Delbart FC, Tempe J (1983) Further extension of the opine concept: plasmids in *Agrobacterium rhizogenes* cooperate for opine degradation. Mol Gen Genet 19:204–214

Phelep M, Petit A, Martin L, Duhoux E, Tempe J (1991) Transformation and regeneration of a nitrogen-fixing tree, *Allocasuarina verticillata* Lam. Biotechnol 9:461–466

Popa G, Cornea C P, Brezeanu A (2006) Influence of different *Agrobacterium rhizogenes* strains on hairy roots induction in *Eustoma grandiflorum*. Roum Biotechnol Lett 11(1):2587–2592

Porter J (1991) Host range and implications of plant infection by *Agrobacterium rhizogenes*. Crc Cr Rev Plant Sci 10:387–421

Pradel H, Dumke-Lehmann U, Dietrich B, Luckner M (1997) Hairy root cultures of *Digitalis lanata*. Secondary metabolism and plant regeneration. J Plant Physiol 151:209–215

Prinsen E, Bytebier B, Hernalsteens JP, De Greef J, Van Onckelen H (1990) Functional expression of *Agrobacterium tumefaciens* T-DNA onc-genes in *Asparagus* crown gall tissues. Plant Cell Physiol 31:69–75

Prinsen E, Chriqui D, Vilaine F, Tepfer M, Van Onckelen H (1994) Endogenous phytohormones in tobacco plants transformed with *Agrobacterium rhizogenes* pRi T_L-DNA genes. Plant Physiol 144:80–85

Putalun W, Udomsin O, Yusakul G, Juengwatanatrakul T, Sakamoto S, Tanaka H (2010) Enhanced plumbagin production from in vitro cultures of *Drosera burmanii* using elicitation. Biotechnol Lett 32:721–724

Quandt HJ, Pühler A, Broer I (1993) Transgenic root nodules of *Vicia hirsuta* a fast and efficient system for the study of gene expression in indeterminate-type nodules. Mol Plant Microbe Interact 6:699–706

Ramsay G, Kumar A (1990) Transformation of *Vicia faba* cotyledon and stem tissues *Agrobacterium rhizogenes*: infectivity and cytological studies. J Exp Bot 41:841–847

Rao AQ, Bakhsh A, Kiani S, Shahzad K, Shahid AA, Husnain T, Riazuddin S (2009) The myth of plant transformation. Biotechnol Adv 27:753–763

Rao SR, Ravishankar G (2002) Plant cell cultures: chemical factories of secondary metabolites. Biotechnol Adv 20:101–153

Ream W (2002) *Agrobacterium* genetics. In: Streips UN, Yasbin RE (eds) Modern Microbial Genetics, 2nd edn. Wille-Liss Inc., New York, pp 323–348

Rech EL, Golds TJ, Husnain T, Vainstein MH, Jones B, Hammat N, Mulligan BJ, Davey MR (1989) Expression of a chimaeric kanamycin resistance gene introduced into the wild soybean *Glycine canescens* using a cointegrate Ri plasmid vector. Plant Cell Rep 8:33–36

Remeeus PM, van Bezooijen J, Wijbrandi J, van Bezooijen J (1998) In vitro testing is a reliable way to screen the temperature sensitivity of resistant tomatoes against *Meloidogyne incognita*. In: Proceedings of 5th international symposium on crop protection, Universiteit Gent Belgium, vol 63, pp 635–640

Riker AJ, Banfield WM, Wright WH, Keitt GW (1930) Studies on infectious hairy root of nursery apple trees. J Agric Res 41:507–540
Rinallo C, Mittempergher L, Frugis G, Mariotti D (1999) Clonal propagation in the genus Ulmus: improvement of rooting ability by *Agrobacterium rhizogenes* T-DNA genes. J Hortic Sci Biotechnol 74:502–506
Rossi L, Hohn B, Tinland B (1996) Integration of complete transferred DNA units is dependent on the activity of virulence E2 protein of *Agrobacterium tumefaciens*. Proc Natl Acad Sci USA 93:126–130
Rugh CL (2001) Mercury detoxification with transgenic plants and other biotechnological breakthroughs for phytoremediation. In Vitro Cell Dev Biol Plant 37:321–325
Rugini E, Mariotti D (1991) *Agrobacterium rhizogenes* T-DNA genes and rooting in woody species. Acta Hort 300:301–307
Rugini E, Pellegrineschi A, Mencuccini M, Mariotti D (1991) Increase of rooting ability in the woody species kiwi (*Actinidia deliciosa* A. Chev.) by transformation with *Agrobacterium rhizogenes rol* genes. Plant Cell Rep 10:291–295
Rugini E, Muganu M, Gutiérrez-Pesce P E, Lolletti D (1996) Comportamento vegeto-produttivo di alcune specie fruttifere transgeniche per il T-DNA e geni *rol* di *Agrobacterium rhizogenes*. Convegno SIGA, Workshop Organismi geneticamente modificati e resistenze genetiche, Bologna, pp 55–57
Ryder MH, Tate ME, Kerr A (1985) Virulence properties of strains of *Agrobacterium* on the apical and basal surfaces of carrot root discs. Plant Physiol 77:215–221
Saha P, Chakraborti D, Sarkar A, Dutta I, Basu D, Das S (2007) Characterization of vascularspecific RSs1 and *rol*C promoters for their utilization in engineering plants to develop resistance against hemipteran insect pests. Planta 226:429–442
Saito K, Yamazaki M, Anzai H, Yoneyama K, Murakoshi I (1992) Transgenic herbicide-resistant *Atropa belladonna* using an Ri plasmid vector and inheritance of the transgenic trait. Plant Cell Rep 11:219–224
Satheeshkumar K, Jose B, Sonia EV, Seeni S (2009) Isolation of morphovariants through plant regeneration in *A. rhizogenes* induced hairy root cultures of *Plumbago rosea* L. Indian J Biotechnol 8:435–441
Savka MA, Ravillion B, Noel GR, Farrand SK (1990) Induction of hairy roots on cultivated soybean genotypes and their use to propagate the soybean cyst nematode. Phytopathology 80(5):503–508.
Schmulling T, Schell J, Spena A (1988) Single genes from *Agrobacterium rhizogenes* influence plant development. EMBO J 7:2621–2629
Schmülling T, Fladung M, Grossmann K, Schell J (1993) Hormonal content and sensitivity of transgenic tobacco and potato plants expressing single *rol* genes of *Agrobacterium rhizogenes* T-DNA. Plant J 3:371–382
Schröder G, Waffenschmidt S, Weiler E, Schröder J (1984) The region of Ti plasmid codes for an enzyme synthesizing indole-3-acetic acid. Eur J Biochem 138:387–391
Scorza R, Zimmerman TW, Cordts JM, Footen KJ (1994) Horticultural characteristics of transgenic tobacco expressing the *rol*C gene from *Agrobacterium*. J Amer Soc Hort Sci 119(5):1091–1098
Sentoku N, Sato Y, Matsuoka M (2000) Overexpression of rice OSH genes induces ectopic shoots on leaf sheaths of transgenic rice plants. Dev Biol 220:358–364
Sevon N, Oksman-Caldentey KM (2002) *Agrobacterium rhizogenes* mediated transformation: root cultures as a source of alkaloids. Planta Med 68:859–868
Shahin EA, Sukhapinda K, Simpson RB, Spivey R (1986) Transformation of cultivated tomato by a binary vector in *Agrobacterium rhizogenesis*: transgenic plants with normal phenotypes harbor binary vector T-DNA but no Ri-plasmid T-DNA. Theor Appl Genet 72:770–777
Shen WH, Petit A, Guern J, Tempe J (1988) Hairy roots are more sensitive to auxin than normal roots. Proc Natl Acad Sci 85:3417–3421
Shen WH, Davioud E, David C, Barbier-Brygoo H, Tempe J, Guern J (1990) High sensitivity to auxin is a common feature of hairy root. Plant Physiol 94:554–560

Shin DI, Podila GK, Huang Y, Karnosky DF (1994) Transgenic larch expressing genes for herbicide and insect resistance. Can J For Res 4:2059–2067

Shinde AN, Malpathak N, Fulzele PD (2009) Enhanced production of phytoestrogenic isoflavones from hairy root cultures of *Psoralea corylifolia* L. using elicitation and precursor feding. Biotechnol Bioprocess E 14:288–294

Shkryl YN, Veremeichik GN, Bulgakov VP, Tchernoded GK, Mischenko NP, Fedoreyev SA, Zhuravlev YN (2008) Individual and combined effects of the *rol*A, B and C genes on anthraquinone production in Rubia cordifolia transformed calli. Biotechnol Bioeng 100(1):118–125

Shoja HM (2010) Contribution to the study of the *Agrobacterium rhizogenes* plast genes, *rol*B and *rol*C, and their homologs in *Nicotiana tabacum*. Universite de Strasbourg, France

Sinha N, Williams R, Hake S (1993) Overexpression of the maize homeobox gene, KNOTTED-1, causes a switch from determinate to indeterminate cell fates. Genes Dev 7:787–795

Sinkar V, Pythoud F, White F, Nester E, Gordon M (1988) *rol*A locus of the Ri plasmid directs developmental abnormalities in transgenic plants. Genes Dev 2:688–697

Slightom JL, Durand-Tardif M, Jouanin L, Tepfer D (1986) Nucleotide sequence analysis of TL-DNA of *Agrobacterium rhizogenes* agropine type plasmid. J Biol Chem 261:108–121

Smigocki AC, Hammerschlag FA (1991) Regeneration of plants from peach embryo cells infected with a shooty mutant strain of *Agrobacterium*. J Amer Soc Hort Sci 116:1092–1097

Smulders M JM, Croes AF, Kemp A, Hese KM, Harren F, Wullems GJ (1991) Inhibition by ethylene of auxin promotion of flower bud formation in tobacco explants is absent in plants transformed by *Agrobacterium rhizogenes*. Plant Physiol 96:1131–1135

Spano L, Costantino P (1982) Regeneration of plants from callus cultures of roots induced by *Agrobacterium rhizogenes* on tobacco. Z Pflanzenphysiol 106:87–92

Spano L, Mariotti D, Cardarelli M, Branca C, Costantino P (1988) Morphogenesis and auxin sensitivity of transgenic tobacco with different complements of Ri T-DNA. Plant Physiol 87(2):479–483

Specq A, Hansen G, Vaubert D, Clerot D, Heron JN, Tempe J, Brevet J (1994) Studies on hairy root T-DNA: regulation and properties of ORF13 from *Agrobacterium rhizogenes* 8196. In Plant Pathogenic Bacteria, Versailles, pp 465–468

Spena A, Schmulling T, Konct C, Schell JS (1987) Independent and synergistic activity of *rol*A, B and C loci in stimulating abnormal growth in plants. EMBO J 206(13):3891–3899

Spiral J, Thierry C, Paillard M, Petiard V. (1993) Obtention de plantules de *Coffea canephora* Pierre (Robusta) transformées par *Agrobacterium rhizogenes*. C R Acad Sci Hebd Seances Acad Sci 316:1–6

Sretenovic-Rajicic T, Ninkovi S, Miljus-Dukic J, Vinterhalter B, Vinterhalter D (2006) *Agrobacterium rhizogenes*-mediated transformation of *Brassica oleracea* var. *sabauda* and *B. oleracea* var. *capitata*. Biol Plant 50:525–530

Stieger PA, Meyer AD, Kathmann P, Fründt C, Niederhauser I, Barone M, Kuhlemeier C (2004) The orf13 T-DNA Gene of *Agrobacterium rhizogenes* confers meristematic competence to differentiated cells. Plant Physiol 135(3):1798–1808

Stiller J, Nasinec V, Svoboda S, Nemcova B, Machackova T (1992) Effects of agrobacterial oncogenes in kidney vetch (*Anthyllis vulneraria* L.). Plant Cell Rep 11:363–367

Stiller J, Martirani L, Tuppale S, Chian RJ, Chiurazzi M, Gresshoff PM (1997) High frequency transformation and regeneration of transgenic plants in the model legume *Lotus japonicus*. J Exp Bot 48:1357–1365

Stummer BE, Smith SE, Langridge P (1995) Genetic transformation of *Verticordia grandis* (Myrtaceae) using wild-type *Agrobacterium rhizogenes* and binary *Agrobacterium* vectors. Plant Sci 111:51–62

Sudha CG, Obul Reddy B, Ravishankar GA, Seeni S (2003) Production of ajmalicine and ajmaline in hairy root cultures of *Rauvolfia micrantha* Hookf, a rare and endemic medicinal plant. Biotechnol Lett 25:631–636

Suginuma C, Akihama T (1995) Transformation of gentian with *Agrobacterium rhizogenes*. Acta Hort 392:153–160

Sun, L-Y, Monneuse M-O, Martin-Tanguy J, Tepfer D (1991) Changes in flowering and accumulation of polyamines and hydroxycinnamic acid-polyamine conjugates in tobacco plants transformed by theA locus from the Ri T_L-DNA of *Agrobacterium rhizogenes*. Plant Sci 80:145–146

Suzuki M (1989) SPXX, a frequent sequence motif in gene regulatory proteins. J Mol Biol 207:61–84

Suzuki H, Fowler T, Tierney M (1993) Deletion analysis and localization of SbPRP1, a soybean cell wall protein gene, in roots of transgenic tobacco and cowpea. Plant Mol Biol 21:109–119

Suzuki K, Tanaka N, Kamada H, Yamashita I (2001) Mikimopine synthase (mis) gene on pRi1724. Gene 263:49–58

Swain SS, Rout KK, Chand PK (2012) Production of Triterpenoid Anti-cancer Compound Taraxerol in *Agrobacterium*-Transformed Root Cultures of Butterfly Pea (*Clitoria ternatea* L.). Appl Biochem Biotechnol 168(3):487–503

Talano MA, Agostini E, Medina MI, Reinoso H, Tordable Mdel C, Tigier HA, de Forchetti SM (2006) Changes in lignosuberization of cell walls of tomato hairy roots produced by salt treatment with the release of a basic peroxidase. J Plant Physiol 163:740–749

Tanaka N, Takao M, Matsumoto T (1994) *Agrobacterium rhizogenes* mediated transformation and regeneration of *Vinca minor* L. Plant Tiss Cult Lett 11:191–198

Tanaka N, Fujikawa Y, Aly MAM, Saneoka H, Fujita K, Yamashita I (2001) Proliferation and *rol* gene expression in hairy root lines of Egyptian clover (*Trifolium alexandrinum* L.). Plant Cell Tiss Org 66:175–182

Tao R, Handa T, Tamura M, Sugiura A (1994) Genetic transformation of Japanese persimmon (*Diospyros kaki* L.) by *Agrobacterium rhizogenes* wild type strain A4. J Jap Soc Hort Sci 63:283–289

Taylor BH, Amasino RM, White EW, Gordon MP (1985) T-DNA analysis of plants regenerated from hairy root tumors. Mol Gen Genet 201: 554–557

Taylor NJ, Fauquet CM (2002) Microparticle bombardment as a tool in plant science and agricultural biotechnology. DNA and Cell Biol 21(12):963–977

Tempe J, Petit A, Farrand SK (1984) Induction of cell proliferation by *Agrobacterium tumafaciens* and *A. rhizogenes*: a parasite's point of view D.P.S Verma and T. Hohn. Genes involved in microbe-plant interactions. Springer-Verlag, New York, pp 271–286

Tepfer D (1983) The potential uses of *Agrobacterium rhizogenes* in the genetic engineering of higher plants: nature got there first. In: Lurquin P, Kleinhofs A (eds) Genetic engineering in eukaryotes. Plenum, New York, pp 153–164

Tepfer D (1984) Transformation of several species of higher plants by *Agrobacterium rhizogenes*: sexual transmission of the transformed genotype and phenotype. Cell 37:959–967

Tepfer D (1990) Genetic transformation using *Agrobacterium rhizogenes*. Physiol Plant 79:140–146

Thimmaraju R, Venkatachalam L, Bhagyalakshmi N (2008) Morphometric and biochemical characterization of red beet (*Beta vulgaris* L.) hairy roots obtained after single and double transformations. Plant Cell Rep 27:1039–1052

Thomas MR, Rose RJ, Nolan KE (1992) Genetic transformation of *Medicago truncatula* using *Agrobacterium* with genetically modified Ri and disarmed Ti plasmids. Plant Cell Rep 11:113–117

Thomashow LS, Reeves S, Thomashow MF (1984) Crown gall oncogenesis: evidence that a T-DNA gene from the *Agrobacterium* Ti plasmid pTiA6 encodes an enzyme that catalyzes synthesis of indoleacetic acid. Proc Natl Acad Sci 81:5071–5075

Tomilov A, Tomilova N, Yoder JI (2007) *Agrobacterium tumefaciens* and *Agrobacterium rhizogenes* transformed roots of the parasitic plant *Triphysaria versicolor* retain parasitic competence. Planta 225(5):1059–1071

Trovato M, Mauro ML, Costantino P, Altamura MM (1997) The *rol*D gene from *Agrobacterium rhizogenes* is developmentally regulated in transgenic tobacco. Protoplasma 197:111–120

Trovato M, Maras B, Linhares F, Constantino P (2001) The plant oncogene *rol*D encodes a functional ornithine cyclodeaminase. Proc Natl Acad Sci 98:13449–13453

Trulson AJ, Simpson RB, Shahin EA (1986) Transformation of cucumber (*Cucumis sativus* L.) plants with *Agrobacterium rhizogenes*. Theor Appl Gene 73:11–15

Turgut Kara N, Ari S (2008) In vitro plant regeneration from embryogenic cell suspension culture of *Astragalus chrysochlorus* (*Leguminoseae*). Afr J Biotechnol 7(9):1250–1255

Turgut Kara N, Ari S (2010) The optimization of voltage parameter for tissue electroporation in somatic embryos of *Astragalus chrysochlorus* (Leguminosae). Afr J Biotechnol 9(29):4584–4588

Tzfira T, Citovsky V (2000) From host recognition to T-DNA integration: the function of bacterial and plant genes in the *Agrobacterium*-plant cell interaction. Mol Plant Pathol 11(4):201–212

Tzfira T, Yarnitzky O, Vainstein A, Altman A (1996). *Agrobacterium rhizogenes*-mediated DNA transfer in *Pinus halepensis* Mill. Plant Cell Rep 16:26–31

Umber M, Clement B, Otten L (2005) The T-DNA oncogene A4- orf8 from *Agrobacterium rhizogenes* strain A4 induces abnormal growth in tobacco. Mol Plant Microbe Interact 18:205–211

Uozumi N, Kobayashi T (1997) Artificial seed production through hairy root regeneration. In: Doran PM (ed) Hairy roots: culture and applications. Harwood Academic Publishers, Amsterdam, pp 113–122

Uozumi N, Ohtake Y, Nakashimada Y, Morikawa Y, Tanaka N, Kobayashi T (1996) Efficient regeneration from GUS-transformed Ajuga hairy root. J Ferm Bioeng 81:374–378

Vain P (2007) Thirty years of plant transformation technology development. Plant Biotechnol J 5:221–229

van Altvorst AC, Bino RJ, van Dijk AJ, Lamers AMJ, Lindhout WH, Van Der Mark F, Dons JJM (1992) Effects of the introduction of *Agrobacterium rhizogenes rol* genes on tomato plant and flower development. Plant Sci 83:77–85

van de Velde W, Mergeay J, Holsters M, Goormachtig S (2003) *Agrobacterium rhizogenes*-mediated transformation of *Sesbania rostrata*. Plant Sci 165:1281–1288

van der Salm TPM, Van Der Toorn CJG, Bouwer R, Haenisch ten Cate CH, Dons HJM (1997) Production of *rol* gene transformed plants of *Rosa hybrida* L. and characterization of their rooting ability. Mol Breed 3:39–47

van Onckelen H, Prinsen E, Inze D, Rudelsheim P, van Lijsebettens M, Follin A, Schell J, van Montagu M, De Greef J (1986) *Agrobacterium* T-DNA gene codes for tryptophan 2-monooxygenase activity in tobacco crown gall cells. FEBS Lett 198:357–360

Vansuyt G, Vilaine F, Tepfer M, Rossingnol M (1992) *rol*A modulates the sensitivity to auxin of the protontranslocationcatalyzed by the plasmamembrane H^+-ATPase in transformed tobacco. FEBS Lett 298:89–92

Veena V, Taylor CG (2007) *Agrobacterium rhizogenes*: recent developments and promising applications. In Vitro Cell Dev Biol Plant 43:383–403

Vilaine F, Casse-Delbart F (1987) A new vector derived from Agrobacterium rhizogenes plasmids: a micro-Ri plasmid and its use to construct a mini-Ri plasmid. Gene 55(1):105–14

Vilaine F, Charbonnier C, Casse-Delbart F (1987) Further insight concerning the T_L-region of the Ri plasmid of *Agrobacterium rhizogenes* strain A4: transfer of a 1.9 kb fragment is sufficient to induce transformed roots on tobacco leaf fragments. Mol Gen Genet 210:111–115

Vinterhalter B, Orbović V, Vinterhalter D (1999) Transgenic root cultures of *Gentiana punctata* L. Acta Soc Bot Pol 4:275–280

Visser RGF, Hesseling-Meinders A, Jacobsen E, Nijdam H, Witholt B, Feenstra WJ (1989) Expression and inheritance of inserted markers in binary vectors carrying *Agrobacterium rhizogenes* transformed potato (*Solanum tuberosum* L.). Theor Appl Genet 78:705–14

Walton NJ, Belshaw NJ (1988) The effect of cadaverine on the formation of anabasine from lysine in hairy root cultures of *Nicotiana hesperis*. Plant Cell Rep 7:115–118

Wang CY, Chiao MT, Yen PJ, Huang WC, Hou CC, Chien SC, Yeh KC, Yang WC, Shyur LF, Yang NS (2006) Modulatory effects of *Echinacea purpurea* extracts on human dendritic cells: a cell- and gene-based study. Genomics 88:801–808

Ward DV, Zambryski P (2001) The six functions of *Agrobacterium* VirE2. Proc Natl Acad Sci USA 98:385–386

Weising K, Kahl G (1996) Natural genetic engineering of plant cells: the molecular biology of crown gall and hairy root disease. World J Microbiol Biotechnol 2:327–351

Weller SA, Stead DE (2002) Detection of root mat associated *Agrobacterium* strains from plant material and other sample types by post-enrichment TaqMan PCR. J Appl Microbiol 92:118–126

Weller SA, Stead DE, Young JPW (2005) Induction of root-mat symptoms on cucumber plants by *Rhizobium*, but not by *Ochrobactrum* or *Sinorhizobium*, harbouring a cucumopine Ri plasmid. Plant Pathol 54:799–805

White LO (1972) The taxonomy of the crown gall organism *Agrobacterium tumefaciens* and its relationship to *Rhizobia* and to other *Agrobacterium*. J Gen Microbiol 72:565–574

White FF, Ghidossi G, Gordon MP, Nester EW (1982) Tumor induction by *Agrobacterium rhizogenes involves* the transfer of plasmid DNA to the plant genome. Proc Natl Acad Sci USA 79:3193–3197

White FF, Taylo BH, Huffman GA, Gordon MP, Nesterr EW (1985) Molecular and genetic analysis of the transferred DNA regions of the root-inducing plasmid of *Agrobacterium rhizogenes*. J Bacteriol 164(1):33–44

Willems A, Collins MD (1993) Phylogenetic analysis of *Rhizobia* and *Agrobacteria* based on 16S rRNA gene sequences. Intl J Syst Bacteriol 43:305–313

Willmitzer L, Sanchez-Serrano J, Buschfeld E, Schell J (1982) DNA from *Agrobacterium rhizogenes* is transferred to and expressed in axenic hairy root plant tissues. Mol Gen Genet 186:16–22

Willmitzer L, Dhaese P, Schreier PH, Schmalenbach W, Van Montagu M, Schell J (1983) Size, location and polarity of T-DNA-encoded transcripts in nopaline crown gall tumors, common transcripts in octopine and nopaline tumors. Cell 32(4):1045–1056

Woese CR, Gupta R, Hahn CM, Zillig W, Tu J (1984) The phylogenetic relationships of three sulfur-dependent *archaebacteria*. Syst Appl Microbiol 5:97–105

Yadav NS, Van Der Leyden J, Bennett DR, Barnes WM, Chilton M-D (1982) Short direct repeats flank the T-DNA on a nopaline Ti plasmid. Proc Natl Acad Sci 79:6322–6326

Yamada T, Palm CJ, Brooks B, Kosuge T (1985) Nucleotide sequences of the *Pseudomonas savastanoi* indole acetic acid gene show homology with *Agrobacterium tumefaciens* T-DNA. Proc Natl Acad Sci 82:6522–6526

Yamazaki M, Son L, Hayashi T, Morita N, Asamizu T, Mourakoshi I, Saito K (1996) Transgenic fertile *Scoparia dulcis* L., a folk medicinal plant, conferred with a herbicide-resistant trait using an Ri binary vector. Plant Cell Rep 15:317–321

Yang DC, Choi YE (2000) Production of transgenic plants via *Agrobacterium rhizogenes*-mediated transformation of *Panax ginseng*. Plant Cell Rep 19(5):491–496

Yasuda H, Tada Y, Hayashi Y, Jomori T, Takaiwa F (2005) Expression of the small peptide GLP-1 in transgenic plants. Transgenic Res 14(5):677–684

Yibrah HS, Grönroos R, Lindroth A, Franzén H, Clapham D, von Arnold S (1996) *Agrobacterium rhizogenes*-mediated induction of adventitious rooting from *Pinus contorta* hypocotyls and the effect of 5-azacytidine on transgene activity. Transgenic Res 5:75–85

Yokoyama R, Hirose T, Fujii N, Aspuria ET, Kato A, Uchimiya H (1994) The *rol*C promoter of *Agrobacterium rhizogenes* Ri plasmid is activated by sucrose in transgenic tobacco plants. Mol Gen Genet 244:15–22

Yoshimatsu K, Shimomura K (1992) Transformation of opium poppy (*Papaver somniferum* L.) with *Agrobacterium rhizogenes* MAFF 03-01724. Plant Cell Rep 11:132–136

Yusibov VM, Steck TR, Gupta V, Gelvin SB (1994) Association of single-stranded transferred DNA from *Agrobacterium tumefaciens* with tobacco cells. Proc Natl Acad Sci USA 91:2994–2998

Zambryski P, Joos H, Genetello C, Leemans J, Van Montagu M, Schell J (1983) Ti-plasmid vector for the introduction of DNA into plant cells without alteration of their normal regeneration capacity. EMBO J 2:2143–2150

Zdravkovic-Korac S, Muhovski Y, Druart PH, Calic D, Radojevic LJ (2004) *Agrobacterium rhizogenes*-mediated DNA transfer to *Aesculus hippocastanum* L. and the regeneration of transformed plants. Plant Cell Rep 22:698–704

Zhan XC, Jones DA, Kerr A (1988) Regeneration of flax plants transformed by *Agrobacterium rhizogenes*. Plant Mol Biol 11:551–559

Zhu JP, Oger M, Schrammeijer B, Hooykaas PJJ, Farrand SK, Winans SC (2000) The bases of crown gall tumorigenesis. J Bacteriol 182:3885–3895

Ziemienowicz A, Merkle T, Schoumacher F, Hohn B, Rossi L (2001) Import of *Agrobacterium* T-DNA into plant nuclei: two distinct functions of VirD2 and VirE2 proteins. Plant Cell 13:369–383

Zuker A, Tzfira T, Scovel G, Ovadis M, Shklarman E, Itzhaki H (2001) *rol*C-transgenic carnation with improved agronomic traits: quantitative and qualitative analyses of greenhouse-grown plants. J Am Soc Hortic Sci 126:13–18

Zupan JR, Zambryski P, Citovsky V (1996) *Agrobacterium* VirE2 protein mediates nuclear uptake of single-stranded DNA in plant cells. Proc Natl Acad Sci 93:2392–2397

Chapter 2
Bioinformatic Tools in Crop Improvement

L. F. De Filippis

Abstract Bioinformatic resources and web databases are essential for the most effective use of genetic, proteomic, metabolomic and phenome information important in increasing agricultural crop productivity. Innovations in web based platforms for omics based research, and application of such information has provided the necessary platform to promote molecular based research in model plants, as well as important crop plants. Combinations of multiple omics web based sites and integration of outcomes is now an important strategy to identify molecular systems promoting comparative genomics, the biological properties in many species, and to accelerate gene discovery and functional analyses. The review details recent advances in plant omics data acquisition sites, together with relevant databases and advance molecular technology under clear biological categories. The information is set out under the molecular biology divisions of; DNA based resources and sequencing, RNA and variation analysis, proteomics, structural proteins, and post-translation modifications, metabolomics, phenome and plant comparative analyses. Tables of relevant web sites are presented under similar headings for convenience, and the application of bioinformation data is reviewed in light of the possible use of these resources for crop improvement. Finally, a long list of future perspectives and research still to be attempted is detailed, which in the fullness of time should enable the full potential of bioinformatics and use in crop improvement programs to be achieved.

Introduction

Sustainable agricultural production and food security are two important issues of concern in response to population increase, environmental degradation and climate change (Brown and Funk 2008; Turner et al. 2009). According to the United Nations, the world population increases by 70–75 million people annually, an aver-

L. F. De Filippis (✉)
Centre for Environmental Sustainability (CENS), Department of Environmental Sciences,
University of Technology, Broadway/Sydney NSW 2007,
P O Box 123, Sydney, Australia
e-mail: lou.defilippis@uts.edu.au

K.R. Hakeem et al. (eds.), *Crop Improvement*, DOI 10.1007/978-1-4614-7028-1_2, age of more than two persons every second; and over 95% of these will live in developing countries (De Filippis 2012). It will be difficult satisfying the needs of this growing population and avoid serious food shortages or even famine from the limited arable land and natural resources available. These factors combined have already resulted in food deficiency and malnutrition, which have become serious health problems. Additionally, recent increased demand for biofuel crops has created a new market for agricultural commodities, causing even more stress on food security (Ozturk et al. 2006; Ozturk 2010; Hakeem et al. 2012). In order to try to resolve these problems and increase crop yields, breeding plants based on a better molecular understanding of gene function, and on the regulatory mechanisms involved in crop production (Pinstrup-Andersen and Cohen 2000; Takeda and Matsuoka 2008) appears to be necessary. Plant molecular biology continues to progress, and important gene sequences and their function have been described; many of which are related to crop yields (production), crop quality (protein and carbohydrate), and tolerance to biotic and abiotic stresses (De Filippis 2012). There are legal, social and political barriers to the full potential use of crop biotechnology and transgenic plants, nevertheless advances in these fields have lead to improvements in agriculture and human life. One vital tool of biotechnology is 'bioinformatics', which is commonly used to genetically type and identify genotypic and phenotypic changes in plants, and this information is important for improvement in performance of crop plants (Ahmad et al. 2011).

The complete genome sequence of the mustard plant *Arabidopsis thaliana* has been available to scientists since 2000 (International Arabidopsis Genome Initiative 2000; Somerville and Dangl 2000). Similarly, the rice (*Oryza sativun* cv. *japonica*) complete genome sequence has been documented since 2005 (International Rice Genome Sequencing Project 2005; Itoh et al. 2007; Hakeem et al. 2012). The rice genome sequencing project in particular with its molecular methods and DNA markers on chromosomes, introduced important developments in mapping populations and chromosome marker resources, which accelerated the isolation of agronomically important quantitative trait loci (QTLs) in crop breeding programs (Ashikari et al. 2005; Konishi et al. 2006; Ma et al. 2006; Kurakawa et al. 2007; Ma et al. 2007; Zhang et al. 2007).

Each biological element that can been measured, can also be represented in a typical plant cell, tissue and organ at various molecular and/or morphological levels, or in other words a conceptual model with layers ranging from the 'genome' to the 'phenome'; a model called 'omic space' (Fig. 2.1) (Toyoda and Wada 2004). Advances in each 'omics' research area have become essential for investigations of gene function and structure, and the type of phenotypic changes present in plants. A schematic presentation of relevant 'omics' resource is shown in Fig. 2.1, together with the current status of available areas of research from Arabidopsis, rice, soybean, corn and *Brassica*; just to cite a few examples. Some of these advances have included improved methods for gene expression, gene modifications, molecular breeding, plant genome and proteome interactions, and metabolite profiling. Large volumes of information in biological resources, mass identification of mutant lines and full-length cDNAs, and the publication of this information in web-based data

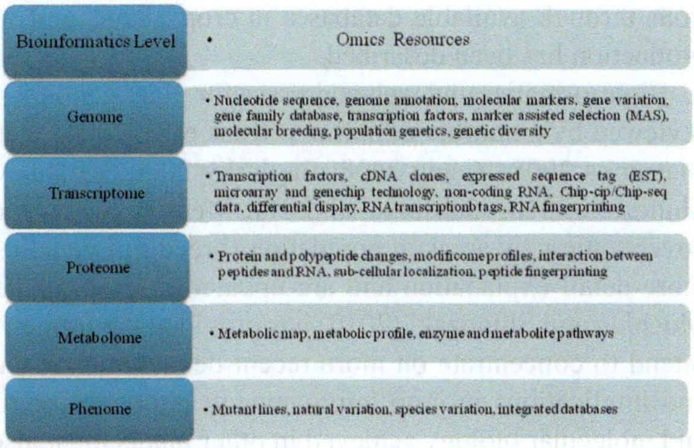

Fig. 2.1 A conceptual model called 'omic space' with layers ranging from the 'genome' to the 'phenome'. (After Toyoda and Wada 2004)

banks have been available for some time (Brady and Provart 2009; Kuromori et al. 2009; Seki and Shinozaki 2009).

Bioinformatic information and web sites have become important for crop scientists in gene data mining, and linking this knowledge to its biological significance (Mochida and Shinozaki 2010). However there needs to be a note of caution. As genomic and proteomic knowledge expands, new forms of electronic data becomes available to help interpret results. Biological data is notoriously variable (even unreliable at times) and 'noisy' in electronic form, due to living systems being complex and measurement and analysis technologies are often imperfect. In my experience two approaches for reducing 'noise' and help reliability of this type of data are required; aggregation and visualisation. Firstly, when combined, multiple forms of evidence become more and more accurate than for example a single source of data, simply because each replicate form of the data reduces overall uncertainty. Secondly, the human mind is an outstanding data analysis tool. It can absorb textual data rather poorly, but it can assimilate visual information in great detail, and the mind can process visual data efficiently to help identify common trends and themes (Cline and Kent 2009).

In this chapter, we provide an overview of the many web-based resources available for use in 'omics' plant research, with particular emphasis on recent progress related to crop species and crop improvement. Therefore we describe DNA and RNA sequence-related resources, molecular markers, whole genome sequencing, protein coding and non-coding transcripts, and provide molecular technology updates. We then review resources important for genetic map-based approaches such as QTL analyses and population genetic (diversity) studies. We also describe the current status of resources and some technologies for transcriptomics, proteomics and metabolomics; however some of these research areas are more comprehensively described in other chapters of this book. We then review molecular developments in each 'omics' field, as well as instances of their combined uses in investigations of particular crop systems. Mutant genotypes for use in 'phenome' research will be discussed, and the integration of 'omics' data between plant species in comparative genomics is dealt with. Throughout this review we provide examples of applications through available databases in crop plants, and where improvement in crop production has been described.

Bioinformatics and web addresses for plant genomics and proteomics have been reviewed by a number of authors (Rose et al. 2004; Sterck et al. 2007; Takeda and Matsuoka 2008; Zhang 2008; Baginsky 2009; Varshney et al. 2009; Mochida and Shinozaki 2010; Jackson et al. 2011; Memon 2012), and this review will basically cover some new areas in population (breeding) genetics, and topics which require more detail explanation and are updated in crop plants. The excellent review by Mochida and Shinozaki (2010) has provided the framework for this review, and we intend to concentrate on more recent developments, and focus on bioinformation and implications in crop improvement; although the technology, instrumentation and molecular biology achieved in other plants must also be covered.

DNA Based Sequence Resources

Genome Sequencing Projects

Initially, the publication and accumulation of nucleotide sequences for model plants only provided fundamental information, however now these base sequences form the fundamentals of research in functional plant genetics in applied species such as crops and domestic animals. Furthermore, DNA sequence data continues to be central in providing the genomic basis for accelerating molecular level understanding of basic biological mechanisms, and the application of such information to crops. In this section, we describe recently developed plant sequencing advancements. Species-specific nucleotide sequences are now providing information related to phenotypic characters, even when based on genome comparative analyses from the few model plants available (Cogburn et al. 2007; Flicek et al. 2008; Paterson 2008; Tanaka et al. 2008).

The genome sequence of *Arabidopsis thaliana* is now used as a model species in plant molecular biology mainly because of its small size, short generation time and high efficiency of transformation. The genome sequence of rice (*Oryza sativa*), including *japonica* and *indica* (an important staple food and a model monocotyledon) has also been used for comparative studies. These two plants still provide the only model plant systems to date, however several genome sequencing projects involving other plants have been completed, and many others are in progress; these are detailed in Table 2.1. Listed below are six of the most important web-based sites for DNA based genome sequencing and annotation projects, their purpose and their URL are detailed in Table 2.2.

NCBI—BioProject

The NCBI site provides genome sequences and information for many plant species (Viridiplantae) designed to facilitate comparative genomic studies amongst the

Table 2.1 List of plant species in which partial or whole genomes have been sequenced. (Data extracted from the following internet sites: http://www.ncbi.nlm.nih.gov/genomes/PLANTS/PlantList.html; http://www.arabidopsis.org/portals/genAnnotation/other_genomes/index.jsp; http://www.ildis.org/)

Division	Class	Species
Non Vascular	Algae	Chlamydomonas reinhardtii Chlorella variabilis Coccomyxa sp. Cyanidioschyzon merolae Ectocarpus siliculosus Micromonas pusilla Micromonas sp. Ostreococcus lucimarinus Ostreococcus tauri Volvox carteri Zostera marina
	Bryophytes	Physcomitrella patens Selaginella moellendorffii
Vascular	Dicotyledons	Amborella trichopoda Aquilegia sp. Arabidopsis lyrata Arabidopsis thaliana Arachis hypogaea Asclepias syriaca Beta vulgaris Boechera holboellii Brassica napus Brassica napa Brassica rapa Caffea caneophora Cajanus cajan Cannabis sativa Capsella rubella Carica papaya Castanea mollissima Citrullus lanatus Citrus clementine Corchorus olitorius Cucumis sativus Eucalyptus grandis Fragaria vesca Glycine max Gossypium hirsutum Gossypium raimonddi Hordeum vulgare Jatropha curcas Lactuca sativa Linum usitatissinun Lotus japonicas Malus domestica Manihot esculenta Medicago truncatula Mimulus guttatus Phaseolus vulgaris Pinus taeda Populus tremula Ricinus communis Theobroma cacao Populus nigra Populus trichocarpa Prunus avium Prunus persica Pyrus bretschneideri Rubus idaeus Salix purpure Solanum lycopersicum Solanum pimpinellifolium Solanum tuberosum Spirodella polyrhiza Thellungiella parvula Vitis vinifera
	Monocotyledons	Brachypodium distachyon Elaeis guineensis Miscanthus giganteus Musa acuminate malaccensis Oryza sativa Oryza glaberrima Panicum hallii Panicum virgatum Phoenix dactylifera Seratia italic Sorghum bicolor Triticum aestivum Zea mays

many other records of plants there. The current version consists of documentation in at least 115 different plants with partial sequences, and about 40,000 Expressed Sequence Tags (ESTs). It also contains separate sites and resources for other web based tools, data banks and other web servers, including agronomically important crops for food or fruit, medicinal plants, a number of green algae, pathogenic bacteria and fungi, viruses and animals. It will be important to become familiar and navigate through this very important site.

Phytozome

The site includes genome sequences and data sets for various crop species designed to facilitate comparative genomic studies amongst other green plants. The current version consists of 31 plant species wholly or partially sequenced, and is set-up into 10 evolutionary significant nodes.

Table 2.2 Integrative databases for DNA, Gene Sequences and Population Genetics analysis in plants

Database Name	Plant Species/Purpose	URL
Home—BioProject—NCBI	Multi-Purpose site; Over 1000 genomes; plants, animals, bacteria, fungi, virus	http://www.ncbi.nlm.nih.gov/sites/entrez?db=bioproject
Phytozome v8.0: Details	Over 31 species of plants; some software	http://www.phytozome.net/Phytozome_info.php
Gramene	Over 29 species of mainly monocots	http://www.gramene.org/
BLAST: Basic Local Alignment Search Tool	Multi-Purpose site for genome comparison; plants, animals, bacteria, fungi, virus	http://blast.ncbi.nlm.nih.gov/Blast.cgi
GrainGenes Class Browser: Marker	Triticeae and Avena site; nearly 200 species	http://wheat.pw.usda.gov/cgi-bin/graingenes/browse.cgi?class=marker
PlantGDB—Resource Plant Comparative Genomics	Multi Plant site; 15 dicot, 7 monocot, 3 other plant species	http://www.plantgdb.org/
TreeView	Phylogenetic tree software	http://taxonomy.zoology.gla.ac.uk/rod/treeview.html
Rod Page	Population genetics, gene diversity software	http://taxonomy.zoology.gla.ac.uk/rod/rod.html
Software	Softwere site; BLAST, sequence alignment	http://evolve.zoo.ox.ac.uk/Evolve/Software.html
LALIGN Server	Sequence alignment software	http://www.ch.embnet.org/software/LALIGN_form.html
PopGene	Population genetics software	http://www2.unil.ch/popgen/softwares/fstat.htm
Arlequin 3.11	Population genetics program	http://cmpg.unibe.ch/software/arlequin3/
ANU Bot Zool	Population genetics statistics, software	http://www.anu.edu.au/BoZo/GenAlEx/genalex_download_6_1.php
SOPH U AB	Population genetics programs	http://www.soph.uab.edu/ssg/linkage/population
Francis Yeh	PopGen software	http://www.ualberta.ca/~fyeh/
IBD	Gene and Distance relations statistics	http://www.bio.sdsu.edu/pub/andy/IBD.html
PRIMER-E	Population software and gene diversity statistics and indices	http://www.primer-e.com/

Gramene

An information resource established as a portal for grass species and grass genomics; including genome sequence information. The current version provides data on

24 plants; including 12 wild and domesticated rice genomes. An organelle data bank is also available from this site (Sect. 4.5).

NCBI—Entrez

Tracks over 800 whole genome projects from biological organisms, and the 115 species of Viridiplantae; including agronomically important crops for food and fruit, medicine and a number of green algae. The Entrez database can be accessed through the home page of NCBI.

NCBI—BLAST

One of the most important sites and tools available, in determining base similarities between nucleotide sequences in databanks. It also contains protein searches and queries. It includes searches for translated nucleotide sequences, conserved domains, multiple alignment tools, evolutionary relationships, and can be applied to all organisms, or limited to specific plants.

GrainGenes and PlantGDB

GrainGenes is a specific database for Triticeae and *Avena* genes, markers, maps and germplasm. PlantGDB contains sequences and a search engine linked to NCIB BLAST for 15 Dicotyledon, 7 Monocotyledon and 3 other plant species. It contains more limited information than the NCBI site over most plants, but is especially useful for agricultural grain species.

DNA Sequencing and UltraHigh-Throughput

Genome sequence information aids researchers in identify genes and gene families, including the identification of coding or non-coding regions, regulatory genes, and repetitive sequences within the genome (e.g. simple sequence repeats—SSRs); all of these are important in molecular biology. This type of information has become primary material for the design of genome advancements, such as microarrays, tilling arrays or molecular and chromosome markers, and these methods are important in whole plant genomic sequencing (Sect. 2.3 below). Pyrosequencing, massive parallel DNA sequencing and single molecule sequencing are adaptations of existing methods, which have become available in recent years (Margulies et al. 2005; Ansorge 2009). These new technologies have provided researchers with new methods to addressed web information in an entirely different way, and ongoing innovations in next-generation sequencing technology (Sect. 2.3), and the release of new genome sequenced plants (listed in Table 2.1) is expected to accelerate the use of the DNA based web-information considerably in crop plants.

Whole Genome Sequencing

Information obtained from whole-genome sequencing in plants allows attempts at chromosome-scale genetic comparisons, thereby identifying conserved genetic areas, which can facilitate identification and documentation of similar genomic sequences in related plant species (Haas et al. 2004; De Bodt et al. 2005). Whole-genome comparisons identifying chromosomal duplication of alleles among related species for example can provide comparative evolutionary histories and diversification of species in ecology, taxonomy and plant breeding (Paterson et al. 2009; Schnable et al. 2009). Next-generation sequencing will allow identification of even more fundamental diversity and variation in genes amongst and between individuals, strains and/or populations. Single nucleotide polymorphisms (SNPs) have been central to these advancements, and SSR fragments have been shown to map consistently in many non-sequenced plant species; a capability that is of immensely important in genetic research. A genome re-sequencing project to identify whole-genome sequence variations in 1001 strains (accessions) of Arabidopsis is in progress. On completion this data will become an important resource for future genetics and population studies to identify alleles associated with phenotypes and diversity across entire plant species (http://1001genomes.org/) (Weigel and Mott 2009). In the same way a high-throughput method for genotyping recombinations in populations of rice, using whole-genome resequencing data generated by the Illumina Genome Analyzer has already been initiated (Huang et al. 2009).

Molecular (DNA) Markers

Identification and location of available molecular DNA markers have contributed significantly to marker-assisted studies and selection (MAS) in plant breeding, and in a wider range of research, including species identification and evolution. Genetic markers constructed to cover the complete genome may allow identification of individual genes associated with complex traits by QTL analysis, and the identification of genetic diversity and induced variations (Feltus et al. 2004; Varshney et al. 2005; Caicedo et al. 2007). Genome sequencing and large-scale EST databanks (Sect. 3) have become important for the construction of molecular markers, and a number of genome-wide rice DNA polymorphic markers have been constructed based on co-alignment between *japonica* and *indica* rice ESTs (Han and Xue 2003; Shen et al. 2004). Computer assisted EST-base single-nucleotide polymorphisms (SNPs) and/or EST-SNP markers for the purpose of identifying sequence-tagged sites (STS) has progressed for numerous species; including the crop plants of barley, wheat, maize, melon, *Brassica*, common bean, sunflower, potato, citrus and grapevine (Mullins et al. 2006; Torada et al. 2006; Jaillon et al. 2007; Heesacker et al. 2008; Kota et al. 2008; Talon and Gmitter 2008; Blair et al. 2009; Deleu et al. 2009; Kaur et al. 2009; Li et al. 2009).

Some molecular markers identified this way allow the indirect selection of interesting genotypes (i.e. breeding lines in crops), and these cultivars constitute an essential tool for the development of marker-assisted selection (MAS) in plant breeding. The use of DNA markers (and indirectly EST markers from RNA) for direct selection offers greater potential gains in breeding for QTL and traits with low heritability, and these can be the most difficult to work with in crop breeding. However these low heritability traits are also amongst the most interesting and the most difficult to develop.

When a locus has many variants, or alleles, it is referred to as being polymorphic. Mutation(s) at a number of loci generate multiple alleles, most of which are eliminated from the population by genetic drift or breeding selection. Only a small number of alleles are incorporated into the population by chance or selection. Most polymorphisms can be genetically straightforward, with two alleles directly determining two versions of the same protein (gene), however, some can be highly complex, with multiple, related genes in a complex system of metabolic differences. Crop breeders have known the complexity of multiple alleles for decades. However with the advent of molecular markers, genetic diversity and other forms of genetic structure in breeding populations is possible. Listed in Table 2.2 are the most important web-based sites for DNA markers and some of the population statistics programs and web resources commonly in use. Molecular markers fall into a number of types listed below, each having positive and negative features, and careful consideration is required before they are adopted in any type of research (Hoang et al. 2009; De Filippis 2012).

Restriction Fragment Length Polymorphism (RFLP)

RFLP requires hydrolysis of probe DNA from samples. RFLP can provide high quality data but has severe restrictions on throughput because large amounts of DNA are required, and because it is not based on amplification of the target DNA via the polymerase chain reaction (PCR).

Random Amplified Polymorphic DNA (RAPD)

RAPD is a method based on PCR but uses arbitrary short primers (10 bases long) to identity plant DNA regions. No knowledge of the genome is needed, but by the same token markers can target many places on the genome. Results can be inconsistent and only dominant genes can be identified.

Simple Sequence Repeats (SSR)

SSR are high quality and consistent DNA markers, but they are the most expensive to develop. SSR markers require extensive band sequencing data for each marker developed, and often the markers are species and even cultivar specific. However they are molecular markers of choice in crop plants.

Amplified Fragment Length Polymorphism (AFLP)

AFLP requires enzymatic degradation of DNA and careful fragment separation, where only a sub-fraction of the population genetic data is sampled by PCR. It can provide too much information at any time. It is more technically demanding and information can be difficult to interpret. It produces very good high quality data, which is suitable for high output sources and automation.

Single Nucleotide Polymorphism (SNP)

SNP relies on the fact that the vast majority of differences in eukaryotic organisms are surprising point mutations in their DNA. So there are a vast number of polymorphisms that are SNPs. The biggest advantage is automation and techniques that do not require electrophoresis to separate fragments. However it does require DNA sequencing which can be costly. SNPs are becoming more and more important as molecular markers for genome information and advancement in crop plants.

Expressed Sequence Tags (EST)

ESTs require cDNA synthesis from RNA, and therefore are the only markers listed which are based on RNA. Preferences for this method should be for crop species where there is already extensive sequencing, and part or full EST data present (Sect. 3).

NCBI—Plant Markers

A genetic marker web database that contains molecular markers such as SNP, SSR and conserved ortholog set cosmid (COS) markers and primers from various plant resources (Heesacker et al. 2008).

GrainGenes

Theweb site for Triticeae genomics, and provides considerable detail of DNA markers and chromosome linkage map data on wheat, barley, rye and oat (Carollo et al. 2005).

Gramene

A database for plant comparative genomics providing gene information and some genetic linkage maps for 29 monocotyledon (grass) species, using some of the more important and more commonly used genetic markers detailed above (Liang et al. 2008; Ware 2007).

Genetic Diversity and Population Genetics Analysis

Comprehensive discussion of the genetic and statistical analyses employed in population genetics is beyond the scope of this review, but I refer you to the following books and reviews (Clark and Gorley 2001; Conte et al. 2008; Wall et al. 2008; Barnholtz-Sloan and Tiwari 2009; Pu et al. 2009). Population gene family data are usually produced by computational procedures including a first step that conducts an all-against-all sequence similarity analysis or matrix, and then a second step in building clusters of inter- and intra- population analysis parameters, by methods such as Markov Clustering (MCL), multi dimensional scaling (MDS), and principle component analysis (PPO); using programs like PRIMER and Arlequin (Table 2.1). Advanced software statistics (GenAlEx—Table 2.2) can yield indices and information that are useful for further statistical analysis and phylogenetic studies using Analysis of Co-Variance (ANCOVA), Analysis of Similarity and Analysis of Variance (ANOVA). Listed in Table 2.2 are a number of important web-based sites for population genetics analysis and computation; and for ease their purpose and URL are also presented there.

RNA Variation Resources

EST and cDNA

Expressed Sequence Tag (ESTs) are determined by partial sequencing of randomly picked gene transcripts that have originated from isolated RNA and converted to cDNA (Adams et al. 1993). Since cDNA and EST collections can be easily generated regardless of chromosome and gene complexity, this method has been applied not only to model plants, but also to a number of crop species with large genomes; due mainly to polyploidy and/or to the number of repetitive sequences. Because EST data collected from cDNA libraries of an organism consists of redundant sequences from the same gene locus or RNA target, it is often necessary to perform EST grouping by metabolic and/or functional units. Then these groups are further consolidated into alignment sequences for each transcript before further analysis (Ewing et al. 1998; Huang and Madan 1999; Masoudi-Nejad et al. 2006). The comprehensive and rapid accumulation of cDNA clones, together with mass data sets of their sequence tags have become important resources for functional genomics (Boguski et al. 1993). ESTs derived from tissues in a range of developmental stages or under various kinds of stress could significantly facilitate discovery of new genes and their function. For example, large-scale expression analysis, genome comparative DNA sequences and the design of expressed gene-specific molecular markers and probes for microarrays have only been possible with extensive EST data (Zhang et al. 2004; Kawaura et al. 2006; Mochida et al. 2006). Listed below are a number of important web-based sites for RNA analysis and ESTs; and for ease their purpose and URL are detailed in Table 2.3.

TriMEDB

The Triticeae Mapped EST database (TriMEDB) provides information regarding mapped cDNA motifs, which are related to barley and wheat sequences (Mochida et al. 2008); a similar database, TriFLDB has much the same information (Mochida et al. 2009b).

NCBI—dbEST (Expressed Sequence Tag)

There are over 63 million ESTs in the NCBI dbEST databank, a most important public domain EST database that includes a number of crop plant species (Boguski et al. 1993). The data sets obtained from representative transcripts can be used as unified transcript and sequence data, in line with other web sites below (Lee et al. 2005; Close et al. 2007; Duvick et al. 2008).

NCBI—UniGene

Identifies transcripts from the same locus as expressed in different types of tissue, age or health status. Most importantly it reports not only on ESTs, but also on related proteins and clone resources.

TIGR

Plant transcript assemblies and gene indices web site. The databank relies on EST/cDNA sequences linked to GenBank from NCBI. New releases and new plant databases are documented regularly.

Table 2.3 Integrative databases for RNA and Expressed Sequence Tag (EST) analysis in plants

Database Name	Plant Species/Purpose	*Uniform Resource Locator* (URL)
TriMEDB:Triticeae Mapped EST DataBase ver.2.0	Triticeae EST database; over 6000 sequences	http://trimedb.psc.riken.jp/index.pl
NCBI dbEST	EST of the extensive databank from NCBI; multi species	http://www.ncbi.nlm.nih.gov/dbEST/
Home—UniGene—NCBI	Unigene (EST) comparison site from NCBI; multi species	http://www.ncbi.nlm.nih.gov/unigene
TIGR Plant Transcript Assemblies	TA and EST of many plants, including conifers, algae, monocots and dicots	http://plantta.jcvi.org/
DFCI—Plant Gene Indices	Plant gene indices (similar to TA) of 60 important agricultural plants	http://compbio.dfci.harvard.edu/tgi/plant.html
HarvEST Home Page	EST database of 10 of the most important agricultural/horticultural plants	http://harvest.ucr.edu/
HARVEST-BLAST.ORG	HarvEST Blast search software from NCBI	http://www.harvest-blast.org/
Plant MicroRNA Database	Plant micro RNA website and database of some commercial plants	http://bioinformatics.cau.edu.cn/PMRD/
Diversity Arrays Technology Pty Ltd (DArT P/L) | Diversity Arrays Technology	Analysis of molecular diversity; SNP, SSR, AFLP, RFLP, methylation	http://www.diversityarrays.com/index.html
Home | Affymetrix	GeneChip mi RNA arrays information—Affymetrix	http://www.affymetrix.com/estore/
miRNA Array | Noncoding RNA | Affymetrix	GeneChip mi RNA information and species	http://www.affymetrix.com/estore/browse/products.jsp?productId=131473#1_1
GeneChip Medicago Genome Array | Affymetrix	Medicago GeneChip array and its symbiont	http://www.affymetrix.com/browse/products.jsp?productId=131472#1_1
GoldenGate Genotyping Assay—A flexible, pre-optimized assay	Site for Golden Gate genotype assay	http://www.illumina.com/technology/goldengate_genotyping_assay.ilmn
Illumina—Assay Technology	Golden Gate genotype technology information	http://www.illumina.com/company/assay_technology.ilmn
Illumina Genotyping	High-throughput SNP genotyping—Illumina	http://dnatech.genomecenter.ucdavis.edu/illumina.html
CLC Biology	Multi-purpose gene work bench; with downloads	http://www.clcbio.com/index.php?id=354

Plant Gene Index

The web site documents a number of animal, plant, protist and fungi species. The site contains a number of web tools, and contains cDNA and genechip data from a number of crop plants.

HarvEST

HarvEST software is available for 10 important agricultural crops. It originated as an EST database with software linked to gene function, microarray design and SNP identification. It is also available as a HarvEST BLAST search engine from an alternative site present on the homepage.

cDNA (Full Length)

Partial cDNAs are useful for rapidly documenting and cataloguing targeted genes, but they are not used or suitable for further study of gene function. This is because the most popular method for preparing a cDNA library does not provide the full-length cDNA that includes the capped site sequences. The biotinylated cap trap method using a thermostabilised reverse transcriptase is one method for constructing full-length cDNA-enriched libraries suitable for studies of gene function; and these have become invaluable for life science projects (Maeda et al. 2006; Tanaka et al. 2008; Yamasaki et al. 2008a). The sequences derived from full-length cDNAs can also help in identifying transcribed regions in completed or draft genomes in other plants. In Arabidopsis and rice, full-length cDNA sequences have been used to identify genomic structural features, such as transcription start sites (TSSs) and transcriptional genes and variant alleles in metabolic activity (Iida et al. 2004; Itoh et al. 2007; Yamamoto et al. 2009). In species for which we have draft genomes, such as *Physcomitrella*, soybean and poplar, full-length cDNA clones have been used to help consolidate genomic (gene) and chromosome structure and function; and this should also greatly contribute to discovery of new genetic information (Nanjo et al. 2007; Ralph et al. 2008a; Umezawa et al. 2008).

Full-length cDNA libraries have contributed to functional analysis using over-expressors in reverse genetics. The full-length cDNA overexpressor (FOX) gene hunting system, which uses full-length cDNA from transgenic plants as overex-pressors, has introduced another approach to high-throughput analysis of functional genes associated with phenotypic traits (Ichikawa et al. 2006; Fujita et al. 2007; Kondou et al. 2009). Full-length enriched cDNA libraries have been constructed for non-sequenced crop or forestry species, such as wheat (*Triticum aestivum*), barley (*Hordeum vulgare*), cassava (*Manihot esculenta*), Japanese cedar (*Cryptomeria japonica*), Sitka spruce (*Picea sitchensis*) and Lotus (Sato et al. 2008; Kawaura et al. 2009; Sato et al. 2009); as well as for plant species showing specific characteristics such as salt tolerance in salt cress (*Thellungiella halophila*) and selenium accumulation in a number of species (De Filippis 2010). Full-length cDNA libraries serve as primary sequence resources for designing microarray probes, and as cloned resources for genetic engineering to improve crop efficiency (Sakurai et al. 2007; Futamura et al. 2008; Ralph et al. 2008b; Taji et al. 2008). Because of the important function of full-length cDNA resources in 'omics' data and information, it is essential to establish relevant information resources that provide gateways to these sites, and to integrate this to related data sets derived from other 'omics' field. A comprehensive computational tool to achieve this total integration is the CLC Biology workbench system (Table 2.3).

small RNA (sRNA) Information

In plants, sRNAs, which can include microRNAs (miRNAs), short interfering RNAs (siRNAs) and trans-acting siRNAs (ta-siRNAs) play important roles in epigenetic processes, and may control gene activities involved in plant development and homeostasis (Ruiz-Ferrer and Voinnet 2009). These RNA molecules are important regulatory resources that should be detailed, and their expression analysed using the most recent next-generational genomic methods (Nobuta et al. 2007; Chellappan and Jin 2009). In maize, sRNAs in the wild type and in the isogenic *mop1-1* loss-of-function mutant were analysed by deep sequencing using Illumina's sequencing-by-synthesis (SBS) technology to characterise possible regulatory roles for maize sRNA (Nobuta et al. 2008). In poplar, expressed sRNAs from leaves and vegetative buds were also studied using high throughput Roche 454 pyrosequencing, and genes in similar families of miRNA identified, including novel new genes (Barakat et al. 2007). Nucleotide sequencing of *Brachypodium* sRNAs have also been performed, resulting in the identification of miRNAs involved in low temperature stress tolerance (Zhang et al. 2009a). The plant miRNA database (PMRD) is a useful information resource on plant miRNA, and is available on the web site of PMRD (Zhang et al. 2009b). The Affymetrix GeneChip miRNA Arrays web site is also important for such studied. Listed in Table 2.3 are important web-based sites and URL for databanks often used in regulatory plant micro RNA research.

Variation Analysis Platforms

High-throughput polymorphic analyses are important for studying genome-wide genotyping using a hybridization-based SNP molecular marker methods, which have been used to analyse Arabidopsis ecotypes and rice strains, and the data identified variation patterns for each species. The Arabidopsis 1,001 project and genome-wide variation study is one of the few available information web sites so far containing this information. Therefore, the demand is rapidly increasing for high throughput and cost-effective platforms for comprehensive variation analysis studies (also called 'variome' analysis). Complete genome resequencing information are already being realized as a direct solution to 'variome' analysis in species whose reference genome sequence data are already available; the URL list for the web sites used for variation analysis are detailed below and URL listed in Table 2.3.

DArT P/L

Diversity Array Technology (DArT) is a high throughput genotyping system that was developed based on microarrays (Jaccoud et al. 2001; Wenzl et al. 2007). In various crop species such as wheat, barley and sorghum, DArT markers have been used with other conventional molecular markers to construct dense genetic maps and/or to perform genetic association studies (Crossa et al. 2007; Peleg et al. 2008; Mace et al. 2009). This genotype technology web site describes DNA variations using SNP, as well as other molecular markers; and also described is information on DNA methylation.

Gene Chip Array

In barley and wheat, *Affymetrix* GeneChip Arrays have been used to show nucleotide polymorphisms based on the differential hybridization of GeneChip probes (Rostoks et al. 2005; Bernardo et al. 2009). The Affimetrix site includes a specific *Medicago* chip and a miRNA site. The *Illumina* GoldenGate Assay allows the simultaneous analysis of up to 1,536 SNPs in 96 samples, and has been used to analyse genotypes of segregating populations in order to construct genetic maps of SNPs in crops such as barley, wheat and soybean (Hyten et al. 2008; Akhunov et al. 2009; Close et al. 2009).

SNP Genotyping

The University of California, Davis site contains a number of options for high throughput genotyping. It operates a commercial division where SNP genotyping can be performed using Illumina technology and GoldenGate Assay.

Analysis of Variation

Comprehensive gene family data sets are usually produced by computer programs, including a sequence similarity search, and then a step for building clusters of EST families by methods such as Markov Clustering (MCL), multi dimensional scaling (MDS) and principle component analysis (PPO) (Table 2.1). Advanced statistics can yield data sets that are useful for further variation analysis using gene markers, as well as phylogenetic studies, using Analysis of Co-Variance and Phylogeny programs (see also Sect. 2.5). Listed in Table 2.1 are a number of web-based sites for genetic variation analysis and computation; and for ease their purpose and URL are also detailed there.

Transcriptome Resources

Comprehensive, high-throughput analysis of gene expression, also called 'transcriptome' analysis, is a good approach to screen targeted genes, predict gene function and discover *cis*-regulatory motifs. Hybridization-based methods, such as that used in microarrays and GeneChips have been well established now for acquiring large-scale gene expression profiles from various species. The recent rapid accumulation of data containing large-scale gene expression profiles, and comparison of this data to large repositories in genetic databanks have provided large amounts of information now available in the public domain. This public data is an efficient and valuable resource for many secondary uses, such as co-expression of genes and comparative genomic studies. Furthermore, as next-generation DNA sequencing applications and deep sequencing of short fragments of expressed RNAs and sRNAs become common, they become important tools to use in both genome-sequenced and non-sequenced species (Harbers and Carninci 2005; de Hoon and Hayashizaki 2008). Listed below are the most important web-based sites for microchip and microarray analysis; their purposes and their URL are detailed in Table 2.4.

Sequence Tag Based Transcriptomics

Documentation of large-scale sequence ESTs from cDNA libraries was an early approach in developing transcriptome data. The alternative is to use ESTs that are randomly sequenced in an unbiased cDNA library, which are classified into clusters of transcriptional units using sequence-clustering and/or other assembly methods. The abundance of each transcript unit expressed in each tissue is then estimated by counting the number of ESTs with identifiers for each cDNA library and/or each sequence cluster. The same methodological principles have been applied in human and mouse, and a form of 'organism map' to determine the transcriptome in various tissues and organs has been realised (Hishiki et al. 2000; Kawamoto et al. 2000; Ogasawara et al. 2006). There are no impediments in similar methods and approaches being use in plant and crop transcriptomics.

Table 2.4 Integrative databases for Microarray and Microchip technology and analysis in plants

Database Name	Plant Species/Purpose	*Uniform Resource Locator* (URL)
Serial Analysis of Gene Expression	Serial analysis of gene expression (SAGE) information page	http://www.sagenet.org/
Serial Analysis of Gene Expression-SAGE™ | Life Technologies	SAGE differential expression of genes	http://www.invitrogen.com/site/us/en/home/Products-and-Services/Applications/Sequencing/Capillary-Electrophoresis-Sequencing/SAGE-Sequencing.html
Microarray, SAGE and other gene expression databases | HSLS	Microarray, SAGE and gene expression data bases; multi-species	http://www.hsls.pitt.edu/obrc/index.php?page=gene_expression_databases
Serial Analysis of Gene Expression	Applications for SAGE; multi purpose	http://www.sagenet.org/findings/index.html
GermSAGE | Home	Germ cell line SAGE; mainly animal species	http://germsage.nichd.nih.gov/germsage/home.html
Bioinformatics resources	Ultimate web site for SAGE; applications, information, books etc	http://www.bioinformatics.fr/resources.php?tag=sage
MAGE—Workgroups—FGED	Microarray Analysis of Gene Expression; MAGE site	http://www.mged.org/Workgroups/MAGE/mage.html
CEA Direction des sciences du vivant—iBiTec-S: Adaptation du SAGE pour micro-quantités	SADE; SAGE adaptation for downsized extracts (microquantities)	http://www-dsv.cea.fr/dsv/instituts/institut-de-biologie-et-de-technologies-de-saclay-ibitec-s/unites/service-de-biologie-integrative-et-genetique-moleculaire-sbigem/laboratoire-de-physiogenomique-lpg/adaptation-du-sage-pour-micro-quantites
Next-Gen Sequence Databases	Next-generation sequences databases; some commercial crops	http://mpss.udel.edu/
PlantPromoterDB	Genome sequences and regulatory elements for Rice and Arabidopsis	http://ppdb.agr.gifu-u.ac.jp/ppdb/cgi-bin/index.cgi

UniGene

The digital differential display (DDD) tool, a component of the NCBI's UniGene database has been applied in large-scale cDNA projects for various species, including crop plants (Mochida et al. 2003; Fei et al. 2004; Sterky et al. 2004; Zhang et al. 2004). Although this approach, coupled with cDNA clone resources has facilitated gene discovery and expression profiling, various disadvantages including high cost and limited resolution due to the need for large-scale sequencing are still present.

SAGE

Serial analysis of gene expression (SAGE) is a method based on sequencing of short-read cDNA tags. SAGE allows the identification of a large number of transcripts present in tissues, and enables quantitative comparison of transcriptomes (Velculescu et al. 1995). Sequencing of selected clones from the SAGE library allows the efficient collection of transcript sequencetags (TSS). Complete genome sequencing or large-scale ESTs data are required, and because few plants fall into this category, very few crop plants have approached the genetics of crop improvement in this manner.

SAGE Derivatives

Some derivatives of the original SAGE protocol (MAGE, SADE, microSAGE, miniSAGE, longSAGE, superSAGE, deepSAGE, GermSAGE, 5′ SAGE) have been developed to improve the utility of SAGE (Hashimoto et al. 2004; Anisimov 2008). For example, superSAGE is an improved version of SAGE, and this method has been applied to quantitative gene expression of both rice infected host cells and their pathogen (Matsumura et al. 2003). The development of superSAGE tags has also been used to design probes directly for oligonucleotide based microarrays in plants (Matsumura et al. 2008).

Parallel Signature Sequencing (MPSS)

Massive parallel signature sequencing (MPSS) is uses as an alternate method to quantify gene expression levels, and generate short sequence tags using a microbead array (Brenner et al. 2000). The database of MPSS includes information from *Arabidopsis*, rice, grape, soybean, *Medicago*, maize, *Brachypodium* and *Magnaporthe grisea* (the rice blast fungus) (Nakano et al. 2006).

MPSS 2 and Plus (*Arabidopsis* profiling)

A new MPSS method has been used to perform genome-scale expression profiling of sRNAs in *Arabidopsis* and rice (Lu et al. 2006; Nobuta et al. 2007). The MPSS2 method has recently been used for quantitative analysis of the 5′ end of transcripts, which are then coupled to the cap-trap method for full-length cDNA cloning; it is reported to be applicable to a number of different plants.

High Density Mapping of TSS

The method has been applied to perform high-density mapping of TSS in Arabidopsis, and identify genome-scale presence of plant promoters (Yamamoto et al. 2009). The data set of Arabidopsis CT-MPSS tags is accessible from the PPDB site, a plant promoter database that provides promoter information and motifs for Arabidopsis and rice (Yamamoto and Obokata 2008).

Hybridisation-Based Platforms

DNA microarrays had their beginning with Brown's research at Stanford University in 1995 (Schena et al. 1995), and since then, microarray and DNA chip-related technologies have advanced rapidly; and their application has expanded to a wide variety of life science disciplines. DNA microarray or GeneChip analysis are designed to acquire comprehensive data of the molecular abundance of each molecule in a given sample, based on its simultaneous hybridization with a large population of synthetic (DNA or cDNA) oligonucleotide species; usually immobilized on a glass slide or on a silicon chip. With the recent and rapid increase in the number of sequenced plant species, the availability of DNA microarrays have also increased for transcriptome analysis. For example, Seki and co-workers designed a custom DNA microarray that uses 7,000 full-length cDNA clones of Arabidopsis as probes, and then successfully screened genes in response to abiotic stresses using a two-colour method (Seki et al. 2002a). With the recent increase in commercially available DNA microarrays, laboratories are able to use a particular DNA microarray design to obtain transcriptome data from many experiments, and in so doing accumulate comprehensive information on organism-specific transcriptional data. Gene expression analysis and gene chip technology, URL web sites and the use of such technology are detailed in Table 2.5.

DNA Microarray Types

DNA microarrays can be classified into two types: (i) the 'spotting' type, which was developed at Stanford University; and (ii) the 'on-chip synthesis' type based on manufactured probes by a number of molecular biology companies. The spotting type was commonly used during the early years of transcriptome research. This method entailed preparing DNA microarrays by spotting a cDNA solution onto a glass slide, but spot microarrays are being replaced quickly by on-chip synthesis type.

Table 2.5 Integrative databases for Gene Expression analysis and technology in plants

Database Name	Plant Species/Purpose	*Uniform Resource Locator* (URL)
Roche NimbleGen | Array Synthesis	Roche platform; information on MAS digital technology	http://www.nimblegen.com/company/technology/index.html
Roche NimbleGen	Information website for biochip technology	http://www.biochipnet.com/node/146
Gene Expression Omnibus (GEO) Main page	GEO navigation and browser software—NCBI	http://www.ncbi.nlm.nih.gov/geo/
AtGenExpress—Weigel World	Microarray data and web information for plant species	http://www.weigelworld.org/resources/microarray/AtGenExpress/
TAIR—Gene Expression	Micrarray data and gene expression, and stress factors in Arabidopsis	http://arabidopsis.org/portals/expression/microarray/ATGenExpress.jsp
AtGenExpress JPN (Arabidopsis Gene Expression profile data base) | Database Registry | RIKEN	Arabidopsis gene expression profile data base from RIKEN	https://database.riken.jp/sw/en/AtGenExpress_JPN_(Arabidopsis_Gene_Expression_profile_data_base)/crib151s2rib151s85i/
ATTED-II: Home	Site for the co-regulation gene functions; mainly in Arabidopsis and rice	http://atted.jp/
GENEVESTIGATOR—The gene expression search engine	Site for the search engine—Geneinvestigator; biomedical, plant biology	https://www.genevestigator.com/gv/
The BAR and other Data Analysis Tools for Plant Biology	Bio Array resource for plant biology; number of crop plants	http://bar.utoronto.ca/welcome.htm
http://www.bar.utoronto.ca/3DDI/	3D data display initiative for large scale dataset display	http://3ddi.org/
http://www.arexdb.org/	Arabidopsis gene expression data base web site	http://www.arexdb.org/
Yale Rice Project	Whole genome transcription profile for rice cells, tissues	http://bioinformatics.med.yale.edu/riceatlas/
Tiling Array, Tiling primers & probes, design primers for tiling, amplify whole genome tiling arrays	Design whole genome primers for use in array design, including DNA methylation	http://www.premierbiosoft.com/dnamicroarray/tiling_array.html
Tiling Arrays, Library File Updates | Affymetrix	GeneChip tilling array, software and analysis	http://www.affymetrix.com/support/technical/libraryfileupdatesmain.affx
Arabidopsis thaliana—Tiling Array Express—Weigel World	Arabidopsis tilling array express; genome browser and used for visual cross platform comparison	http://www.weigelworld.org/resources/microarray/at-tax

Table 2.5 (continued)

Database Name	Plant Species/Purpose	*Uniform Resource Locator* (URL)
Publication Detail	Arabidopsis whole genome tilling array and expression and transcript identification	http://www.arabidopsis.org/servlets/TairObject?type=publication&id=501727375
ChIP-Seq	On line analysis tool; cross-referencing numerous datasets and gene expression profiles	http://ccg.vital-it.ch/chipseq/
MACS—Model-based Analysis for ChIP-Seq	Model based analysis for ChIP Seq datasets and software	http://liulab.dfci.harvard.edu/MACS/
SOLiD® ChIP-Seq | Life Technologies	Chromatin immunoprecipitation (ChIP) sequencing kits protocols	http://www.invitrogen.com/site/us/en/home/Products-and-Services/Applications/epigenetics-noncoding-rna-research/Chromatin-Remodeling/Chromatin-Immunoprecipitation-ChIP/SOLiD-ChIP-Seq-.html
Chromatin Immunoprecipitation sequencing (ChIP-Seq)—Data	ChIP sequencing technology; chromatin immunoprecipitation	http://www.illumina.com/technology/chip_seq_assay.ilmn

On-Chip Transcriptome Research

The on-chip *in situ* oligo synthesis-based method is a light-directed chemical process that combines solid-phase synthesis with photolithographic techniques. This method was initially employed with the Affymetrix-manufactured GeneChip Array system. In the Affymetrix GeneChip system, a known gene or potentially expressed sequence is represented on the chip by up to 20 unique oligonucleotide probes. The 'On-Chip' technology is fast becoming more readily available and more cost effective.

High Density DNA Array

Roche NimbleGen and Agilent Technology offer platforms to manufacture high-density DNA arrays based on Roche's patented Maskless Array Synthesizer (MAS) technology, and on a non-contact industrial inkjet printing process, both of which are also used for *in situ* oligonucleotide synthesis.

NCBI Gene Expression Omnibus (GEO)

NCBI's Gene Expression Omnibus (GEO) and the European Bioinformatics Institute (EBI)'s ArrayExpress are important web sites serving as primary archives of transcriptome data in the public domain (Parkinson et al. 2007; Barrett et al. 2009).

There are cross-linkages on this web site to several focused databases that provide calculated transcriptome data with user-friendly interfaces.

AtGenExpress

AtGenExpress is a multinational project designed to uncover the transcriptome of *A. thaliana*. The data sets collected in AtGenExpress have been one of the most comprehensive resources for the Arabidopsis transcriptome described to date (Kilian et al. 2007; Goda et al. 2008).

ATTED II

ATTED II is a database that provides co-expression analysis calculated from publicly available Arabidopsis ATH1 GeneChip data (Obayashi et al. 2007, 2009). Co-expression analysis generated from collected transcriptomes has aided gene expression studies in plants.

GeneInvestigator

Genevestigator, which is a reference expression database and meta-analysis system, also provides summary information from hundreds of microarray experiments on various organisms, including Arabidopsis, barley and soybean, with an easyinterface forresults (Zimmermann et al. 2004).

Electronic Fluorescence Pictograph (eFP)

The electronic fluorescent pictograph (eFP) browser provides gene expression patterns collected from *Arabidopsis*, poplar, *Medicago*, rice and barley via a user-friendly program (Winter et al. 2007). The Bio-Array resources on this site, however, directs most of the information to *Arabidopsis*, and the 3DDI site provides three dimensional display data designed to generate extendible models for plants.

AREX

The Arabidopsis Gene Expression Database AREX is a web site that provides data of high-resolution gene expression patterns of root tissues in *Arabidopsis* (Birnbaum et al. 2003; Brady et al. 2007).

RICEATLAS

The RICEATLAS is a database housing rice transcriptome data covering various types of tissues using laser capture microdissection (Jiao et al. 2009). It is housed within the search engine of the Yale Virtual Centre for Cellular Expression Profiling of Rice web address, and in the TIGR site.

Tilling Arrays

Tilling arrays are high-density oligonucleotide probes covering the genome in a particular organism, and therefore are a platform for analysing expressed regions throughout the genome. The method is effective in discovering novel genes and elucidating their structure. For example, Seki and co-workers performed transcriptome analysis in Arabidopsis under abiotic stress using whole-genome tilling array and described a number of antisense transcripts induced by stress (Matsui et al. 2008).

At-TAX

Arabidopsis thaliana Tilling Array Express (At-TAX) is a whole-genome tilling array resource for developmental expression analysis and transcript identification (Laubinger et al. 2008; Zeller et al. 2009). The utility of tilling arrays has been extended recently by coupling this platform with an immunoprecipitation method (see Sects. 3.6.13 and 3.6.14 below).

MADS

The MADS domain uses a chromatin immunoprecipitation (ChIP) method that has been used to identify transcriptional regulator sites for somatic embryogenesis, when coupled with the Affymetrix tilling array for Arabidopsis. This rather complex method to use and master, nevertheless found approximately 2,000 sites of regulation (Zheng et al. 2009).

ChIP–Seq

A methylcytosine immunoprecipitation (mCIP) method, in combination with Arabidopsis tilling array, can map comprehensive DNA methylation sites; and is often referred to as 'methylome' data (Zhang et al. 2006). Sequencing of co-precipitated DNA and a protein using next generation sequencing, 'ChIP-seq', has also become an important experimental approach (Park 2009). MACS and SOLiD ChIP-Seq web sites contain information and commercial kits and alternative analysis using the ChIP-Seq platform. A chromatin ChIP-Seq assay is also available from Illumina using the chromatin immunoprecipitation approach (Farrer et al. 2009).

Protein (Proteomics)

Genome sequencing projects for several organisms have been completed, but proteome analysis, which is the detailed investigation of the function, modification network and 3D structure of proteins, has gained increase attention (De Filippis and Magel 2012; Memon 2012). Large-scale proteome information can be an important resource for a better understanding of protein function in cellular systems, which are controlled primarily by polypeptides and proteins. Protein dynamic properties reflect cell and organ differences in terms of growth, development and response to biotic and abiotic stress. The primary objective of plant proteomics has traditionally been to simply identify all (or most of) the proteins in cells, organs and tissues. Recent rapid technical advances in proteomics (e.g. protein separation and purification, advances in mass spectrometry and methodological development in protein quantification and identification) have lead scientists into the second stage of proteomics, including quantitative proteomics, subcellular proteomics, modifications of proteins and protein-metabolite interactions (Rose et al. 2004; Rossignol et al. 2006; Baginsky 2009; Yates et al. 2009).

Resources in Proteomics

The different Web-accessible plant proteome-related databases are summarized on the proteomics subcommittee home web page of the Multinational Arabidopsis Steering Committee (MASCP); under the heading of 'Proteomic Databases and Resources'. A summary of the information in various basic and advance proteomics sites and databases is given in Table 2.6, including their relevant URL.

Proteome Profiling

A typical experimental research pathway for protein profiling can be summarized as; (i) sample preparation of impure plant protein mixtures, (ii) separation and purification of the proteins, (iii) detection of proteins and/or polypeptides, and (iv) identification of fractionated or ionised proteins and polypeptides. Various technical advances for each step of the process above have greatly increased the overall performance, efficiency and cost effectiveness of plant proteomics research (Rose et al. 2004; Jorrin-Novo et al. 2009; Uauy et al. 2009).

Table 2.6 Integrative databases for Proteomics and Protein (Polypeptide) analysis in plants

Database Name	Plant Species/Purpose	*Uniform Resource Locator* (URL)
Matrix Science—Mascot	Home page for peptide mass fingerprint database, searches and peptide identification	http://www.matrixscience.com/search_form_select.html
Entrez Databases	Entrez database and tools; a multi-purpose proteomics site	http://www.ncbi.nlm.nih.gov/About/tools/restable_mol.html
Main Page—Mascp	Arabidopsis proteomic multinational sub-committee	http://www.masc-proteomics.org/mascp/index.php/Main_Page
World-2DPAGE Constellation: SWISS-2DPAGE	ExPASy Swiss protein 2D PAGE; important web site for documentation and software	http://world-2dpage.expasy.org/swiss-2dpage/
The World-2DPAGE Constellation	ExPASy Swiss protein 2D PAGE site; important repository and search engine software	http://world-2dpage.expasy.org/
Kazusa Genome Resources	Genome and proteomic resources from Miyakogusa.jp	http://genome.kazusa.or.jp/
Cyanobase	CyanoBase; resources for cyanobacteria	http://genome.kazusa.or.jp/cyanobase
Rhizobase	RhizoBase; resources for rhyzobacteria	http://genome.kazusa.or.jp/rhizobase/
MudPIT	Multidimensional protein identification technology	http://www.proteome.soton.ac.uk/mudpit.htm
FTICR/MS Technology	Fourier transform mass spectrometry (FTMS) derived from ion cyclotron resonance (ICR) spectrometry	http://www.wangnmr.com/FTICR_MS_technology.htm
DIGE	Protein abundance changes using dige multi-fluorescent dye	http://www.med.uc.edu/proteomics/dige.htm
DECODON—Differential in gel electrophoresis (DIGE)—image analysis—Delta2D	Generic proteomics site for information on methodologies	http://www.decodon.com/Solutions/Delta2D/digeAnalysis.html
Proteomics services: 2D DIGE, iTRAQ, MS, and Phosphorylation sites	Generic proteomics site for information on methodologies	http://www.appliedbiomics.com/
3Dye™ 2D DIGE Kits	Kits and uses for 3Dye 2D DIGE	http://www.lumiprobe.com/p/2d-dige-kits
Isotope-Coded Affinity Tags (ICAT) Methodology	The ICAT methodology and its uses—an interactive site	http://www.bio.davidson.edu/courses/genomics/ICAT/ICAT.html
ICAT, Quantitation, Quantification, Isotope Coded Affinity Tag Technology	ICAT uses in relative quantitative proteomics, especially regulation	http://www.creative-proteomics.com/ICAT.htm

Table 2.6 (continued)

Database Name	Plant Species/Purpose	*Uniform Resource Locator* (URL)
SILAC—Stable isotope labeling by amino acids in cell culture	Stable isotope labeling by amino acid in cell culturing for MS quantitative proteomics	http://www.silac.org/
Super-SILAC Technology for Quantitative Proteomics in Neoplasms	Peptide fingerprint analysis method used in conjunction with MS ionization	http://medgadget.com/2011/02/supersilac_technology_for_quantitative_proteomics_in_neoplasms.html
Duke Proteomics Core Facility—Protein Quantitation—IGSP	Facilities and methodology for protein quantification using methods described in Table 8	http://www.genome.duke.edu/cores/proteomics/services/protein-quantitation/

Sample Preparation

Sample preparation is perhaps one of the most important steps in proteomics research. Methods that use trichloroacetic acid (TCA) and acetone are still the most commonly used procedures for protein precipitation and separation from other metabolites in plant mixtures (Song et al. 2006; Wang et al. 2006). An alternate method using phenol and ammonium acetate/methanol is also popular for plant separation. Sample purification effectively improves protein detection and increases proteome separation in subsequent steps by reducing interference in samples. The use of different reagents to separate proteins by their different solubilities and membrane associations can be effective in reducing the complexity of proteins in fractions, and to enrich rare proteins in samples. Modifications of these basic procedures can be used to separate membrane associated proteins in large amounts from the remainder of the soluble protein fraction (Agrawal et al. 2005; Barjaktarovic et al. 2007).

Chromatography (SDS-PAGE)

Sequential treatment of the isolated proteins on various inert substrates is a common method for separating protein samples based on a combination of solubility, molecular mass and isoelectric point. At one time only one-dimensional SDS–PAGE had wide use in separating complex proteins, based only on their molecular mass, but now this approach has limited application. In contrast, high-resolution separation of proteins by two-dimensional gel electrophoresis (2-DE), which uses isoelectric focusing (IEF) in the first dimension and SDS–PAGE in the second dimension, is a more effective and exact technique. The recent development of the immobilized pH gradient (IPG)-IEF strips for use in the first dimension has improved reproducibility and resolution. The 2-DE method has been widely accepted in proteomics for a range of important crop and bacterial species containing massive cell walls (Chen and Harmon 2006; Herbert et al. 2006; Berkelman and Stenstedt 2002). Databases housing 2-DE information have been developed and released to the public.

For example, the Swiss Institute of Bioinformatics SWISS-2DPAGE database, Expasy and the Kazusa DNA Research Institute Cyano2Dbase are important. Different chromatography separation methods, such as gel filtration chromatography, ion exchange chromatography and affinity chromatography can also be used, but are used less often in plants (Wu et al. 2005; Frolich and Arnold 2006).

Identification Methods

To identify each protein or polypeptide found in a sample, peptide mass fingerprinting has been widely employed. Currently the most efficient method available consists of two steps; (i) enzymatic digestion of well separated proteins excised from gels into smaller peptides, and (ii) accurate mass measurements (fingerprints) of the peptide fractions using mass spectrometry (MS). Various 'in-gel' digestion methods, and modifications of these have been used to separate protein samples using 2-DE, and further developments of these methods will continue to play an important role in proteomics. The MS equipment generally consists of a source to ionise samples and a mass spectrometer(s) to detect the ionized samples. The matrix-assisted laser desorption ionization (MALDI) method is used in combination with time of flight (TOF) MS (as MALDI-TOF-MS), or the electrospray ionization (ESI) method is used in combination with quadrupole (Q) or ion trap (IT) MS. More recent developments in MS procedures, such as Q-TOF MS, IT-TOF MS or MALDI Q-TOF MS, have become popular. Furthermore, ion fragmentation by collision-induced dissociation (CID) using tandem MS such as Q-TOF MS or post-source decay (PSD) using MALDI-TOF MS have been used to determine more correctly peptide amino acid sequences. Eventually though, to identify the target proteins obtained, peptide mass fingerprint data are searched against a database of theoretically predicted masses of known amino acid sequences (Hirano et al. 2004; Newton et al. 2004). To aid in the correct identification of proteins and polypeptides, pI and molecular mass information from gels are used to check the accuracy of MS identification fingerprint data (De Filippis and Magel 2012).

Shotgun Proteomics

Conventional gel electrophoresis-based separation is by far the most common method used, however a gel-free separation method has been used from time to time; sometimes being referred to as a 'shotgun proteomics' approach. In this gel-free method, the protein mixture is directly digested into peptides and separated by one of the separation and identification methods just described.

MudPIT

The multidimensional protein identification technology (MudPIT) consists of a combination of different separation methods described before, in atypical 'shotgun approach'. MudPIT is especially suitable for the analysis of proteins that are difficult to separate by 2-DE, as well as for high-volume analysis by automated analytical instruments now in common use (Yates et al. 2009).

FT-ICR MS

Fourier transformation ion cyclotron resonance mass spectrometry (FT-ICR MS) possesses high resolution, high sensitivity, high dynamic range and high mass measurement accuracy. The high resolution and precision of FT-ICR MS allows researchers to carry out 'top-down proteomics', similar to a 'shotgun approach' in which an intact protein mixture is analysed directly, without separation and/or purification of the proteins (Bogdanov and Smith 2005).

Quantitative Proteomics

Quantification of the abundance of proteins identified is important for a better understanding of protein dynamics and kinetics, in response to cellular activities and environmental changes. A quantitative proteome approach also plays a crucial role in the discovery of key proteomic changes, including expression, repression, interaction and modification of proteins that are associated with genetic variations and/ or phenotypic changes in organisms (Gstaiger and Aebersold 2009).

DIGE

Difference gel electrophoresis (DIGE) is now a popular method for differential display of proteins for quantitative protein comparisons. In DIGE, protein samples are labelled with different fluorescent dyes before 2-D electrophoresis, enabling accurate determination of differences in protein abundance between samples (Rossignol et al. 2006). This method is effective in minimising and even negatinggel to gel variation while significantly increasing accuracy and reproducibility of samples. There are a number of commercial suppliers of 2D DIGE based gels available and ready for proteomic studies.

iCAT (iTRAQ)

Isotope-coded affinity tags (ICATs) and isobaric tags for relative and absolute quantification and comparison in basic regulation of proteins (iTRAQ), are other important methods in proteomics.

SILAC

Stable isotope labelling with amino acids in cell cultures (SILAC) is a method used for protein differential display using stable isotope labelling (Jorrin-Novo et al. 2009). Super SILAC technology is available from suppliers, and well suited for plant cell and tissue culture comparisons.

MS-MS Analysis (Differential Isotopes)

Using a single MS/MS analysis, corresponding peptides from each sample are differentially detected based on mass shifts caused by the different isotopes used; and this type of analysis allows comparison of relative protein and polypeptide abundance between samples.

LC-MS/MS

Recently, label-free quantitative techniques have been developed to facilitate high-throughput comparisons specific for proteomic expression. For example, label-free quantification in the proteomes from each of two samples are separately analysed using liquid chromatography (LC)-MS/MS. Then, each MS spectrum is aligned to calculate relative protein abundance and changes based on ion intensity differences, such as peptide peak areas or peak heights in the chromatograms. Finally, MS/MS analysis is used to identify the peptides of interest (Gstaiger and Aebersold 2009).

Subcellular Proteomics (Organelles)

Accurate and quick proteome analysis of cell organelles has become very important for understanding the various enzymatic activities within cell organelles, the compartmentalisation of metabolites and metabolic pathways, cellular logistics such as protein targets, their movement and regulation; and it has become very important to understand proteomic dynamics at the subcellular level (Andersen and Mann 2006; Chen and Harmon 2006; Baginsky 2009). A number of different approaches listed below have been applied to analyse the proteome of organelles and subcellular compartments of plant cells. Studies so far have included cell organelles and compartments like the chloroplasts, etioplasts, amyloplasts, chromoplasts, mitochondria, vacuoles, plasma membrane, nucleus, peroxisomes, cytosolic ribosomes and cell walls (Baginsky 2009).

Chloroplasts, Mitochondia and Organelles

Proteomic analyses of chloroplasts, mitochondria and othercell fractions have been carried out to determine detailed localisation of proteins in sub-cellular compartments. Methods for organelle isolation and purification are already available, and are essential in the initial steps before protein is isolated and identified from them (Sakai et al. 2004; Holy and Perkins 2009).Techniques for quantitative proteomics, such as ICAT and iTRAQ described above, are then applied to resolve quantitative data on the proteome in each organelle or cell compartment. In *Arabidopsis*, rice and algae, differential proteome profiles of the plant plasma membranes were obtained, and used to identify different proteins expressed in response to environmental stresses such as cold, salt stress and bacterial elicitors (Benschop et al. 2007; Katz et al. 2007; Cheng et al. 2009; Minami et al. 2009). Listed below are the most important web-based sites for sub-cellular proteomics, their purpose and the URL are detailed in Table 2.7. Several databases below provide subcellular proteome information.

Rice Proteome Database

The rice proteome database is a 2-DE image information base for rice that contains data from various tissues, as well as subcellular compartments and organelles (Komatsu 2005).

Plant Organelle Database and GOBASE

These two web sites detail numerous external links to plant organelles (O'Brien et al. 2009).

NASC Proteomics Database

The Nottingham *Arabidopsis* Stock Centre (NASC) Proteomics database is also useful for both cellular and organelle data mining of proteins.

SUBA

The Sub-cellular location database for *Arabidopsis* proteins (SUBA) provides sub-cellular proteome analytical and energy data on proteins in subcellular compartments (Dunkley et al. 2006).

Table 2.7 Integrative databases for Sub-cellular Proteomics and Protein Modification analysis in plants

Database Name	Plant Species/Purpose	*Uniform Resource Locator* (URL)
Enzyme Database—BRENDA	A comprehensive protein and enzyme information search and retrevel data system; an important proteomics site	http://www.brenda-enzymes.org/
Rice Proteome Database	Rice proteome database; protocols and tools including sub-cellular proteins and modifications of proteins	http://gene64.dna.affrc.go.jp/RPD/
Proteomics Database for Arabidopsis data	Arabidopsis NASC) proteome database; protocols and tools including sub-cellular and modifications of proteins	http://proteomics.arabidopsis.info/
SUBA II	Search engine for Arabidopsis proteins, including phosphoproteins and protein modifications	http://suba.plantenergy.uwa.edu.au/
My 2D-PAGE—map viewer	Soybean proteome database; 2D protocols and tools including sub-cellular and modifications of proteins	http://proteome.dc.affrc.go.jp/cgi-bin/2d/2d_view_map.cgi
PhosPhAt 3.0	Protein phosphorylation database mainly using MS, and predictor software	http://phosphat.mpimp-golm.mpg.de/
P3DB—Plant Protein Phosphorylation DataBase	Plant protein phosphorylation database for 6 plants, but contains over 11,000 phosphoproteins and 32,000 phosphosites	http://www.p3db.org/
GOBASE—The Organelle Genome Database	Gene databank for cell orgamelles and sub-cellular compartmentation	http://gobase.bcm.umontreal.ca/

Soybean Proteome Database

The soybean proteome database also provides 2-DEdata for various tissues, as well as for subcellular compartments and organelles (Sakata et al. 2009).

Gramene

A database for plant comparative genomics and proteomics providing information on monocotyledon (grass) plant species, and is cross linked to the GenBank protein and organelle databases (Ware 2007).

Post-Translation Protein Modification

Modificon

Modificon research is when a comprehensive approach is used to investigate various kinds of post-translational protein modifications, which can play an important role in our current understanding of proteomics. Modificon data reports and identifies modified proteins, and elucidates and coordinates the role of each protein modification with its associated biological action (Kwon et al. 2006). In this regards protein phosphorylation is one of the most critical key regulatory process we have discovered that can control the expression of many regulatory proteins. The list of web sites and URL for protein modification structure and analysis are given in Table 2.7, including their URL.

Protein Phosphorylation

Protein phosphorylation is an important regulatory step in most signalling pathways, and is a widespread protein modification step affecting most basic cellular processes in eukaryotic organisms (Schmelzle and White 2006). Advances in MS-based technologies, accompanied by phosphopeptide enrichment techniques have allowed researchers to perform high-volume, large-scale *in vivo* phosphorylation site mapping. So far, several different plant phosphoproteome studies have been reported (Nuhse et al. 2004; de la Fuente van Bentem et al. 2006; Benschop et al. 2007; Nuhse et al. 2007; Sugiyama et al. 2008). For example, the proteome-wide mapping of *in vivo* phosphorylation sites in *Arabidopsis* have recently been achieved, and some other preliminary studies have been completed on other plants (Chitteti and Peng 2007; Barjaktarovic et al. 2009).

LC MS/MS

A primary method for proteome-wide mapping of *in vivo* phosphorylation sites in Arabidopsis, and this has recently been achieved by using complementary phosphopeptide enrichment methods, coupled to high-accuracy LC-MS/MS with a Finnigan LTQ-Orbitrap (Sugiyama et al. 2008).

PhosPhAT

The *Arabidopsis* Protein Phosphorylation Site Database (PhosPhAt) provides information on *Arabidopsis* phosphorylation sites, which have been identified in MS by different research groups.

P3DB

The Plant Protein Phosphorylation Database (P3DB) is an information resource for plant phosphoproteome research, andprovides a resource for protein phosphorylation data and detail information from multiple plant species (Gao et al. 2009).

Ubiquitination

Ubiquitination of protein is another one of the post-translational modifications occurring in most eukaryotic cells, including plants. Protein ubiquitination is another regulatory mechanism that controls protein localization and activity. Several large-scale analyses of protein ubiquitination sites in plants have been reported (Maor et al. 2007; Manzano et al. 2008; Igawa et al. 2009).

Anti-Ubiquitin Antibody

In Arabidopsis, affinity purification using an anti-ubiquitin antibody, and the subsequent use of MS/MS analyses has been performed to identify ubiquitinated proteins (Igawa et al. 2009).

Structural Proteomics

Large data sets of protein 3D structures are also important as information resources for elucidating relationships between protein function and structure, or for analysing the active sites and their molecular identity in various protein complexes. To deal with the technical aspects, methodology and interpretation of protein structure in more detail is beyond the scope of this review, and I refer you to some recent reviews on this topic (Yan and Chen 2005; Wlodawer et al. 2008). However listed below in summary form are some of the most important features for structural proteomics and web-based sites for protein structural projects, their purpose and URL are detailed in Table 2.8.

ISGO

The International Structural Genomics Organization (ISGO) site (Stevens et al. 2001) is designed to facilitate co-operation to determine protein structures using existing instrumentation in a number of international laboratories for structural proteomics, and have identified many DNA-binding domains (DBDs) of plant-specific transcription factors (TF) (Yamasaki et al. 2004; Yamasaki et al. 2008b).

Table 2.8 Integrative databases for Structural Proteomics analysis in plants

Database Name	Plant Species/Purpose	*Uniform Resource Locator* (URL)
International Structural Genomics Organization	ISGO main web site: Reports, Activities and Publications	http://www.isgo.org/
Welcome to RSGI (RIKEN)	RIKEN Structural Genomic Proteomic Initiative—Home page	http://www.rsgi.riken.jp/
RSGI (RIKEN Structural Genomics/Proteomics Initiative) structure | Integrated Database of Protein | SciNetS	Integrated database for protein structure, including Arabidopsis	https://database.riken.jp/sw/links/en/cria266s4i/6p
Download | (RIKEN Structural Genomics/ Proteomics Initiative) Somantic Ontology Class Database | RDF/ OWL	Download web page for most of the RIKEN proteomic software	http://biolod.org/class/cria266s4i/RSGI_RIKEN_Structural_Genomics_Proteomics_Initiative_structure
Home : PSI-Nature Structural Biology Knowledgebase	Web site to keep informed on structural genomics and biology	http://www.sbkb.org/
The Protein Structure Initiative: achievements and visions for the future—F1000 Biology	Web site for research articles and methods on structural genomics and structural biology	http://f1000.com/reports/b/4/7
PSI Pilot Phase Fact Sheet—National Institute of General Medical Sciences	NIH web site for proteomics fact sheets and protein information in mainly the health sciences	http://www.nigms.nih.gov/Research/FeaturedPrograms/PSI/Background/PilotFacts.htm
RCSB Protein Data Bank—RCSB PDB	Portal and search engine –Biological Macromolecular Structures and Resouces	http://www.pdb.org/pdb/home/home.do
CLC bio: Integrated 3D molecule viewer	Integrated workbench for analyzing 2D gel data and MS peptide fingerprinting profiles	http://www.clcbio.com/index.php?id=500
I-TASSER server for protein structure and function prediction	Complete function and structure prediction for over 1,00,000 proteins from international sources linked to the NCBI protein databanks	http://zhanglab.ccmb.med.umich.edu/I-TASSER/
DisEMBL 1.5—Predictors of intrinsic protein disorder	Intrinsic protein disorder prediction web site linked to SWISS protein identification	http://dis.embl.de/
GTOP database	Genomes TO Protein structure and function site; unfortunately contains no plant data, but has interesting bacteria and virus	http://spock.genes.nig.ac.jp/~genome/gtop.html

Table 2.8 (continued)

Database Name	Plant Species/Purpose	*Uniform Resource Locator* (URL)
CATH: Protein Structure Classification Database—Prof. Orengo's Bioinfomatics Group at UCL, London, UK	Proteins classified into structural domains and superfamilies; useful web site once protein identity is known even with plants	http://www.cathdb.info/
SCOP: Structural Classification of Proteins	Contains proteins in fold domains and superfamilies; useful in assessing protein structural similarities across species	http://scop.mrclmb.cam.ac.uk/scop/index.html
PANTHER—Classifications of Genes and Proteins	Browser and search engine for proteins, and divides them into ontogeny and superfamilies	http://www.pantherdb.org/

RIKEN (RSGI)

The RIKEN Structural Genomics/Proteomics Initiative (RSGI) in Japan is an important and key centre for structural genomic and proteomic analysis. BioLOD.org is another of the RSGI sub-web sites for structural proteomics. The RIKEN SGPI has solved over 2,700 protein structures, including 33 from Arabidopsis that appear on another important web site, the PDB site listed below.

SBKB from PSI Nature

Keeps you informed about advances in structural biology and structural proteomics. It is an easy to navigate web site and includes information on other proteomic research done at other web addresses.

PSI and F1000

The Protein Structure Initiative (PSI) in the USA, and the structural genomics centres of Europe (Yokoyama et al. 2000) are important sites. The PSI has promoted large-scale attempts to determine the 3D structure of proteins and deposition into a database. In 2005, the PSI shifted to its second phase, known as PSI-2 'information phase' which was to solve more challenging structures such as protein complexes and folding, and to identify membrane associated proteins (Fox et al. 2008).

PDB

The number of solved protein structures appearing in the protein data bank (PDB) is fast growing, and has been one of the most popular resources for biomolecular structural data; the PDB site has had a dramatically increased in available data deposited during the past decade (Kouranov et al. 2006).

CLC BIO

CLC has an integrated 3D molecular viewer for determining structures of proteins. It has a fully navigational and integrated 3D viewer tool for use. It has file systems and graphics that are compatible with publication of the 3D structures directly.

Cell-Free System (Wheat Germ Embryo)

Although methodological problems still exist in structural and analytical proteomics, some new methods and computational advances have played important roles in cellular protein determination and identification. One major bottleneck has been the production of proteins as they may be present in living organism, in 3D structure. Most researchers in structural proteomics have used *Escherichia coli* cells for protein production in automated methods, as an approximation of the cell-based 3D protein structure. However *Escherichia coli* is a bacterium and prokaryote, and doubts have been expressed as to the protein folds and structures assembled in a prokaryote being relevant to eukaryotic organisms. Cell-free expression systems have also been used mainly as a method to address several limitations of cell-based methods, such as protein quality and quantity, and high throughput issues. The wheat germ embryo cell-free system has been developed as a eukaryotic cell-free system to overcome such problems, and has the advantage of producing multi-domain proteins (Endo and Sawasaki 2003, 2006). For example, a comparative study of protein production from 96 *Arabidopsis* open reading frames (ORFs) demonstrated marked differences in protein profiles between the wheat germ cell system and the cell free system (Tyler et al. 2005).

NMR Methods Plus CP/MAS/DD

The technology and platform of NMR spectroscopy has played an important role in structural proteomics. High-resolution multidimensional solid-state NMR methods used in combination with cross polarization (CP), magic angle spinning (MAS) and dipolar decoupling (DD) are now becoming the methods of choice for structural analysis in NMR equipment (Castellani et al. 2002; McDermott 2009). Recent improvements in NMR include a cryoprobe for improved sensitivity, a micro-coil probe for sample mitigation, and a flow-probe designed to shorten preparation time.

X-Ray Crystallography

X-ray crystallography has also been used to determine the protein structures of almost 90 % of protein entries in the PDB database. The third generation X-ray synchrotrons have become essential for macromolecular crystallography (MX) of large proteins and protein complexes (Samatey et al. 2001). For example, the synchrotron SPring-8 of RIKEN in Japan has been used to determine the structures of important membrane proteins and large complex proteins; such as Ca^{2+}-ATPase, rhodopsin and flagellin (Palczewski et al. 2000; Samatey et al. 2001; Toyoshima et al. 2003).

Informationand Web Resources in Structural Proteomics

Bioinformatics and related databases are therefore important tools for advancing the study of structural proteomics. The methods used in computational prediction of protein 3D structures are readily available and described by Zhang (2008) and Zhang (2009b). Free modeling, which is sometimes called '*de novo*' modelling is used to predict the 3D structure of proteins, and the web-based sites for information on structural proteomics, their purpose and URL are detailed in Table 2.8.

I-TASSER

A number of web servers and computational tools for free and/or template-based protein modelling have recently been made available; for example, the I-TASSER internet server.

CASP

I-TASSER is often used in Critical Assessment of Techniques for Protein Structure Prediction (CASP) (Zhang 2009a). This template-based modelling method is a comparatively new method for matching proteins using evolutionarily related proteins of known structure as a template.

Swiss Model Server (DisEMBL)

There are many web services and tools (e.g. Swiss Institute of Bioinformatics SWISS-MODEL server) to support template-based modelling (Schwede et al. 2003). The Intrinsic Protein Disorder prediction program, which is linked to the Swiss Model server, is one such piece of software.

GTOP

Databases housing previously predicted structures from amino acid sequences by template-based modelling for a wide range of species exist on this site; the Genome TO Protein structure and function (GTOP) database. Data in this site is obtained by the application of various computational tools for structural prediction from sequences of amino acids and genomes (Fukuchi et al. 2009).

CATH/SCOP

The databases for structure-related protein classification, as typified by CATH and the Structural Classification of Proteins (SCOP) data sites, have provided important clues to the relationship between proteins, protein function and protein evolution (Greene et al. 2007; Andreeva et al. 2008).

PANTHER

A database for protein families based on conserved protein domains, the Protein Analysis Through Evolutionary Relationships (PANTHER) is an important site for classifying proteins into families. The database contains the evolution and function of proteins, and the resources are used in genome-wide identification of genes (Mi et al. 2005; Wilson et al. 2007; Finn et al. 2008; Wilson et al. 2008).

Plant-Metabolism (Metabolomics)

Metabolomics is referred to as an understanding of metabolism primarily based on comprehensive and multidimensional approaches, by taking advantage of various analytical instruments and bioinformation available to identify metabolites. Metabolomic data can include individual and multiple assessments of metabolites, and to quantify particular metabolites in order to provide advantages over chemical phenotype analysis. The plant metabolome is complex enough for an individual plant, but it is even more challenging for comparison between plants (Bino et al. 2004; von Roepenack-Lahaye et al. 2004). Therefore, plant metabolomics is a great analytical challenge, but nevertheless it is important to an understanding of plant growth and development. Metabolomics has the ability to improve knowledge of plant cell systems, to engineer molecular breeding and to improve the productivity and function of plants in areas like stress tolerance, pharmaceutical production, food quality, biomaterials and energy (Trethewey 2004; Oksman-Caldentey and Saito 2005; Fernie and Schauer 2009). In this section, we include metabolomic analytical sites for plants, metabolic profiling, instrumentation and their applications with respect to other 'omics' research. We also describe plant metabolomics related computational tools and databases commonly in use.

Instruments in Metabolomics

Many technological advances have recently been made in instrumentation related to metabolomics. Metabolomics data starts with the acquisition of metabolic fingerprints using these analytical instruments (Fiehn et al. 2000; Roessner et al. 2001; Fernie et al. 2004). Methods for sample separation, such as gas chromatography (GC), high-performance or ultra performance liquid chromatography (LC) and capillary electrophoresis (CE) are used, in conjunction with various types of MS. CE-MS is especially effective, because it is a highly-sensitive method for separating and analysing biological molecules (Ramautar et al. 2009). Detailed below, are the URL web pages for explanation, methodology and instruments used in metabolomics, and outlined in Table 2.9.

QMS and TOF MS

QMS is especially usefull in metabolomics. TOF MS are also well regarded for use in metabolomics. Triple Q (QqQ) MS (a tandem-type MS) and Q-TOF (a hybrid type MS) are also in common use.

FT-ICR MS

Methods that do not involve pre-separation of samples can be used, e.g. FT-ICR MS, are often employed, allowing for MS analysis even on quite crude plant preparations (Werner et al. 2008).

NMR Based Methods

NMR-based methods are used in metabolomic analysis (Dixon et al. 2006; Schripsema 2009). These methods can be broadly classified into solution NMR and insoluble or solid-state NMR, according to sample solubility. High-resolution (hr)-MAS techniques can generate suitable metabolic fingerprints from insoluble samples and solid-state preparations (Bertocchi and Paci 2008). One-dimensional NMRis where protons (1 H) are observed (1 H-NMR), however more detailed analyses and metabolite identification and/or flux analysis can be obtained with other nuclei, particularly 13 C and 15 N (Kikuchi et al. 2004; Sekiyama and Kikuchi 2007). Processed sets of data are subsequently used to identify metabolites corresponding to each spectrum signal by searching against standard compound databases. In non-target analyses, spectrum data sets that include spectra of unknown compounds are subjected to statistical analyses, such as multivariate analysis (Sect. 5.1.8 below), to mine data for biological significance (Tikunov et al. 2005). In target analyses, spec-

Table 2.9 Integrative databases for Metabolomics and Instrumentation analysis in plants

Database Name	Plant Species/Purpose	*Uniform Resource Locator* (URL)
GC/MS Web Site	Web site for explanation of gas chromatography and mass spectrometry	http://www.scientific.org/tutorials/articles/gcms.html
Comprehensive Pesticide Residue Analysis by LC/MS/MS Using an Ultra Aqueous C18 Column/Foods, Flavor—Restek Chromatography	Agricultural, food, flavor and fragrance identity of compounds via LC/MS/MS—a commercial company site	http://www.restek.com/Technical-Resources/Technical-Library/Foods-Flavors-Fragrances/fff_A020
Lipid Analysis with GC-MS, LC-MS, FT-MS—Metabolomics Fiehn Lab	Metabolite (lipid) identity of compounds via LC/MS/MS—another commercial site	http://fiehnlab.ucdavis.edu/staff/kind/Metabolomics/LipidAnalysis/
FTIR Analysis	Industrial application and compound identity—FTIR	http://www.semlab.com/ftir.html
FTIR Analysis—Fourier Transform Infrared Spectroscopy	Web site explaining FTIR analysis and identity	http://www.ides.com/articles/testing/2008/FTIR_Analysis.asp
NMR Analysis, Processing and Prediction	Explanation of NMR analysis	http://nmr-analysis.blogspot.com.au/
NMR Analysis	NMR analysis, methodology and equipment requirements	http://www.intertek.com/analysis/nmr/
Metabolome Analysis Service by CE-MS	Various applications for CE-MS in drugs, toxicology, disease, blood, cell and tissue; including possible use in plants	http://www.infinitebio.com/products/HMT_Metaborolomics_Analysis/index.html
QMS 403 C Aëolos®—Quadrupole Mass Spectrometer-NETZSCH	New dimensions and equipment for gas analysis; the QMS 403; the quadruple MS instrument	http://www.netzsch-thermal-analysis.com/en/products/detail/pid,33.html
Evolved Gas Analysis-EGA (QMS, TG-GC-MS, FTIR, SKIMMER)	Quadruple MS and standard MS analysis, especially coupled to FTIR; pulse thermal analysis	http://www.netzsch-thermal-analysis.com/en/products/evolved-gas-analysis/
GEN | Application Notes: MALDI-TOF MS Analysis of Nanotrap Enriched Low Molecular Weight Protein—Human Serum example	MALDI-TOF MS; used to select Biomarkers which are often masked by high abundant proteins, and/or may be susceptible to proteolysis	http://www.genengnews.com/application-notes/maldi-tof-ms-analysis-of-nanotrap-enriched-low-molecular-weight-proteins-from-human-serum/22/
NMR Spectroscopy	Web site to explain Proton NMR spectroscopy and multi uses	http://www2.chemistry.msu.edu/faculty/reusch/VirtTxtJml/Spectrpy/nmr/nmr1.htm

Table 2.9 (continued)

Database Name	Plant Species/Purpose	*Uniform Resource Locator* (URL)
SOMA | Self Organizing Map Application Site	A web tool used in DNA and protein fragment classification	http://soma.arb-silva.de/
Recent advances from the F1000 Biology Reports	Research article on the web describing recent advances in metabolomics and greener pastures (ie in plant biology)	http://f1000.com/reports/b/2/7/

trum data sets that are associated with particular compounds are used as metabolic profiles for each compound.

Statistics (PCA-HC-SOM)

Data analysis is most important in determination of biological significance in metabolomics. Statistical analyses using multivariate analysis, such as principal component analysis (PCA), hierarchical clustering analysis (HCA) and self-organization mapping (SOM), are typically used to classify samples and/or metabolites (Hirai et al. 2004; Jonsson et al. 2004; Matsuda et al. 2009). The visualization of metabolic profiles on metabolic maps is also used in combination with other 'omics' methods, which can include gene expression profiles of particular genes encoding the enzymes involved in those particular pathways (Thimm et al. 2004; Tokimatsu et al. 2005).

Metabolite Profiling Plants

The systematic collection of metabolite profiles is the initial step in metabolomics. This step can be performed with various instruments (described above) capable of high turnover and simultaneously many measurements. Comprehensive metabolic profile data can contribute to the understanding of the cellular system in response to changes in intracellular and extracellular metabolites. Furthermore, the changes in metabolic pathways associated with genetic variations can be evaluated as chemical fingerprints to identify genes involved in metabolism. A number of studies of metabolic profiling in plant species have been performed that have resulted in the release of results; unfortunately these results are mostly present on web databases. Listed below are the most important web-based sites for metabolite and product analysis, their purpose and URL are detailed in Table 2.10.

Table 2.10 Integrative databases for Metabolite and Products analysis in plants

Database Name	Plant Species/Purpose	*Uniform Resource Locator* (URL)
Metabolomic Tool Kit—GARNet	Arabidopsis web site predicting metabolites and enzymes; software 'omics viewer is particularly usefull	http://www.garnetcommunity.org.uk/resources/metabolomic-tool-kit
Metabolomics Bioinformatics	Search engine for metabolomics and protein	http://appliedbioinformatics.wur.nl/moto/
Database—KOMICS	A series of databases for plant transcriptome and metabolome analysis	http://www.kazusa.or.jp/komics/en/en-database.html
ARMec Repository	Database for mass spectrometry identity	http://www.armec.org/MetaboliteLibrary/
CSB.DB: GMD	Golm metabolome database for some model organisms, including Arabidopsis	http://csbdb.mpimp-golm.mpg.de/csbdb/gmd/gmd.html
MS2T	Phytochemical spectral data for proteomic research; Arabidopsis, rice, wheat, soybean, brassica, allium	http://prime.psc.riken.jp/lcms/ms2tview/ms2tview.html
PRIMe LC-MS Branch	Mass spectrometry of plant fractionation databank of secondary metabolites	http://prime.psc.riken.jp/lcms/
AtMetExpress LCMS	Mass spectrometry data of different growth stage and metabolites in Arabidopsis	http://prime.psc.riken.jp/lcms/AtMetExpress/
PRIMe: Simple BL-SOM	Self-organising map and statistics tool to detail genes/metabolites into a series of rows for analysis	http://prime.psc.riken.jp/?action=blsom_index
TAIR—Home Page	Arabidopsis information on genetic maps, markers, sequencing and expression	http://www.arabidopsis.org/
KEGG PATHWAY Database	Collection of manually drawn pathway maps for molecular interactions	http://www.genome.jp/kegg/pathway.html
Plant metabolic pathway database (PMN/Plant Cyc) home page	Broad network of plant metabolic pathways; including literature, analysis, proteins, enzymes	http://www.plantcyc.org/
Pathway Tools Information Site	Comprehensive systems biology software and bioinformatics databank	http://bioinformatics.ai.sri.com/ptools/
Home—MapMan	Software for analysis of RNA and sequences	http://mapman.gabipd.org/web/guest
Home—KaPPA—Portal	Web based databank for visual uploaded 'omics' metabolic pathway data	http://kpv.kazusa.or.jp/en/

Table 2.10 (continued)

Database Name	Plant Species/Purpose	*Uniform Resource Locator* (URL)
PRIMe: Platform for RIKEN Metabolomics	Web based service for transcriptomics and metabolomics in life science	http://prime.psc.riken.jp/
International Potato Center	Potato centre web site which includes the potato databank	http://cipotato.org/

Plant Metabolomics

In the case of Arabidopsis, an NSF-funded multi-institutional project aimed at fully developing the metabolomics database, 'Plantmetabolomics', has recently been documented through GARNet. The site contains a tool kit of resources for metabolic data analysis and comparison.

Solanaceae (Tomato/Potato)

Several databases for Solanaceae and other plants are available (also web sites below) at the Plant Research International at Wageningen, Netherlands homepage. The Metabolome Tomato Database (Moco et al. 2006) and potato database CIPOTATO were developed here (Mullins et al. 2006).

KOMICS

The KOMICS (Kazusa-omics) database collects annotation data of metabolite peaks detected by LC-FT-ICR-MS, and the web site contains a representative metabolome data set for the model tomato cultivar, 'Micro-tom' (Iijima et al. 2008).

ARMeC Repository Project

The Armec Repository Project provides metabolome data on the potato, and serves as a data repository for metabolite peaks detected by ESI-MS.

Golm Metabolome Database

The Golm Metabolome Database (GMD) provides public access to custom mass spectral libraries and metabolite profile data, and contains additional tools and base information (Kopka et al. 2005).

MS2T/PRIMe

The MS/MS spectral tag (MS2T) libraries at the Riken Metabolomics (PRIMe) website provides access to libraries of phytochemical LC-MS2 spectra obtained from various plant species.

LC-ESI-Q TOF/MS

By using the automatic MS2 retreval function of LC-ESI-QTOF/MS (Matsuda et al. 2009), MS2T/PRIMe and LC-ESI-Q TOF/MS databases can play crucial roles as information resources and repositories of large amount of new data. They also serve as tools for further integration of metabolic profiles containing comprehensive data acquired from other 'omics' research (Akiyama et al. 2008).

Combined Approaches in Metabolomics

Metabolome research can help in the understanding of metabolic activity at the cell, tissue and organ levels, and achieving profiles complementary to other 'omics' research; which can aid genetic variation studies. At present these combination of methods have been demonstrated mostly in the model plant Arabidopsis by utilising the many other 'omics' web sites and resources that currently exist for this model plant; including whole-genome sequencing, large-scale transcriptome data and related expression data, bioresources from mutants, and full-length cDNA clones. The experimental scheme for systematic information retrieval of gene-to-metabolite data through a combination of transcriptome and metabolome resources has been demonstrated by Saito's group at the RIKEN Plant Science Centre (Saito et al. 2008). The URL list and resources below are also listed in Table 2.10.

map (BL-SOM)

Data containing transcriptome and metabolome changes in Arabidopsis under stress conditions induced by deficiency of sulfur and nitrogen were analysed using batch-learning, self-organizing map (BL-SOM) analysis, which enabled the identification of genes involved in glucosinolate biosynthesis (Hirai et al. 2004). An integrated approach with an activation-tagged mutant line with overexpressed genes wasd used to identify genes involved in anthocyanin biosynthesis (Tohge et al. 2005).

ATTED II Database

Co-expression data of the *Arabidopsis* transcriptome has been provided by the ATTED-II database. Investigation of key genes involved in specific metabolic pathways, and metabolome analysis was used with mutant lines (Obayashi et al. 2009). The ATTED-II database has identified genes involved in lipid metabolism, and UDP-glucose pyrophosphorylase 3 (UGP3) as an essential requirement in the first step of sulfolipid biosynthesis (Okazaki et al. 2009). Co-expression analysis was used to identify genes related to flavonoid biosynthesis, and the role of two key and important flavonoid pathway genes *UGT78D3* and *RHM1* (Yonekura-Sakakibara et al. 2008).

Results in Metabolite Profiling

The metabolic pathways that act in response to cold and dehydration conditions in *Arabidopsis* were investigated by metabolome analysis using MS and microarray analysis of overexpressors in genes encoding transcriptional factors (Maruyama et al. 2009). Metabolomic profiling was also used to investigate chemical phenotypic changes between wild-type Arabidopsis and a knockout mutant of the *NCED3* gene under dehydration. The metabolic data was integrated with transcriptome data to reveal ABA-dependent regulatory pathways (Urano et al. 2009). Metabolome profiling has also been used to evaluate chemical phenotypes of natural variations and/or segregating populations in plant ecology and plant breeding. Analysis between metabolic expression and genomic diversity will enable the discovery of more key genes involved in differences between metabolic and phenotype expression of plants (Schauer et al. 2008; Fu et al. 2009).

Metabolite QTL (mQTL)

Metabolite QTL (mQTL) analysis using segregated populations has been applied to various plant species such as *Arabidopsis*, poplar and tomato in a popular 'forward genetics' approach (Morreel et al. 2006; Schauer et al. 2006; Lisec et al. 2008; Rowe et al. 2008; Schauer et al. 2008).

Metabolic and Genomic Diversities

The recent availability of data for genome-wide variation acquired by high-throughput genotyping methods, including high volume resequencing, has provided some details of genes association with nucleotide variation and phenotypic changes; especially ibn relation to the identification of key genes that play significant roles in evolutionary histories and phylogeny (Sect. 2.5). Attempts to mine correlative patterns between metabolic and genomic diversities have recently been applied to sesame and rice using seeds of natural variant phenotypes (Laurentin et al. 2008; Mochida et al. 2009a).

Information Resources for Metabolomics

Various information resources related to metabolomics have played a central role not only in metabolome research but also in synergistic integration of data with other 'omics' information through a variety of sites; such as the web sites listed below and their URL are detailed in Table 2.10.

TAIR

The web site of metabolome resources at TAIR provides a summarized list of web hyperlinks to resources that facilitate metabolome research around the world, and other web pages and sites.

KEGG

A set of biological pathway maps is available via the Kyoto Encyclopaedia of Genes and Genomes (KEGG), using a popular database for general information on life sciences called the KEGG PATHWAY Database (Kanehisa and Goto 2000; Kanehisa et al. 2008).

PMN

The Plant Metabolic Network (PMN) is a collaborative project that aims to build plant metabolic pathway databases for plants. It contains a number of other important web tools described below.

Plant Cyc

PlantCyc, is a comprehensive plant biochemical pathway database that contains information from available literature, and computational analyses and tools for genes, enzymes, compounds, reactions and pathways involved in primary and secondary plant metabolism.

AraCyc/PoplarCyc

AraCyc and PoplarCyc are also available at the PMN web site, which provides a review of information about metabolic pathways in Arabidopsis and poplar respectively (Mueller et al. 2003). There are also metabolic pathway databases for several other plant species generated by PMN researchers. MapMan is a tool to project 'omics' data onto metabolic pathways (Thimm et al. 2004).

KaPPA-View

KaPPA-View is a web-based analysis tool that can be used to superimpose transcriptome and metabolome data onto plant metabolic pathway maps (Tokimatsu et al. 2005).

PRIMe

PRIMe is a web-based service that provides data of metabolites measured by multidimensional NMR spectroscopy, GC-MS, LC-MS and CE-MS, in an integrated approach to analysis of comprehensive data within the metabolome and transcriptome search engines present there (Akiyama et al. 2008).

Plant Phenotype (Phenome) Analysis

Analysis of mutants is an effective approach for investigation of gene structure and function (Springer 2000; Stanford et al. 2001). Collections of mutant lines are also essential bioresources for accelerating forward and reverse genetics in plants. The available mutant resources for phenome analysis in plant species have been well described in a recent review by Kuromori et al. (2009). The demand for comprehensive collection of mutants and related information has increased dramatically encouraged by the high-throughput and genome-wide phenome analysis in plants (Alonso and Ecker 2006). Listed below are six of the most important web-based sites for phenome analysis and insertion mutants, their purpose and their URL are detailed in Table 2.11.

Insertion Mutants

Insertion mutant resources with index data that document the inserted gene mutation position have become extremely beneficial resources to investigate functional analysis of genes that can be up-regulated by promoters, or disrupted, down-regulated and silenced by reverse genetics.

Table 2.11 Integrative databases for Phenotype and Genotype Tagging analysis in plants

Database Name	Plant Species/Purpose	*Uniform Resource Locator* (URL)
SNPs between Nipponbare IRGSP build4—Koshihikari	Generic browser for SNP in plants, animals and micoorganisms, and chromosome locations	http://koshigenome.dna.affrc.go.jp/cgi-bin/gbrowse/koshig/?help=general
Rice Genome Annotation Project	Sequence annotation of the 12 rice chromosomes	http://rice.plantbiology.msu.edu/
OryzaSNP @ MSU	Rice SNP identification and sequence; over 150,000 high quality SNP	http://oryzasnp.plantbiology.msu.edu/
OryGenesDB: database for rice reverse genetics	Rice database and information on Flanking Sequence Tags (FST)	http://orygenesdb.cirad.fr/
FOX Hunting | RIKEN SciNetS	Full-length cDNA library for plant expression studies by repeated transformation	http://nazunafox.psc.database.riken.jp/sw/en/FOX_Hunting_/ria61i/
Rice FOX Database	Rice full-length cDNA overexpressed Arabidopsis mutant database website	http://ricefox.psc.riken.jp/
Tilling	Targeting induced local lesions in genomes website; wide application in plant gene function and breeding	http://tilling.ucdavis.edu/index.php/Main_Page
TILLING Genes To Improve Soybeans	Soybean targeting induced local lesions in genomes site	http://www.ars.usda.gov/is/ar/archive/jul05/genes0705.htm
About RevGen UK	Platform and data for gel based reverse genetics in Lotus, Medicago, Brassica	http://revgenuk.jic.ac.uk/about.htm
TILLING DATABASE	Platform and databank for phenotype recording in Pea, Tomato and Brachypodium	http://urgv.evry.inra.fr/UTILLdb
CATMA Website	Complete Arabidopsis transcriptome microarray page for gene specific tag (GST)	http://www.catma.org/
http://www.agrikola.org/	Systematic RNAi knockouts in Arabidopsis GST from the CATMA consortium	http://www.agrikola.org/index.php?o=/agrikola/%20html/index
GO:0009616 virus induced gene silencing	Virus induced gene silencing wbsite; including cross-links	http://www.ebi.ac.uk/QuickGO/GTerm?id=GO:0009616
fioreDB: Database for Flower Bio-engineering by CRES-T system	Gene silencing technology and database for genes encoding crop, flower factors	http://www.cres-t.org/fiore/public_db/index.shtml

T-DNA-tagged (SNP)

Transferred DNA-tagged (T-DNA-tagged) lines and transposon-tagged lines have recently become popular for investigation of insertion mutants in plants, and short nucleotide polymorphisms (SNPs) (Krysan et al. 1999). The two component maize transposon, *Activator* (*Ac*)/*Dissociation* (*Ds*), is a popular system for the production of transposon-based insertion mutations, and generation of mutants with a high proportion of single-copy insertions (SNP) (Long et al. 1993). In rice, the endogenous retrotransposon *Tos17*, which can be activated under controlled conditions, is also available for the study of the insertion mutant lines of *japonica* rice cultivars (see under SNP markers for rice).

Nipponbare

Nipponbare (Miyao et al. 2003, 2007), is the web resource that provides information on the rice *Tos17* mutant rice lines with flanking sequences of insertion.

Maize Enhancer/Suppressor Mutator

The maize *Enhancer/Suppressor Mutator* (*En/Spm*) element has also been used as an effectivetool for the study of functional genomics in maize cultivars (Kumar et al. 2005).

Enhanced Trap (ET)/Gene Trap (GT)

The enhancer trap (ET) and the gene trap (GT) genetic constructs have been coupled with T-DNA and *Ac/Ds* transposons, and can facilitate discovery of genes adjacent to promoter or enhancer activities sites in plants (Sundaresan et al. 1995; An et al. 2005).

OryGenes DB

OryGenesDB is a database that integrates information of available insertion mutant resources in rice (Droc et al. 2009). There are a number of resources for insertion mutant populations with insertion site index and tagged data available for various other plant species (Kuromori et al. 2009)

Activation Tagging

Activation tagging (AT) is another popular method for generating gain-of-function mutant lines. The method uses T-DNA or a transposable element containing the cauliflower mosaic virus 35S enhancer (Weigel et al. 2000). Transcriptional activation of genes near the insertion aids in novel phenotype identification and identifies genes that are redundant and/or essential for survival. AT resources have now been used to isolate genes from Arabidopsis, rice, petunia and tomato (Kakimoto 1996; Zubko et al. 2002; Mathews et al. 2003; Mori et al. 2007). Recently, AT systems using transposons of maize *En/Spm* or *Ac/Ds* have been developed in *Arabidopsis*, rice and soybean (Weigel et al. 2000; An et al. 2005: Schneider et al. 2005; Qu et al. 2008; Kuromori et al. 2009). Web-based sites for genotype tagging and mutant lines, and their purpose and URL are detailed in Table 2.11.

Fox Hunting

The FOX gene hunting system has been developed as an efficient gain-of-function system that combines a transformation algorithm with large-scale information from full-length cDNA clones (Ichikawa et al. 2006). The system can be applied across plant species, like the development of a set of full-length rice cDNA clones aimed at '*in planta*' high-throughput screening of rice functional genes; but with Arabidopsis set as the host reference species (Nakamura et al. 2007; Kondou et al. 2009). Similar results (overexpressors using cDNA) have been achieved in tobacco (Lein et al. 2008).

Chemical and Physical Mutagenesis

Spontaneous and induced mutations are the major source of most of the existing genetic variation in plants, and are commonly used in plant breeding. The occurrence of spontaneous mutations in nature is relatively rare and difficult to identify because they can be recessive, or are deleterious and quickly eliminated. Increasing the rate of mutation (ie induced mutations) can provide additional sources of variant genotypes important in plant breeding. Mutagenic agents include alkylating agents, ethyl methanesulfonate (EMS), sodium azide and methylnitrosourea (MNU), or X-ray and UV-light, fast-neutrons, ion-beam irradiationand nuclear (alpha, beta, gamma rays) radiation. Alkylating agents that react with DNA to change nucleotide sequences produce relatively few useful point mutations. However, the absorption of ionising radiation produces more complex DNA and structural chromosomal changes, and are considered the mutagenic agents of choice in plant breeding applications, however most of these mutagenic agents have been used to generate mutant populations for many years now; these mutant lines have been particularly useful in forward genetics studies in various plant species (Hoang et al. 2009; Uauy et al. 2009; Jackson et al. 2011).

Mutations can occur in tissue cultured plantlets and the process can be rapid; this process is sometimes called 'somaclonal variation' and these plants have also found value in plant breeding. In recent years the use of tissue culture in combination with radiation induced mutations have resulted in a number of desired genotypes in rice, and a number of these have been used directly or indirectly in breeding programs (Hoang et al. 2009; Rahman et al. 2012). Some web sites and URL for developing mutant lines, analysis and use in crop breeding are detailed in Table 2.11.

TILLING

Targeting induced local lesions in genomes (TILLING) was developed as a reverse-genetics tool that provides an allelic series of induced point mutations in genes of importance (Till et al. 2004, 2006). Because high-throughput TILLING permits the rapid and low-cost recovery of induced point mutations in populations of chemically mutagenised individuals, the methods employed have had a level of acceptance and have yielded informationon various animal and crop plant species.

EcoTILLING

The TILLING methodology can also be used to explore allelic variations that appear in natural populations; this technology is called EcoTILLING (Comai et al. 2004; Wang et al. 2006). Several laboratory sites have established TILLING and/or EcoTILLING centers for researches in the public domain (Barkley and Wang 2008). TILLING mutagenic lines and projects in rice, tomato, soybean and Arabidopsis have been performed at the University of California, Davis Genome Centre.

RevGenUK

RevGenUK at the John Innes Centre provides TILLING service for population studies and information in *Medicago trunculata, Lotus japonica* and *Brassica rapa.*

UTILLdb/ INRA

UTILLdb of INRA is another database for TILLING populations of pea and tomato that provide an interface to search for TILLed crop lines based on phenotype descriptions.

Gene Silencing Techniques

Although insertion mutagenesis is an effective method for generating loss-of-function mutants, it has severe limitations in use with redundant genes and lethal mutations. To overcome these limitations, methods to interrupt gene expression have been developed and applied to the functional analysis of plant genes. Some web sites and URL for gene silencing are detailed in Table 2.11.

RNAi

RNA interference (RNAi) is a method for RNA-mediated gene silencing by sequence-specific degradation of homologous mRNA, triggered by double-stranded RNA (dsRNA); also known as post-transcriptional gene silencing (PTGS) (Waterhouse et al. 1998; Chuang and Meyerowitz 2000).

ihp RNA

Constitutive expression of an intron-containing self-complementary hairpin RNA (ihpRNA) has been another method for silencing target genes in plants. With demands for conditional silencing of target genes (the most useful silencing in genetics is that which results in prevention of plant regeneration or embryonic lethality), RNAi systems using chemical-inducible Cre/LoxP recombination or a promoter of heat shock inducible genes have been developed (Guo et al. 2003; Masclaux et al. 2004).

CATMA

In *Arabidopsis*, the Complete *Arabidopsis* Transcriptome MicroArray (CATMA) project has been initiated to design and produce high-quality gene-specific and gene silencing sequence tags (GSTs) covering most of the *Arabidopsis* genome.

AGRIKOLA

Using the GST data set of the CATMA project, the *Arabidopsis* Genomic RNAi Knock-out Line Analysis (in AGRIKOLA) project has also been started, with the goal of systematically analysing *Arabidopsis* genes by RNAi interference (Hilson et al. 2003). The *Medicago trunculata* RNAi database is also available on this web site as an information resource for RNAi-based gene silencing.

VIGS

Virus-induced gene silencing (VIGS) is a derivative method of the ones above that takes advantage of the plant RNAi-mediated antiviral defence mechanism, via RNA interference. The VIGS system was used to assess the function of almost 5000 random *Nicotiana benthamiana* cDNAs in disease resistance (Lu et al. 2003a, b).

CRES-T

The chimeric repressor silencing technology (CRES-T) system was developed as a novel method for gene silencing, and a plant specific repression domain that act as a repressor in transgenic plants; and these can inhibit the expression of target genes (Hiratsu et al. 2003). The CRES-T system has been applied to Arabidopsis in order to analyse their biological function, and to obtain transgenic plants with agronomically preferable traits. An associated database, FioreDB (Mitsuda and Ohme-Takagi 2009) also provides gene silencing information through chimeric repression in flowers.

Plant Comparison Genomics and Databases

The accumulation of nucleotide sequences for many of the agricultural crop species and domestic animals, will allow us to perform genome-wide comparative analyses with the aim of discovering new and important genes involved in phenotypic expression (Sato and Tabata 2006; Itoh et al. 2007; Neale and Ingvarsson 2008). The accumulation of genomic resources derived from various species, such as the extensive collection of cDNAs, ESTs and data from whole-genome sequencing, should facilitate sharing of information about gene function between model plants and other less described crop plants. In time this will also accelerate molecular and cellular systems related to agronomically important traits. A number of information resources for plant genomics comparisons and data exchange on the web have appeared, along with appropriate analytical tools. Here we highlight integrative databases promoting complete plant comparative genomics that we have not described previously. The URLs of each integrative database in plant genomics are shown in Table 2.12.

Plant Portal Information

TAIR

TAIR is one of the most popular and integrated information resources in plants, and although mentioned before it plays important roles as a portal in Arabidopsis research (Swarbreck et al. 2008).

SIGnAL

The Salk Institute Genomic Analysis Laboratory (SIGnAL) is also an information resource that integrates various data sets of whole plant'omics' results, again mainly related to *Arabidopsis*.

Table 2.12 Integrative databases for Comparison Genomics Database analysis in plants

Database Name	Plant Species/Purpose	*Uniform Resource Locator* (URL)
SIGnAL: Salk Institute Genomic Analysis Laboratory Home Page	Salk insertion sequence database; Rice, Arabidopsis included, also Human data	http://signal.salk.edu/
RARGE- RIKEN Arabidopsis Genome Encyclopedia	Site for downloading a number of important web browsers and search engines for Arabidopsis	https://database.riken.jp/sw/en/RIKEN_RARGE_Promoter/ria12i/
WebGBrowse	Generic genome browser used for display along reference sequences	http://webgbrowse.cgb.indiana.edu/cgi-bin/webgbrowse/uploadData
GBrowse.org Home Page	Portal for entry to software, repository, and browsers	http://www.gbrowse.org/index.html
Generic Model Organism Database Project—Generic Genome Browser at SourceForge.net	A free and easy to use database and browser in a number of different languages	http://sourceforge.net/projects/gmod/files/Generic%20Genome%20Browser/
Sol Genomics Network	Solonaceae website for QTL and molecular breeding; mainly tomato	http://nar.oxfordjournals.org/content/early/2010/10/08/nar.gkq866.full
SoyBase.org	SoyBase and soybean breeders toolbox; molecular and physiological data present	http://soybase.org/
MaizeGDB	Maize informatics; molecular and physiological data present	http://www.maizegdb.org/
Plant Genome Duplication Database	Database to catalogue whole plant genomes; focus on 30 agronomic flowering plants	http://chibba.agtec.uga.edu/duplication/
GRASSIUS: Plant Genome Project	Systems approach to identify regulatory netwirks in grasses	http://grassius.org/plantgenome.html
GRASSIUS: About	Home Page for extensive web resource for gene expression, and regulation in grasses	http://grassius.org/about.html
GRASSIUS: GrassTFDB: Transcription Factor Database	Transcriptional factors database for grasses; maize, sugarcane, sorghum and rice	http://grassius.org/grasstfdb.html
LegumeTFDB	Digital library and database for legume transcriptional factors	http://dl.acm.org/citation.cfm?id=1707758
SoybeanTFDB	Digital library and database for legume transcriptional factors	http://soybeantfdb.psc.riken.jp/

RARGE

The RIKEN *Arabidopsis* Genome Encyclopaedia (RARGE) is a portal site providing a gateway for access to comprehensive 'omics' data and/or bioresources (Sakurai et al. 2005). The site also house cross-referenced data between each described gene and its description, such as full-length cDNA clones, gene mutants, gene expression patterns, homologous genes and phenotypic expression.

Gbrowse

Web browsers are commonly used to visualize genes along with genome sequences and associated information, genome browsers such as Gbrowse have been very often used (Donlin 2007).

Gramene

Gramene is a popular site for integrated rice information and plant comparative genomics in grasses. Gramene offers integrated genome associated data including sequences and molecular markers, but is also an important QTL database for breeding in Gramineae species (Ware 2007; Liang et al. 2008).

Sol Genomics

The integration of a number of resources have recently been completed for various individual plant species. The Sol genomics network is a portal site for Solanaceae species that includes information on the tomato and potato genome sequencing projects (Mueller et al. 2005; Mullins et al. 2006).

Soybase

SoyBase is a resource portal and repository site for genomic soybean research and it includes released whole-genome sequence data for this important crop plant.

Maize GDB

The MaizeGDB is the community database for biological information about *Zea mays*, and includes genetic and genomic data sets and other related information on maize (Lawrence et al. 2004).

Genome-wide Comparison in Plants

The completion of a number of genome sequencing projects in plants has increased information on genome-scale comparative analyses and data that facilitate identification of conserved and/or characteristic properties between plant species. Using model and inferred proteome data deduced from sequenced genomes of plants has enabled several efforts to completed and construct comprehensive gene families in other species for comparison. The aim of establishing platforms to verify and compare gene content and elucidating the process of gene duplication and functional diversification among species is on the way (Sterck et al. 2007). Web-based sites for genome-wide comparison in plants, their purpose and URL are detailed in Table 2.11.

Markov Clustering and Multi Dimensional Scaling

Comprehensive gene family data sets are usually produced by computational procedures including a step that conducts an all-against-all sequence similarity search matrices, and then a step for building clusters of protein families by methods such as Markov Clustering (MCL), multi dimensional scaling (MDS) (see Sects. 2.5, 3.4 and 5.1; programs and statistical URL sites are listed in Table 2.1).

PGDD

Correlated gene arrangements among taxa, along with chromosomal location, also known as synteny and collinearity, have become valuable sites for inference of shared ancestry amongst genes, and for transfer of knowledge from one species to another related species (Tang et al. 2008a). The plant genome duplication database (PGDD) provides a data set of intra-genome or cross-genome syntenic relationships identified throughout genome-sequenced plant species at present (Tang et al. 2008b).

Databases for Plant Genomics

Databases housing sets of genomic information and annotations of cross-reference species are now essential for a better understanding of the biology of plants, in particular gene families and/or particular cellular processes. In plants, the genome-wide identification of genes encoding transfer factors (TFs) in Arabidopsis have been reported, and comparisons with other organisms revealed important information (Riechmann et al. 2000; Guo et al. 2008). Further integration of such data must be performed, establishing an integrative, knowledge-based resource across related plant species in terms of comparative genomics of regulation processes. A number of general and specific web-based crop plant databases, their purpose and URL are detailed in Table 2.11.

GRASSIUS

GRASSIUS provides the first step toward building a comprehensive platform for integration of data, tools and resources in comparative regulatory genomics across the grass species (Yilmaz et al. 2009).

Grass TFDB

The Grass Transcription Factor Database (GrassTFDB) of GRASSIUS houses integrated information on a number of important crop species, and the species listed below in Sect. 7.3.3 are all available.

Crop Plant Specific Databases

GRASSIUS contains MaizeTFDB, RiceTFDB, SorghumTFDB, SugarcaneTFDB and Brachypodium TFDB.

Legume TFDB

The LegumeTFDB provides predicted TF encoding genes in the genome sequences of three major legume species: soybean, *L. japonicus* and *M. trunculata*; and is an extended version of the Soybean TFDB site below.

Soybean TFDB

SoybeanTFD is aimed at a more integrated knowledge base for legume TFs, providing a public resource for comparative genomics of the TFs of legumes, non-legume plants and even other organisms (Mochida et al. 2009c, 2010).

Conclusions and Future Perspective

High-throughput DNA technological advances have provided new opportunities to develop collections of related genomic and proteomic information for crop plants, never before available to plant and agricultural scientists. Such comprehensive approaches can provide an excellent starting point for integrating knowledge and allowing the comparison among, and within crop species. This approach promises to be an efficient way to discover new information on genes and their function that are important in crop improvement. However there is much research still to be done!

Pyrosequencing procedures, massive parallel DNA sequencing and single molecule sequencing are now becoming normal and readily available to scientists. Additionally, these new technologies have provided researchers with new avenues to addressed web information at the entire genome level in the fields of inter-species and intra-species comparative genomics and evolutionary genetics. In evolutionary and population genetics, knowledge of genetic diversity, natural and induced variations and population structure are not only important in ecology, but are also important to the breeding and biodiversity of crop plants in agriculture.

New and efficient procedures for whole-genome *de novo* sequencing in crop plants is perhaps one of the most anticipated innovations for next-generation sequencing and related applications. Although, to date, this approach has been realized only in bacteria, a number of tentative attempts are being made to realise this advancement in higher plants, and this must be developed further in crop species.

Breeding crop plants for QTL and traits with low heritability can be the most interesting and the most difficult to work with in breeding programs, but marker assisted selection (MAS) is improving all the time and stable, specific more informative molecular markers must be developed for crop plants. In crop plants especially these traits with low heritability can be important in improving yield, quality and production, and despite being difficult to master must be addressed.

Expressed Sequence Tags (ESTs) derived from different and specific tissues, including tissues from organisms in a range of developmental stages or under biotic and abiotic stress could significantly facilitate gene discovery, and this information is vital in the design of molecular markers and probes for microarrays and gene-chip studies. Full-length cDNA libraries have contributed to function alanalysis by creating overexpressors used in reverse genetics, and this approach is only beginning to be applied in crop plants. However for results to be easily interpreted, reverse genetics must be used in conjunction with comparative analyses of 'modificon' events among plants.

RNA interference, includingsRNAs, microRNAs (miRNAs), short interfering RNAs (siRNAs) and trans-acting siRNAs (ta-siRNAs) are playing important roles as crucial components of epigenetic processes, and gene networks involved in plant development and homeostasis. More information on identification and expression of interfering RNA molecules is necessary, especially by using next-generational genomic technologies in crop plants.

Measurement in abundance of transcripts expressed in cells, tissues and organs can now be estimated from sequence clusters. These methodological principle have been applied in human and mouse, in the form of a type of 'tissue or organ map' to derive the transcriptome in organs; and such 'plant tissue and organ' mapping could be instrumental as reference material for increased crop production.

Differential display methods are now cost effective and produce large amounts of data. Serial analysis of gene expression (SAGE) and derivatives of this method can genetically define host cells and their pathogens simultaneously in crop plants, and additionally superSAGE tags have been used to design probes directly for oligo microarrays in plants and pathogen alike. Microarray and DNA chip-related technologies have advanced rapidly and their application has expanded to a wide variety of crop science disciplines. Today commercially available DNA microarrays are becoming easier and cheaper to obtain, which promises increased availability of 'transciptome' information in all plants

Functional proteomics, including quantitative proteomics, subcellular proteomics and various modifications of proteins and polypeptides, and protein-protein and protein-DNA interactions are new and essential technologies to develop in crop plants. The MALDI or ESI MS methods are still popular and important, but more recent MS procedures must be considered and used in crop plants to obtain high resolution, high sensitivity, high dynamic range and high mass measurement accuracy.

Difference gel electrophoresis (DIGE), isotope-coded affinity tags (ICATs) and isobaric tags for relative and absolute quantitation (iTRAQ) of proteins will enable a better insight into basic regulation of proteins and polypeptides. Stable isotope labelling with amino acids in cell culture (SILAC) combined with MS-MS Analysis (using differential isotope analysis) are further improvements in protein determination important to crop plant proteomics.

Cell organelle analysis of their proteome and metabolite contents are essential for understanding the various enzymatic activities within cell organelles, the compartmentalisation of metabolic pathways, protein targeting, trafficking and regulation, and polypeptide dynamics for many of the crop plants still to be investigated. Modificon research can be aimed to identify modified proteins, through protein phosphorylation and ubiquitination, and although these important studies are just beginning they are becoming more and more important to our understanding of plant metabolism.

Systematic collection of metabolite profiles and advances in instrumentation like NMR analysis will create new information in metabolomics. Chemical phenotype identification are important to develop in crop plants, which can be used to identify genes involved in particular metabolic pathways and cellular processes. This type of information is important to integrate with other 'omics' research, such as profiles of the transcriptome and proteome, now mainly available in Arabidopsis.

It is essential in crop breeding to associate and determine molecular, metabolic, genomic and proteomic diversity of species, cultivars and breeding lines in those plants. This will increase our knowledge and identify many of the genes involved in phenotypic changes, and would also aid in the identification of genetic associations, providing a strong basis for better utilisation of marker assisted selection (MAS), and the use of more stable and informative molecular markers in plant breeding.

T-DNA-tagged lines have emerged as a popular mutant resource, due to the rapid generation of large-scale mutant populations, but are not readily available in many crop plants. Activation tagging (AT) is a popular method for generating gain-of-function mutants important in breeding, and a better knowledge across related breeding stocks and species in terms of comparative genomics. Computer assembly into gene families can verify gene content and elucidates gene duplication (polyploidy).

Chemical mutagenic agents and physical mutagens should be utilised more often to obtain a comprehensive set of crop mutant lines, and then employ these to target induced local lesions in genomes (ie TILLING) as a tool for general reverse-genetics, and RNA interference (RNAi) to screen for gene silencing and gene inactivation. Gene silencing and virus-induced gene silencing (VIGS) can be used in crop plants to discover key genes involved in phenotypic differences.

References

Adams MD, Soares MB, Kerlavage AR, Fields C, Venter JC (1993) Rapid cDNA sequencing (expressed sequence tags) from a directionally cloned human infant brain cDNA library. Nat Genet 4:373–380

Ahmad P, Ashraf M, Younis M, Hu X, Kumar A, Akram NA, Al-Qurainy F (2011) Role of transgenic plants in agriculture and biopharming. Biotech Advances 30:525–540

Agrawal GK, Yonekura M, Iwahashi Y, Iwahashi H, Rakwal R (2005) System, trends and perspectives of proteomics in dicot plants. Part I: technologies in proteome establishment. J Chromatogr B 815:109–123

Akhunov E, Nicolet C, Dvorak J (2009) Single nucleotide polymorphism genotyping in polyploid wheat with the Illumina GoldenGate assay. Theor App Genet 119:507–517

Akiyama K, Chikayama E, Yuasa H, Shimada Y, Tohge T, Shinozaki K et al (2008) PRIMe: a Web site that assembles tools for metabolomics and transcriptomics. In Silico Biol 8:339–345

Alonso JM, Ecker JR (2006) Moving forward in reverse: genetic technologies to enable genome-wide phenomic screens in Arabidopsis. Nat Rev Genet 7:524–536

An G, Jeong DH, Jung KH, Lee S (2005) Reverse genetic approaches for functional genomics of rice. Plant Mol Biol 59:111–123

Andersen JS, Mann M (2006) Organellar proteomics: turning inventories into insights. EMBO Rep 7:874–879

Andreeva A, Howorth D, Chandonia JM, Brenner SE, Hubbard TJ, Chothia C et al (2008) Data growth and its impact on the SCOP database: new developments. Nucleic Acids Res 36:D419–D425

Anisimov SV (2008) Serial analysis of gene expression (SAGE): 13 years of application in research. Curr Pharm Biotechnol 9:338–350

Ansorge WJ (2009) Next-generation DNA sequencing techniques. Nat Biotechnol 25:195–203

Ashikari M, Sakakibara H, Lin S, Yamamoto T, Takashi T, Nishimura A et al (2005) Cytokinin oxidase regulates rice grain production. Science 309:741–745

Baginsky S (2009) Plant proteomics: concepts, applications, and novel strategies for data interpretation. Mass Spectrom Rev 28:93–120

Barakat A, Wall PK, Diloreto S, Depamphilis CW, Carlson JE (2007) Conservation and divergence of microRNAs in *Populus*. BMC Genomics 8:481

Barjaktarovic Z, Nordheim A, Lamkemeyer T, Fladerer C, Madlung J, Hampp R (2007) Time course of changes in amounts of specific proteins upon exposure to hyper-g, 2-D clinorotation, and 3-D random positioning of *Arabidopsis* cultures. J Exper Botany 58:4357–4363

Barjaktarovic Z, Schultz W, Madlung J, Fladerer C, Nordheim A, Hampp R (2009) Changes in the effective gravitational field strength affect the state of phosphorylation of stress related proteins in callus of *Arabidopsis thaliana*. J Exper Botany 60:779–789

Barkley NA, Wang ML (2008) Application of TILLING and EcoTILLING as reverse genetic approaches to elucidate the function of genes in plants and animals. Curr Genomics 9:212–226

Barnholtz-Sloan JS, Tiwari HK (2009) Population Genetics. In: Krawetz S (ed) Bioinformatics for Systems Biology. Humana Press (Springer Science Business Media)

Barrett T, Troup DB, Wilhite SE, Ledoux P, Rudnev D, Evangelista C et al (2009) NCBI GEO: archive for high-throughput functional genomic data. Nucleic Acids Res 37:D885–D890

Benschop JJ, Mohammed S, O'Flaherty M, Heck AJ, Slijper M, Menke FL (2007) Quantitative phosphoproteomics of early elicitor signaling in Arabidopsis. Mol Cell Proteomics 6:1198–1214

Berkelman T, Stenstedt T (2002) 2-D Electrophoresis: Principles and methods. Amersham Bioscience AB, Uppsala

Bernardo AN, Bradbury PJ, Ma H, Hu S, Bowden RL, Buckler ES et al (2009) Discovery and mapping of single feature polymorphisms in wheat using Affymetrix arrays. BMC Genomics 10:251

Bertocchi F, Paci M (2008) Applications of high-resolution solidstate NMR spectroscopy in food science. J Agric Food Chem 5:9317–9327

Bino RJ, Hall RD, Fiehn O, Kopka J, Saito K, Draper J et al (2004) Potential of metabolomics as a functional genomics tool. Trends Plant Sci 9:418–425

Birnbaum K, Shasha DE, Wang JY, Jung JW, Lambert GM, Galbraith DW et al (2003) A gene expression map of the Arabidopsis root. Science 302:1956–1960

Blair MW, Torres MM, Giraldo MC, Pedraza F (2009) Development and diversity of Andean-derived, gene-based microsatellites for common bean (*Phaseolus vulgaris* L). BMC Plant Biol 9:100

Bogdanov B, Smith RD (2005) Proteomics by FTICR mass spectrometry: top down and bottom up. Mass Spectrom Rev 24:168–200

Boguski MS, Lowe TM, Tolstoshev CM (1993) dbEST—database for 'expressed sequence tags'. Nat Genet 4:332–333

Brady SM, Provart NJ (2009) Web-queryable large-scale data sets for hypothesis generation in plant biology. Plant Cell 21:1034–1051

Brenner S, Johnson M, Bridgham J, Golda G, Lloyd DH, Johnson D et al (2000) Gene expression analysis by massively parallel signature sequencing (MPSS) on microbead arrays. Nat Biotechnol 18:630–634

Brown M, Funk CC (2008) Climate. Food security under climate change. Science 319:580–581

Caicedo AL, Williamson SH, Hernandez RD, Boyko A, Fledel-Alon A, York TL et al (2007) Genome-wide patterns of nucleotide polymorphism in domesticated rice. PLoS Genet 3:1745–1756

Carollo V, Matthews DE, Lazo GR, Blake TK, Hummel DD, Lui N et al (2005) GrainGenes 2.0. an improved resource for the smallgrains community. Plant Physiol 139:643–651

Castellani F, van Rossum B, Diehl A, Schubert M, Rehbein K, Oschkinat H (2002) Structure of a protein determined by solid-state magic-angle-spinning NMR spectroscopy. Nature 420:98–102

Chellappan P, Jin H (2009) Discovery of plant microRNAs and short-interfering RNAs by deep parallel sequencing. Methods Mol Biol 495:121–132

Chen S, Harmon AC (2006) Advances in plant proteomics. Proteomics 6:5504–5516

Cheng Y, Qi Y, Zhu Q, Chen X, Wang N, Zhao X et al (2009) New changes in the plasma-membrane-associated proteome of rice roots under salt stress. Proteomics 9:3100–3114

Chitteti BR, Peng Z (2007) Proteome and phosphoproteome differential expression under salinity stress in rice (*Oryza sativa*) root. J Proteome Res 2007:1718–1727

Chuang CF, Meyerowitz EM (2000) Specific and heritable genetic interference by double-stranded RNA in Arabidopsis thaliana. Proc Natl Acad Sci USA 97:4985–4990

Clark KR, Gorley RN (2001) Primer version 5.2.7 user manual/tutorial. Plymouth Marine Laboratory. PRIMER-E Ltd, Plymouth, UK

Cline MS, Kent WJ (2009) Understanding genome browsing. Nature Biotechnol 27:153–155

Close TJ, Wanamaker S, Roose ML, Lyon M (2007) HarvEST: an EST database and viewing software. Methods Mol Biol 406:161–178

Close TJ, Bhat PR, Lonardi S, Wu Y, Rostoks N, Ramsay L et al (2009) Development and implementation of high-throughput SNP genotyping in barley. BMC Genomics 10:582

Cogburn LA, Porter TE, Duclos MJ, Simon J, Burgess SC, Zhu JJ et al (2007) Functional genomics of the chicken-a model organism. Poult Sci 86:2059–2094

Comai L, Young K, Till BJ, Reynolds SH, Greene EA, Codomo CA et al (2004) Efficient discovery of DNA polymorphisms in natural populations by Ecotilling. Plant J 37:778–786

Conte MG, Gaillard S, Lanau N, Rouard M, Perin C (2008) GreenPhylDB: a database for plant comparative genomics. Nucleic Acids Res 36:D991–D998

Crossa J, Burgueno J, Dreisigacker S, Vargas M, Herrera-Foessel SA, Lillemo M et al (2007) Association analysis of historical bread wheat germplasm using additive genetic covariance of relatives and population structure. Genetics 177:1889–1913

De Bodt S, Maere S, Van de Peer Y (2005) Genome duplication and the origin of angiosperms. Trends Ecol Evol 20:591–597

De Filippis LF (2010) Biochemical and molecular aspects in phytoremediation of selenium. In: Ashraf M, Ozturk M, Ahmad MSA (eds) Plant Adaptation and Phytoremediation. Springer Science – Business Media

De Filippis LF (2012) Breeding for biotic stress tolerance in plants. In: Asharaf M, Ozturk M, Ahmad MSA, Aksoy A (eds) Crop Production for Agricultural Improvement. Springer Science – Business Media

De Filippis LF, Magel E (2012) Identification of biochemical differences between the sapwoos and transition zone in *Robinia pseudoacacia* L. by differential display of proteins. Z Holzforschung 66:543–549

de Hoon M, Hayashizaki Y (2008) Deep cap analysis gene expression (CAGE): genome-wide identification of promoters, quantification of their expression, and network inference. Biotechniques 44:627–628

de la Fuente van Bentem S, Anrather D, Roitinger E, Djamei A, Hufnagl T, Barta A et al (2006) Phosphoproteomics reveals extensive in vivo phosphorylation of Arabidopsis proteins involved in RNA metabolism. Nucleic Acids Res 34:3267–3278

Deleu W, Esteras C, Roig C, Gonzalez-To M, Fernandez-Silva I, Gonzalez-Ibeas DD et al (2009) A set of EST-SNPs for map saturation and cultivar identification in melon. BMC Plant Biol 9:90

Dixon RA, Gang DR, Charlton AJ, Fiehn O, Kuiper HA, Reynolds TL et al (2006) Applications of metabolomics in agriculture. J Agric Food Chem 54:8984–8994

Donlin MJ (2007) Using the Generic Genome Browser (GBrowse). Curr Protoc Bioinformatics Chapter 9:Unit 9

Dunkley TP, Hester S, Shadforth IP, Runions J, Weimar T, Hanton SL et al (2006) Mapping the Arabidopsis organelle proteome. Proc Natl Acad Sci USA 103:6518–6523

Duvick J, Fu A, Muppirala U, Sabharwal M, Wilkerson MD, Lawrence CJ et al (2008) PlantGDB: a resource for comparative plant genomics. Nucleic Acids Res 36:D959–D965

Endo Y, Sawasaki T (2003) High-throughput, genome-scale protein production method based on the wheat germ cell-free expression system. Biotechnol Adv 21:695–713

Endo Y, Sawasaki T (2006) Cell-free expression systems for eukaryotic protein production. Curr Opin Biotechnol 17:373–380

Ewing B, Hillier L, Wendl MC, Green P (1998) Base-calling of automated sequencer traces using phred. I. Accuracy assessment. Genome Res 8:175–185

Farrer RA, Kemen E, Jones JD, Studholme DJ (2009) De novo assembly of the *Pseudomonas syringae* pv. syringae B728a genome using Illumina/Solexa short sequence reads. FEMS Microbiol Lett 291:103–111

Fei Z, Tang X, Alba RM, White JA, Ronning CM, Martin GB et al (2004) Comprehensive EST analysis of tomato and comparative genomics of fruit ripening. Plant J 40:47–59

Feltus FA, Wan J, Schulze SR, Estill JC, Jiang N, Paterson AH (2004) An SNP resource for rice genetics and breeding based on subspecies *indica* and *japonica* genome alignments. Genome Res 14:1812–1819

Fernie AR, Schauer N (2009) Metabolomics-assisted breeding: a viable option for crop improvement? Trends Genet 25:39–48

Fernie AR, Trethewey RN, Krotzky AJ, Willmitzer L (2004) Metabolite profiling: from diagnostics to systems biology. Nat Rev Mol Cell Biol 5:763–769

Fiehn O, Kopka J, Dormann P, Altmann T, Trethewey RN, Willmitzer L (2000) Metabolite profiling for plant functional genomics. Nat Biotechnol 18:1157–1161

Finn RD, Tate J, Mistry J, Coggill PC, Sammut SJ, Hotz HR et al (2008) The Pfam protein families database. Nucleic Acids Res 36:D281–D288

Flicek P, Aken BL, Beal K, Ballester B, Caccamo M, Chen Y et al (2008) Ensembl 2008. Nucleic Acids Res 36:D707–D714

Fox BG, Goulding C, Malkowski MG, Stewart L, Deacon A (2008) Structural genomics: from genes to structures with valuable materials and many questions in between. Nat Methods 5:129–132

Frölich T, Arnold GJ (2006) Proteome research based on modern liquid chromatography-tandem mass spectrometry: Separation, identification and quantification. J Neural Transmission 133:973–994

Fu J, Keurentjes JJ, Bouwmeester H, America T, Verstappen FW, Ward JL et al (2009) System-wide molecular evidence for phenotypic buffering in Arabidopsis. Nat Genet 41:166–167

Fujita M, Mizukado S, Fujita Y, Ichikawa T, Nakazawa M, Seki M et al (2007) Identification of stress-tolerance-related transcriptionfactor genes via mini-scale Full-length cDNA Over-eXpressor (FOX) gene hunting system. Biochem Biophys Res Commun 364:250–257

Fukuchi S, Homma K, Sakamoto S, Sugawara H, Tateno Y, Gojobori T et al (2009) The GTOP database in 2009: updated content and novel features to expand and deepen insights into protein structures and functions. Nucleic Acids Res 37:D333–D337

Futamura N, Totoki Y, Toyoda A, Igasaki T, Nanjo T, Seki M et al (2008) Characterization of expressed sequence tags from a full-length enriched cDNA library of *Cryptomeria japonica* male strobili. BMC Genomics 9:383

Gao J, Agrawal GK, Thelen JJ, Xu D (2009) P3DB: a plant protein phosphorylation database. Nucleic Acids Res 37:D960–D962

Goda H, Sasaki E, Akiyama K, Maruyama-Nakashita A, Nakabayashi K, Li W et al (2008) The AtGenExpress hormone and chemical treatment data set: experimental design, data evaluation, model data analysis and data access. Plant J 55:526–542

Greene LH, Lewis TE, Addou S, Cuff A, Dallman T, Dibley M et al (2007) The CATH domain structure database: new protocols and classification levels give a more comprehensive resource for exploring evolution. Nucleic Acids Res 35:D291–D297

Gstaiger M, Aebersold R (2009) Applying mass spectrometrybased proteomics to genetics, genomics and network biology. Nat Rev Genet 10:617–627

Guo HS, Fei JF, Xie Q, Chua NH (2003) A chemical-regulated inducible RNAi system in plants. Plant J 34:383–392

Guo A, He K, Liu D, Bai S, Gu X, Wei L et al (2005) DATF: a database of Arabidopsis transcription factors. Bioinformatics 21:2568–2569

Guo AY, Chen X, Gao G, Zhang H, Zhu QH, Liu XC et al (2008) PlantTFDB: a comprehensive plant transcription factor database. Nucleic Acids Res 36:D966–D969

Hakeem KR, Ahmad A, Iqbal M, Gucel S, Ozturk M (2011) Nitrogen efficient rice genotype can reduce nitrate pollution. Environ Sci Pollut Res 18:1184–1193

Hakeem K, Ozturk M, Memon AR (2012) Biotechnology as an aid for crop improvement to overcome food shortage. In: Ashraf et al (eds) Crop production for agricultural improvement. Springer

Haas BJ, Delcher AL, Wortman JR, Salzberg SL (2004) DAGchainer: a tool for mining segmental genome duplications and synteny. Bioinformatics 20:3643–3646

Han B, Xue Y (2003) Genome-wide intraspecific DNA-sequence variations in rice. Curr Opin Plant Biol 6:134–138

Harbers M, Carninci P (2005) Tag-based approaches for transcriptome research and genome annotation. Nat Methods 2:495–502

Hashimoto S, Suzuki Y, Kasai Y, Morohoshi K, Yamada T, Sese J et al (2004) 5′-end SAGE for the analysis of transcriptional start sites. Nat Biotechnol 22:1146–1149

Hayashizaki Y (2003) RIKEN mouse genome encyclopedia. Mech Ageing Dev 124:93–102

Heesacker A, Kishore VK, Gao W, Tang S, Kolkman JM, Gingle A et al (2008) SSRs and INDELs mined from the sunflower EST database: abundance, polymorphisms, and cross-taxa utility. Theor Appl Gene 117:1021–1029

Herbert BR, Grinyer J, McCarthy JT, Isaacs M, Harry E et al (2006) Improved 2-DE of microorganisms after acid digestion. Electrophoresis 27:1630–1640

Hilson P, Small I, Kuiper MT (2003) European consortia building integrated resources for Arabidopsis functional genomics. Curr Opin Plant Biol 6:426–429

Hirai MY, Yano M, Goodenowe DB, Kanaya S, Kimura T, Awazuhara M et al (2004) Integration of transcriptomics and metabolomics for understanding of global responses to nutritional stresses in *Arabidopsis thaliana*. Proc Natl Acad Sci USA 101:10205–10210

Hirano H, Islam N, Kawasaki H (2004) Technical aspects of functional proteomics in plants. Phytochemistry 65:1487–1498

Hiratsu K, Matsui K, Koyama T, Ohme-Takagi M (2003) Dominant repression of target genes by chimeric repressors that include the EAR motif, a repression domain, in *Arabidopsis*. Plant J 34:733–739

Hishiki T, Kawamoto S, Morishita S, Okubo K (2000) BodyMap: a human and mouse gene expression database. Nucleic Acids Res 28:136–138

Holy J, Perkins E (2009) Structure and function of the nucleus and cell organelles. In: Krawetz S (ed) Bioinformatics for Systems Biology. Humana Press (Springer Science – Business Media)

Hoang TML, De Filippis LF, Le XT (2009) Salt tolerance and screening for genetic changes in rice mutants after gamma irradiation using RAPD and microsatellite (RAMP) markers. Open J Plant Sci 2:12–18

Huang X, Madan A (1999) CAP3: A DNA sequence assembly program. Genome Res 9:868–877

Huang X, Feng Q, Qian Q, Zhao Q, Wang L, Wang A et al (2009) High-throughput genotyping by whole-genome resequencing. Genome Res 19:1068–1076

Hyten DL, Song Q, Choi IY, Yoon MS, Specht JE, Matukumalli LK et al (2008) High-throughput genotyping with the GoldenGate assay in the complex genome of soybean. Theor Appl Genet 116:945–952

Ichikawa T, Nakazawa M, Kawashima M, Iizumi H, Kuroda H, Kondou Y et al (2006) The FOX hunting system: an alternative gain-of-function gene hunting technique. Plant J 48:974–985

Igawa T, Fujiwara M, Takahashi H, Sawasaki T, Endo Y, Seki M et al (2009) Isolation and identification of ubiquitin-related proteins from Arabidopsis seedlings. J Exp Bot 60:3067–3073

Iida K, Seki M, Sakurai T, Satou M, Akiyama K, Toyoda T et al (2004) Genome-wide analysis of alternative pre-mRNA splicing in Arabidopsis thaliana based on full-length cDNA sequences. Nucleic Acids Res 32:5096–5103

Iijima Y, Nakamura Y, Ogata Y, Tanaka K, Sakurai N, Suda K et al (2008) Metabolite annotations based on the integration of mass spectral information. Plant J 54:949–962

Imanishi T, Itoh T, Suzuki Y, O'Donovan C, Fukuchi S, Koyanagi KO et al (2004) Integrative annotation of 21,037 human genes validated by full-length cDNA clones. PLoS Biol 2:e162

International Rice Genome Sequencing Project (2005) The map-based sequence of the rice genome. Nature 436:793–800

International Arabidopsis Genome Initiative (2000) Analysis of the genome sequence of the flowering plant *Arabidopsis thaliana*. Nature 408:796–815

Itoh T, Tanaka T, Barrero RA, Yamasaki C, Fujii Y, Hilton PB et al (2007) Curated genome annotation of *Oryza sativa* ssp. Japonica and comparative genome analysis with *Arabidopsis thaliana*. Genome Res 17:175–183

Jaccoud D, Peng K, Feinstein D, Kilian A (2001) Diversity arrays: a solid state technology for sequence information independent genotyping. Nucleic Acids Res 29:E25

Jackson SA, Iwata A, Lee S-H, Schmutz J, Shoemaker R (2011) Sequencing crop genomes: approaches and applications. New Pytol 191:915–926

Jaillon O, Aury JM, Noel B, Policriti A, Clepet C, Casagrande A et al (2007) The grapevine genome sequence suggests ancestral hexaploidization in major angiosperm phyla. Nature 449:463–467

Jiao Y, Tausta SL, Gandotra N, Sun N, Liu T, Caly NK et al (2009) A transcriptome atlas of rice cell types uncovers cellular, functional and developmental hierarchies. Nat Genet 41:258–263

Jonsson P, Gullberg J, Nordstrom A, Kusano M, Kowalczyk M, Sjostrom M et al (2004) A strategy for identifying differences in large series of metabolomic samples analyzed by GC/MS. Anal Chem 76:1738–1745

Jorrin-Novo JV, Maldonado AM, Echevarria-Zomeno S, Valledor L, Castillejo MA, Curto M et al (2009) Plant proteomics update (2007–2008): second-generation proteomic techniques, an ap-

propriate experimental design, and data analysis to fulfill MIAPE standards, increase plant proteome coverage and expand biological knowledge. J Proteomics 72:285–314
Kakimoto T (1996) CKI1, a histidine kinase homolog implicated in cytokinin signal transduction. Science 274:982–985
Kanehisa M, Goto S (2000) KEGG: Kyoto encyclopedia of genes and genomes. Nucleic Acids Res 28:27–30
Kanehisa M, Araki M, Goto S, Hattori M, Hirakawa M, Itoh M et al (2008) KEGG for linking genomes to life and the environment. Nucleic Acids Res 36:D480–D484
Katz A, Waridel P, Shevchenko A, Pick U (2007) Salt-induced changes in the plasma membrane proteome of the halotolerant alga *Dunaliella salina* as revealed by blue native gel electrophoresis and nano-LC-MS/MS analysis. Mol Cell Proteomics 6:1459–1472
Kaur S, Cogan NO, Ye G, Baillie RC, Hand ML, Ling AE et al (2009) Genetic map construction and QTL mapping of resistance to blackleg (*Leptosphaeria maculans*) disease in Australian canola (*Brassica napus* L.) cultivars. Theor Appl Genet 120:71–83
Kawamoto S, Yoshii J, Mizuno K, Ito K, Miyamoto Y, Ohnishi T et al (2000) BodyMap: a collection of 3′ESTs for analysis of human gene expression information. Genome Res 10:1817–1827
Kawaura K, Mochida K, Yamazaki Y, Ogihara Y (2006) Transcriptome analysis of salinity stress responses in common wheat using a 22k oligo-DNA microarray. Funct Integr Genomics 6:132–142
Kawaura K, Mochida K, Enju A, Totoki Y, Toyoda A, Sakaki Y et al (2009) Assessment of adaptive evolution between wheat and rice as deduced from full-length common wheat cDNA sequence data and expression patterns. BMC Genomics 10:271
Kigawa T, Yabuki T, Yoshida Y, Tsutsui M, Ito Y, Shibata T et al (1999) Cell-free production and stable-isotope labeling of milligram quantities of proteins. FEBS Lett 442:15–19
Kikuchi J, Shinozaki K, Hirayama T (2004) Stable isotope labelling of *Arabidopsis thaliana* for an NMR-based metabolomics approach. Plant Cell Physiol 45:1099–1104
Kilian J, Whitehead D, Horak J, Wanke D, Weinl S, Batistic O et al (2007) The AtGenExpress global stress expression dataset: protocols, evaluation and model data analysis of UV-B light, drought and cold stress responses. Plant J 50:347–363
Komatsu S (2005) Rice proteome database: a step toward functional analysis of the rice genome. Plant Mol Biol 59:179–190
Kondou Y, Higuchi M, Takahashi S, Sakurai T, Ichikawa T, Kuroda H et al (2009) Systematic approaches to using the FOX hunting system to identify useful rice genes. Plant J 57:883–894
Konishi S, Izawa T, Lin SY, Ebana K, Fukuta Y, Sasaki T, Yano M (2006) An SNP caused loss of seed shattering during rice domestication. Science 312:1392–1396
Kopka J, Schauer N, Krueger S, Birkemeyer C, Usadel B, Burgmüller E et al (2005) GMD@CSB.DB: the Golm Metabolome Database. Bioinformatics 21:1635–1638
Kota R, Varshney RK, Prasad M, Zhang H, Stein N, Graner A (2008) EST-derived single nucleotide polymorphism markers for assembling genetic and physical maps of the barley genome. FunctIntegr Genomics 8:223–233
Kouranov A, Xie L, de la Cruz J, Chen L, Westbrook J, Bourne PE et al (2006) The RCSB PDB information portal for structural genomics. Nucleic Acids Res 34:D302–D305
Krysan PJ, Young JC, Sussman MR (1999) T-DNA as an insertional mutagen in Arabidopsis. Plant Cell 11:2283–2290
Kumar CS, Wing RA, Sundaresan V (2005) Efficient insertional mutagenesis in rice using the maize En/Spm elements. Plant J 44:879–892
Kurakawa T, Ueda N, Maekawa M, Kobayashi K, Kojima M, Nagato Y et al (2007) Direct control of shoot meristem activity by a cytokinin-activating enzyme. Nature 445:652–655
Kuromori T, Takahashi S, Kondou Y, Shinozaki K, Matsui M (2009) Phenome analysis in plant species using loss-of-function and gain-of-function mutants. Plant Cell Physiol 50:1215–1231
Kwon SJ, Choi EY, Choi YJ, Ahn JH, Park OK (2006) Proteomics studies of post-translational modifications in plants. J Exp Bot 57:1547–1551

Laubinger S, Zeller G, Henz SR, Sachsenberg T, Widmer CK, Naouar N et al (2008) At-TAX: a whole genome tiling array resource for developmental expression analysis and transcript identification in *Arabidopsis thaliana*. Genome Biol 9:R112

Laurentin H, Ratzinger A, Karlovsky P (2008) Relationship between metabolic and genomic diversity in sesame (*Sesamum indicum* L.). BMC Genomics 9:250

Lawrence CJ, Dong Q, Polacco ML, Seigfried TE, Brendel V (2004) MaizeGDB, the community database for maize genetics and genomics. Nucleic Acids Res 32:D393–D397

Lee Y, Tsai J, Sunkara S, Karamycheva S, Pertea G, Sultana R et al (2005) The TIGR Gene Indices: clustering and assembling EST and known genes and integration with eukaryotic genomes. Nucleic Acids Res 33:D71–D74

Lein W, Usadel B, Stitt M, Reindl A, Ehrhardt T, Sonnewald U et al (2008) Large-scale phenotyping of transgenic tobacco plants (*Nicotiana tabacum*) to identify essential leaf functions. Plant Biotechnol J 6:246–263

Li F, Kitashiba H, Inaba K, Nishio T (2009) A *Brassica rapa* linkage map of EST-based SNP markers for identification of candidate genes controlling flowering time and leaf morphological traits. DNA Res 16:311–323

Liang C, Jaiswal P, Hebbard C, Avraham S, Buckler ES, Casstevens T et al (2008) Gramene: a growing plant comparative genomics resource. Nucleic Acids Res 36:D947–D953

Lisec J, Meyer RC, Steinfath M, Redestig H, Becher M, Witucka-Wall H et al (2008) Identification of metabolic and biomass QTL in *Arabidopsis thaliana* in a parallel analysis of RIL and IL populations. Plant J 53:960–972

Liu X, Lu T, Yu S, Li Y, Huang Y, Huang T et al (2007) A collection of 10,096 indica rice full-length cDNAs reveals highly expressed sequence divergence between *Oryza sativa indica* and *japonica* subspecies. Plant Mol Biol 65:403–415

Long D, Martin M, Sundberg E, Swinburne J, Puangsomlee P, Coupland G (1993) The maize transposable element system Ac/Ds as a mutagen in Arabidopsis: identification of an albino mutation induced by Ds insertion. Proc Natl Acad Sci USA 90:10370–10374

Lu R, Malcuit I, Moffett P, Ruiz MT, Peart J, Wu AJ et al (2003a) High throughput virus-induced gene silencing implicates heat shock protein 90 in plant disease resistance. EMBO J 22:5690–5699

Lu R, Martin-Hernandez AM, Peart JR, Malcuit I, Baulcombe DC (2003b) Virus-induced gene silencing in plants. Methods 30:296–303

Lu C, Kulkarni K, Souret FF, MuthuValliappan R, Tej SS, Poethig RS et al (2006) MicroRNAs and other small RNAs enriched in the Arabidopsis RNA-dependent RNA polymerase-2 mutant. Genome Res 16:1276–1288

Ma JF, Tamai K, Yamaji N, Mitani N, Konishi S, Katsuhara M et al (2006) A silicon transporter in rice. Nature 440:688–691

Ma JF, Yamaji N, Mitani N, Tamai K, Konishi S, Fujiwara T et al (2007) An efflux transporter of silicon in rice. Nature 448:209–212

Mace ES, Rami JF, Bouchet S, Klein PE, Klein RR, Kilian A et al (2009) A consensus genetic map of sorghum that integrates multiple component maps and high-throughput Diversity Array Technology (DArT) markers. BMC Plant Biol 9:13

Maeda N, Kasukawa T, Oyama R, Gough J, Frith M, Engström PG et al (2006) Transcript annotation in FANTOM3: mouse gene catalog based on physical cDNAs. PLoS Genet 2:e62

Manzano C, Abraham Z, Lopez-Torrejon G, Del Pozo JC (2008) Identification of ubiquitinated proteins in Arabidopsis. Plant Mol Biol 68:145–158

Maor R, Jones A, Nuhse TS, Studholme DJ, Peck SC, Shirasu K (2007) Multidimensional protein identification technology (MudPIT) analysis of ubiquitinated proteins in plants. Mol Cell Proteomics 6:601–610

Margulies M, Egholm M, Altman WE, Attiya S, Bader JS, Bemben LA et al (2005) Genome sequencing in microfabricated high-density picolitre reactors. Nature 437:376–380

Marques MC, Alonso-Cantabrana H, Forment J, Arribas R, Alamar S, Conejero V et al (2009) A new set of ESTs and cDNA clones from full-length and normalized libraries for gene discovery and functional characterization in citrus. BMC Genomics 10:428

Maruyama K, Takeda M, Kidokoro S, Yamada K, Sakuma Y, Urano K et al (2009) Metabolic pathways involved in cold acclimation identified by integrated analysis of metabolites and transcripts regulated by DREB1A and DREB2A. Plant Physiol 150:1972–1980

Masclaux F, Charpenteau M, Takahashi T, Pont-Lezica R, Galaud JP (2004) Gene silencing using a heat-inducible RNAi system in Arabidopsis. Biochem Biophys Res Commun 321:364–369

Masoudi-Nejad A, Tonomura K, Kawashima S, Moriya Y, Suzuki M, Itoh M et al (2006) EGassembler: online bioinformatics service for large-scale processing, clustering and assembling ESTs and genomic DNA fragments. Nucleic Acids Res 34:W459–W462

Mathews H, Clendennen SK, Caldwell CG, Liu XL, Connors K, Matheis N et al (2003) Activation tagging in tomato identifies a transcriptional regulator of anthocyanin biosynthesis, modification, and transport. Plant Cell 15:1689–1703

Matsuda F, Yonekura-Sakakibara K, Niida R, Kuromori T, Shinozaki K, Saito K (2009) MS/MS spectral tag-based annotation of non-targeted profile of plant secondary metabolites. Plant J 57:555–577

Matsui A, Ishida J, Morosawa T, Mochizuki Y, Kaminuma E, Endo TA et al (2008) Arabidopsis transcriptome analysis under drought, cold, high-salinity and ABA treatment conditions using a tiling array. Plant Cell Physiol 49:1135–1149

Matsumura H, Reich S, Ito A, Saitoh H, Kamoun S, Winter P et al (2003) Gene expression analysis of plant host–pathogen interactions by SuperSAGE. Proc Natl Acad Sci USA 100:15718–15723

Matsumura H, Kruger DH, Kahl G, Terauchi R (2008) SuperSAGE: a modern platform for genome-wide quantitative transcript profiling. Curr Pharm Biotechno 9:368–374

McDermott A (2009) Structure and dynamics of membrane proteins by magic angle spinning solid-state NMR. Annu Rev Biophys 38:385–403

Memon AR (2012) Transcriptomics and proteomics analysis of root nodules of modern legume plants. In: Asharaf M, Ahmad MSA, Ozturk M, Aksoy A (eds) Crop Production for Agricultural Improvement. Springer Science – Business Media

Mi H, Lazareva-Ulitsky B, Loo R, Kejariwal A, Vandergriff J, Rabkin S et al (2005) The PANTHER database of protein families, subfamilies, functions and pathways. Nucleic Acids Res 33:D284–D288

Minami A, Fujiwara M, Furuto A, Fukao Y, Yamashita T, Kamo M et al (2009) Alterations in detergent-resistant plasma membrane microdomains in *Arabidopsis thaliana* during cold acclimation. Plant Cell Physiol 50:341–359

Ming R, Hou S, Feng Y, Yu Q, Dionne-Laporte A, Saw JH et al (2008) The draft genome of the transgenic tropical fruit tree papaya (*Carica papaya* Linnaeus). Nature 452:991–996

Mitsuda N, Ohme-Takagi M (2009) Functional analysis of transcription factors in Arabidopsis. Plant Cell Physiol 50:1232–1248

Miyao A, Tanaka K, Murata K, Sawaki H, Takeda S, Abe K et al (2003) Target site specificity of the Tos17 retrotransposon shows a preference for insertion within genes and against insertion in retrotransposon-rich regions of the genome. Plant Cell 15:1771–1780

Miyao A, Iwasaki Y, Kitano H, Itoh J, Maekawa M, Murata K et al (2007) A large-scale collection of phenotypic data describing an insertional mutant population to facilitate functional analysis of rice genes. Plant Mol Bio 63:625–635

Mochida K, Shinozaki K (2010) Genomics and bioinformatics resources for crop improvement. Plant Cell Physiol 51:497–523

Mochida K, Yamazaki Y, Ogihara Y (2003) Discrimination of homoeologous gene expression in hexaploid wheat by SNP analysis. Mol Genet Genomics 270:371–377

Mochida K, Kawaura K, Shimosaka E, Kawakami N, Shin-I T, Kohara Y et al (2006) Tissue expression map of a large number of expressed sequence tags and its application to in silico screening of stress response genes in common wheat. Mol Gen Genomics 276:304–312

Mochida K, Saisho D, Yoshida T, Sakurai T, Shinozaki K (2008) TriMEDB: a database to integrate transcribed markers and facilitate genetic studies of the tribe Triticeae. BMC Plant Biol 8:72

Mochida K, Furuta T, Ebana K, Shinozaki K, Kikuchi J (2009a) Correlation exploration of metabolic and genomic diversities in rice. BMC Genomics 10:568

Mochida K, Yoshida T, Sakurai T, Ogihara Y, Shinozaki K (2009b) TriFLDB: a database of clustered full-length coding sequences from Triticeae with applications to comparative grass genomics. Plant Physiol 150:1135–1146

Mochida K, Yoshida T, Sakurai T, Yamaguchi-Shinozaki K, Shinozaki K, Tran LS (2009c) In silico analysis of transcription factor repertoire and prediction of stress responsive transcription factors in soybean. DNA Res 16:353–369

Mochida K, Yoshida T, Sakurai T, Yamaguchi-Shinozaki K, Shinozaki K, Tran LS (2010) LegumeTFDB: an integrative database of *Glycine max, Lotus japonicus* and *Medicago truncatula* transcription factors. Bioinformatics 26:290–291

Moco S, Bino RJ, Vorst O, Verhoeven HA, de Groot J, van Beek TA et al (2006) A liquid chromatography–mass spectrometry-based metabolome database for tomato. Plant Physiol 141:1205–1218

Mori M, Tomita C, Sugimoto K, Hasegawa M, Hayashi N, Dubouzet JG et al (2007) Isolation and molecular characterization of a Spotted leaf 18 mutant by modified activation-tagging in rice. Plant Mol Biol 63:847–860

Morreel K, Goeminne G, Storme V, Sterck L, Ralph J, Coppieters W et al (2006) Genetical metabolomics of flavonoid biosynthesis in Populus: a case study. Plant J 47:224–237

Mueller LA, Zhang P, Rhee SY (2003) AraCyc: a biochemical pathway database for Arabidopsis. Plant Physiol 132:453–460

Mueller LA, Solow TH, Taylor N, Skwarecki B, Buels R, Binns J et al (2005) The SOL Genomics Network: a comparative resource for Solanaceae biology and beyond. Plant Physiol 138:1310–1317

Mullins E, Milbourne D, Petti C, Doyle-Prestwich BM, Meade C (2006) Potato in the age of biotechnology. Trends Plant Sci 11:254–260

Nakamura H, Hakata M, Amano K, Miyao A, Toki N, Kajikawa M et al (2007) A genome-wide gain-of function analysis of rice genes using the FOX-hunting system. Plant Mol Biol 65:357–371

Nakano M, Nobuta K, Vemaraju K, Tej SS, Skogen JW, Meyers BC (2006) Plant MPSS databases: signature-based transcriptional resources for analyses of mRNA and small RNA. Nucleic Acids Res 34:D731–D735

Nanjo T, Sakurai T, Totoki Y, Toyoda A, Nishiguchi M, Kado D et al (2007) Functional annotation of 19,841 *Populus nigra* full length enriched cDNA clones. BMC Genomics 8:448

Neale DB, Ingvarsson PK (2008) Population, quantitative and comparative genomics of adaptation in forest trees. Curr Opin Plant Biol 11:149–155

Newton RP, Brenton AG, Smith CJ, Dudley E (2004) Plant proteome analysis by mass spectrometry: principles, problems, pitfalls and recent developments. Phytochemistry 65:1449–1485

Nobuta K, Venu RC, Lu C, Belo A, Vemaraju K, Kulkarni K et al (2007) An expression atlas of rice mRNAs and small RNAs. Nat Biotechnol 25:473–477

Nobuta K, Lu C, Shrivastava R, Pillay M, De Paoli E, Accerbi M et al (2008) Distinct size distribution of endogeneous siRNAs in maize: evidence from deep sequencing in the mop1–1 mutant. Proc Natl Acad Sci USA 105:14958–14963

Nuhse TS, Stensballe A, Jensen ON, Peck SC (2004) Phosphoproteomics of the Arabidopsis plasma membrane and a new phosphorylation site database. Plant Cell 16:2394–2405

Nuhse TS, Bottrill AR, Jones AM, Peck SC (2007) Quantitative phosphoproteomic analysis of plasma membrane proteins reveals regulatory mechanisms of plant innate immune responses. Plant J 51:931–940

Obayashi T, Kinoshita K, Nakai K, Shibaoka M, Hayashi S, Saeki M et al (2007) ATTED-II: a database of co-expressed genes and cis elements for identifying co-regulated gene groups in Arabidopsis. Nucleic Acids Res 35:D863–D869

Obayashi T, Hayashi S, Saeki M, Ohta H, Kinoshita K (2009) ATTED-II provides coexpressed gene networks for Arabidopsis. Nucleic Acids Res 37:D987–D991

O'Brien EA, Zhang Y, Wang E, Marie V, Badejoko W, Lang BF, Burger G (2009) GOBASE: an organelle genome database. Nucleic Acids Res 37:D946–D950

Ogasawara O, Otsuji M, Watanabe K, Iizuka T, Tamura T, Hishiki T et al (2006) BodyMap-Xs: anatomical breakdown of 17 million animal ESTs for cross-species comparison of gene expression. Nucleic Acids Res 34:D628–D631

Okazaki Y, Shimojima M, Sawada Y, Toyooka K, Narisawa T, Mochida K et al (2009) A chloroplastic UDP-glucose pyrophosphorylase from Arabidopsis is the committed enzyme for the first step of sulfolipid biosynthesis. Plant Cell 21:892–909

Oksman-Caldentey KM, Saito K (2005) Integrating genomics and metabolomics for engineering plant metabolic pathways. Curr Opin Biotechnol 16:174–179

Ozturk M (2010) Agricultural residues and their role in bioenergy production. Version steele 24 May 09. Proceedings-second consultation AgroResidues-Second expert consultation 'The utilization of agricultural residues with special emphasis on utilization of agricultural residues as biofuel', Cairo Egypt, November 2007, pp 31–43

Ozturk M, Ergin M, Akçiçek E (2006) Ecology, sustainability, and stewardship. Proc.of the 13th IAS science conference on "Energy for Sustainable Development" and "Science for the Future of the Islamic World and Humanity", Kuching/Sarawak, Malaysia (2003). In: Ergin M, Zou'bi MR (eds) Islamic World Academy of Sciences (IAS). National Printing Press, Amman, pp 261–278

Palczewski K, Kumasaka T, Hori T, Behnke CA, Motoshima H, Fox BA et al (2000) Crystal structure of rhodopsin: a G proteincoupled receptor. Science 289:739–745

Park PJ (2009) ChIP-seq: advantages and challenges of a maturing technology. Nat Rev Genet 10:669–680

Parkinson H, Kapushesky M, Shojatalab M, Abeygunawardena N, Coulson R, Farne A et al (2007) ArrayExpress-a public database of microarray experiments and gene expression profiles. NucleicAcids Res 35:D747–D750

Paterson AH (2008) Genomics of sorghum. Int J Plant Genomics 2008:362–451

Paterson AH, Bowers JE, Bruggmann R, Dubchak I, Grimwood J, Gundlach H et al (2009) The *Sorghum bicolor* genome and the diversification of grasses. Nature 457:551–556

Peleg Z, Saranga Y, Suprunova T, Ronin Y, Roder MS, Kilian A et al (2008) High-density genetic map of durum wheat × wild emmer wheat based on SSR and DArT markers. Theor Appl Genet 117:103–115

Pinstrup-Andersen P, Cohen M (2000) Modern biotechnology for food and agriculture: Risks and opportunities for the poor. In: Persley GJ, Lantin MM (eds) Agricultural biotechnology and the poor. Consultative Group on International Agricultural Research, Washington, DC

Qu S, Desai A, Wing R, Sundaresan V (2008) A versatile transposonbased activation tag vector system for functional genomics in cereals and other monocot plants. Plant Physiol 146:189–199

Rahman U, Shaheen T, Ashraf M, Zafar Y (2012) Bridging genomic and classic breeding approaches for improving crop productivity. In: Asharaf M, Ozturk M, Ahmad MSA, Aksoy A (eds) Crop Production for Agricultural Improvement. Springer Science – Business Media

Ralph SG, Chun HJ, Cooper D, Kirkpatrick R, Kolosova N, Gunter L et al (2008a) Analysis of 4,664 high-quality sequence-finished poplar full-length cDNA clones and their utility for the discovery of genes responding to insect feeding. BMC Genomics 9:57

Ralph SG, Chun HJ, Kolosova N, Cooper D, Oddy C, Ritland CE et al (2008b) A conifer genomics resource of 200,000 spruce (*Picea* spp.) ESTs and 6,464 high-quality, sequence-finished full-length cDNAs for Sitka spruce (*Picea sitchensis*). BMC Genomics 9:484

Ramautar R, Somsen GW, de Jong GJ (2009) CE-MS in metabolomics. Electrophoresis 30:276–291

Riechmann JL, Heard J, Martin G, Reuber L, Jiang C, Keddie J et al (2000) Arabidopsis transcription factors: genome-wide comparative analysis among eukaryotes. Science 290:2105–2110

Roessner U, Luedemann A, Brust D, Fiehn O, Linke T, Willmitzer L et al (2001) Metabolic profiling allows comprehensive phenotyping of genetically or environmentally modified plant systems. Plant Cell 13:11–29

Rose JCK, Bashir S, Giovannoni JJ, Jahn MM, Saravanan RS (2004) Tackling the plant proteome: Practicalapproaches, hurdles and experimental tools. Plant J 39:715–733

Rossignol M, Peltier JB, Mock HP, Matros A, Maldonado AM, Jorrin JV (2006) Plant proteome analysis: a 2004–2006 update. Proteomics 6:5529–5548
Rostoks N, Borevitz JO, Hedley PE, Russell J, Mudie S, Morris J et al (2005) Single-feature polymorphism discovery in the barley transcriptome. Genome Biol 6:R54
Rowe HC, Hansen BG, Halkier BA, Kliebenstein DJ (2008) Biochemical networks and epistasis shape the *Arabidopsis thaliana* metabolome. Plant Cell 20:1199–1216
Ruiz-Ferrer V, Voinnet O (2009) Roles of plant small RNAs in biotic stress responses. Annu Rev Plant Biol 60:485–510
Saito K, Hirai MY, Yonekura-Sakakibara K (2008) Decoding genes with co-expression networks and metabolomics-'majority report by precogs'. Trends Plant Sci 13:36–43
Sakai A, Takano H, Kuroiwa T (2004) Organelle nuclei in higher plants: structure, composition, function and evolution. Int Rev Cytol 238:59–118
Sakata K, Ohyanagi H, Nobori H, Nakamura T, Hashiguchi A, Nanjo Y et al (2009) Soybean proteome database: a data resource for plant differential 'omics'. J Proteome Res 8:3539–3548
Sakurai T, Plata G, Rodriguez-Zapata F, Seki M, Salcedo A, Toyoda A et al (2007) Sequencing analysis of 20,000 full-length cDNA clones from cassava reveals lineage specific expansions in gene families related to stress response. BMC Plant Bio 7:66
Samatey FA, Imada K, Nagashima S, Vonderviszt F, Kumasaka T, Yamamoto M et al (2001) Structure of the bacterial flagellar protofilament and implications for a switch for supercoiling. Nature 410:331–337
Sato S, Tabata S (2006) *Lotus japonicus* as a platform for legume research. Curr Opin Plant Bio 9:128–132
Sato S, Nakamura Y, Kaneko T, Asamizu E, Kato T, Nakao M et al (2008) Genome structure of the legume, *Lotus japonicus*. DNA Res 15:227–239
Sato K, Shin IT, Seki M, Shinozaki K, Yoshida H, Takeda K et al (2009) Development of 5006 full-length cDNAs in barley: a tool for accessing cereal genomics resources. DNA Res 16:81–89
Schauer N, Semel Y, Roessner U, Gur A, Balbo I, Carrari F et al (2006) Comprehensive metabolic profiling and phenotyping of interspecific introgression lines for tomato improvement. Nat Biotechnol 24:447–454
Schauer N, Semel Y, Balbo I, Steinfath M, Repsilber D, Selbig J et al (2008) Mode of inheritance of primary metabolic traits in tomato. Plant Cell 20:509–523
Schena M, Shalon D, Davis RW, Brown PO (1995) Quantitative monitoring of gene expression patterns with a complementary DNA microarray. Science 270:467–470
Schmelzle K, White FM (2006) Phosphoproteomic approaches to elucidate cellular signaling networks. Curr Opin Biotechnol 17:406–414
Schnable PS, Ware Doulton RS, Stein JC, Wei F, Pasternak S et al (2009) The B73 maize genome: complexity, diversity, and dynamics. Science 326:1112–1115
Schneider A, Kirch T, Gigolashvili T, Mock HP, Sonnewald U, Simon R et al (2005) A transposon-based activation-tagging population in Arabidopsis thaliana (TAMARA) and its applicationin the identification of dominant developmental and metabolic mutations. FEBS Let 579:4622–4628
Schripsema J (2009) Application of NMR in plant metabolomics: techniques. problems and prospects. Phytochem Anal 21:14–21
Schwede T, Kopp J, Guex N, Peitsch MC (2003) SWISS-MODEL: an automated protein homology-modeling server. Nucleic Acids Res 31:3381–3385
Seki M, Shinozaki K (2009) Functional genomics using RIKEN Arabidopsis thaliana full-length cDNAs. J Plant Res 122:355–366
Seki M, Narusaka M, Ishida J, Nanjo T, Fujita M, Oono Y et al (2002a) Monitoring the expression profiles of 7000 Arabidopsis genes under drought, cold and high-salinity stresses using a full-length cDNA microarray. Plant J 31:279–292
Seki M, Narusaka M, Kamiya A, Ishida J, Satou M, Sakurai T et al (2002b) Functional annotation of a full-length *Arabidopsis* cDNA collection. Science 296:141–145
Sekiyama Y, Kikuchi J (2007) Towards dynamic metabolic network measurements by multi-dimensional NMR-based fluxomics. Phytochemistry 68:2320–2329

Shen YJ, Jiang H, Jin JP, Zhang ZB, Xi B, He YY et al (2004) Development of genome-wide DNA polymorphism database for map-based cloning of rice genes. Plant Physiol 135:1198–1205
Somerville C, Dangl J (2000) Genomics. Plant biology in 2010. Science 290:2077–2078
Song J, Braun G, Bevis E, Doncaster K (2006) A simple protocol for protein extraction of recalcitrant fruit tissues suitable for 2-D electrophoresis and MS analysis. Electrophoresis 27:3144–3151
Springer PS (2000) Gene traps: tools for plant development and genomics. Plant Cell 12:1007–1020
Stanford WL, Cohn JB, Cordes SP (2001) Gene-trap mutagenesis: past, present and beyond. Nat Rev Gene 2:756–768
Sterck L, Rombauts S, Vandepoele K, Rouze P, Van de Peer Y (2007) How many genes are there in plants (… and why are they there)? Curr Opin Plant Biol 10:199–203
Sterky F, Bhalerao RR, Unneberg P, Segerman B, Nilsson P, Brunner AM et al (2004) A *Populus* EST resource for plant functional genomics. Proc Natl Acad Sci USA 101:13951–13956
Stevens RC, Yokoyama S, Wilson IA (2001) Global efforts in structural genomics. Science 294:89–92
Sugiyama N, Nakagami H, Mochida K, Daudi A, Tomita M, Shirasu K et al (2008) Large-scale phosphorylation mapping reveals the extent of tyrosine phosphorylation in Arabidopsis. Mol Syst Biol 4:193
Swarbreck D, Wilks C, Lamesch P, Berardini TZ, Garcia-Hernandez M, Foerster H et al (2008) The Arabidopsis Information Resource (TAIR): gene structure and function annotation. Nucleic Acids Res 36:D1009–D1014
Taji T, Sakurai T, Mochida K, Ishiwata A, Kurotani A, Totoki Y et al (2008) Large-scale collection and annotation of full-length enriched cDNAs from a model halophyte, *Thellungiella halophila*. BMC Plant Biol 8:115
Takeda S, Matsuoka M (2008) Genetic approaches to crop improvement: responding to environmental and population changes. Nat Rev Genet 9:444–457
Talon M, Gmitter FG Jr (2008) Citrus genomics. Inter J Plant Genomics 2008:1–17
Tanaka T, Antonio BA, Kikuchi S, Matsumoto T, Nagamura Y, Numa Y et al (2008) The Rice Annotation Project Database (RAP-DB): 2008 update. Nucleic Acids Res 36:D1028–D1033
Tang H, Bowers JE, Wang X, Ming R, Alam M, Paterson AH (2008a) Synteny and collinearity in plant genomes. Science 320:486–488
Tang H, Wang X, Bowers JE, Ming R, Alam M, Paterson AH (2008b) Unraveling ancient hexaploidy through multiply-aligned angiosperm gene maps. Genome Res 18:1944–1954
Thimm O, Blasing O, Gibon Y, Nagel A, Meyer S, Kruger P et al (2004) MAPMAN: a user-driven tool to display genomics data sets onto diagrams of metabolic pathways and other biological processes. Plant J 37:914–939
Tikunov Y, Lommen A, de Vos CH, Verhoeven HA, Bino RJ, Hall RD et al (2005) A novel approach for nontargeted data analysis for metabolomics. Large-scale profiling of tomato fruit volatiles. Plant Physiol 139:1125–1137
Till BJ, Reynolds SH, Weil C, Springer N, Burtner C, Young K et al (2004) Discovery of induced point mutations in maize genes by TILLING. BMC Plant Biol 4:12
Till BJ, Colbert T, Codomo C, Enns L, Johnson J, Reynolds SH et al (2006) High-throughput TILLING for Arabidopsis. Methods Mol Biol 323:127–35
Tohge T, Nishiyama Y, Hirai MY, Yano M, Nakajima J, Awazuhara M et al (2005) Functional genomics by integrated analysis of metabolome and transcriptome of Arabidopsis plants over-expressing an MYB transcription factor. Plant J 42:218–235
Tokimatsu T, Sakurai N, Suzuki H, Ohta H, Nishitani K, Koyama T et al (2005) KaPPA-view: a web-based analysis tool for integration of transcript and metabolite data on plant metabolic pathway maps. Plant Physiol 138:1289–1300
Torada A, Koike M, Mochida K, Ogihara Y (2006) SSR-based linkage map with new markers using an intraspecific population of common wheat. Theor Appl Gene 112:1042–1051
Toyoda T, Wada A (2004) Omic space: coordinate-based integration and analysis of genomic phenomic interactions. Bioinformatics 20:1759–1765

Toyoshima C, Nomura H, Sugita Y (2003) Crystal structures of Ca^{2+}-ATPase in various physiological states. Ann NY Acad Sci 986:1–8

Trethewey RN (2004) Metabolite profiling as an aid to metabolic engineering in plants. Curr Opin Plant Biol 7:196–201

Turner WR, Oppenheimer M, Wilcove DS (2009) A force to fight global warming. Nature 462:278–279

Tyler RC, Aceti DJ, Bingman CA, Cornilescu CC, Fox BG, Frederick RO et al (2005) Comparison of cell-based and cell-free protocols for producing target proteins from the *Arabidopsis thaliana* genome for structural studies. Proteins 59:633–643

Uauy C, Paraiso F, Colasuonno P, Tran RK, Tsai H, Berardi S, Comai L, Dubcovsky J (2009) A modified TILLING approach to detect induced mutations in tetraploid and hexaploid wheat. BMC Plant Biol 9:115–129

Umezawa T, Sakurai T, Totoki Y, Toyoda A, Seki M, Ishiwata A et al (2008) Sequencing and analysis of approximately 40,000 soybean cDNA clones from a full-length-enriched cDNA library. DNA Res 15:333–346

Urano K, Maruyama K, Ogata Y, Morishita Y, Takeda M, Sakurai N et al (2009) Characterization of the ABA-regulated global responses to dehydration in Arabidopsis by metabolomics. Plant J 57:1065–1078

Varshney RK, Graner A, Sorrells ME (2005) Genomics-assisted breeding for crop improvement. Trends Plant Sci 10:621–630

Varshney RK, Nayak SN, May GD, Jackson SA (2009) Next-generation sequencing technologies and their implications for crop genetics and breeding. Trends Biotechnol 27:522–530

Velculescu VE, Zhang L, Vogelstein B, Kinzler KW (1995) Serial analysis of gene expression. Science 270:484–487

Wall PK, Leebens-Mack J, Muller KF, Field D, Altman NS, de Pamphilis CW (2008) PlantTribes: a gene and gene family resource for comparative genomics in plants. Nucleic Acids Res 36:D970–D976

Wang DK, Sun ZX, Tao YZ (2006) Application of TILLING in plant improvement. Yi Chuan Xue Bao 33:957–964

Ware D (2007) Gramene: a resource for comparative grass genomics. Methods Mol Biol 406:315–330

Waterhouse PM, Graham MW, Wang MB (1998) Virus resistance and gene silencing in plants can be induced by simultaneous expression of sense and antisense RNA. Proc Natl Acad Sci USA 95:13959–13964

Weigel D, Mott R (2009) The 1001 genomes project for *Arabidopsis thaliana*. Genome Biol 10:107

Weigel D, Ahn JH, Blazquez MA, Borevitz JO, Christensen SK, Fankhauser C et al (2000) Activation tagging in Arabidopsis. Plant Physiol 122:1003–1013

Wenzl P, Raman H, Wang J, Zhou M, Huttner E, Kilian A (2007) A DArT platform for quantitative bulked segregant analysis. BMC Genomics 8:196

Werner E, Heilier JF, Ducruix C, Ezan E, Junot C, Tabet JC (2008) Mass spectrometry for the identification of the discriminating signals from metabolomics: current status and future trends. J Chromatogr B 871:143–163

Wilson D, Madera M, Vogel C, Chothia C, Gough J (2007) The SUPERFAMILY database in 2007: families and functions. Nucleic Acids Res 35:D308–D313

Wilson D, Charoensawan V, Kummerfeld SK, Teichmann SA (2008) DBD-taxonomically broad transcription factor predictions: new content and functionality. Nucleic Acids Res 36:D88–D92

Winter D, Vinegar B, Nahal H, Ammar R, Wilson GV, Provart NJ (2007) An 'electronic fluorescent pictograph' browser for exploring and analyzing large-scale biological data sets. PLoS ONE 2:e718

Wlodawer A, Minor W, Dauer Z, Jaskolski M (2008) Protein crystallography for non-crystallographers, or how to get the best (but not more) from published macromolecular structures. FEBS J 275:1–21

Wu WW, Wang G, Baek SJ, Shen R-F (2005) Comparative study of three proteomic quantitative methods, DIGE, cICAT, and iTRAQ, using 2D gel- or LC- MAL:DI TOF/TOF. J Proteome Res 5:651–658

Yamamoto YY, Obokata J (2008) ppdb: a plant promoter database. Nucleic Acids Res 36:D977–D981

Yamamoto YY, Yoshitsugu T, Sakurai T, Seki M, Shinozaki K, Obokata J (2009) Heterogeneity of Arabidopsis core promoters revealed by high-density TSS analysis. Plant J 60:350–362

Yamasaki K, Kigawa T, Inoue M, Tateno M, Yamasaki T, Yabuki T et al (2004) Solution structure of the B3 DNA binding domain of the *Arabidopsis* cold-responsive transcription factor RAV1. Plant Cell 16:3448–3459

Yamasaki C, Murakami K, Fujii Y, Sato Y, Harada E, Takeda J et al (2008a) The H-Invitational Database (H-InvDB), a comprehensive annotation resource for human genes and transcripts. Nucleic Acids Res 36:D793–D799

Yamasaki K, Kigawa T, Inoue M, Watanabe S, Tateno M, Seki M et al (2008b) Structures and evolutionary origins of plant-specific transcription factor DNA-binding domains. Plant Physiol Bioche 46:394–401

Yan W, Chen S (2005) Mass spectrometry-based quantitative proteomic profiling. Brief Funct Genome Proteome 4:1–12

Yates JR, Ruse CI, Nakorchevsky A (2009) Proteomics by mass spectrometry: approaches, advances, and applications. Annu Rev Biomed Eng 11:49–79

Yilmaz A, Nishiyama MY Jr, Fuentes BG, Souza GM, Janies D, Gray J et al (2009) GRASSIUS: a platform for comparative regulatory genomics across the grasses. Plant Physiol 149:171–180

Yonekura-Sakakibara K, Tohge T, Matsuda F, Nakabayashi R, Takayama H, Niida R et al (2008) Comprehensive flavonol profiling and transcriptome coexpression analysis leading to decoding gene-metabolite correlations in Arabidopsis. Plant Cell 20:2160–2176

Zeller G, Henz SR, Widmer CK, Sachsenberg T, Ratsch G, Weigel D et al (2009) Stress-induced changes in the *Arabidopsis thaliana* transcriptome analyzed using whole-genome tiling arrays. Plant J 58:1068–1082

Zhang Y (2008) Progress and challenges in protein structure prediction. Curr Opin Struct Biol 18:342–348

Zhang Y (2009a) I-TASSER: fully automated protein structure prediction in CASP8. Proteins 77(Suppl 9):100–113

Zhang Y (2009b) Protein structure prediction: when is it useful? Curr Opin Struct Biol 19:145–155

Zhang H, Sreenivasulu N, Weschke W, Stein N, Rudd S, Radchuk V et al (2004) Large-scale analysis of the barley transcriptome based on expressed sequence tags. Plant J 40:276–290

Zhang X, Yazaki J, Sundaresan A, Cokus S, Chan SW, Chen H et al (2006) Genome-wide high resolution mapping and functional analysis of DNA methylation in Arabidopsis. Cell 126:1189–1201

Zhang J, Xu Y, Huan Q, Chong K (2009a) Deep sequencing of *Brachypodium* small RNAs at the global genome level identifies microRNAs involved in cold stress response. BMC Genomics 10:449

Zhang Z, Yu J, Li D, Liu F, Zhou X, Wang T, Ling Y, Su Z (2009b) PMRD: plant microRNA database. Nucleic Acids Res 38:D806–D814

Zheng Y, Ren N, Wang H, Stromberg AJ, Perry SE (2009) Global identification of targets of the Arabidopsis MADS domain protein AGAMOUS-Like 15. Plant Cell 21:2563–2577

Zhu QH, Guo AY, Gao G, Zhong YF, Xu M, Huang M et al (2007) DPTF: a database of poplar transcription factors. Bioinformatics 23:1307–1308

Zimmermann P, Hirsch-Hoffmann M, Hennig L, Gruissem W (2004) GENEVESTIGATOR. *Arabidopsis* microarray database and analysis toolbox. Plant Physiol 136:2621–2632

Zubko E, Adams CJ, Machaekova I, Malbeck J, Scollan C, Meyer P (2002) Activation tagging identifies a gene from *Petunia hybrida* responsible for the production of active cytokinins in plants. Plant J 29:797–808

Chapter 3
Crop Improvement Through Plant Tissue Culture

Sumiya Jamsheed, Saiema Rasool, Shivani Koul, Mohamed Mahgoub Azooz and Parvaiz Ahmad

Abstract Plant tissue culture has emerged as a powerful and cost-effective tool for the crop improvement. Tissue culture is alternatively called cell, tissue and organ culture through *in vitro* condition. It can be employed for large scale propagation of disease free clones and gene pool conservation. Agricultural industry has applied immensely *in vitro* propagation approach for large scale plant multiplication of elite superior varieties. As a result, hundreds of plant tissue culture laboratories have come up worldwide, especially in the developing countries due to cheap labour costs. Tissue culture has been exploited to create genetic variability from which crop plants can be improved and to increase the number of desirable germplasms available to the plant breeder. The selection of somaclonal variations appearing in the regenerated plants may be genetically stable and useful in crop improvement. Available methods for the transfer of genes could significantly simplify the breeding procedures and overcome some of the agronomic and environmental problems, which otherwise would not be achievable through conventional propagation methods. Transgenic crops resistant to pests, insects, diseases and other abiotic stresses is a great achievement in the field of plant biotechnology. This article is cover *in vitro* propagation and role of biotechnology in crop improvement.

P. Ahmad (✉)
Department of Botany, GDC Anantnag, Jammu and Kashmir 192101, India
e-mail: parvaizbot@yahoo.com

S. Jamsheed · S. Rasool · S. Koul
Department of Botany, Faculty of Science, Jamia Hamdard, New Delhi 110062, India
e-mail: sumiyajkhan@gmail.com

S. Rasool
e-mail: saimu083@yahoo.co.in

S. Koul
e-mail: shivanee_kaul@rediffmail.com

M. M. Azooz
Department of Biological Sciences, Faculty of Science, King Faisal University, Hofuf, Saudi Arabia

Department of Botany, Faculty of Science, South Valley University, 83523 Qena, Egypt
e-mail: azzozm@yahoo.com

K. R. Hakeem et al. (eds.), *Crop Improvement,* DOI 10.1007/978-1-4614-7028-1_3,

Introduction

Plant biotechnology involves a number of technologies and the techniques, methods and strategies involved in in-vitro culture are only a part of it. Plant tissue culture has now become a major component of this applied branch. Advances made in molecular biology can be manifested in plants through plant tissue culture. This technique is new and has changed the scenario of plant science (Hussain and Hasnain 2012). Apart from the conventional methods of pollination and cross fertilization, there are a number of methods for producing genetically modified plants. Last 20 years have witnessed a number of developments in this field. However, the modern molecular biological techniques are still under way to make a broad based development on crop improvement to raise a selected plant to the stage of cultivar release (Hussain et al. 2011). A number of recalcitrant crops are now able to regenerate using the techniques of in-vitro culture utilizing cells or calli or protoplasts in this process and each such explants can be used in genetic transformation (Davey et al. 2005). Now a days the application of tissue culture to various branches of plant science like plant breeding, horticulture, forestry, industrial production of compounds and conservation of ever depleting natural genetic resources has been the focal point of research (Roy et al. 2011).

Advancement in the techniques of protoplast, cell, tissue and organ culture and regeneration of whole plants has resulted in the development of tissue culture as a technology (Thorpe 2012). The technique has advanced rapidly over the years due to extensive investigations into problems related to basic and applied aspects of plants. The phenomenon of growth, metabolism, differentiation and morphogenesis can be well understood by the knowledge of tissue culture (Karkonen et al. 2011). Plant tissue culture is of great interest to molecular biologists, plant breeders and industrialists. The methods of tissue culture have been used as an important aid to conventional methods of plant improvement. They have been used for the production of genetically modified superior clones, ex-situ conservation of germplasm, pathogen free plants as well as in the synthesis of many important secondary compounds (including pharmaceuticals). The advantages offered by tissue culture in agriculture and general plant biotechnology have well been witnessed by many research labs and industries (Mustafa et al. 2011).

The conventional breeding programmes can be complemented by biotechnology and expedite the crop improvement programmes. A large number of centres are involved in studies involving *in vitro* culture and selection, micropropagation, embryo rescue, genetic transformation, marker assisted characterization and DNA fingerprinting worldwide. Micropropagation protocols and somatic embryogenesis has been achieved in a number of important genera. Germplasm screening has become successful due to the techniques of *in vitro* selections for antibiotic tolerance and fungal toxin resistance. *Agrobacterium tumefaciens* mediated transformation has been established in a number of cereal, fruit and vegetable crops (Azria and Bhalla 2011; Pons et al. 2012). A number of fruit ripening genes have been cloned and transferred into plants and DNA fingerprinting for genetic diversity analysis has been conducted on many crop species (Shi and Zhang 2012, Chandel et al. 2010). The technique of tissue culture helps in crop improvement through different approaches like breeding, wide hybridization, haploidy, somaclonal variation and micropropagation.

Approaches for Crop Improvement

Distant Hybridisation

Crop improvement can be achieved by the methods of genetic transfer, whether via single gene, through genetic engineering, or multiple genes, through conventional breeding or tissue culture techniques. In angiosperms fertilization depends upon a number of factors which include, transfer of pollen grains from anthers to stigma, germination of pollen grains to produce a pollen tube, penetration of pollen tube to the stigma and the style to reach the ovule. The pollen tube discharge triggers and the two sperm nuclei then fuse with their respective partners i.e., the egg cell and the secondary nucleus which results in the formation of the embryo and endosperm respectively. However, this phenomenon can be stopped at any stage, resulting in a barrier to hybridization and thus, the inhibition of gene transfer between the two plants. However, in case of distant crosses involving individuals of different species or genus, a number of barriers have to be overcome for hybridization to take place. These barriers include pre-fertilization barriers like failure of pollen tube to germinate or poor pollen tube growth which can be overcome by *in vitro* fertilization (Dresselhaus et al. 2011) and post-fertilization barriers such as no endosperm development which may be overcome by embryo, ovule or pod culture. Protoplast fusion has been successful in producing the desired hybrids in plants where fertilization can't be induced by *in vitro* treatments (Ingram 2010).

In Vitro Fertilization

In vitro fertilization has been used to obtain both interspecific and intergeneric hybrids by overcoming physiological based self-incompatibility. A large number of plant species have been obtained via pollination of pistils and ovules (Palanivelu and Preuss 2006). This range includes agricultural crops, such as tobacco, rice, corn, clover, poppy and cotton. Wide hybridization, pollination with abortive pollen, delayed pollination, and physicochemical treatment of the ovary can be used to produce haploids (Islam and Tuteja 2012). *In vitro* fertilization studies have helped to understand many important phenomena regarding pollination and fertilization. A glycoprotein, TTS was purified from tobacco stylar tissue, which supports pollen tube growth (Hancock et al. 2005)

Embryo Culture

Poor endosperm development in wide hybridization results in embryo abortion during post-zygotic events. The in-vitro embryo culture technique has been successful in overcoming the major barrier and solving the problems of low seed set, seed dormancy, slow seed germination, including embryo growth in the absence of a symbiotic partner and the production of monoploids (Lin et al. 2011). The breeding cycle of a number of ornamentals like roses, orchids and banana and *Colocasia* has been reduced (Henning et al. 2004). A number of interspecific and intergeneric hybrids of a number of agriculturally important crops have been produced, including cotton, barley, tomato, rice, jute, *Hordeum* x *Secale*, *Tripsacum*x *Zea* and some *Brassicas* (Sanei et al. 2010; Tommonaro et al. 2012). At least seven Canadian barley cultivar were obtained from the material selected from doubled haploids originating through the bulbosum method of cross-fertilization and embryo rescue (Munoz-Amatrian et al. 2009). Monoploid wheat varieties have also been produced by this technique (Zhang et al. 2008). An *in vitro* spikelet culture system was developed to check the fruit set and early fruit development in rice crop. The cultured ovary of pollinated spikelets developed into fruits with an embryo and endosperm. While unpollinated spikelets when cultured on a medium containing 2,4-D, developed parthenocarpic fruits (Uchiumi and Okamoto 2010). An in-vitro protocol was developed for the culture of immature embryos of *Medicago truncatula* that permits their development in a way comparable to that observed in plants (Ochatt 2011).

Protoplast Fusion

When hybrid plants cannot be produced by conventional breeding, the role of protoplast fusion comes into play which acts as a means of creating unique hybrid plants. Protoplasts can be obtained from a number of crop species (Wang et al. 2011). However, while protoplasts of any two plants can be fused by chemical or physical means, production of unique somatic hybrid production is limited by the ability to generate the fused product and sterility in the interspecific hybrids rather than the production of protoplasts. *Nicotiana* is the best example of the use of protoplast fusion to improve crop production. The somatic hybrid products of a chemical fusion of protoplasts have been produced with modified alkaloid content and disease resistance of commercial tobacco cultivars (Patel et al. 2011).

The genetic components needed for taxol synthesis in *Taxus chinensis var. mairei* were transferred to a more tractable plant *Bupleurum scorzonerifolium*. RAPD data of the hybrid genome confirmed the presence of 82.4–96.8% genome *of B. scorzonerifolium* and 4.6–13.9% from the donor i.e., *T. chinensis* (Zhang et al. 2011).

Haploids

The significance of haploids in genetics and plant breeding has been realized for a long time. However, their exploitation remained restricted because of the extremely low frequency with which they occur in nature (usually 0.001–0.01 %). Spontaneous production of haploids usually occurs through the process of parthenogenesis (embryo development from an unfertilized egg). However, they reproduce the characters of the male parent alone suggesting their origin through ovule androgenesis (embryo development inside the ovule by the activity of the male nucleus alone). Haploid plants have the gametophytic number of chromosomes (Atanassov et al. 1995; Zapata Arias et al. 1995). The production of haploids in tomato has been tried and is still at a poor stage of development. The process of early embryogenesis from isolated microspores and the disruption of normal meiotic development and change of developmental fate towards callus proliferation, morphogenesis and plant regeneration have been shown in tomato by using light and electron microscopy (Segui-Simaro and Nuez 2007). For cell culture studies and breeding in flax, haploid and double haploid material and homozygous lines need to be produced. Anther culture has proved to be the most successful method producing doubled haploid lines in flax (Obert et al. 2009).

A fast and cheap method to obtain pure or homozygous lines is a priority for hybrid seed production in important crop plants. Pure lines can be produced traditionally by inbreeding and selection techniques, which are time consuming and costly. Alternatively, it has become possible through a biotechnological approach to accelerate the production of homozygous lines i.e., the induction of androgenesis to generate double haploid plants. Androgenesis reduces this process to a single generation, which implies time and cost saving. Due to these advantages, androgenic doubled haploids are the choice in a number of important crop plants where the methodology is well set up. In *solanaceae* family, crops like eggplant and pepper anther cultures are used for doubled haploid production and recent advances in the knowledge of embryo development are opening new ways to achieve the final goal of an efficient protocol in recalcitrant species (Segui-Simaro et al. 2011).

Gynogenesis is the phenomenon of production of whole plants from the unfertilized ovules. In Gentian (*Gentiana triflora*, *G. scabra* and their hybrids) an ornamental flower, unfertilized ovules were cultured in a medium containing a high concentration of sucrose (Doi et al. 2011) results in production of young plantlets. Although the embryos showed genotypic variation, all the genotypes gave response. The ovules collected from flower buds just before anthesis showed higher response. The dark culture condition also gave more number of haploid embryos as in 16-hour light condition.

Somaclonal Variation

The phenotypic variation of plants regenerated from cell culture is referred to as somaclonal variation. Apart from the mutant cell lines and plants obtained as a result of mutations many variants have been obtained through the tissue culture process cycle itself. The somaclonal variants may be genetic or epigenetic and are usually observed in the regenerated plantlets (Velker et al. 2012). These are dependent on the natural variations in a population of cells. Somaclonal variation may be due to pre-existing genetic differences in the cells or variations induced by tissue culture. The variation may be created through several types of mutations like inversion, deletion, duplication, gene amplification or de-amplification, by the activation of transposable element, point mutations, or re-activation of silent genes in multigene families (Karp 1994). Many aspects of the mechanism of somaclonal variation remain undefined, however in rice transposition of retrotransposons is one of the main causes of somaclonal variation (Pistelli et al. 2012). Somaclonal variation may also be induced in cultures via tissue culture process as reported in *Saintpaulia sp.* (Sato et al. 2011). A number of somaclonal variations observed in *in vitro* raised regenerants have been found to be of agricultural and horticultural significance. Some of such important alterations include chloroplast and chromoplast physiology, growth and development of the plant, seed yield, morphology of the flower and leaf, production of essential oils, fruit solids and disease resistance. Many of the important crops including wheat, rice, oats, maize, sugarcane, alfalfa, tobacco, tomato, potato, oil seed rape and celery have been observed with such variations (Karp 1994). This variation can also be obtained from gametic tissue.

One of the most important advantages of somaclonal variation is the induction of more genetic variability in economically important crops (Schellenbaum et al. 2008). *In vitro* selection of such somaclonal variants or rapid plant screening methods will be valuable. Enhancement in some somaclonal variants has been reported under *in vitro* conditions that include resistance to diseases, pathotoxins and herbicides or tolerance to different stress conditions (Zebrowska 2010).

Micropropagation

Propagation of Plants

Almost all types of plants can now be regenerated into plantlets from explants or callus. Thus, majority of the plant species have now well established micropropagation protocols and at present among the different techniques of plant tissue culture technology micropropagation is of widest use (Loyola-Vargas and Ochoa Alejo 2012). At present there are a number of tissue culture firms involved in *in vitro* multiplication, elimination of pathogens, storage of germplasm, genetic manipulation and plant-breeding programs (Ding et al. 2008). Micropropagation plays a major role in crop improvement. However, there are several limitations to the use of this technique. Up to 70 % of the production costs of micropropagation are required to fulfill the cost of labour needed to transfer tissue repeatedly between vessels and for asepsis. Tissue culture laboratory is greatly affected by the problems of vitrification, acclimatization and contamination (Doran 2009). A large number of desirable economic traits are lost in the tissue-cultured products due to genetic variation in cultured lines, such as polyploidy, aneuploidy and mutations. Enhancing axillary-bud breaking, production of adventitious buds and somatic embryogenesis are the three methods of micropropagation. In the latter two methods, differentiated structures arise directly or indirectly from callus. Axillary bud breaking produces very less number of plantlets as the number of shoots cultured is affected by the number of axillary buds cultured. However, it is the most widely used method in commercial micropropagation as it produces the most true to type plantlets. Adventitious budding is advantageous as bud primordial may be formed on any part of the inoculums (Brown and Thorpe 1996). Unfortunately, largest number of plantlets can be produced by somatic embryogenesis but only limited numbers of species respond to it. The use of bioreactors helps in large scale production of somatic embryos and their delivery in the form of synthetic seeds.

Synthetic Seeds

Synthetic or artificial seed is analogous to a zygotic seed and may be defined as a somatic embryo encapsulated inside a coating (Redenbaugh 1993). Synthetic seeds may be of different types: somatic embryos in a coating of water gel, dried and coated somatic embryos, suspended in a fluid, and buds encapsulated in a water gel. The use of synthetic seeds is advantageous over the traditional micropropagation protocols as it may have a cost of saving, as the labour intensive step of transferring plants from *in vitro* to field conditions. Other applications of synthetic seeds include the male sterile and parental line maintenance, for hybrid crop production and the preservation and multiplication of woody plants that have long juvenile developmental phase (Marimuthu et al. 2011). However, before the widespread of this technology, somaclonal variation has to be minimized. The production of high quality embryos at large scale must be perfected in the desired species and the micropropagation protocols will have to be cost-effective compared with existing seed or technologies.

Pathogen Eradication

The crop plants multiplying through vegetative propagation are generally infected with pathogens. Although in many cases the presence of the pathogen may not be obvious, but the yield or quality may be substantially reduced due to infection. *In vitro* culture has helped in increasing the yield of many crop plants. In China, virus free potatoes, produced by *in vitro* culture gave higher yields than the normal field plants with increase up to 150 % (Meiyalaghan et al. 2011). Seeds are responsible for only 10 % of viral infection, therefore, careful propagation from seeds can eliminate most of the viruses. The viruses are not distributed uniformly in the plant and the apical meristems are usually free from viruses (Wang and Valkonen 2008). The culture of the apical meristem coupled with chemo or thermo-therapy, have been used to produce virus-free material for micropropagation (Cantrill et al. 2005).

Germplasm Preservation

Conservation of germplasm, under *in vitro* storage can be done by slow growth conditions i.e., at low temperature or by fortifying the medium with growth retarding compounds. This is usually done by cryopreservation or by desiccated synthetic seeds (Silva et al. 2012). All such technologies depend on reducing or stopping growth and metabolic activity. However, the limiting factors are lack of a common method suitable for all species and genotypes, the expenditure and somaclonal variations and non-intentional cell-type selection in the preserved material e.g., cell divisions resulting in aneuploidy at low temperature or non-optimal conditions giving a particular cell type a selective growth advantage (Hajari et al. 2011).

Plant Tissue Culture, a Back Bone to Genetically Modified (GM) Crops

The expression of heterosis via hybrid vigour could be realized by performing pollination under controlled conditions as this leads to the development of new generations that performed better in the field than either of the parents and the progeny of the subsequent generation. Manipulation of the genetic makeup is one of the major activities in plant breeding, thus, a availability of genetic diversity is the prerequisite in plant breeding (Shefferson and Roach 2012). And the role of biotechnology and tissue culture comes into play in this area for creating genetic diversity and manipulating genetic variability. Although there is still lot of integration required most of the plant biotechnology and plant breeding programmes, but field trials of transgenics have now become much more common. Therefore, the modern technologies have revolutionized the advancements in the crop improvement programmes and it has been predicted for more than a decade (Ranganath 2011).

Various plant species are being modified genetically, either by vector dependent e.g., *Agrobacterium* or vector independent, which includes biolistic, micro-injection and liposome methods (Tagaki et al. 2011; She et al. 2012). In majority of the cases, tissue culture technology has been used to recover the modified cells or tissues. In fact, plant tissue culture techniques have played a major and important role in the development of genetic engineering. Plant tissue culture has helped in achieving many great successes in the field of transgenics. Davis and Reznikov (1992) have set milestones in the field of plant biotechnology by using a range of protoplast, microscope, tissue and organ culture protocols in many crop plants. Development of efficient transformation methods can enable the possibility of obtaining transgenic events that are devoid of marker gene/s upfront. Eggplant, an economically important vegetable crop does form only a non-significant percentage of agricultural production as it is susceptible to a number of pathogens with bacterial and fungal wilts being most devastating. A crop improvement approach has been developed which involves an *Agrobacterium* mediated transformation from two types of egg plants, a *Solanum melongena* L. breeding line, and *S. melongena* L. Black egg plant (Todaro et al. 2011). The use of leaf derive callus for the generation of stably transformed maize plants has been reported (Ahmadabadi et al. 2007; Chen et al. 2010). This method combines both conventional breeding and biotechnological techniques. Genetic transformation using *Agrobacterium tumefaciens* mediated transformation has been reported in mango, an important fruit crop (Krishna and Singh 2007; Singh et al. 2011). A technique for developing transgenics without the use of any selectable marker gene has been developed in peanut by taking advantage of high and consistent transformation potential of this crop (Bhatnagar et al. 2010). The technique of genetic engineering especially gene transfer depends upon plant tissue culture for the foreseeable future (Budzianowska 2009).

Besides this backcloth, a number of transgenic crops have been developed and a few are being grown in many parts of the world and these crops are either herbicide-tolerant or insect resistant.

Insect Resistance

For 21st century, the demands of sustainable agriculture were fulfilled by the genetically engineered crop resistance to insect pests that offers the prospective user-friendly environment and consumer friendly method of crop production. These biotech crops are genetically modified with *Bacillus thuringiensis* (Bt) endotoxins for insect resistance and the development of Bt crops stands as one of the most outstanding successes of transgenic plant biotechnology (Table 3.1). The Bt is a strong biological insecticide, which comprises of crystal protein endotoxin that is produced by some stains of soil bacterium *Bacillus thuringiensis* (Goudar et al. 2012). The Bt crystal (cry) genes are toxic to lepidopterans (Cohen et al. 2000), dipterans (Andrews et al. 1987) and coleopterans (Herrnstadt et al. 1986). Bt cry protein is toxic to insects (BANR 2000) and non-toxic to humans and animals. The Bt toxin gene was first discovered by Ishiwaki in 1901, in diseased silkworms and the first Bt toxin gene was cloned in 1981 (Jain et al. 2007). In 1986, the field test of transgenic tobacco expressing Bt toxin was performed. Moreover, in 1988 the first GM plant of japonica rice was produced and after that indica rice in 1990 (Ahmad et al. 2012). Now the biotech crops are grown globally including soybean, maize, cotton, canola, squash, papaya, sugar beet and tomato and the global biotech crop area derives soybean, corn, cotton and canola (Brookes and Barfoot 2011). The available data this time showed that biotech crops like soybean accounted the largest share (52 %) followed by corn (30 %) and canola (5 %). 16.7 million farmers across 29 countries (10 industrialized and 19 developing countries) planted 160 million hectares of biotech crops in 2011 and 90 % were small and resource poor farmers in developing countries (James 2010; Ahmad et al. 2012). However the combination of transgenic expression and improved protein stability has resulted even death of the Bt-resistant insects (Chougule and Bonning 2012; Kota et al. 1999). But now a days other insecticidal proteins like lectins, antibodies, protease inhibitors, wasp

Table 3.1 Transgenic plants expressing genes for insect and disease resistance

Plant	Gene	Resistance to	Reference
Potato	*Cry1Ab*	Potato tuber moth	Kumar et al. (2010)
Rice	*Cry1Ab*	Lepidopteron	Qi et al. 2009
Tobacco	Magi6 peptide	*Spodoptera frugiperda*	Hernández-Campuzano et al. 2009
Rice (Indica, Basmati)	*Cry1Ac, Cry2A*	YSB[a]	Bashir et al. (2005)
Rice (Indica, Minghuli 63)	*Cry2A*	YSB	Chen et al. (2005)
Rice (Indica, Minghuli 63)	*Cry1Ac, Cry2A, Cry9c*	YSB and Asiatic rice borer	Chen et al. 2008
Rice (Elite Vietnamese)	Fused gene, *Cry1Ab-1B* and hybrid Bt gene, *Cry1A/Cry1Ac*	YSB	Ho et al. (2006)
Indica pusa basmati 1, Japonica, Tainung 67	Potato proteinase inhibitor 2 (Pin 2)	YSB	Bhutani et al. (2006)
Indica basmati 370	*Cry1Ac, Cry2A*	YSB	Riaz et al. (2006)
Rice (Korean varieties) P-I, P-II, P-III	*Cry1Ab*	YSB	Kim et al. (2008)
Rice (Zhuxian B)	Sbti + GNA	Leaf folder + BPH	Li et al. (2005)
Indica rice	*Cry1Ab, Cry1Ac, gna*	YSB	Ramesh et al. (2004)
Indica rice	*Cry1Ab, Cry1Ac*	YSB	Alcantara et al. (2004)
Indica rice	*Cry1Ac, Cry2A, gna*	Lepidopteron insects	Rahman et al. (2007)
Indica rice	Chitinase + β-1,3-glucanase genes	*Rhizoctonia solani*	Sridevi et al. (2008)
Rape	*hrf2* gene encoding $\text{harpin}_{\text{xooc}}$ protein	*Sclerotinia sclerotinorium*	Ma et al. (2008)
Tobacco	p35 gene from baculovirus *Autographa californica*	TMV[a]	Wang et al. (2008)
Japonica	*Pi-d2*	Rice leaf blast and neck blast	Chen et al. (2010)
Tobacco	*GbTLP1*	*Verticillium dahliae*	Munis et al. (2010)
Potato	*StPUB17* (UND/PUB/ARM) repeat type gene	*Phytophthora infestans*	Ni et al. (2010)
Potato	*RB* resistance gene	*Potato late blight*	Liu and Halterman (2009)
Wheat	Ta-Tlp (thaumatin-like protein gene)	Powdery mildew and *Fusarium* head blight	Xing et al. (2008)

[a] *YSB* yellow stem borer, *TMV* tobacco mosaic virus

and microbial insecticides, spider toxins and insect peptide hormone have been discovered (Whetstone and Hammock 2007; Van Damme 2008). For instance, bacterium *Photorhabdus luminescens* produces photorhabdus toxin, which is an alternative to Bt for transgenic production. And the combination of photorhabdus toxins and Bt toxins in transgenic crops can be used to fight insect resistance. US based company Monsanto with India's Maharastra Hybrid Seeds Company (Mahyco) has recently developed a Bt eggplant (*Solanum melongena*) by incorporating a crystal gene (Cry1Ac) from *B. thuringiensis* (Krattiger 2010; Cotter 2011).

Disease Resistance

Diseases caused by bacteria, fungi, viruses and nematodes are responsible for damaging the crop plants, i.e phytopathogens have been threatening human life through the loss of crop production. For example the production of potato has been threatened by several fungal diseases and microbial pathogens resulting in 20% yield loss (Walter et al. 2011). Even though in potato the *Botrytis cinerea* is not the main disease, it may cause other serious diseases like *Fusarium solani* and *Phytophthora infestans*. But the transgenic approaches have used the genes encoding pathogenesis related (PR) proteins that confer the resistance to fungal pathogens (Hoshikawa et al. 2012; Gao et al. 2000) (Table 3.1). These PR proteins have antimicrobial properties against many fungal and bacterial pathogens. Transgenic approaches provide a powerful tool for the development of disease resistant crops (Melchers and Stuiver 2000; Rommens and Kishor 2000; Ellis et al. 2000). One approach involves the viral gene expression that interferes with the completion of life cycle of viruses. Powell-Abel et al. 1986 discovered that the coat proteins for TMV (tobacco mosaic virus) expressed in host plant interfered with the replication and plants expressing the TMV coat protein gene were resistant to TMV infection. This approach is now widely used to protect the crops from a large number of viruses (Mundembe et al. 2009). In 1992, china was the first country to commercialize these virus resistant transgenic crops (Brookes and Barfoot 2012; James 1997). After this virus-resistant tomato, squash and watermelon plants were also produced (Meeusen 1996). In oilseed rape, overexpression of tomato chitinase gene with a strong promoter gene has resulted increased resistance fungal attack in plants (i.e pathogens such as *Cylindrosporium concentricum* and *Phoma lingam)* (Grison et al. 1996). Anand et al. (2003) reported that wheat plants co-expressing a chitinase and β-1, 3-glucanase genes obtained to *Fusarium graminearum*. And under the green house and field conditions, transgenic potato plants expressing alfalfa antifungal peptide (alfAEP) showed strong defense activity against fungal pathogen, *Verticillium dahliae* (Gao et al. 2000). Thionins (PR proteins) are toxic to phytopathogens by attacking the cell membrane to increase their permeability and cause the death of the fungal cell due to leakage of proteins, nucleotides and other components (Chan et al. 2005; Hoshikawa et al. 2012). For example, overexpressed gamma-thionin gene from wasabi (*Eutrema wasabi*) in transgenic rice enhanced resistance to rice blast disease caused by *Magnaporthe grisea* (Kanzaki et al. 2002). Even in transgenic potato plants, potato thionin (snaking-1) gene expressed enhanced resistance to *Rhizoctonia solani* and *Erwinia cartovora* (Almasia et al. 2008). The wasabi thionin gene in transgenic potato plants had antifungal activity against gray mold (*Botrytis cinerea*)

(Khan et al. 2006; Hoshikawa et al. 2012). Another disease that adversely affect the barely and wheat production is *Fusarium* head blight (FHB). *Fusarium* produced trichothecene mycotoxin deoxynivalenol (DON) contamination with food is a great risk for humans and animals because trichothecenes are cytotoxins for eukaryotic cell. Recently, Di et al. (2010) has verified that expression of an N-terminal fragment of yeast L3 (L3Δ) in wheat showed reduction in disease ruthlessness and improved level of DON in transgenic wheat kernel when compared to non-transgenic wheat plants. Trichothecenes play multiple roles in the cell. They inhibit the protein synthesis (McLaughlin et al. 2009; Di et al. 2010). Critical role of trichothecene mycotoxin (tcmI) in the protein synthesis (Grant et al. 1976), which encodes the ribosomal protein L3 in yeast was observed by McLaughlin et al. (2009). Overexpression of RPL3 gene in transgenic plants induces resistance to trichothecene mycotoxin deoxynivalenol (DON) (McLaughlin et al. 2009).

Herbicide Resistance

Effective weed control has become one of the most important procedures in cropping operations to ensure good quality harvest. But the required mechanical weed control practices are now viewed as unsatisfactory and have been replaced by chemical weed control using herbicides. Herbicides are vicious for most of the plants because they function by disrupting the essential processes like photosynthesis, pigment biosynthesis, mitosis or essential amino acid biosynthesis (Mulwa and Wanza 2006). Now a day these herbicides have been replaced by new chemicals like glyphosate that is environmental friendly as it is degraded rapidly by soil micro-organisms (Day 1996). Glyphosate a highly translocated foliar herbicide was discovered in 1970 (Franz et al. 1996). Glyphosate is a non specific herbicide like many other herbicides and kills the green plants, therefore it can be used prior to seed emergence. Plants expressing herbicide tolerance accounted for 71 % of all the transgenic crops were grown worldwide in 1998 and 1999 (James 1999). Herbicide tolerant soybean, corn, cotton and canola represents the major transgenic products (James 1999; Liu 1999). Scientists in 1983, at Monsanto and Washington University isolated the common soil bacteria, *Agrobacterium tumefaciens* strain CP4, which is highly tolerant to glyphosate because its EPSPS (5-enolpyruvylshikimate-3-phosphate synthase) is less sensitive to inhibition by glyphosate than EPSPS found in plants (Watrud et al. 2004). In 1986, they successfully inserted the CP4 EPSPS gene into the plant genome and obtained the glyphosate resistant (GR) plants. GR soybean was commercialized within 10 years. Initial GR crops were the most quickly adopted technology in the history of agriculture (James 2007).

Recently, herbicide resistant *Amaranthus palmeri* by expressing glyphosate-insensitive herbicide target site gene, 5-enolpyruvylshikimate-3-phosphate synthase (EPSP) that is involved in the shikimate cycle where it catalyzes the reversible addition of the enolpyruvyl moiety of phosphor-enolpyruvate to shikimate 3-phosphate was developed (Gaines et al. 2010). In the western and central Africa considerable loss of maize was observed by a parasitic weed *Striga hermonthica*. Menkir et al.

(2010) incorporated an imidazolinone resistance (IR) XA17 gene into some maize lines that confers resistance to imazaquin and nicosulfuron herbicides. These IR-maize lines showed resistance to the *Striga hermonthica* weed and the yield loss was minimized to a considerable level. The expression of bar gene responsible for resistance to herbicides in sweet potato was demonstrated by Zang et al. (2009).

Approaches that have been used to generate herbicide-tolerant crops are: (1) Decrease the sensitivity of the plants to the herbicide by modifying the sensitivity of the target enzyme and (2) engineer a herbicide detoxifying pathway in to the plant (Simoens and Van Montagu 1995). For instance glyphosate and acifluoren tolerance is included in first approach. Transgenic plants tolerant to the herbicide acifluorfen, which inhibits chlorophyll biosynthesis, have been produced through overexpression of the target enzyme involved in chlorophyll biosynthesis (Lermontova and Grimm 2000; Ahmad et al. 2012). Second approach includes the resistance to glufosinate and bromoxynil. To enhance the metabolism of these herbicides various genes were introduced and the active compound is converted to products are non-toxic to the crop (Haumann 1997; Ahmad et al. 2012). Now the critics of herbicide resistant crops fear the over use of herbicides and the development of herbicide resistant weeds. But the herbicide resistant weeds can be controlled by rotating the crops with different transgenic modes of action. Various environmentally friendly herbicides and their corresponding resistant genes are available that makes crop rotation practices possible.

Abiotic Stress

Abiotic stress (Salinity, drought, temperature, UV Radiations etc.) has been found to have negative impact on the crop production. The crop loss due to these abiotic stresses is responsible for the enormous economic loss worldwide (Ahmad et al. 2008, 2010, 2012).

Conventional breeding techniques have been used to improve the crop production but much success has not been achieved in generating stress tolerant plants. Plant biologists have developed transgenic technology to generate stress tolerant plants and to improve the crop production. Modification of the biochemical pathways through the transgenic approach and overexpression of the stress tolerant genes have great success in achieving the target of generating stress tolerant plants. For further information see Table 3.2.

Conclusions and Future Perspectives

Plant tissue culture has emerged as an inescapable tool with possibilities for complementing the conventional methods in plant breeding and crop improvement. These techniques have proved successful and are now being used globally for the ex-situ conservation of the plants including crop plants. Plant cell/tissue culture is a rapidly

Table 3.2 Some promising genes that can be expressed in plants for abiotic stress tolerance

Gene and gene product	Plant	Resistance to	Reference
betA (Choline dehydrogenase)	Tobacco	Salinity and low temperature	Holmstrom et al. (2000)
BADH1 (Betaine aldehyde dehydrogenase)	Tomato	Salinity	Jia et al. (2002)
EctA, ectB, ectC	Tobacco	Salinity	Nakayama et al. (2000)
OstA, OstB (Trehalose-6-P synthase, Trehalose-6-P phosphatase)	Tobacco rice	Salt, drought	Garg et al. (2002)
TPS and TPP (Trehalose synthesis)	Arabidopsis	Drought, salt, temperature	Miranda et al. (2007)
TPP1 (Trehalose synthesis)	Rice	Salt and cold	Ge et al. (2008)
TPS1 (Trehalose synthesis)	Alfalfa	Drought, salt, temperature	Suarez et al. (2009)
WCOR15 (cold induced gene)	Tobacco	Freezing	Shimamura et al. (2006)
AtOAT (Ornithine amino transferase)	Rice	Drought and salt	Jang et al. (2003)
pdc1 (Pyruvate decarboxylase overexpression)	Rice	Submergence tolerance	Minhas and Grover (1999)
pdc1 and pdc2 (Pyruvate decarboxylase overexpression)	Arabidopsis	Hypoxic stress survival	Ismond et al. (2003)
ppo (Polyphenol oxidases suppression)	Tomato	Drought	Thipyapong et al. (2004)
SAMDC (polyamine synthesis)	Tobacco	Drought, salinity, *Verticillium, Fusarium wilts*	Waie and Rajam (2003)
SPDS (Spermidine synthase)	Arabidopsis	Salinity	Bagni et al. (2006)
P5CS (Δ^1-pyrroline-5-carboxylate synthase)	Bean	Drought, salt and cold	Chen et al. (2009)
P5CS (Δ^1-pyrroline-5-carboxylate synthase)	Potato	Salt	Hmida-Sayari et al. (2005)
P5CS (Δ^1-pyrroline-5-carboxylate synthase)	Wheat	Drought	Vendruscolo et al. (2007)
adc (polyamine synthesis)	Rice	Drought	Capell et al. (2004)
Osm1 to *Osm4* (Osmotin protein accumulation)	Strawberry	Salt and drought	Husaini and Abdin (2008)
ME-leaN4 (Lea protein)	Lettuce	Salt	Park et al. (2005)
Os LEA3-1 (Lea protein)	Rice	Drought	Xiao et al. (2007)
HVA1 (group 3 LEA protein gene)	Mulberry	Salt and drought	Lal et al. (2008)
BhLEA1, LEA2 (LEA protein)	Tobacco	Drought	Liu et al. (2009)
HAL3 (FMN-binding protein)	Arabidopsis	Salt and osmotic tolerance	Espinosa-Ruiz et al. (1999)
HAL1	Arabidopsis	Salt	Ellul et al. (2003)

Table 3.2 (continued)

Gene and gene product	Plant	Resistance to	Reference
HAL1	Watermelon	Salt	Yang et al. (2001)
OsDREB1A	Arabidopsis	Drought, salt and cold tolerance	Dubouzet et al. (2003)
DREB1A (Transcription factor)	Paspalum grass	Drought	James et al. (2008)
DREB1A (Transcription factor)	Tobacco	Salt	Cong et al. (2008)
DREB1A, DREB2A (Transcription factor)	Arabidopsis	Drought and cold	Maruyama et al. (2009)
OsNAC10 (Transcription factor)	Rice	Drought	Jeong et al. (2010)
OsSMCP1 (Transcription factor)	Arabidopsis	Salt	Yokotani et al. (2009)
Osmyb4 (Cold induced transcription factor)	Apple	Drought and cold tolerance	Pasquali et al. (2008)
A1fin1 (Transcription factor)	Alfalfa	Salt	Winicov (2000)
OrbHLH2 (Transcription factor)	Arabidopsis	Salt and osmotic stress	Zhou et al. (2009)
OsWRKY45 (Transcription factor)	Arabidopsis	Drought	Qiu and Yu (2009)
Tsi1 (EREBP/AP2 DNA binding motif)	Tobacco	Salt and pathogen	Park et al. (2001)
CBF1 (DREB1B)	Tomato	Drought	Hsieh et al. (2002)
CBF4	Arabidopsis	Drought	Haake et al. (2002)
ABF3/ABF4	Arabidopsis	Drought	Kang et al. (2002)
AtMYC2/AtMYB2	Arabidopsis	Drought	Abe et al. (2003)
ZPT2-3 (Cys2/His2-type Zinc-finger protein)	Petunia	Drought	Sugano et al. (2003)
CpMYB10	Arabidopsis	Drought and salt	Villalobos et al. (2004)
FeSOD (Superoxide dismutase)	Tobacco	Salt and oxidative stress	Van Camp et al. (1996)
MnSOD	Arabidopsis	Oxidative stress	Wang et al. (2004)
MnSOD	Rice	Oxidative stress	Tanaka et al. (1999)
Glutathione-S-transferase/ glutathione peroxidase	Tobacco	Salt and cold	Roxas et al. (1997)
KatE (Catalase)	Tobacco	Salt and oxidative stress	Al-Taweel et al. (2007)
DHAR1 (Dehydroascorbate reductase)	Arabidopsis	Salt tolerance	Ushimaru et al. (2006)
AtALDH3 (Aldehyde dehydrogenase)	Arabidopsis	Drought, salt and oxidative stress	Sunkar et al. (2003)
MsALR (Aldose/aldehyde reductase)	Alfalfa	Drought and heavy metal	Oberschall et al. (2000)
Ascorbate peroxidise	Tobacco	Drought and salt	Badawi et al. (2004)
GlyI and GlyII (Glyoxylase)	Tobacco	Salt	Yadav et al. (2005)
OsCDPK (calcium dependent protein kinase)	Rice	Drought and salt	Saijo et al (2000)

Table 3.2 (continued)

Gene and gene product	Plant	Resistance to	Reference
Cnb1 (Calcineurin)	Tobacco	Salt	Pardo et al (1998)
DnaK (Heat shock proteins)	Tobacco	Salt	Sugino et al. (1999)
AtHsp 17.6A (Small heat shock protein)	Arabidopsis	Drought and salt	Sun et al (2001)
AtGSK1	Arabidopsis	Drought and Salt	Piao et al (2001)
AtNDPK2 (Nucleotide diphosphate kinase)	Arabidopsis	Salt, cold, methyl viologen	Moon et al. (2002)
AtNHX1 (Vacuolar Na^+/H^+ antiporter)	Tomato	Salt	Zhang and Blumwald (2001)
AtNHX1 (Vacuolar Na^+/H^+ antiporter)	Mustard	Salt	Zhang et al. (2001)
AtNHX1 (Vacuolar Na^+/H^+ antiporter)	Rice	Salt	Ohta et al. (2002)
SOS1 (Plasma membrane Na^+/H^+ antiporter)	Arabidopsis	Salt	Shi et al. (2003)
AVP1 (K^+/Na^+ transport regulation)	Arabidopsis	Drought and salt	Gaxiola et al. (2001)
CaXTH3 (Xyloglucan endotransglucosylase)	Arabidopsis	Drought and salt	Cho et al. (2006)
ZmOPR1 (12-Oxo-Phytodie-noic acid reductases)	Arabidopsis	Osmotic and salt stress	Gu et al. (2008)
SPCP2 (papain-like cysteine protease)	Arabidopsis	Salt and drought	Chen et al. (2010)
W6 (Ethylene responsive factor gene)	Tobacco	Salt tolerance	Lu et al. (2010)
TSRF1 (Ethylene responsive factor)	Rice	Drought	Quan et al. (2010)
TERF2/LeERF2 (Ethylene responsive factor)	Tomato tobacco	Freezing	Zhang and Huang (2010)
StPUB17 (UND/PUB/ARM) repeat type gene	Potato	Salt tolerance	Ni et al. (2010)

developing technology which holds promise of restructuring agricultural, horticultural and forestry practices. Cultured explants undergo frequent genetic changes which are expressed at biochemical or molecular level. The genetic variability expressed in regenerated plants can be transmitted to the progeny through sexual or vegetative propagation. Rapid advances in plant genetic engineering have made it possible to improve endogenous metabolic pathways and/or bestow foreign functions. Utilization of plants, however, is disturbed by environmental stresses such as drought, high salinity, low temperature, and pathogens, which either directly cause cell death or inhibit growth by disarranging the intracellular water balance. The application of tissue culture technology as a central tool or as an adjunct to other methods, including recombinant DNA techniques, is at the vanguard in plant modification and improvement for agriculture, horticulture and forestry.

References

Abe H, Urao T, Ito T, Seki M, Shinozaki K, Yamaguchi-Shinozaki K (2003) Arabidopsis AtMYC2 (b-HLH) and AtMYB2 (MYB) function as transcriptional activators in abscisic acid signalling. Plant Cell 15:63–78

Ahmad P, Sarwat M, Sharma S (2008) Reactive oxygen species, antioxidants and signaling in plants. J Plant Biol 51(3):167–173

Ahmad P, Jaleel CA, Salem MA, Nabi G, Sharma S (2010) Roles of Enzymatic and non-enzymatic antioxidants in plants during abiotic stress. Crit Rev Biotechnol 30(3):161–175

Ahmad P, Ashraf M, Younis M, Hu X, Kumar A, Akram NA, Al-Qurainy F (2012) Role of transgenic plants in agriculture and biopharming. Biotechnol Adv 30:524–540

Ahmadabadi M, Ruf S, Bock R (2007) A leaf-based regeneration and transformation system for maize (*Zea mays* L.). Transgenic Res 16:437–448

Alcantara EP, Aguda RM, Curtiss A, Dean DH, Cohen MB (2004) *Bacillus thuringiensis* δ-endotoxin binding to brush border membrane vesicles of rice stem borers. Arch Insect Biochem Physiol 55:169–177

Almasia NI, Bazzini AA, Hopp HE, Vazquez-Rovere C (2008) Overexpression of snakin-1 gene enhances resistance to *Rhizoctonia solani* and *Erwinia carotovora* in transgenic potato plants. Mol Plant Pathol 9:329–338

Al-Taweel K, Iwaki T, Yabuta Y, Shigeoka S, Murata N, Wadan AA (2007) Bacterial transgene for catalase protects translation of D1 protein during exposure of salt-stressed tobacco leaves to strong light. Plant Physiol 145:258–265

Anand A, Zhou T, Trick HN, Gill BS, Bockus WW, Muthukrishnan S (2003) Green-house and field testing of transgenic wheat plants stably expressing genes for thaumatin-like protein, chitinase and glucanase against *Fusarium graminearum*. J Exp Bot 54:1101–1111

Andrews RW, Fausr R, Wabiko MH, Roymond KC, Bulla LA (1987) Biotechnology of Bt: a critical review. Biotechnol 6:163–232

Atanassov A, Zagorska N, Boyadjiev P, Djilianov D (1995) *In vitro* production of haploid plants. World J Microbiol Biotechnol 11:400–408

Azria D, Bhalla PL (2011) Agrobacterium-mediated transformation of Australian rice varieties and promoter analysis of major pollen allergen gene, Ory s 1. Plant Cell Rep 30:1673–1681

Badawi GH, Kawano N, Yamauchi Y, Shimada E, Sasaki R, Kubo A, Tanaka K (2004) Overexpression of ascorbate peroxidase in tobacco chloroplasts enhances the tolerance to salt stress and water deficit. Physiol Plant 121:231–237

Bagni N, Ruiz-Carrasco K, Franceschetti M, Fornalè S, Fornasiero RB, Tassoni A (2006) Polyamine metabolism and biosynthetic gene expression in arabidopsis thaliana under salt stress. Plant Physiol Biochem 44:776–786

BANR (Board on Agriculture and Natural Resources) (2000) Genetically modified pest protected plant: science and regulation. Committee on genetically modified pest protected plants. National Research Council, National Academy Press, Washington, p 292

Bashir K, Husnain T, Fatima T, Riaz N, Makhdoom R, Riazuddin S (2005) Novel indica basmati line (B-370) expressing two unrelated genes of *Bacillus Thuringiensis* is highly resistant to two lepidopteran insects in the field. Crop Protection 24:870–879

Bhatnagar M, Prasad K, Bhatnagar-Mathur P, Narasu ML, Waliyar F, Sharma KK (2010) An efficient method for the production of marker-free transgenic plants of peanut (*Arachis hypogaea* L.). Plant Cell Rep 29:495–502

Bhutani S, Kumar R, Chauhan R, Singh R, Choudhury VK, Choudhury JB (2006) Development of transgenic indica rice plants containing potato proteinase inhibitor 2 gene with improved defense against yellow stem borer. Physiol Mol Biol Plant 12:43–52

Brookes G, Barfoot P (2011) Global impact of biotech crops: environmental effects 1996–2009. Lands Biosci 2:34–49

Brookes G, Barfoot P (2012) GM crops: global socio-economic and environmental impacts 1996–2010. Dorchester, UK

Brown DCW, Thorpe TA (1996) Crop improvement through tissue culture. World J Microbiol Biotechnol 11:409–415

Budzianowska A (2009) *In-vitro* cultures of tobacco and their impact on development of plant biotechnology. Biotechnol Adv 25:223–243

Cantrill LC, Overall RL, Goodwin PB (2005) Changes in macromolecular movement accompany organogenesis in thin cell layers of *Torenia fournieri*. Planta 222:933–946

Capell T, Bassie L, Christou P (2004) Modulation of the polyamine biosynthetic pathway in transgenic rice confers tolerance to drought stress. PNAS 101:9909–9914

Chan YL, Prasad V, Sanjaya C, Chen KH, Liu PC, Chan MT, Cheng CP (2005) Transgenic tomato plants expressing an Arabidopsis thionin (i2.1) driven by fruit-inactive promoter battle against phytopathogenic attack. Planta 221:386–393

Chandel G, Datta K, Datta SK (2010) Detection of genomic changes in transgenic Bt rice populations through genetic fingerprinting using amplified fragment length polymorphism (AFLP). GM Crops 1:327–336

Chen H, Tang W, Xu C, Li X, Lin Y, Zhang Q (2005) Transgenic indica rice plants harboring a synthetic *cry2A** gene of *Bacillus thuringiensis* exhibit enhanced resistance against lepidopteran rice pests. Theor Appl Genet 111:1330–1337

Chen H, Zhang G, Zhang Q, Lin Y (2008) Effect of transgenic Bacillus thuringiensis rice lines on mortality and feeding behavior of rice stem borers (Lepidoptera: Crambidae). J Econ Entomol 101:182–189

Chen HJ, Su CT, Lin CH, Huang GJ, Lin YH (2010) Expression of sweet potato cysteine protease *SPCP2* altered developmental characteristics and stress responses in transgenic *Arabidopsis* plants. J Plant Physiol 167:838–847

Chen JB, Wang SM, Jing RL, Mao XG (2009) Cloning the PvP5CS gene from common bean (*Phaseolus vulgaris*) and its expression patterns under abiotic stresses. J Plant Physiol 166:12–19

Chen ZY, Brown RL, Damann KE, Clevel TE (2010) PR10 expression in maize and its effect on host resistance against *Aspergillus flavus* infection and aflatoxin production. Mol Plant Pathol 11:69–81

Cho SK, Kim JE, Park JA, Eom TJ, Kim WT (2006) Constitutive expression of abiotic stress-inducible hot pepper *CaXTH3*, which encodes a xyloglucan endotransglucosylase/hydrolase homolog, improves drought and salt tolerance in transgenic *Arabidopsis* plants. FEBS Letters 580:3136–3144

Chougule NP, Bonning BC (2012) Toxins for transgenic resistance to hemipteran pests. Toxins 4:405–429

Cohen BM, Gould F, Bentur JC (2000) Bt rice: practical steps to sustainable use. Int Rice Res 25:4–10

Cong L, Zheng HC, Zhang YX, Chai TY (2008) *Arabidopsis DREB1A* confers high salinity tolerance and regulates the expression of GA dioxygenases in tobacco. Plant Sci 174:156–164

Cotter J (2011) Genetically engineered (GE) Bt egg-plant (talong): health risks, environmental impacts and contamination from field trials. Greenpeace Breifing, GRL-TN-02. http://gaia-health.com/articles451/000460-greenpeacebt.pdf

Davey MR, Anthony P, Power JB, Lowe KC (2005) Plant protoplasts: status and biotechnological perspectives. Biotechnol Adv 23:131–171

Davis J, Reznikoff WS (1992) Milestones in biotechnology. Classic papers on genetic engineering. Boston Butterworth Heinemann, Boston

Di R, Blechl A, Dill-Macky R, Tortora A, Tumer NE (2010) Expression of a truncated form of yeast ribosomal protein L3 in transgenic wheat improves resistance to fusarium head blight. Plant Sci 178:374–380

Ding F, Jin S, Hong N, Zhong Y, Cao Q, Yi G, Wang G (2008) Vitrification-cryopreservation, an efficient method for eliminating *Candidatus liberobacter* asiaticus, the citrus Huanglongbing pathogen, from in-vitro adult shoot tips. Plant Cell Rep 27:241–250

Doi H, Yokoi S, Hikage T, Nishihara M, Tsutsumi KI, Takahata Y (2011) Gynogenesis in gentians (Gentiana triflora, G. scabra): production of haploids and doubled haploids. Plant Cell Rep 30:1099–1096

Doran PM (2009) Application of plant tissue cultures in phytoremediation research: incentives and limitations. Biotechnol Bioeng 103:60–76

Dresselhaus T, Lausser A, Marton ML (2011) Using maize as a model to study pollen tube growth and guidance, cross-incompatibility and sperm delivery in grasses. Ann Bot 108:727–737

Dubouzet JG, Sakuma Y, Ito Y, Kasuga M, Dubouzet EG, Miura S, Seki M, Shinozaki K, Yamaguchi-Shinozaki K (2003) *OsDREB* genes in rice, (*Oryza sativa* L.) encode transcription activators that function in drought, high salt and cold responsive gene expression. Plant J 33:751–763

Ellis J, Dodds P, Pryor T (2000) Structure, function and evolution of plant disease resistance genes. Curr Opin Plant Biol 3:278–284

Ellul P, Rios G, Atares A, Roig LA, Serrano R, Moreno V (2003) The expression of the *Saccharomyces cerevisiae* HAL1 gene increases salt tolerance in transgenic watermelon (*Citrullus lanatus* (Thunb.) Matsun. & Nakai.). Theor Appl Genet 107:462–469

Espinosa-Ruiz A, Belles JM, Serrano R, Culianez-Macia FA (1999) *Arabidopsis thaliana* AtHAL3: a flavoprotein related to salt and osmotic tolerance and plant growth. Plant J 20:529–539

Franz JE, Mao MK, Sikorski JA (1996) Glyphosate: a unique global pesticide. American Chemical Society, Washington, p 653

Gaines TA, Zhang W, Wang D, Bukun B, Chisholm ST, Shaner DL, Nissen SJ, et al. (2010) Gene amplification confers glyphosate resistance in *Amaranthus palmeri*. Proc Nat Acad Sci USA 107:1029–1034

Gao AG, Hakimi SM, Mittanck CA, Wu Y, Woerner BM, Stark DM, Shah DM, Liang J, Rommens CMT (2000) Fungal pathogen protection in potato by expression of a plant defensin peptide. Nat Biotechnol 18:1307–1310

Garg AK, Kim JK, Owens TG, Ranwala AP, Choi YD, Kochian LV, Wu RJ (2002) Trehalose accumulation in rice plants confers high tolerance levels to different abiotic stresses. PNAS 99:15898–15903

Gaxiola R, Li J, Undurraga S, Dang LM, Allen GJ, Alper SL, Fink GR (2001) Drought and salt tolerant plants result from overexpression of the AVP1 H + pump. Proc Nat Acad Sci USA 98:11444–11449

Ge LF, Chao DY, Shi M, Zhu MZ, Gao JP, Lin HX (2008) Overexpression of the trehalose-6-phosphate phosphatase gene OsTPP1 confers stress tolerance in rice and results in the activation of stress responsive genes. Planta 228:191–201

Goudar G, Alagawadi GAR, Krishnaraj PU, Basavana Goud K (2012) Characterization of *Bacillus thuringiensis* isolates of Western Ghats and their insecticidal activity against diamond back moth (*Plutella xylostella* L.). Karnataka J Agric Sci 25(2):199–202

Grant PG, Schindler D, Davies JE (1976) Mapping of trichodermin resistance in *Saccharomyces cerevisiue*: a genetic locus for a component of the 60s ribosomal subunit. Genetics 83:667–673

Grison R, Grezes-Besset B, Schneider M, Lucante N, Olsen L, Leguay JJ, Toppan A (1996) Field tolerance to fungal pathogens of *Brassica napus* constitutively expressing a chimeric chitinase gene. Nat Biotechnol 14:643–646

Gu D, Liu X, Wang M, Zheng J, Hou W, Wang G, Wang J (2008) Overexpression of *ZmOPR1* in *Arabidopsis* enhanced the tolerance to osmotic and salt stress during seed germination. Plant Sci 174:124–130

Haake V, Riechmann JLCD, Pineda O, Thomashow MF, Zhang JZ (2002) Transcription factor CBF4 is a regulator of drought adaptation in Arabidopsis. Plant Physiol 130:639–648

Hajari E, Berjak P, Pammenter NW, Watt MP (2011) A novel means for cryopreservation of germplasm of the recalcitrant-seeded species, Ekebergia capensis. Cryo Letters 32:308–316

Hancock CN, Kent L, McClure BA (2005) The stylar 120 kDa glycoprotein is required for S-specific pollen rejection in *Nicotiana*. Plant J 43:716–723

Haumann BE (1997) Bioengineered oilseed acreage escalating. Inform 8:804–811

Henning L, Gruissem W, Grossniklaus U, Kohler C (2004) Transcriptional programs of early reproductive stages in arabidopsis. Plant Physiol 135:1765–1775

Hernández-Campuzano B, Suárez R, Lina L, Hernández V, Villegas E, Corzo G, Iturriaga G (2009) Expression of a spider venom peptide in transgenic tobacco confers insect resistance. Toxicon 53:122–128

Herrnstadt C, George GS, Edward WR, David L (1986) A new strain of *Bacillus thuringiensis* with activity against coleopteran insects. Nat Biotechnol 4:305–308

Hmida-Sayari A, Gargouri-Bouzid R, Bidani A, Jaoua L, Savouré A, Jaoua S (2005) Overexpression of Δ^1-pyrroline-5-carboxylate synthetase increases proline production and confers salt tolerance in transgenic potato plants. Plant Sci 169:746–752

Ho NH, Baisakh N, Oliva N, Datta K, Frutos R, Datta SK (2006) Translational fusion hybrid *Bt* genes confer resistance against yellow stem borer in transgenic elite vietnamese rice (*Oryza sativa* L.) Cultivars. Crop Sci 46:781–789

Holmstrom KO, Somersalo S, Manda A, Palva TE, Welin B (2000) Improved tolerance to salinity and low temperature in transgenic tobacco producing glycine betaine. J Exp Bot 51:177–185

Hoshikawa K, Ishihara G, Takahashi H, Nakamura I (2012) Enhanced resistance to gray mold (*Botrytis cinerea*) in transgenic potato plants expressing thionin genes isolated from Brassicaceae species. Plant Biotechnol 29:87–93

Hsieh TH, Lee JT, Charng YY, Chan MT (2002) Tomato plants ectopically expressing Arabidopsis CBF1 show enhanced resistance to water deficit stress. Plant Physiol 130:618–626

Husaini AM, Abdin MZ (2008) Overexpression of tobacco osmotin gene leads to salt stress tolerance in strawberry (*Fragaria x ananassa* Duch.) plants. Indian J Biotechnol 7:465–472

Hussain A, Hasnain S (2012) Comparative assessment of the efficacy of bacterial and cyanobacterial phytohormones in plant tissue culture. World J Microbiol Biotechnol 28:1459–1466

Hussain AM, Rashid Z, Mir RU, Aquil B (2011) *In vitro* cultures of tobacco and their impact on development of plant biotechnology. GM Crops 2:150–162

Ingram GC (2010) Family life at close quarters: communication and constraint in angiosperm seed development. Protoplasma 247:195–214

Islam SM, Tuteja N (2012) Enhancement of and rogenesis by abiotic stress and other pretreatments in major crop species. Plant Sc 182:134–144

Ismond KP, Dolferus R, Pauw MD, Dennis ES, Good AG (2003) Enhanced low oxygen survival in arabidopsis through increased metabolic flux in the fermentative pathway. Plant Physiol 132:1292–1302

Jain M, Chengalrayan K, Abouzid A, Gallo M (2007) Prospecting the utility of a/MI/mannose selection system for the recovery of transgenic sugarcane (Saccharum spp. hybrid) plants. Plant Cell Rep 26:581–590

James C (1997) Global status of transgenic crops in 1997. Int Service Acquisition Agric biotechnol Appl 5:1–30

James C (1999) Global review of commercialized transgenic crops in 1999. Int Service Acquisition Agric Biotechnol Appl 12:1–7

James C (2007) Global Status of Commercialized Biotech/GM Crops. http://www.isaaa.org/resources/publications/briefs/

James C (2010) A global overview of biotech (GM) crops, adoption, impact and future prospects. GM Crops 1:8–12

James VA, Neibaur I, Altpeter F (2008) Stress inducible expression of the DREB1A transcription factor from xeric, *Hordeum spontaneum* L. in turf and forage grass (*Paspalum notatum* Flugge) enhances abiotic stress tolerance. Transgenic Res 17:93–104

Jang IC, Oh SJ, Seo JS, Choi WB, Song SI, Kim CH, Kim YS, Seo HS, Choi YD, Nahm BH, Kim JK (2003) Expression of a bifunctional fusion of the *Escherichia coli* genes for trehalose-6-phosphate synthase and trehalose-6-phosphate phosphates in transgenic rice plants increases trehalose accumulation and abiotic stress tolerance without stunting growth. Plant Physiol 131:516–524

Jeong JS, Kim YS, Baek KH, Jung H, Ha SW, Choi YD, Kim M, Reuzeau C, Kim JK (2010) Root specific expression of *OsNAC10* improves drought tolerance and grain yield in rice under field drought conditions. Plant Physiol 153:185–197

Jia GX, Zhu ZQ, Chang FQ, Li YX (2002) Transformation of tomato with the BADH gene from atriplex improves salt tolerance. Plant Cell Rep 21:141–146

Kang J, Choi H, Im M, Kim SY (2002) *Arabidopsis* basic leucine zipper proteins that mediate stress-responsive abscisic acid signaling. Plant Cell 14:343–357

Kanzaki H, Nirasawa S, Saitoh H, Ito M, Nishihara M, Terauchi R, Nakamura I (2002) Overexpression of the wasabi defensin gene confers enhanced resistance to blast fungus (*Magnaporthe grisea*) in transgenic rice. Theor Appl Genet 105:809–814

Karkonen A, Santanen A, Iwamoto K, Fukuda H (2011) Plant tissue cultures. Methods Mol Biol 715:1–20

Karp A (1994) Origins, causes and uses of variation in plant tissue cultures. In: Thorpe TA (ed) Plant cell and tissue culture: methods and applications in agriculture. New York Academic Press, pp 181–211

Khan RS, Nishihara M, Yamamura S, Nakamura I, Mii M (2006) Transgenic potatoes expressing wasabi defensin peptide confer partial resistance to gray mold (*Botrytis cinerea*). Plant Biotechnol 23:179–183

Kim S, Kim C, Li W, Kim T, Li Y, Zaidi MA, Altosaar I (2008) Inheritance and field performance of transgenic Korean Bt rice lines resistant to rice yellow stem borer. Euphytica 164:829–839

Kota M, Daniell H, Varma S, Garczynski SF, Gould F, Moar WJ (1999) Overexpression of the *Bacillus thuringiensis* (Bt) Cry2Aa2 protein in chloroplasts confers resistance to plants against susceptible and Bt-resistant insects. Proc Natl Acad Sci USA 96:1840–1850

Krattiger A (2010) Intellectual property, commercial needs and humanitarian benefits: must there be a conflict? New Biotechnol 27:573–577

Krishna H, Singh SK (2007) Biotechnological advances in mango (*Mangifera indica* L.) and their future implication in crop improvement: a review. Biotechnol Adv 25:223–243

Kumar M, Chimote V, SinghR, Mishra GP, Naik PS, PandeySk, Chakrabarti SK (2010) Development of Bt transgenic potatoes for effective control of potato tuber moth by using *cry1Ab* gene regulated by GBSS promoter. Crop Prot 29:121–127

Lal S, Gulyani V, Khurana P (2008) Overexpression of *HVA1* gene from barley generates tolerance to salinity and water stress in transgenic mulberry (*Morus indica*). Transgenic Res 17:651–663

Lermontova I, Grimm B (2000) Overexpression of plastidic protoporphyrinogen IX oxidase leads to resistance to the diphenyl-ether herbicide acifluorfen. Plant Physiol 122:75–84

Li G, Xu X, Xing H, Zhu H, Fan Q (2005) Insects resistance to nilaparvata lugens and cnephalocrosis medinalis in transgenic indica rice and the inheritance of gna + sbtitransgenes. Pest Manag Sci 61:390–396

Lin M, Wu Q, Zheng S, Tian H (2011) Tissue culture and polyploidy induction of morinda officinalis. Zhongguo Zhong Yao Za Zhi 36:2325–2328

Liu K (1999) Biotech crops: products, properties, and prospects. Food Technol 53:42–48

Liu X, Wang Z, Wang L, Wu R, Phillips J, Deng X (2009) *LEA* 4 group genes from the resurrection plant *Boea hygrometrica* confer dehydration tolerance in transgenic tobacco. Plant Sci 176:90–98

Liu Z, Halterman D (2009) Analysis of proteins differentially accumulated during potato late blight resistance mediated by the *RB* resistance gene. Physiol Mol Plant Pathol 74:151–160

Loyola-Vargas VM, Ochoa-Alejo N (2012) An introduction to plant cell culture: the future ahead. Methods Mol Biol 877:1–8

Lu Y, Liu P, Xu ZS, Zhang RY, Liu L, Li LC, Chen M, Ye XG, Chen YF, Ma YZ (2010) Overexpression of *W6* gene increases salt tolerance in transgenic tobacco plants. Acta Agron Sinica 34:984–990

Ma L, Huo R, Gao Xw, He D, Shao M, Wang Q (2008) Transgenic Rape with *hrf2* Gene Encoding $Harpin_{Xooc}$ Resistant to *Sclerotinia sclerotinorium*. Agricul Sci China 7:455–461

Marimuthu MP, Jolivet S, Ravi M, Pereira L, Davda JN, Cromer L, Wang L, Noque F, Chan SW, Siddiqi I, Mercier R (2011) Synthetic clonal reproduction through seeds. Science 331:876

Maruyama K, Takeda M, Kidokoro S, Yamada S, Sakuma Y, Urano K, Fujita M, Yoshiwara K, Matsukura S, Morishita Y, Sasaki R, Suzuki H, Saito K, Shibata D, Shinozaki K, Yamaguchi-Shinozaki K (2009) Metabolic pathways involved in cold acclimation identified by integrated analysis of metabolites and transcripts regulated by DREB1A and DREB2A. Plant Physiol 150:1972–1980

McLaughlin JE, Bin-Umer MA, Tortora A, Mendez N, McCormick S, Tumera NE (2009) A genome-wide screen in saccharomyces cerevisiae reveals a critical role for the mitochondria in the toxicity of a trichothecene mycotoxin. Proc Natl Acad Sci 106:21883–21888

Meeusen RL (1996) Commercialization of transgenic seed products: two case studies. In: Collins GB, Shepherd RJ (eds) Engineering plants for commercial products and applications. NY Acad Sci, New York, pp 172–176

Meiyalaghan S, Barrell PJ, Jacobs JM, Conner AJ (2011) Regeneration of multiple shoots from transgenic potato events facilitates the recovery of phenotypically normal lines: assessing a cry9Aa2 gene conferring insect resistance. BMC Biotechnol 11:93

Melchers LS, Stuiver MH (2000) Novel genes for disease-resistance breeding. Curr Opin Plant Biol 3:147–152

Menkir A, Chikoye D, Lum F (2010) Incorporating an herbicide resistance gene into tropical maize with inherent polygenic resistance to control Striga hermonthica (Del.) Benth. Plant Breed 129:385–392

Minhas D, Grover A (1999) Transcript levels of genes encoding various glycolytic and fermentation enzymes change in response to abiotic stresses. Plant Sci 146:41–51

Miranda JA, Avonce N, Suarez R, Thevelein JM, Dijck PV, Iturriaga G (2007) A bifunctional TPS–TPP enzyme from yeast confers tolerance to multiple and extreme abiotic-stress conditions in transgenic *Arabidopsis*. Planta 226:1411–1421

Moon H, Baek D, Lee B, Prasad DT, Lee SY, Cho MJ, Lim CO, Choi MS, Bahk J, Kim MO, Hong JC, Yun DJ (2002) Soybean ascorbate peroxidase suppresses Bax-induced apoptosis in yeast by inhibiting oxygen radical generation. Biochem Biophys Res Commun 290:457–462

Mulwa RMS, Mwanza LM (2006) Biotechnology approaches to developing herbicide tolerance/selectivity in crops. Afr J Biotechnol 5(5):396–404

Mundembe R, Matibiri A, Sithole-Niang I (2009) Transgenic plants expressing the coat protein gene of cowpea aphid-borne mosaic potyvirus predominantly convey the delayed symptom development phenotype. Afr J Biotechnol 8:2682–2690

Munis MFH, Tu L, Deng F, Tan J, Xu L, Xu S, LongL, Zhang X (2010) A thaumatin-like protein gene involved in cotton fiber secondary cell wall development enhances resistance against *Verticillium dahliae* and other stresses in transgenic tobacco. Biochem Biophy Res Commu 393:38–44

Munoz-Amatrian M, Svensson Jt, Castillo AM, Close TJ, Valles MP (2009) Microspore embryogenesis: assignment of genes to embryo formation and green vs. albino plant production. Funct Integr Genomics 9:311–323

Mustafa NR, de Winter W, van Irene F, Verportee R (2011) Initiation, growth and cryopreservation of plant cell suspension cultures. Nat Protoc 6:715–742

Nakayama H, Yoshida K, Ono H, Murooka Y, Shinmyo A (2000) Ectoine, the compatible solute of *Halomonas elongata*, confers hyperosmotic tolerance in cultured tobacco cells. Plant Physiol 122:1239–1247

Ni X, Tian Z, Liu J, Song B, Li J, Shi X, Xie C (2010) StPUB17, a novel potato UND/PUB/ARM repeat type gene, is associated with late blight resistance and NaCl stress. Plant Sci 178:158–169

Oberschall A, Deak M, Török K, Saa L, Vass I, Kovacs I, Feher A, Dudits D, Horvath GV (2000) A novel aldose/aldehyde reductase protects transgenic plants against lipid peroxidation under chemical and drought stresses. Plant J 24:437–446

Obert B, Zackova Z, Samaj J, Pretova A (2009) Doubled haploid production in Flax (*Linum usitatissimum L.*). Biotechnol Adv 27:371–375

Ochatt SJ (2011) Immature seeds and embryos of *Medicago truncatula* cultured *in vitro*. Methods Mol Biol 710:39–52

Ohta M, Hayashi Y, Nakashima A, Hamada A, Tanaka A, Nakamura T, Hayakawa T (2002) Introduction of a Na^+/H^+ antiporter gene from *Atriplex gmelini* confers salt tolerance to rice. FEBS Lett 532:279–282

Palanivelu R, Preuss D (2006) Distinct short-range ovule signals attract or repel Arabidopsis thaliana pollen tubes *in vitro*. BMC Plant Biol 6:1–7

Pardo JM, Reddy MP, Yang S, Maggio A, Huh GH, Matsumoto T, Coca MA, Paino-D'Urzo M, Koiwa H, Yun DJ, Watad AA, Bressan RA, Hasegawa PM (1998) Stress signaling through Ca2+/calmodulin dependent protein phosphatase calcineurin mediates salt adaptation in plants. Proc Natl Acad Sci 95:9681–9686

Park BJ, Liu Z, Kanno A, Kameya T (2005) Increased tolerance to salt- and water-deficit stress in transgenic lettuce (*Lactuca sativa* L.) by constitutive expression of LEA. Plant Growth Regul 45:165–171

Park JM, Park CJ, Lee SB, Ham BK, Shin R, Paek KH (2001) Overexpression of the tobacco Tsi1 gene encoding an EREBP/AP2-type transcription factor enhances resistance against pathogen attack and osmotic stress in tobacco. Plant Cell 13:1035–1046

Pasquali G, Biricolti S, Locatelli F, Baldoni E, Mattana M (2008) *Osmyb4* expression improves adaptive responses to drought and cold stress in transgenic apples. Plant Cell Rep 27:1677–1686

Patel D, Power JB, Anthony P, Badakshi F, Pat Heslop-Harrison JS, Davey MR (2011) Somatic hybrid plants of *Nicotiana* x sanderae (+) N. debneyi with fungal resistance to *Peronospora tabacina*. Ann Bot 108:809–819

Piao HL, Lim JH, Kim SJ, Cheong GW, Hwang I (2001) Constitutive overexpression of AtGSK1 induces NaCl stress responses in the absence of NaCl stress and results in enhanced NaCl tolerance in *Arabidopsis*. Plant J 27:305–314

Pistelli L, Iacona C, Miano D, Crilli M, Colao MC, Mensuali-Sodi A, Muleo R (2012) Molecular spectrum of somaclonal variation in regenerated rice revealed by whole genome sequencing. Plant Cell Physiol 53:256–264

Pons E, Peris JE, Pena L (2012) Field performance of transgenic citrus trees: assessment of the long-term expression of uidA and nptII transgenes and its impact on relevant agronomic and phenotypic characteristics. BMC Biotechnol 12:41

Powell-Abel P, Nelson RS, De B, Hoffman H, Rogers SG, Fraley RT, Beachy RN (1986) Delay of disease development in transgenic plants that express the tobacco mosaic virus coat protein gene. Science 232:738–743

Qi YB, YeSH, Lu YT, Jin QS, Zhang XM (2009) Development of marker-free transgenic *Cry1Ab* rice with lepidopteran pest resistance by *Agrobacterium* mixture-mediated co-transformation. Rice Sci 16:181–186

Qiu Y, Yu D (2009) Overexpression of the stress-induced *OsWRKY45* enhances disease resistance and drought tolerance in *Arabidopsis*. Environ Exp Bot 65:35–47

Quan R, Hu S, Zhang Z, Zhang H, Zhang Z, Huang R (2010) Overexpression of an ERF transcription factor *TSRF1* improves rice drought tolerance. Plant Biotech J 8:476–488

Rahman M, Rashid H, Shahid AA, Bashir K, Husnain T, Riazuddin S (2007) Insect resistance and risk assessment studies of advanced generations of basmati rice expressing two genes of *Bacillus thuringiensis*. Elect J Biotech 10(2):240–251

Ramesh S, Nagadhara D, Reddy VD, Rao KV (2004) Production of transgenic indica rice resistant to yellow stem borer and sap-sucking insects, using super-binary vectors of *Agrobacterium tumefaciens*. Plant Sci 166:1077–1085

Ranganath RM (2011) Developmental switches that hold the key to a revolution in crop biotechnology. Nat Rev Genet 12:224

Redenbaugh K (1993) Synseeds: applications of synthetic seeds to crop improvement. CRC Press, Boca Ratan FL

Riaz N, Husnain T, Fatima T, Makhdoom R, Bashir K, Masson L, Altosaar I, Riazuddin S (2006) Development of Indica basmati rice harboring two insecticidal genes for sustainable resistance against lepidopteran insects. South African J Bot 72:217–223

Rommens CM, Kishore GM (2000) Exploiting the full potential of disease-resistance genes for agricultural use. Curr Opin Biotechnol 11:120–125

Roxas VP, Smith Jr RK, Allen ER, Allen RD (1997) Overexpression of glutathione-S-transferase/glutathione peroxidase enhances the growth of transgenic tobacco seedlings during stress. Nat Biotechnol 15:988–991

Roy AK, Malaviya DR, Kaushal P (2011) Generation of interspecific hybrids of trifolium using embryo rescue techniques. Methods Mol Biol 710:141–151

Saijo Y, Hata S, Kyozuka J, Shimamoto K, Izui K (2000) Overexpression of a single Ca^{2+} dependent protein kinase confers both cold and salt/drought tolerance on rice plants. Plant J 23:319–327

Sanei M, Pickering R, Fuchs J, Banei Moghaddam AM, Dziurlikowska A, Houben A (2010) Interspecific hybrids of hordeum marinum ssp. marinum x H. bulbosum are mitotically stable and reveal no gross alterations in chromatin properties. Cytogenet Genome Res 129:110–116

Sato M, Hosokawa M, Doi M (2011) Somaclonal variation is induced *De Novo* via the tissue culture process: a study quantifying mutated cells in *Saintpaulia*. Plos One 6:e23541

Schellenbaum P, Mohler V, Wenzel G, Walter B (2008) Variation in DNA methylation patterns of grapevine somaclones (*Vitis vinifera* L.). BMC Plant Biol 8:78

Segui-Simaro JM, Nuez F (2007) Embryogenesis induction, callogenesis and plant regeneration by *in vitro* culture of tomato isolated microspores and whole anthers. J Exp Bot 58:1119–1132

Segui-Simaro JM, Corral-Martinez P, Parra-Vega V, Gonzalez-Garcia B (2011) Androgenesis in recalcitrant solanaceous crops. Plant Cell Rep 30:765–778

She M, Yin G, Li J, Li X, Du L, Ma W, Ye X (2012) Efficient regeneration potential is closely related to auxin exposure time and catalase metabolism during the somatic embryogenesis of immature embryos in *triticum aestivum* L. Mol Biotechnol 4:213–217

Shefferson RP, Roach DA (2012) The triple helix of plantago lanceolata: genetics and the environment interact to determine population dynamics. Ecology 93:793–802

Shi H, Lee BH, Wu SJ, Zhu JK (2003) Overexpression of a plasma membrane Na^+/H^+ antiporter gene improves salt tolerance in Arabidopsis thaliana. Nat Biotechnol 21:81–85

Shi HY, Zhang YX (2012) Pear ACO genes encoding putative 1-aminocyclopropane-1-carboxylate oxidase homologs are functionally expressed during fruit ripening and involved in response to salicylic acid. Mol Biol Rep 12:19–26

Shimamura C, Ohno R, Nakamura C, Takumi S (2006) Improvement of freezing tolerance in tobacco plants expressing a cold-responsive and chloroplast-targeting protein WCOR15 of wheat. J Plant Physiol 163:213–219

Silva Rde C, Camillo J, Scherwinski-Pereira JE (2012) A method for seedling recovery in jatropha curcas after cryogenic exposure of the seeds. Rev Biol Trop 60:473–482

Simoens C, Van Montagu M (1995) Genetic engineering in plants. Hum Reprod Update 1:523–542

Singh RK, Ali SA, Nath P, Sane VA (2011) Activation of ethylene-responsive p-hydroxyphenylpyruvate dioxygenase leads to increased tocopherol levels during ripening in mango. J Exp Bot 62:3375–3385

Sridevi G, Parameswari C, Sabapathi N, Raghupathy V, Veluthambi K (2008) Combined expression of chitinase and β-1,3-glucanase genes in *indica* rice (*Oryza sativa* L.) enhances resistance against *Rhizoctonia solani*. Plant Sci 175:283–290

Suarez R, Calderon C, Iturriaga G (2009) Enhanced tolerance to multiple abiotic stresses in transgenic alfalfa accumulating trehalose. Crop Sci 49:1791–1799

Sugano S, Kaminaka H, Rybka Z, Catala R, Salinas J, Matsui K, Ohme-Tkagi M, Takatsuji H (2003) Stress-responsive zinc finger gene ZPT2–3 plays a role in drought tolerance in petunia. Plant J 36:830–841

Sugino M, Hibino T, Tanaka Y, Nii N, Takabe T, Takabe T (1999) Overexpression of DnaK from a halotolerant cyanobacterium *Aphanotece halophytica* acquires resistance to salt stress in transgenic tobacco plants. Plant Sci 146:81–88

Sun W, Bernard C, van de Cotte, Van Montagu M, Verbruggen N (2001) *At-HSP17. 6A*, encoding a small heat-shock protein in *Arabidopsis*, can enhance osmo-tolerance upon overexpression. Plant J 27:407–415

Sunkar R, Bartels D, Kirch HH (2003) Overexpression of a stress inducible aldehyde dehydrogenase gene from *Arabidopsis thaliana* in transgenic plants improves stress tolerance. Plant J 35:452–464

Tagaki K, Nishizawa K, Hirose A, Kita A, Ishimoto M (2011) Manipulation of saponin biosynthesis by RNA interference-mediated silencing of β-amyrin synthase gene expression in soybean. Plant Cell Rep 30:1835–1846

Tanaka K, Hibino T, Hayashi Y, Tanaka A, Kishitani S, Takabe T, Yokota S, Takabe T (1999) Salt tolerance of transgenic rice overexpressing yeast mitochondrial Mn-SOD in chloroplasts. Plant Sci 148:131–138

Thipyapong P, Melkonian J, Wolfe DW, Steffens JC (2004) Suppression of polyphenol oxidase increases stress tolerance in tomato. Plant Sci 167:693–703

Thorpe T (2012) History of plant tissue culture. Methods Mol Biol 877:9–27

Todaro A, Cavallaro R, Argento S, Branca F, Spagna G (2011) Study and characterization of polyphenol oxidase from egg plant (*Solanum melongena L.*). J Agric Food Chem 59:11244–11248

Tommonaro G, de prisco R, Abbamondi Gr, Marzocco S, Saturnino C, Poli A, Nicolaus B (2012) Evaluation of antioxidant properties, total phenolic content, and biological activities of new tomato hybrids of industrial interest. J Med Food 15:483–489

Uchiumi T, Okamoto T (2010) Rice fruit development is associated with an increased IAA content in pollinated ovaries. Planta 232:579–592

Ushimaru T, Nakagawa T, Fujioka Y, Daicho K, Naito M, Yamauchi Y, Nonaka H, Amako K, Yamawaki K, Murata N (2006) Transgenic arabidopsis plants expressing the rice dehydroascorbate reductase gene are resistant to salt stress. J Plant Physiol 163:1179–1184

Van Camp W, Capiau K, Van Montagu M, Inze D, Slooten L (1996) Enhancement of oxidative stress tolerance in transgenic tobacco plants overproducing Fe-superoxide dismutase in chloroplasts. Plant Physiol 112:1703–1714

Van Damme EJM (2008) Plant lectins as part of the plant defense system against insects. In: Schaller A (ed) Induced plant resistance to herbivory. Springer Dordrecht, Netherlands, pp 285–307

Velker BA, Denomme MM, Mann MR (2012) Embryo culture and epigenetics. Methods Mol Biol 912:399–421

Vendruscolo EC, Schuster I, Pileggi M, Scapim CA, Molinari HB, Marur CJ, Vieira LG (2007) Stress induced synthesis of praline confers tolerance to water deficit in transgenic wheat. J Plant Physiol 164:1367–1376

Villalobos MA, Bartels D, Iturriaga G (2004) Stress tolerance and glucose insensitive phenotypes in arabidopsis overexpressing the *CpMYB10* transcription factor gene. Plant Physiol 135:309–324

Waie B, Rajam MV (2003) Effect of increased polyamine biosynthesis on stress responses in transgenic tobacco by introduction of human *S*- denosylmethionine gene. Plant Sci 164:727–734

Walter RS, Rosemary L, Gary DF, Weingartner DP (2011) Compendium of potato diseases (APS compendium of plant disease series). APS Press, The American Phytopathological Society, St. Paul

Wang GX, Tang Y, Yan H, Sheng XG, Hao WW, Zhang L, Lu K, Liu F (2011) Production and characterization of interspecific somatic hybrids between brassica oleracea var. botrytis and B. nigra and their progenies for the selection of advanced pre-breeding materials. Plant Cell Rep 30:1811–1821

Wang QC, Valkonen JP (2008) Elimination of two viruses which interact synergistically from sweet potato by shoot tip culture and cryotherapy. J Virol Methods 154:135–145

Wang W, Vinocur B, Shoseyov O, Altman A (2004) Role of plant heat shock proteins and molecular chaperones in the abiotic stress response. Trends Plant Sci 9:244–252

Wang Z, Song J, Zhang Y, Yang B, Wang Y, Chen S (2008) Mechanism analysis of broad-spectrum disease resistance induced by expression of anti-apoptotic *p35* gene in tobacco. Chinese J Biotech 24:1707–1713

Watrud LS, Lee EH, Fairbrother A, Burdick C, Reichman JR, Bollman M, Storm M, King G, Van de Water PK (2004) Evidence for landscape-level, pollen-mediated gene flow from genetically modified creeping bentgrass with CP4 EPSPS as a marker. Proc Natl Acad Sci USA 101:14533–14538

Whetstone PA, Hammock BD (2007) Delivery methods for peptide and protein toxins in insect control. Toxicon 49:576–596

Winicov I (2000) Alfin1 transcription factor overexpression enhances plant root growth under normal and saline conditions and improves salt tolerance in alfalfa. Planta 210:416–422

Xiao B, Huang Y, Tang N, Xiong L (2007) Overexpression of a LEA gene in rice improves drought resistance under field conditions. Theor Appl Genet 115:35–46

Xing LP, Wang HZ, Jiang ZN, Ni JL, Cao AZ, Yu L, Chen PD (2008) Transformation of wheat thaumatin-like protein gene and analysis of reactions to powdery mildew and *Fusarium* head blight in transgenic plants. Acta Agronomica Sinica 34:349–354

Yadav SK, Singla-Pareek SL, Reddy MK, Sopory SK (2005) Transgenic tobacco plants over-expressing glyoxylase enzymes resist an increase in methylglyoxal and maintain higher reduced glutathione levels under salinity stress. FEBS Lett 579:6265–6271

Yang SX, Zhao YX, Zhang Q, He YK, Zhang H, Luo HA (2001) L1 mediate salt adaptation in *Arabidopsis thaliana*. Cell Res 11:142–148

Yokotani N, Ichikawa T, Kondou Y, Maeda S, Iwabuchi M, Mori M, Hirochika H, Mastsui M, Oda K (2009) Overexpression of a rice gene encoding a small C2 domain protein *OsSMCP1* increases tolerance to abiotic and biotic stresses in transgenic Arabidopsis. Plant Mol Biol 71:391–402

Zang N, Zhai H, Gao S, Chen W, He S, Liu Q (2009) Efficient production of transgenic plants using the bar gene for herbicide resistance in sweet potato. Sci Hort 122:649–653

Zapata-Arias FJ, Torrizo LB, Ando A (1995) Current developments in plant biotechnology for genetic improvement, the case of rice (*Oryza sativa* L.). World J Microbiol Biotechnol 11:393–399

Zebrowska JI (2010) *In vitro* selection in resistance breeding of strawberry (Fragaria x ananassa duch.). Commun Agric appl Biol Sci 75:699–704

Zhang L, Chen Q, Yuan Z, Xiang Z, Zheng Y, Liu D (2008) Production of aneuhaploid and euhaploid sporocytes by meiotic restitution in fertile hybrids between durum wheat langdon chromosome substitution lines and *Aegilops tauschii*. J Genet Genomics 35:617–623

Zhang F, Wang P, Ji D, Kang G, Xiang F (2011) Asymmetric somatic hybridization between bupleurum scorzonerifolium willd and taxus chinensis var. mairei. Plant Cell Rep 30:1857–1864

Zhang HX, Blumwald E (2001) Transgenic salt tolerant tomato plants accumulate salt in foliage but not in fruit. Nat Biotechnol 19:765–768

Zhang HX, Hodson JN, Williams JP, Blumwald E (2001) Engineering salt-tolerant *Brassica* plants: characterization of yield and seed oil quality in transgenic plants with increased vacuolar sodium accumulation. Proc Natl Acad Sci USA 98:12832–12836

Zhang Z, Huang R (2010) Enhanced tolerance to freezing in tobacco and tomato overexpressing transcription factor TERF2/LeERF2 is modulated by ethylene biosynthesis. Plant Mol Bio 73:241–249

Zhou J, Li F, Wang JL, Ma Y, Chong K, Xu YY (2009) Basic helix-loop-helix transcription factor from wild rice (*OrbHLH2*) improves tolerance to salt- and osmotic stress in *Arabidopsis*. J Plant Physiol 166:1296–1306

Chapter 4
Mutagenesis—A Potential Approach for Crop Improvement

Rajib Roychowdhury and Jagatpati Tah

Abstract Global environmental dissociative changes are now in steady state. Its negative impacts were gradually imposed on a wide range of crops and thus crop improvement was hindered as well. Given this challenge, existing and new, appropriate technologies need to be integrated for global crop improvement. Among the different present approaches, mutagenesis and mutation breeding and the isolation of improved or novel phenotypes in conjunction with conventional breeding programmes can result in mutant varieties endowed with new and desirable variation of agrometrical traits. Induced mutations and its related technologies play very well in this ground and this overall strategy helps to trace the crop genetic diversity along with its biodiversity maintenance. Such induced mutagenesis, a crucial step in crop improvement programme, is now successful in application due to the advancement and incorporation of large-scale selection techniques, micropropagation and other *in vitro* culture methods, molecular biology tools and techniques in modern crop breeding performance. Time to time, different mutation techniques and their application processes are changing significantly; in this perspective, insertional mutagenesis and retrotransposons are taking more supports for mutational tagging and new mutation generation. For details investigation on plant structure and function, mutagenic agents and their precise role are much essential as it can produce mutants with some phenotypic changes. Functional genomics studies make the ultimatum platform on this field of study where few crop plants were used for mutational experimentation on some prime agronomic traits till now. This is a prerequisite step and is applying on diverse crop for further improvement. High throughput DNA technologies for mutation screening such as TILLING (Targeting Induced Limited

R. Roychowdhury (✉)
Department of Biotechnology, Visva-Bharati, Santiniketan 731235, West Bengal, India
e-mail: rajibroychowdhurybu@gmail.com

J. Tah
Botany Department (UGC-CAS), The University of Burdwan,
Burdwan 713104, West Bengal, India
e-mail: jagatpati_tah@yahoo.com

K. R. Hakeem et al. (eds.), *Crop Improvement,* DOI 10.1007/978-1-4614-7028-1_4,

Lesions IN Genomes), high-resolution melt analysis (HRM), ECOTILLING etc. are the key techniques and resources in molecular mutation breeding. Molecular mutation breeding will significantly increase both the efficiency and efficacy of mutation techniques in crop breeding. Such modern and classical technologies are using for the development of mutation induction with the objective of using a set of globally important crops to validate identified relevant novel techniques and build these into modular pipelines to serve as technology packages for induced crop mutations. Thus, mutation assisted plant breeding will play a crucial role in the generation of 'designer crop varieties' to address the uncertainties of global climate variability and change, and the challenges of global plant-product insecurity.

Introduction

Mutation breeding is the purposeful application of mutations in plant breeding area. It offers good prospects for the domestication of promising underutilized wild species, for agricultural or horticultural uses as well as for improving adaptation of recently introduced crops to unsuitable environments. Mutagenesis has remained popular for close to a century because of its simplicity, technical and economic viability, applicability to all plant species and usability at small or large scales (Siddiqui and Khan 1999). More than 2,000 plant varieties that contain induced mutations have been officially released for cultivation either directly as new varieties or used as parents to derive new varieties without the regulatory restrictions faced by genetically modified material (Maluszynski et al. 2000; Waugh et al. 2006). The main strategy in mutation-based breeding has been to upgrade the well-adapted plant varieties by improving a few desirable major yield and quality traits (Ahloowalia et al. 2004; Wilde et al. 2012). Besides, the increased yield and enhanced quality of the novel varieties included several other components such as subsequent use for breeding, improved harvest index from heterosis in hybrid cultivars, response to increased agronomic inputs, and consumer preference.

Plant breeding categorized into three sub-types as mutation breeding, recombination breeding and transgenic breeding has the potential of generating variation and selection of target lines. In case of mutation breeding, the basic fundamental and the unique feature is the generation of new mutated alleles. The key steps includes analysis of difference in the sensitivity of different genotypes and plant tissues to different mutations often measured using lethal doses (LD), generation of genetic chimeras after mutagenic treatment and analysis of their effect on transmission of mutated alleles and segregation in the subsequent generation and also often the recessive nature of induced mutations. This knowledge is important for establishing proper doses and modes of mutagenic treatment. Apart from this, the knowledge can also be employed for the planning of methodology of harvesting and growing second mutant (M_2) populations from first mutant (M_1) generations (Table 4.1). Like any other scientific innovative technology, mutation breeding has its advantages and limitations. The advantages being creation of new genetic al-

Table 4.1 Three important plant breeding strategies

Breeding Methods	Source of genetic variation	Transmission, expression and inheritance	Nature of gene action	Breeding generations
Mutation breeding	New alleles artificially and randomly created from endogenous genes	Induced mutations subject to diplontic and haplontic selection	Mostly recessive alleles	About 2–3 generations
Recombinant breeding	Recombination of gene alleles from parental varieties	No selective transmission; co-segregation of closely linked alleles	Dominant, recessive alleles, and QTLs	About 10 generations
Transgenic breeding	Insertion of new genes or modification of endogenous genes	Expression of transgenes subject to position effect or silencing	Mostly dominant alleles	About 3 generations

QTL-Quantitative trait loci

leles that do not persist in germplasm pools and the induction of new gene alleles for a commercial variety such that new varieties carrying desired mutation alleles can be directly used as a commercial variety. Also, the limited genetic changes of any single plant of a mutated population and the often recessive nature enable breeders to develop a new variety in a short breeding cycle. The limitation being its limited power in generating the dominant alleles which might be desired; its less effectiveness than cross breeding for a trait needs for a combination of multiple alleles, such as tolerance to abiotic stresses. The low mutation frequency requires growing and screening a large population for selection of desired mutants at a reasonable confidence. This becomes very expensive for traits that have to be evaluated through laborious phenotypic analysis (Roychowdhury 2011; Roychowdhury and Tah 2011).

The knowledge of the extent to which the desirable characters with economic values are heritable is a prerequisite for any crop improvement programme (Roychowdhury and Tah 2011b; Roychowdhury et al. 2011a). Breeders have continually retained their interest in the grouping of the germplasm and the pedigree of selected cultivars since the information might be particularly helpful in effective breeding strategy determination (Ali et al. 2011). For this purpose, inducible mutation, using chemical or physical mutagens, is a suitable source of producing variation through mutation breeding procedure (Domingo et al. 2007; Roychowdhury and Tah 2011a) which can produce several improved mutant varieties with high demanding economic value (Din et al. 2004). From implicational point of view, it is quite possible to induce gene-mutation artificially with the help of some potent chemical mutagens to create any new variation in crops.

Time to time, the spectrum of available mutation techniques has also significantly increased. Important practical results have been achieved through the use of chronic irradiation from gamma-field or by irradiation with heavy ion beam or by chemical mutagenesis. *In vitro* cultured somaclonal variations have been proved to be useful for creating variation in many characters, especially the ones that can be selected under *in vitro* cultural environments. Classical insertional mutagenesis and more recently retro-transposons have become almost irreplaceable tools in generating and tagging of new mutations for crop improvement. Potent mutagenic agents ('mutagens' that cause mutation) can be used to produce the 'morphological mutants' that are essential for dissecting plant structure, functions and their regulation (Maluszynski 1999). The traditional mission of mutation breeding technology deals with the development of new and desired variation(s) through breeding programs for overall crop improvement that has recently been significantly spread widely. Induced mutations can play an important role in the conservation and preservation of crop biodiversity. Induced mutations and related advance technologies are important not only for extending genetic diversity of major crops but also are an important additional source of biodiversity enhancement of neglected and local crops.

Mutation and Mutagens

A mutation is a sudden heritable change in the DNA of a living cell, not caused by the common phenomena of genetic segregation or genetic recombination. Mutations may occur in nature without intentional human intervention, and are said to be spontaneous. Spontaneous mutations may result from the activity of mobile genetic elements (transposons) that can move around to different positions within the genome of a single cell and affect the activity of the gene in which they are inserted (Wessler 2006). Mobile genetic elements affect the gene function through various mechanisms. Retro-transposons, for example, move in the genome by being transcribed to RNA and then back to DNA by reverse transcriptase, while DNA transposons move directly from one position to another within the genome using a transposase enzyme to 'cut and paste' them within the genome, causing spontaneous mutations (Kidwell 2005). Most spontaneous mutations occur in very low frequencies (10^{-6}) of an individual gene. Moreover, not all phenotypically observed variation refers to genetic changes. At the same time, not all the spontaneous changes in the DNA ultimately result in permanent changes of the DNA. Even if such changes would be permanent, they may not always result in visible or detectable effects (Ranel 1989). For example, there may be latent adaptive mutations in African nightshade that help the plants to survive in the wild, but these are not known so far. Besides, spontaneous mutations depend on chance and make breeding programmes considerably slow. Although selection for economically useful spontaneous mutants still takes place with some level of success (Ahloowalia et al. 2004; Wilde et al. 2012), the purposeful induction of a specifically desired mutation at a specific time and place, and in a selected genotype for a selected purpose is a much more attractive option.

Mutation Induction

A physical or chemical agent that changes the genetic information (usually DNA) of an organism and thus increasing the number of mutations above the natural background level is called a mutagen (Fig. 4.1). DeVries (1905) suggested the use of radiation to induce mutations. The discovery that X-ray induced mutations in *Drosophila melanogaster* (Muller 1927) and in *Hordeum vulgare* (Stadler 1928) led to the use of radiation-induced mutations for changing plant traits by plant breeders and geneticists. Auerbach and Robson (1946) reported the use of chemicals such as mustard gas to be highly mutagenic. Since then a number of agents have been discovered that can increase the frequency of artificially induced mutations. The main mutagens available for induction of mutations include UV radiation, electromagnetic waves such as X-rays, γ-rays and cosmic rays; fast moving particles such as α-particles, β-particles and neutrons; and chemical agents such as, alkylating agents, acridines, azides, hydroxyl amides, etc. In general, ionizing radiations such as X-rays and γ-rays are preferred over chemical mutagens because of their ease of application, good penetration and reproducibility, high mutation frequency and less disposal problems. The X-rays are obtained from X-ray machines by bombarding tungsten or molybdenum with electrons in a vacuum, whereas γ-rays are obtained from radioisotopes like ^{60}Co and ^{137}Caesium in the γ-chamber. The UV-radiations possess limited tissue penetrating ability [low linear energy transfer or in short LET] and cause relatively little damage except after prolonged exposure as a result of which their use is restricted to pollen grains (Kovacs and Keresztes 2002).

The application of this phenomenon has come a long way to become a real tool, not only in crop breeding but also in basic research on the plant genome, its structure and function. Breeders were the first to recognize the potential of induced mutations through analogy with spontaneous mutants, often selected as new plant types in many crops, from cereals to apples, not to mention ornamental and decorative plants. Many mutants with desired traits were selected in the second or third generation after mutagenic treatment and subsequently released as new cultivars after agronomic evaluation in regional and national trials. These or other mutants developed with mutations in desired traits, even though not released as new cultivars, have been used in cross-breeding programmes as a source of particular alleles, often allelic to the spontaneous ones, but in a desired genotype. Among them were sources for characters such as short stature and lodging resistance; disease resistance; oil quality; and increased nitrogen fixation. These mutated genes are especially valuable as the best currently grown cultivar was usually selected for mutagenic treatment. A desired mutation in a good genetic background is a very attractive component in breeding programmes. This approach is much simpler and faster than crossing with an exotic source, and it is one of the main reasons for the

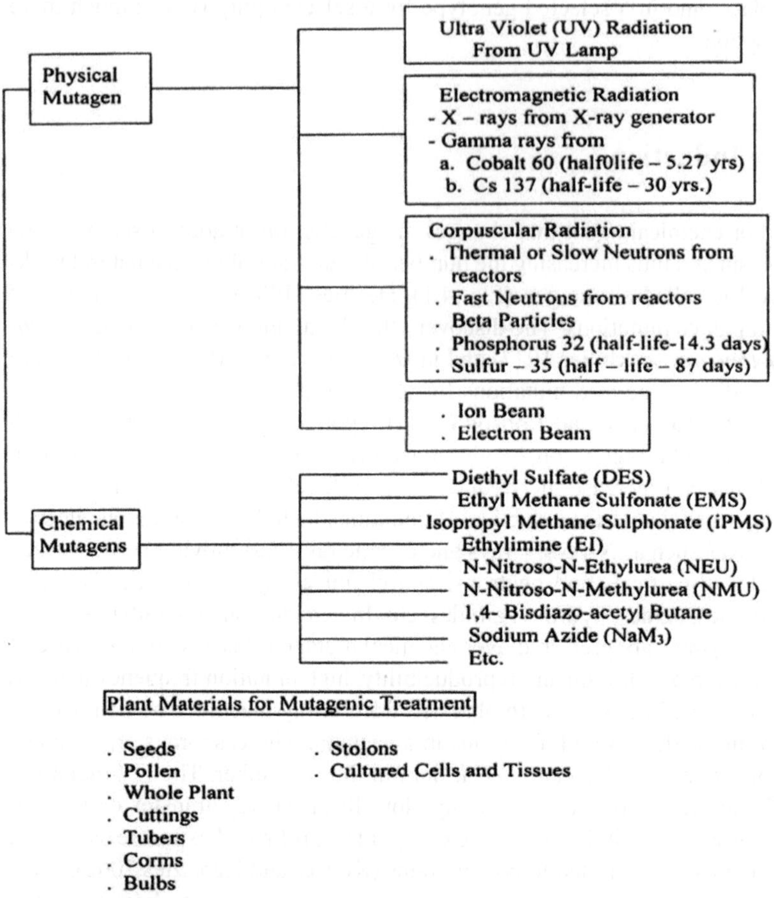

Fig. 4.1 Different kinds of physical and chemical mutagens that are mostly used and selected plants parts for mutagenic treatment/induction

wide use of mutated alleles in the breeding of numerous species. Mutation induction raises the natural mutation rates 10–100 fold, expanding the opportunity to isolate a higher number of mutants in a limited space. Today, induced mutations are ideal for augmenting natural variation in germplasm and as an alternative to hybridization and recombination in plant breeding. Mutations provide new starting material for the production of new cultivars and on the other hand they offer excellent tools for identifying new genes, for studying the nature of genes and their way of controlling biochemical pathways (Micke et al. 1990). The genetic variation from mutagenesis is different from that existing in germplasm collections or obtainable from crossing as it is not yet selected by nature or man and thus contains traits which were not favored during evolution or previous plant breeding activities. Besides, genes for a desired trait may not be fit or may be tightly linked with undesirable genes so that recombination through hybridization is rare or impossible. For example, genes causing pollen abortion cannot be transmitted gametophytically to future generations; consequently homozygous plants with all aborted pollen are lost in the cause of evolution.

Mutation and mutation breeding is a tool and being used to study the nature and function of genes which are the building blocks and basis of plant growth and development, thereby producing raw materials for genetic improvement of economic crops (Adamu and Aliyu 2007). Mutation induction offers significant increase in crop production (Kharkwal and Shu 2009) and the possibility of inducing desired attributes that either cannot be found in nature or have been lost during evolution. Treatment with mutagens alters genes or breaks chromosomes. Gene mutations occur naturally as errors in DNA replication. Most of these errors are repaired but some may pass to the next cell division to become established in the plant offspring as spontaneous mutations. Gene mutations without phenotypic expressions are usually not recognized. Consequently, genetic variation appears rather limited and breeders have to resort to mutation induction (Novak and Brunner 1992; Kozgar et al. 2012).

Mutagenic agents have been used to induce useful phenotypic variations in plants for more than seventy decades (Vasline et al. 2005; Roychowdhury et al. 2011). During the past 70 years, more than 2,543 mutant cultivars from 175 plant species including ornamentals, cereals, oilseeds, pulses, vegetables, fruits and fibers have been officially released in 50 countries all over the world (Maluszynski et al. 2000; Chopra 2005). Chemical mutagenesis (the non-GMO approach) is an approach to create mutation in plants for their improvement of potential agronomic traits. In any mutation breeding programme, selection of an effective and efficient mutagen is very essential to produce high frequency of desirable mutation. Many chemical mutagens have been employed for obtaining useful mutants in various crop species (Singh and Singh 2001a; Roychowdhury and Tah 2011c; Roychowdhury et al. 2012a). However the various workers emphasizes that artificial induction of mutation by colchicine (COL), ethyl methane sulphonate (EMS), sodium azide (SA), maleic hydrazide (MH) provides tool to overcome the limitations of variability in plants and induces specific improvement without disturbing their better attributes (Mensah and Obadoni 2007; Islam 2010; Roychowdhury 2011; Roychowdhury and Tah 2011; Roychowdhury et al. 2011b; Gnanamurthy et al. 2012). It might be considered that, these chemical induced growth abnormalities were mainly due to cell death and suppression of mitosis at different exposures. Several factors such as properties of mutagens, duration of treatment, pH, pre- and post-treatment, temperature and oxygen concentrations, etc. influence the effect of mutagens. The dose of a mutagen applied is an important consideration in any mutagenesis programme. Generally, it was observed that higher the concentrations of the mutagen greater the biological damage. To enhance the mutagenic effectiveness and efficiency and especially the metabolite, more knowledge about the effect of time, pH value, temperature, seed soaking and various concentrations are required (Khan et al. 2009; Roychowdhury and Tah 2011a, 2011c; Roychowdhury et al. 2012a). Crop plants offers many opportunities exploitation of mutations, recombination and of increasing genetic variability in quantitatively inherited agronomic characters. Induced mutations are also useful when it is desired to improve easily identifiable characters.

Initial studies on induced mutations were mainly directed to finding optimum combination of mutagen and dose to elicit the best response. Both physical and chemical mutagens were tested in various crop species such as wheat, barley, rice, tobacco, corn, Brassica, fruit crops and vegetables. These studies helped to initiate large-scale mutation breeding experiments for various practical applications (Chopra 2005). Since various physical and chemical mutagens are known to act in different ways to cause DNA lesions, combined effects of mutagens were investigated. In wheat, combined treatment with UV and X-rays showed dose-dependent effects. UV pretreatment of seeds reduced the frequency of mutations at low doses of X-rays (11–16 kr) but increased it at high doses of 22–30 kr (Swaminathan and Natarajan 1959; Bansal et al. 1962). In barley, treatment with S-2 aminoethylisothiouronium bromide hydrobromide (AET) was tested as both pre- and post-treatment with X-rays. Frequency of chromosome aberrations and chlorophyll mutations registered a significant drop when AET treatment was followed by X-ray irradiation. On the other hand, post x-ray treatment of AET caused a slight drop in chromosome aberrations (Chopra et al. 1965). Similarly, combined treatment of two chemical mutagens, ethyl methane sulfonate (EMS) and hydroxyl amine (HA), was investigated in wheat. Data of chlorophyll and viable mutations indicated that EMS is a potent mutagen in *Triticum dicoccum* but HA is a weak mutagen. But when HA was administered after EMS treatment, there was a significant drop in mutation frequency indicating that HA may be involved in mutational repair process (Chopra and Swaminathan 1966). Studies with *Drosophila* showed that formaldehyde, which is not mutagenic in female flies, could enhance mutation frequency when administered following X-ray treatment. This suggested that formaldehyde might be blocking some DNA repair process (Mahajani and Chopra 1973).

Mutagens and Their Doses

One of most crucial requirements for a successful breeding programme is the selection of an effective and efficient doze of a mutagen for mutagenizing the starting material. Historically, the effectiveness of a mutagen has been measured in terms of biological effect that it produces. It is, however, desirable to establish a relationship between the observed biological effect to a well-defined and easily measurable physical quantity characterizing the amount of radiation or chemical mutagen responsible for that effect (Roychowdhury 2011). Therefore, the mutagenic effect in biological targets is commonly and conveniently described in terms of dose-effect relationships. In quantitative radiation biology, the 'simple dose' (D) is the amount of energy absorbed per mass of irradiated matter at the point of interest. The special unit of D is rad (1 tad 100 erg/g $= 10^{-2}$ joule/kg), expressed in terms of time as rad/h, rad/min and rad/s. Thus, among others, changes in radiation doses and duration of exposure of biological material to the irradiation are important parameters of physical mutations. In case of chemical agents, the dose of treatment is determined based on several parameters viz., (i) concentration, (ii) duration of treatment and (iii) tem-

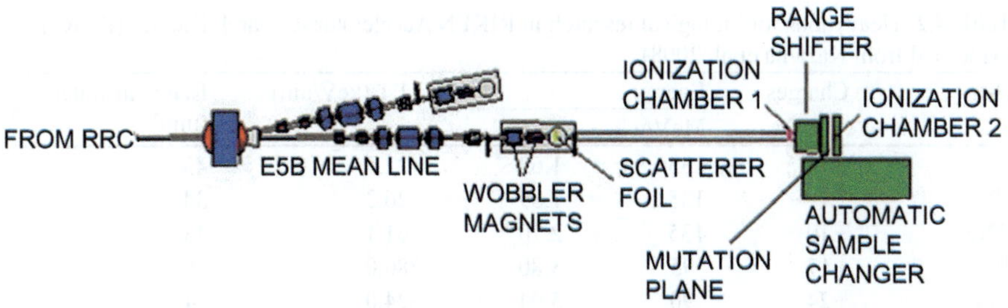

Fig. 4.2 The schematic view of E5B beam line. (RRC = RIKEN Ring Cyclotron)

perature during treatment. The concentration of chemical mutagen is determined by the per cent strength of the chemical in the solvent (distilled water). The volume of treatment solution is also equally important so as to provide each seed or organ an opportunity to absorb the same number of moles of mutagen. It is generally accepted that a treatment giving 30–40 % growth reduction is likely to give an optimal mutation yield in crops. The treatment duration must provide an opportunity for hydration and full penetration through the treated tissue of the mutagen. Long treatment is advisable, but it can be shortened by using pre-soaked seeds. However, the treatment duration also depends on the hydrolytic rate of the mutagen. For a short period, a high concentration is used after pre-soaking at high temperatures. The temperature of mutagenic solution greatly influences the mutagenic process. When there is no published information of mutation dose in a particular crop, we often resort to LD_{50} (Lethal Dose-50), which is a common parameter to decide the effective doses of both physical and chemical mutagens (Albokari et al. 2012). Thus, LD_{50} is a dose, which results in 50 % mortality of treated seeds (Roychowdhury 2011). With ionizing radiations, a dose which restricts survival to 50 % (LD_{50}) or growth to 50 % (GR_{50}) is a good treatment.

Ion Beam Mutagenesis

Application of ion beams for mutation induction was started with low-energy ions in China in the late 1980s and with heavy ions in Japan in the early 1990s. While ion beam technology has been used for food crop improvement in China, it has been more extensively used for floriculture plants in Japan. Ion beams as a mutagen are different from other physical mutagens such as gamma or X-rays in that they not only involve energy transfer (as gamma or X-rays), but also mass deposition and charge exchange (Hase et al. 2012); hence could result in complex DNA damage and changes that are not found when gamma or X-rays are used (high percentage of double strand breaks and subsequent chromosome aberrations). Ion beams are produced by particle accelerators, i.e. cyclotrons. Figure 4.2 is a schematic view of the E5B beam line available in the RIKEN Accelerator Research Facility (RARF), Japan.

Table 4.2 Heavy ions for biological research in RIKEN Accelerator Research Facility (RARF). (Modified from Kazama et al. 2008)

Heavy ions	Charges	Energy		LET (keV/μm)	Range in water (mm)
		MeV/u	GeV		
^{12}C	+6	135	1.62	22.5	43
^{14}N	+7	135	1.89	26.3	34
^{20}Ne	+10	135	2.70	61.1	23
^{40}Ar	+17	95	3.80	280.0	8
^{56}Fe	+24	90	5.04	624.0	4

Typical heavy ions used for irradiation on biological samples are neon-20, nitrogen-14, carbon-12, lithium-7, argon-40, iron-56 (Table 4.2).

They have different energy levels and linear energy transfer (LET), ionization densities which correlate to the complexity of DNA damage, and different ranges of penetration (Fig. 4.3).

LET is the energy deposited to target material when an ionizing particle passes through it. Once an accelerated particle encounters any substance, it gradually loses its own energy (i.e., the same amount of energy is transferred to the substance causing damage.) and eventually stops at the point where the maximum energy loss is observed (Fig. 4.4). In this figure, an ionizing particle gradually loses its own energy as it slows down in the target material. LET refers to this energy loss, which is deposited to the material. In this cartoon, LET is represented by wavy lines. LET reaches its maximum just before the ionizing particle stops. Immediately after this peak, LET plunges to zero. LET is usually expressed in kilo electron volts per micrometer (keV/mm), which represents the average amount of energy lost per unit distance. Ion beams have a relatively high LET (around 10–1,000 keV/μm or higher), while X-rays, γ-rays and electrons have low LETs (around 0.2 keV/μm). Therefore, ion beams are able to cause more severe damage to living cells than other forms radiation, resulting in the high relative biological effectiveness (Blakely 1992; Lett 1992). It is possible to modulate the treatment of plant material with one species of ion at different LETs by passing the ions through a combination of absorbers—since changes in the LET of ion species occur as they pass through matter (Kazama et al. 2008).

Studies have shown that the biological effect of ion beam radiation is dependent on absorption doses and LET values but independent of ion species (Kazama et al. 2008), which means that the treatment of carbon-12 would produce similar biological effect on rice seeds as neon-20 if the same dose (say 50 Gy) and same LET (say 30 keV/μm) is applied. DNA double strand breaks are believed to be the most important consequence of ion beam radiation. Very complex repair mechanisms have been unveiled but are prone to errors due to double-strand breaks and lead to deletions, insertions, inversions and translocations. Studies on the mutant gene alleles induced by ion beam radiation showed that most mutations are deletions and that the size of DNA deletion is LET-dependent. Most complex DNA damage caused by the intricate set of effects of heavy ion beams (HIB) escapes the repair efforts

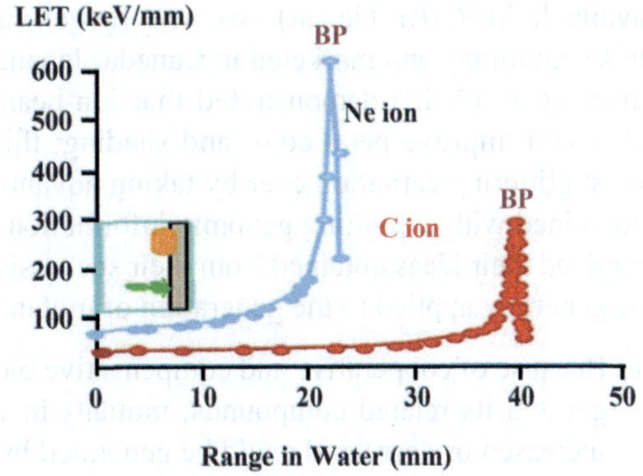

Fig. 4.3 A beam with sufficient energy penetrates a plantlet and/or plant tissue with rather low and uniform LET and it will then drastically increase towards the end of the track that is known as the Bragg peak (BP). (Modified from Kazama et al. 2008)

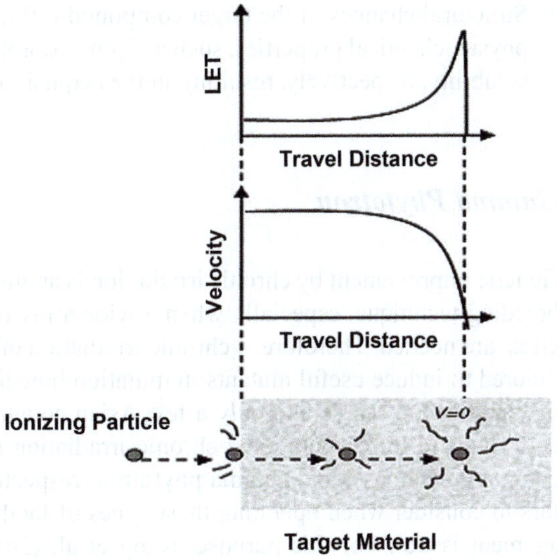

Fig. 4.4 Conceptual diagram of LET

and thus is described as more biologically effective and mutagenic than X-rays and gamma rays. A wider mutation spectrum and less collateral physiological damage (i.e. effect on plant survival and growth) is commonly reported for ion beam radiation as compared to other mutagens, which is considered an important advantage. In China, 23 new rice and wheat mutant varieties have been bred using ion beam technology and released for large scale commercial production (more than 1 million ha per annum). The wheat variety 'Wanmai 54' displayed excellent resistance to head scab disease and rust disease and recorded the highest yield in the national new wheat variety yield trial (2003–2007), with yield increases over control variety of 7–10.6%. In Japan, ion beam technology has been used for generating mutants for a vast number of plant species by various researchers; for example, a consortium of more than 90 user groups was established to utilize the ion beam technology available in RARF (Japan). Six new flower varieties have been developed using this technology and marketed in Canada, Japan, USA and the EU since 2002. Okamura et al. (2012) demonstrated that ion-beam radiation mediated breeding can alter and improve petal color and shading; this leads to the success creating the most glittering carnation ever by taking advantage of new mutagenesis techniques combined with exploiting genomic information. Here, Nakayam et al. (2012) summarized their ideas obtained from their successive ion-beam mutant studies that can be generally applied to the generation of mutants as follows:

1. Because of cooperative and compensative biosynthetic regulation between a target and its related compounds, mutants in which the target compounds either increased or decreased could be generated by ion-beam irradiation.
2. When multiple compounds are concerned in the expression of one phenotype, different types of mutants occur among the same phenotype.
3. Structural changes of the target compound influence the physical, chemical and physicochemical properties, such as light-absorption, co-pigmentation effect and solubility, respectively, resulting in the acquisition of a novel phenotype.

Gamma Phytotron

Genetic improvement by chronic irradiation is another important option of mutation breeding technique, especially when a wide array of mutants and minimal growth arrest are needed. Therefore, a chronic irradiation of living plant materials has been favored to induce useful mutants in mutation breeding. Unfortunately, these kinds' of facilities are scarce and only a few Asian countries including Japan, Malaysia, and Thailand have operational chronic irradiation facilities such as gamma field, gamma greenhouse, and gamma phytotron, respectively. There are still many factors to consider when operating these types of facilities such as security and management issues. For this purpose, Kang et al. (2010) constructed a new gamma phytotron which can be occupied with living pot plants or cultured callus during long periods of chronic irradiation at lower doses. The ionizing source is ^{60}Co with the radioactivity strength of about 400 curies. The facility consists of an irradiation room, a non-irradiation room, a glasshouse for acclimation, an operating room, and an office. The total area of the irradiation room is about 104.16 m^2. The target plant materials for a gamma ray irradiation can be arranged from 2 m (612.9 m Gy/h) to 7 m (60.1 m Gy/h) from the ^{60}Co source at present.

For safety reasons, the building, where the ^{60}Co source is located, is surrounded by concrete walls with 1.2 m depth and a twofold lead shielded door between the control room and the irradiation room. Moreover, the irradiation room is equipped with two CCD camera systems, which enable an inner situation check of the control room. The irradiation room and non-irradiation control room have automatic control systems for various ranges of temperature (15–35 °C), humidity (50–80 %) and light condition (maximum 30,000 lx), which can be finely setup according to

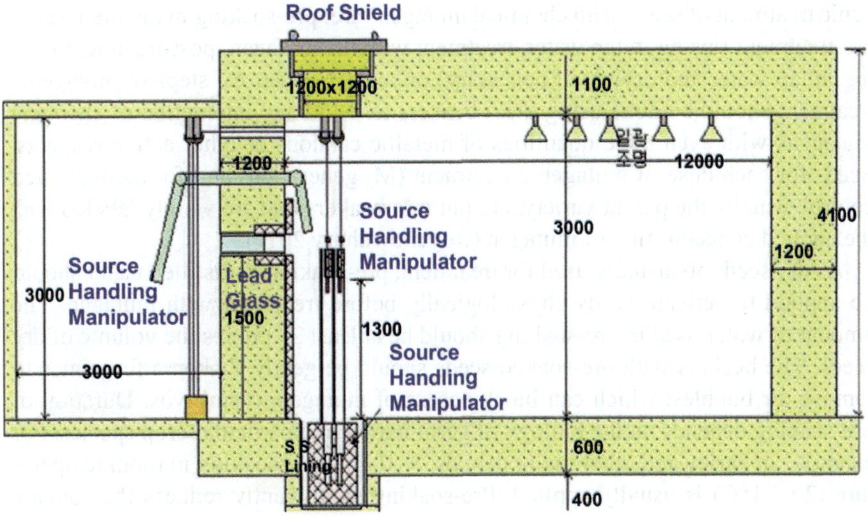

Fig. 4.5 The diagrammatic view of the gamma ray source and associated apparatus in gamma phytotron. (Modified from Kang et al. 2010)

the growth conditions of targeted plants (Fig. 4.5). The difference of mutagenic effects of the acute and chronic irradiation can be compared using the same treatment dose. It is expected that the heavy applications of the chronic gamma phytotron will be extended to various crop plants, which are eventually provided to domestic and global communities for mutation breeding and fundamental research.

Chemical Mutagenesis

The use of chemical mutagens is also very simple and can be done in any biological laboratory with basic equipment. However, it should be kept in mind that most chemical mutagens are also strong carcinogens. For this reason, all steps of mutagenic treatment should be carried out wearing gloves and under a Biohazard flow-hood. These safety conditions are not necessary for treatment with sodium azide, which is a very powerful mutagen, but only for a limited number of species, including barley, rice, maize, oat, sorghum, sesame, jute and soybean. Numerous chemical mutagens have been successfully used for crop improvement.

The mutagenic action of a chemical mutagen induces somatic and genetic effects in a treated cell, tissue or organ. After treatment of seeds, only unrepaired damage to the DNA in initial cells of the sporogenic layer (germ line cells) are transferred as mutations to the next generation. Other mutations in somatic cells of the embryo, including mitotic chromosomal aberrations, together with toxic action of a mutagen on all components of cytosol, affect plant growth and development, and are called the 'somatic effect' of the mutagen. The steps generally followed in mutagenic treatment of seeds with chemical mutagens are: pre-soaking in distilled water, pre-treatment rinsing in tap water, treatment with the mutagen, post-treatment rinsing in tap water and drying (if necessary) on a filter paper. All steps of mutagenic treatment should be done using glass beakers to avoid any interaction of chemical mutagens with even trace quantities of metallic cataions or other active reagents. Seeds for each dose of mutagenic treatment (M_0 generation) and for the untreated control, usually the parent variety, are put into beakers that are visibly labeled with the applied concentration of mutagen (Roychowdhury 2011).

As dry seeds are usually used for treatment, pre-soaking in distilled water should be applied to activate seeds physiologically before treatment with mutagen. The amount of water used in pre-soaking should be at least 2–3 times the volume of dry seeds. The beakers with pre-soaked seeds should be gently shaken a few times to remove air bubbles, which can block access of mutagen to embryos. Duration of pre-soaking depends on the biology of germination of a particular crop species. For example, in barley and other major cereals, 8–10 h of pre-soaking in room temperature (20–24 °C) is usually applied. Pre-soaking significantly reduces the somatic effect of chemical mutagen (Roychowdhury and Tah 2011a). Short washing, 2–3 times in room-temperature tap water should be applied after soaking to remove water-soluble substances leaching from the seed. Such prepared seeds are ready for mutagenic treatment. It is advisable to use three doses of mutagen for a large-scale field experiment. This is especially desired for regions with very variable and unpredictable weather conditions during the growing period of mutagenetically treated material. Drought, cold and heat can significantly modify the somatic effect of a mutagen and influence the final effect of treatment. The concentration of mutagen, its duration and temperature of treatment are understood under the term 'dose' in chemical mutagenesis. A temperature of mutagenic solution of 22–24°C is most often applied for the seed treatment of various crop species. The use of other temperatures is also possible. However, it should be noted that the increased temperature will significantly shorten the half-life of chemical mutagen and generate products of hydrolysis that can increase undesired somatic effect of a mutagen. This is especially relevant to treatment with mutagens such as diethyl sulfate (DES) or ethyl methane sulphonate (EMS). To obtain equal penetration of a mutagen through the cells of a seed embryo, it is necessary to treat seeds in a water solution of the mutagen for 3–5 h. Similar to the pre-soaking, the treatment should be done with a significant surplus of mutagenic solution, some 2–3 times the volume of the dry seeds. In cereals, about 1–1.5 ml of mutagenic solution is applied per seed. The concentration of the mutagen should be considered, together with duration of the treatment. A shorter treatment time with higher concentration of mutagen can increase somatic effects and could be insufficient to penetrate equally all cells in the plant material. A gentler treatment requires a lower concentration but longer period of application.

Extensive post-treatment rinsing several times in room-temperature tap water is necessary to stop action of the mutagen and to remove its residues from the surface of the seeds. To facilitate sowing, the treated seeds can be dried on filter paper under a fume hood. However, too intensive drying, especially with increased air temperature, can enhance somatic effects of the mutagen. Surface-dry seeds are ready for sowing and are termed the M_1 generation. In a well-organized laboratory, pre-soaking is done overnight and mutagenic treatment in the early morning. This allows the M_1 seeds to be sown the same day. Should this be impossible, due to prolonged pre-soaking or mutagenic treatment, the mutagen treated seeds, after brief drying, can be kept in a refrigerator at a temperature of around 6–8 °C. Some mutagens are active in a particular acidity of a treatment solution. This is the case for sodium azide, which is a very efficient mutagen in several species if applied at low pH. For this reason, sodium azide is dissolved in a phosphate buffer at pH 3 and this solution is used for treatment.

Mutation Breeding Strategy

A sequential strategy is essential for any mutation breeding steps where mutagenic induction and its mutagenesis are much helpful for autogamous crops than the cross-pollinating one. This is due to several problems regarding the incorporation, selection and maintenance of recessive mutations in crop plant, many plant breeding problems in the cross-pollinating species, sometime many handling based problem for the existing variability. Where the lack of variability exists for specific and simply inherited traits, the basis of choosing between induced mutations and hybridization is essentially the same in self- and cross-fertilizing species. However, the genetic consequences of the failure of recessive imitations to express in cross-fertilizing systems without forced selling or sib-mating must be taken into consideration in assessing the cost of such ventures. The efficiency of mutation breeding, more than any other breeding method is dependent on the effectiveness with which useful variants can be recognized in M_2 or M_3 generation. The first step in the mutation breeding selection process is to reduce the population of potential variants to a sufficiently small fraction to permit more detailed analysis and evaluation.

Probability of Obtaining Mutants

Mutations occur more or less at random and for mutagenesis-derived populations, unlike segregating populations derived from cross-breeding, there is no clue as to the kind or magnitude of genetic change. A particular gene can be expected to mutate once in about 10,000 mutagen treated cells, provided that an effective treatment was given. On an average, it appears that tail about five cells of the embryonic shoot apex may become part of the germ line and thus relate to the next generation. This would mean that for seed propagated species the M_2 generation of 2,000 mutagen treated (M_1) plants has to be examined in order to have reasonable chance of finding mutation in a particular gene. To extend the simplified calculation further, one might assume that plant genome possesses $10–100 \times 10^3$ relevant genes. Based upon the single locus mutation rate estimate of 1×10^{-4} this would mean that in an experiment using 2,000 M_1 plants, there would $10–100 \times 10^{-3}$ mutations or 1–10 per treated cell. Of these, some may be easily recognizable, others not; few usable, most of them not. Admittedly such estimates cannot be very accurate but the old, of magnitude should be acceptable and, therefore, be taken into account (Roychowdhury 2011).

Size of M_1 Population

The M_1 population is the plants that are generated from mutagenized seeds or other propagules. Determination of target population size in M_1 is the most crucial component of mutation breeding. The target should be fixed so as to allow high number of mutation events, yet the population size should be manageable by the breeder. It is obvious that the population size will depend on the inheritance pattern of the gene. If the mutation is monogenic recessive, the probability of recovering a mutant phenotype will be higher than for a trait controlled by more than one gene. In mathematical terms, if 'n' is the number of mutation events in the M_1 generation after treatment, and 'P' is the probability of occurrence of at least one mutation, then-

$$n = \frac{\log(1-P)}{\log(1-\mu)}, \text{ where } \mu \text{ is the rate of mutation}$$

With a mutation frequency of 1×10^{-4}, 'n' equals to 46,520 for a monogenic recessive trait. As there are two alleles, the number may be reduced to 23,260, indicating that about 25,000 plants are to be grown to obtain a useful mutation in M_1 generation. In practice, ten times of the size has to be considered, because the mutation produced may be useful or undesirable. Mutation breeding is an input intensive process. It is therefore advisable to select mutagens with high mutation frequency, so that M_1 generation size can be reduced.

While planting M_1 population, it is suitable to divide the whole seed lot in different small sections for ease of screening and analysis of chimera. It is to be remembered that germ-line mutations take place only in the initial cells of embryo, so depending on the nature of the species, products of initial divisions should be screened. For example, cereals like rice, wheat, barley, oat etc. produce multiple tillers. Those tillers that generates first (primary tillers) have maximum chance to carry mutation. In case of tuber crops like potato, the mutation may be present in any of the stems arising from different discs of a tuber, so each of them has equal chance to give rise to a mutation. Obviously, here the segregation pattern of mutation will depend on both the number of stems as well as the ploidy level of potato, which is an autotetraploid crop. Mutation breeding in polyploidy crop is more difficult than a diploid crop due to therecombination and segregation problem.

Genetically, a mutant plant in M_1 should be heterozygous, because during treatment, only one allele is affected by one mutation. Probability of occurrence of a mutation in both the alleles simultaneously is product of individual probabilities of mutation and therefore, is extremely low. It is not possible to identify a recessive mutation in M_1 stage; only dominant mutations can be identified. However, due to occurrence of this expression is also sectorial, or may not be observable. A breeder should attempt to screen mutations in M_2 generation, where the mutation will segregate generating homozygotes for recessive or dominant alleles. The M_1 plants should not be allowed to cross pollinate, because recombination will lead to generation of new variability that will be difficult to separate from effects of mutation (Roychowdhury 2011).

M_1 Generation Maintenance

The treated seeds need to be handled with care. The seeds treated, with physical mutagens can be stored before sowing. However, the seeds treated with chemical mutagens should be washed thoroughly and be planted as soon as possible. If the seeds cannot be plowed soon for various reasons such as weather or long transport, the seeds should be dried in shade to a moisture content of about 13 % as soon as possible without causing any damage to the seeds. Soil conditions can have considerable influence on survival and growth of the M_1. Nitrogen fertility of the soil should be normal or slightly sub-normal to limit excessive vegetative growth. However, other nutrients should be at optimum levels. The time of sowing should be slightly later (2 or 3 weeks) than normal so as to reduce excessive vegetative growth. The purpose of isolation of the M_1 is to avoid the introduction of genetic variability other than that induced with the mutagenic treatment. Most mutagen treatments will induce some pollen sterility increasing the amount of out crossing. M_1 population should be planted 75–100 m apart from the parental or other genotypes of the same crop species. If the crop is frequently insect pollinated, as with some legumes, the required isolation may he greater and other means of isolation may be required. Mutagen treated M_1 materials normally flower over a longer period than the control materials. A slightly later sowing of M_1 material than the parental genotype will permit separation of flowering times. When mutation breeding is practiced with a limited number of M_1 types of a crop, it may be possible to grow M_1 treatments side by side since F_1 hybrids may not occur or could easily be recognized by means of marker traits such as flower, plant or spike colour. Mechanically isolation can be achieved by bagging spikes in cereals using plastic or paper bags to prevent cross-pollination and bird damage. M_1 generation can also be grown in a green house or in the screen enclosures to achieve control over pollination by insects. Methods of harvesting the M_1 populations will depend on the pattern of ontogenetic development in the species, the methods of screening and the generation to be screened for mutants. In case the seed yield from each branch is reasonably adequate, it is suggested that each primary branch may be harvested separately. In case of cereals, individual plant or spike can be harvested (Roychowdhury 2011; Roychowdhury et al. 2011c, 2012).

Management of M_2 Population

Sowing of M_2 generation depends upon the method of harvesting of M_1 generation. Two methods of sowing M_2 generation can be followed. Firstly, M_1 plant to row, where all seeds produced from a single plant are grown in row. The success of its use will depend, to a large extent, on how well the branching has been controlled because it tends to dilute the yield of M_2 mutants. Second method is of M_1 spike or branch to row, which offers the greatest precision with regard to the origin of a mutant when the material treated is genetically homogeneous as regards the non-mutant allele and when outcrossing is controlled.

Mainly three types of screening/selection techniques can be employed for the selection of mutants in M_2 and subsequent generation viz. visual, mechanical/physical and other methods (Roychowdhury et al. 2012). Visual screening is the most effective and efficient method for identifying mutant phenotypes. Visual selection often is the prime basis for selecting for disease resistance, earliness, plant height, colour changes, ion-shattering, adaptation to soil, climate, growing period etc. Mechanical or physical selection can be used very efficiently for seed size, shape, weight, density, etc., using appropriate sieving machinery. In other category, chemical, biochemical, physiological, physio-chemical like screening procedures may be needed for selecting certain types of mutants. Low alkaloid content mutants can he selected using colorimetric tests. Colorimetric, chromatographic or electrophoresis techniques may be used to select isolate protein variants.

Propagation and Evaluation of Mutants

When a mutant appears promising, seed multiplication for extensive field testing is necessary. The mutant, the mother strain and other varieties with which it is intended to compare, should produce comparable seed properties for the basic trial seed. Mutants of vegetatively propagated plants can be multiplied by the usual method for the crop in question such as cuttings, grafting, budding, layering, bulbs, tubers etc. The methods of testing mutants in comparative trials are essentially the same as for any other newly developed strain. It is intended to find whether the mutant promises to become a variety surpassing the value of the mother strain and of the best available variety (a) in at least one property or (b) by a better combination of different characters; mutants of growth rhythm, growth habit, structure and yield components should be tested in a wide range of environments such as locations, soil, water and nutrient conditions, seed rates, planting, distances, sowing dates etc.

Mutagenesis and Genetic Variability

Diversifying the limited genetic variability for agronomic traits of interest, especially yield and its associate attributes and developing new crop cultivars are much demanding in this modern era (Roychowdhury et al. 2012). Due to lack of sufficient natural variability, the mutation breeding performance in crop species can significantly accelerate many breeding endeavors, which have proven difficult with classical breeding procedures (Roychowdhury 2011; Roychowdhury et al. 2011). Various metrical attributes like seed weight, number of branches, leaves, flowers, leaf area, etc., are very much complex in nature because it is governed by polygenes and greatly influenced by environmental factors (Roychowdhury et al. 2011, 2011c, 2012). This may raise breeder's concern, since the genetic organization provides the base for crop enhancement of environmental adaptation, yield and other associated attributes. The presence of adequate genetic variability between treatments of a cultivar is critically important (Fasoula and Fasoula 2002). Moreover, the genetic progress in a breeding program is actually dependent on the variation in the present gene pool (Dreisigacker et al. 2004) associated with the magnitude of several genetic parameters.

It is a powerful and effective tool in the hands of plant breeders for self-pollinating crops having narrow genetic base as well as for cross-pollinating crops (Micke 1988). The success of any breeding program depends to a large extent on the amount of genetic variability present in the population. The role of mutation breeding in increasing the genetic variability for desired traits in various crop plants have been proved beyond doubt by a number of scientists (Tah 2006; Adamu and Aliyu 2007; Khan and Goyal 2009; Kozgar et al. 2011; Mostafa 2011; Kozgar et al. 2012). Wide spectrum of genetic variability has been induced using both physical and chemical mutagens in order to utilize it in crop improvement and inheritance studies (Patil 1966; Ashri 1970; Gowda et al. 1996). Induced mutations have been used to generate genetic variability and have been successfully utilized to improve yield and yield components of various crops like *Oryza sativa* (Singh et al. 1998), *Dianthus caryophyllus* (Roychowdhury and Tah 2011b), *Solanum melongena* (Roychowdhury et al. 2011c), *Cicer arietinum* (Kozgar et al. 2012), *Vicia faba* (Ismail et al. 1977), *Vigna radiata* (Wani and Khan 2006; Roychowdhury et al. 2012), *Vigna unguiculata* (Mensah and Akomeah 1992), *Cajanus cajan* (Srivastava and Singh 1996), *Vigna mungo* (Singh and Singh 2001b) and *Lens culinaris* (Khan et al. 2006). These reports show that mutagenesis is a potential tool to be employed for crop improvement.

Overall variability must be partitioned into heritable and non-heritable components with the aid of genetic parameters such as genotypic and phenotypic coefficients of variation, heritability and genetic advance (Ariyo 1987; Roychowdhury and Tah 2011b; Roychowdhury et al. 2011a, 2011c, 2012). Genetic variability studies provide basic information regarding the genetic properties of the population based on which breeding methods are formulated for further improvement of the crop. These studies are also helpful to know about the nature and extent of variability that can be attributed to different causes, sensitive nature of the crop to environmental influences, heritability of the characters and genetic advance that can be realized in practical breeding. Progress in any crop improvement venture depends mainly on the magnitude of genetic variability and heritability present in the source material. The extent of variability is measured by genotypic coefficient of variance (GCV) and phenotypic coefficient of variance (PCV) which provides information about relative amount of variation in different characters. Hence, to have a thorough comprehensive idea, it is necessary to have and analytical assessment of metrical components. Since heritability is also influenced by environment, the information on heritability alone may not help in pin pointing characters enforcing selection. Nevertheless, the heritability estimates in conjunction with the predicted genetic advance will be more reliable (Johnson et al. 1955). Heritability gives the information on the magnitude of inheritance of quantitative traits, while genetic advance will be helpful in formulating suitable selection procedures. Thus such studies permit an effective screening of large plant population leading to generate demanding mutant lines. Therefore, an investigative attempt is essential for estimation the extent of various genetic parameters like variability, heritability and genetic advance in mutagen treated crop lines for some common agronomically important metrical traits and establishing a suitable breeding procedure, except expensive available molecular breeding methods along with developing a high quality and better yielding new crop germplasm to increase its diversity and sustainable agro-economical market demand by analyzing the stability capability and effect of mutagens for character improvement (Roychowdhury et al. 2011c).

Introduce of analysis of variance (ANOVA) revealed either significant or non-significant differences (at 1% and/or 5% level of probability) amongst the crop genotypes for all the analyzed traits. When highly significant differences were found, the crop genotypes showed a wide range of variation for all the characters studied. After analysis of variance (ANOVA) for each character, *F*-value was calculated. The wide range of *F*-values provides bright scope to select superior and suitable genotypes to be incorporated in the breeding programmes for further crop improvement. It is also important to note the value of coefficient of variation (CV) and critical difference (CD) values whose significant numerals indicate that the crop cultivar is suitable for its respective locational field where prevailing environmental effects were favorable. The higher CD value indicates higher stability in that environment (Roychowdhury 2011).

Substantial differences within phenotypic (PCV) and genotypic (GCV) co-efficient of variation were needed to mark for all the studied attributes. The least difference between PCV and GCV indicating phenotypic variability is reliable measure of genotypic variability. The PCV was higher than the corresponding GCV for all time the traits which might be due to the interaction of the genotypes with the environment to some degree or other denoting environmental factors influencing the expression of these characters. Close correspondence between PCV and GCV for the characters implied their relative resistance to prevailing environmental variation. Higher values of PCV and GCV indicate the presence of high degree of variability and better scope for improvement. However, low values have indicated narrow range of variation for these characters and provides very least scope for selection. This also described that genetic factors were predominantly responsible for expression of these attributes and selection could be made effectively on the basis of phenotypic performance (Roychowdhury et al. 2011c).

The heritability estimates indicate the relative amount of estimates have been found to be satisfactory tools for selection based on phenotypic performance. The high estimates of heritability suggested that selection based on phenotypic performance would be more effective. However, heritability values alone may not provide clear predictability of the breeding value. Heritability in conjugation with genetic advance over mean (GAM) and/or genetic gain is more effective and reliable in predicting the resultant effect of selection (Roychowdhury and Tah 2011b; Roychowdhury et al. 2011c, 2012). High heritability combined with high genetic gain indicates less influence of environment in expression of these characters; and prevalence of additive gene action in their inheritance (Panse 1957). Hence, these metrical traits require simple selection in breeding programmes. High heritability with moderate genetic gain indicates that the characters were governed by additive gene interaction. High heritability coupled with low genetic gain indicating non-additive gene action; hence heterosis breeding would be recommended for that trait.

Mutational Analysis of Plant Structure and Function

As a prerequisite for functional genomics, mutational analysis of the most important characters that determine the plant productivity should be considered for the most important crops. Germplasm collection and maintenance is necessary for the recovery of various crop mutants. Rice, maize, barley, mung bean, carnation, tomato are the only positive examples of crop mutant germplasm conservation. In all these collections, the number of mutants with described and characterized mutations of genes responsible for plant productivity or for other agronomically important and desirable characters for breeding is exceptionally low.

According to Brown and Peters (1996), during investigation on mouse genomics for dissection of basic pathophysiological mechanisms, the first defined term '**Phenotype gap**' depicts that many mouse mutations are extremely valuable for the investigation of human diseases and for identification of the critical genes involved in human pathologies. This 'phenotype gap' concept can easily be extended to genetic investigation of plant species as a basic component of the mutational analysis of any crop plants. The phenotype gap will reflect the gulf between available mutant resources and the full range of phenotypes of an investigated plant species.

In *Arabidopsis thaliana*, having a very low amount of DNA per haploid genome, it seems that the phenotype gap is also very wide. Probably only 1.8 % of visible markers have been described in which 167 genes are expected per megabase (Mb) and an average number of identified visible markers of 3 per Mb have been identified (Vizir et al. 1994). The genome of rice (*Oryza sativa*), barley (*Hordeum vulgare*) and wheat (*Triticum aestivum*) contain about 4, 37 and 115 times more DNA

(DNA amounts correspond to genome length) than *A. thaliana*. This indicates the amount of work that should be done to fill in the 'phenotype gap' in higher plants, especially the major crop species. Because of the small rice genome size, large scale mutational work should be initiated with this crop. The synteny of cereal genomes can help in the use of mutated genes in other cereals.

To narrow the 'phenotype gap' in crop plant species, it is necessary to expand mutant resources in breadth and depth (Brown and Peters 1996) by recovering mutations at new loci and recovering further mutations at known mutated loci. Closing the 'phenotype gap' requires efficient mutagenesis protocols and sensitive screening methods. As current mutagenesis has a great number of mutagenic agents such as various types of radiation, chemical mutagenesis, *in vitro* conditions, insertional mutagenesis, and activation of retro-transposons, the efficient and sensitive screening method is still the most limiting factor for isolation of a particular mutation.

There are many misconceptions related to frequency of specific locus mutations. Most probably, underestimation of the frequency of mutations induced by radiation or chemical mutagens leads to a very critical assessment of their usefulness in generating desired genetic variability and diversity in plants. It has generally been accepted, from the last 3 decades, that the average frequency of induced mutations is approximately on the level of 1×10^{-6}. This figure ignores the data related to the level of spontaneous mutations which have almost similar level for higher eukaryotes (Drake et al. 1998). Consequently, too high of mutagen doses have often been used, which induced too many mutations in the nucleus of each treated cell. The generative progeny that develop from this cell segregate for many characters that may negatively influence agronomically important characters, such as adaptability and yield potential. As a result, due to the use of high doses, many mutants were selected in mutated populations but most frequently with significant modifications in parental genetic background that made their usefulness in breeding programs highly questionable. The effectiveness of mutational strategies was also compromised by improper handling of successive mutated generations due to misunderstanding of the genetic consequences of chimeric structure of first mutant generation (M_1) plants and the adoption of '**diplontic selection**' concept. In reality, the frequency of mutations at numerous loci is much higher, as is indicated by the frequency of mutants in the second mutant generation (M_2) of some crop species.

Recent developments of gene transfer technology have enormous promise for improvement of plant productivity; however, there is a lack of available new genes which can be transferred to current high-yielding varieties and further significantly increase yield. In other words, there are no genes that have been identified which can contribute to world crop production. Borlaug (1997) referred to these genes as 'master genes' and concluded that Biotechnology may be a new window through which to search for new 'master genes' for high yield potential by eliminating the confounding effects of other genes. Therefore until new master genes are discovered, alternative solutions for crop improvement must be pursued. Further increases in crop yield may involve breeding for improved root systems. Breeding programs for high yield and adaptability have only indirectly selected favorable root systems.

Nevertheless, such an approach cannot replace a gene recombination breeding program that focuses on such characters of the root system as the dynamics of soil penetration; seminal, adventitious, and secondary root numbers; total root length; weight, number, and distribution of root hairs; and many physiological characters that directly influence plant productivity. Phosphorus uptake in barley can be almost doubled by increasing root hair density and length (Gahoonia and Nielsen 1997). The availability of phosphorous, zinc and other nutrients in poor soils as well as water and nitrogen nutrition depend also on mycorrhiza associations (Barker et al. 1998). Mycorrhizal fungi transfer assimilated carbon between tobacco plants (Muller and Dulieu 1998). These examples indicate an urgent need for further development of selection methods and for studies of the inheritance of characters related to the root system structure, function, and their linkage with other plant characters. Mutational analysis of selected root characters in breadth and depth would be the most desired approach, especially since a high frequency of induced mutations has been observed in relation to the root characters. More than 3.3 % of progenies of barley M_1 plants have indicated mutation in root system characters after combined seed treatment with sodium azide (NaN_3) and N-methyl-N-nitroso-urea (MNH) according to the mutagenic treatment method described by Szarejko and Maluszynski (1980). Mutant lines selected in M_3 generation indicated mutations related to root hairs, number and length of seminal roots, rootlessness, and abnormal root tip development.

Root mutants, described in maize, were obtained after mutagenic treatment with EMS and mutator MU (Feix et al. 1997). Mutants with unusual gravitropism behavior, aberrant lateral root formation, premature root degradation, and with lack of crown and brace roots were described in mutated generations. The genetic analysis of mutants indicated that the formation of the various root types and classes is controlled by different genes. Mutational analysis has been demonstrated as a powerful tool to dissect signaling pathways for plant defense responses (Dangl et al. 1996; Yang et al. 1997). There are also several examples of the use of mutational analysis to define the physical size, organization, and the sequence complexity of the major cluster of pathogenesis-related genes or the fine gene structure, e.g. downy mildew resistance genes in lettuce (Anderson et al. 1996) and for the *Mlo* locus for powdery mildew resistance in barley (Buschges et al. 1997).

Induced mutations in rice, especially for semi-dwarfness and earliness, are most often used to demonstrate the fastest way to obtain these characters in genotypes where crosses can modify particular characters such as adaptability, aroma, taste as well as requirements of local markets. More recently, mutation techniques have also been used to generate mutants with particular requirements related to quality characters where a rapid selection method is available. Very useful mutants have been obtained for fatty acid composition in rapeseed (Kott 1995), canola (Wong and Swanson 1991), flax (Dribnenki et al. 1996), soybean (Schnebly et al. 1995), cuphea (Knapp and Tagliani 1991), camelina (Vollmann et al. 1997), for grain quality in rice (Kumamaru et al. 1997) and for amylose-free starch in potato (Leij et al. 1991).

Officially Released Mutant Varieties

Of a total of 1,847 accessions of the FAO/IAEA Mutant Varieties Database (http://www-mvd.iaea.org), crop species are represented by 1,357 officially released mutant cultivars and ornamental and decorative plant species by 490 mutant varieties. Crop mutant cultivars were mainly developed in seed propagated plant species (1,284 entries), whereas vegetatively propagated crops are represented by only 73 varieties. Among the cereals (869 mutant varieties), rice (333) ranks first, followed by barley (261), bread wheat (147), maize (49), durum wheat (25), and others (54). Most of the rice mutant varieties (67.6 %) were released as '**direct mutants**', i.e. direct seed multiplication of selected mutants and their subsequent distribution to farmers. In addition, some mutants such as 'Reimei' (Japan) and 'Calrose 76' (United States) were successfully used in extensive crossbreeding programs. Semi-dwarfness (129 varieties) and earliness (117 varieties) were the most often selected characters from the treated populations. The list of improved characters also contains traits desired for increasing sustainability in rice production, i.e. tolerance to cold (13) salinity (6), and photoperiod insensitivity (3). The vast majority (201) of the directly released rice mutant varieties was induced with physical and only 25 with chemical mutagens. Radiation was applied in 199 cases and laser mutagenesis only in the development of two mutant varieties. Among the radiation sources, gamma rays were used in 199 cases, including 37 varieties developed by chronic gamma irradiation, followed by 14 with X-rays, 9 with neutrons, and 3 varieties with other sources of radiation. Methyl- and ethyl-nitroso-urea (12) as well as ethyl methane sulphonate or EMS (9) was most commonly used as chemical mutagens to induce mutations for breeding new varieties. According to the database, the mutant rice varieties were officially released in 26 countries. The seven top countries are: China (117), Japan (46), India (31), United States (23) and Vietnam (14). The economic impact of rice mutant varieties has been reviewed by Rutger (1992).

Achievements Through Mutagenesis

Several achievements in crop improvement through mutation breeding: mutation breeding efforts to date have resulted into two major outcomes—improved varieties that are directly used as variety for commercial cultivation and new genetic stocks with improved characters or with better combining ability. Although development of new cultivars has been the primary objective of mutation breeding, the genetic stocks developed can have numerous applications in plant breeding, from being used as a donor parent in conventional breeding programme or as a parent in hybrid breeding programme. Apart from these, mutation research itself has also a very different objective, i.e., mapping of genes. The technique of identification of a gene by knockdown of the phenotypic expression through induced mutagenesis is a major component of today research on molecular genetics and genomics. Specific sequences like transposons can be randomly inserted into a genotype through genetic crossing, which when inserted within a gene, blocks its transcription, thereby causing loss of phenotype. The gene then can be identified by using the sequences of transposon, a technique known as transposon tagging. A variety of other techniques including RNA interference, gene trap, activation tagging and virus induced gene silencing are based on the same principle of knocking out the phenotype through inactivation either at DNA or RNA level to establish gene-phenotype relationship. Discussion of such techniques is beyond the scope of discussion and will primarily concentrate on the application of mutation induction for crop improvement purpose only. International Atomic Energy Association (IAEA) has categorized its mutant variety database of 3,220 (as on December 2008) varieties according to six breeding methods namely:

1. Development of commercial cultivars through direct mutagenesis of genotype (2,738 genotypes),
2. Development of variety using mutant line as one of the parents in crossing programme (375 genotypes),
3. Development of variety through crossing of two mutants (28 genotypes),
4. Development of hybrid variety using mutant genotype as one parent (26 genotypes),
5. Development of variety through mutation of segregating generation (53 genotypes).

The first category includes 273R varieties and involves all the released variety through mutagenesis of seed, vegetative propagules and cultured tissues. Among the other classes, more success have resulted from using mutant line as parent in breeding programme as well as mutagenesis of breeding nurseries. Country wise, China ranks first in development of new varieties through induced mutagenesis and is well ahead of other countries in numbers. Many of these mutant varieties have been developed in rice, the principal food crop of India, China and other some Asian countries, through induced mutagenesis of seed as well as anther culture. Major commercial mutant varieties of China have been developed in rice, wheat, maize, *Capsicum*, cotton, tomato and groundnut. India ranks second after China, developing about 240 mutant varieties of different crops through direct mutagenesis of which major varieties have been developed for rice, wheat, barley, pearl millet, jute, groundnut, soybean, chickpea, mung bean, cowpea, black gram, sugarcane, chrysanthemum, rose and *Dahlia*. Additionally, about 50 varieties have been developed through using mutant lines in breeding programme. Indian mutation breeding programme became successful in the sixties with development of mutant varieties in wheat and rice and thereafter flourished in the next decades where new mutants have been developed in about 60 agricultural and horticultural crops. The major methods of mutation breeding in India involve irradiation with gamma rays, X-rays and treatment with EMS. Major Indian Institutes involved in mutation breeding are Bhabha Atomic Research Centre (BARC) in Mumbai, Indian Agricultural Research

Institute (IARI) in New Delhi and National Botanical Research Institute (NBRI) in Lucknow. Besides, some State Universities like The University of Burdwan (BU), Punjab Agricultural University (PAU), Tamil Nadu Agricultural University (TNAU) and crop specific National Research Centers have contributed a lot in mutation breeding programme in India.

Functional Genomics Approach

In plants, the two most common methods for producing reduction-of-function mutations are antisense RNA suppression (Finnegan et al. 1996) and insertional mutagenesis (van Houwelingen et al. 1998; Speulman et al. 1999). However, antisense RNA suppression requires considerable effort for any given target gene before knowing whether it will work, and insertional mutagenesis occurs at a low frequency per genome. However, its efficacy is not yet clear; for example, epigenetic phenotypes can be variegated and unpredictable (Que and Jorgensen 1998). Because these techniques rely either on *Agrobacterium* T-DNA vectors for transmission or on an endogenous tagging system, their usefulness as general reverse genetics methods is limited to very few plant species. Moreover, these techniques produce a very limited range of allele types. Therefore, as the amount of sequence data grows for Arabidopsis and other organisms, it is important to develop genome-scale reverse genetic strategies that are automated, broadly applicable, and capable of creating the wide range of mutant alleles that is needed for functional analysis.

Targeting specific loci is especially attractive when only a few genes of interest exist. Of the targeting methods for plants, posttranscriptional gene silencing (PTGS) is becoming increasingly popular (Waterhouse et al. 1998), replacing the less reliable antisense suppression methods that have been used for years. PTGS takes advantage of the innate RNAi system that is found in most eukaryotes, in which double-stranded RNAs are processed into 22–25-bp pieces (Baulcombe 2002) that can diffuse out of cells through plasmodesmata and the vascular system and into other cells. These pieces then target homologous transcripts for degradation and even can target genes for DNA methylation (Matzke et al. 2001). Reports of success using this method are encouraging (Chuang and Meyerowitz 2000), although the efficiency of silencing can vary, so results may be unpredictable. Furthermore, throughput is limited by the needs to engineer a construct for each gene of interest and to individually transform plants with each one. PTGS may be the best way to simultaneously target multiple closely related genes in a family.

Homologous recombination may be the most desirable strategy for targeted mutagenesis and has been routine in some microbial organisms such as *E. coli* and yeast for decades (Struhl et al. 1979). However, this technique has thus far been difficult or infeasible in multicellular eukaryotes, which have less-active homologous recombination systems. In a few model organisms, including mammals (Capecchi 2000), flies (Rong and Golic 2000), and the moss *Physcomytrella* (Schaefer 2001), there has been substantial progress in developing targeting by homologous recombination. However, in higher plants, homologous recombination is not yet efficient enough for practical use (Wang et al. 2001). Another potentially powerful targeting method uses chimeric RNA/DNA hybrid oligonucleotides to introduce base changes, insertions, or deletions (Rice et al. 2000), although, to our knowledge, the broad applicability of this method remains to be demonstrated.

Nucleotide sequence variation is a major determinant of heritable phenotypic difference in plant genomes. Variation can either be natural, from divergent populations, or induced through treatment with mutagens (Till et al. 2007a). There are several methods used in discovery mutations, which are natural or induced through treatment with mutagens in the genomes. TILLING and ECOTILLING are closely related methods that are useful in the rapid detection of small mutations and natural polymorphisms, respectively.

TILLING: A Best Screening Tool for Mutant Plant Population

Genetic variation is a powerful resource that humans have exploited over the millennia to advance biological knowledge and generate the crops and horticultural varieties that have become so much a part of everyday life. In recent years, the availability of genomic sequences from many plant species and the development of a wide array of molecular-genetic technologies have enhanced our ability to detect or engineer such variation at specific genetic loci (reverse genetics), greatly expanding our capacity for both probing gene function and genetic engineering. McCallum et al. (2000a) have introduced a new reverse genetic strategy that combines the high density of point mutations provided by traditional chemical mutagenesis with rapid mutational screening to discover induced lesions. TILLING (*T*argeting *I*nduced *L*ocal *L*esions *I*N *G*enomes) combines chemical mutagenesis (Koornneef et al. 1982) with a sensitive mutation detection instrument.

The TILLING strategy utilizes traditional mutagenesis followed by high throughput mutation discovery (Mccallum et al. 2000b; Colbert et al. 2001). The main steps in TILLING are mutagenesis, the development of a non-chimeric population, preparation of a germplasm stock, DNA extraction and sample pooling, screening the population for induced mutations, and the validation and evaluation of mutants (Fig. 4.6). The methods required for each step can be applied to many species, making the TILLING process broadly applicable. Mutants discovered by TILLING can be used for gene-function studies and can be introduced into breeding programs.

In a pilot experiment, DNA from a collection of EMS-mutagenized Arabidopsis plants was pooled, subjected to PCR amplification, and screened for mutations using denaturing HPLC (DHPLC). DHPLC detects mismatches in heteroduplexes created by melting and annealing of heteroallelic DNA. Among the lesions detected were base transitions causing missense and nonsense changes that can be used for phenotypic analyses.

TILLING is suitable for any organism that can be heavily mutagenized, even those that lack genetic tools. Starting with a homozygous population is desirable,

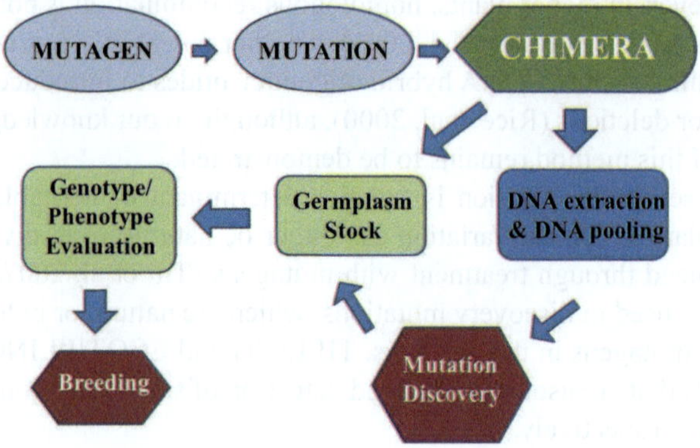

Fig. 4.6 A TILLING chart for gene function analysis and developing new crop varieties. A mutagenized population is prepared using a mutagen that primarily causes small lesions (single nucleotide polymorphisms, or insertions/deletions) randomly throughout the genome. Many mutagenic treatments produce a chimeric plant in the first generation. Chimeras are dissolved and a structured population is typically developed. A germplasm stock is prepared for long term storage of mutant lines, and DNA is extracted from each individual mutant. DNAs are pooled and the library of samples is screened for induced mutations in selected regions of target genes. Candidate mutants are removed from the germplasm stock and further characterized genotypically and phenotypically. Individuals or lines exhibiting the desired characteristics can be incorporated into the breeding program

because DHPLC will detect polymorphisms. Nevertheless, this strategy can be applied to species and hybrids that cannot be practically homozygosed: we and others have detected rare polymorphisms in a heteroallelic background using DHPLC. The general applicability of TILLING makes it appropriate for genetic modification of crops, and there may be agricultural interest in producing phenotypic variants without introducing foreign DNA of any type into a plant's genome.

TILLING consists of several major steps: development of a mutagenized population, DNA preparation and pooling, and mutation discovery (Fig. 4.7). At first, random mutations are induced in genomes by using chemical mutagens; seeds are mutagenized by treatment with ethyl methane sulfonate (EMS). The resulting M_1 plants are self-fertilized to get the M_2 individuals which are used to prepare DNA samples for mutational screening. DNA is extracted from test samples. The DNA samples are pooled and arrayed into microtiter plates. Screening for mutations begins with PCR amplification of a target fragment of up to 1.5 kb using gene-specific infra-red dye-labeled primers. The forward primer is 5′-end labeled with a fluorescent dye that is detected at 700 nm (IRDye 700) and the reverse primer is labeled with the IRDye 800 nm (Till et al. 2006). These PCR products are denatured and re-annealed to allow the formation of mismatches, or heteroduplexes, which represent naturally occurring single nucleotide polymorphisms (SNPs) and induced SNPs. Samples, are then incubated with a single-strand specific nuclease to digest mismatched base pairs. For mismatch-specific cleavage, several enzymes, including S1 nuclease (Howard et al. 1999) and T4 endonuclease VII (Youil et al. 1996)

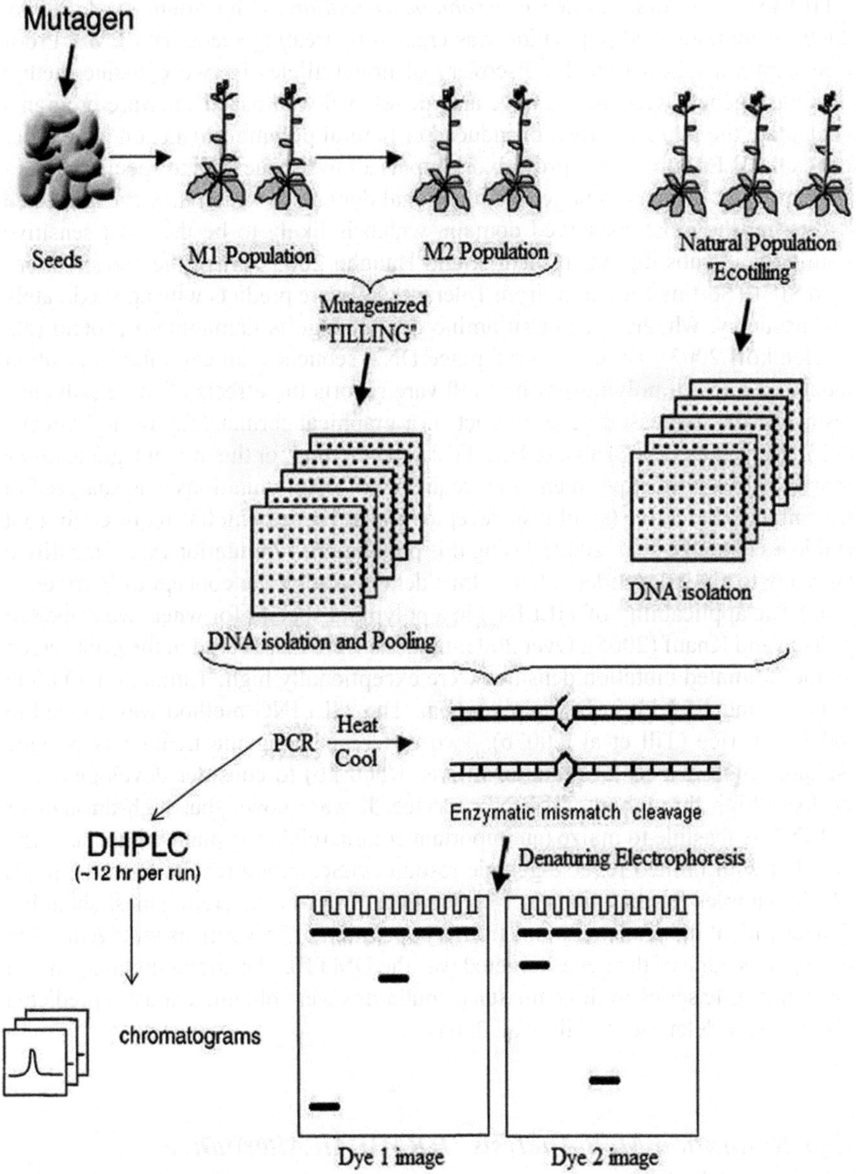

Fig. 4.7 Schematic diagram of the overall TILLING and ECOTILLING strategy for plants. (Modified from Simsek and Kacar 2010)

have been used. After cleavaging, DNA samples are purified from buffer components and then each sample is loaded onto a denaturing polyacrylamide gel. Cleaved heteroduplexes produced two smaller molecular weight products, one labeled with IRDye 700 and the other with IRDye 800, whose sizes added up to the size of the full length product (Till et al. 2007b).

TILLING was first applied to *Arabidopsis thaliana* (McCallum et al. 2000a, 2000b). A mutagenized population was created by treating seeds with EMS. Proof of concept was shown by the discovery of novel alleles in two cytosine methyl transferase genes. Researchers have also developed web based software programs to calculate the putative effect of induced or natural polymorphisms on gene function. CODDLE (http://www.proweb.org/input) allows requestors to specifically design their PCR primers to target the functional domain in which they are interested or to target the most-conserved domain, which is likely to be the most sensitive to amino-acid substitutions (Gilchrist and Haughn 2005). Also, the conservation-based SIFT (Sorting Intolerant from Tolerant) software predicts with approximately 75% accuracy, whether or not an amino acid change is damaging a protein (Ng and Henikoff 2003). By using a reference DNA sequence, an exon/intron position model and a list of polymorphisms, software reports the effects of these polymorphisms on the expressed gene product in a graphical format (Taylor and Greene 2003). Perry et al. (2003) adapted the TILLING method for the model legume *Lotus japonicas*. In a pilot experiment, the frequency of point mutations was analyzed in the symbiosis defective (symbiosis receptor kinase) gene, which is required for root symbioses (Stracke et al. 2002). Using this population, 17 mutations were identified that relate to six independent alleles, thus demonstrating the concept of Perry et al. (2003) The applicability of TILLING in a polyploid species for wheat was reported by Slade and Knauf (2005). Over 200 mutations were discovered in the pilot screen and the estimated mutation densities were exceptionally high: 1 mutation/40 kb in tetraploid and 1/24 kb in hexaploid wheat. The TILLING method was applied to model crop rice (Till et al. 2007b). Two different mutagenic treatments provide a suitably high density of mutations (over 1/500 kb) to consider development of rice for a high throughput TILLING service. It was shown that high-throughput TILLING is feasible to maize (an important commercial crop plant with a large genome but with limited reverse-genetic resources).Screening results from the pools of DNA samples for mutations in 1 kb segments from 11 different genes, obtaining 17 independent induced mutations from a population of 750 pollens mutagenized in maize plants. One of the genes targeted was the DMT102chromomethylase gene, in which an allelic series of three missense mutations were obtained and are predicted to be strongly deleterious (Till et al. 2004).

High-Resolution Melt Analysis (HRM): An Alternative Screening Platform

Although Cel1-based TILLING is very efficient for detecting mutations in large (1–2 kb) exon-rich amplicons from target genes, it is less productive when used to screen genes with multiple small exons separated by larger introns, as mutations in introns, except those at splice junctions, rarely affect gene function. High-resolution melt analysis (HRM) has been established as an alternative screening platform for such targets. HRM depends on the loss of fluorescence from intercalating dyes bound

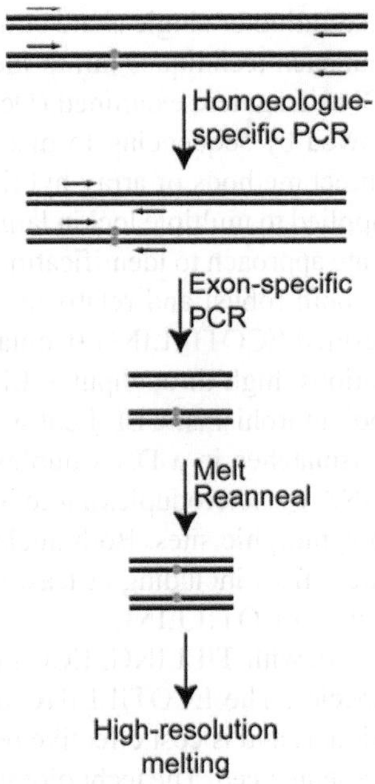

Fig. 4.8 Schematic of PCR and heteroduplex production in high-resolution melt analysis (HRM)

to double-stranded DNA during thermal denaturation (Ririe et al. 1997). Accurate control of temperature and continuous monitoring of fluorescence in instruments allows detection of single base mismatches in amplicons up to 500 bp. The method has been used both for genotyping and SNP discovery in medical genetics (Zhou et al. 2004, 2005), and SNP genotyping in plants has been demonstrated. Mutation scanning by HRM in hexaploid wheat requires a two-step amplification process, first, using homeologue-specific primers to amplify a larger amplicon containing several coding regions, followed by HRM analysis using primers specific for each exon or part thereof; a simple flowchart is shown in Fig. 4.8. As the melt analysis following PCR is extremely rapid, the throughput of this technique is equal to or greater than that of Cel1-based TILLING and is, arguably, easier to establish (Parry et al. 2009).

Ecotilling

The genomes of individuals within a single species contain significant genetic variation that has arisen from spontaneous mutation. The vast majority of this diversity is in the form of single nucleotide changes commonly referred to as simple nucleotide polymorphisms (SNPs). Such naturally occurring SNPs are of great interest to scientists because they are useful as genetic markers in mapping, breeding and genotyping and can provide information concerning gene structure, linkage disequilibrium, population structure or adaptation. A number of different techniques for identifying SNPs have been developed. Some of these detect differences in denaturation or single stand structure that result from changes in nucleotide sequence but such techniques fail to identify the number or position of mutations within the DNA fragment examined (DeFrancesco and Perkel 2001) so detection must be followed by sequencing to distinguish between different polymorphisms. The more direct methods of array hybridization or sequencing are currently expensive when applied to multiple loci in large numbers of individuals. TILLING provides an alternate approach to identification of naturally occurring SNPs in large populations that is both robust and relatively inexpensive. This application of TILLING has been termed ECOTILLING (Comai et al. 2004). Allowing of forceful discovery of mutations, high throughput TILLING technology is ideal for the detection of natural polymorphisms: CEL I cut with partial efficiency, allowing the display of multiple mismatches in a DNA duplex. Therefore, interrogating an unknown homologous DNA by heteroduplexing to a known sequence reveals the number and position of polymorphic sites. Both nucleotide changes and small insertions and deletions are identified, including at least some repeat number polymorphisms. This method is called ECOTILLING.

As with TILLING, ECOTILLING is general, and should be applicable to most species. The ECOTILLING allows the rapid detection of variation in many individuals and is cost effective because only one individual for each haplotype need to be sequenced. The technology is applicable to any organism including those that are heterozygous and polyploid (Comai et al. 2004). Sixty-three novel SNPs were identified in 9 target genes, for 41 tree accessions. The ECOTILLING method also was applied to sugarcane (*Saccharum* sp.), a complex polyploidy species, as a model to develop and test new protocols for high throughput ECOTILLING using capillary electrophoresis (Eliot et al. 2008). If SNPs in a population occur relatively rarely (less than one polymorphic individual per pool), the DNA of up to eight such individuals can be pooled, as is done in TILLING. However, when most individuals within a population differ at one or more base pairs in any given specific target sequence, 8-fold pooling will complicate genotyping. For this reason, in highly heterozygous species, the genomic DNA from each individual is usually pooled only with DNA from a reference individual for which the target has been sequenced. In addition, to detect those loci that were heterozygous prior to pooling, unpooled genomic DNA from an individual is Ecotilled separately.

ECOTILLING detects the number and relative position of all SNP's, including point mutations, and small insertions and deletions, within a target sequence in each individual tested. Thus both the spectrum of natural variation within the target sequence and the distribution of that variation throughout the population can be established. If knowledge of the specific nucleotide changes is required then DNA sequencing must be done following ECOTILLING (Fig. 4.7). However, since the number of different genotypes will normally be much smaller than the number of individuals examined, the target DNA from only a few representative individuals will need to be sequenced to establish the exact array of genotypes thus reducing the cost of SNP detection relative to direct sequencing (Simek and Novoselovi 2012). The efficacy of ECOTILLING has been demonstrated by two recent studies involving representative plants from different ecotypes of *Arabidopsis thaliana* (Comai et al. 2004) and different sub-populations of the poplar *Populus trichocarpa* (Gilchrist and Haughn 2005). Both studies were effective in rapidly identifying numerous target sequence SNP's within the populations examined. The poplar study provided information on population heterozygosity and linkage disequilibrium, identified a conserved potential regulatory domain in an intron and generated ecotype and species specific markers for genotyping. The fact that *A. thaliana* is a small inbreeding annual while *P. trichocarpa* is a large out-breeding perennial underscore the universality of ECOTILLING.

Conclusion and Future Perspectives

The spectrum of available mutation techniques has significantly increased; as a result, following the recent trend in the release of crop mutant varieties in some countries, the number of officially released mutant varieties listed in the FAO/IAEA Mutant Varieties Database will exceed exponentially year after year. The conventional breeding method takes several years to develop a new cultivar/variety from wild species. Induced mutagenesis and its breeding approaches are potential tools and being highly used in crops to improve their quality and quantitative yield traits. Mutagenic induction is much easy to apply on crops and inexpensive to develop desired agronomical traits, high yield, stress tolerance properties and resistant ability in various crops. Developing genetically novel germplasm becomes more feasible concurrent with the enhancement of breeding techniques, genomics, molecular manipulations and genetic engineering. The cost effectiveness of applying new mutation associated technologies (mutation breeding) and trained manpower would be of paramount importance for crop improvement. In classical mutation breeding, induced mutations are embedded in mutants that are either directly or indirectly (through crosses with other varieties) used for developing a new variety, whereby it is rather difficult to trace the mutated genes in subsequent breeding. The mutant plant species can be easily selected from some conventional simple screening and by PCR and non PCR based techniques. Therefore, it should be applied on various crops. It is now possible to tag mutated genes, pyramid them into a single elite breeding line, and follow them up in subsequent breeding programs. In view of the significance of conventionally induced mutants in functional genomics, there is great opportunity ahead for us in the era of genomics.

Now, several modern and classical technologies are using for the development of mutation induction with the objective of using a set of globally important crops to validate identified relevant novel techniques and build these into modular pipelines to serve as technology packages for induced crop mutations. Thus, mutation assisted plant breeding will play a crucial role in the generation of 'designer crop varieties' to address the uncertainties of global climate variability and change, and the challenges of global food insecurity.

References

Adamu AK, Aliyu H (2007) Morphological effects of sodium azide on tomato (*Lycopersicon esculentum* Mill.). Sci World J 2(4):9–12

Ahloowalia BS, Maluszynski M, Nichterlein K (2004) Global impact of mutation derived varieties. Euphytica 135:187–204

Albokari MMA, Alzahrani SM, Alsalman AS (2012) Radiosensitivity of some local cultivars of wheat (*Triticum aestivum* L.) to gamma irradiation. Bangladesh J Bot 41(1):1–5

Ali Z, Xu ZL, Zhang DY, He XL, Bahadur S, Yi JX (2011) Molecular diversity analysis of eggplant (*Solanum melongena*) genetic resources. Genet Mol Res 10(2):1141–1155

Anderson PA, Okubara PA, Arroyo-Garcia R, Meyers BC, Michelmore RW (1996) Molecular analysis of irradiation-induced and spontaneous deletion mutants at a disease resistance locus in *Lactuca sativa*. Mol General Genet 251:316–325

Ariyo OJ (1987) Variation and heritability of fifteen characters in Okra (*Abelmoschus esculentus* L.). Tropic Agric 45:67–71

Ashri A (1970) A dominant mutation with variable penetrance and expressivity induced by diethyl sulfate in peanuts (*Arachis hypogaea* Linn.). Mutation Res 9:473–480

Auerbach C, Robson JM (1946) Chemical production of mutations. Nature 157:302

Bansal HC, Chopra VL, Swaminathan MS (1962) Effect of ultraviolet pre- and post-treatment on yield of mutations by X-rays in wheat. Indian J Genet 22:162–166

Barker SJ, Tagu D, Delp G (1998) Regulation of root and fungal morphogenesis in mycorrhizal symbioses. Plant Physiol 116:1201–1207

Baulcombe D (2002) RNA silencing. Curr Biol 12:R82–R84

Blakely EA (1992) Cell inactivation by heavy charged particles. Radiat Environ Biophys 31:181–196

Borlaug NE (1997) Feeding a world of 10 billion people: the miracle ahead. Plant Tiss Cul Biotech 3(3):119–127

Brown SDM, Peters J (1996) Combining mutagenesis and genomics in the mouse—closing the phenotype gap. Trends Genet 12(11):433–435

Buschges R, Hollrichter K, Panstruga R, Simons G, Wolter M, Frijters A, van Daelen R, von der Lee T, Diergaarde P, Groenendijk J, Topsch S, Vos P, Salamini F, Schulze-Lefert P (1997) The barley *Mlo* gene: a novel control element of plant pathogen resistance. Cell 88:695–705

Capecchi MR (2000) How close are we to implementing gene targeting in animals other than the mouse? Proc Natl Acad Sci U S A 97:956–957

Chopra VL (2005) Mutagenesis: investigating the process and processing the outcome for crop improvement. Curr Sci 89(2):353–359

Chopra VL, Swaminathan MS (1966) Mutagenic efficiency of individual and combined treatments of ethyl-methane sulfonate and hydroxyl amine in emmer wheat. Indian J Genet 26:59–62

Chopra VL, Kapoor ML, Swaminathan MS (1965) Effects of pre- and post-treatments with S-2-aminoethyllisothiouronium bromide hydrobromide on the frequency of chromosome aberrations and chlorophyll mutations induced by X-rays in barley. Indian J Exp Biol 3:123–125

Chuang CF, Meyerowitz EM (2000) Specific and heritable genetic interference by double-stranded RNA in *Arabidopsis thaliana*. Proc Natl Acad Sci U S A 97:4985–4990

Colbert T, Till BJ, Tampa R et al (2001) High-throughput screening for induced point mutations. Plant Physiol 126:480–484

Comai L, Young K, Reynolds SH, Codomo C, Enns L, Johnson J, Burtner C, Henikoff JG, Grene EA, Till BJ, Henikoff S (2004) Efficient discovery of nucleotide polymorphisms in populations by ECOTILLING. Plant J 37:778–786

Dangl JL, Dietrich RA, Richberg MH (1996) Death don't have no mercy: cell death programs in plant-microbe interaction. Plant Cell 8:1793–1807

DeFrancesco L, Perkel JM (2001) In search of genomic variation. Scientist 15:24–26

DeVries H (1905) Species and varieties: their origin by mutation. The Open Court Publishing Company, Chicago

Din R, Qasim M, Ahmad K (2004) Radio sensitivity of various wheat genotypes in M1 generation. Int J Agric Biol 6:898–900

Domingo C, Andres F, Talon M (2007) Rice cv. Bahia mutagenized population: a new resource for rice breeding in the Mediterranean basin. Span J Agric Res 5:341–347

Drake JW, Charlesworth B, Charlesworth D, Crow JF (1998) Rates of spontaneous mutation. Genetics 148:1667–1686

Dreisigacker S, Zhang P, Warburton ML, Van Ginkel M (2004) SSR and pedigree analyses of genetic diversity among CIMMYT wheat lines targeted to different mega environments. Crop Sci 44:381–388

Dribnenki JCP, Green AG, Atlin GN (1996) Linola TM 989 low linolenic flax. Can J Plant Sci 76:329–331

Eliot F, Cordeiro G, Bundock PC, Henry RJ (2008) SNP discovery by ECOTILLING using capillary electrophoresis. Plant Genotyping 2: SNP Technol 5:78–87

Fasoula VA, Fasoula DA (2002) Principles underlying genetic improvement for high and stable crop yield potential. Field Crops Res 75:191–209

Feix G, Hochholdinger F, Wulff D (1997) Genetic analysis of root formation in maize. In: Developmental pathways in plants: biotechnological implications. The Hebrew University of Jerusalem, Rehovot, p 10

Finnegan EJ, Peacock WJ, Dennis ES (1996) Reduced DNA methylation in *Arabidopsis thaliana* results in abnormal plant development. Proc Natl Acad Sci U S A 93:8449–8454

Gahoonia TS, Nielsen NE (1997) Variation in root hairs of barley cultivars doubled soil phosphorous uptake. Euphytica 98:177–182

Gilchrist EJ, Haughn GW (2005) TILLING without a plough: a new method with applications for reverse genetics. Curr Opin Plant Biol 8:211–215

Gnanamurthy S, Dhanavel D, Chidambaram ALA (2012) Frequency in germination studies of chlorophyll mutants in effectiveness and efficiency using chemical mutagens. Int J Curr Life Sci 2(3):23–27

Gowda MVC, Nadaf HL, Sheshagiri R (1996) The role of mutation in intraspecific differentiation of groundnut (*Arachis hypogaea* L.). Euphytica 90:105–113

Hase Y, Akita Y, Kitamura S, Narumi I, Tanaka A (2012) Development of an efficient mutagenesis technique using ion beams: Toward more controlled mutation breeding. Plant Biotechnol 29:193–200

Howard JT, Ward J, Watson JN, Roux KH (1999) Heteroduplex cleavage analysis using S1 nuclease. Biotech 27:18–19

Islam SMS (2010) The effect of colchicine pretreatment on isolated microspore culture of wheat (*Triticum aestivum* L.). Aus J Crop Sci 4(9):660–665

Ismail MA, Heakal MY, Fayed A (1977) Improvement of yield through induced mutagenesis in broad beans. Indian J Genet Plant Breed 36(3):347–350

Johnson HW, Robinson HF, Comstock RE (1955) Estimation of genetic and environmental variability in soybean. Agron J 47:314–318

Kang SY, Kim JB, Lee GJ, Kim DS (2010) Gamma phytotron: a new chronic gamma irradiation facility. Plant Mut Rep 2(2):50–51

Kazama Y, Saito H, Yamamoto YY, Hayashi Y, Ichida H, Ryuto H, Fukunishi N, Abe T (2008) LET-dependent effects of heavy-ion beam irradiation in *Arabidopsis thaliana*. Plant Biotechnol 25:113–117

Khan S, Goyal S (2009) Improvement of mungbean varieties through induced mutations. Afr J Plant Sci 3:174–180

Khan S, Wani MR, Parveen K (2006) Sodium azide induced high yielding early mutant in lentil. Agric Sci Digest 26(1):65–66

Khan S, Al-Qurainy F, Anwar F (2009) Sodium azide: a chemical mutagen for enhancement of agronomic traits of crop plants. Environ Int J Sci Tech 4:1–21

Kharkwal MC, Shu QY (2009) The role of induced mutations in world food security. Shu QY (eds) In: Induced plant mutations in the genomics era. Food and Agriculture Organization of the United Nations, Rome, pp 33–38

Kidwell MG (2005) Transposable elements. In: Gregory TR (eds) The evolution of the genome. Elsevier, San Diego, pp 165–221

Knapp SJ, Tagliani LA (1991) Two medium chain fatty acid mutants of *Cuphea viscosissima*. Plant Breed 106:338–341

Koornneef M, Dellaert LW, Van Der Veen JH (1982) EMS- and radiation-induced mutation frequencies at individual loci in *Arabidopsis thaliana* (L.) Heynh. Mutat Res 93:109–123

Kott L (1995) Production of mutants using the rapeseed doubled haploid system. In: Induced mutations and molecular techniques for crop improvement. IAEA, Vienna

Kovacs E, Keresztes A (2002) Effect of gamma and UV-BIC radiation on plant cells. Micron 33:199–210

Kozgar MI, Goyal S, Khan S (2011) EMS induced mutational variability in *Vigna radiata* and *Vigna mungo*. Res J Bot 6:31–37

Kozgar MI, Khan S, Wani MR (2012) Variability and correlations studies for total iron and manganese contents of chickpea (*Cicer arietinum* L.) high yielding mutants. American J Food Tech 7:437–444

Kumamaru T, Sato H, Satoh H (1997) High-lysine mutants of rice, *Oryza sativa* L. Plant Breed 116:245–249

Leij FR, Visser RGF, Oosterhaven K, Kop DAM, Jacobsen E, Feenstra WJ (1991) Complementation of the amylose-free starch mutant of potato (*Solanum tuberosum*) by the gene encoding granule-bound starch synthase. Theor Appl Genet 82:289–295

Lett JT (1992) Damage to cellular DNA from particulate radiations, the efficacy of its processing and the radiosensitivity of mammalian cells. Emphasis on DNA double strand breaks and chromatin breaks. Radiat Environ Biophys 31:257–277

Mahajani S, Chopra VL (1973) Interaction of formaldehyde and X-rays on mutation production in *Drosophila melanogaster*. Curr Sci 42:473–474

Maluszynski M (1999) Crop germplasm enhancement through mutation techniques. Proceedings of the International Symposium on Rice Germplasm Evaluation and Enhancement Arkansas, Stuttgart, pp 74–83

Maluszynski M, Nichterlein K, van Zanten L, Ahloowalia BS (2000) Officially released mutant varieties—the FAO/IAEA Database. Mut Breed Rev 12:1–12

Matzke M, Matzke AJM, Kooter JM (2001) RNA: guiding gene silencing. Science 293:1080–1083

McCallum CM, Comai L, Greene EA, Henikoff S (2000a) Targeting Induced Local Lesions in Genomes (TILLING) for plant functional genomics. Plant Physiol 123:439–442

McCallum CM, Comai L, Grene EA, Henikoff S (2000b) Targeted screening for induced mutations. Nat Biotechnol 18:455–457

Mensah JK, Akomeah PA (1992) Mutagenic effects of hydroxylamine and streptomycin on the growth and seed yield of cowpea (*Vigna unguiculata* L. Walp). Legume Res 15(1):39–44

Mensah JK, Obadoni O (2007) Effects of sodium azide on yield parameters of groundnut (*Arachis hypogea* L.). Afr J Biotechnol 6:20–25

Micke A (1988) Genetic improvement of grain legumes using induced mutations: an overview. In: Improvement of grain legume production using induced mutations. IAEA, Vienna, pp 1–51

Micke A, Donini B, Maluszynski M (1990) Induced mutations for crop improvement. Mut Breed Rev 7:41

Mostafa GG (2011) Effect of sodium azide on the growth and variability induction in *Helianthus annuus* L. Int J Plant Breed Genet 5:76–85

Muller HJ (1927) Artificial transmutation of the gene. Science 66:84–87

Muller J, Dulieu H (1998) Enhanced growth of non-photosynthesizing tobacco mutants in the presence of a mycorrhizal inoculum. J Exp Bot 49(321):707–711

Nakayama M, Tanikawa N, Morita Y, Ban Y (2012) Comprehensive analyses of anthocyanin and related compounds to understand flower color change in ion-beam mutants of cyclamen (*Cyclamen* spp.) and carnation (*Dianthus caryophyllus*). Plant Biotechnol 29:215–221

Ng PC, Henikoff S (2003) SIFT: predicting amino acid changes that affect protein function. Nucleic Acids Res 31:3812–3814

Novak FJ, Brunner H (1992) Plant breeding: induced mutation technology for crop improvement. IAEA Bull 4:25–32

Okamura M, Umemoto N, Onishi N (2012) Breeding glittering carnations by an efficient mutagenesis system. Plant Biotechnol 29:209–214

Panse VG (1957) Genetics of quantitative characters in relation to plant breeding. Indian J Genet Plant Breed 17:318–328

Parry MAJ, Madgwick PJ, Bayon C, Tearall K, Hernandez-Lopez A, Baudo M, Rakszegi M, Hamada W, Al-Yassin A, Ouabbou H, Labhilili M, Phillips AL (2009) Mutation discovery for crop improvement. J Exp Bot 60(10):2817–2825

Patil SH (1966) Mutations induced in groundnut by X-rays. Indian J Genet 26A:334–348

Perry JA, Wang TL, Welham TJ, Gardner S, Pike JM, Yoshida S, Parniske M (2003) A TILLING reverse genetics tool and a web accessible collection of mutants of the legume Lotus japonicus. Plant Physiol 131:866–871

Que Q, Jorgensen RA (1998) Homology-based control of gene expression patterns in transgenic petunia flowers. Dev Genet 22:100–109

Ranel C (1989) The nature of spontaneous mutations. Mutation Res 212:3342

Rice MC, May GD, Kipp PD, Parekh H, Kmiec EB (2000) Genetic repair of mutations in plant cell-free extracts directed by specific chimeric oligonucleotides. Plant Physiol 123:427–438

Ririe KM, Rasmussen RP, Wittwer CT (1997) Product differentiation by analysis of DNA melting curves during the polymerase chain reaction. Anal Biochem 245:154–160

Rong YS, Golic KG (2000) Gene targeting by homologous recombination in Drosophila. Science 288:2013–2018

Roychowdhury R (2011) Effect of chemical mutagens on carnation (*Dianthus caryophyllus* L.): a mutation breeding approach. LAP Lambert Academic Publishing, Germany, pp 1–276

Roychowdhury R, Tah J (2011) Mutation breeding in *Dianthus caryophyllus* L. for economic traits. Electronic J Plant Breed 2(2):282–286

Roychowdhury R, Tah J (2011a) Assessment of chemical mutagenic effects in mutation breeding programme for M1 generation of Carnation (*Dianthus caryophyllus*). Res Plant Biol 1(4):23–32

Roychowdhury R, Tah J (2011b) Genetic variability study for yield and associated quantitative characters in mutant genotypes of *Dianthus caryophyllus* L. Int J Biosci 1(5):38–44

Roychowdhury R, Tah J (2011c) Chemical mutagenic action on seed germination and related agrometrical traits in M1 *Dianthus* generation. Curr Bot 2(8):19–23

Roychowdhury R, Bandyopadhyay A, Dalal T, Tah J (2011) Biometrical analysis for some agro-economic characters in M1 generation of *Dianthus caryophyllus*. Plant Archives 11(2):989–994

Roychowdhury R, Tah J, Dalal T, Bandyopadhyay A (2011a) Selection response and correlation studies for metrical traits in mutant carnation (*Dianthus caryophyllus* L.) genotypes. Cont J Agric Sci 5(3):6–14

Roychowdhury R, Sultana P, Tah J (2011b) Morphological architecture of foliar stomata in M2 carnation (*Dianthus caryophyllus* L.) genotypes using scanning electron microscopy (SEM). Electronic J Plant Breed 2(4):583–588

Roychowdhury R, Roy S, Tah J (2011c) Estimation of heritable components of variation and character selection in eggplant (*Solanum melongena* L.) for mutation breeding programme. Cont J Biol Sci 4(2):31–36

Roychowdhury R, Datta S, Gupta P, Tah J (2012) Analysis of genetic parameters on mutant populations of mungbean (*Vigna radiata* L.) after ethyl methane sulphonate treatment. Not Sci Biol 4(1):137–143

Roychowdhury R, Alam MJF, Bishnu S, Dalal T, Tah J (2012a) Comparative study for chemical mutagenesis on seed germination, survivability and pollen sterility in M1 and M2 generations of *Dianthus*. Plant Breed Seed Sci 65(1):29–38

Rutger JN (1992) Impact of mutation breeding in rice—a review. Mut Breed Rev 8:1–24

Schaefer DG (2001) Gene targeting in *Physcomitrella patens*. Curr Opin Plant Biol 4:143–150

Schnebly SR, Fehr WR, Welke GA, Hammond EG, Duvick DN (1995) Inheritance of reduced and elevated palmitate in mutant lines of soybean. Crop Sci 34:829–833

Siddiqui BA, Khan S (1999) Breeding in crop plants: mutations and *in vitro* mutation breeding. Kalyani, India, p 402

Simek R, Novoselovi D (2012) The use of reverse genetics approach in plant genomics. Poljoprivreda 18(1):14–18

Simsek O, Kacar YA (2010) Discovery of mutations with TILLING and ECOTILLING in plant genomes. Sci Res Essays 5(24): 3799–3802

Singh J, Singh S (2001a) Induced mutations in basmati rice (*Oryza sativa* L.). Diamond Jub Symp, New Delhi, p 212

Singh M, Singh VP (2001b) Genetic analysis of certain mutant lines of Urdbean for yield and quality traits in M4 generation. Indian J Pulses Res 14(1):60–62

Singh S, Richharia AK, Joshi AK (1998) An assessment of gamma ray induced mutations in rice (*Oryza sativa* L.). Indian J Genet Plant Breed 58(4):455–463

Slade AJ, Knauf VC (2005) TILLING moves beyond functional genomics into crop improvement. Transgenic Res 14:109–115

Speulman E, Metz PL, van Arkel G, te Lintel Hekkert B, Stiekema WJ, Pereira A (1999) A two-component enhancer-inhibitor transposon mutagenesis system for functional analysis of the Arabidopsis genome. Plant Cell 11:1853–1866

Srivastava A, Singh VP (1996) Induced high yielding Pigeon pea mutants. Mut Breed Newsletter 42:8–9

Stadler LJ (1928) Mutation in barley induced by X-rays and radium. Science 67:186–187

Stracke S, Kistner C, Yoshida S, Mulder L, Sato S, Kaneko T, Tabata S, Sandal N, Stougaard J, Szczyglowski K, Parniske M (2002) A receptor-like kinase required for both bacterial and fungal symbiosis. Nature 417:959–962

Struhl K, Stinchcomb DT, Scherer S, Davis RW (1979) High-frequency transformation of yeast: autonomous replication of hybrid DNA molecules. Proc Natl Acad Sci U S A 76:1035–1039

Swaminathan MS, Natarajan AT (1959) Effect of ultraviolet pretreatment on yield of mutations by X-rays in wheat. Science 130:1407–1409

Szarejko I, Maluszynski M (1980) Analysis of usefulness of sodium azide in plant breeding. Acta Biol 9:60–66

Tah PR (2006) Induced macromutation in mungbean [*Vigna radiata* (L.) Wilczek]. Int J Bot 2:219–228

Taylor NE, Greene EA (2003) PARSESNP: a tool for the analysis of nucleotide polymorphisms. Nucleic Acids Res 31:3808–3811

Till BJ, Burtner C, Comai L, Henikoff S (2004) Mismatch cleavage by single-strand specific nucleases. Nucleic Acids Res 32:2632–2641

Till BJ, Zerr T, Bowers E, Grene EA, Comai L, Henikoff S (2006) High throughput discovery of rare human nucleotide polymorphisms by ECOTILLING. Nucleic Acids Res 34:99

Till BJ, Comai L, Henikoff S (2007a) TILLING and ECOTILLING for crop improvement. In: Genomics-assisted crop improvement, vol 1. Genomics approaches and platforms. Springer, Germany, pp 333–349

Till BJ, Cooper J, Tai TH, Colowit P, Greene EA, Henikoff S, Comai L (2007b) Discovery of chemically induced mutations in rice by TILLING. BMC Plant Biol 7:19

Van Houwelingen A, Souer E, Spelt K, Kloos D, Mol J, Koes R (1998) Analysis of flower pigmentation mutants generated by random transposon mutagenesis in *Petunia hybrida*. Plant J 13:39–50

Vasline YA, Vennila S, Ganesan J (2005) Mutation—an alternate source of variability. UGC national seminar on present scenario in plant science research. Department Botany, Annamalai University, Annamalainagar, pp 42

Vizir IY, Anderson ML, Wilson ZA, Mulligan BJ (1994) Isolation of deficiencies in the Arabidopsis genome by gamma irradiation of pollen. Genetics 137:1111–1119

Vollmann J, Damboeck A, Baumgartner S, Ruckenbauer P (1997) Selection of induced mutants with improved linolenic acid content in camelina. Fett/Lipid 99(10):357–361

Wang HX, Viret JF, Eldridge A, Perera R, Signer ER et al (2001) Positive-negative selection for homologous recombination in *Arabidopsis*. Gene 272:249–55

Wani MR, Khan S (2006) Estimates of genetic variability in mutated populations and the scope of selection for yield attributes in *Vigna radiata* (L.) Wilczek. Egypt J Biol 8:1–6

Waterhouse PM, Graham MW, Wang MB (1998) Virus resistance and gene silencing in plants can be induced by simultaneous expression of sense and antisense RNA. Proc Natl Acad Sci U S A 95:13959–13964

Waugh R, Leader DJ, McCallum N, Caldwell D (2006) Harvesting the potential of induced biological diversity. Trends Plant Sci 11:71–79

Wessler SR (2006) Transposable elements and the evolution of eukaryotic genomes. Proc Natl Acad Sci U S A 103:17600–17601

Wilde HD, Chen Y, Jiang P, Bhattacharya A (2012) Targeted mutation breeding of horticultural plants. Emir J Food Agric 24(1):31–41

Wong RSC, Swanson E (1991) Genetic modification of canola oil: high oleic acid canola. In: Haberstroh C, Morris CE (eds) Fat and cholesterol reduced food. Gull, Houston, pp 154–164

Yang Y, Shah J, Klessig DF (1997) Signal perception and transduction in plant defense responses. Genes Dev 11:1621–1639

Youil R, Kemper B, Cotton RG (1996) Detection of 81 of 81 known mouse beta-globin promoter mutations with T4 endonuclease VII—the EMC method. Genomics 32:431–435

Zhou L, Vandersteen J, Wang L, Fuller T, Taylor M, Palais B, Wittwer CT (2004) High-resolution DNA melting curve analysis to establish HLA genotypic identity. Tissue Antigens 64:156–164

Zhou LM, Wang L, Palais R, Pryor R, Wittwer CT (2005) High resolution DNA melting analysis for simultaneous mutation scanning and genotyping in solution. Clinical Chem 51:1770–1777

Chapter 5
Role of Bio-fertilizers in Crop Improvement

Majeed–ul-Hassan Chesti, Tabasum N. Qadri, Asiya Hamid, Javed Qadri, Mohamed Mahgoub Azooz and Parvaiz Ahmad

Abstract Bio-fertilizers are cost effective, ecofriendly and renewable source of plant nutrients to supplement chemical fertilizers in sustainable agricultural system. Bio-fertilizers are preparations containing living cells of efficient nitrogen fixing, P-solubilizing/mobilizing or cellulose decomposing microorganisms, which when applied to seed or inoculated into the soil enhance availability of nutrients to plants either working symbiotically/asymbiotically or through solubiization of soil nutrients such as phosphorus or decomposition of complex materials. Bio-fertilizers are gaining impetus due to the growing emphasis on maintenance of soil health, curtail the environmental pollution and cut down on the use of chemicals in agriculture. Bio-fertilizers are also ideal input for reducing the cost of cultivation and for practicing organic farming. In the present context of very high cost of chemical fertilizers, the bio-fertilizers assume special significance.

M.-ul-H. Chesti (✉) · J. Qadri
Regional Agricultural Research Station, SKUAST-J, Tandwal, Rajouri 185131, Jammu and Kashmir, India
e-mail: chestimh2008@yahoo.co.in

J. Qadri
e-mail: qadrijavid2008@gmai.com

T. N. Qadri · A. Hamid
Department of Botany, GDC Budgam, Srinagar, Jammu and Kashmir, India
e-mail: tabasumqadri@yahoo.co.in

A. Hamid
e-mail: hameed901@gmail.com

M. M. Azooz
Department of Botany, Faculty of Science, South Valley University, 83523 Qena, Egypt
e-mail: azzozm@yahoo.com

Department of Biological Sciences, Faculty of Science, King Faisal University, Al Ahsa, Saudi Arabia

P. Ahmad
Department of Botany, GDC Anantnag, Srinagar, Jammu and Kashmir, India
e-mail: parvaizbot@yahoo.com

K. R. Hakeem et al. (eds.), *Crop Improvement,* DOI 10.1007/978-1-4614-7028-1_5,

Introduction

Bio-fertilizers are microbial inoculants of bacteria, algae, fungi that augment the availability of nutrients to the plants. Use of bio-fertilizers, in contrast to chemical fertilizers, accounts economical and ecological benefits to farmers (Brahmaprakash and Sahu 2012). Different types of microorganisms show the potential of converting essential soil nutrients which are in unavailable form to available form with the help of biological activity biological process such as nitrogen fixation and solubilization of rock phosphate (Rokhzadi et al. 2008). Bio-fertilizers improve plant growth, protect plants from amelioration of toxic effect in soils, root pest and disease control, improved water usage and soil fertility (Halim 2009; El-yazeid et al. 2007; Badawi et al. 2011; Mader et al. 2011; Mohammadi and Sohrabi 2012).

In addition they get engaged in symbiotic as well as associative microbial activities with higher plants Tiwari et al. 2003. These being an economical and safer source of plant nutrition for increasing the agricultural production, improve soil fertility and are called mini fertilizer factories (Vyas et al. 2008). The microorganisms form root nodules in leguminous plants by colonizing roots of legumes. Nitrogen fixation, phosphorus solublization and phytohormone production abilities have been observed and result in enhancement of agricultural productivity, e.g. *Rhizobium* for legumes (grain, fodder) (Ali et al. 2010) plant growth promoting rhizobacteria (PGPR) for cereals (wheat, rice, grasses etc.), *Azolla* for rice ecosystem, and actinomycetes (*Frankia* spp.) (Zhang et al. 2012), for forest trees (Danso et al. 1992). These microorganisms also have the ability to convert atmospheric nitrogen to plant usable form and can provide up to 200 kg N/ha/crop. Besides nitrogen, phosphorus is an essential element for crop production. Another group of bacteria which play important role in stimulating growth of plants are plant growth-promoting rhizobacteria (PGPR), they in addition to stimulating growth of plants also control plant pathogens, and pest infestation i.e they act as bio-fertilizers as well as biopesticides and ought to have meticulous consideration for agricultural purposes (Lugtenberg and Kamilova 2009). PGPR colonizes the rhizosphere, i.e, around the root, and even in the intercellular spaces of root.

Advantages of Bio-fertilizers over Chemical Fertilizers

Uses of microbial products have various advantages over traditional chemicals for agricultural purposes (Mahdi et al. 2010). These products have been commended safer than many of the chemicals that are used, they can fix atmospheric nitrogen in nodules of leguminous plants and soil and make it available to the plants and increase the fertility of soil (Shankar et al. 2012). They solublize the insoluble forms of phosphorous and again make is available to the plants (Hashemabadi et al. 2012), they also produce hormones which promote the growth of rhizosphere in addition to these properties they also help in mineralization of soil by decomposing the organic matter (Mahdi et al. 2010).

Above all neither toxic exudates of these microbes, nor microbes themselves are accumulated in the food chain, self-replication of microbes curtails the need for repeated application and target organisms rarely build up resistance as is observed when chemical agents are used to get rid of the pests detrimental to plant development (Mahdi et al. 2010).

Agricultural land deprived of essential nutrients gets impoverished after long term cultivation, to provide or nourish the soil nutrient content under conventional farming system, farmers use apply elevated doses of chemical fertilizers which in turn contaminate the ecosystem. Thus to implement the agricultural land a balanced and accountable use of organic agriculture is required. The principles of organic farming also outline the similar concepts where the soil health and biodiversity is built up to sustain the plant growth in longer term (Mahdi et al. 2010). Various beneficial microbes and their products found in rhizosphere are useful to plants by means of promoting growth or by acting as bio-control agents or both and are termed as Plant Growth-Promoting Rhizobacteria (PGPR) (Akhtar et al. 2012; Faramarzi et al. 2012). Rosenblueth and Martinez (2006) described several endophytic bacteria from different plant species mainly belonging to genera *Rhizobium, Azospirillum, Bacillus, Pseudomonas, Azotobacter, Burkholderia, Herbaspirillum, etc.* play beneficial roles e.g. endophytic N-fixation, increased P-uptake, improve photosynthesis and plant vigor, tolerance to biotic as well as abiotic stresses and in addition to these properties they act as insecticides and help in phytoremediation of polluted soils. Bio-fertilizers application can be used on crops prior to planting i.e. directly to soil, as a side dressing or as a foliar spray because it does not pollute and it adds humus to the soil (Raj 2007; Venkatashwarlu 2008). Co-inoculation of some *Pseudomonas* and *Bacillus* strains along with effective *Rhizobium* spp. is shown to stimulate chickpea growth, nodulation and nitrogen fixation (Mohammadi et al. 2010). Findings of Mohammadi et al. (2010) showed that the highest sugar, protein, starch contents, nodule weight and seed nitrogen, potassium, phosphorus of chickpea were obtained from combined application of phosphate solubilizing bacteria, *Rhizobium* and *Trichoderma* fungus. The Bio-fertilizers fix nitrogen in the soil that benefits the plant to overcome the nutritional stress. Appropriate doses of phosphorus, potassium, zinc, iron, molybdenum and cobalt along with fertilizers mitigate the stress and the legume starts responding directly to the nutrient. Usually most of the nitrogen fixed passes directly into the plant whereas some of it gets leaked into the soil for non-legume plant. However, after death, decay of these legumes by micro-organisms nitrogen eventually returns to the soil.

Types of Bio-fertilizers

A variety of recognized microorganisms for nitrogen fixation are also used such as *Azorhizobium caulinodans* and is effectively utilized in rice and maize. Likewise *Acetobacter and Sinorhizobium* have been used for sugarcane and soybean crop. Respectively microbes like *Thiobacillus and Thiooxidans* are known for sulphur and iron oxidization.

Nitrogen Fixing Bio-fertilizers

The nitrogen fixing bacteria are of two types' i.e., biological nitrogen fixation (symbiotic) and non-symbiotic nitrogen fixation (free living). The former develops as an association with crop plants through formation of nodules in their roots while as free living bacteria can fix atmospheric nitrogen without association with plants.

Biological Nitrogen Fixation

Atmosphere contains approximately 70 % of N which is not readily available form and therefore is not consumed by living organisms. It can be made available with the help of chemical or biological processes, though chemical nitrogen fertilizers are relatively expensive (Zilli et al. 2004). Living organisms utilize nitrogen in the form of ammonia to synthesize proteins, nucleic acids, amino acids, and other nitrogen-containing compounds for the maintenance of life. The process of conversion of inert N_2 to biologically important NH_3 with the help of bacteria is called biological Nitrogen Fixation. The nitrogen fixation is done by the bacteria, and the NH_3 produced is absorbed by the plant.

This biological reduction of nitrogen to ammonia is performed only by some prokaryotes and is a highly oxygen-sensitive process. The Biological nitrogen fixation includes diverse range of diazotrophic soil microbes belonging to aerobes (*Azotobacter, Beijerinckia, Drexia*), facultative anaerobes (*Clostridium, Pseudomonas, Rhizobium*), heterotrophs (*Klebsiella, Enterobacter*), phototrophs (*Anabaena, Nostoc, Azosprillium*) The most competent nitrogen fixers establish a symbiosis with higher plants in which the energy for nitrogen fixation and, in general, the oxygen protection system in particular are provided by the plant partner. In these symbiotic relationships prokaryotic partnership is provided by soil bacteria *Rhizobium* in leguminous plants and *Frankia* bacteria in actinorhizal symbiosis. Biological Nitrogen Fixation confers tremendous amount of NH_3 to natural ecosystems.

Rhizobium

Rhizobium is a gram-negative, free living organism present in soil, once it comes in contact with specific legume crop, nitrogen fixation starts and this rhizobium-legume association is of significant environmental and agricultural importance in view of the fact that it accounts for an estimated 180 million tons biological nitrogen fixation per year (Postgate 1982). *Rhizobium* invades the root hairs of the legumes by forming nodules. First time, bacterium capable of fixing nitrogen was isolated from nodules of a legume in 1888 by Beijirinck from Holland. Later on this bacterium was reported in Bergey's Manual of Determinative Bacteriology under the genus *Rhizobium. Rhizobium* has the ability to fix the atmospheric N in symbiotic state only. Rhizobium also exists as an endosymbiotic N fixing microorganism associated with root of legumes. It enters into plants through the root system then it forms nodule. The name Rhizobium was established by Frank in 1889. Seven distinct species of rhizobium has so far being discovered on the basis of "Cross Inoculation Group Concept" and more than twenty cross-inoculations groups have been established so far. Out of this, merely seven are most important. One group of rhizobia are very slow growing and are known as Bradyrhizobium while as other group is rhizobia is fast growing and is known as Rhizobium. Both slow growing as well as fast growing rhizobia has ability to fix atmospheric nitrogen. They create a symbiotic association with legumes and some non-legumes like Parasponia. Rhizobium legume symbiosis is very host-specific process and it fixes N in particular host plant only, this host specifity is mediated by plant compounds such as flavonoids (Goethals et al. 1992). Flavonoid activates the nod genes present in Rhizobium. The communication of rhizobia and legumes begins with signal exchange and recognition of the symbiotic partners, which is followed by attachment of the rhizobia to the plant root hairs. After infection, the root hair starts deforming, and the bacteria invades the plant by a newly produced infection thread growing through it at the same time, cortical cells which are mitotically activated, give rise to the nodule primordium. Infection threads grow toward the primordium, and the bacteria are then released into the cytoplasm of the host cells, surrounded by a plant derived peribacteroid membrane (PBM) (Van Workum et al. 1998). In the course of process the nodule primordium develops into a mature nodule, while the bacteria differentiate into their endosymbiotic form that is called as the bacteroid. The effective nodules are filled with pink sap called leghaemoglobin pigment. Leghaemoglobin regulates the supply of oxygen to the bacteria and helps the activity of nitrogenase enzyme and other regulatory enzymes (Choudhury and Kennedy 2004). The nitrogenase is responsible for reduction of nitrogen to ammonia in the process of nitrogen fixation. Bacteroids, together with the surrounding PBMs, are called symbiosomes. When symbiosomes are developed, bacteria synthesize nitrogenase, which catalyzes the reduction of nitrogen (Mylona et al. 1995). The product of nitrogen fixation, ammonia, is then exported to the plant. The plant provides all immediate nutrients and energy for the bacteria and just in a week small bead like structures i.e., nodules are formed. The root nodules act as a micro fermentor for biological N fixation where they can convert atmospheric N into ammonia. Rhizobium is able to induce the shoot and root growth in rice plants. (Yanni and El-Fattah 1999). Nodules occur in many shapes such as in *Alfalfa* and clover, nodules are fingerlike, round in *Lentil*, palm shaped in *Cicer*, though the entire nodule is generally less than 1/2 in in diameter during favorable conditions. Since the Nitrogen fixed is not free so the plant must contribute a considerable amount of energy in the form of photosynthates and other essential nutritional factors which are important for the bacteria. Rhizobium plays a key role and is the maximum researched bio-fertilizer (Mishra and Dadhich 2010). Currently the legume-rhizobia symbiosis has been extended to economically essential food crops or cereals and certain rhizobia that are competent of crack entry into ruptured epidermis during emergence of lateral rootlets in cereal crops (Kalia and Gupta 2002) the process can be improved by the addition of phytohormones (Kannaiyan et al. 2001) or use of signal chemicals (Amutha and Kannaiyan 2000).

Classification of *Rhizobium* Bio-fertilizers

1. *Rhizobium leguminosarum*
2. *Biovarphaseoli Phaseolus* (Bean)
3. *Biovarviceae Vicea* (Vetch)
4. *Biovartrifolii Trifolium* (Berseem)
5. *Rhizobium meliloti Melilotus* (Senji) *Rhizobium loti*
6. *Bradyrhizobium japonicum Glycine* (Soybean)
7. *Bradyrhizobium species Lupinus* (Lupin), *Vigna, Cicer*

Recently two more genera have been included in the family Rhizobiaceae. They are ***Sinorhizobium*** and ***Azorhizobium*** which are nodulating the Soybean and Dhaincha **(*Sesbania*)**, respectively. ***Azorhizobium caulinodans*** were isolated from the stem nodules of *Sesbania rostrata* but can also colonise and produce nodules in rice roots. *Azorhizobium caulinodans* also capable of fixing nitrogen in the free living state (Mandon et al. 1998).

Blue Green Algae (BGA)/Cyanobacteria

Blue-green algae or cyanobacteria are photosynthetic prokaryotes capable of fixing nitrogen with the help of enzyme nitrogenase. They are generally aquatic, small organisms visible as a single cell or large accumulation of cells (colonies) or strings of cells i.e. trichomes under microscope, sometimes accumulations are so large that they can be seen with a naked eye. Another name for blue-green algae is cyanophytes, cyanobacteria and most recently cyanoprokaryotes. As far as vegetative structure is considered they are resemble algae and other free living bodies. Their requirements for light, nutrients and carbon dioxide are similar. Certain types of blue-green algae have tiny gas vesicles in their cells that help to regulate buoyancy or get submerged under water in response to light fluctuations and availability of nutrient. BGA include *Anabaena, Nostoc, Plectonema, Syctonema, Calothrix, Aulosira, Tolypothrix*. Among these commercially available representatives are cultures of *Anabaena, Nostoc, Tolyphorix and Aulosira*. The blue-green alga (*Anabaena azollae*) shows a symbiotic association with *Azolla* (aquatic fern) and also fixes atmospheric nitrogen. BGA has shown to be associated with the *Azolla* present in ventral pore along the dorsal lobe of each vegetative leaf. This endophyte fixes atmospheric nitrogen and remains inside the tissue of the water fern in addition to its use in utilization in paddy fields. BGA fixes atmospheric nitrogen in semi aquatic ecosystem and takes part in photosynthetic activity. *Azolla* is a fast growing water fern and has ability to double its weight within a week. *Azolla* being rich in organic manure mineralizes the soil nitrogen rapidly and is made available to the plants. It is a protein rich feed to fish and poultry. BGA besides nitrogen fixation also synthesizes and releases growth stimulating substances viz., auxin and amino compounds that enhance the growth of rice plants. Algae can be multiplied in the paddy field by broadcasting the inoculants at the rate of about 10 kg/ha. It has been observed that incorporation of *Azolla-Anabaena* to paddy field increases rice yields and addition of dried *Azolla filiculoides* at the rate of 93 kg N/ha has increased a rice yield upto 70 %, the increase obtained with an equivalent amount of ammonium sulphate (Anitha and Kannaiyan 1999). In a field experiment the cyanobacteria was used to degrade coir pith with the help of lignolytic enzyme. (Malliga et al. 1996) and produced cyanopith, it can be used as bio-fertilizer to improve the crop productivity (Jha and Prasad 2005). Coir pith contains high lignin (31 %), cellulose (27 %), content (Bhat et al. 2003) and carbon nitrogen ratio (C/N) of 104:1 (Palaniappan 2005). Manoharan et al. 2011 used cynopith as bio-fertilizers on *Amaranthus dubius* that increases the growth of *Amaranthus*.

Azospirillum

Azospirillum is microaerophilic, free living, non-symbiotic, loosely associative nitrogen fixing bacteria and it establishes a close association with various plants mainly with C_4 maize, sorghum, sugarcane, ray grass, *Amaranthus* etc. This microorganism fixes atmospheric N and makes it available for plants in asymbiotic manner (Steenhoudt and Vanderleyden 2006). *Azospirillum* grows in the rhizosphere of the plants or occasionally penetrates into the root tissues but is not able to produce any visible nodule or out growth on the root tissue but grows intracellularly (Saikia et al. 2007). This association is due to the ability of the microbe to use malic acid, an organic acid formed for capturing CO_2 as a carbon source. It also secretes various phytohormones which include gibberellins, cytokinins, auxins and affect development and morphology of root by increasing root length, number of root hair cells, lateral roots. Azosprillium also secretes iron-chelating siderphores that help in the sequestering of iron sufficient for plant growth (Romerheld and Marshner 1986). A free living nitrogen fixing bacteria was for the first time reported by Beijerinck in 1925 under the name of *Spirillum lipoferum* and later on renamed this organism as *Azospirillum* (nitrogen fixing Spirillum) in 1978. *Azosprillium* is one of recognized dominant soil microbe and is able to fix about 10–40 kgN/ha. *The Azosprillium* inoculation improves vegetative growth of the plants (Naderifar and Daneshian 2012). Till date only four species of *Azospirillum* have been identified which include *A. lipoferum, A. brasilense, A. amazonense, A. iraquense*. Among these species only *A. brasilense and A. lipoferum* are very common in Indian soils. Inoculation of vegetable crops with *Azospirillum* has resulted in yield enhancement. The field experiment of *Azosprillium* with maize was examined and was confirmed that this association benefits enzyme activating glutamine synthetase and glutamine synthetase in the leaves of paranodulated maize plants. Bhaskara Rao and Charyulu (2005) studied the association of *A. lipoferum* inoculated to foxtail millet plant in combination with N fertilizer and demonstrated the increase in plant growth level, dry weight of shoot and root over when compared with control plants.

Maize plants inoculated with *Azospirillum* showed high rate of photosynthesis and stomatal conduction leading to high yield compared to control plants (Kumar and Bhaskara Rao 2012). Rice plant inoculated with *A. brasilense* at population rate of 8×10^{-7}/g of dry weight under field conditions showed a yield of 1.6–10.5 g plant-1 (Mirza et al. 2000; Malik et al. 2002). *Azospirillum* population enhances the uptake of P and NH_4 compounds in rice plants (Murty and Ladha 1988).

Azotobactor

Azotobactor is *a* gram-negative, aerobic, heterotrophic, rod shaped nitrogen fixing bacteria present in alkaline and neutral soils (Lakshmi-narayana 1993). They are free living organism present in soil, water and also in association with some plants (Gandora et al. 1998; Martyniuk and Martyniuk 2003). Various species of *Azotobacter* are *A. agilis, A. chrococcum, A. beijerinckii, A. vinelandii, A. ingrinis.* Among these *Azotobacter*, *Azotobactor chrococcum* is the most commonly found in arable soils of India. In addition to its capability to fix atmospheric nitrogen (20–40 Kg N/ha) for different crops, it can also produce various growth promoting substances *viz.,* auxins, and gibberellins cytokinins, indole acetic acid including vitamins and antibiotics, which control plant pathogens and help to maintain soil fertility. Azotobacter produces slime like substances which help in aggregation of soil particles. Many strains of *Azotobactor* exhibit fungicidal properties against certain species of fungus. Various crop plants like rice, maize, cotton, sugarcane, pearl millet, vegetable and some plantation crops show response to *Azotobacter.* Occurrence of organic matter in uncultivated soil promotes its multiplication and nitrogen fixing capacity. Field experiments carried out on *Azotobacter* under different agro-climatic conditions pointed out that Azotobacter is suitable when inoculated with seed or seedling of crop plants like onion, brinjal, tomato and cabbage. *Azotobacter* being heaviest among breathing organism and requires a large amount of organic carbon for its growth. Although it is poor competitor for nutrients in soil but it enhances plant growth through nitrogen fixation, release of growth promoting substances, and fungicidal substances. It improves seed germination and plant growth. N fixation process which is highly sensitive to O_2, *Azotobacter* have special mechanism against O_2 it reduces the concentration of O_2 in the cells (Shank Yu et al. 2005). Nitrogenase enzyme is also sensitive to O_2, but is supposed that the extreme respiration role of *Azotobacter* utilizes free O_2 within the cells and protects the nitrogenase (Kumar and Bhaskara Rao 2012). *Azotobacter* species have various types of nitrogenases viz., molybdenum–iron nitrogenase, vanadium–iron nitrogenase (Robson et al. 1986; Narula et al. 2000). *Azotobacter* requires carbon source for their energy (Kanungo et al. 1997) and is capable of fixing 10 mg N/g of carbohydrates in field conditions. *Azotobacter is* believed to be one of the significant bio-fertilizer for rice and other cereals, it can be applied by seed dipping and seedling root dipping methods (Kannaiyan et al. 1980; Kannaiyan 1999; Ruttimann et al. 2003; Singh et al. 1999). *Azotobacter* can also able to enhance the growth in wheat crop (Kader et al. 2002).

Non-legume-Frankia Symbiosis

Frankia a genus of actinomycetes, is a free, gram's positive nitrogen fixing bacterium that lives in soil and develops symbiotic interaction with various trees and shrubs forming symbiotic nodules (Verghese and Misra 2002). There are about 264 species belonging to 25 genera which take part in Frankia symbiosis. The Frankia is of fundamental and ecological interests for diverse reasons that include its wide distribution, its ability to fix nitrogen, differentiate specialized cell for nitrogen fixation (Verghese 2002). These specialized cells are called sporangium and vesicles and in addition to it can nodulate non-leguminous trees by forming root nodules, such *as Casuarina, Alnus, Dansea, Myrica, Elaeagnus* (Dawson et al. 2005; Franche et al. 2009). In wastelands fertility of soil can be improved by growing such non-leguminous plants in nitrogen deficient soils. In the process of nodulation, *Frankia* develops as little lateral swelling on roots and subsequently develops into new lobes at their apices forming cluster coralloid structure (Duhoux et al. 2001). Inoculation of *Frankia* enhances growth, nodulation, nitrogenase activity of nodule and nodule dry weight of *Casuarina* and *Alnus* plants. They live in the soil and have a symbiotic relationship with certain woody angiosperms, called actinorhizal plants. *Frankia sp.* produces three types of cells: sporangiospores, hyphae, and diazo-vesicles (Tjepkema et al. 1980), these diazo-vesicles are spherical, thick walled, lipid-enveloped cellular structures responsible for providing sufficient nitrogen to the host plant during symbiosis. *Frankia* enter into plants by root hair infection, nodules formed on lateral roots with cortical cylinder of vascular tissue (Ganesh et al. 1994). *Frankia* supplies almost total nitrogen needed by host plant and thus can establish a nitrogen-fixing symbiosis with host plants where nitrogen is the limiting factor for plant development. Therefore, actinorhizal plants colonize and often prosper in soils that are low in combined nitrogen (Benson and Silvester 1993). Symbiotic interaction of this category adds a large quantity of new nitrogen to numerous ecosystems such as temperate forests, dry chaparral, sand dunes, mine wastes etc. They also assist in creating and transporting certain root hormones, controlling pathogens and nematodes, water retention, mineral uptake, root exploration and resource sharing (Benson and Silvester 1993). *Frankia* specifically fixes nitrogen in the air and produces molecules that other plants can use. *Frankia* is said to be responsible for 15 % of the biologically fixed nitrogen in the world (Trujillo 2008).

Plant Growth Promoting Rhizobacteria

An assemblage of rhizobacteria (bacteria on rhizosphere) to facilitate beneficial effect on plant growth is referred to as plant growth promoting rhizobacteria or PGPR (Schroth and Hacock 1981). PGPR belong to several genera, e.g., *Alcaligenes*, *Agrobacterium Arthrobacter*, Azotobacter, Actinoplanes, *Bacillus*, *Bradyrhizobium*, *Amorphosporangium*, *Pseudomonas* sp., *Enterobacter*, *Rhizobium*, *Erwinia*, *Cellulomonas*, *Streptomyces Flavobacterium*, and *Xanthomonas* (Weller 1988). In a recent study it was found that PGPR covers a wide range of plant species. In all successful plant microbe interactions, the capability to colonize plant habitats is important. Single bacterial cells can affix to surfaces and, after repeated cell divisions and proliferation, dense aggregates are formed which are commonly referred to as macro colonies or biofilms (Mohammadi and Sohrabi 2012). Steps of colonization include attraction, recognition, adherence, invasion (only endophytes and pathogens), colonization and growth, and several strategies to establish interactions (Nihorimbere et al. 2011). There is crosstalk between plant roots and soil microbes. Plants roots initiate crosstalk by producing signals that are recognized by the microbes, which in turn produce signals that initiate colonization (Berg 2009). PGPR reach root surfaces by active motility facilitated by flagella and are guided by chemotactic responses. This implies that PGPR capability highly depends either on their abilities to take advantage of a specific environment or on their abilities to adapt to changing conditions or plant species (Nihorimbere et al. 2011). Habibi et al. (2011) strongly recommended that use of bio-fertilizers (combined strains) in addition with organic and chemical fertilizers have resulted in the maximum grain yield and oil yield in medicinal pumpkin. They revealed that 50 % of required nitrogen and phosphorus fertilizers might be replaced by bio and organic fertilizers, since bio and organic fertilizers improve the efficiency of recommended nitrogen and phosphorus fertilizers and reduced the cost of chemical fertilizers and also prevent the environment pollution from extensive application of chemical fertilizers. de Freitas et al. (1993) demonstrated that inoculation of beans with *Rhizobium. leguminosarum* and *Pseudomonas putida* increased the number of nodules and acetylene reduction activity (ARA) significantly. A significant positive effect on grain yield and ARA in roots of barley was obtained due to combined inoculation of nitrogen fixer's *A. lipoferum*, *Arthrobacter mysorens* and the phosphate solubilizing strain *Agrobacterium radiobacter* by Belimov et al. (1995). Radhakrishnan (1996) revealed that inoculation of *Azospirillum* and phosphor-bacteria resulted in higher root biomass and more bolls in cotton. Findings of Mohammadi (2010) showed that inoculation of bio-fertilizers (PSB + *Trichoderma* fungi) + application of FYM had a great influence on canola growth, height and grain yield when compared to control treatment. Findings of Mohammadi et al. (2011) showed that application of bio-fertilizers had a significant effects on nutrient uptake of chickpea combined application of Phosphate solubilizing bacteria and *Trichoderma harzianum* produced the highest leaf P content and grain P content. Capacity of *Bacillus* sp. to produce organic acid such as gluconic, citric and fumaric acids under P-limiting conditions may increase the solubility of poorly soluble phosphorus (Mohammadi and Sohrabi 2012).

Phosphorus Solubilizing/Mobilizing Microorganisms (PSM)

Phosphorous makes about 0.2 % of the plant on dry weight basis. It has distinct role in plant metabolism which includes cell division, cell development, photosynthesis, breakdown of sugars, nuclear transport within plants, and transfer of genetic characteristics from one generation to another generation and regulation of metabolic pathway (Rodriguez and Fraga 1999). The plant obtains their phosphate requirements from the soil pool. It occurs in soil as inorganic phosphate, produced by weathering of rocks that is unable to be utilized by the plants (Lee et al. 2005) or as organic phosphate derived from decaying plant, animal or microorganisms (Rodriguez and Fraga 1999). About 15–20 % of applied phosphorus is recovered from the crops and rest gets fixed in the soil and is not readily available to the plants. A group of morphologically different microorganisms which have the property of solubilizing the fixed phosphorous by producing organic acids and enzymes and to make them easily available to the crops are known as Phosphorous Solublising Microorganisms (PSM). They include diverse species of *Bacillus, Aspergillus, Pseudomonas, Penicillium, Agrobacterium, Achromobacter, Burkholderia, Aerobacter, Erwinia, Micrococcus, Flavobacterium and Trichoderma*. These organisms solubilize the fixed soil phosphorus thereby releasing the citrate and water soluble phosphorus so as to help in mineralizing organic phosphate compounds that are present in the organic wastes (Rodriguez and Fraga 1999). These microorganisms have the property to bring phosphate solublization by secreting organic acids such as propionic acid, lactic acid, formic acid, acetic acid, succinic acids etc. these acids lower the pH and help to dissolve the phosphate bound (Rodriguez and Fraga 1999). They also produce growth promoting substances e.g. IAA, GA etc. experiments conducted in field conditions in India have shown to replace 20–50 kg P_2O_5/ha in different crops due to PSM's inoculation (Vora and Shelat 1996, 1998, 1999). Improvement in seed germination by application of PSB has been reported by Sharma et al. (2007) in *Cicer arietinum*. Various horticultural plants and vegetables were successfully inoculated with P-solubilizing bio-fertilizers to obtain higher yields (Khan et al. 2010; Velineni and Brahmaprakash 2011). Field experiments demonstrated that P-solubilizing bio-fertilizers in addition to improving the growth and quality of crops, also reduced) the usage of chemical or organic fertilizers significantly (Young 1990; Chang and Young 1992a, b; Young et al. 1998a, b; Young and Chen 1999; Chang and Young 1999; Young et al. 2000; Liu and Young 2001; Young et al. 2003). Phosphate solubilizing bacteria has the capacity to convert inorganic unavailable phosphorus form to soluble forms like HPO_4^{2-} and $H_2PO_4^-$ with the help of processes like organic acid production, chelation and ion exchange reactions and make them available to plants (Chang and Yang 2009; Banerjee et al. 2010). Naturally occurring rhizospheric phosphorus solubilizing microorganism (PSM) has a long history and dates back to 1903 (Khan et al. 2007). Alam et al. (2002) pointed out that bacteria are more effective in phosphorus solubilization than fungi. Among the whole microbial population in soil, phosphate solubilizing bacteria (PSB) comprise 1–50 %, whereas phosphorus solubilizing fungi (PSF) are only 0.1–0.5 %. (Chen et al. 2006). Number of phosphorous solubilizing bacteria amongst total PSM in north Iranian soil was around 88 % (Fallah 2006). Microorganisms concerned in phosphorus acquirement include mycorrhizal fungi and PSMs (Fankem et al. 2006). Among the soil bacterial communities, effective phosphate solubilizers ectorhizospheric strains from *Pseudomonas* and *Bacilli*, and endosymbiotic rhizobia have been described as (Igual et al. 2001). Strains from bacterial genera *Pseudomonas, Bacillus, Rhizobium* and *Enterobacter* along with *Penicillium* and *Aspergillus* fungi are the most influential P solubilizers (Whitelaw 2000). *B. circulans, Bacillus megaterium, B. subtilis, B. sircalmous, B. polymyxa, Enterobacter and Pseudomonas striata*, can be referred as the most important strains (Subbarao 1988; Kucey et al. 1989). A fungus *Arthrobotrys oligospora* is also found to have the ability to solubilize the phosphate rocks (Duponnois et al. 2006). Increased high percentage of PSM is concentrated in the rhizosphere, and they are metabolically more active than from other sources (Vazquez et al. 2000). By and large, 1 g of fertile soil contains about 101–1010 bacteria, and their live weight may exceed 2,000 kg ha^{-1}. Soil bacteria can be cocci (sphere, 0.5 μm), bacilli (rod, 0.5–0.3 μm) or spiral (1–100 μm) shapes. Bacilli are common in soil, where as spirilli are very rare in natural environments (Baudoin et al. 2002). The PSB are cosmopolitan and vary in forms and population in diverse soils. Their population depends upon the physical and chemical properties organic content and phosphorous content of soil and cultural activities (Kim et al. 1998). Maximum populations of PSB are found in agricultural and rangeland soils (Yahya and Azawi 1998). In north of Iran, the PSB count ranged from 0 to 107 cells g^{-1} soil, with 3.98 % population of PSB among total bacteria (Fallah 2006). Mineralization and solubilization potential for organic and inorganic phosphorus, are also shown by bacterial populations (Hilda and Fraga 1999; Khiari and Parent 2005). Phosphorus solubilizing activity is determined by the capacity of microbes to liberate metabolites such as organic acids, which through their hydroxyl and carboxyl groups chelate the cation bound to phosphate, than are transformed to soluble forms (Sagoe et al. 1998). Various microbial processes/mechanisms including organic acid production and proton extrusion are used in Phosphate solubilization. (Surange 1995; Dutton and Evans 1996; Nahas 1996). A wide range of microbial P solubilization mechanisms exist in nature and much of the global cycling of insoluble organic and inorganic soil phosphates is attributed to bacteria and fungi (Banik and Dey 1982). Whitelaw (2000) suggested that Phosphorus solubilization is also carried out by a large number of saprophytic bacteria and fungi acting on sparingly soluble soil phosphates, mainly by chelation-mediated mechanisms. Phosphate solubilizing microorganisms secrete organic acids and enzymes that act on insoluble phosphates and convert it into soluble form, thus, proving P to plants (Ponmurugan and Gopi 2006). Inorganic P is solubilized by the action of organic and inorganic acids secreted by PSB in which hydroxyl and carboxyl groups of acids chelate cations (Al, Fe, Ca) and decrease the pH in basic soils (Kpomblekou and Tabatabai 1994; Stevenson 2005). The PSB dissolve the soil P through production of low molecular weight organic acids mainly gluconic and ketogluconic acids (Goldstein 1995; Deubel et al. 2000), in addition to lowering the pH of rhizosphere. The pH of rhizosphere is lowered through biotical production of proton/bicarbonate release (anion/cation balance) and gaseous (O_2/CO_2) exchanges. Phosphorus solubilization ability of PSB has direct correlation with pH of the medium. In addition to phosphorous solublization ability of PSB, they also can improve plant growth by enhancing the availability of other trace element such as iron (Fe), zinc (Zn), etc. Gull et al. (2004) suggested that PSB can solubilize the fixed soil P and applied phosphates resulting in higher crop yields. According to Goenadi et al. (2000) direct application of phosphate rock is usually ineffective in the short time period of most annual crops. Gyaneshwar et al. (2002) suggested that acid producing microorganisms are able to increase the solubilization of phosphatic rock. The PSB in conjunction with single super phosphate and rock phosphate reduce the P dose by 25 and 50 %, respectively (Sundara et al. 2002). *Pseudomonas striata* and *Bacillus polymyxa* solubilized 156 and 116 mg P L^{-1}, respectively (Rodríguez and Fraga 1999). *Pseudomonas fluorescens* solubilized 100 mg P L^{-1} containing $Ca_3(PO_4)^2$ or 92 and 51 mg P L^{-1} containing $AlPO_4$ and $FePO_4$, respectively (Henri et al. 2008).

Mycorrhiza

Mycorrhizae are mutualistic associations between fungi and plant roots. The host plant gets mineral nutrients from mycorrhizal fungi, while as the fungus partener is provided with photosynthetic products from the host plant (Jakobsen et al. 2002). Fungi become integrated into the root structure, or fungi lives in close association with plant roots. Fungal hyphae may live on the external surface of roots (ectomycorrhizal) or may invade root cells (endomycorrhizal). Mycorrhiza belong to fungi kingdom Basidiomycetes, Ascomycetes and Zygomycetes. Mycorrhizal fungi, and fungi generally, have a strong influence on soil structure (Rillig and Mummey 2006). Their hyphal strands help to hold soil aggregates together, and they also excrete organic substances that help cement the aggregates (Rillig and Mummey 2006). Hyphae conduct water and immobile nutrients (like P) to roots despite disruption of capillary water flow in soil. Of the many types of mycorrhizal association the most important association which are economically as well as ecologically importance are: ectomycorrhizal associations, and the endomycorrhizal association of the vesicular-arbuscular (VA) type (Rillig and Mummey 2006). In case of ectomycorrhizal associations, the fungi attack the cortical region of the host root devoid of piercing cortical cells. Ectomychorrhizae are recognized to occur in the families of Salicaceae, Fagaceae, Pinaceae, Betulaceae, Tiliaceae, Juglandaceae and Ceasalpinionideae. The ectomycorrhizal roots lack root hairs and are covered by a sheath of fungal hypae which almost looks like host tissue. This tissue is called Pseudoparenchamatous sheath. Hyphae from this sheath enter into the cortex and remain in the outer cortical region to form a network called Hartig's net (Alizadeh 2011). The nutrients absorbed by the hyphae are transported to the plant with the help of this Hartig's net. Infection of host plants by ectomycorrhizal fungi frequently leads to changes in feeder roots that are apparent to the naked eye but in case of endomycorrhizal associations of the VA type, the fungi penetrate the cortical cells and form clusters of delicately divided hyphae known as *arbuscules* in the cortex (Alizadeh 2011). They also form vesicles, which are membrane-bound organelles of varying shapes, inside or outside the cortical cells. Arbuscules are supposed to be the sites where resources are exchanged among the host plant and the fungi (Alizadeh 2011). Vesicles in general serve as storage space but when they are old they can serve as reproductive structures. Vesicles and arbuscules, together with large spores, comprise the diagnostic features of the VA mycorrhizas. Most ectomycorrhizal fungi belong to several genera within the class Basidiomycetes, while some belong to the zygosporic Zygomycetes and Ascomycetes. On the other hand, AM fungi belong to six genera within the azygosporous zygomycetes

Vesicular Arbuscular Mycorrhiza (VAM)

VAM are common, ancient and most fascinating class of fungi which proves to be very beneficial to plants (Alizadeh 2011). VAM is an endotrophic mycorrhiza formed by aseptate phycomycetous fungi. They produce an interconnected network of hyphae between cortical cells that extend to the soil and hence absorb various nutrients and water (Sally et al. 2011) VAM forms an association with various crop plants which include monocot, dicot, annual or perennial crops. The use of VAM enhances growth of plants in less fertile soils besides application of FYM and cereal-legume crop rotations. Whereas, application of chemicals mostly fungicide suppresses its existence.

Mycorrhiza enhances the feeding areas of the plant root as the hyphae spreads around the roots. It also mobilizes the nutrients particularly phosphorous that are present in organic or inorganic form in soil and translocate it to plants with the help of extensive mycelium. In addition to translocation of phosphorous to plant it also stores the nutrients and removes the toxic substances for example, phenolics which otherwise hinder nutrient availability in addition to this it also provides protection against other fungi and nematodes. VAM also assists in transfer of nutrients other than phosphorus, like zinc and sulfur Cu (copper), K (potassium), Al (aluminum), Mn (manganese), Fe (iron) and Mg (magnesium) from the soil to the plant roots. They act as intracellular obligate fungal endo-symbiont by penetrating the root cortex (Alizadeh 2011). In addition they possess vesicles intended for storage of nutrients and arbuscular for transferring these nutrients into root system as well as enhances water absorption. However, in ecto-mycorrhiza, the hyphae cover both outside and within the root in the intercellular spaces of epidermis and cortex. Trees are usually found to be infected with ectomycorrhiza, they increase the tolerance of plants against drought and salt stress, increase the photosynthetic activity of plants, higher chlorophyll content, higher leaf water potential restored capacity (Wang 1989, 1998). VAM helps in soil conservation and soil aggregation, increase the resistance of plants against root-pathogens, increases habitatrestoration (Dodd 2000).

Conclusions and Future Prospects

Bio-fertilizers increase crop productivity by increasing availability or uptake of nutrients through solubilization or increased absorption stimulation of plant growth with the help of hormonal action or antibiosis, or by decomposition of organic residues. Moreover, bio-fertilizers also help to reduce the use of chemical fertilizers which in turn reduces the amount and cost of chemical fertilizers and thus prevents the environment pollution from extensive application of chemical fertilizers.

To get better productivity of agricultural lands and to maintain this productivity, the integrated approach to determine the most favorable plant-microorganism interaction is important. The bio-fertilizers are thought to be more expensive and show unpredictable performance. Besides, the effect on the crops is slow, compared to chemical fertilizers. In order to get potential benefit from bio-fertilizers in commercial agriculture, consistency in their performance is to be improved. Special care such as mode of application on crops and to keep them effective for extensive use is needed. As bio-fertilizers contain living organisms, their concert therefore depends on environment surrounding them, Short shelf life, lack of suitable carrier materials, susceptibility to high temperature, problems in transportation and storage of bio-fertilizers are major bottlenecks that are at a standstill and have to be solved in order to acquire efficient inoculation. The main criteria to take into consideration in making of bio-fertilizers are microbes' growth profile, types and optimum condition of organism, and formulation of inoculums, methods of application and storage of the product are all critical to the success for a sustainable agriculture.

References

Abdul Halim NB (2009) Effects of using enhanced biofertilizer containing N-fixer bacteria on patchouli growth. Thesis. Faculty of chemical and natural resources engineering university Malaysia, Pahang. p. 145

Akhtar A, Hisamuddin, Robab MI, Abbasi, Sharf R (2012) Plant growth promoting rhizobacteria: an overview. J Nat Prod Plant Resour 2(1):19–31

Alam S, Khalil S, Ayub N, Rashid M (2002) *In vitro* solubilization of inorganic phosphate by phosphate solubilizing microorganism (PSM) from maize rhizosphere. Intl J Agric Biol 4:454–458

Ali B, Sabri AN, Hasnain S (2010) Rhizobacterial potential to alter auxin content and growth of *Vigna radiata* (L.). World J Microbiol Biotechnol 26:1379–1384

Alizadeh O (2011). Mycorrhizal symbiosis. Adv Studies Biol 6(3):273–281

Amutha K, Kannaiyan S (2000) Effect of kinetin and inoculation of *Azorhizobium cauulinodans* on induction of lateral rootlets and paranodules in rice. J Microb World 2:1–7

Anitha KG, Kannaiyan S (1999) Carrier based immobilized cyanobacterial cultures as biofertilizer for rice. Paper presented in the seminar on modern trends in plant sciences, AVC College, Mannampandal, Tamil Nadu, India

Badawi FSHF, Biomy, AMM, Desoky AH (2011) Peanut plant growth and yield as influenced by co-inoculation with *Bradyrhizobium* and some rhizo-microorganisms under sandy loam soil conditions. Ann Agric Sci 56:17–25

Banerjee S, Palit R, Sengupta C, Standing D (2010) Stress induced phosphate solubilization by Arthrobacter sp. and Bacillus sp. Isolated from tomato rhizosphere. Aust J Crop Sci 4(6):378–383

Banik S, Dey BK (1982) Available phosphate content of an alluvial soil as influenced by inoculation of some isolated phosphate solubilizing microorganisms. Plant Soil 69:353–364

Baudoin E, Benizri E, Guckert A (2002) Impact of growth stages on bacterial community structure along maize roots by metabolic and genetic finger printing. Appl Soil Ecol 19:135–145

Belimov AA, Kojemiakov AP, Chuvarliyeva CV (1995) Interaction between barley and mixed cultures of nitrogen fixing and phosphate-solubilizing bacteria. Plant Soil 173:29–37

Benson DR, Silvester WB (1993) Biology of Frankia strains, actinomycete symbionts of actinorhizal plants. Microbiol Rev 57(2):293–319

Berg G (2009) Plant-microbe interactions promoting plant growth and health: perspectives for controlled use of microorganisms in agriculture. Appl Microbiol Biotech 84:11–18

Bhaskara Rao KV, Charyulu PBBN (2005) Evaluation of effect of inoculation of *Azospirillum* on the yield of *Setaria italic* (L.). Afr J Biotech 4(9):989–995

Bhat A, Bhat A, Narayan P (2003) Chromatographic analysis of phenolic and study of kalson lignin biodegraded, coir pith using *Pleurotus Sajorcaju*. BSc, Dissertation, University of Kerala, Alappuzha.

Brahmaprakash GP, Sahu PK (2012) Bio-fertilizers for sustainability. J Ind Inst Sci 92:37–62

Chang CH, Yang SS (2009) Thermo-tolerant phosphate-solubilizing microbes for multi-functional biofertilizer preparation. Biores Technol 100:1648–1658

Chang FP, Young CC (1992a) Ecology of vesicular-arbuscular mycorrhizal fungi in tea garden soils. Tai Tea Res Bull 11:63–77

Chang FP, Young CC (1992b) Effects of VA mycorrhizal fungi and phosphorus-solubilizing bacteria inoculated on growth of tea cuttings in plastic bag. Tai Tea Res Bull 11:79–89

Chang FP, Young CC (1999) Studies on soil inoculation with P-solubilizing bacteria and P fertilizer on P-uptake and quality of tea. Soil Environ 2:35–44

Chen YP, Rekha PD, Arunshen AB, Lai WA, Young CC (2006) Phosphate solubilizing bacteria from subtropical soil and their tri-calcium phosphate solubilizing abilities. Appl Soil Ecol 34:33–41

Choudhury ATMA, Kennedy IR (2004) Prospects and potentials for systems of biological nitrogen fixation in sustainable rice production. Biol Fert Soils 39:219–227

Danso SKA, Bowen GD, Snaginga N (1992) Biological nitrogen fixation in trees in agro-ecosystems. Plant and Soil 141:177–196

Dawson JO, Gottfried GJ, Urbana IL, Hahn D (2005). Occurrence, structure, and nitrogen-fixation of root nodules of actinorhizal Arizona alder. USDA forest service proceedings RMRS-P-36. pp. 75–79

De Freitas JR, Gupta VVS, Germida JJ (1993). Influence of Pseudomonas syringae R 25 and P. putida R 105 on the growth and N2 fixation (ARA) of pea (*Pisum sativum* L.) and field bean (*Phaseolus vulgaris*). Biol Fertil Soils 16:215–220

Deubel A, Gransee A, Merbach W (2000) Transformation of organic rhizodeposites by rhizoplane bacteria and its influence on the availability of tertiary calcium phosphates. J Plant Nutr Soil Sci 163:387–392

Dodd JC (2000) The role of arbuscular mycorrhizal fungi in agro- and natural ecosystems. Outlook Agric 29:55–62

Duhoux E, Rinaudo G, Diem HG, Auguy F, Fernandez D, Bogusz D, Franche C, Dommergues Y, Huguenin B (2001) Angiosperm Gymnostoma trees produce root nodules colonized by arbuscular mycorrhizal fungi related to Glomus. New Phytol 149:115–125

Duponnois R, Kisa M, Plenchette C (2006) Phosphate solubilizing potential of the nemato fungus Arthrobotrys oligospora. J Plant Nutr Soil Sci 169:280–282

Dutton MV, Evans CS (1996) Oxalate production by fungi: Its role in pathogenicity and ecology in the soil environment. Can J Microbiol 42:881–895

El-Yazeid AA, Abou-Aly HA, Mady MA, Moussa SAM (2007) Enhancing growth, productivity and quality of squash plants using phosphate dissolving microorganisms (bio phosphor) combined with boron foliar spray. Res J Agric Biol Sci 3(4):274–286

Fallah A (2006) Abundance and distribution of phosphate solubilizing bacteria and fungi in some soil samples from north of Iran. 18th World Congress of soil science, July 9

Fankem H, Nwaga D, Deubel A, Dieng L, Merbach W, Etoa FS (2006) Occurrence and functioning of phosphate solubilizing microorganisms from oil palm tree rhizosphere in Cameroon. Afr J Biotech 5:2450–2460

Faramarzi A, Pourgorban MA, Ansari MH, Taghizadeh R (2012) The effects of plant growth promoting rhizobacteria (PGPR) inoculation on the yield and yield components of grain corn (*Zea mays* L.) in Astara. Iran J Food Agricul Environ 10(1):299–301

Franche C, Kristina Lindstrom K, Elmerich C (2009) Nitrogen-fixing bacteria associated with leguminous and non-leguminous plants. Plant Soil 321:35–59

Gandora V, Gupta RD, Bhardwaj KKR (1998) Abundance of *Azotobacter* in great soil groups of North-West Himalayas. J Ind Soc Soil Sci 46(3):379–383

Ganesh G, Misra AK, Chapelon C, Normand P (1994) Morphological and molecular characterization of *Frankia sp.* isolates from nodules of *Alnus nepalensis Don*. Arch Microbiol 161:152–155

Goenadi DH, Siswanto G, Sugiarto Y (2000) Bioactivation of poorly soluble phosphate rocks with a phosphorus-solubilizing fungus. Soil Sci Soc Am J 64:927–932

Goethals K, Van Montagu M, Holsters M (1992) Conserved motifs in a divergent nod box of *Azorhizobium caulinodans* ORS571 reveals a common structure in promoters regulated by LysR-type proteins. Proc Natl Acad Sci USA 89:1646–1650

Goldstein AH (1995) Recent progress in understanding the molecular genetics and biochemistry of calcium phosphate solubilization by Gram-negative bacteria. Biol Agri Hort 12:185–193

Gull M, Hafeez FE, Saleem M, Malik KA (2004) Phosphorus uptake and growth promotion of chickpea by co-inoculation of mineral phosphate solubilizing bacteria. Aust J Exp Agric 44:623–628

Gyaneshwar P, Kumar GN, Parekh LJ, Poole PS (2002) Role of soil microorganisms in improving P nutrition of plants. Plant Soil 245:83–93

Habibi A, Heidari G, Sohrabi Y, Badakhshan H, Mohammadi K (2011) Influence of bio, organic and chemical fertilizers on medicinal pumpkin traits. J Med Plant Res 5(23):5590–5597

Hashemabadi D, Zaredost F, Ziyabari MB, Zarchini M, Kaviani B, Solimandarabi MJ, Torkashvand AM, Zarchini S (2012) Influence of phosphate bio-fertilizer on quantity and quality features of marigold (*Tagetes erecta* L.). Australian J Crop Sci 6(6):1101–1109

Henri F, Laurette NN, Annette A, John Q, Wolfgang M, François-Xavier E, Dieudonne E (2008) Solubilization of inorganic phosphates and plant growth promotion by strains of Pseudomonas fluorescens isolated from acidic soils of Cameroon. Afr J Microbiol Res 2:171–178

Hilda R, Fraga R (1999) Phosphate solubilizing bacteria and their role in plant growth promotion. Biotechnol Adv 17:319–339

Igual JM, Valverde A, Cervantes E, Velazquez E (2001) Phosphate-solubilizing bacteria asinoculants for agriculture: use of updated molecular techniques in their study. Agronomie 21:561–568

Jakobsen I, Smith SE, Smith FA (2002) Function and diversity of arbuscular Mycorrhizae in carbon and mineral nutrition. In: van der Heijden MGA, Sanders I (eds) Mycorrhizal ecology. Ecological studies, vol 157. Springer, Berlin

Jha MN, Prasad AN (2005) Efficiency of new inexpensive cyanobacteria biofertilizer including its self life. World J Microb Biotechnol 22(1):73–79

Kader MA, Miar MH, Hoque MS (2002) Effects of *Azotobacter* inoculants on the yield and Nitrogen uptake by Wheat. J Biol Sci 2(4):259–261

Kalia A, Gupta RP (2002) Nodule induction in non-legumes by rhizobia. Ind J Microbiol 42:183–193

Kannaiyan S (1999) Biological fertilizers for rice production. In: Kannaiyan S (ed) Bioresources technology for sustainable agriculture, Associate Publishing Company, New Delhi, pp 1–29

Kannaiyan S, Govindarajan K, Lewin HD (1980) Effect of foliar spray of *Azotobacter chroococcum* on rice crop. Plant Soil 56:487–490

Kannaiyan S, Kumar K, Gopalswamy G (2001). Stem nodulating Sesbania rostrata as a bioresource for rice production, induction of nodulation and endophytic N2 fixation in rice. Madras Agric J 88:547–560

Kanungo PK, Panda D, Adhiya TK, Ramakrishnan B, Rao VR (1997) Nitrogenase activity and nitrogen fixing bacteria assoiciated with rhizosphere of rice cultivars. J Sci Food Aric 73:485–488

Khan MS, Zaidi A, Ahemad M, Oves M, Wani PA (2010). Plant growth promotion by phosphate solubilizing fungi: current perspective. Arch Agron Soil Sci 56:73–98

Khan MS, Zaidi A, Wani PA (2007) Role of phosphate-solubilizing microorganisms in sustainable agriculture—a review. Agron Sustain Dev 27:29–43

Khiari L, Parent LE (2005) Phosphorus transformations in acid light-textured soils treated with dry swine manure. Can J Soil Sci 85:75–87

Kim KY, Jordan D, Mc Donald GA (1998) Effect of phosphate-solubilizing bacteria and vesicular-arbuscular mycorrhizae on tomato growth and soil microbial activity. Biol Fertil Soils 26:79–87

Kpomblekou K, Tabatabai MA (1994) Effect of organic acids on release of phosphorus from phosphate rocks. Soil Sci 158:442–453

Kucey RMN, Janzen HH, Leggett ME (1989) Microbially mediated increases in plant-available phosphorus. Ad Agron 42:199–228

Kumar SR, Bhaskara Rao KV (2012) Biological nitrogen fixation: a review. IJALS 1:1–9

Lee KD, Bai Y, Smith D, Han HS, Supanjani S (2005) Isolation of plant growth promoting endophytic bacteria from bean nodule. Res J Agric Biol Sci 1(3):232–236

Lugtenberg B, Kamilova F (2009) Plant-growth-promoting rhizobacteria. Ann Rev Microbiol 63:541–546

Mader P, Kaiser F, Adholeya A, Singh R, Uppal HS, Sharma AK, Srivastava R, Sahai V, Aragno M, Wiemken A, Johri BN, Fried PM (2011) Inoculation of root microorganisms for sustainable wheat-rice and wheat-black gram rotations in India. Soil Biol & Biochem. 43:609–619

Mahdi SS, Hassan GI, Samoon SA, Rather HA, Showkat AD, Zehra B (2010) Bio-fertilizers in organic agriculture. J Phytol 2(10):42–54

Malik KA, Mirza MS, Hassan U, Mehnaz S, Rasul G, Haurat J, Bally R, Normand P (2002) The role of plant-associated beneficial bacteria in rice-wheat cropping system. In: Kennedy IR, Choudhury ATMA (eds) Biofertilisers in action. Rural Industries Research and Development Corporation, Canberra, pp 73–83

Malliga PL, Uma L, Subramanian G (1996) Lignolytic activity of the cyanobacterium *Anabaena Azolla* ML2 and the value of coir waste as a carrier for BGA bio-fertilizers. Microbios 86:175–183

Mandon K, Michel-Reydellet N, Encarnacion S, Kaminski PA, Leija A, Cevallos MA, Elmerich C, Mora J (1998) Poly-β-hydroxybutyrate turnover in *Azorhizobium caulinodans* is required for growth and affects *nifA* expression. J Bacteriol 180:5070–5076

Manoharan G, Chitra Devi K, Malliga P (2011) Effect of cyanopith and cyanospray biofertilizer on *Amaranthus Dubius*. Int J Env Sci 2(1):352–360

Martyniuk S, Martyniuk M (2003) Occurrence of *Azotobacter Spp.* In Some Polish Soils. J Environ Stud 12(3):371–374

Mirza MS, Rasul G, Mehnaz S, Ladha JK, Ali S, Malik KA (2000) Beneficial effects of inoculated nitrogen-fixing bacteria on rice. International Rice Research Institute, Los Baos, Philippines, pp 191–204

Mishra BK, Dadhich SK (2010) Methodology of nitrogen biofertilizer production. J Adv Dev Res 1(1):3–6

Mohammadi K (2010) Ecophysiological response of canola (*Brassica napus* L.) to different fertility systems in crop rotation. Ph.D thesis. Agronomy Department. Tarbiat Modares University, Tehran, Iran. p. 354

Mohammadi K, Ghalavand A, Aghaalikhani M, Heidari GR, Sohrabi Y (2011) Introducing the sustainable soil fertility system for chickpea (Cicer arietinum L.). Afr J Biotech 10(32):6011–6020

Mohammadi K, Ghalavand A, Aghaalikhani M, Sohrabi Y, Heidari GR (2010) Impressibility of chickpea seed quality from different systems of increasing soil fertility. Electronic J Crop Prod 3(1):103–119

Mohammadi K, Sohrabi Y (2012) Bacterial bio-fertilizers for sustainable crop production: a review. J Agri Bio Sci 7:307–316

Murty MG, Ladha JK (1988) Influence of *Azospirillum* inoculation on the mineral uptake and growth of rice under hydroponic conditions. Plant Soil 108:281–285

Mylona P, Pawlowski K, Bisseling T (1995) Symbiotic nitrogen fixation. Plant Cell 7:869–885

Naderifar M, Daneshian J (2012) Effect of different nitrogen and bio-fertilizers effect on growth and yield of *Brassica napus* L. Intl J Agri Crop Sci 4(8):478–482

Nahas E (1996) Factors determining rock phosphate solubilization by microorganism isolated from soil. World J Microb Biotech 12:18–23

Narula N, Kumar V, Behl RK, Deubel A, Granse A, Merbach W (2000) Effect of P-solubilizing A. chroococcum on NPK uptake in P responsive wheat genotypes grown under green house conditions. J Plant Nutr Soil Sci 163:93–398

Nihorimbere V, Ongena M, Smargiassi M, Thonart P (2011) Beneficial effect of the rhizosphere microbial community for plant growth and health. Biotechnol Agron Soc Environ 15(2):327–337

Palaniappan P (2005) Effect of coir waste based basal cyanobacterial bio-fertilizers and foliar spray on cowpea *(Vigna unigiculata*). M.Sc., Dissertation, Bharathidasa University, Tiruchirappalli, Tamil Nadu, India

Ponmurugan P, Gopi C (2006) In vitro production of growth regulators and phosphatase activity by phosphate solubilizing bacteria. Afr J Biotechnol 5:348–350

Postgate JR (1982) The fundamentals of nitrogen fixation. Cambridge University Press, Cambridge

Radhakrishnan KC (1996) Role of bio-fertilizers in cotton productivity. National seminar biofertilizer production problem and constraints, TNAU, Coimbatore, Jan 24–25

Raj SA (2007) Bio-fertilizers for micronutrients. Biofertilizer Newsletter (July), pp 8–10

Rillig MC, Mummey DL (2006) Mycorrhizas and soil structure. New Phytol 171:41–53

Robson RL, Eady RR, Richardson TH, Miller RW, Hawkins M, Postgate JR (1986) London. Nature 322:388–390

Rodrguez H, Fraga R (1999) Phosphate solubilizing bacteria and their role in plant growth promotion. Biotech Adv 17:319–339

Rokhzadi A. Asgharzadeh A, Darvish F, Nourmohammadi G, Majidi E (2008) Influence of plant growth-promoting rhizobacteria on dry matter accumulation and yield of chickpea (*Cicer arietinum* L.) under field condition. Am-Euras J Agric Environ Sci 3(2):253–257

Romerheld V, Marshner H (1986) Evidence for az specific uptake system for iron physiderophores in roots of grasses. Plant Physiol 80:175–180

Rosenblueth M, Martinez RE (2006) Bacterial endophytes and their interactions with hosts. Mol Plant Microb In 9(8):827–837.

Ruttimann JC, Rubio LM, Dean DR Ludden PW (2003) VnfY is required for full activity of the vanadium-containing dinitrogenase in *Azotobacter vinelandii.* J Bacteriol 185:2383–2386

Sagoe C, Ando T, Kouno K, Nagaoka T (1998) Relative importance of protons and solution calcium concentration in phosphate rock dissolution by organic acids. Soil Sci Plant Nutr 44:617–625

Saikia SP, Jain V, Khetarpal S, Aravind S (2007) Dinitrogen fixation activity of *Azospirillum brasilense* in maize (*Zea mays*). J Curr sci 93(9/10):1296–1300

Schroth MN, Hancock JG (1981) Selected topics in biological control. Ann Rev Microbiol 35:453–476

Shankar PV, Vishwas PS, Pathade GR (2012) Comparative use of chemical and biological fertilizers and pesticides and their health effects on farmers in Satara tahsil, Maharashtra. Universal J Environ Res Technol 2:174–178

Sharma K, Dak G, Agrawal A, Bhatnagar M, Sharma R (2007) Effect of phosphate solubilizing bacteria on germination of *Cicer arietinum* seeds and seedling growth. J Her Med Toxicol 1(1):61–63

Singh MS, Devi RKT, Singh NI (1999) Evaluation of methods for *Azotobacter* application on the yield of rice. Ind J Hill Farm 12:22–24

Steenhoudt O, Vanderleyden J (2006) *Azospirillum*, a free-living nitrogen-fixing bacterium closely associated with grasses: genetic, biochemical and ecological aspects. FEMS Microbiol Review 24:487–506

Stevenson FJ (2005) Cycles of Soil: Carbon, Nitrogen, Phosphorus, Sulfur, Micronutrients. Wiley, New York

Subbarao NS (1988) Phosphate solubilizing microorganism. In: Biofertilizer in agriculture and forestry. Regional Biofert. Dev. Centre, Hissar, India. pp. 133–142

Sundara B, Natarajan V, Hari K (2002) Influence of phosphorus solubilizing bacteria on the changes in soil available phosphorus and sugarcane yields. Field Crops Res 77:43–49

Tjepkema JD, Ormerod W, Torrey JG (1980) Vesicle formation and acetylene reduction activity in *Frankia* sp. CP11 cultured in defined nutrient media. London. Nature 287:633–635

Van Workum WAT, Van Slageren S, Van Brussel AAN, Kijne JW (1998) Role of exopolysaccharides of Rhizobium leguminosarum bv. viciae as host plant-specific molecules required for infection thread formation during nodulation of Vicia sativa. Mol Plant Microbe Interact 11:1233–1241

Vazquez P, Holguin G, Puente M, Cortes AE, Bashan Y (2000) Phosphate solubilizing microorganisms associated with the rhizosphere of mangroves in a semi arid coastal lagoon. Biol Fertil Soils 30:460–468

Velineni S, Brahmaprakash GP (2011) Survival and phosphate solubilizing ability of *Bacillus megaterium* in liquid inoculants under high temperature and desiccation stress. J Agr Sci Tech 13:795–802

Venkatashwarlu B (2008) Role of bio-fertilizers in organic farming: organic farming in rain fed agriculture. Central institute for dry land agriculture, Hyderabad. pp. 85–95

Verghese S, Misra AK (2002) *Frankia*–actinorhizal symbiosis with special reference to host-microsymbiont relationship. Curr Sci 83(4):404–408

Vora MS, Shelat HH (1996) Isolation and screening of micro-organisms for dissolving tricalcium Phosphate. J Madras Agric 83:353–354

Vyas RV, Shelat HN, Vora MS (2008) Bio-fertilizers techniques for sustainable production of major crops for green revolution in Gujarat—an overveiw. Green Farm 1(10–11):68–72

Wang CL (1998) Response of *Phaseolus angularis*. Weight inoculated with arbuscular mycorrhizal fungi under drought stress. J Agri Assoc China 181:92–101

Weller DM (1988) Biological control of soil borne plant pathogens in the rhizosphere with bacteria. Ann Rev Phytopathol 26:379–407

Whitelaw MA (2000) Growth promotion of plants inoculated with phosphate solubilizing fungi. Adv Agron 69:99–151

Yahya A, Azawi SKA (1998) Occurrence of phosphate solubilizing bacteria in some Iranian soils. Plant Soil 117:135–141

Yanni YG, El-Fattah FKA (1999) Towards integrated biofertilization management with free living and associative dinitrogen fixers for enhancing rice performance in the Nile delta. Symbiosis 27:319–331

Young CC (1990) Effects of phosphorus-solubilizing bacteria and vesicular-arbuscular mycorrhizal fungi on the growth of tree species in subtropical-tropical soils. Soil Sci Plant Nut 36:225–231

Young CC, Juang TC, Chao CC (1988) Effect of Rhizobium and VA mycorrhiza inoculation on nodulation, symbiotic nitrogen fixation and soybean yield in subtropical-tropical fields. Biol Pert Soils 6:165–169

Young CC, Chang CH, Chen LF, Chao CC (1998a) Studies on the properties of nitrogen fixation and ferric phosphate solubilizing bacteria isolated from a Taiwan soil. J Chinese Agri Chem Soci 36:201–210

Young CC, Hung LJ, Chen LF (1998b) Studies on the exudate and characteristics of P-solubilization with P-solubilizing rhizobia for four green manures. Soil Environ 1:7–17

Young CC, Chen LG, Chen LF (2000) Effects of nitrogen fertilization on loss of nutrient in four-Taiwan soils. J Agricul Assoc China 1:1–14

Young CC, Lai WA, Shen FT, Hung MH, Huang WS, Arun AB (2003) Exploring the microbial potentiality to augment soil fertility in Taiwan. In the Proceedings of 6th ESAFS inter-national conference: soil management technology on low productivity and degraded soils, Taipei, Taiwan. pp. 25–27

Zhang J, Wang M, Cao Y, Wu L, Xu S (2012) Iron and zinc accumulation trend in a Japonica rice grains after anthesis. Afr J Agric Res 7:1312–1316

Zilli JE, Valisheski RR, Filho FR, Nerves MP, Rumianek NG (2004) Assessment of cowpea rhizobium diversity in Cerrado area of northeastern Brazil. Braz J Microbiol 35(4):281–287

Chapter 6
Plant-Microorganism Interactions: Effects on the Tolerance of Plants to Biotic and Abiotic Stresses

Muriel da Silva Folli-Pereira, Lydice Sant'Anna Meira-Haddad, Cristina Maria Nobre Sobral de Vilhena da Cruz Houghton and Maria Catarina Megumi Kasuya

Abstract Arbuscular mycorrhizal fungi (AMF) establish mutualistic symbiosis (arbuscular mycorrhiza—AM), with the roots of most species of terrestrial plants, acting as a bridge between the soil and plants. AMF are critical in the establishment and adaptation of plants in locations severely disturbed. They affect also the physico-chemical properties of substrate and act for the formation and maintenance of soil structure, acting in the aggregation of soil particles. The AM occurs in the roots of most plants, promoting improvements in the growth and development of plants and increase in tolerance and/or plant resistance to various adverse environmental agents and can also be used as a potential biological control agent of plant diseases. The different responses of plants to this symbiosis can be assigned to the functional diversity of AM, depending of the interaction between AMF, plants and environmental conditions. The establishment and functioning of MAs during stress conditions involves a complex process of recognition and development, concurrently at physiological, biochemical and molecular changes in both symbionts. In addition, mycorrhizal colonization of roots has a significant impact on the gene expression of several plants that encode proteins presumably involved in tolerance to stress. In this context, whereas the AMF are essential in the establishment and

M. C. Megumi Kasuya (✉) · M. da Silva Folli-Pereira
Departamento de Microbiologia, Universidade Federal de Viçosa, 36570-000 Viçosa, Minas Gerais, Brasil
e-mail: mkasuya@ufv.br

M. da Silva Folli-Pereira
e-mail: mfolli@gmail.com

L. Sant'Anna Meira-Haddad
Centro de Ciências Agrárias, Ambientais e Biológicas,
Universidade Federal do Recôncavo da Bahia, Cruz das Almas, BA, Brasil
e-mail: lydicemeira@yahoo.com.br

C. M. N. S. de Vilhena da Cruz Houghton
Faculdade de Ciências, Universidade de Lisboa, 1749-016, Lisboa, Portugal
e-mail: cristinacruzhoughton@gmail.com

K. R. Hakeem et al. (eds.), *Crop Improvement,* DOI 10.1007/978-1-4614-7028-1_6,

adaptation of plants on disturbed sites, this chapter will be covered the molecular and physiological mechanisms of the association MA, responsible for this adaptation and greater tolerance of plants to biotic and abiotic stresses.

Introduction

Plants are constantly exposed to various abiotic and biotic stresses, such as radiation, temperature, water, minerals, plants, animals, and microorganisms, which alter the biosynthesis and development of plants as a consequence of oxidative burst (Gill and Tuteja 2010). During aerobic metabolism, molecular oxygen is partially reduced, generating transient intermediates that are highly reactive and damaging to the cell. This partial reduction of O_2 produces reactive oxygen species (ROS), including the superoxide anion (O_2), the hydroxyl radical (OH^-), and hydrogen peroxide (H_2O_2). Further conversions can occur among these molecules, transforming them into even more reactive species (Hajiboland and Joudmand 2009; Gill and Tuteja 2010; Abdel Latef and Chaoxing 2011). These ROS cause morphological, physiological and molecular alterations, resulting in plant damage and eventual death. However, plants have enzymatic and non-enzymatic defence systems to minimise the effects of ROS (Arfaoui et al. 2007; Gill and Tuteja 2010).

The microhabitat of the rhizosphere is a specialised ecosystem, where microbial populations are highly favoured, allowing the growth, development and multiplication of these microorganisms. Plants can form beneficial associations with these microorganisms, acquiring tolerance and/or resistance to biotic and abiotic stress factors (Dimkpa et al. 2009; Singh et al. 2011)

Bacteria and fungi are the most numerous inhabitants of the rhizosphere. Plant growth-promoting rhizobacteria (PGPR) are beneficial bacteria that colonise the root system of plants and promote growth through various mechanisms including producing plant growth regulators, increasing the cycling and availability of soil nutrients, functioning as pathogen biocontrol agents and conferring tolerance and/or resistance to biotic and abiotic stresses (del Mar Alguacil et al. 2009). Arbuscular mycorrhizal fungi (AMF) are common inhabitants of soil that form mutualistic associations within the root systems of a large number of agricultural plant species, conferring benefits to host plants distributed throughout various habitats (Smith et al. 2010). Arbuscular mycorrhizae improve the nutritional status of plants, facilitate plant adaptation to different ecosystems, and increase plant tolerance to biotic and abiotic stress factors, and they are also considered to be biocontrol agents (Singh et al. 2012)

Plants that associate with microorganisms become more tolerant to stress conditions. Several mechanisms have been described to explain this greater tolerance (Pozo et al. 2010). The physiological changes induced in the plant through relationships with symbionts prevent pathogenic attacks and activate defence mechanisms. Phytoalexins, pathogenesis-related proteins, and agents for the lignification of the cell wall have been reported in mycorrhizal plants in areas far from infection sites, thus indicating the occurrence of systemic resistance. AMF play an important role in enhancing the absorption of water and nutrients, conferring an enhanced hydration and nutritional status to the plant (Pozo et al. 2010; Folli-Pereira et al. 2012).

The benefits of microorganisms might help plants in terms of tolerance to several biotic and abiotic stress factors. Thus, it is necessary to understand each symbiosis to ensure the best utilisation of the benefits to plants.

AMF and Growth Promoter Bacteria as Potential Factors Involved in Plant Tolerance to Stress

Plant growth promoting rhizobacteria (PGPRs) are bacteria that increase plant growth through interactions with plant roots. PGPRs represent a functionally active portion of the soil biota present inside the plant root, in the rhizosphere, or on the rhizoplane. The literature indicates their potential as a component of the sustainable management of agricultural systems, conferring characteristics such as greater resistance to biotic and abiotic stress conditions to the host and promoting plant growth, leading to their widespread biotechnological use. PGPRs can enhance growth of plants through a variety of mechanisms: (a) production of plant hormones; (b) controlling pathogens; and (c) increasing tolerance to stress conditions (Kohler et al. 2006, 2009; Jalili et al. 2009; del Mar Alguacil et al. 2009).

The mechanisms of plant growth promotion include direct actions, such as biological nitrogen fixation, production of plant growth regulator hormones, and the solubilisation of inorganic phosphate, and indirect actions, such as biological controls, production of siderophores and allelochemicals and the induction of local and systemic resistance. The use of PGPRs in biotechnology has intensified with the production of antibiotics and other bioactive molecules and the application of these microorganisms in bioremediation processes and transgenic techniques.

These mechanisms are potentially applicable in the field for the quantitative and qualitative improvement of agricultural production. The use of these microorganisms in agriculture depends on knowledge of their diversity, plant-bacteria interaction mechanisms, and the ability to maintain, manipulate, and modify beneficial populations under field conditions.

PGPRs, especially those belonging to the fluorescent *Pseudomonas* group, have been well studied but have a disadvantage relative to *Bacillus*. *Bacillus* have developed a resistance structure, called the endospore, which is produced under adverse conditions and enables the bacterium to withstand these conditions. Once the conditions change in favour of the microorganism, its reproduction cycle allows it to express its beneficial characteristics to the host plant.

Arbuscular mycorrhizal associations in turn have important consequences for nutrient cycling in soil, providing plants with essential nutrients such as phosphorus when they are scarce or if they have low mobility in the soil solution. In exchange, photosynthetic carbon is transported to the soil through the transfer of sugar from the root to the AMF, which subsequently translocates this carbon as lipids and sugars into the external mycelium distributed throughout the soil (Bago et al. 2003). The symbiosis between plants and AMF also results in the reduction of physiological losses through stress (Munier-Lamy et al. 2007) and consequently faster growth, leading to economy input and a reduction of environmental contamination (Huang et al. 2009). Moreover, these fungi can act as potential biological control agents, reducing the effects and damage from plant pathogens through indirect means or increased nutrition and plant resistance (Meira-Haddad 2008; Folli-Pereira et al. 2012). AMF also play an important role in the aggregation of soil particles.

PGPRs have been traditionally used as inducers of systemic resistance to diseases in plants. Currently, new PGPRs have been proposed for use in agricultural crops. Studies concerning the specific interactions of symbiotic microorganisms and pathogens have demonstrated the complexity of rhizospheric interactions involving both mycorrhizal and non-mycorrhizal fungi, beneficial and pathogenic bacteria, the plant and the soil. It is no longer possible to study only the isolated microorganism without considering the complexity of its habitat. The association of PGPRs with other microorganisms such as AMF is both economical and practical.

PGPRs colonise plant roots and promote the development of AMF through the enhanced absorption of P and N (Artursson et al. 2006; Richardson et al. 2009). However, there are limited data concerning the PGPR hyphal inoculation of AMF (Hartmann et al. 2009).

The strength of the physiological phases of bacterial binding to AMF hyphae varies, including a weak electrostatic binding in the first stage, followed by a strong bond in the second stage, which is related to the production of cellulose and other extracellular bacterial products (Artursson et al. 2006). Indeed, mutant bacteria are unable to produce these products in the presence of the AMF hyphae (Artursson et al. 2006). Some bacterial strains such as *Pseudomonas* spp. can colonise both plant roots and AMF hyphae, suggesting that the mechanisms of the related processes could be relatively similar.

The association of gram-positive bacteria with AMF is high compared with gram-negative bacteria, although this relationship has not been verified (Artursson et al. 2005). The significance of these interactions is due to the synergistic interaction of some important PGPRs, including the gram-positive *Bacillus spp*., with AMF (Francis et al. 2010). The enzymes of soil bacteria and AMF can also influence the decomposition of organic matter in the soil (de Boer et al. 2005).

There are a large number of bacteria, including PGPR and *Rhizobium*, which promote the activity and development of AMF (Frey-Klett et al. 2007; Richardson et al. 2009). These mycorrhizal helper bacteria are usually fungus-specific (Rillig et al. 2005) and promote the growth of specific AMF during symbiosis with the host plant. This specificity has been attributed to the size and surface rugosity of the spore (Bharadwaj et al. 2008). Thus, Frey-Klett et al. (2007) proposed the term "mycorrhizal auxiliary bacteria" to describe a broader concept than "mycorrhizal helper bacteria" (MHB), which includes the beginning of the arbuscular mycorrhizal symbiosis and the effects of MHB on the biocontrol of other species in terms of specificity. MHB may influence the germination of spores, affect the spore wall (de Boer et al. 2005), produce stimulants such as CO_2 (Carpenter-Boggs et al. 1995), or affect the absorption of P by the AMF (Ruiz-Lozano and Bonfante 2000).

Moreover, the AMF might compete for nutrients from the soil, resulting in affecting changes within the bacterial community of the rhizosphere. The association of some bacteria with AMF is specific (Artursson et al. 2005), suggesting that fungal exudates stimulate communication between bacteria and AMF (Artursson et al. 2006). Indeed, some genera of bacteria, including *Arthrobacter* and *Bacillus*, were most commonly observed in the hyphosphere or within soil around AMF hyphae, while *Pseudomonas* spp. were more distributed in the rhizosphere of *Sorghum bicolour* (Artursson et al. 2005).

The adhesion of PGPRs to AMF is determined by the formation of biofilms (Seneviratne et al. 2009). MHB, which primarily includes *Bacillus* and *Pseudomonas*, can affect the functions of the AMF, influencing root permeability, root exudation, AMF colonisation of the host root, and phytohormone production, thereby mitigating the adverse effects of the environment on hyphal growth and stimulating the growth of root hairs in plants. Some strains of rhizobia are also able to affect the pre-symbiotic stage of fungi, influencing spore germination and hyphal growth (Frey-Klett et al. 2007).

The symbiosis between AMF and plants contributes to the stability of soil aggregates, including soils with high salinity (Caravaca et al. 2005). Stability is initiated through macroaggregates (>250 mm), which tangle hyphae and deposit organic substances that assist in the subsequent stability of soil aggregates. A key factor in the contribution of these fungi to the stabilisation of saline soils is the production of glomalin, a glycoprotein that acts as an insoluble glue to stabilise aggregates (Gadkar and Rillig 2006).

The influence of AMF on plant growth has been attributed to bacteria associated with the mycorrhizosphere (Larsen et al. 2009). The production of exopolysaccharides (EPSs) in response to adverse environmental conditions such as drought can contribute to soil aggregation, leading to increased water retention in the rhizosphere (Kaci et al. 2005), which can eventually affect the growth of AMF in these soils. The effectiveness of the PGPR inoculation of plants leads to soil stabilisation and promotes soil fertility (Kohler et al. 2006). The study of the antagonistic and synergistic effects of different microbial inoculants when co-inoculated is a crucial step in the development of efficient host-microorganism combinations. The inoculation with rhizobacteria, alone or in combination with AMF, improves the physical properties of the soil, even under saline stress.

Inoculation with AMF is an effective method of increasing the capacity of host plants to establish and address stress situations, such as nutrient deficiency, drought and soil disturbance (Caravaca et al. 2003). In fact, several authors have indicated that AMF inoculation stimulates the absorption of water and nutrients, especially N and P (Jeffries et al. 2003; Folli-Pereira et al. 2012) in the plant or enhances the aggregation of eroded soils (Caravaca et al. 2002) to improve seedling performance. In return, the mycorrhizal plants provide fungus with photosynthetic C, which is delivered into the soil through fungal hyphae. Thus, the formation of mycorrhizae can affect the microbial population in the rhizosphere directly or indirectly through changes in exudation patterns or fungal exudates. Moreover, soil microorganisms might affect the formation and function of arbuscular mycorrhiza (AM). Growth-promoting mycorrhizal helper bacteria are known to stimulate the mycelial growth of AMF or improve the establishment of the mycorrhizal association (Toro et al. 1997).

The combined inoculation of beneficial microorganisms in the soil rhizosphere reduces the need for agricultural chemicals that are harmful to the environment; consequently, these microorganisms are gaining more attention for establishing sustainable agroecosystems. Indeed, microorganisms are active at the soil-plant interface, where microcosm systems such as the rhizosphere are developed (Cordier et al. 2000). Carbon flows are essential to the functioning of the rhizosphere. Many microbial interactions are responsible for key environmental processes, such as biogeochemical nutrient cycles and the maintenance of plant health and soil quality.

The effectiveness of microorganisms as modifiers of soil fertility and facilitators of plant development has been verified through the analysis of alterations in nutritional *status* and plant development. The combined inoculation of selected microorganisms in the rhizosphere has been recommended to maximise the growth and nutrition of plants. The study of the antagonistic and synergistic effects of different microbial inoculants when co-inoculated is a crucial step for the development of effective microorganism-host combinations. It has been reported that the double inoculation of *Glomus intraradices* and *Bacillus subtilis* promotes the establishment of the AMF and increases the plant biomass and P accumulation (Toro et al. 1997).

Inoculation with both growth-promoting bacteria and AMF produced decreased Na and increased K absorption in lettuce leaves, increasing the salinity tolerance of plants (Kohler et al. 2009). The PGPR strain *Pseudomonas mendocina* produces exopolysaccharides (Kohler et al. 2006) that bind to cations, including Na, thereby decreasing the content of Na available for absorption by plants.

AMF might influence bacterial communities in the soil, including PGPRs that are involved in soil aggregation through the exudation of carbon derived from photosynthesis in the mycorrhizosphere. However, the mechanisms underlying changes in the soil matrix and their significance for soil aggregation are poorly understood. Unlike AMF, which exert a strong influence on the scale of macroaggregates, rhizobacteria directly influence the formation and stabilisation of microaggregates. Thus, the AMF-mediated alteration of prokaryotic communities could indirectly influence aggregation processes at smaller scales than macroaggregates. In drought conditions, the formation of aggregates in the soil and the consequent soil stabilisation are essential for the increased accumulation of water in the soil, which consequently increases plant productivity during water stress. In addition, PGPRs and AMF produce phytohormones that contribute to increased development and root growth, and plant roots contribute to the stability of soil aggregates directly through the root "material" and indirectly through the stimulation of microbial activity in the rhizosphere.

Glomalin in Soil: The Importance of the Soil-Plant-Microorganism System

AMF are critical for the establishment and adaptation of plants in severely disturbed locations, including those contaminated with heavy metals (Vallino et al. 2006). They also affect the physicochemical characteristics of the substrate and contribute to the formation and maintenance of soil structure through the aggregation of soil particles, hyphal exudates, and residues. Moreover, mycorrhizal fungi produce glomalin, a protein extracted from soil as glomalin-related soil protein (GRSP) (Rillig 2004), which plays a key role in the stability of the soil (Bedini et al. 2009).

GRSP is an alkali-soluble protein material related to AMF (Rillig 2004; Nichols and Wright 2006), whose biochemical nature has not been elucidated. As fungal hyphae are shed (Driver et al. 2005), GRSP is transferred to the soil as a complex of repeat monomer structures connected by hydrophobic interactions (Nichols 2003), which bind to soil particles and stabilise aggregates (Rillig and Mummey 2006). In addition, GRSP contains bound iron (0.04–8.8 %) (Nichols 2003), but it does not contain phenolic compounds such as tannins (Rillig et al. 2001).

Glomalin contains approximately 60 % carbohydrates, comprising N linked to oligosaccharides. It also contains Fe, which is insoluble in water. Glomalin exhibits high hydrophobicity, which might contribute to the initiation of soil aggregation. The amount of immunoreactive glomalin extracted from the soil is directly proportional to soil aggregate stability in various regions of the world. Glomalin was detected in large amounts in many soils (Nichols 2003), which has been attributed to the fact that AMF colonise the roots of approximately 80 % of vascular plant species and have a global distribution. Large "pools" of glomalin might result from their high persistence in soil (Rillig et al. 2001).

Soil aggregates have also become an important protective environment for AMF hyphae. In degraded soils in recovery, improved aggregation is accompanied by an increased amount of colonised fine roots and hyphae that influence the geometric diameter of the aggregates. Because well-aggregated soils are less affected by erosion and more favourable for plant development, the effects of AMF on aggregation contribute to agricultural productivity and sustainability and to the conservation and functionality of natural ecosystems.

C losses in the soil result from leaching and erosion (Rillig et al. 2006). Stable soil structural units (aggregates) provide resistance to erosion. The importance of AMF in reducing erosion losses is related to their role in soil aggregation (Rillig 2004) and consequently in nutrient cycling through the reduction of carbon leaching in soils (Rillig et al. 2006).

Cations are bound to GRSP in quantities that vary in different soils (Nichols 2003; Chern et al. 2007). Recently, González-Chávez et al. (2004) clearly showed an increased binding capacity of GRSP to heavy metals (MTs) (Cu, Pb and Cd). Based on the results of his investigation, it has been suggested that this sequestration could be important for biostabilisation in soils contaminated with MTs. Bedini et al. (2009) showed that the amounts of Cu, Ni, Pb and Co bound to GRSP were, respectively 2.3, 0.83, 0.24, 0.24% of the total amount of MTs present in contaminated soil, thereby reducing the bioavailability of toxic elements and, consequently, plant stress. Vodnik et al. (2008) showed that GRSP represented 21.2% of the organic matter in soil contaminated with MTs, which was positively correlated with the concentrations of Pb and Zn in the soil; notably, the amount of lead bound to GRSP ranged from 0.69 to 23.4 mg g^{-1} DW GRSP, which represented 0.8–15.5% of the total Pb in the soil.

Wright et al. (1996) hypothesised that AMF secrete glomalin into the soil, which helps in soil aggregation. This model was directly based on the observed correlation between the GRSP concentrations with the stability of soil aggregates in water. The increase of soil aggregation would benefit both the host and associated AMF, justifying the energy "cost" of glomalin production. Experimental evidence, though obtained in an artificial manner, suggested that relations between the production of glomalin, soil aggregation and the enhancement of extraradicular AMF hyphae growth might indeed exist (Bedini et al. 2010). However, AMF also appear to produce GRSP in soils where organic matter is not the primary binding agent in the soil, and GRSP and soil aggregation are not correlated (Rillig et al. 2003). This finding suggests that the promotion of soil aggregation might not be the primary function of glomalin. In addition, the AMF communities and many other groups of soil biota profit from an improved soil structure (Niklaus et al. 2003), which makes it unlikely that the promotion of soil aggregation is the primary function of glomalin.

Using an *in vitro* sterile culture system, Driver et al. (2005) showed that most (80%) of the glomalin was contained in the fungal mycelium, rather than in the liquid growth medium. It is unclear if this result translates from the artificial aqueous culture system to the soil environment, or if it applies to fungi across the spectrum of AMF species. However, if it does, it suggests that a primary function of glomalin may be in the living fungus. Indeed, the putative function of glomalin is homologous to that of heat shock proteins. Based on these observations, Purin and Rillig (2007) proposed a new model for glomalin function. This model has the following key components: (a) glomalin primarily functions as chaperone in the cell. It is known that certain chaperones have the ability to act as a signal, resulting in greater thermotolerance and control of spore viability; (b) in the context of soil aggregation, the environmental function of glomalin is secondary to its primary physiological function.

There are few reports of heat shock proteins (Hsp) that act as chaperones in Glomeromycota, other than glomalin. Using the AMF species *G. intraradices*, Porcel et al. (2006) showed the expression of the small Hsp 30 improved plant tolerance to water stress.

Functional Diversity of AMF as a Determinant in the Ability of AMF to Increase Plant Tolerance to Stress Conditions

Arbuscular mycorrhizal fungi form the most common mutualistic relationship in nature with the roots of approximately 80% of terrestrial plants, with a presumed origin of approximately 460 million years ago (INVAM 2012). The intimacy of mycorrhizal associations provides a seamless morphological and physiological integration, resulting in increased functional compatibility. Fungal hyphae act as an extension of the plant root system, conferring increased absorption of water and nutrients to plants, while the plant provides the fungus with photo-assimilates, allowing it to complete its life cycle, which only occurs in the presence of the host in the case of AMF (Smith and Read 2008). Although this symbiosis is often considered to be mutualistic because the AMF receive carbon from the plant, the net effect on the plant capacity varies from mutualistic to parasitic (Kiers and van der Heijdenet 2006), depending on the ecological conditions and plant-fungus combinations.

Spores, fragments of colonised roots and the extraradicular mycelium of soils are the primary potential sources of inoculum, contributing the colonisation of plants. The relative contribution of each type of propagule to the colonisation of plant roots is difficult to determine.

Colonising ability (Avio et al. 2006) is used to describe the ability of AMF to propagate inside the plant roots. As such, it should be considered to be a measure of the "*steady state*", differing from the level of colonisation observed in a particular segment of the root at a given time. The dynamic colonisation process requires a continuous signal exchange during the growth of hyphae and roots. Different AMF can colonise a particular host species at the same level, whereas the symbiotic effectiveness, measured as the growth response, can vary substantially (Smith et al. 2004). Abiotic and biotic factors influence the symbiotic effectiveness between the two partners at the organismal and cellular level. At the *community level*, abiotic factors such as the availability of soil nutrients; the micro and macroclimate, including light and moisture (Staddon et al. 2003); and biotic factors such as community composition (Klironomos et al. 2000) indirectly influence symbiotic effectiveness. Interactions with pathogens and parasites affect carbon gain at the community level and the organismal levels.

Because of the lack of evidence for "taxonomic specificity", the different symbiotic responses of the host plant to the various AMF isolates suggest the existence of a "functional specificity" (Finlay 2004). This specificity is related to the balance between benefits and costs of the fungus for the host, which is sometimes attributed to differences in the colonisation degree or the efficiency of nutrient transport between fungus and plant. There may be a preferential fungus-plant association at a certain stage of plant development, which is modulated by the physiology and ecology of the plant through mechanisms of evolutionary convergence between symbionts (Pawlowska 2004). Therefore, a functional mycorrhiza results from the seamless morphological and physiological integration of partners, reflecting complex biochemical, genetic and physiological interactions and relationships that depend on the nature of the soil and the environment.

Although there is no evidence for host-specific AMF, there is evidence of functional specificity when considering the effects of these fungi on host plants (Pouyu-Rojas et al. 2006). As fungal isolates may vary depending on environmental changes, it is important to evaluate occurrence and functional diversity as critical factors in the structure of the plant community and ecosystem productivity (O'Connor et al. 2002).

Knowing the AMF community structure of a certain environment or biome and evaluating the functional diversity of these symbionts are critically important when trying to exploit the potential of these fungi. The functional diversity of arbuscular mycorrhizae (AMs) has often been defined in terms of responses in plant growth, which can range from negative to positive, depending on the particular plant-fungus combination and environmental conditions (Johnson et al. 1997). This functional diversity can be measured using the colonisation rate, absorption of nutrients and plant growth effects. Plants respond differently to different AMF, and these responses are observed both among the AMF of different species and among isolates of the same species (Munkvold et al. 2004; Smith et al. 2004).

Pouyu-Rojas et al. (2006) suggested the existence of selectivity and differentiated symbiotic compatibility, with preferred combinations in the formation of AMs and variable responses depending on the mycorrhizal genotype involved in the fungus-plant relationship. Some studies have shown that a given AMF species originates from the same soil and colonises different plant species with distinct sporulation patterns (Eom et al. 2000). In some cases, AMF that promote the host growth in one plant species could inhibit growth in another (Smith and Read 2008), and this beneficial or parasitic relationship depends on the fungus-plant combination and environmental conditions (Johnson et al. 1997; Smith and Read 2008).

Additionally, individual species of AMF can vary greatly in their response to the growth of different plant species, and variations can occur both among AMF isolates belonging to different species and isolates of the same species (Munkvold et al. 2004; Smith et al. 2004). Consequently, the presence or absence of certain AMF species influences structural changes in the population (Klironomos et al. 2000). For example, increasing the diversity of these fungi in the soil (Rillig 2004) influences the diversity, structure and productivity of the plant community (Heijden et al. 2004), in experimental studies performed in greenhouses and in natural ecosystems.

Inoculation with different AMF species differentially alters the growth and coexistence of different plant species (Heijden et al. 2003), and increasing the species richness of these fungi increases the diversity and productivity of plants. Santos (2008) verified the influence of AMF species richness in the soil community on the initial growth of tree species native to Brazil. The results of this study showed that the benefits of increased AMF richness are greater when plants are grown in complex communities with a considerable amount of competition. However, those studies used a single individual as a representative of each AMF species, and each culture was initiated from a single spore. Therefore, these results cannot indicate whether intraspecific variation could account for the differences observed between isolates of the same species.

Hart and Klironomos (2002) showed that variation in plant growth was greater among plants inoculated with different AMF species than among those inoculated with different isolates of the same species. However, this finding does not prove that the variability within isolates is not ecologically important. More recently, considerable variability has been observed within AMF species. Munkvold et al. (2004) demonstrated that large differences in plant growth and P absorption were observed within AMF species, showing the importance of the ecological potential of variability within AMF species. Hart and Reader (2002), tested the effect of 21 AMF isolates on plant growth, and showed that AMF families also differ in the benefits conferred to host plants, although there is great variability within and between AMF species and genera. These studies show that there is considerable functional diversity in AMF and that variability within an AMF species can be greater than between different AMF species or genera. This functional diversity is important for individual plant growth and the composition of plant communities. Thus, it is clear that there is great functional diversity in AMF and that the increase in the diversity of these fungi in soil (Rillig 2004) influences the diversity and productivity of the plant community.

It is unclear whether plants utilise this diversity to select efficient AMF or AMF combinations that are more beneficial in terms of the stimulation of their growth (Heijden et al. 2004). Thus, it is important to determine whether increasing the AMF diversity in soil influences plants and which plant-fungus combinations occur preferentially and effectively. Moreover, it would be important to determine whether the inoculation of plants with a mixture of AMF reconstitutes the AMF community observed in nature. Studies like these can be used to monitor plant performance and reveal whether the diversity of the AMF species in plant roots is linked to functional diversity.

Almost all data on the variability of the functions or functional diversity of AMF were obtained from experiments in which plants have been inoculated with an AMF isolate and the plant growth or total P absorption was measured. Such experiments are not entirely relevant to field situations when more than one AMF species is generally present in a single root system (Jansa et al. 2003). Currently, the challenge is to establish mixed communities using different AMF species to assess whether plants are able to select efficient AMF or AMF combinations that are complementary in their functions. However, such studies are complex because of the difficulty to identify the AMF that are colonising the roots, which becomes a limiting factor for understanding the control of these relationships. The consequences of the simultaneous colonisation of a plant by functionally different AMF have been little explored.

If a plant is colonised by AMF species that are complementary in their functions, for example the absorption of nutrients from different soil "pools", they can be more beneficial to the plant as a mixture than any one species separately (Alkan et al. 2006; Gustafson and Casper 2006). Johnson et al. (2004) showed that the diversity of AMF in the roots of *Plantago lanceolata* was positively correlated with the concentrations of P and N in the shoots of the plants. However, other studies indicate that the maximum benefits to the plants could be achieved with a single efficient AMF species and that increasing the mycorrhizal diversity would not result in greater benefits for the plants. According to Santos (2008), increasing AMF diversity in the community present in the soil can increase the chances for the establishment of a fungus species that is more efficient for plant growth. Thus, it is important to characterise the community structure of AMF in a particular environment or ecosystem and assess the functional diversity of these symbionts to establish whether there is a relationship between the AMF diversity and benefits to plants.

Physiological Aspects of the Arbuscular Mycorrhizal Association in the Plant Tolerance to Stress

The establishment of AMs confers the plants with a range of benefits, primarily through the extraradicular mycelium of the fungus that facilitates the absorption of nutrients from areas located beyond the depletion zone of the roots, particularly phosphorus, and it increases the availability and translocation of nutrients to cortex cells in the plant roots. Other relevant effects of AMF are increased plant resistance to pathogens of the root system and water absorption capacity. In soil, they favour aggregate formation and stability, not only through the physical action of the mycelium but also through the action of glomalin. Through the enhancement of the hydric and nutritional status of plants, AMF can contribute the increased tolerance to environmental stress conditions.

Thus, the symbiosis between plants and AMF also results in the reduction of losses by stresses (Munier-Lamy et al. 2007) and consequently faster growth, with economic inputs and the reduction of environmental contamination (Huang et al. 2009). Furthermore, these fungi can act as potential biological control agents, reducing the effects or damage caused by plant pathogens through indirect means, enhanced plant nutrition, or increased resistance in the root system.

The plant response to colonisation by AMF depends on the severity and frequency of drought, and other soil conditions. AMF can affect the growth and productivity of the host plant under conditions of high and low humidity (Borowicz 2010). Thus, symbiosis could increase plant responses to moderate water deficit through various mechanisms, including increased water absorption from the soil through the hyphae (Augé et al. 2003), alteration of hormonal levels causing changes in stomatal conductance (Augé et al. 2008), increased leaf turgor and osmotic potential reduction (Wu et al. 2006), and improved nutrition of the host plant (Chen et al. 2005).

Mycorrhizal plants develop a root system uses carbon more efficiently. Consequently, these plants convert larger quantities of photosynthates in the root development to increase their absorption capacity (Neocleous and Vasilakakis 2007).

The chlorophyll concentration in the leaves is an important physiological index for determining the degree of photosynthesis in plants. AMF can increase the chlorophyll concentration in the leaves. Indeed, mycorrhizal plants growing under stress conditions possess greener leaves, suggesting that salt interferes with the synthesis of chlorophyll (Colla et al. 2008). Mycorrhizal inoculation also increases the absorption of phosphorus and magnesium and reduces the sodium content in the plant, which in turn increases the chlorophyll content and consequently improves the overall performance of mycorrhizal plants under stress conditions (Sheng et al. 2008).

Plants associated with AMF often have greater resistance to saline stress, perhaps with greater consistency than the stress due to drought. Salinity negatively affects the formation and functioning of the mycorrhizal symbiosis (Sheng et al. 2008). Studies indicate that AMF can increase plant growth and nutrient absorption, reduce losses in productivity under salinity conditions and improve the tolerance to salinity (Hajiboland et al. 2010). The colonisation of plant roots by some AMF is reduced in the presence of NaCl (Giri et al. 2007), potentially due to the direct effect of NaCl on the fungi (Juniper and Abbott 2006), indicating that salinity can inhibit the formation of mycorrhiza (Sheng et al. 2008).

Many researchers have reported that AMF increases the ability of plants to address saline stress (Jahromi et al. 2008) because of the enhanced absorption of nutrients by the plants (Asghari et al. 2005) and the ionic equilibrium (Giri et al. 2007), which protects enzymatic activity (Rabie and Almadini 2005), and facilitation of water absorption. However, there are few studies concerning the influence of mycorrhizal inoculation on photosynthesis and water relations during saline stress. Some reports indicate that mycorrhizal colonisation can improve the relative water content in squash leaves (Colla et al. 2008), hydric potential and photosynthesis of maize plants (Sheng et al. 2008), and chlorophyll concentration in the leaves of various plant species (Sannazzaro et al. 2006; Colla et al. 2008).

Recent findings suggest that glomalin might indirectly influence the storage of carbon in the soil through the stabilisation of soil aggregates (Zhu and Miller 2003) and soil stability. The stability of soil aggregates is one of the most important properties to control plant growth in arid and semi-arid environments through the control of the soil-plant hydric status.

The establishment of mycorrhizal associations results in increased tolerance of plants to environmental stresses (Tang et al. 2009). However, little is known about the physiological and molecular mechanisms responsible for this greater tolerance. Increased activity and induction of new isoenzymes that participate in the antioxidant system in inoculated plants allow the plant to tolerate excess superoxide radicals generated during the prevalence of stress conditions.

Salinity induces oxidative stress in plants (Hajiboland and Joudmand 2009). Plant cells contain an array of protection mechanisms and repair systems that can minimise the occurrence of oxidative damage caused by reactive oxygen species (ROS) (Abdel Latef and Chaoxing 2011). The induction of enzymes that eliminate ROS such as superoxide dismutase (SOD), catalase (CAT), peroxidase (POD) and ascorbate peroxidase (APX) is the most common mechanism for detoxifying the ROS synthesised during a stress response.

Information about the response of the antioxidant defence system under conditions of stress in mycorrhizal plants is contradictory: an increase, lack of change, and even a decrease in SOD, CAT, APX and POD activity were reported in mycorrhizal soy (Porcel et al. 2003) subjected to hydric stress and tomatoes subjected to salinity (He et al. 2007; Hajiboland et al. 2010).

Under conditions of hydric deficit, plants attempt to maintain their water content by accumulating compatible, non-toxic solutes such as proline and glycine betaine, which do not interfere with the normal physiological processes of the plant (Ma et al. 2006; Zhang et al. 2008). The accumulation of these solutes is a sensitive physiological index of plants in response to salt and other stresses (Peng et al. 2008). For plants to survive under conditions of salinity and water, adjusting the leaf osmotic potential is important and requires intracellular osmotic balance. Thus, under hydric and salinity stresses, plants accumulate some organic solutes (proline, soluble sugars, glycine betaine, among others) and inorganic ions to maintain greater osmotic adjustment (Yang et al. 2009). The presence of AMF in the roots could modify the osmotic potential of the leaves and influence the carbohydrate composition and proline level.

Proline is the most common compatible osmolyte in plants and plays an important role in increasing the adaptation of plants to drought and salinity (Hasegawa et al. 2000). In addition to osmotic adjustment, this molecule has other proposed functions in osmotically stressed plant tissues: it maintains and protects the integrity of the plasma membrane (Hincha and Hagemann 2004), acts as a source of carbon and nitrogen, and eliminates hydroxyl radicals. Proline accumulation in mycorrhizal plants subjected to drought has been reported, and the variable effects of mycorrhizal colonisation on the levels of proline in plants under saline stress have been observed. However, to date, there is little information concerning the influence of colonisation by arbuscular mycorrhizal fungi on this accumulation (Sannazzaro et al. 2007).

Despite the accumulation of proline induced in plants under stress (Andrade et al. 2009), evidence of the effects of mycorrhizal symbiosis on the levels of proline or soluble amino acids are scarce or null under stress conditions. Andrade et al. (2010) observed that soluble amino acids and the proline content of the leaves of mycorrhizal and non-mycorrhizal bean plants increased in response to the addition of Cu to soil, suggesting a stress response similar to the excess of this metal in the soil. However, proline accumulation in the leaves of mycorrhizal plants showed a more pronounced increase in response to Cu in the soil when compared to homologous non-mycorrhizal plants, indicating a possible role of this amino acid in the response to Cu toxicity in mycorrhizal plants, which exhibited greater biomass accumulation than non-mycorrhizal plants.

Moreover, AMF increase the vigorousness of the root system and stimulate the production of hormones by plants (Yao et al. 2005). Thus, the increase in plant tolerance to hydric and/or saline stress might be related to the increased expression of genes in response to stress.

Glycine betaine acts as a protective non-toxic osmolyte during periods of drought in many organisms, including algae, bacteria, large plants and animals (Treberg and

Driedzic 2007). It is synthesised at high levels in many plant species in response to several types of environmental stresses, acting not only as an osmoprotectant but also as a stabiliser of proteins and membranes (Oishi and Ebina 2005). This compound appears to be a critical determinant of stress tolerance. Its accumulation is induced under stress conditions, and this accumulation is correlated with the tolerance level (Wu et al. 2008).

The role of arbuscular mycorrhizae in attenuating the stress caused by heavy metals in plants growing in contaminated soils has been recognised (Göhre and Paszkowski 2006). Improvement of the nutritional status and reduced or altered metal absorption are among the greatest benefits related to mycorrhizae on host plants under heavy metal stress (Andrade et al. 2008). AMF can alter the concentration of metals in plants through the immobilisation of metal in the cell wall of intra or extraradicular hyphae, metal chelation through compounds secreted by AMF such as glomalin (Vodnik et al. 2008), or metallic compartmentalisation in fungal cells. Thus, these fungi act as a filter for metal, reducing local concentrations in the soil and creating a suitable environment for plant growth in soils contaminated with metals (Göhre and Paszkowski 2006).

The mycorrhizal association can alter the metal absorption in plants (Andrade et al. 2008), with reports of both the increase and reduction of metal concentrations in plant tissues. As a consequence of physiological alterations, mycorrhizal plants perform better under metal stress conditions (Paradi et al. 2003). At the molecular level, the expression of some genes related to plant tolerance to heavy metals is altered through arbuscular mycorrhizal symbiosis. However, the global mechanisms by which the fungi reduce the phytotoxicity of the metal in plants have not been fully elucidated, and the results f some studies show conflicting results, depending on specific plant/fungal species/metal interactions.

The literature has reported several detoxification mechanisms in plants, but the mechanisms associated with AMF vary among plant species. Variation is also observed for the metal used, applied concentration, plant organ, and duration of the exposure (Gratão et al. 2008).

Abiotic Stress and its Influence on the AMF Community in the Soil

Mycorrhizae are complex symbioses formed by several components that determine the colonisation rate, incidence of propagules and the effects and functions of the symbiosis for plants and ecosystems. The primary components of this system are the fungus, the plant, and the environment (soil), which have strong interrelationships and interdependences.

Arbuscular mycorrhizae (AMs) are of widespread occurrence in superior plants, and AMF are prevalent among fungi normally found in the rhizosphere or among root colonisers. It is estimated that most plant species (approximately 250,000 species) are capable of forming AMs. Therefore, this association has widespread occurrence, except in plants that are members of the following families: Brassicaceae, Amarantaceae, Comelinaceae, Juncaceae, Proteaceae, Poligonaceae, Cyperaceae and Chenopodiaceae. Approximately 87 % of the Cruciferae (Brassicaceae), 67 % of the Chenopodiaceae, 37 % of the Poligonaceae and 4 % of the legumes do not form AMs. Surveys conducted in various regions of the world confirm that AMs are much more abundant than ectomycorrhizae and occur in most Phanerogams (97 %), including almost all species of agronomic and pastoral interest and forest species native to the tropics.

The richness of AMF varies greatly, and two to 33 species per ecosystem have been identified. Although several studies have been conducted, the wealth, diversity and symbiotic potential of AMF populations in Brazilian ecosystems have not yet been sufficiently studied. The occurrence of AMF in the country includes surveys conducted in various crops and non-cultivated ecosystems. Many of them reveal the richness of the species, with many of them that have not yet been identified (approximately 20 % of the species observed).

The cultivation of soil and the imposition of environmental stress cause major changes in the structure of fungal communities through changes in the distribution and dominance of the species. These effects are due to biotic and abiotic changes in the edaphic environment, such as changes in the vegetation (roots) and chemical properties of the soil, especially in the components of acidity, availability of nutrients, water, salinity and heavy metal contamination. Propagules of these fungi are present in almost all soils, and the type of vegetation and environment determine the occurrence and degree of root colonisation. AMF have reduced occurrence or are absent in soils that are fumigated, severely disturbed by erosion, subject to mining, in areas of civil construction, under long fallow period or flooding, and those cultivated for long periods with non-host species and high concentrations of environmental pollutants.

The presence of heavy metals at toxic concentrations in the soil greatly influences the AMF. The excess metals reduce spore germination, mycelia growth, degree of colonisation and sporulation of these fungi, causing a significant impact on their ecology and diversity (Klauberg-Filho et al. 2005). Despite these effects, more than 30 species of AMF have been identified in contaminated soils worldwide and some at high frequencies, such as *Paraglomus occultum*, *G. clarum*, *G. intraradices* and *Scutellospora pellucida*, in addition to abundant colonisation and sporulation. Even at high concentrations of toxic metals, increased colonisation rates and spore densities have been reported (Gaur and Adhoeya 2004); however, in soils contaminated with Cd, Zn, Cu and Pb, the species richness decreases with the increasing concentration of these metals in the soil (Klauberg-Filho et al. 2002). The presence of heavy metals inhibits spore germination and mycelial growth, reducing the mycorrhizal colonisation of plants. Several studies provide evidence of the different AMF behaviours in relation to excess metals in the soil, and several isolates are known to be tolerant to multiple metal contaminants in the soil. Considering the importance of these fungi in the ecology of plants, isolates tolerant to heavy metals are of great interest in the revegetation of areas degraded by the accumulation of these elements.

The "arable" layer of the soil is where the absorbing roots of plants are concentrated, becoming the primary habitat and reservoir of AMF propagules in ecosystems. Any factor impacting this layer will exert a great influence on the AMF community. Weissenhorn et al. (1993) and Weissenhorn et al. (1994) evaluated the tolerance of isolates of *Glomus mosseae* obtained from adjacent areas that were polluted or not polluted with heavy metals (Cd and Zn), and also in relation to a reference isolate maintained in the laboratory. Germination tests showed that the isolates obtained from contaminated areas showed greater tolerance to heavy metals than the isolates from adjacent uncontaminated areas. This result demonstrates that different isolates of the same "species" are functionally distinct and suggest that AMF have the ability to adapt to anthropogenic changes.

The AMF responses to heavy metals are diverse at the fungus species level (Hildebrandt et al. 2007). For example, *Glomus etunicatum* was more sensitive to Cd, Pb and Zn than *G. intraradices* (Pawlowska and Charvat 2004), and the *G. mosseae* isolated from soils polluted with heavy metals was more tolerant to Cd than the same species isolated from a non-polluted substrate (Weissenhorn et al. 1994). An adequate understanding of the AMF community under stress by heavy metals could contribute to the recognition of the interactions between fungi and heavy metals and future revegetation or phytoremediation of regions polluted by heavy metals (Hildebrandt et al. 2007). Studies indicate that many species of plants growing well in areas polluted by a single heavy metal, such as *Fragaria vesca*, *Viola calaminaria*, *Veronica rechingeri*, *Solidago giante*, *Thymus polytrichus* and *Thlaspi praecox*, were colonised by various AMF and AMF isolates that can positively act to regulate plant resistance to heavy metal stress (Zarei et al. 2008; Sonjak et al. 2009).

Studies conducted with AMF in preserved and disturbed areas show the importance of these fungi in the studied areas (Silva et al. 2005). Silva et al. (2005) identified 15 species of AMF in an area of preserved caatinga and an area degraded by copper mining and observed a strong reduction in plant diversity and AMF species community; the community was quantitatively and qualitatively affected by mining activity. In areas of high salinity, Yano-Melo et al. (2003) identified 21 taxa of AMF, especially *G. mosseae* and *G. intraradices*, which favoured sporulation in the first cycle of multiplication in a trap culture and decreased from the second cycle.

Salinity stress negatively affects the formation and function of mycorrhizal symbiosis by inhibition of spore germination, plant colonisation and formation of new spores (Juniper and Abbott 2006; Giri et al. 2007; Abdel Latef and Chaoxing 2011). Other environmental factors such as soil water content, concentrations of available phosphorus, organic matter content in the soil, soil pH and vegetation coverage affect levels of colonisation in plants through AMF. AMF in wetland habitats are periodically exposed to anaerobic conditions and high salinity in soils (Bohrer et al. 2004; Carvalho et al. 2004). Depending on the AMF species, soil salinity levels can affect spore production and germination (Carvalho et al. 2004). The presence of heavy metals in toxic concentrations in the soil also exert great influence on the AMF, and the excess metal reduces spore germination, mycelial growth, degree of colonisation, and sporulation of these fungi, which might have a significant impact on their ecology and diversity (Klauberg-Filho et al. 2005).

Because mycorrhizae are compartmentalised biological systems, they are influenced by the effects of the environment and countless edaphic factors of each component that directly or indirectly regulate the formation, operation, and occurrence of AMs. The components and controlling factors have constant and intense interactions, such that a change in any of these factors influences the occurrence of mycorrhizae and AMF propagules.

Establishment of AMF in Extreme Temperature Conditions

Arbuscular mycorrhizal fungi (AMF) may respond to high temperature conditions by changing their morphology, modifying their external environment, or adapting their internal metabolism, although the degree of phenotypic plasticity might vary. Because AMF obtain carbon from autotrophic host plants, fungi can also be exposed to stress through changes in carbon allocation from the host plant. Results obtained from the refinement and application of molecular identification methods in recent years has revealed that the degree of host specificity by some mycorrhizal fungi might be greater than expected. This result implies that the availability of compatible roots influences the survival of the fungus and changes in species composition in plant communities. Restricting the supply of assimilates from the compatible host root could limit the growth of certain fungi in rehabilitation areas. Therefore, in many situations, mycorrhizal colonisation appears to be more dependent on the host plant than on the temperature (Hawkes et al. 2008), and normally high temperatures such as 35 and 40 °C show no significant effects on mycorrhizal development (Zhu et al. 2010).

There are few studies that examine in detail the factors that affect the survival of specific AMF in their natural habitats. Instead, the effects of physical-chemical factors, especially temperature, on plants are widely reported.

Few plant species survive under continuous temperatures above 45 °C. Both photosynthesis and respiration are inhibited at supra-optimal temperatures. However, as the temperature increases, the photosynthetic rate decreases more rapidly than the respiration.

The structure and stability of cell membranes are important during high temperature stress. The excessive fluidity of lipid membranes at elevated temperatures is correlated with the loss of physiological function. In some species, the acclimatisation to high temperatures is associated with the increased saturation of fatty acids in the lipids, which makes membranes less fluid. The strength of hydrogen bonds and electrostatic interactions between the polar groups of the aqueous phase of the membrane decreases, which results in a stronger association between integral proteins of the membrane and its lipid phase. Thus, high temperatures modify the composition and structure of membranes, resulting in the loss of ions and the inhibition of metabolic processes such as photosynthesis and respiration.

One aspect common to fungi and plants when subjected to high temperature stress is the generation of reactive oxygen species. The uncontrolled accumulation of ROS generates oxidative stress and can cause membrane lipid peroxidation, inactivation of enzymes containing SH groups, and RNA and DNA damage. ROS, particularly the superoxide radical (O_2^-) and hydrogen peroxide (H_2O_2), are generated in the cytoplasm, chloroplasts, mitochondria, peroxisomes, and apoplast.

In microorganisms, particularly AMF, the components of antioxidant systems are not well known. It is known, however, that enzymes such as catalase, peroxidise, and superoxide dismutase participate in the decomposition of ROS. Historically, research has shown that the establishment of mycorrhizal associations result in increased plant tolerance to adverse environmental factors, although many of the effects are attributed to enhanced plant nutrition associated with the AMF. The increased activity and induction of new isoenzymes that participate in the antioxidant system in mycorrhizal plants confers tolerance to excess superoxide radicals generated during the prevalence of stress conditions (Costa 2003). In arbuscular mycorrhizal associations (*Trifolium pratense–G. mosseae*), there is an increase in the activity and synthesis of new SOD isoenzymes induced through symbiosis (Palma et al. 1993).

Through the involvement of oxidative stress enzymes against oxidative damage caused by the increased production of ROS during stress conditions, AMF can increase the capacity to resist oxidative and environmental stresses in the plant, conferring increased tolerance to ROS, although the role of these enzymes in mycorrhizae is little elucidated.

Arbuscular Mycorrhizal Fungi in Plants Tolerance to Nematode Attacks

The protection of plants against abiotic stress caused by pathogens in soil can also be attributed to AMF (Moraes et al. 2004; Meira 2004; Hol and Cook 2005; Elsen et al. 2008; Meira-Haddad 2008; Vos et al. 2012).

AMF can be considered to be biocontrol agents (Azcón-Aguilar and Barea 1996; Pozo et al. 2002) and have received much attention for promoting resistance and/or tolerance, decreasing the incidence and severity of plant diseases, and reducing the number of soil pathogens (Cordier et al. 1998; Hol and Cook 2005; Borges et al. 2007; Meira-Haddad 2008; Vos et al. 2012).

According to Azcón-Aguilar and Barea (1996), AMF promotes the following mechanisms for biological control of plant diseases: improving the nutritional status of the host plant, compensation for damage caused by the pathogen, competition for the site of infection and colonisation site, anatomical and morphological changes in the root system of the host, changes in the microbial population of the rhizosphere, and activation of systemic and localised defence mechanisms (Pozo et al. 2002).

Several studies have demonstrated that AMF affect the reproduction of nematodes by reducing oviposition, the number of individuals in the roots of infected plants, and the number of galls and by increasing plant tolerance to pathogen attack through reductions in their development (Elsen et al. 2003; de La Peña et al. 2006; Jaiti et al. 2008; Meira-Haddad 2008).

Populations of *Pratylenchus coffeae, Radophulus similis, Meloidogyne javanica*, among others, were reduced by *G. mosseae* and *G. intraradices* when associated with banana roots (Pinochet et al. 1996; Elsen et al. 2003). The mycorrhizal fungi might affect nematode reproduction by reducing galls and eggs and inhibiting penetration (Siddiqui and Mahmood 1995; Meira-Haddad 2008; Vos et al. 2012). These diverse effects indicate that this interaction is specific, and the plant genotype, nematode species, fungal isolates, and changes in the environment could explain these different responses (Siddiqui and Mahmood 1995; Hol and Cook 2005; Borges et al. 2007; Jaiti et al. 2008).

Using the root compartmentalisation system of two banana cultivars, it was possible to demonstrate the direct effect of AMF on *M. incognita* (Meira-Haddad 2008). When AMF and nematodes were inoculated in the same compartment, a reduction in the number of eggs in the cv. Prata-anã and both the number of eggs and galls in the cv. FHIA 01 was observed (Meira-Haddad 2008). Studies have shown that plants colonised by AMF have few galls containing few females, and the nematodes are smaller in size. Because of this characteristic, they need more time to develop to adulthood (Diedhiou et al. 2003; Freire et al. 2007; de La Peña et al. 2006).

Cells of the root system with arbuscules were not infected with the pathogens, and its proliferation was reduced in the mycorrhizal roots and also in parts of the root system, demonstrating that bioprotection is directly linked to AMF root colonisation. Cytomolecular studies have shown that the systemic and localised protective effect induced of AMF colonisation involves the accumulation of defence molecules in combination with the elicitation reaction of the host cell wall. Modifications of the cell wall associated with localised resistance and the formation of papillae characterises systemic resistance to *P. parasitica* in mycorrhizal tomato plants (Cordier et al. 1998).

The changes in the root system caused by the AMF promote the vigorous growth of the plant, thus reducing the negative effect of the pathogen (Siddiqui and Mahmood 1995). The reduction of plant growth due to the establishment of nematodes in roots is lower when colonised by mycorrhizal fungi (Cofcewicz et al. 2001).

The success of the plant defence system against pathogens depends primarily on the recognition of invasion by the pathogen in the initial stages for the activation of defence response cascades. Plants in mycorrhizal symbiosis undergo biochemical, physiological and molecular alterations related to the plant defence system for the establishment of symbiosis (Garcia-Garrido and Ocampo 2002). However, the plant defence responses are limited, transient, and restricted to specific cells; however, the reactions share similarities with the physiological reactions observed during colonisation by pathogens (Garcia-Garrido and Ocampo 2002; Lambais et al. 2003). The mycorrhizal colonisation acts as the primary system of plant defence to pathogen attack (Elsen et al. 2008).

The physiological changes in the plant caused by the symbionts prevent pathogen attack and activate defence mechanisms because proteins related to pathogenesis, phytoalexin production, and cell wall lignification have been reported in mycorrhizal plants at regions far from the infection sites, indicating the occurrence of systemic resistance (Cordier et al. 1998; Pozo et al. 2002; Selosse et al. 2004).

The successful establishment of mycorrhiza is essential for the control of nematodes and has a negative effect on the reproduction of these organisms (Cordier et al. 1998; Elsen et al. 2003; Vos et al. 2012).

The bioprotector effect of the AMF to plant pathogens might be related to the induction of localised or systemic resistance (Cordier et al. 1998; Pozo et al. 2002; Elsen et al. 2008). When colonised by AMF, plants exhibit biochemical, physiological and molecular alterations related to the plant defence system as symbiosis is established (Garcia-Garrido and Ocampo 2002; De Gara et al. 2003; Selosse et al. 2004). However, the plant defence responses are limited, transient, and restricted to specific cells, but the plant reactions have physiological similarities with the reactions observed during colonisation by pathogens (Garcia-Garrido and Ocampo 2002; Lambais et al. 2003). Mycorrhizal colonisation acts as the primary system of plant defence against pathogens (Elsen et al. 2008). Plants with higher antioxidant activities are more tolerant to different stresses, and mycorrhizae increase the activity of antioxidant enzymes, such as peroxidise, catalase and superoxide dismutase (Costa 2003; Lambais et al. 2003; Meira 2004; Arfaoui et al. 2007).

Peroxidase is an enzyme that is transiently induced and subsequently suppressed during mycorrhizal colonisation. The peroxidases catalyse the oxidative polymerisation of phenylpropanoids for the production of lignin and are involved in cross-linking the proteins of the cell wall, thus contributing to increased rigidity (Siddiqui and Mahmood 1995; Hol and Cook 2005; del Río et al. 2006). Consequently, hydrogen peroxide plays an important role in strengthening the cell wall and the systemic induction of defence genes (del Río et al. 2006).

Phenylalanine ammonia lyase is a key enzyme in the phenylpropanoid pathway, which is responsible for the deamination of L-phenylalanine to form *trans*-cinnamic acid and ammonia. *Trans*-cinnamic acid is incorporated in different phenolic compounds to produce phytoalexins, which are antimicrobial compounds that are closely related to the resistance of plants to pathogens (Wuyts et al. 2006; Arfaoui et al. 2007).

The study of the mechanisms involved in the AMF bioprotection to nematodes has been limited due to obligatory biotrophism and parasitism of both. AMF can be considered to be biological control agents; however, the diversity of responses to the combination AMF-nematode and plants is unique. Generalisations are hindered because these interactions are dependent on the host, nematode species, AMF species and combinations of nematode and AMF initial inoculum density, soil fertility and nematode inoculum.

Conclusions and Future Perspectives

The various interactions that occur between plant roots and microorganisms in the soil are of importance to ecosystems. The understanding of these interactions can greatly benefit agriculture through the manipulation of populations of common microorganisms that inhabit the soil associated with roots, which is a promising area of research that can be an effective option for increasing plant tolerance to biotic and abiotic stresses. Beneficial microorganisms that promote plant growth and confer a protective effect against soil pathogens are considered to be of great value to production systems.

Despite of more than 120 years, since the first descriptions and hypotheses about the functionality of the mycorrhizal associations, it is suspected that the deeper impact of this symbiosis is yet to be revealed.

The effort by the potentiation of AMF in the field, as well as by the generation of related techniques, demand studies incorporating multiplication protocols of AMF. Implies consider this component in long-term studies that seek to detect not only its impact on the growth and development of a plant, but about the magnitude of its contribution to global events and structure of plant communities.

With the perspective opened by molecular techniques, there are the opportunity to understand mechanisms of evolution of plant species and the symbiosis. It remains to researchers in AM extend your range of research in a multidisciplinary effort, even because, without this approach, it will not possible to understand the full dimension of this formidable symbiosis.

The AMF and the symbiotic association process require an interaction between root and fungi, so far not clarified with regard to the mechanism for the recognition of symbiotic partners and interaction, or the moment from which it is recognized as an association.

The symbiosis between plants and AMF results in the reduction of losses by stresses and consequently faster growth, with economic inputs and the reduction of environmental contamination. Furthermore, AMF can act as potential biological control agents, reducing the effects or damage caused by plant pathogens through indirect means, enhanced plant nutrition, or increased resistance in the root system.

Little is known about the physiological and molecular mechanisms responsible for greater tolerance of mycorrhizal plants. The knowledge that one of the physiological responses to biotic and abiotic stress in plants mediated by AM consists in the increase of the activity of an oxidative stress set of enzymes (SOD, APX, POD), in the production of compounds with antimicrobial activity (phenols, quinones, *phytoalexins*) and the activation of enzymes that catalyze the reactions for the production of compounds that act as chemical or physical barriers. Furthermore, increased activity and induction of new isoenzymes that participate in the antioxidant system in inoculated plants can allow the plant to tolerate excess superoxide radicals generated during the prevalence of stress conditions.

Know the community structure of the AMF of a particular environment or biome, and evaluate the functional diversity of these symbionts are of fundamental importance when you want to explore the potential of these fungi to increase the plants tolerance to biotic and abiotic stress. There are few works that examine in detail the factors that affect the survival of AMF in their natural habitats. On the contrary, the effects of physical and chemical factors on plants are widely reported.

Studies on the factors that may regulate the establishment and functioning of AM and changes biochemical, physiological and molecular in both symbionts during biotic and abiotic stress conditions have been made in recent years. However, the elucidation of these mechanisms is still far from being completed. The limited information about the genetics of the AMF and the difficulties encountered in carrying out these studies, which is hampered by the obligatory symbiotic relationship that is required and by the complexity of fungal genomics, has contributed to limiting the knowledge of that symbiosis.

References

Abdel Latef AAH, Chaoxing H (2011) Effect of arbuscular mycorrhizal fungi on growth, mineral nutrition, antioxidant enzymes activity and fruit yield of tomato grown under salinity stress. Sci Horticul 127:228–233

Alkan N, Gadkar V, Yarden O, Kapulnik Y (2006) Analysis of quantitative interactions between two species of arbuscular mycorrhizal fungi, *Glomus mosseae* and *G. intraradices*, by real-time PCR. Appl Environ Microbiol 72(6):4192–4199

Andrade SAL, Gratão PL, Silveira APD, Schiavinato MA, Azevedo RA, Mazzafera P (2009) Zn uptake, physiological response and stress attenuation in mycorrhizal jack bean growing in soil with increasing Zn concentrations. Chemosphere 75:1363–1370

Andrade SAL, Gratão PL, Azevedo RA, Silveira APD, Schiavinato MA Mazzafera P (2010) Biochemical and physiological changes in jack bean under mycorrhizal symbiosis growing in soil with increasing Cu concentrations. Environ Exp Bot 68:198–207

Andrade SAL, Silveira APD, Jorge RA, de Abreu MF (2008) Cadmium accumulation in sunflower plants influenced by arbuscular mycorrhiza. Int J Phytorem 10:1–13

Arfaoui A, El hadrami A, Mabrouk Y, Sifi B, Boudabous A, El hadrami I, Daayf F, Chérif M (2007) Treatment of chickpea with *Rhizobium* isolates enhances the expression of phenylpropanoid defense-related genes in response to infection by *Fusarium oxysporum* f. sp. *ciceris*. Int J Phytorem 45:470–479

Artursson V, Finlay RD, Jansson JK (2005) Combined bromodeoxyuridine immunocapture and terminal restriction fragment length polymorphism analysis highlights differences in the active soil bacterial metagenome due to *Glomus mosseae* inoculation or plant species. Environ Microbiol 7:1952–1966

Artursson V, Finlay RD, Jansson JK (2006) Interactions between arbuscular mycorrhizal fungi and bacteria and their potential for stimulating plant growth. Environ Microbiol 8:1–10

Asghari H, Marschner P, Smith S, Smith F (2005) Growth response of *Atriplex nummularia* to inoculation with arbuscular mycorrhizal fungi at different salinity levels. Plant Soil 273:245–256

Augé RM, Moore JL, Cho K, Stutz JC, Sylvia DM, Al-Agely A, Saxton AM (2003) Relating dehydration resistance of mycorrhizal *Phaseolus vulgaris* to soil and root colonization by hyphae. J Plant Physiol 160:1147–1156

Augé RM, Toler HD, Sams CE, Nasim G (2008) Hydraulic conductance and water potential gradients in squash leaves showing mycorrhiza-induced increases in stomatal conductance. Mycorrhiza 18:115–121

Avio L, Pellegrino E, Bonari E, Giovannetti M (2006) Functional diversity of arbuscular mycorrhizal fungal isolates in relation to extraradical mycelial networks. New Phytol 172:347–357

Azcón-Aguilar C, Barea JM (1996) Arbuscular micorrizas and biological control of soil-borne plant pathogens—an overview of the mechanisms involved. Mycorrhiza 6:457–464

Bago B, Pfeffer PE, Abubaker J, Jun J, Allen JW, Brouillette J, Douds DD, Lammers PJ, Shachar-Hill Y (2003) Carbon export from arbuscular mycorrhizal roots involves the translocation of carbohydrate as well as lipid. Plant Physiol 131:1496–1507

Bedini S, Pellegrino E, Avio L, Pellegrini S, Bazzoffi P, Argese E, Giovannetti M (2009) Changes in soil aggregation and glomalin-related soil protein content as affected by the arbuscular mycorrhizal fungal species *Glomus mosseae* and *Glomus intraradices*. Soil Biol Biochem 41:1491–1496

Bedini S, Turrini A, Rigo C, Argese E, Giovannetti M (2010) Molecular characterization and glomalin production of arbuscular mycorrhizal fungi colonizing a heavy metal polluted ash disposal island, downtown Venice. Soil Biol Biochem 42:758–765

Bharadwaj DP, Lundquist PO, Persson P, Alstrom S (2008) Evidence for specificity of cultivable bacteria associated with arbuscular mycorrhizal fungal spores (multitrophic interactions in the rhizosphere). FEMS Microbiol Ecol 65:310–322

Bohrer EK, Friese CF, Amon JP (2004) Seasonal dynamic of arbuscular mycorrhizal fungi in differing wetland habitats. Mycorrhiza 14:329–337

Borges AJS, Trindade AV, Matos AP, Peixoto MFS (2007) Redução do mal-do-panamá em bananeira-maçã por inoculação de fungo micorrízico arbuscular. Pesqui Agropecu Bras 42:35–41 (in Portuguese)

Borowicz VA (2010) The impact of arbuscular mycorrhizal fungi on strawberry tolerance to root damage and drought stress. Pedobiologia 53:265–270

Caravaca F, Hernández MT, García C, Roldán A (2002) Improvement of rhizosphere aggregates stability of afforested semi-arid plant species subjected to mycorrhizal inoculation and compost addition. Geoderma 108:133–144

Caravaca F, Barea JM, Palenzuela J, Figueroa D, Alguacil MM, Roldán A (2003) Establishment of shrub species in a degraded semiarid site after inoculation with native or allochthonous arbuscular mycorrhizal fungi. Appl Soil Ecol 22:103–111

Caravaca F, Alguacil MM, Azcón R, Parlade J, Torres P, Roldán A (2005) Establishment of two ectomycorrhizal shrub species in a semiarid site after *"in situ"* amendment with sugar beet, rock phosphate and *Aspergillus niger*. Microbiol Ecol 49:73–82

Carpenter-Boggs L, Loynachan TE, Stahl PD (1995) Spore germination of *Gigaspora margarita* stimulated by volatiles of soil-isolated actinomycetes. Soil Biol Biochem 27:1445–1451

Carvalho LM, Correia PM, Martins-Loucao A (2004) Arbuscular mycorrhizal fungal propagules in a salt marsh. Mycorrhiza 14:165–170

Chen X, Tang JJ, Zhi GY, Hu SJ (2005) Arbuscular mycorrhizal colonization and phosphorus acquisition of plants: effects of coexisting plant species. Appl Soil Ecol 28:259–269

Chern EC, Tsai DW, Ogunseitan OA (2007) Deposition of glomalin-related soil protein and sequestered toxic metals into watersheds. Environ Sci Technol 41:3566–3572

Cofcewicz ET, Medeiros CAB, Carneiro RMDG, Pierobom CR (2001) Interação dos fungos micorrízicos arbusculares *Glomus etunicatum* e *Gigaspora margarita* e o nematóide das galhas *Meloidogyne javanica* em tomateiro. Fitopatol Bras 26:65–70 (in Portuguese)

Colla G, Rouphael Y, Cardarelli M, Tullio M, Rivera CM, Rea E (2008) Alleviation of salt stress by arbuscular mycorrhizal in zucchini plants grown at low and high phosphorus concentration. Biol Fert Soils 44:501–509

Cordier C, Pozo MJ, Barea JM, Gianinazzi S, Gianinazzi-Pearson V (1998) Cell defense responses associated with localized and systemic resistance to *Phytophthora parasitica* induced in tomato by an arbuscular mycorrhizal fungus. Mol Plant Microbe Interact 11:1017–1028

Cordier C, Lemoine MC, Lemanceau P, Gianinazzi-Pearson V, Gianinazzi S (2000) The benefical rhizosphere: a necessary strategy for microplant production. Acta Hortic 530:259–268

Costa FA (2003) Micorrizas arbusculares e enzimas do estresse oxidativo em raízes de cenoura transformada e em morangueiro. Doctoral thesis, Universidade Federal de Viçosa, Minas Gerais, Brazil 92p (in Portuguese)

de Boer W, Folman LB, Summerbell RC, Boddy L (2005) Living in a fungal world: impact of fungi on soil bacterial niche development. FEMS Microbiol Rev 29:795–811

De Gara L, Pinto MC, Tommasi F (2003) The antioxidant systems vis-a-vis reactive oxygen species during plant-pathogen interaction. Plant Physiol Bioch 41:863–870

De La Peña E, Echeverría SR, Van Der Putten WH, Freitas H, Moens M (2006) Mechanism of control of root-feeding nematodes by mycorrhizal fungi in the dune grass *Ammophila arenaria*. New Phytol 169:829–840

del Mar Alguacil M, Kohler J, Caravaca F, Roldán A (2009) Differential effects of *Pseudomonas mendocina* and *Glomus intraradices* on lettuce plants physiological response and aquaporin PIP2 gene expression under elevated atmospheric CO_2 and drought. Microbiol Ecol 58:942–951

Del Río LA, Sandalio LM, Corpas FJ, Palma JM, Barroso JB (2006) Reactive oxygen species and reactive nitrogen species in peroxisomes. production, scavenging, and role in cell signaling. Plant Physiol 141:330–335

Diedhiou PM, Hallmann J, Oerke EC, Dehne HW (2003) Effects of arbuscular mycorrhizal fungi and a non-pathogenic *Fusarium oxysporum* on *Meloidogyne incognita* infestation of tomato. Mycorrhiza 13:199–204

Dimkpa C, Weinand T, Asch F (2009) Plant-rhizobacteria interactions alleviate abiotic stress conditions. Plant Cell Environm 32:1682–1694

Driver JD, Holben WE, Rillig MC (2005) Characterization of glomalin as a hyphal wall component of arbuscular mycorrhizal fungi. Soil Biol Biochem 37:101–106

Elsen A, Beeterens R, Swennen R, De Waele D (2003) Effects of an arbuscular mycorrhizal fungus and two plant-parasitic nematodes on *Musa* genotypes differing in root morphology. Biol Fert Soils 38:367–376

Elsen A, Gervasio D, Swennen R, De Waele D (2008) AMF-induced biocontrol against plant parasitic nematodes in *Musa* sp.: a systemic effect. Mycorrhiza 18:251–256

Eom AH, Hartnett DC, Wilson GWT (2000) Host plant species effects on arbuscular mycorrhizal fungal communities in tallgrass prairie. Oecologia 122(3):435–444

Finlay RD (2004) Mycorrhizal fungi and their multifunctional roles. Mycologist 18:91–96

Folli-Pereira MS, Meira-Haddad LS, Rassol N, Otoni WC, Kasuya MCM (2012) Development of mycorrhized vitroplants of *Jatropha curcas* L. at different rooting stages. Plant Biotech Rep. DOI: 10.1007/s11816–012-0232–5

Francis I, Holsters M, Vereecke D (2010) The Gram-positive side of plant--microbe interactions. Environ Microbiol 12:1–12

Freire ES, Campos VP, Dutra MR, Rocha FS, Silva JRC, Pozza EA (2007) Infectividade de juvenis do segundo estádio de *Meloidogyne incognita* em tomateiro após privação alimentar em solo e água em diferentes condições. Summa Phytopathol 33:270–274 (in Portuguese)

Frey-Klett P, Garbaye J, Tarkka M (2007) The mycorrhiza helper bacteria revisited. New Phytol 176:22–36

Gadkar V, Rillig M (2006) The arbuscular mycorrhizal fungal protein glomalin is a putative homolog of heat shock protein. FEMS Microbiol Ecol 263:93–101

Garcia-Garrido JM, Ocampo JA (2002) Regulation of the plant defence response in arbuscular mycorrhizal symbiosis. J Exp Bot 53:1377–1386

Gaur A, Adholeya A (2004) Prospects of arbuscular mycorrhizal fungi in phytoremediation of heavy metal contaminated soils. Curr Sci 86(4):528–534

Gill SS, Tuteja N (2010) Reactive oxygen species and antioxidant machinery in abiotic stress tolerance in crop plant. Plant Physiol Biochem 48:909–930

Giri B, Kapoor R, Mukerji KG (2007) Improved tolerance of *Acacia nilotica* to salt stress by arbuscular mycorrhiza, *Glomus fasciculatum* may be partly related to elevated K/Na ratios in root and shoot tissues. Microbial Ecol 54:753–760

Göhre V, Paszkowski U (2006) Contribution of the arbuscular mycorrhizal symbiosis to heavy metal phytoremediation. Planta 223:1115–1123

González-Chávez C, Carrillo-González R, Wright SF, Nichols KA (2004) The role of glomalin, a protein produced by arbuscular mycorrhizal fungi in sequestering potentially toxic elements. Environ Poll 130:317–323

Gratão PL, Monteiro CC, Antunes AM, Peres LEP, Azevedo RA (2008) Acquired tolerance of tomato (*Lycopersicon esculentum* cv. Micro-Tom) plants to cadmium-induced stress. Ann Appl Biol 153:321–333

Gustafson DJ, Casper BB (2006) Differential host plant performance as a function of soil arbuscular mycorrhizal fungal communities: experimentally manipulating co-occurring *Glomus* species. Plant Ecol 183(2):257–263

Hajiboland R, Aliasgharzadeh A, Laiegh SF, Poschenrieder C (2010) Colonization with arbuscular mycorrhizal fungi Improves salinity tolerance of tomato (*Solanum lycopersicum* L.) plants. Plant Soil 331:313–327

Hajiboland R, Joudmand A (2009) The K/Na replacement and function of antioxidant defense system in sugar beet (*Beta vulgaris* L.) cultivars. Acta Agr Sc B-S P 59:246–259

Hart MM, Klironomos JN (2002) Diversity of arbuscular mycorrhizal fungi and ecosystem functioning. In: van der Heijden MGA, Sanders IR (eds) Mycorrhizal Ecology. Ecological Studies, vol 157. Springer-Verlag, Heidelberg, pp 225–242

Hart MM, Reader RJ (2002) Host plant benefit from association with arbuscular mycorrhizal fungi: variation due to differences in size of mycelium. Biol Fert Soils 36:357–366

Hartmann A, Schmid M, Van Tuinen D, Berg G (2009) Plant-driven selection of microbes. Plant Soil 321:235–257

Hasegawa PM, Bressan RA, Zhu JK, Bohnert HJ (2000) Plant cellular and molecular responses to high salinity. Annu Rev Plant Physiol 51:463–499

Hawkes CV, Hartley IP, Ineson P, Fitter AH (2008) Soil temperature affects carbon allocation with arbuscular mycorrhizal networks and carbon transport from plant to fungus. Glob Change Biol 14:1181–1190

He Z, He C, Zhang Z, Zou Z, Wang H (2007) Changes of antioxidative enzymes and cell membrane osmosis in tomato colonized by arbuscular mycorrhizae under NaCl stress. Colloids Surfaces B Biointerfaces 59:128–133

Heijden MGA, van der Scheublin TR, Brader A (2004) Taxonomic and functional diversity in arbuscular mycorrhizal fungi—is there any relationship? New Phytol 164(2):201–204

Heijden MGA, van der Wiemken A, Sanders IR (2003) Different arbuscular mycorrhizal fungi alter coexistence and resource distribution between co-occurring plants. New Phytol 157(3):569–578

Hildebrandt U, Regvar M, Bothe H (2007) Arbuscular mycorrhiza and heavy metal tolerance. Phytochemistry 68:139–146

Hincha DK, Hagemann M (2004) Stabilization of model membranes during drying by compatible solutes involved in the stress tolerance of plants and microorganisms. Biochem J 383(2):277–283

Hol WHG, Cook R (2005) An overview of arbuscular mycorrhizal fungi--nematode interactions. Basic Appl Ecol 6:489–503

Huang H, Zhang S, Wu N, Luo L, Christie P (2009) Influence of *Glomus etunicatum*/ *Zea mays* mycorrhiza on atrazine degradation, soil phosphatase and dehydrogenase activities, and soil microbial community structure. Soil Biol Biochem 41:726–734

INVAM (2012) Intrepretation of Morphology. In: International Culture Collection of VA Mycorrhizal Fungi. http://invam.caf.wvu.edu/fungi/taxonomy/concepts/concepts.htm

Jahromi F, Aroca R, Porcel R, Ruiz-Lozano JM (2008) Influence of salinity on the in vitro development of *Glomus intraradices* and on the *in vivo* physiological and molecular responses of mycorrhizal lettuce plants. Microbial Ecol 55:45–53

Jaiti F, Meddich A, El Hadrami I (2008) Effectiveness of arbuscular mycorrhizal fungi in the protection of date palm (*Phoenix dactylifera* L.) against bayoud disease. Physiol Mol Plant P 71:166–173

Jalili F, Khavazi K, Pazira E, Nejati A, Asadi RH, Rasuli SH, Miransari M (2009) Isolation and characterization of ACC deaminase producing fluorescent pseudomonas, to alleviate salinity stress on canola (*Brassica napus* L.) growth. J Plant Physiol 166:667–674

Jansa J, Mozafar A, Kuhn G, Anken T, Ruh R, Sanders IR, Frossard E (2003) Soil tillage affects the community structure of mycorrhizal fungi in maize roots. Ecol Appl 13(4):1164–1176

Jeffries P, Gianinazzi S, Perotto S, Turnau K, Barea JM (2003) The contribution of arbuscular mycorrhizal fungi in sustainable maintenance of plant health and soil fertility. Biol Fert Soils 37:1–16

Johnson D, Vandenkoornhuyse PJ, Leake JR, Gilbert L, Booth RE, Grime JP, Young JPW, Read DJ (2004) Plant communities affect arbuscular mycorrhizal fungal diversity and community composition in grassland microcosms. New Phytol 161(2):503–515

Johnson NC, Graham JH, Smith FA (1997) Functioning of mycorrhizal associations along the mutualism--parasitism continuum. New Phytol 135:575–585

Juniper S, Abbott LK (2006) Soil salinity delays germination and limits growth of hyphae from propagules of arbuscular mycorrhizal fungi. Mycorrhiza 16:371–379

Kaci Y, Heyraud A, Barakat M, Heulin T (2005) Isolation and identification of an EPS-producing Rhizobium strain from arid soil (Algeria): characterization of its EPS and the effect of inoculation on wheat rhizosphere soil structure. Res Microbiol 156:522–531

Kiers ET, Van Der Heijden GA (2006) Mutualistic stability in the arbuscular mycorrhizal symbiosis: exploring hypotheses of evolutionary cooperation. Ecology 87:1627–1636

Klauberg-Filho O, Siqueira JO, Moreira FMS (2002) Fungos micorrízicos arbusculares em solos de área poluída com metais pesados. Rev Bras Cienc Solo 6:125–134 (in Portuguese)

Klauberg-Filho O, Siqueira JO, Moreira FMS, Soares CRFS, Silva S (2005) Ecologia, função e potencial de aplicação de FMAs em condições de excesso de metais pesados. Tópicos em ciência do solo. Soc Bras Cienc Solo 4:85–144 (in Portuguese)

Klironomos JN, McCune J, Hart M, Neville J (2000) The influence of arbuscular mycorrhizae on the relationship between plant diversity and productivity. Ecol Lett 3:137–141

Kohler J, Caravaca F, Carrasco L, Roldan A (2006) Contribution of *Pseudomonas mendocina* and *Glomus intraradices* to aggregate stabilization and promotion of biological fertility in rhizosphere soil of lettuce plants under field conditions. Soil Use Manage 22:298–304

Kohler J, Caravaca F, Roldan A (2009) Effect of drought on the stability of rhizosphere soil aggregates of Lactuca sativa grown in a degraded soil inoculated with PGPR and AM fungi. Appl Soil Ecol 42:160–165

Lambais MR, Ríos-Ruiz WF, Andrade RM (2003) Antioxidant responses in bean (*Phaseolus vulgaris*) roots colonized by arbuscular mycorrhizal fungi. New Phytol 160:421–428

Larsen J, Cornejo P, Barea JM (2009) Interactions between the arbuscular mycorrhizal fungus *Glomus intraradices* and the plant growth promoting rhizobacteria *Paenibacillus polymyxa* and *P. macerans* in the mycorrhizosphere of *Cucumis sativus*. Soil Biol Biochem 41:286–292

Ma QQ, Wang W, Li YH, Li DQ, Zou Q (2006) Alleviation of photoinhibition in drought-stressed wheat (Triticum aestivum) by foliar-applied glycinebetaine. J Plant Physiol 163(2):165–175

Meira LS (2004) Atividade de enzimas do estresse oxidativo em morangueiro micropropagado e inoculado com fungos micorrízicos arbusculares durante a aclimatização. Master's thesis, Universidade Federal de Viçosa, Minas Gerais, Brazil 37p (in Portuguese)

Meira-Haddad LS (2008) Efeito de fungos micorrízicos arbusculares sobre o parasitismo do nematóide das galhas em plantas de bananeira micropropagadas. Doctoral thesis, Universidade Federal de Viçosa, Minas Gerais, Brazil 134p (in Portuguese)

Moraes RM, Andrade Z, Bedir E, Dayan FE, Lata H, Khan I, Pereira AMS (2004) Arbuscular mycorrhizal improves acclimatization and increases lignan content of micropropagated mayapple (*Podophyllum peltatum* L.). Plant Sci 166:23–29

Munier-Lamy C, Deneux-Mustin S, Mustin C, Merlet D, Berthelin J, Leyval C (2007) Selenium bioavailability and uptake as affected by four different plants in a loamy clay soil with particular attention to mycorrhizae inoculated ryegrass. J Environ Radioactiv 97:148–158

Munkvold L, Kjoller R, Vestberg M, Rosendahl S, Jakobsen I (2004) High functional diversity within species of arbuscular mycorrhizal fungi. New Phytol 164:357–364

Neocleous D, Vasilakakis M (2007) Effects of NaCl stress on red raspberry (*Rubus idaeus* L. 'Autumn Bliss'). Sci Hortic-Amsterdam 112(3):282–289

Nichols K (2003) Characterization of glomalin-a glycoprotein produced by arbuscular mycorrhizal fungi. Ph D Dissertation. University of Maryland, College Park, Maryland, pp 285

Nichols KA, Wright SF (2006) Carbon and nitrogen in operationally defined soil organic matter pools. Biol Fert Soils 43:215–220

Niklaus PA, Alphei J, Ebersberger D, Kampichler C, Kandeler E, Tscherko D (2003) Six years of in situ CO2 enrichment evoke changes in soil structure and soil biota of nutrient-poor grassland. Glob Change Biol 9:585–600

O'Connor, PJ, Smith SE, Smith FA (2002) Arbuscular mycorrhizas influence plant diversity and community structure in a semiarid herbland. New Phytol 154:209–218

Oishi H, Ebina M (2005) Isolation of cDNA and enzymatic properties of betaine aldehyde dehydrogenase from *Zoysia tenuifolia*. J Plant Physiol 162:1077–1086

Palma JM, Longa MA, del Río LA, Arines J (1993) Superoxide dismutase in vesicular arbuscular mycorrhizal red clover plants. Physiol Plantarum 87:77–83

Paradi I, Bratek Z, Lang F (2003) Influence of arbuscular mycorrhiza and phosphorus supply on polyamine content, growth and photosynthesis of *Plantago lanceolata*. Plant Biology 46:563–569

Pawlowska TE (2004) Genetic processes in arbuscular mycorrhizal fungi. FEMS Microbiol Lett 251:185–192

Pawlowska TE, Charvat I (2004) Heavy-metal stress and developmental patterns of arbuscular mycorrhizal fungi. Appl Environ Microbiol 70:6643–6649

Peng YL, Gao ZW, Gao Y, Liu GF, Sheng LX, Wang DL (2008) Eco-physiological characteristics of alfalfa seedlings in response to various mixed salt-alkaline stresses. J Integr Plant Biol 50:29–39

Pinochet J, Calvet C, Camprubi A, Fernandez A (1996) Interactions between migratory endoparasitic nematodes and arbuscular mycorrhizal fungi in perennial crops: a review. Plant Soil 185:183–190

Porcel R, Aroca R, Cano C, Bago A, Ruiz-Lozano JM (2006) Identification of a gene from the arbuscular mycorrhizal fungus *Glomus intraradices* encoding for a 14-3-3 protein that is up-regulated by drought stress during the AM symbiosis. Microbial Ecol 53:575–582

Porcel R, Barea JM, Ruiz-Lozano JM (2003) Antioxidant activies in mycorrhizal soybean plants under drought stress and their possible relationship to the process of nodule senescence. New Phytol 157:135–143

Pouyu-Rojas E, Siqueira JO, Santos JGD (2006) Compatibilidade simbiótica de fungos micorrízicos arbusculares com mudas de espécies arbóreas tropicais. Rev Bras Cienc Solo 30(3):413–424 (in Portuguese)

Pozo MJ, Cordier C, Dumas-Gaudot E, Gianinazzi S, Barea JM, Azcón-Aguilar C (2002) Localized versus systemic effect of arbuscular mycorrhizal fungi on defense responses *Phytophthora* infection in tomato plants. Jou Exp Bot 53:525–534

Pozo MJ, Jung SC, Lopez-Rãez JA, Azcón-Aguilar C (2010) Impact of arbuscular mycorrhizal symbiosis on plant response to biotic stress: the role of plant defence mechanisms. Arbuscular mycorrhizas: Physiol Funct 3:193–207

Purin S, Rillig MC (2007) The arbuscular mycorrhizal fungal protein glomalin: limitations, progress, and new hypothesis for its function. Pedobiologia 51:123–130

Rabie GH, Almadini AM (2005) Role of bioinoculants in development of salt-tolerance of *Vicia faba* plants. Afr J Biotechnol 4:210–222

Richardson A, Barea J-M, McNeill A (2009) PRIGENT-COMBARET C. Acquisition of phosphorus and nitrogen in the rhizosphere and plant growth promotion by microorganisms. Plant Soil 321:305–339

Rillig MC (2004) Arbuscular mycorrhizae, glomalin and soil quality. Can J Soil Sci 84:355–363

Rillig MC, Lutgen ER, Ramsey PW, Klironomos JN, Gannon JE (2005) Microbiota accompanying different arbuscular mycorrhizal fungal isolates influence soil aggregation. Pedobiology 49:251–259

Rillig MC, Hoye AT, Carran A (2006) Minimal direct contribution of arbuscular mycorrhizal fungi to DOC leaching in grassland through losses of glomalin-related soil protein. Soil Biol Biochem 38:2967–2970

Rillig MC, Mummey DL (2006) Mycorrhizas and soil structure. New Phytol 171:41–53

Rillig MC, Ramsey PW, Morris S, Paul EA (2003) Glomalin, an arbuscular-mycorrhizal fungal soil protein, responds to land-use change. Plant Soil 253:293–9

Rillig MC, Wright SF, Nichols KA, Schmidt WF, Torn MS (2001) Large contribution of arbuscular mycorrhizal fungi to soil carbon pools in tropical forest soils. Plant Soil 233:167–177

Ruiz-Lozano JM, Bonfante P (2000) Intracellular *Burkholderia* of the arbuscular mycorrhizal fungus *Gigaspora margarita* possesses the vacB gene, which is involved in host cell colonization by bacteria. Microbiology Ecol 39:137–144

Sannazzaro AI, Echeverría M, Albertó EO, Ruiz OA, Menéndez AB (2007) Modulation of polyamine balance in *Lotus glaber* by salinity and arbuscular mycorrhiza. Plant Physiol Bioch 45:39–46

Sannazzaro AI, Ruiz OA, Alberto EO, Menendez AB (2006) Alleviation of salt stress in *Lotus glaber* by *Glomus intraradices*. Plant Soil 285:279–287

Santos JGD (2008) Riqueza de fungos micorrízicos arbusculares no solo e no crescimento inicial de espécies arbóreas nativas. Doctoral thesis, Universidade Federal de Lavras, Minas Gerais, Brazil 120p (in Portuguese)

Selosse MA, Baudoin E, Vandenkoornhuyse P (2004) Symbiotic microorganisms, a key for ecological success and protection of plants. C R Biol 327:639–648

Seneviratne G, Thilakaratne RMMS, Jayasekara APDA, Seneviratne KACN, Padmathilake KRE, De Silva MSDL (2009) Developing beneficial microbial biofilms on roots of non-legumes: a novel biofertilizing technique. In: Khan MS et al (eds) Microbial strategy for crop improvement. Springer-Verlag, Berlin, pp 51–61

Sheng M, Tang M, Chen H, Yang B, Zhang F, Huang Y (2008) Influence of arbuscular mycorrhizae on photosynthesis and water status of maize plants under salt stress. Mycorrhiza 18:287–296

Siddiqui Z A, Mahmood I (1995) Role of plant symbionts in nematode management: a review. Bioresour Technol 54:217–226

Silva GA, Trufem SFB, Saggin Júnior OJ, Mais LC (2005) Arbuscular mycorrhizal fungi in a semiarid copper mining area in Brazil. Mycorrhiza 15(1):47–53

Singh LP, Gill SS, Tuteja N (2011) Unraveling the role of fungal symbionts in plant abiotic stress tolerance. Plant Signal Behav 6:175–191

Singh AK, Hamel C, DePauw RM, Knox RE (2012) Genetic variability in arbuscular mycorrhizal fungi compatibility supports the selection of durum wheat genotypes for enhancing soil ecological services and cropping systems in Canada. Can J Microbiol 58:293–302

Smith SE, Read DJ (2008) Mycorrhizal Symbiosis, 3rd edn. Academic Press, London

Smith SE, Facelli E, Pope S, Smith FA (2010) Plant performance in stressful environments: Interpreting new and established knowledge of the roles of Arbuscular Mycorrhizas. Plant Soil 326:3–20

Smith SE, Smith FA, Jakobsen I (2004) Functional diversity in arbuscular mycorrhizal (AM) symbioses: the contribution of the mycorrhizal P uptake pathway is not correlated with mycorrhizal responses in growth and total P uptake. New Phytol 162:511–524

Sonjak S, Beguiristain T, Leyval C, Regvar M (2009) Temporal temperature gradient gel electrophoresis (TTGE) analysis of arbuscular mycorrhizal fungi associated with selected plants from saline and metal polluted environments. Plant Soil 314:25–34

Staddon PL, Thompson K, Jakobsen I, Grime JP, Askew AP, Alastair Fitter AH (2003) Mycorrhizal fungal abundance is affected by long-term climatic manipulations in the field. Global Change Biol 9:186–194

Tang M, Chen H, Huang JC, Tian ZQ (2009) AM fungi effects on the growth and physiology of *Zea mays* seedlings under diesel stress. Soil Biol Biochem 41(5):936–940

Toro M, Azcón R, Barea JM (1997) Improvement of arbuscular mycorrhizal development by inoculation of soil with phosphate solubilizing rhizobacteria to improve rock phosphate bioavailability (32P) and nutrient cycling. Appl Environ Microbiol 63:4408–4412

Treberg JR, Driedzic WR (2007) The accumulation and synthesis of betaine in winter skate (*Leucoraja ocellata*). Comp Biochem Physiol A: Mol Integr Physiol 147(2):475–83

Vallino M, Massa N, Lumini E, Bianciotto V, Berta G, Bonfante P (2006) Assessment of arbuscular mycorrhizal fungal diversity in roots of *Solidago gigantea* growing in a polluted soil in Northern Italy. Environ Microbiol 8:971–983

Vodnik D, Grčmana H, Mačeka I, Van Elterenb JT, Kovačevičc M (2008) The contribution of glomalin-related soil protein to Pb and Zn sequestration in polluted soil. Sci Total Environ 392:130–136

Vos C, Claerhout S, Mkandawire R, Panis B, De Waele D, Elsen A (2012) Mycorrhiza-induced resistance in banana acts on nematode host location and penetration. Soil Biol Biochem 47:60–66

Weissenhorn I, Glashoff A, Leyval C, Berthelin J (1994) Differential tolerance to Cd and Zn of arbuscular mycorrhizal (Am) fungal spores isolated from heavy metal-polluted and unpolluted soils. Plant Soil 167:189–196

Weissenhorn I, Leyval C, Berthelin J (1993) Cd-tolerant arbuscular mycorrhizal (Am) fungi from heavy-metal polluted soils. Plant Soil 157:247–256

Wright SF, Upadhyaya A (1996) Extraction of an abundant and unusual protein from soil and comparison with hyphal protein of arbuscular mycorrhizal fungi. Soil Sci 161:575–586

Wu QS, Zou YN, Xia RX (2006) Effects of water stress and arbuscular mycorrhizal fungi on reactive oxygen metabolism and antioxidant production by citrus (*Citrus tangerine*) roots. Eur J Soil Biol 42:166–172

Wu WQ, Su XY, Xia Y, Wang YS, Luan LJ (2008) An, The *Suaeda liaotungensis* kitag betaine aldehyde dehydrogenase gene improves salt tolerance of transgenic maize mediated with minimum linear length of DNA fragment. Euphytica 159:17–25

Wuyts N, Waele D, Swennen R (2006) Activity of phenylalanine ammonia-lyase, peroxidase and polyphenol oxidase in roots of banana (*Musa acuminata* AAA, cvs Grande Naine and Yangambi Km5) before and after infection with *Radopholus similis*. Nematology 8:201–209

Yang CW, Xu HH, Wang LL, Liu J, Shi DC, Wang GD (2009) Comparative effects of salt-stress and alkali-stress on the growth, photosynthesis, solute accumulation, and ion balance of barley plants. Photosynthetica 47(1):79–86

Yano-Melo AM, Trufem SFB, Maia LC (2003) Arbuscular mycorrhizal fungi in salinized and surrounded areas at the São Francisco submedium valley, Brazil. Hoehnea 30(2):79–87

Yao Q, Zhu HH, Chen JZ (2005) Growth responses and endogenous IAA and iPAs changes of litchi (*Litchi chinensis* Sonn.) seedlings induced by arbuscular mycorrhizal fungal inoculation. Sci Hortic 105:145–151

Zarei M, König S, Hempel S, Nekouei MK, Savaghebi G, Buscot F (2008) Community structure of arbuscular mycorrhizal fungi associated to Veronica rechingeri at the Anguran zinc and lead mining region. Environ Pollut 156:1277–1283

Zhang F-L, Niu B, Wang Y-C, Chen F, Wang S-H, Xu Y, Jiang L-D, Gao J, Wu J, Tang L, Jia Y-J (2008) A novel betaine aldehyde dehydrogenase gene from *Jatropha curcas,* encoding an enzyme implicated in adaptation to environmental stress. Plant Sci 174:510–518

Zhu XC, Song FB, Xu HW (2010) Influence of arbuscular mycorrhiza on lipid peroxidation and antioxidant enzyme activity of maize plants under temperature stress. Mycorrhiza 20:325–332

Zhu YG, Miller RM (2003) Carbon cycling by arbuscular mycorrhizal fungi in soil-plant systems. Trends Plant Sci 8:407–409

Chapter 7
Biotic Stress and Crop Improvement: A Wheat Focus Around Novel Strategies

Alvina Gul Kazi, Awais Rasheed and Abdul Mujeeb-Kazi

Abstract Currently much of the wheat genetic variability is obtained through conventional crop improvement methods involving land races and normal varieties. Hence, the germplasm base available in the form of cultivars is becoming increasingly narrow and the need for widening the gene pool is essential in view of the emerging biotic and abiotic stresses due to global warming and climate change. Major abiotic constraints that have surfaced are due to increased salinity, water logging, drought and heat. Biotic stresses of emphasis here additionally contribute to the crops productivity situation. To counter these maladies a broad genetic base is essential to have on hand and its implementation a dire need forming the focus of this communication. New and useful genetic variations exist in the wild uncultivated wheat progenitor species that can be utilized for the enhancement of the existing wheat breeding pools and improve yield stability. These genetic variations can be harnessed through a combination of conventional breeding methods coupled with interspecific, intraspecific and intergeneric hybridization approaches popularly known as "wide crossing" that independently and cumulatively augment the available genetic variability for wheat improvement.

Diploid wheat progenitors (2n=2x=14) A, B, and D are the constituents of bread wheat (*Triticum aestivum* L) offering extensive diversity that contributes to crop improvement by providing novel allelic enrichment. A and D genome dip-

---

A. G. Kazi (✉)
Atta-ur-Rahman School of Applied Biosciences,
National University of Sciences and Technology (NUST),
Islamabad, Pakistan
e-mail: alvina_gul@yahoo.com

A. Rasheed
Department of Plant Sciences, Quaid-i-Azam University,
Islamabad, Pakistan
e-mail: awaispbg@gmail.com

A. Mujeeb-Kazi
National Institute of Biotechnology and Genetic Engineering (NIBGE),
Faisalabad, Pakistan
e-mail: m.kazi@cgiar.org

K. R. Hakeem et al. (eds.), *Crop Improvement,* DOI 10.1007/978-1-4614-7028-1_7,

loids belong to the "primary" gene pool and the B(S) genome to the "secondary" pool. Exploiting these diploids requires skills of developing user friendly genetic stocks commonly known as "synthetic hexaploids (SH)". The stocks are produced by combining durum wheat cultivars (2n=4x=28) with each diploid thus generating hexaploids that are genomically AABBDD, AABBAA and AABBBB(SS). All stocks cytologically are expected to be 2n=6x=42 and major resources and provide unique allelic diversity for wheat improvement.

Biotic stresses of significance vary according to location and our major ones are the three rusts, karnal bunt with upcoming concern prevailing for powdery mildew, barley yellow dwarf and the new emergence of spot blotch. Progress to combat these stresses has be driven in tandem with locational priorities and these dictates have shifted global and national focus among the rusts to stem rust with the threat of race UG99's spread linked with a local races presence. Thus diversity for exploitation has extended beyond the diploid relatives to include tertiary gene pool resources where most notable mention is of the diploid *Thinopyrum bessarabicum* that has the potential to address multiple stress factors and will be elucidated in an agglomerated manner to embrace various accessional sources as they relate to the major biotic stresses resistance management.

Introduction

Currently much of the wheat genetic variability is obtained through conventional crop improvement methods involving land races and normal varieties. Hence, the germplasm base available in the form of cultivars is becoming increasingly narrow and the need for widening the gene pool is essential in view of the emerging biotic and abiotic stresses due to global warming and climate change. Major abiotic constraints that have surfaced are due to increased salinity, water logging, drought and heat (Mujeeb-Kazi et al. 2008a). Biotic stresses of emphasis here additionally contribute to the crops productivity situation. To counter these maladies a broad genetic base is essential to have on hand and its implementation a dire need forming the focus of this communication. New and useful genetic variations exist in the wild uncultivated wheat progenitor species that can be utilized for the enhancement of the existing wheat breeding pools and improve yield stability. These genetic variations can be harnessed through a combination of conventional breeding methods coupled with interspecific, intraspecific and intergeneric hybridization approaches popularly known as "wide crossing" that independently and cumulatively augment the available genetic variability for wheat improvement.

Diploid wheat progenitors (2n=2x=14) A, B, and D are the constituents of bread wheat (*Triticum aestivum* L) offering extensive diversity that contributes to crop improvement by providing novel allelic enrichment. A and D genome diploids belong to the "primary" gene pool and the B(S) genome to the "secondary" pool. Exploiting these diploids requires skills of developing user friendly genetic stocks commonly known as "synthetic hexaploids (SH)" (Mujeeb-Kazi et al. 1996a). The stocks are produced by combining durum wheat cultivars (2n=4x=28) with each diploid thus generating hexaploids that are genomically AABBDD, AABBAA and AABBBB(SS). All stocks cytologically are expected to be 2n=6x=42 and major genetic resources that provide unique allelic diversity for wheat improvement.

Biotic stresses of significance vary according to location and our major ones are the three rusts, karnal bunt with upcoming concern prevailing for powdery mildew, barley yellow dwarf and the new emergence of spot blotch (Mujeeb-Kazi et al. 2008b). Progress to combat these stresses has be driven in tandem with locational priorities and these dictates have shifted global and national focus among the rusts to stem rust with the threat of race UG99's spread linked with a local races presence (Mirza et al. 2010). Thus diversity for exploitation has extended beyond the diploid relatives to include tertiary gene pool resources where most notable mention is of the diploid *Thinopyrum bessarabicum* that has the potential to address multiple stress factors and will be elucidated in an agglomerated manner to embrace various accessional sources as they relate to the major biotic stresses resistance management. This communication covers our major biotic stresses, address a few of international importance using strategies that embrace diverse means of introgressing genes integrating technologies that add to the efficiency of pre-breeding and breeding to deliver outputs that are expected to form a conduit to food security. The overall theme is captioned "wide crossing".

Major credit for motivating research on the course of wide crosses goes to Kruse 1967, 1969, 1973, 1974 following the events of 1891 (Rimpau) and 1904 (Farrar). Initial impetus was derived from the wheat/barley findings of Kruse 1974 that paved the way for significant cytogenetical outputs by Islam et al. (1981) and followed by some additional digressions with other *Triticeae* members Sharma and Gill (1983), Mujeeb-Kazi and Kimber (1985), Mujeeb-Kazi et al. (1987, 1989), Wang (1989), Jiang et al. (1994), Sharma (1995). All these reports have centered on "intergeneric hybridization" and considered complex for realizing alien genetic transfers. Parallel to these efforts since mid-1980s emerged the era of exploiting of close relatives particularly the diploid wheat progenitor *Aegilops tauschii* (2n=2x=14, DD) via direct crossing (Alonso and Kimber 1984; Gill and Raupp 1987) or via bridge crossing (Mujeeb-Kazi and Hettel 1995; Mujeeb-Kazi et al. 1996a). Both these strategies will be highlighted but for handling practical outputs for key biotic stresses only those at a priority in our perception shall be considered that others could modify according to their desires.

Wheat Production

The productivity levels across varied environments are separated into the irrigated and rainfed regimes of cultivation. Stress constraints vary as well and these are locational holding their specific priority profiles. Often diversity to address a trait is present in conventional sources but when limitations prevail then unique sources are tapped. Over the past few decades this emphasis on suing novel genetic resources has increased and the benefits have also resulted (Mujeeb-Kazi et al. 2008b). Maximum benefits on a practical scale has come from the closely related wild progenitors like the D genome diploid grass *Ae. tauschii* that has resulted in varieties in China, Spain, Afghanistan, Ecuador and an abundance of advanced varietal candidate lines globally possessing biotic stress resistances and high yield levels (David Bonnett, Personal Communication). Specifying the contribution of special resources towards biotic stresses a brief consideration covers spot blotch, *Septoria tritici*, Karnal bunt, *Fusarium graminearum,* powdery mildew, yellow rust and stem rust, substantiated by inputs from other resources that are more divergent are outputs for spot blotch and stem rust.

Germplasm for Combating Biotic Stresses

The wheat family members are distributed within three gene pools; primary, secondary and tertiary (Jiang et al. 1994) and their utilization ranges from a relative ease to complex based upon genetic distance and genetic affinity traversing from perfect homology to genomic homeology. Details to elucidate the species distribution and range are provided in Dewey (1984), Kimber and Feldman (1987).

a. The conventional resource. The globally available accessions in *ex situ* gene banks number approximately 800,000 of which 3% are of wild wheats (Valkoun 2001). Those categorized as conventional wheats, land races and grouped into winter, spring and facultative fall in this section from which special mention will be made of some land races and naturally originated wheats that have been in extensive use since mid-1970s that carry the spontaneous translocation T1BL.1RS. National land races approach 1,000 in Pakistan of which 112 are widely studied and have been evaluated for various parameters. They are further unique diversity candidates to exploit for wheat improvement.
b. The unique wild/exotic gene pool resources including the diploid progenitors of the primary and secondary gene pool plus various tetraploids with their contribution of derived usable stocks which address biotic stresses. In extensive use have been the D genome followed by the A and in very limited use so far the B(S) genomes. Capturing interest also are the tetraploids *Triticum dicoccum*, *T. dicoccoides*, *T. carthlicum* and the hexaploid *T. spelta*.
c. The tertiary gene pool species with their diversity made user friendly and specifically targeted for a biotic stress of current global significance and a potent threat to wheat production will be addressed. The contributions from this pool have been tabulated and reported by Mujeeb-Kazi et al. (1987, 1989, 2008a), Sharma and Gill (1983), Sharma (1995), Friebe et al. (1996), Mujeeb-Kazi (2003, 2005, 2006a), Trethowan and Mujeeb-Kazi (2008), Ogbonnaya et al. (2013) elucidating that almost all major stresses encountered in wheat cultivation plots can benefit from alleles of value present in these resources.

Germplasm Choice for Wheat Improvement

The range of available genetic resources for improving wheat is enormous but if the time span for the delivery of final products that translate into varieties is the measure then the priority would follow intraspecific, interspecific and intergeneric.

At the intraspecific level *T. dicoccum/T. dicoccoides* followed by *T. carthlicum* take lead for both durum and bread wheat improvement with *T. spelta* being a good candidate for bread wheat.

At the interspecific level based upon homology the D genome heads the list along the strategy of bridge crossing and direct crossing. For additive variation A genome accessions come next and lastly usage of the B(S) from the Sitposis section that is rarely exploited for applied goals of wheat production.

At the intergeneric level most of the tertiary gene pool species are genomically far removed from the wheat A, B and D genomes. Hence parental choice of the alien resource is paramount. If trait is present across various ploidy levels then the preference of the species to be exploited would be the lower ploidy. Using such a strategy the salt tolerance gene transfers from the diploids *Thinopyrum bessarabicum* or *Th. elongatum* are preferred over the tetraploid source *T. junceum* or the decaploid *Th. ponticum*. It is fortuitous that the diploid *Th. bessarabicum* is also resistant to the UG99 pathotypes and thus for this paper it shall be discussed in detail.

Hybrid Production

Hybridization varies from the conventional to the radical combinations and thus can be rather easy to highly complex. In general where polyploidy levels differ, the higher polyploidy parent serves as the female and after crossing seed is set that may mature and be shriveled or would have to have its embryo excised from after 10 to 15 days and give plantlets that are self-sterile requiring induced doubling or backcrossing for further use in pre-breeding or breeding.

Pre-Breeding

Of late this term has been used to justify program organization in some international output projections. Unfortunately assigning the pre-breeding category to those crosses where F1 embryo rescue is not needed needs to be separated and fitted into the conventional hybridization category well within the work arsenal of any wheat breeder. Intraspecifc crosses have across many decades being handled by breeders where the classic examples are of the bread wheat/ durum wheat combinations forming the pentaploid and their further exploitation. This remains a 100% field operation and calling it pre-breeding in our perception is inappropriate. Thus all crosses using the AABB tetraploids onto wheat or AABBDD hexaploids with AABB tetraploids that do not require embryo culture and laboratory assistance to handle self-fertility or chromosome doubling should fall under the breeder's domain of conventional breeding and not categorized as "pre-breeding".

Wide hybridization returning to Triticale production, the classic wheat/barley hybrids then additional Triticeae species that require embryo rescue and media preparation skills, special plantlet regeneration strategies, handling care of the hybrid that is self-sterile with its advance via amphiploidy or backcrossing to affect alien transfers is the true categorization of the term "pre-breeding" and has been in operation since the start of such efforts highlighted in reports of Rimpau (1891), Farrar (1904), Kruse (1967, 1969), Islam et al. (1978), Sharma and Gill (1983), Mujeeb-Kazi and Kimber (1985), Mujeeb-Kazi et al. (1987, 1989, 1996a, 2013), Sharma (1995) and Mujeeb-Kazi and Hettel (1995). Outputs from such diverse crossings require infrastructure and skills that are hard to find within the conventional breeding professionals and requires expertise and controlled environment facilities that permits adopting special tools for F1 crossing to be done, compatibility to be harnessed, tissue culture to rescue the embryos, cytology to validate the hybrid, growing conditions that promote vigorous growth, amphiploid induction and/ or backcrossing before the BC or the amphiploid could be manipulated to deliver advanced lines that have the potential to become varietal materials. These facets are true "pre-breeding" steps that has been not been mentioned and the term loosely used to chart program structures in developing world programs fostered by international funding.

Breeding

An efficient protocol being followed in CIMMYT has been limited backcrossing coupled with selected modified bulk for breeding program efficiency. Where novel diversity becomes a donor of alleles the same protocol has been utilized effectively. With the focus on bread wheat improvement these steps are as follows:

a. Bread wheat parent as the female crossed by the novel tetraploids to produce pentaploid F_1 hybrids that upon limited backcrossing are advanced mediated by selected bulk and ultimate cytology to generate hexaploid euploids ($2n=6x=42$, AABBDD)
b. Similar crossing as in (a) of bread wheat by synthetic hexaploids to result in F_1's that are ABD AAB, ABD ABB(S) or ABD ABD to yield BC1's and upon advance deliver euploids with the 42 complement.

Often with the D genome hexaploids the F1 hybrids exhibit necrosis this knowledge of the necrotic genes is important to overcome this limitation. Also the tough rachis requires a sizeable F2 population to allow for selecting free threshing derivatives. Even though major efforts are on limited backcrossing use of F1 top-crossing in wide crossing has been exploited significantly (Mujeeb-Kazi and Asiedu 1990).

Mediated in this breeding scheme is achieving homozygosity as early as possible and often on the segregating F3 population where each selected individual is made homozygous via the detached tiller strategy (Riera-Lizarazu and Mujeeb-Kazi 1990) that uses haploids from crossing the selections by maize.

The Interspecific Contribution to Biotic Stresses: Some Crucial Steps in Outputs Generation to Combat Biotic Stresses

Bridge Crosses

The genetic stocks used for identifying resistance are known as synthetic hexaploids and these are of the three wheat genomes. According to their usage level they genomically rank as the D, A and B with their hexaploids being AABBDD, AABBAA and AABBSS with all being 2n=6x=42. These have been elaborately discussed in literature (Mujeeb-Kazi 2003, 2006a; Mujeeb-Kazi et al. 2004, 2008b, Ogbonnaya et al. 2013).

The Synthetic Production Across A, B (S) and D Genome Diploids, Maintenance and Their Utilization Mode

Standard conventional procedures of crossing using *T. turgidum* cultivars as the female parent, embryo rescue, seedling differentiation, hybrid seedling development and induction of the doubled fertile product (amphiploid/ synthetic hexaploid) permit harnessing of the various stocks that are 2n=6x=42 (AABBAA, AABBSS and AABBDD). Lesser has been the use of *T. dicoccum* and *T. dicoccoides* as the female parental source. No special modifications are needed but subtle variations do enhance the frequency of the outputs.

The variations are in:

a. use of early pollination or bud pollination,
b. giving cold treatment of 6C to the plated embryos for 2–3 weeks and
c. delaying the colchicine treatment until the 6 leaf stage around a vigorously growing hybrid plantlet or slightly later.

The synthetics that result are in limited amount as to seed resource and these C-0 stocks require increase that is done on each seed after its cytological validation having the 42 complement. This is essential since aneuploidy does prevail (Mujeeb-Kazi and Hettel 1995). Such seed increases for each synthetic allows for building up each entries seed quantity allowing for materials to be screened for biotic and abiotic stresses, conduct molecular diagnostics and distribute globally. Extensively in use has been the D genome synthetics and their elite sub-sets 1 plus 2. Lesser targeted usage also happens and special requests are met from the euploid reserve holdings. One significant provision was of a synthetic hexaploid wheat stock to Cornell University wheat program where after being crossed with the cultivar "Opata" the famous ITMI population of 150 RIL's was developed . This population was the conduit for developing the wheat microsatellite map (Roder et al. 1998). D genome synthetic production has been of worldwide interest but large numbers have been associated with the CIMMYT wide cross program where close to 1,200 spring and winter habits synthetics have been produced (Mujeeb-Kazi et al. 2008b) and these numbers are increasing (Bonnett, Personal Communication with A Mujeeb-Kazi). In depth details of the global inputs of various laboratories towards the D genome stocks have been recently reviewed by Ogbonnaya et al. (2013) where the current status of their practical utilization has also been elucidated.

The close relative genetic diversity beyond the D genome has exploited accessions of the A diploid resources *T. boeoticum*, *T. monococcum* and *T. urartu* and to a very limited extent the B(S) genome diploid *Ae. speltoides* of the Sitopsis section. The A and the B(S) genome stocks have yet to be widely utilized.

The use of the SH route is categorized as bridge crossing as upon crossing selected SH s with bread wheat for its improvement all three genomes are contributors and allelic richness is harnessed from all three of the SH genomes, i.e., A, B and D allowing for intraspecific and interspecific coverage to occur where the intraspecific portion simulates the bread wheat/ durum wheat pentaploid breeding protocol.

Direct Crosses

The most efficient technique for exploiting *Ae. tauschii* variability for bread wheat improvement is to achieve direct transfers from resistant/tolerant *Ae. tauschii* accessions to bread wheat. The methodology rapidly produces improved BC1 derivatives with the six genomes (AABBDD), five of which (AABBD) resemble the elite bread wheat cultivar used in the cross (Fig. 7.1). Aneuploidy in the BC does surface and thus recovery of euploids ($2n=6x=42$) requires cytology. Advantages of direct crossing have been elucidated by Cox et al. (1990) and has tremendous potential to go beyond the D genome into the A genome diploids. For a targeted approach screening of the genomic resource is important for practical wheat productivity goals. Alonso and Kimber (1984), Cox et al. (1990, 1991) and Gill and Raupp (1987) unequivocally placed priority on direct *Ae. tauschii* crossing with bread wheat cultivars. Based on the transfer of stem rust resistance from *Ae. tauschii* to the cultivar Chinese Spring Alonso and Kimber (1984) determined that one backcross onto the F1 hybrids reinstated 92 % of the genotype of the recurrent parent.

Where there are constraints to screening of the *Ae. tauschii* accessions, screening the synthetic hexaploids derived from *T. turgidum*/*Ae. tauschii* is an alternative particularly where the durum parent is susceptible and thus the SH resistance is

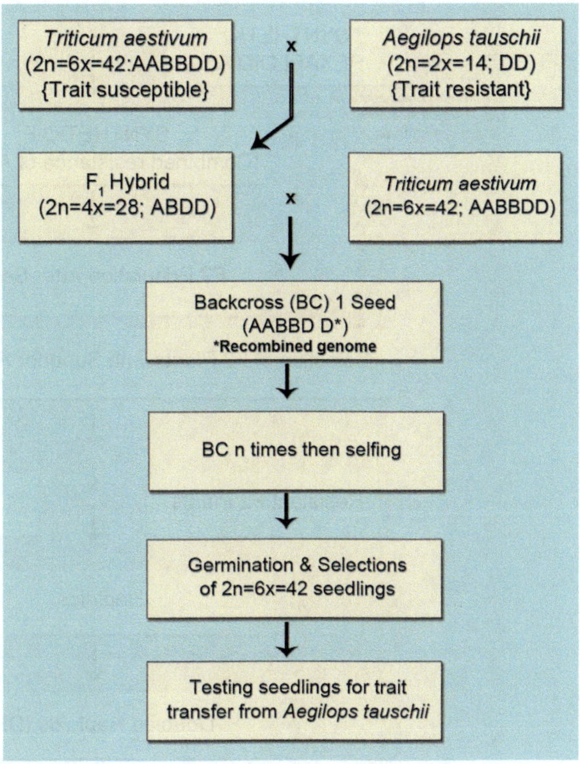

Fig. 7.1 Protocol for resistant trait transfer from desired *Ae. tauschii* accession into an elite but susceptible *T. aestivum* cultivar by "direct crossing"

attributed to the *Ae. tauschii* accession. Therefore, the accession can be selected from such synthetics for a trait and used in direct crosses.

Innovative Use of Resistance Pyramiding within "D" Genome Synthetics

Underutilized but with a high potential is the modus operandi is combining resistances identified in two divergent D genome synthetics for a biotic stress that adds efficiency to pre-breeding/breeding. Divergent synthetics with resistance contribution from two accessions are first crossed and their F1's superior resistance performance than either parent detected. The F1 upon selfing generates a segregating F2 population from which superior plants with resistance resembling the F1 and better than either parent in the cross are selected. Representative tillers detached and doubled haploids produced, which upon further seed increase and screening will have value additive of the two synthetic parents in the DH derivatives (Fig. 7.2). Integrated in this scheme could be the DNA profile of both SHs of the cross to ensure that DNA polymorphism is prevalent extended further to genes in either parent that

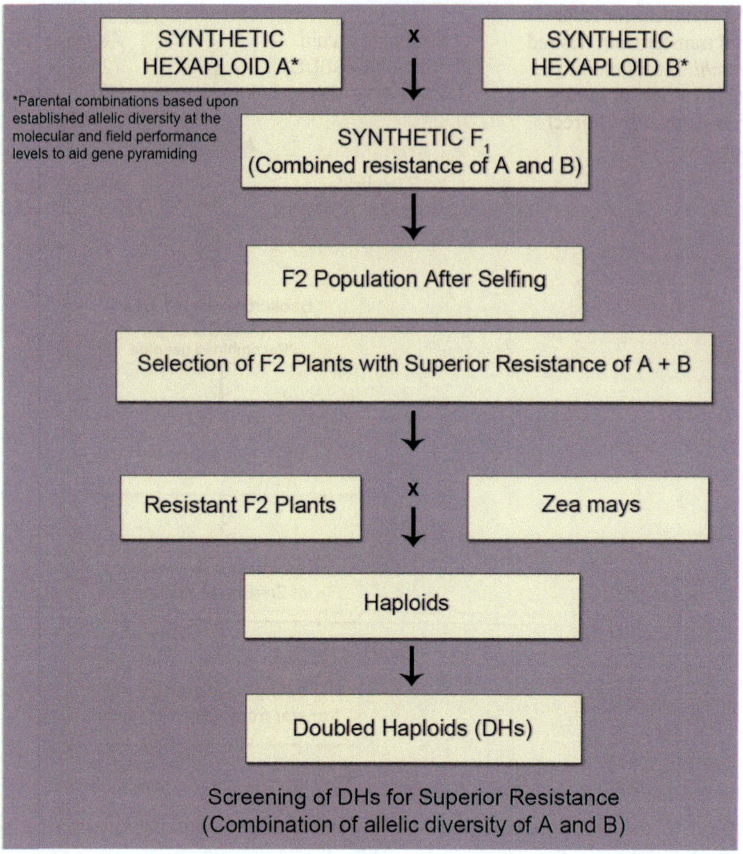

Fig. 7.2 Protocol for pyramiding resistance value of polymorphic synthetic hexaploid wheats for adding to breeding efficiency

could be combined via marker technology. Hence in the offing exists a genetic resource area that could combine desired SHs for enhancing recombination breeding efficiency and for which the entire over a 1,000 D genome SH's of the D genome and a couple hundred of the A genome await exploitation.

Additional Diversification from the A and B (S) Genomes

With production and maintenance protocols similar to those of the D genome diploid accession SH derivatives, uniqueness of the A and B (S) genome synthetics resides essentially in widening the variability resource that allows for novel allelic richness to become user friendly and provide the wheat crop additional genes that would be additive for resistance durability. The generated resource can not only enrich durum breeding programs but can also complement the efforts with bread wheat improvement akin to what pentaploid breeding (2n=5x=35, AABBD) efforts provide due to bread/durum crossings. This diversification comes from numerous accessions of the diploid progenitors and is extended to cover the tetraploids not being extensively used in the current breeding programs. The allelic richness that becomes available to exploit does have a down side and needs voicing. In case where genetic diversity is distributed in the two ploidy level crops by combining and shuffling their genes, a fear exists that the spectrum of divergence may be reduced or eliminated thus narrowing the base making germplasm derivatives of breeding programs prone to greater susceptibility occurrences.

The Practical Output Contribution of Novel Genomic Diversity

Apart from the conventional mode of wheat improvement is the newer trend where close relatives of the primary gene pool are exploited for improvement programs and categorized as interspecific breeding. The focus here is on the D genome that could easily be extended to the A genome diploids and novel AB tetraploids. Though direct crossing has greater precision this group's emphasis has been on adding diversity to wheat across its three genomes for global handling of complex stresses. Hence bridge crossing has been favored and has contributed on the applied scale as evidenced from global varietal releases, registration of stocks and generation of pre-bred materials. Information has been captured from research conducted up to 2004 in CIMMYT Mexico and reported in Annual Wheat Newsletter of 2004 with additional outputs from similar resource materials but environmentally removed to Islamabad, Pakistan from 2005 until the present having the group leader as the common lead person throughout.

The most significant practical usage of the D genome synthetics for biotic stresses to date have been for Fusarium head blight, Septoria leaf blotch, Cochliobolus spot blotch, karnal bunt, yellow, leaf and stem rust and powdery mildew. The strategy has been to identify resistance first in the primary synthetic hexaploids and then exploit those in crosses onto elite wheat cultivars that need such improvement. Each of these biotic stresses are briefly discussed:

Fusarium Head Scab

The internationally recognized Sumai 3 has over the past few decades stood as the resistant standard for scab resistance and if the emphasis is on Type2 resistance then acceptance of up to 15% infection is considered as resistant since the central inoculated florets get damaged, thus immunity is never present. The extensive testing done in Mexico by the CIMMYT Wheat Wide Crossing program researchers allowed interspecific and intergeneric cross combination products to define several entries that were similar to or better than Sumai 3; better in the sense that the inoculated florets in the middle of the spike did not damage the seed formation completely as happens with Sumai 3. Two groups with potent resistance were selected and one set is of the D genome synthetics (Zaharieva et al. 2003) and another that has pyramided resistance from a intergeneric combination combined with a primary synthetic. The former SH group formed a sub-set with 35 primaries and the latter a few sister lines with extended details to cover all 4 categories of evaluation from type 1 to 4 and then evaluated for multiple stresses (Mujeeb-Kazi et al. 2004)

Septoria Leaf Blotch

Septoria leaf blotch (*Bipolaris sorokiniana*) limits wheat production in high rainfall areas across 10.4×10^6 hectares globally. Disease scoring is of a double-digit scale from 1-1 (resistant) to 9-9 (susceptible) and recorded over the three grain development stages (watery, milky, doughy). The D genome synthetics proved to be superior for their resistance levels with numerous giving scores between 1-1 and 3-1 compared to the bread wheat cultivars that exhibited a susceptible trend with scores between 4-1 and 9-9. The identification of resistance in unique SH sources led breeders to exploit the germplasm in their wheat improvement efforts. The derivatives allowed for selection of good agronomic plant progenies with high levels of resistance and also led to germplasm registrations (Mujeeb-Kazi et al. 2000, 2001a). Similar resistance was also observed in the A genome synthetics (AABBAA) where the score range of 1-1–2-1 in abundance were superior to the levels seen for their durum parents that had the best at 4-1 to poor forms reaching 8-9. Recently, Aggarwal et al. (2011) developed a SCAR marker for detection of spot blotch in leaf and field soil which is suggested to play a key role in effective management of this disease.

Cochliobolus Spot Blotch

Spot blotch affects wheat crops across several environments from Latin America, Asia and southeast Asia with Bangladesh being represented as a major disease location. Its presence in the wheat crop cycle of 2009–2010 in lower Punjab of Pakistan was alarming causing a leading cultivar "Bhakkar" to be banned from further planting in that area. The screening site for this disease is the most severe in Mexico at the location "Poza Rica" where field screening under natural conditions allowed numerous entries to be selected with resistance that included genetic stocks of various genomes and their advanced pre-bred derivatives. The best resistant lines selected were later called "Mayoor" and "Sabuf" leading to crop registration (Mujeeb-Kazi et al. 1996b). Subsequently when focus shifted to exploitation of close wheat progenitors, the D genome diversity at the basic primary level and for the advanced derivatives became significant for wheat breeding efforts. From the Mayoor and Sabuf test scores of 9-2–9-4 (9-9 susceptible) screening and a grain finish between 1 and 3 (5 poor blemished grain) the derivatives and D genome stocks utilizing *Ae. tauschii* produced selected products that scored 9-2 with a grain finish of 1–2. These were far superior than Mayoor and Sabuf and utilized via gene pyramiding options. Mayoor was hybridized with a synthetic combination TKSN1081/Ae. tauschii (222) to yield superior spot blotch resistant derivatives which also possess multiple disease resistance that covers scab (type 1–4), Septoria and karnal bunt. Both Mayoor and Sabuf have been further utilized in the development of molecular mapping populations with susceptible wheat Flycatcher and Ciano (Mujeeb-Kazi et al. 2004) as follows:

1. Mayoor//TKSN 1081/*Ae tauschii* (222)/3/Flycatcher with 171 doubled haploids
2. Sabuf/3/Bacanora//Ceta/*Ae. tauschii*(895)/4/Flycatcher with 125 doubled haploids
3. Sabuf/3/Bacanora//Ceta/*Ae. tauschii*(895)/4/Ciano with 102 doubled haploids.

The above populations are a conduit for molecular studies involving QTL mapping aspects and testing internationally.

Karnal Bunt

This soil borne disease (*Neovossia indica*) is a tremendous quarantine concern as once the soil is infected its occurrence is un-ending. The national level of acceptance of grain for consumption is set for 3 % infection. Grains shipped across nations spread the disease if they carry it and thus caution is exercised to wash seed, treat it and stringently apply quarantine testing to provide seed disbursement that is virtually 0 % infectious. Disease free sites are used for seed increase as in Mexico by CIMMYT. However some countries are lax as to the seed been tested across wheat cultivation zones and danger prevails as to its spread. Diversity for resistance exists and breeding efforts are preferred. Durum cultivars are generally field resistant with bread wheats open to high susceptibility levels. Search for resistant resources has been a priority for the past 2–3 decades. Since swift outputs have been sought the choice of the D genome has had a high priority and fortuitously the SH wheats derived from durum/*Ae. tauschii* combinations gave outputs where immunity was abundant in the materials tested. Stringent testing under controlled testing made the field resistant durums susceptible with their SH products remaining immune (Villareal et al. 1996); an unequivocal proof that D genome accessions carried the resistance. This led to the identification of user friendly developed stocks and their registration (Mujeeb-Kazi et al. 2001b). Further the SH wheats upon crossing to susceptible elite bread wheats gave derivatives that had resistance transferred into the selections (Mujeeb-Kazi et al. 2006a). The identification of QTLs underlying the resistance to karnal bunt has been carried out using disease screening in multiple-environment data and it was established that two major QTLs reside on chromosomes 3BL and 2DS which ultimately reduced the disease spread (Sukhwinder-Singh et al. 2012).

Rusts (Leaf, Stem and Yellow)

Of the total 215 million hectares area planted to bread and durum wheat globally about 44 % (95 million hectares) are in Asia. Sixty-nine million hectares are in China, India and Pakistan. Most of the farmers are classed as "poor" or "small" farmers and hence food security plus production stability are of significant importance. Stem rust has been under control since the green revolution times of the mid-sixties. Leaf rust (caused by *Puccinia triticina*) and yellow rust (caused by *Puccinia strifor-mis*) however have the potential to affect production levels up to 60 and 43 million hectares respectively in Asia if susceptible cultivars were grown (Singh et al. 2004). Though fungicidal applications offer control their use is an added cost to farmers besides being unsafe environmentally. Hence growing resistant cultivars is the most effective and efficient control strategy (Rizwan et al. 2008). These major stresses have to be simultaneously addressed.

Rusts have remained very dynamic pathogens that have consistently existed as a major wheat-breeding objective globally. Currently major attention is given to stem rust and yellow rust is a close second that should not be a reason to look at leaf rust with complacency. The conventional picture details numerous genes for all three rusts in the Wheat Rust compilation (McIntosh et al. 2003). Genes have been identified from within conventional wheat cultivars and also from alien species. The resource of the D genome *Ae. tauschii* is of interest for this presentation and its contributions to yellow rust and stem rust shall be elucidated. An in depth coverage of leaf rust has been made by Dubin and Brennan (2009) in the IFPRI 2020 Vision Initiative Report hence it is not covered here.

Stripe or Yellow Rust

Stripe rust (yellow rust), caused by *Puccinia striiformis* f. sp. *tritici*, is an important foliar disease of wheat. It occurs in wheat growing areas of temperate, moist and cool regions in all continents except Antarctica (Chen 2005). Its wider prevalence is a global threat to wheat production inflicting about 30–100 % grain losses besides affecting the quality of grain and forage (Chen 2005). In China, India and Pakistan; the top wheat producers in Asia where 59.3 million hectares are under wheat cultivation, stripe rust prevails in 24.8 million hectares i.e., ~ 40 % of wheat grown area (Singh et al. 2004). The deployment of stripe rust resistant genes is the most effective method to protect wheat productivity and several stripe rust resistant genes have been deployed successfully in the past. So far 84 *Yr* genes have been designated in wheat out of which 36 genes have temporary designations (McIntosh et al. in MacGenes 2010). Additionally, 52 QTLs have also been identified conferring resistance to stripe rust in bread and durum wheats (McIntosh et al. in MacGenes 2010).

Utilization of genetic resistance and its incorporation in wheat demands genetic resources with enormous potential. These genetic resources have been categorized as wild relatives, elite cultivars and landraces (Bux et al. 2012a; Kazi et al. 2012; Arif et al. 2012). Wild progenitors of wheat possess abundant unutilized genetic diversity. There are several stripe rust resistance genes derived from wild relatives like *Yr5* from *T. spelta* (Kema 1992), *Yr8* from *Ae. comosa* (Riley et al. 1968), *Yr9* from *Secale cereale* (Zeller 1973), *Yr28* from *Ae. tauschii* (Singh et al. 1998), *Yr37* from *Ae. kotschyi* (Marais et al. 2005), *Yr38* from *Ae. sharonensis* (Marais et al. 2006), *Yr40* from *Ae. geniculata* (Kuraparthy et al. 2007) and *Yr42* from *Ae. neglecta* (Marias et al. 2009). Recently, Ren et al. (2012) tagged a *Yr* gene in synthetic hexaploid line C110 and designated the gene as *YrC110*. Unfortunately, no resistance gene to stripe rust has been identified and transferred to wheat from the A-genome diploid progenitors *T. monococcum and T. urartu.*

Stem Rust

Stem rust (*Puccinia graminis tritici*) resistance got high attention after the new race TTKSK (UG99) emerged in Uganda in 1999 (Pretorius et al. 2000). Stem rust per se has the potential to devastate wheat in all continents (Dubin and Brennan 2009; Hodson 2011) The subsequent spread of TTKSK in that region soon found it to attack wheat in Kenya, Ethiopia, Yemen reaching up to Iran throwing the SE Asia region in jeopardy as the migration trends could take the pathogen into Pakistan and beyond. CIMMYT wheat breeding program had been utilizing the D genome synthetics for various other attributes and advanced derivatives screened in Kenya gave encouraging resistant results. Selections were high yielding and also UG99 resistant (Singh et al. 2011a, b) show promise and the danger from its spread somewhat reduced. Global alliances and funding have alleviated the hazard from this new race and also from its mutant forms. The threat of the race and its lineage has been substantiated (Singh et al. 2008).

In Pakistan exists a local race of stem rust that has shown virulence and is an added concern for that region. We have advocated that screening against this race should be a priority and from the resistant selections further evaluations be made in Kenya to ensure that final selections have resistance against both forms (i.e. UG99, mutants, and the local Pakistan race).

Resistance gene *Sr2* provided stability to wheat varieties with the release of Yaqui 50 in Mexico and other *Sr2* carrying wheats released since then stabilized the stem rust situation in Mexico and other countries where semi-dwarf wheats got adopted. When present alone *Sr2* gene confers slow rusting that is inadequate under heavy disease pressure but does provide satisfactory resistance when it is in combination with other minor genes. Identifying/developing adapted resistant cultivars in a relatively short time and replacing the susceptible cultivars before rust migrates into our terrain is the strategy to mitigate potential losses. Although several genes will provide resistance to the race UG99 the long-term strategy should focus on rebuilding the "*Sr2* complex" to achieve long-term durability. The complex to be built will involve an assemblage of slow rusting gene *Sr2* with other unknown additive genes of similar nature (Singh et al. 2006). Addressing the target swiftly has been considered very crucial as migratory paths present a gloomy picture for wheat production should adequate resistance not be incorporated in regional wheat varieties (Hodson et al. 2005, Reynolds and Borlaug 2006). To add to the swiftness would be efficient tools (Mujeeb-Kazi et al. 2006b; Randhawa et al. 2009) as an integral means to drive the gene transfers (Mago et al. 2005) and give allelic output stability (Mujeeb-Kazi 2003, 2005, 2006). The allelic diversity from unique genetic resources will also be a significant aid (Coghlan 2006; Rizwan et al. 2007; Simonite 2006).

Hence, the availability of broad-based genetic variability is a pre-requisite for having a sound and successful wheat improvement program. Genomic diversity is one unique option available and the maximum ease that permits exploitation of this resource comes from the D and A genome diploids of the primary Triticeae gene pool that have generated via pre-breeding the synthetic hexaploid germplasm (Mujeeb-Kazi 2003, 2006). Synthetic hexaploids created by crossing *Triticum turgidum* with *Aegilops tauschii* tap the desirable genes present in the wild D genome diploid species (Trethowan and Mujeeb-Kazi 2008; Trethowan and Van-Ginkel 2009). These synthetic hexaploid wheats have been used as an intermediary for transferring resistance genes from the wild D genome ancestor to cultivated wheat. As both synthetic hexaploid and bread wheat varieties have the same genomic constitution with perfect homology they can be readily inter-crossed.

Several varieties with UG99 and its lineage lines have been released in various countries based upon data gathered from Kenya and Ethiopia (Joshi et al. 2011) and as early as 2006 in CIMMYT a targeted program to increase yield and possess stem and yellow rust resistance got actively moving with its superior products obtained as reported by Singh et al. (2011b). It was encouraging to see that in the various promising high yielding and resistant lines identified a significant number had unique genetic resources in their pedigrees that included several of *Ae. squarrosa* (Syn. *Ae. tauschii*) and *Thinopyrum acutum*. Almost all possessed APR genes and others that contributed. The contribution to yield from alien resources has been well demonstrated by the varietal release in Sichuan, China of Chaunmai 42 that reported a yield enhancement of 22.7% over the earlier cultivar Chaunmai 107 (Yang et al. 2009).

In initial stages has been the contribution with basic research aspects where significant contributions related to UG99 pathotypes resistance has been reported. This has come through screening of various accessions of the AA diploid progenitor wheats (Rouse and Jin 2011), the D genome *Ae. tauschii* accessions (Rouse et al. 2011) and the tertiary gene pool diploid *Th. bessarabicum* promise (Xu et al. 2009). All the above have opened up avenues for global researchers to embark on volatile "pre-breeding" programs where from the closely related forms of the AA and DD resource pay offs will be swifter while those from the E^bE^b genome transfers more time consuming due to the genetic distance of that diploid resource but still needed as translocations from this source may be additive to what has recently been reported by Qi et al. (2011) with *Dasypyrum villosum* (*Sr52*). A new dimension to the wild progenitor usage also exists in land race genetic diversity resources that are worthy of exploiting as done recently by Pretorius et al. (2012) for the South African land races where *Sr2, Sr24* and *Sr34* were detected via marker diagnostics.

A Major Anticipated Contribution of the Tertiary Gene Pool using Th. bessarabicum

Th. bessarabicum is a diploid salt tolerant grass species with the E^b genome and a preferred alien diversity resource extensively used by wide cross researchers since early 1980's in England, USA and Mexico. It has been combined with *T. aestivum* cv Chinese Spring to yield an amphiploid that is an octoploid ($2n=8x=56$, $AABBDDE^bE^b$). All seven addition lines have since been produced from the initial five were studied by Zhang et al. 2002. The initial addition line set was shared with R.R-C Wang by A. Mujeeb-Kazi from CIMMYT, Mexico. Very recently the complete set of seven was developed by Kazi 2011. The germplasm provided to Xu by R.R-C Wang in USA has been screened by Xu et al. 2009 for evaluating each addition line ($2n=6x+42+2=44$; 1J $\{E^b\}$ to 7 $\{E^b\}$) and the amphiploid for UG99 resistance. Results have shown considerable promise for resistance to race TTKSK (known as UG99 or TTKS). Disease scores of 0, 1 and 2 were considered as low infection types and the data obtained by Xu et al. 2009 suggests that the amphiploid, and addition lines 1J, 2J, 6J and 7J ($J=E^b$) fell in the low infection category. Underway has been a program to introgress the value of resistance from the alien diploid species into wheat using the *Ph* gene manipulation strategy exploiting the recessive ph genetic stock developed by late ER Sears. This effort was initiated to promote wheat alien translocations from the amphiploid source (Mujeeb-Kazi 2006a) and also initiated in a targeted manner from each disomic addition line of *Th. bessarabicum* (Kazi 2011). Several Robertsonian and subtle homologous translocations have been identified by C-banding and euploid progeny ($2n=6x=42$) recovered (Kazi 2011). Additional resources studies by Xu et al. 2009 included *Th. intermedium, Th. elongatum, Th. ponticum, Elynus recticetus, Ae. caudate* and *Ae. speltoides* which suggest the value of this removed tertiary gene pool for harnessing diversity of value.

Powdery Mildew

Wheat powdery mildew, caused by *Blumeria graminis* (DC.) E.O. Speer f. sp. *Tritici* Em. Marchal (*Bgt*) = *Erysiphe graminis* DC. *Ex Merat* f. sp. *Tritici* Em. Marchal, is one of the most devastating diseases of common wheat occurs in many areas, including China, Germany, Japan, Russia, United Kingdom, South and West Asia, North and East Africa, and the Southeastern United States. Yield losses range from 13–34 % due to this disease (Griffey et al. 1993). To date, 41 loci with more than 60 genes/alleles for resistance to powdery mildew have been identified and located on 18 different chromosomes in bread wheat. 29 resistance genes/alleles have been tagged with different types of molecular markers. The desirable type of resistance to powdery mildew is called adult plant resistance (APR), which retards infection, growth and reproduction of the pathogen in adult plants but not in seedlings. It is also called "slow mildewing" and "partial resistance". APR in wheat cultivar Knox and its derivatives remained effective against powdery mildew infection during the 20 years in which these cultivars were grown commercially (Shaner 1973). Another cultivar, Massey, which is a derivative of Knox62, was developed and released in 1981 (Starling et al. 1984) and still has effective powdery mildew resistance in adult plants. Common sources of *Pm* genes are different species within the primary, secondary and tertiary gene pools. Many of the resistance genes were introduced from ancestral and other wild species related to common wheat such as *Triticum monococcum,* close relative of the A genome progenitor *Triticum uratu*, the B genome progenitor *Aegilops speltoides*, and the D genome progenitor *Ae. Tauschii* (Hsam and Zeller 2002). They reported a total of 22 resistance alleles at ten loci in common wheat indicating that *Pm* genes may still be found in cultivated wheat. Earlier studies by Mains (1933) identified that the wild wheat relatives *T. monococcum* (AA genomes), *T. dicoccum* (AABB), and *T. timopheevi* (AAGG) are the sources of resistance genes to powdery mildew as early as 1933. Screening of old wheat cultivars, landraces and related species for resistance to powdery mildew started in the 1930's (Hsam and Zeller 2002). *Pm* genes were identified in many different, widely distributed wheat cultivars and landraces. *Pm5a* and *Pm5b*, followed by *Pm2, Pm6,* and *Pm8* are the most common in Europe, Asia and Mediterranean cultivars. *Pm3a* is commonly found in wheat cultivars grown in diverse geographical locations including the Balkans, Japan, China and the US. *Pm3c* was identified in Germany, while *Pm3d* was found in several European countries and China. *Pm4a* has been used in commercial wheat cultivars in Germany and China. A number of commercially grown cultivars have been found to have *Pm* gene combinations (Heun and Fischbeck 1987). The best known cultivars are Normandie with *Pm1, Pm2,* and *Pm9,* Maris Huntsman with *Pm2* and *Pm6*, Kronjuvel with *Pm4b* and *Pm8*, and 623/65 with *Pm4b* and *Pm8* (Liu et al. 1999). Gene transfer from species within the primary gene pool of *Triticum* with homologous chromosomes to wheat can be done directly by hybridization, recombination and backcrossing.

Ae. tauschii has proved to be an important source of resistance against powdery mildew. Earlier Gill et al. (1986) screened 60 *Ae. tauschii* accessions to four different *Bgt* isolates and identified 11 highly resistant and 20 moderately resistant accessions. Hsam and Zeller (2002) transferred two resistance genes, *Pm2* and *Pm19*, from *Ae. tauschii* to common wheat. Two germplasm lines, NC96B-GTD3 and NC97BGTD7, were developed which carries resistance genes *Pm34* and *Pm35* against PM (Murphy et al. 1998; Mirinda et al. 2006, 2007). Similarly, *Pm25* has been transferred from diploid *T. monococcum* to common wheat germplasm NC96BGTA5 (Shi et al. 1998; Murphy et al. 1998). In tetraploid wild emmer wheat (*T. dicoccoides*), seven PM resistance genes viz. *Pm16, Pm26, Pm3, Pm31, Pm36, Pm41* and *pm42* have been identified and transferred to wheat (Rong et al. 2000; Liu et al. 2002; Hsam and Zeller 2002; Hua et al. 2009). Recently, Rafiq et al. (2012) identified several A- (*T. monococcum*; *T. urartu*) and S-genome (*Ae. speltoi-

des) derived amphiploids resistant to PM. Both diploid and tetraploid parental lines were proposed to carry resistance genes against PM.

The other resistance sources against powdery mildew include *Pm12 (*6B) and Pm32 (1B) from *Ae. speltoides* (Jia 1996; Hsam et al. 2003), *Pm29* (7D) from *Ae. geniculata* (Zeller et al. 2002), *Pm34* and *Pm35* (5DL) from *Ae. tauschii* (Miranda et al. 2006, 2007; Qiu et al. 2006), *Pm39* from *Ae. umbellulata* (Zhu et al. 2006), and some undesignated genes from *Ae. longissima*, *Ae. searsii*, *Ae. umbellulata* (Buloichik et al. 2008), *Ae. comosa* (Bennett 1984) and *Ae sharonesis* (Olivera et al. 2007). From *T. monococcum*, *Pm25* and three temporarily designated genes, *Pm2026, Mlm3033* and *Mlm80*, have been introduced in wheat (McIntosh et al. 2010).

Molecular Diagnosis for Host Resistance

So far, six genes providing resistance against diseases in wheat have been successfully cloned (Liu et al. 2012). Diagnostic markers having ability to capture allelic variation have been developed for a major powdery mildew resistance locus, *Pm3*, and *Lr34/Yr18/Pm38* locus providing broad spectrum resistance against leaf rust, stripe rust and powdery mildew (Tommasini et al. 2006; Lagudah et al. 2009; Miedner et al. 2012). The cloning followed by sequencing of the adjacent untranslated regions of *Pm3* resistance alleles helped in development of seven allele-specific molecular markers which successfully discriminated allelic variants at the *Pm3* locus (Tommasini et al. 2006). In multiple studies, these markers identified desirable alleles in the US and European wheat cultivars (Peusha et al. 2008; Chen et al. 2009; Lillemo et al. 2010; Mohler et al. 2011). However, their use for identification of alleles in Chinese wheats is restricted due to the susceptibility of the all seven alleles in China. The major locus *Lr34/Yr18/Pm38* provides durable resistance against many diseases and its usage is encouraged worldwide. The functional markers developed by Lagudah et al. (2009) are very simple to apply due to their easy resolution on agarose gels. This marker provided positive association between stripe rust and marker data on a wide array of wheat lines developed at CIMMYT (Wu et al. 2010). In another study on Chinese landraces, this marker showed the presence of *Lr34/Yr18/ Pm38* allele in 82.1 % genotypes. However, 25 % of these genotypes were found susceptible to stripe rust in field (Wu et al. 2010). The susceptibility of landraces with positive *Lr34/Yr18/ Pm38 a*llele is proposed due to the presence of inhibitor genes or absence of a functional gene that is essential in the biosynthetic pathway for the expression of Lr34/Yr18/Pm38. For more than a decade, the T1BL.1RS translocation has been widely used in global wheat breeding programs. Several agronomic features and resistance to diseases are associated with this translocation, although its resistances to diseases have been overcome in many locations. Therefore, it is important to identify the T1BL.1RS translocation in wheat breeding. Functional markers based on the rye secalin gene on 1RS were successfully applied in breeding (Liu et al. 2008). Another important stripe rust resistance gene, *Yr17* located on chromosome 2NS of *T. ventricosum* Tausch. has been translocated to the short arm of bread wheat chromosome 2AS (Helguera et al. 2003), and this chromosomal segment also conferred resistance to leaf rust (*Lr37*) and stem rust (Sr38). The *Lr19* gene, originated from decaploid *Th. ponticum*, was transferred into durum wheat, and widely conferring resistance to leaf rust in wheat (Gennaro et al. 2009). *Ae. tauschii* Cosson was the donor of the *Lr21* that is a durable and highly effective leaf rust resistance gene, and it has been incorporated into wheat cultivar and is available for breeding (Talbert et al. 1994). The leaf rust resistance gene *Lr47* confers resistance to a wide spectrum of leaf rust strains. This gene was recently transferred from chromosome 7S of *T. speltoides* Tausch to chromosome 7A of common wheat (Helguera et al. 2000). Leaf rust resistance gene *Lr51*, located within a segment of *T. speltoides* Tausch chromosome 1S, was translocated to the long arm of chromosome 1B of bread wheat, which is resistant to the current predominant races in USA (Helguera et al. 2005). The gene-specific markers *Xucw108* and *Xuhw89* for *Gpc-B1* and *Yr36* originated from chromosome 6BS of *T. turgidum* ssp. dicoccoides. They were identified and validated in a collection of 117 cultivated tetraploid and hexaploid wheat germplasm (Uauy et al. 2006).

Molecular Basis of Disease Resistance

It is important to understand the molecular mechanism of disease resistance to devise sustainable control (Bux et al. 2012b). The earlier findings of Flor (1956) proposing gene-for-gene hypothesis provided basis for predicting the molecular basis of disease resistance. The molecular interpretation of Flor's findings, avirulent (*Avr*) genes encode signal transduction that is perceived by the products of plant *R* genes, are regarded as foundation concept in disease resistance. The *R*-gene/avirulence factor complex is thought to instigate a series of signaling cascades leading to disease resistance. Rapid oxidative bursts, cell wall strengthening, induction of defense gene expression and rapid cell death at the site of infection are the downstream cellular events that confer resistance state (Morel and Dangl 1997). In more elaborated form, the direct or indirect recognition of pathogen by host '*R*' genes lead to a resistance response known as effector-triggered immunity (ETI), which includes localized programmed cell death (PCD), known as the hypersensitive response (HR) which ultimately restrict pathogen growth (Dangl et al. 1996). *R* gene proteins serve to recognize pathogen effectors either through direct interaction or as guards for target molecules and are known to confer resistance to bacteria, viruses, nematodes, oomycetes, insects, and biotrophic fungi (Martin et al. 2003). Genes for resistance, their protein products, and underlying mechanism are being investigated (Hammond-Kosack and Jones 1996). Sufficient progress has been made in these aspects that will facilitate developing effective control strategies.

Most of the *R* genes cloned so far, revealed a nucleotide-binding site (NBS) and a leucine-rich-repeat (LRR) region. These are the most abundant types in plant species (Meyers et al. 1998). These plant *R* genes encode proteins that have a putative amino-terminal signaling domain, a nucleotide binding site and a series of carboxy-terminal leucine repeats (Meyers et al. 2005). Two different types of NBS-LRR proteins have been reported. One major class has an amino-terminal TIR (Toll/interleukin receptor) domain also called TIR-NBS-LRR or TNL proteins. Other class includes the genes which encode an amino-terminal coiled-coiled motif (CC-NBS-LRR or CNL proteins). The mechanism of resistance induced by *Pm3a* and its other allelic forms in wheat (Feng et al. 2010) and by *Pb1* against rice blast clearly exemplifies that the single amino acid residue at the final position of the kinase-2 motif is the characteristic of coiled-coil (CC) motif while the trytophan (W) and aspartic acid (D) are the characteristics of TIR-type proteins. The details of the molecular functions of these protein domains and their interacting partners are still being established. However, the consistent identification of this class of proteins across diverse plant species demonstrates that the NBS-LRR genes are a pillar of plant defense against pathogen. The majority of the *R* genes in *Arabidopsis* are TNL genes; however they have not yet been reported in cereals. In rice, about 1,500 NBS coding sequences were analyzed and not a single sequence was known to have TIR binding domain (Zhou et al. 2004).Three leaf rust resistance genes, *Lr1, Lr10, Lr21* and one powdery mildew resistance gene, *Pmb3b*, are known to have CNL type domain (Feuillet et al. 2003; Huang et al. 2003; Yahiaoui et al. 2004).

Conclusions and Future Prospects

Three areas of prevailing concern that have received enormous attention are global warming, climate change and food security. Associated with these has been population growth with futuristic projections made first at 2025 (8.2 billion) and now for 2050 at 9.2 billion expected to touch 10 billion swiftly by 2055. The three catchy buzz words above have generated many discussion fora, created a lot of debate and delivered numerous "smart" ideas as the outcome of various interactive sessions. However, at the end of the day if we look at national "wheat" based food security it is doubtful if we are better off now in 2012 than we were a few years before. Our productivity is not that earth shaking, average national yields in several countries fluctuate yearly and the yield gap remains as significant as before. The stress constraints are more complex and the resolve to adopt a way forward for achieving promising returns rather feeble. Have we addressed these constraints and can the outputs be quantified? The trend is negative or on the conservative side "STATIC" as to progress. The umbrella cover that integrates research/production activities is non-existent and resources are far from being neatly agglomerated. Taking wheat in Pakistan as an example we still hover around 25 million tons per acre and mean yields are not beyond 25–30 maunds. The per hectare figure is estimated at 2.6 t. Stress constraints still dominate the scene and one wonders if solutions have been found to allow us to see durability with high value. Problems galore relate to the three rusts with stem rust the one abused most for personal projection and professional clout but done with questionable structural quality science. Then come other attributes like Karnal bunt. Minor diseases like mildew, spot blotch, BYDV, aphids, loose smut are additive and the environmental limitations imposed by heat, water and salinity more complex.

Increasing yield per se must receive special attention as we project forward and for that a holistic strategy is vital that embraces both biotic/abiotic stresses complimented by exploiting targeted yield traits. Details to directly increase yield are essential to have and it is prudent to be aware how we can get more per unit area through focused research strategies. Crucial will be "food" security that emanates from gene and varietal deployment. The tools are there in enormous untapped close and distant genomic diversity mines, large spikes, 1,000 grain weight, photosynthetic efficiency, root profile, various phenology parameters and efficient technologies that should be exploited for practical benefit. Advocates of food security must set targets stringently and implement. The vision 2050 is excellent if we prepare ourselves accordingly to combat those challenges by on ground outputs that capture the wide expanses of allele rich diversity resources. Paramount would be holistic strategies that maintain good multidisciplinary balance, embrace integration, exploit the conventional resources, selectively tap the high technology inputs, deliver products that can redefine the output limits that currently exist and have built in durability that can sustain the produce within the national boundaries and also protect us from the vagrancies of pathogenic migrations via environmental means.

Turbulent environmental scenarios will necessitate researchers to harness genes from close and distant species resources distributed within the three Triticeae gene pools and utilize efficient technologies that can add efficiency to delivering varietal output products in a swift manner. Efforts will rely more on the primary gene pool species and accessions but not refrain to exploit those forms that are removed and placed in the tertiary pool. Multiple stresses will be the order of research focus addressed through multidisciplinary research protocols mediated by novel techniques that add efficiency to breeding like limited backcrossing, selected modified bulk, doubled haploidy for rapid homozygosity at earlier segregating generations (preferably F_3), molecular diagnostics (marker assisted selection, gene pyramiding, choice of linked genes and markers) that are allele specific correlated with agronomic practices that maximize production targets that fall under crop management.

True that research is the major factor of providing food security but we can safely say that management can provide faster returns if its cohesive structure is handled astutely to maximize production via "IMPLEMENTATION" of major contributing strategies. Objectivity has to guide progress with "creativity" being the pivotal word in which "diversity" takes the lead for biotic stresses across all wheat genomic levels. Integration of all available technologies shall be essential to cope with the projected population increases that will be dictated by availability of practical outputs (varieties) that have built in stability for agricultural sustainability on farmers fields by agglomeration of genes that offer disease resistance durability that can tackle the volatile dynamic nature of the pathogenic interplay as well as possessing abiotic tolerance for traits that have a static system and thus generally have a longer performance tolerance life. For biotic stresses and use of genomic resources for improvement the diversity combination permutations are excessive and offer abundant options for stabilizing wheat varietal profiles for continued wheat global productivity. The reliance on conventionally available variation within wheat cultivars across its spring, winter and facultative habits will persist, but the course ahead for adding unique variations will increase. The forms of closer D genome affinity or homology receiving greater attention currently (Ogbonnaya et al. 2013) shall be complimented by a significant shift that shall exploit the other two genomes and the tertiary gene pool species (Mujeeb-Kazi et al. 2008b). With all pre-breeding and breeding efforts working in tandem across multiple disciplines aided by molecular diagnostics and efficient crop production technologies that fit the "holistic" production strategies we end this presentation optimistically by projecting that the approximately 2 % annual increase required to feed the growing populace is within our grasp.

References

Aggarwal R, Gupta S, Banerjee S, Singh VB (2011) Development of a SCAR marker for detection of *Bipolaris sorokiniana* causing spot blotch of wheat. Can J Microbiol 57:934–942

Alonso LC, Kimber G (1984) Use of restitution nuclei to introduce alien genetic variation into hexaploid wheat. Z Pflanzenzücht 92:185–189

Arif MA, Bux H, Gul-Kazi A, Rasheed A, Napar AA, Riaz A, Mujeeb-Kazi A (2012) Stripe rust analysis of D-genome synthetic wheats (2n=6x=42; AABBDD) and their molecular diversity. Arch Phytopathol PFL. doi:10.1080/03235408.2012.677492

Bennett FGA (1984) Resistance to powdery mildew in wheat: a review of its use in agriculture and breeding programmes. Plant Pathol 33:279–300

Buloichik AA, Borzyak VS, Voluevich EA (2008) Influence of alien chromosomes on the resistance of soft wheat to biotrophic fungal pathogens. Cytol Genet 42:9–15

Bux H, Ashraf M, Chen XM (2012a) Expression of high temperature adult plant (HTAP) resistance against stripe rust (*Puccinia striiformis* f.sp.*tritici*) in Pakistan wheat landraces. Can J Plant Pathol 34:68–74

Bux H, Ashraf M, Rasheed A, Dipak SP, Kazi AG, Afzaal M (2012b). Molecular basis of disease resistance in cereal crops: an overview. In: Ashraf M, Öztürk M, Ahmad MSA, Aksoy A (eds) Crop production for agriculture improvement. Springer, pp 477–489. doi:10.1007/978-94-007-4116-4_18

Chen XM (2005) Epidemiology and control of stripe rust on wheat (*Puccinia striiformis* f. sp. *tritici*) on wheat. Can J Plant Pathol 27:314–337

Chen YH, Hunger RM, Carver BF, Zhang HL, Yan LL (2009) Genetic characterization of powdery mildew resistance in U.S. hard winter wheat. Mol Breeding 24:141–152

Coghlan A (2006) Synthetic wheat offers hope to the world. New Scientist. On line report on 11 Feb 2006

Cox TS, Hatchett JH, Gill BS, Raupp WJ, Sears RS (1990) Agronomic performance of hexaploid wheat lines derived from direct crosses between wheat and *Aegilops squarrosa*. Plant Breeding 105:271–277

Cox TS, Harrell LC, Chen P, Gill BS (1991) Reproductive behavior of hexaploid/diploid wheat hybrids. Plant Breeding 107:105–118

Dangl JL, Dietrich RA, Richberg MH (1996) Death don't have no mercy: cell death programs in plant-microbe interactions. Plant Cell 8:1793–1807

Dewey DR (1984) The genomic system of classification as a guide to intergeneric hybridization with the perennial Triticeae. In: Gustafsson JP (ed) Gene manipulation in plant improvement, 16th Stadler Genetics Symposium, Plenum Press, New York, pp 209–279

Dubin HJ, Brennan JP (2009) Combating stem and leaf rust of wheat. IFPRI Discussion Paper 00910, November 2009, pp 1–53

Farrar W (1904) Some notes on the wheat "Bobs"; its peculiarities, economic value, and origin. Agric Gaz NSW 15:849–854

Feng DS, Ma X, Lin AL, Wang HG, Tian JC (2010) Isolation of resistance gene analogues to powdery mildew resistance sequences in hexaploid wheat. Biologia Plantarum 54(3):551–555

Feuillet C, Travella S, Stein N, Albar L, Nublat A, Keller B (2003) Map-based isolation of the leaf rust disease resistance gene *Lr10* from the hexaploid wheat (*Triticum aestivum* L.) genome. Proceedings National Academy Sciences 100:15253–15258

Flor H (1956) The complementary genic systems in fl ax and fl ax rust. Adv Genet 8:29–54

Friebe B, Jiang J, Raupp WJ, McIntosh RA, Gill BS (1996) Characterization of wheat-alien translocations conferring resistance to diseases and pests: current status. Euphytica 91:59–87

Gennaro A, Koebner RMB, Ceoloni C (2009) A candidate for *Lr19*, an exotic gene conditioning leaf rust resistance in wheat. Functional Integr Genomics 9:325–334

Gill BS, Raupp WJ (1987) Direct gene transfers from *Aegilops squarrosa* L. to hexaploid wheat. Crop Sci 27:445–450

Gill BS, Raupp WJ, Sharma HC (1986) Resistance in *Aegilops squarrosa* to wheat leaf of rust, wheat powdery mildew, greenbug, and Hessian fly. Plant Dis 70:553–556

Griffey CA, Das MK, Stromberg EL (1993) Effectiveness of adult-plant resistance in reducing grain yield loss to powdery mildew in winter wheat. Plant Dis 77:618–622

Gul-Kazi A, Rasheed A, Mahmood T, Mujeeb-Kazi A (2012) Molecular and morphological diversity with biotic stress resistances of high 1000-grain weight synthetic hexaploid wheats. Pak J Bot 44:1021–1028

Hammond-Kosack KE, Jones JDG (1996) Inducible plant defense mechanisms and resistance gene function. Plant Cell 8:1773–1791

Helguera M, Khan IA, Dubcovsky J (2000) Development of PCR markers for wheat leaf rust resistance gene *Lr47*. Theor Appl Genet 101;625–631

Helguera M, Khan IA, Kolmer J, Lijavetzky D, Zhong-qi L, Dubcovsky J (2003) PCR assays for the *Lr37-Yr17-Sr38* cluster of rust resistance genes and their use to develop isogenic hard red spring wheat lines. Crop Sci 43:1839–1847

Helguera M, Vanzetti L, Soria M, Khan IA, Kolmer J, Dubcovsky J (2005) PCR markers for *Triticum speltoides* leaf rust resistance gene *Lr51* and their use to develop isogenic hard red spring wheat lines. Crop Sci 45:728–734

Heun M, Fischbeck G (1987) Identification of wheat powdery mildew resistant genes by analyzing host-pathogen interactions. Plant Breeding 98:124–129

Hodson DP (2011) Shifting boundaries: challenges for rust monitoring. Euphytica 179:93–104

Hodson DP, Singh RP, Dixon JM (2005) An initial assessment of the potential impact of stem rust (race UG99) on wheat producing regions of Africa and Asia using GIS. Abstracts of the 7th International Wheat Conference, Mar Del Plata, p 142

Hsam SLK, Zeller FJ (2002). Breeding for powdery mildew resistance in common wheat (T. *aestivum* L. em Thell.). In: Belanger RR, Bushnell WR, Dik AJ, Carver TLW (eds) The powdery mildews: a comprehensive treatise, APS Press, St. Paul, pp 219–238

Hsam SLK, Lapochkina IF, Zeller FJ (2003) Chromosomal location of genes for powdery mildew resistance in common wheat (*Triticum aestivum* L. em Thell.). 8. Gene *Pm32* in a wheat-*Aegilops speltoides* translocation line. Euphytica 133:367–370

Hua W, Liu Z, Zhu J, Xie C, Yang T, Zhou Y, Duan X, Sun Q, Liu Z (2009) Identification and genetic mapping of *Pm42*, a new recessive wheat powdery mildew resistance gene derived from wild emmer (*Triticum turgidum* var. *dicoccoides)*. Theor Appl Genet 119: 223–230

Huang L, Brooks SA, Li WL, Fellers JP, Trick HN, Gill BS (2003) Map-based cloning of leaf rust resistance gene *Lr21* from the large and polyploidy genome of bread wheat. Genetics 164:655–664

Islam AKMR, Shepherd KW, Sparrow DHB (1978) Production and characterization of wheat-barley addition lines. Proceedings of 5th International Wheat Genetics Symposium, New Delhi, pp 365–371

Islam AKMR, Shepherd KW, Sparrow DHB (1981) Isolation and characterization of euplasmic wheat-barley chromosome addition lines. Heredity 46:161–174

Jia JZ (1996) RFLP-based maps of the homoeologous group-6 chromosomes of wheat and their application in the tagging of *Pm12*, powdery mildew resistance gene transferred from *Aegilops speltoides* to wheat. Theor Appl Genet 92:559–565

Jiang J, Friebe B, Gill BS (1994) Recent advances in alien gene transfer in wheat. Euphytica 73:199–212

Joshi AK, Azab M, Mossad M, Moselhy M, Osmanzai M, Gelalcha S, Bedada G, Bhatta MR, Hakiom A, Malakar PK, Haque ME, Tiwari TP, Majid A, Kamal MRJ, Bishaw Z, Singh RP, Payne T, Braun HJ (2011) Delivering rust resistant wheat to farmers: a step towards increased food security. Euphytica 179:187–196

Kazi AG (2011) Utilization of *Triticeae* gene pool diversity for wheat improvement. PhD dissertation, Quaid-i-Azam University, Islamabad, pp 1–240

Kema GHJ (1992) Resistance in spelt wheat to yellow rust I. Formal analysis and variation for gliadin patterns. Euphytica 63:207–217

Kimber G, Feldman M (1987) Wild wheat. An introduction. Special Report 353, College of Agriculture, University of Missouri, Columbia, USA

Kruse A (1967) Intergeneric hybrids between *Hordeum vulgare* L. ssp. Distichum (v. Pallas, 2=14) and *Secale cereale* L. (v. Petkus, 2n=14). Royal Veterinary Agricultural College Yearbook 1967. Copenhagen, pp 82–92

Kruse A (1969) Intergeneric hybrids between *Triticum aestivum* L. (v. Koga II. 2=42) and *Avena sativa* L. (v. Stal., 2n=42) with pseudogamous seed formation. Royal Veterinary Agricultural College Yearbook 1969. Copenhagen, pp 188–200

Kruse A (1973) *Hordeum* x *Triticum* hybrids. Hereditas 73:157–161

Kruse A (1974) *Hordeum vulgare* ssp. distichum (var Bomi) x *Triticum aestivum* (var. Koga). An F1 hybrid with generative seed formation. Hereditas 78:319

Kuraparthy V, Chunneja P, Dhaliwal HS, Kaur S, Bowden RL, Gill BS (2007) Characterization and mapping of cryptic alien introgression from *Aegilops geniculata* with new leaf rust and stripe rust resistance genes *Lr57* and *Yr40* in wheat. Theor Appl Genet 114:1379–1389

Lagudah ES, Krattinger SG, Herrera-Foessel S, Singh RP, Huerta-Espino J, Spielmeyer W, Brown-Guedira G, Selter LL, Keller B (2009) Gene-specific markers for the wheat gene *Lr34/Yr18/* Pm38 which confers resistance to multiple fungal pathogens. Theor Appl Genet 119:889–898

Lillemo M, Skinnes H, Brown JKM (2010) Race specific resistance to powdery mildew in Scandinavian wheat cultivars, breeding lines and introduced genotypes with partial resistance. Plant Breeding 129:297–303

Liu Z, Sun Q, Yang T (1999) Development of SCAR markers linked to the *Pm21* gene conferring resistance to powdery mildew in common wheat. Plant Breeding 118:215–219

Liu Z, Sun Q, Ni Z, Nevo E, Yang T (2002) Molecular characterization of a novel powdery mildew resistance gene *Pm30* in wheat originating from wild emmer. Euphytica 123:21–29

Liu C, Yang ZJ, Li GR, Zeng ZX, Zhang Y, Zhou JP, Liu ZH, Ren ZL (2008) Isolation of a new repetitive DNA sequence from *Secale africanum* enables targeting of *Secale* chromatin in wheat background. Euphytica 159:249–258

Liu Y, He Z, Appels R, Xia X (2012) Functional markers in wheat: current status and future prospects. Theor Appl Genet 125:1–10

Mago R, Bariana HS, Dundas IS, Speilmeyer W, Lawrence GJ, Pryor AJ (2005) Development of PCR markers for the selection of wheat stem rust resistance genes *Sr24* and *Sr26* in diverse wheat germplasm. Theor Appl Genet 111:496–504

Mains EB (1933). Host specialization of *Erysiphe graminis* tritici. Proc Natl Acad Sci 19:49–53

Marais GF, McCallum B, Snyman JE, Pretorius ZA, Marais AS (2005) Leaf rust and stripe rust resistance genes *Lr54* and *Yr37* transferred to wheat from *Aegilops kotschyi*. Plant Breeding 124:538–541

Marais F, McCallum B, Marais A (2006) Leaf rust and stripe rust resistance genes derived from *Triticum sharonense*. Euphytica 149:373–380

Marais F, Marais A, McCallum B, Pretorius Z (2009) Transfer of leaf rust and stripe rust resistance genes *Lr62* and *Yr42* from *Aegilops neglecta* Req. ex Bertol. to common wheat. Crop Sci 49:871–879

Martin GB, Bogdanove AJ, Sessa G (2003) Understanding the functions of plant disease resistance proteins. Annu Rev Plant Biol 54:23–61

McIntosh RA, Yamazaki Y, Devos KM, Dubcovsky J, Rogers WJ, Appels R (2003) Catalogue of gene symbols for wheat. In: Pogna NE, Romano M, Pogna EA, Galterio G (eds) Proceedings of the 10th International Wheat Genetics Symposium, Vol. 4, associated CD-Rom, pp 1–34,

McIntosh RA, Yamazaki Y, Dubcovsky J, Rogers J, Morris F, Somers DJ, Appels R, Devos KM (2010) Catalog of gene symbols for wheat. MacGene. http://www.shigen.nig.ac.jp/wheat/komugi/genes/download.jsp. Accessed 25 Mar 2013

Meyers BC, Chin DB, Shen KA, Sivaramakrishnan S, Lavelle DO, Zhang Z, Mitchelmore RW (1998) The major resistance gene cluster in lettuce is highly duplicated and spans several megabases. Plant Cell 10:1817–1832

Meyers BC, Kaushik S, Nandety RS (2005) Evolving disease resistance genes. Curr Opin Plant Biol 8:129–134

Miedner T, Korzun V (2012) Marker-assisted selection for disease resistance in wheat and barley breeding. Phytopathology 102:560–566

Miranda LM, Murphy JP, Marshall D, Leath S (2006) *Pm34*: a new powdery mildew resistance gene transferred from *Aegilops tauschii* Coss. to common wheat (*Triticum aestivum* L.). Theor Appl Genet 113:1497–1504

Miranda LM, Murphy JP, Marshall D, Cowger C, Leath S (2007) Chromosomal location of *Pm35*, a novel *Aegilops tauschii* derived powdery mildew resistance gene introgressed into common wheat (*Triticum aestivum* L.). Theor Appl Genet 114:1451–1456

Mirza JI, Ahmed I, Khanzada KA, Ahmed N, Rattu A, Fayyaz M, Ahmad Y, Harkro AA, Mujeeb-Kazi A (2010) Local stem rust virulence in Pakistan and future breeding strategy. Pak J Bot 43(3):1999–2009

Mohler V, Bauer A, Bauer C, Flath K, Schweizer G, Hartl L (2011) Genetic analysis of powdery mildew resistance in German winter wheat cultivar Cortez. Plant Breeding 130:35–40

Morel J, Dangl JL (1997) The hypersensitive response and the induction of cell death in plants. Cell Death Differ 4:671–683

Mujeeb-Kazi A (2003) Wheat improvement facilitated by novel genetic diversity and in vitro technology. Plant Cell, Tissue and Organ Culture. 13:179–210

Mujeeb-Kazi A (2005) Wide crosses for durum wheat improvement. In: Roya C, Nachit MN, Difonzo N, Araus JL, Pfeiffer WH, Slafer GA (eds) Durum wheat breeding: Current approaches and future strategies. The Haworth Press Inc., pp. 703–743

Mujeeb-Kazi A (2006) Utilization of genetic resources for bread wheat improvement. In: Singh RJ, Jauhar PP (eds) CRC Series, pp 61–97

Mujeeb-Kazi A, Asiedu R (1990) Wide hybridization—potential of alien genetic transfers for *Triticum aestivum* improvement. In: Bajaj YPS (ed) Biotech in agriculture and forestry, Vol 13, Wheat, pp 111–127

Mujeeb-Kazi A, Hettel GP (1995) Utilizing wild grass biodiversity in wheat improvement: 15 years of wide cross research at CIMMYT. CIMMYT Report 2, pp 1–140

Mujeeb-Kazi A, Kimber G (1985) The production, cytology and practicality of wide hybrids in the *Triticeae*. Cereal Res Commun 13:111–124

Mujeeb-Kazi A, Roldan S, Suh DY, Sitch LA, Farooq S (1987) Production and cytogenetic analysis of hybrids between *Triticum aestivum* and some caespitose *Agropyron* species. Genome 29:537–553

Mujeeb-Kazi A, Roldan S, Suh DY, Ter-Kuile N, Farooq S (1989) Production and cytogenetics of *Triticum aestivum* L. hybrids with some rhizomatous *Agropyron* species. Theor Appl Genet 77:162–168

Mujeeb-Kazi A, Rosas V, Roldán S (1996a) Conservation of the genetic variation of *Triticum tauschii* (Coss.) Schmalh. (*Aegilops squarrosa* auct. non L.) in synthetic hexaploid wheats

(*T. turgidum* L. s. lat. X *T. tauschii*; 2n=6x=42, AABBDD) and its potential utilization for wheat improvement. Genet Resour Crop Evol 43:129–134

Mujeeb-Kazi A, Villareal RL, Gilchrist LI, Rajaram S (1996b) Registration of five wheat germplasm lines resistant to *Helminthosporium* leaf blight. Crop Sci 36:216–217

Mujeeb-Kazi A, Gilchrist LI, Villareal RL, Delgado R (2000) Registration of ten wheat germplasm lines resistant to *Septoria tritici* leaf blotch. Crop Sci 40:590–591

Mujeeb-Kazi A, Cano S, Rosas V, Cortes A, Delgado R (2001a) Registration of five synthetic hexaploid wheat and seven bread wheat germplasm lines resistant to wheat spot blotch. Crop Sci 41:1653–1654

Mujeeb-Kazi A, Fuentes-Davila G, Villareal RL, Cortes A, Rosas V, Delgado R (2001b) Registration of 10 synthetic hexaploid wheat and six bread wheat germplasms resistant to karnal bunt. Crop Sci 41:1652–1653

Mujeeb-Kazi A, Delgado R, Cortes A, Cano S, Rosas V, Sanchez J (2004) Progress in exploiting *Aegilops tauschii* for wheat improvement. Annual Wheat Newsletter 50:79–88

Mujeeb-Kazi A., Fuentes-Davilla G, Gul A, Mirza JI (2006a) Karnal bunt resistance in synthetic hexaploid wheats (SH) derived from durum wheat x *Aegilops tauschii* combinations and in some SH x bread wheat derivatives. Cereal Res Commun 34:1199–1205

Mujeeb-Kazi A, Gul A, Ahmad I, Mirza JI (2006b) A simplified and effective protocol for production of bread wheat haploids (n=3x=21, ABD) with some application areas in wheat improvement. Pak J Bot 38:393–406

Mujeeb-Kazi A, Gul A, Ahmad I, Farooq M, Rauf Y, Rahman A, Riaz H (2008a) Genetic resources for some wheat abiotic stress tolerances. In: Ashraf M, Ozturk M, Athar HR (eds) Salinity and water stress: improving crop efficiency, Springer, pp 149–163

Mujeeb-Kazi A, Gul A, Farooq M, Rizwan S, Ahmad I (2008b) Rebirth of synthetic hexaploids with global implications for wheat improvement. Aust J Agric Res 59:391–398

Mujeeb-Kazi A, Kazi AG, Dundas I, Rasheed A, Chen P, Kishi M, Bonnett D, Wang RRC, Bux H, Farrakh S (2013) Genetic diversity for wheat improvement as a conduit for food security. In: Donald Sparks (eds) Advances in Agronomy (in press)

Murphy JP, Leath S, Huynh D, Navarro RA, Shi A (1998) Registration of NC96BGTD1, NC96BGTD2, and NC96BGTD3 wheat germplasm resistant to powdery mildew. Crop Sci 38:570

Ogbonnaya FC, Abdalla O, Mujeeb-Kazi A, Kazi AG, Xu SS, Gosman N, Lagudah ES, Bonnett D, Sorrells ME (2013) Synthetic hexaploid in wheat improvement. In: Janick J (ed) Plant Breeding Reviews (in press)

Olivera PD, Kolmer JA, Anikster Y, Steffenson BJ (2007) Resistance of Sharon goatgrass (*Aegilops sharonensis*) to fungal diseases of wheat. Plant Dis 91:942–950

Peusha H, Enno T, Jakobson I, Tso˜mbalova J, Ingver A, Järve K (2008) Powdery mildew resistance of Nordic spring wheat cultivars grown in Estonia. Acta Agric Scand Sect B-Soil Plant Sci 58:289–296

Pretorius ZA, Singh RP, Wagoire WW, Payne TS (2000) Detection of virulence to wheat stem rust resistanceJ gene *Sr31* in *Puccinia graminis* f. sp. *tritici* in Uganda. Phytopathology 84–203

Pretorius ZA, Jin Y, Bender CM, Herselman L, Prins R (2012) Seedling resistance to stem rust race UG99 and marker analysis for *Sr2, Sr24* and *Sr31* in South African wheat cultivars and lines. Euphytica 186:15–23

Qi LL, Pumphery MO, Friebe B, Zhang P, Qian C, Bowden RL, Rouse MN, Jin Y, Gill BS (2011) A novel Robertsonian translocation event leads to transfer of a stem rust resistance gene (*Sr52)* effective against race UG99 from *Dasypyrum villosum* into bread wheat. Theor Appl Genet 123:159–167

Qiu YC, Sun XL, Zhou RH, Kong XY, Zhang SS, Jia JZ (2006) Identification of microsatellite markers linked to powdery mildew resistance gene *Pm2* in wheat. Cereal Res Commun 34:1267–1273

Rafiq K, Rasheed A, Gul-Kazi A, Bux H, Naz F, Mahmood T, Mujeeb-Kazi A (2012) Powdery mildew resistance in some new wheat amphiploids (2n=6x=42) derived from A- and S-genome diploid progenitors. Plant Genet Resour. doi:10.1017/S1479262112000202

Randhawa HS, Mutti JS, Kidwell K, Morris CF, Chen XM (2009) Rapid and targeted introgression of genes into popular wheat cultivars using marker assisted background selection. PLoS ONE 4(6):1–11 (e5752)

Ren Q, Liu H, Zhang Z, Feng J, Xu S, Zong-jun PU, Zhi-yong XIN (2012) Characterization and molecular mapping of a stripe rust resistance gene in synthetic wheat CI110. J Integr Agric 11:521–527

Reynolds MP, Borlaug NE (2006) Applying innovations and new technologies from international collaborative wheat improvement. J Agr Sci 144:95–110

Riera-Lizarazu O, Mujeeb-Kazi A (1990) Maize (*Zea mays* L.) mediated wheat (*Triticum aestivum* L.) polyhaploid production using various crossing methods. Cereal Res Commun 18:339–345

Riley R, Chapman V, Johnson R (1968) Introduction of yellow rust resistance of *Aegilops comosa* into wheat by genetically induced homoeologous recombination. Nature 217:383–384

Rimpau W (1891) Kreuzungsprodukte landwirtschaftlicher Kulturpflanzen. Landwirtschaftl Jahrb 20:335–371

Rizwan S, Ahmad I, Ashraf M, Sahi GM, Mirza JI, Rattu AR, Mujeeb-Kazi A (2007) New sources of wheat yellow rust (*Puccinia striformis* f. *tritici*) seedling resistance. Pak J Bot 39:1207–1216

Rizwan S, Ahmad I, Ashraf M, Mirza JI, Sahi GM, Rattu AR, Mujeeb-Kazi A (2008) Evaluation of synthetic hexaploid wheats and their durum parents for stripe rust resistance. Rev Mex Fitopatol 25:152–160

Roder MS, Korzun V, Wendehake K, Plaschke J, Tixier M-H, Leroy P, Ganal M (1998) A microsatellite map of wheat. Genetics 149:2007–2023

Rong JK, Millet E, Manisterski J, Feldman M (2000) A new powdery mildew resistance gene: introgression from wild emmer into common wheat and RFLP-based mapping. Euphytica 115:121–126

Rouse MN, Jin Y (2011) Stem rust resistance in A-genome diploid relatives of wheat. Plant Dis 10:1094 (PDIS,-04–10–0260)

Rouse MN, Olson EL, Gill BS, Pumphery MO, Jin Y (2011) Stem rust resistance in *Aegilops tauschii* germplasm. Crop Sci 51:2074–2078

Shaner G (1973). Evaluation of slow-mildewing resistance of Knox wheat in the field. Phytopathology 63:867–872

Sharma HC (1995) How wide can a wide cross be?. Euphytica 82:43–64

Sharma HC, Gill BS (1983) Current status of wide hybridization in wheat. Euphytica 32:17–31

Shi AN, Leath S, Murphy JP (1998) A major gene for powdery mildew resistance transferred to common wheat from wild eikorn wheat. Phytopathology 88:144–147

Simonite T (2006) Ancient genetic tricks shape up wheat; turning back the evolutionary clock offers better crops for dry regions. Nature. On line reporting on Jan. 03, 2006

Singh RP, Nelson JC, Sorrells ME (1998) Mapping *Yr28* and other genes for resistance to stripe rust in wheat. Crop Sci 40:1148–1155

Singh RP, William HM, Huerta-Espino J, Rosewarne G (2004) Wheat rust in Asia: meeting the challenges with old and new technologies. New dimensions for a diverse planet, Proceedings of the 4th International Crop Science Congress, 26 Sep–1 Oct 2004, Brisbane

Singh RP, Hodson DP, Jin Y, Huerta-Espino J, Kinyua MG, Wanyera R, Njau P, Ward RW (2006) Current status, likely migration and strategies to mitigate the threat to wheat production from race Ug99 (TTKS) of stem rust pathogen. CAB reviews: perspectives in Agriculture, Veterinary Science, Nutrition and Natural Resources No. 054. pp 1–13

Singh RP, Hodson DP, Huerta-Espino J, Jin Y, Njau P, Wanyera R, Herrara-Foessel S, Ward R (2008) Will stem rust destroy the world's wheat crop? Adv Agron 98:271–309

Singh RP, Hodson DP, Huerta-Espino J, Jin Y, Bhavani S, Njau P, Herrara-Foessel S, Singh PK, Singh S, Govindan V (2011a) The emergence of UG99 races of the stem rust fungus is a threat to world wheat production. Annu Rev Phytopathol 49:465–481

Singh RP, Huerta-Espino J, Bhavani S, Herrara-Foessel S, Jin Y, Njau P, Singh PK, Velu G, Singh S, Pena RJ, Crossa J (2011b) High yielding CIMMYT spring wheats with resistance to UG99 and other rusts developed through targeted breeding. BGRI Technical Workshop, 13–16 June, St. Paul, pp 98–104

Starling TM, Roane CW, Camper HM (1984). Registration of Massey wheat. Crop Sci 24:1000 (Tenth International Wheat Genetics Symposium, Paestum. pp 772–774)

Sukhwinder-Singh, Hernandez MV, Crossa J, Singh PK, Bains NS (2012) Multi-trait and multi-environment QTL analyses for resistance to wheat diseases. PLoS ONE 7(6):e38008

Talbert LE, Blake NK, Chee PW, Blake TK, Magyar GM (1994) Evaluation of 'sequence-tagged-site' PCR products as molecular markers in wheat. Theor Appl Genet 87:789–794

Tommasini L, Yahiaoui N, Srichumpa P, Keller B (2006) Development of functional markers specific for seven *Pm3* resistance alleles and their validation in the bread wheat gene pool. Theor Appl Genet 114:165–175

Trethowan RM, Mujeeb-Kazi A (2008) Novel germplasm resources for improving environmental stress tolerance of hexaploid wheat. Crop Science 48:1255–1265

Trethowan RM, Van-Ginkel M (2009) Synthetic wheat—an emerging genetic resource. In: Carver B (ed) Wheat science and trade. Chapter 16. Wiley-Blackwell, Iowa, pp 369–386

Uauy C, Distelfeld A, Fahima T, Blechl A, Dubcovsky J (2006) A NAC gene regulating senescence improves grain protein, zinc, and iron content in wheat. Science 314:1298–1301

Valkoun J (2001) Wheat pre-breeding using wild progenitors. Euphytica 119:17–23

Villareal RL, Mujeeb-Kazi A, Fuentes-Davila G, Rajaram S (1996) Registration of four synthetic hexaploid wheat germplasm lines derived from *Triticum turgidum* x *T. tauschii* crosses and resistant to karnal bunt. Crop Sci 36:218–220

Wang RR-C (1989) Intergeneric hybrids involving perennial *Triticeae*. Genet (Life Sci. Advances) 8:57–64

Wu L, Xia XC, Zhu HZ, Li SZ, Zheng YL, He ZH (2010) Molecular characterization of *Lr34/Yr18/Pm38* in 273 CIMMYT wheat cultivars and lines using functional markers. Sci Agric Sin 43:4553–4561

Xu SS, Jin Y, Klindworth DL, Wang RR-C, Cai X (2009) Evaluation and characterization of seedling resistances to stem rust UG99 races in wheat-alien species derivatives. Crop Sci 49:2167–2175

Yahiaoui N, Srichumpa P, Dudler R, Keller B (2004) Genome analysis at different ploidy levels allows cloning of the powdery mildew resistance gene *Pm3b* from hexaploid wheat. Plant J 37:528–538

Yang W, Liu D, Li J, Zhang L, Wei H, Hu X, Zheng Y, He Z, Zou Y (2009) Synthetic hexaploid wheat and its utilization for wheat genetic improvement in China. J Genet Genomics 36:539–546

Zaharieva M, Suenaga K, William HM, Mujeeb-Kazi A (2003) Genetic diversity of synthetic hexaploids with enhanced levels of resistance to *Fusarium* head scab. Annual Wheat Newsletter 49:73–75

Zeller FJ (1973) 1B/1R wheat-rye chromosome substitutions and translocations. In: Sears ER, Sears LMS (eds) Proceedings of the 4th International Wheat Genetics Symposium, Columbia, pp 209–221

Zeller FJ, Kong L, Hartl L, Mohler V, Hsam SLK (2002) Chromosomal location of genes for resistance to powdery mildew in common wheat (*Triticum aestivum* L. em. Thell.). 7. Gene *Pm29* in line Pova. Euphytica 123:187–194

Zhang JY, Li XM, Wang RR-C, Cortes A, Rosas V, Mujeeb-Kazi A (2002) GISH, AFLP, and RAPD characterization of Eb-genome chromosomes in five *Thinopyrum bessarabicum* disomic addition lines of bread wheat. Int J Plant Sci 163:167–174

Zhou T, Wang Y, Chen J, Araki H, Jing Z, Jiang K, Shen J, Tian D (2004) Genome-wide identification of NBS genes in japonica rice reveals significant expansion of divergent non- TIR NBS—LRR genes. Mol Gen Genet 271:402–415

Zhu Z, Zhou R, Kong X, Dong Y, Jia J (2006) Microsatellite marker identification of a *Triticum aestivum-Aegilops umbellulata* substitution line with powdery mildew resistance. Euphytica 150:149–153

Chapter 8
Variability in *Fusarium* species Causing Wilt Disease in Crops: A Transcriptomic Approach to Characterize Dialogue Between Host and Pathogen

Reiaz ul Rehman, Khalid Rehman Hakeem, Inayatullah Tahir, Bilal Ahmad Padder, Mehraj ul Din Shah and Mushtaq Ahmad Teli

Abstract Crops are indispensible for the existence of humans and animals and the commercial importance comes under threat when they are attacked and infected by the pathogens. The crops are constantly under threat and are exposed to various pathogens. Some pathogens are host specific and thus can infect the healthy plant and some of them are opportunists, which gain entry from the wounding site. The most important and devastating among the pathogens are the fungal pathogens and important amongst these is *Fusarium* sp. This genus contains many species attacking diversity of agricultural crops. These are pathogenic to plants and also produce toxins, which affect the animals and humans consuming the plants. This review focuses on the pathogenesis of *Fusarium* sp. causing crop diseases and understanding the host-pathogen interactions including plant defense responses. This article perceives the potential of transcriptomics in association between two-species. The

R. ul Rehman (✉) · K. R. Hakeem
P.G Programme in Bioresources, Department of Botany, University of Kashmir, Srinagar 190006, India
e-mail: reiazrehman@yahoo.co.in

K. R. Hakeem
Faculty of Forestry, Universiti Putra Malaysia, 43400 Serdang, Selangor, Malaysia
e-mail: hakeem.bot@gmail.com

I. Tahir
Department of Botany, University of Kashmir, Srinagar 190006, India
e-mail: inayatullahtahir@gmail.com

B. A. Padder · M. ul Din Shah · M. A. Teli
Division of Plant Pathology, SKAUST-K, Srinagar 191121, India
e-mail: bapadder@rediffmail.com

M. ul Din Shah
e-mail: mehraj547@rediffmail.com

M. A. Teli
e-mail: ahmadm9969@gmail.com

K. R. Hakeem et al. (eds.), *Crop Improvement*, DOI 10.1007/978-1-4614-7028-1_8,

identified association between species (crops and microbes (*Fusarium* sp.) can reveal processes which can be exploited beneficially for applications in biotechnology. Specifically, we address the question how the new knowledge gained from transcriptomic approaches and analyses of interactions between plants and disease causing microbes (*Fusarium*) can be exploited in ways that will ultimately lead to crop improvement by development of crop cultivars that are productive under multiple environmental pressures.

Introduction

The genus *Fusarium* consists of species that attack almost all food crops. Out of 101 most economically important plants, at least 81 are hosts of *Fusarium* sp. (Table 8.1) (Nayaka et al. 2011). The *Fusarium* spp are responsible for various diseases among live-stocks and humans and these diseases are ascribed to mycotoxins produced by these fungi belong to secondary metabolites (Table 8.2). The diseases caused by *Fusarium* spp have had several major economic impacts around the world in time and resulted in loss of billions of dollars. So, due to their economic importance *Fusarium* spp are being used as models for various biological and morphological studies. Plant pathogenic fungi show variability as they are known to contain a considerable number of sub-species/formae specialis or strains, which are almost morphologically identical but they may have quite different infection capacity on their hosts (Nayaka et al. 2011). This variability in pathogenic populations results in variation in host resistance (Leilani et al. 2006). Thus, degree of variations among plant pathogens are must for understanding the pathogen as well as the disease they cause. The amount of pathogen variation may have a direct impact on its biological activity and its role in the environment (Zabalgogeazcoa 2008). The variations may occur at any stage of growth leading to changes in morphology that determine its host range, inoculum potential, infectivity, and virulence (Parker and Gregory 2004). Therefore, it is relevant to monitor pathogen populations for shift in virulence with changes in environmental factors and host cultivar(s). Hence, the variability studies within and between pathogenic populations from different geographical regions is essential for selection of resistant genotypes in breeding programs that aim at developing resistant varieties.

Crop Productivity for Food Security

The Environment has a role in limiting the plant productivity due to biotic and abiotic stresses, e.g. most or all the existing crops can be significantly affected by diseases and have reduced productivity in terms of yield and quality. The prevention of diseases by conventional breeding to yield resistant crops is most effective (Akhond and Machray 2009; Gust et al. 2010) and is said to be the environmentally friendly

Table 8.1 Important *Fusarium* diseases in major food crops

Pathogen	Host plant	Disease
Fusarium spp. *Fusarium graminearum* *Fusarium avenaceum* *Fusarium culmorum* *Fusarium nivale*	Barley	Scab/head blight
Fusarium verticillioides *Fusarium avenaceum* *Fusarium subglutinans* *Fusarium graminearum* *Fusarium culmorum* *Fusarium oxysporum* *Fusarium poae* *Fusarium solani* *Fusarium pallidoroseum*	Corn	Kernel, root and stalk rot, seed rot, seedling blight
Fusarium verticillioides *Fusarium* spp.	Millets	Head mold/top rot
Fusarium verticillioides *Fusarium fujikuroi* *Fusarium proliferatum* *Fusarium spp.*	Paddy	Root rots/seedling blight/bakanae disease
Fusarium verticillioides *Fusarium* spp. *Fusarium subglutinans*	Sorghum	Fusarium head blight, root and stalk rot
Fusarium spp. *Fusarium pseudograminearum* *Fusarium graminearum* *Fusarium avenaceum* *Fusarium culmorum*	Wheat	Crown rot/foot rot, seedling blight, root rot

Fungal names according to http://www.indexfungorum.org/

and responsible approach to disease prevention as opposed to the indiscriminate use of pesticides. The other environmentally sensible approach is to understand and utilize the biotic and abiotic environment of the plant and develop sustainable disease management strategies. However, the development of management strategies requires in-depth knowledge and understanding of the plant-microbe intimate interactions with one another in extremely complex environment leading to different physiological changes within the plant. Further, more information generation and knowledge is required to understand the resource utilization within the plant upon the exposure to the pathogens. The resource management upon exposure to disease in terms of prioritizing themselves for utilizing in defense mechanisms instead of growth and development to develop the sustainable strategies to improve upon health of plants and thus agriculture (Schenk et al. 2012).

Table 8.2 Biochemistry of *Fusarium* mycotoxins

Fusarium mycotoxins	Toxicity mechanism	Toxicity	Examples	Main enzyme
Trichothecenes (biosynthesis: sesquiterpenes)	Inhibition of protein synthesis	Alimentary Toxic Aleukia Akakabi-byo (red mold disease) swine feed refusal. In humans promote neoplasms, cause autoimmune disease	Diacetoxyscirpenol, T-2 toxin, nivalenol, deoxynivalenol (vomitoxin)	Trichodiene synthase (Desjardins and Proctor 2007; Murphy and Armstrong 1995; Berek et al. 2001; Lindsay 1997)
Eumonisins (biosynthesis: polyketides)	Inhibition of sphingolipid metabolism	Carcinogenic: Leukoencephalomalacia (equines) neural tube defects (rodents); pulmonary edema (pigs); neural tube defects esophageal cancer (human)	Fumonisin B1, a propane-1,2,3-tricarboxylic diester of 2-amino-12,16-dimethyl-3,5,10,14,15-pentahydroxyicosane	Fumonisin polyketide synthase (Proctor et al. 2006)
Zearalenones (biosynthesis: nonaketide precursors)	Not acutely toxic	Non-steroidal estrogenic: estrogenic syndromes in swine	Resorcyclic acid lactone	Polyketide synthases (Kim et al. 2005; Gaffoor and Trail 2006)
Beauvercin and enniatins (biosynthesis: N-methylated cyclic hexadepsipeptides)	Non-toxic	Non-ribosomal, cyclic hexadepsipeptides, enniatins: Enniatin A beauvericin	D-2-hydroxyisovaleric acid. L-amino acids: valine, leucine, or isoleucine, phenylalanine	Enniatin synthetase (Member of non-ribosomal peptide synthetase) (Desjardins and Proctor 2007)
Butenolide (4-acetamido-4-hydroxy-2-butenoic acid lactone)	Moderately toxic	Fescue foot (cattle: edema, lameness and gangrenous loss of extremities)		Cytochrome P450 monooxygenase (Harris et al. 2007)
Equisetin N-methyl-2,4-pyrollidone (1-methyl-3-acyl-5-hydroxymethyl-2,4-dione)	Activity against human immunodeficiency virus, gram-positive bacteria	Activity against human immunodeficiency virus		Polyketide synthase-non-ribosomal peptide synthetase (EQIS) (Sims et al. 2005)
Fusarins 2-pyrrolidones (polyketide and TCA)	Mutagenic	Non-ribosomal peptide synthetase epidemiologically with human diseases	Fusarins A, C, D	Hybrid polyketide synthase (Gaffoor et al. 2005)

Plant-Fungi Interaction

The wilting caused by *Fusarium* is result of various factors guided by host-pathogen interaction such as recognition of pathogen by plant root, attachment of pathogen by differentiated structure, penetration of the pathogen to have access to vascular tissue, adaptation of pathogen within the plant system, the proliferation of fungal components (hyphae and microconidia) in xylem vessels and lastly the secretion by the pathogen (proteins and toxins) (Inoue et al. 2002; Di Pietro et al. 2003). The primary defense response by the host includes the production of gums, tyloses and gels (Beckman 1987). Thus, there is the need for understanding host-pathogen association in terms of infection process (at molecular level), which would provide the information about the mechanisms of this interaction. The study of this interaction will unravel the genes involved in different signaling cascades, which would help us in identifying genes involved in resistance as well as susceptibility. This kind of association (*Fusarium*-Tomato) study has been recently reported as a model system for infection process (resistance and susceptibililty) at molecular level (Takken and Rep 2010). Further, some processes related to mechanisms of infection resistance have been determined by molecular techniques such as gene silencing or insertion mutagenesis (Inoue et al. 2002; Di Pietro et al. 2003; Michielse 2009). The knowledge about the infection resistance or susceptibility in agricultural crops would help us in better understanding about the development of strategies in controlling the disease factors such as infection responses and progression. For example, mRNA changes in association (host-pathogen) during the infection process could give information about the resistance and susceptibility processes (Wise et al. 2007).

Use of DNA Based Markers

The advances and development in molecular biology techniques and its applications to genetic analysis has led to better understanding in terms of knowledge which has led to elucidation of behavior and structure of fungal genome. For example, these advances have helped the fungal taxonomists in identification of isolates more rapidly and for determining virulence/toxicity of fungal strains. The advances in molecular biology led to distinction between species (closely related) having little or no similarity at the morphological level (Wulff et al. 2010) or strain identification within species (Chandra et al. 2010). And thus by looking for the variation within DNA sequences, the molecular biology provides the base for precise identification within and between the species. The Polymerase chain reaction (PCR) based assays are being employed widely used for identification, diagnosis and characterization of pathogens (*Fusarium* species) (Doohan et al. 1998). Various PCR based markers are in use (Nayaka et al. 2011), which has led to the understanding of genetic diversity and establishing phylogeny among different *Fusarium* species. The PCR based methods which have been used to generate sequence information include RFLP, AFLP,

RAPD, SCARs, RT-PCR, SNPs, microarrays, pyrosequencing, DNA barcoding and many more and this has led to gene function determination in fungi (Nayaka et al. 2011). The DNA microarrays and next-generation sequencing techniques have helped greatly in genome-wide expression, but in these cases known collection of transcript sequences must be readily available (Roh et al. 2010). However, in comparison with above tools, the method of choice for gene discovery (identifying transcripts) in plant-microbe interaction remains to be cDNA-AFLP (Vuylsteke et al. 2007; Gupta et al. 2009). The cDNA-AFLP technique has been used to study the interaction between the *Fusarium* pathogens and the plant hosts, e.g. *Fusarium* interaction with host *Cucumis melo* through the analysis of differentially expressed genes in the vascular colonization by using the approach of cDNA-AFLP (Szafranska et al. 2008; Sestili et al. 2011). In the species other than model species the utilization of cDNA-AFLP techniques for expression studies is appropriate as it helps in identification of genes in two species during infection and is also important for identifying factors of pathogenesis and virulence and thus would help in developing strategies in controlled manner (Durrant et al. 2000; Guo et al. 2006; Polesani et al. 2008; Wang et al. 2009; Gupta et al. 2010; Zvirin et al. 2010). Under the evolutionary pressure the species of *Fusarium* have modified for better adaptability and thus have colonized in differing ecological systems (Desjardins et al. 1993). These species produce a variety of compounds (bioactive secondary metabolites), which are show toxicity towards plants (Munkvold et al. 1997). These toxins also find their way to animals and humans through the plant products consumed as fodder or food. The *Fusaria* have different strains belonging to many groups which are identical morphologically and thus are difficult to study such as endophytes (Bacon and Hinton 1996), saprophytes (Fracchia et al. 2000), and plant pathogens (Chandra et al. 2008) and thus is the reason for difficulty in establishing taxonomy systems for *Fusarium* species. So, it is required to utilize molecular based methods in differentiating taxa. In this context, in recent times the application of phylogenetic species concept to *Fusarium* systematic has helped in resolving the difficulties in taxonomy. This concept requires several characters such as morphology of species crossing between species and molecular markers in species for it to be statistically powerful (Yli-Mattila et al. 2002). However, the molecular markers (DNA sequence data) are preferred because they provide relevant characters. The molecular data in combination with other morphological characters within *Fusarium* have helped to differentiate species, which were otherwise placed differently (Zeller et al. 2003). It has been noted, based on the DNA sequences that often the strains of *Fusarium* placed in a forma specialis differ significantly and these need not be monophyletic in origin (Baayen et al. 2000). Inferring from this the traits of plant pathogenicity, which have high economic importance, need not to have characters that are evolutionary conserved and thus is part in *Fusarium* species description. Since the arrival of DNA sequencing at the scene it has become an important criterion in diagnostics and distinguishing species. The most commonly used sequences to distinguish *Fusarium* sp. are calmodulin (O'Donnell et al. 2000), TEF (translocation elongation factor 1-α) (Wulff et al. 2010), ITS1 and ITS2 (internally transcribed spacer regions in the ribosomal repeat region) (O'Donnell and Cigelnik 1997), tub2 (β-tubulin)

(O'Donnell et al. 1998), and IGS (intergenic spacer region) (Yli-Mattila and Gagkaeva 2010). However, it has been seen that all these sequences do not equally work well with all species (Nayaka et al. 2011). Further, the proteins such as histone H3, calmodulin, and Tri101 are also used for distinction of species besides (Mule et al. 2005). In short, it is difficult to distinguish and identify *Fusarium* sp. accurately based on old traditional methods because of their morphological similarity and genetic variation. Hence, for the accurate identification and characterization of species DNA-based tools are required. In this direction, SSRs (single sequence repeats) of the EST databases shall provide a way forward in identification and characterization of *Fusarium* sp. These sequences are ubiquitously transcribed which are co-dominant and locus-specific, and also are highly polymorphic, often multi-allelic and finally SSRs are transferrable among species within genera (Power et al. 1996; Morgante et al. 2002; Varshney et al. 2005a, b). Various EST-SSR markers developed from EST databases are used for genotyping in several species of flowering plants (Varshney et al. 2005a) and these have been developed in many plant species like rice (Temnykh et al. 2001) wheat (Eujayl et al. 2002) rye (Hackauf and Wehling 2002) Cotton (Han et al. 2006) Grape (Cordeiro et al. 2001). These EST-SSR markers are gene-tagged associated with the expressed gene and are also linked with alleles of quantitative and qualitative trait locus (QTLs) (Torben et al. 2007). So far, many genomes have been sequenced and thus to the comparative genomics has become an important discipline which helps to extend the information from one species (model) to another unrelated species or between related species having a much complex genome (Gale and Davos 1998). The EST-SSR markers in related species show high level of transferability in comparison to anonymous DNA markers because they are more conserved and thus these markers are useful for comparative genomics, comparative mapping and evolutionary studies across species (Cordeiro et al. 2001; Thiel et al. 2003; Eujayl et al. 2004; Scott et al. 2000; Saha et al. 2004). However, the degree of polymorphism may be limited due to conserved nature of EST-SSRs (Torben et al. 2007) as the transferability across species of the SSR loci within genus is reported above 50 % (Thiel et al. 2003; Eujayl et al. 2004; Peakall et al. 1998; Gaitán-Solís et al. 2002; Dirlewanger et al. 2002) and the transferability across genera of SSRs loci is reported to be poor (Thiel et al. 2003; Peakall et al. 1998; White and Powell 1997; Roa et al. 2000). Amongst the molecular markers the SSRs are widely used for molecular mapping, selection, assessment of genetic diversity, protection of varieties and thus helping to link genotypic and phenotypic variation (Powell et al. 1996; Gupta and Varshney 2000; Varshney et al. 2005a). The SSRs are comprised of tandemly repeated sequences having mono-, di-, tri-, tetra-, penta-, or hexa- nucleotide units (Ellegren 2004) found in coding and non-coding regions in prokaryotic and eukaryotic DNA.

The presence of various characteristics within SSRs such as relative abundance, multi-allelic nature, simple detection, high reproducibility, co-dominant inheritance, multi-allelic nature, and extensive genome coverage make them the ideal molecular markers (Powell et al. 1996). However, the SSRs are known to experience high rate of mutations (reversible length-altering) by replication slippage (transient dissociation of replicating DNA) and unequal crossing over (misaligned reassociation), which leads to variability in number of SSR motifs at a locus (Levinson and Gutman 1987; Richards and Sutherland 1992). SSRs show highest variability among DNA sequences within the genome (Weber 1990), and the rate of mutation and type mainly depends upon the number of repeat motifs (Wierdl et al. 1997). However, the rates of mutations differ among loci, among alleles, and between species (Ellegren 2000). Among the earlier studies the anonymous DNA fragments (isolated from genomic libraries) were utilized for the development of SSR markers. However, recently SSRs are being developed computationally from the sequencing data generated from large-scale EST sequencing projects. The EST-SSR markers are superior and informative than anonymous DNA markers because they are gene-tagged and are associated with the expressed gene and hence linked to quantitative and qualitative trait loci (QTLs) (Andersen and Lübberstedt 2003). The *Fusarium* genomes have been conserved and are well documented, and to compare genetic information from model to related species comparative genomics has become necessary. In short, EST-SSRs derived from expressed genes are conserved with higher transferability to related species in comparison to anonymous DNA markers, thus, are very useful as strong markers for evolutionary studies comparative genomics and comparative mapping across species (Torben 2007).

Fusarium Databases

The comparative analysis among the fungal species is facilitated by the Fusarium database which gives accesses to various sequenced genomes of Fusarium simultaneously. The three most important *Fusarium* spp, which have been studied extensively due to their devastating impact on crops, are *F. oxysporum, F. verticillioides and F. graminearum.* Over the last decade there has been enormous advancement in genomic technology and sequencing of various species has been completed and thus there arose the need for comparative projects. Thus, the Fusarium Comparative project (http://www.broad.mit.edu/annotation/genome/fusarium_group/MultiHome.html) is one such effort which compared the above three species. This project improved gene annotations and identifying non-coding elements. As these species are biologically distinct evolutionary studies were done among these above species and it was found that *F. oxysporum* and *F. verticillioides* genomes have diverged from *F. graminearum*, and thus the comparative of these offers a platform for interaction studies in plant-fungi (pathogenicity, virulence factors and evolution). The comparative genomics among these have highlighted the presence of lineage-specific chromosomes comprising the transposable elements (TEs) and encoding pathogenicity related (PR) genes. These projects have also helped in exploring the genetic composition of these chromosomes among the strains of the above *Fusarium* species. The *F. oxysporum* spp. are ubiquitous plant and soil inhabiting microbes causing the wilt and root rot diseases in over 120 plant species (Michielse and Rep 2009). It has been reported that many of the strains don't show apparent symptoms on the plant and can even protect them from subsequent infections (Alabouvette et al. 2009). Further, some strains have also been identified as pathogenic to humans causing localized or disseminated infections (O'Donnell et al. 2004). The aims and objectives of comparative genomes project has been to make genome sequence data for strains available along with the host specificities. The first among strains of *F. oxysporum* spp was made available in 2007 which caused tomato wilt and on comparison with *F. graminearum* and *F. verticillioides* genomes, and it was discovered that the mobile supernumerary chromosomes contained genes for host specific infection and disease (Ma et al. 2010). Subsequently, 11 more strains of *F. oxysporum* have been sequenced out of which two infected tomato (Gale et al. 2003; Rosewich et al. 1999). The other two of the sequenced strains showed specificity towards crucifers (cabbage yellow disease, radish wilt), *Arabidopsis* (Diener and Ausubel 2005) and the other sequenced strain caused wilt of banana, melon, cotton and pea (Fourie et al. 2011). Some *Fusarium oxysporum* strains causing humans diseases such as strains from human blood (O'Donnell et al. 2004) have been shown to cause necrotic diseases in immune-competent individuals such as outbreak of keratitis in contact lens users (Chang et al. 2006) and may also be disseminating infections in neutropenic patients (Boutati and Anaissie 1997). Furthermore, a strain of *Fusarium oxysporum* has been sequenced which is known to colonize host roots and shown to be biotic component for wilt disease suppression (Fravel et al. 2003). Another important species among *Fusarium* which is distributed worldwide is *Fusarium graminearum* which causes head blight of wheat and barley (O'Donnell et al. 2000, 2004) and it had an economic losses worth of approximately 3 billion dollars in 1990s to U.S. agriculture (McMullen et al. 1997; Windels 2000), further it is becoming a threat due to outbreaks in Canada, Europe, Asia and South America in recent past (Dubin et al. 1997). *Fusarium graminearum* species also causes disease on rice and corn (White 1999; Webster and Gunnell 1992) and results in reduced yield and seed quality and also the infestation makes the food and feed unsuitable due to production of mycotoxins (McMullen et al. 1997). The strains of *Fusarium graminearum* sequenced by International Gibberella zeae Genomics Consortium (IGGR) was PH-1 (NRRL 31084) (Trail and Common 2000) was shown to produce mycotoxins; trichothecenes and zearalenone such as deoxynivalenol and nivalenol toxin (Garvey et al. 2008). Another important species present worldwide is *Fusarium verticillioides* causes kernel and ear rot disease of maize. The abiotic stress conditions such as high temperature, drought along with other biotic factors such as the damage by insect can further amplify and enhance the disease and thus cause reduce crop quality and yield. The negative economic impact of this species causing several animal (Howard et al. 2001; Seefelder et al. 2003; Wilson et al. 1992; Kriek et al. 1981) and human diseases (Seefelder et al. 2003) is due to fumonisin mycotoxins.

Plant-Defense Responses

The evolution of fungi is dependent directly on the development and the spread of green plants and most of fungi are saprophytic and few among them are parasitic. These fungi require nutrients to complete their life cycle, which is provided to them by the living cells. The plants have effective defense mechanisms to limit the spread of parasitic fungi and they have to encounter the primary line of defense provided by cell wall of host plants, which inhibit their penetration to the tissues. The secondary line of defense includes the production of wide range of secondary metabolites or compounds which act as fungicides. These compounds are induced, that is they are produced only after the infection has occurred (e.g. Phytoalexins). The fungal parasites are host specific and some of them even require the alteration of hosts, which may or may not be phylogenetically related for stage wise development on either host. The fungal pathogens have been seen to cause more damage to agriculture (monocultures) than in plant communities rich in species due to high host specificity. To get hold of the plant nutrients the fungal parasites are having at least three strategies. (1) Production of enzymes for cell wall and cuticle breakdown. (2) Production of toxins to either reduce or inhibit the cellular activity. (3) Production of host-specific substances (e.g. phytohormones) and thus disturbing the hormonal equilibrium and thus causing disruption of growth and differentiation of the cells and tissues (e.g. *Gibberella fujikuroi*) produces gibberellins which affect the growth of rice. The literature about pathogenicity of fungi is very extensive and majority of these emphasize on economic considerations, classification, life cycle, symptoms and diagnostic, host range and factors of host resistance. The processes of plants molecular mechanisms after infection have been elucidated in recent past and these can be partially generalized due to variation in possible interaction reactions. The host resistance is not only based on general and unspecific defense mechanisms but also on specific mechanisms during which genetically determined substances are produced which are directed against specific fungal pathogen. Genetic analyses have shown that the host resistance is caused by genes inherited independently of each other, such as the existence of dominant alleles of the respective genes, and resistance genes. It has been shown by microarray analysis that resistance to virus in melon is associated with defense responsive gene expression/induction (Gonzalez-Ibeas et al. 2012a) which led to development of a cost-effective kit for microarrays (Gonzalez-Ibeas et al. 2012b). Further, the silenced lines of melon developed by RNA Interference showed resistance against the viruses (Rodriguez-Hernandez et al. 2012). Plants protect themselves against the diversity of herbivores and microbial pathogens by expressing an array of constitutive and induced defenses rendering the plant an inaccessible or unsuitable food source. The perception of attack and deployment of the induced defenses is primarily mediated by three well-studied defense-signaling pathways that are regulated by jasmonic acid (JA), salicylic acid (SA) and ethylene (ET) (Glazebrook 2005; Howe and Jander 2008; Walling 2009). Herbivores and pathogens introduce a distinct set of elicitors and effectors that are perceived by host plants and these signals allow the plant to tailor its defense response to individual challengers (Glazebrook 2005; Howe and Jander 2008; Walling 2009; McSteen 2008; Zhao et al. 2008a; Stout et al. 2006; Bari and Jones 2009). The SA-regulated defense pathway is activated by biotrophic pathogens (pathogens that invade living plant tissue) and many phloem-feeding insects (Glazebrook 2005; Walling 2009; Kempema et al. 2007; Puthoff et al. 2010; Kusnierczyk et al. 2008; Zarate 2007). Often SA-induced signaling antagonizes JA- and ET-regulated signaling pathways, although exceptions do exist (Mur et al. 2006; Leon-Reyes et al. 2010; Verhage et al. 2010). The suppression of JA/ET-regulated defenses confers susceptibility to many tissue-damaging and phloem-feeding herbivores (Howe and Jander 2008; Zarate 2007; Gao et al. 2007; Pieterse and Dicke 2007) and can influence attraction of natural enemies (Zhang et al. 2009). However, in some plant—herbivore interactions, SA-regulated defenses and/or novel defense-signaling pathways contribute to the plant immune response (Thaler et al. 2010; Bhattarai et al. 2007, 2010). Therefore, the nature of defenses elicited by endosymbiont-like pathogens in their host plants have the potential to profoundly impact the plant's interaction with the insect and/or ability to resist attacks by other pathogens or pests. If an herbivore can circumvent induced plant defenses or plant recognition by vectoring its endosymbiont associate into its host plant during feeding, it may have a selective advantage relative to insects feeding on uninfected plants. Alternatively, effectors from the endosymbiont may circumvent the plant recognition system, compromising plant immune responses and related insect and bacterial resistance in both the JA/ET- and SA-regulated defense pathways. Thus, the endosymbiont's modification of plant defenses could result in a more susceptible host plant for both symbiotic partners.

Understanding Pathogenesis: Role of Systems Biology

The *Fusarium* spp. causes disease to crops (Table 8.1) and the disease effect is huge in terms of economy due to problems in health to animals as well as humans by consuming the contaminated grain (McMullen et al. 1997) with mycotoxins (Garvey et al. 2008). Thus, it becomes necessary to understand pathogenesis (pathogenic genes) and thereby prevent the invasion by these destructive pathogens. The plant pathology discipline describe the pathogenesis genes as those which causes losses or in other words are those which when disrupted causes the reduction of disease symptoms (Idnurm and Howlett 2001). The identification of these genes can be done with either the gene silencing or gene knockout studies (Liu et al. 2010). In *F. graminearum* 49 pathogenic genes have been verified by utilizing the biological methods and then stored in PHI-base database (http://www.phi-base.org/query.php). However, when we consider the genome size of *F. graminearum*, it is realized that compilation of the pathogenesis related genes is a huge uphill task as well as time consuming. To solve this problem, the computational biological methods could provide an alternative for solving this problem, especially after the release of genome sequences in Broad Institute (http://www.broadinstitute.org) (Liu et al. 2010). Hence, for the prediction of pathogenesis genes comparative genomics will be handy and will help in comparison between fungi, which are either pathogenic or non-pathogenic (Zhao et al. 2008b, c). But, recently it has been found that it is difficult to identify pathogenesis genes in *F. graminearum* because there are no unique features in specific genes among pathogenic and non-pathogenic fungi (Liu et al. 2010). On the basis of literature on pathogenicity in model pathogens (Gohre and Robatzek 2008), it can be believed that *F. graminearum* pathogenesis is also involving the networks of proteins and molecules while interacting from itself as well as the ones secreted by the host cells (Liu et al. 2010). Thus based on systems biology approach, the network of molecular interaction in *Fusarium spp.* can provide additional insights into the processes of pathogenesis. Moreover, the protein-protein interaction maps can also provide clues to potential pathogenesis genes and thus pathogenesis procedure (Zhao et al. 2009). The pathogenesis genes are generally expressed differentially before and after invasion so that the pathogen can sustain or pass through the host immune system and thus adapt in the host. Similarly the differentially expressed genes of *F. graminearum* may be pathogenic genes (Liu et al. 2010). However, false positives may be produced by differentially expressed genes while identifying genes involved in disease procedures, as some of these genes are not involved in pathogenesis, despite expression changes during experimentation. However, by integrating the protein interaction and gene expression studies on perturbations (drug, extracellular stimuli) (Zhao et al. 2010) shall be useful in identifying the pathogenesis processes. The approaches in systems biology helps in integrating the protein interaction mapping data and gene expression data and works on assumptions proteins interacting in a network share similar functions and it is termed as "Association rule" (Zhao et al. 2010) and thus are most probably involved in the similar pathways (Zhao et al. 2010). In *F. graminearum* the prediction results have shown that most of the pathogenesis genes belong to G-protein coupled receptors and MAPK signaling pathways (Liu et al. 2010) (Table 8.2).

Transcriptomics: Understanding of Plant—Microbe Interactions

Transcript profiling is a functional genomics tool, which is used most widely and can be conducted along with various genomic tools such as SAGE (serial analysis of gene expression), MPSS (massive parallel signature sequencing) and microarrays (Dilip et al. 2010). The transcriptomics of related species (pathogenic and non-pathogenic) have led to identification of divergence among them in the genome (Wurtzel et al. 2012). The role of regulatory processes in pathogenesis (i.e. the protein abundance) can be compared with reference proteomics as well as transcriptomics (Voge and Marcotte 2012). The transcriptomic approach is necessary for understanding the pathogenesis processes in particular the receptors present in the plants which help in recognizing or perceiving the conserved signatures of pathogens (Schwessinger and Ronald 2012). The responses of plants to the biotic stresses (pathogens) leads to the several changes such as cellular, physiological, biochemical and most importantly the transcriptomic levels. The interactions are very complex and are mediated and regulated by the hormonal signaling pathways (Atkinson and Urwin 2012; Tian et al. 2012). The modulation of defense systems by the plants upon the infection, invasion or any other biotic stress can be studies at the transcriptomic level of interacting host and pathogen (Lee et al. 2012; Tian et al. 2012). The metabolomic and transcriptomic studies could be helpful in

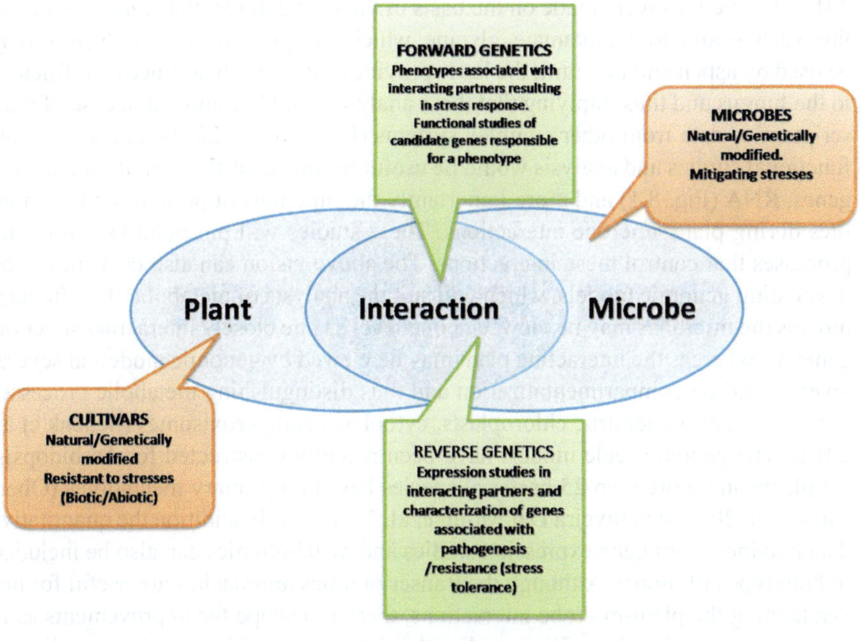

Fig. 8.1 Depicting approaches towards utilizing the interaction of plant-microbe for applications in biotechnology. (Adapted from Schenk et al. 2012)

revealing the mechanisms of antagonism between species infecting the same plant as reported recently among Ustilago maydis and *Fusarium verticillioides* infecting maize (Jonkers et al. 2012). Based on the extensive collection of ESTs, several microarrays have been developed for crops (Torben et al. 2007). The plant expression database of plants and pathogens (PLEXdb, http://www.plexdb.org/index.php) can provide the microarray data of *Fusarium graminearum* (obtained with Affymetrix GeneChip) (Liu et al. 2010). The transcriptomics approach to study the interaction between two- species may lead to the discovery of important genes in plants and *Fusarium* thus leading to characterization, which shall provide disease resistance strategies (Fig. 8.1). The multidisciplinary approaches are required for obtaining a comprehensive systems biology look of the involved processes. The studies of genomics, transcriptomics, proteomics and metabolomics will provide data sets, which shall be integrated using bioinformatics and statistical tools and thus will help to identify important biological processes and thus make prediction models. These approaches are being used to access the information about the associations (beneficial/detrimental) among two or multiple species. For example, the ectomycorrhizal interaction between *Laccaria bicolor* and aspen (*Populus tremuloides*) roots led to the identification of genes expressed significantly, which were later mapped to specific metabolic pathways and hence gave rise to the model of ectomycorrhizal metabolome. Identification of which genes are expressed is done by utilizing the next-generation short-read transcriptomic sequencing data (Larsen et al.

2011). Predictions were made on the basis of the above model that various metabolites such as allantoin, glutamate, glycine, which are synthesized by *L. bicolor* may be used by aspen and in return; the latter provides sugars such as glucose or fructose to the fungus and thus implying that these analyses could be applied in case of transcriptomics data from other complex systems (Larsen et al. 2011). In addition, the functional studies and analysis would be useful for the identification of functions of genes, RNA (Fig. 8.1) and more importantly the functions of proteins and metabolites during plant–microbe interactions. These studies will pin point the important processes that control these interactions. The above vision can also be achieved by developing genomic models, which will suit the analysis of metabolic flux. In these models the microbes may be viewed at one level as one closely interacting super organisms, whereas the interacting plant may be viewed by genomic models at several levels based on compartmentalization and thus distinguishing metabolic processes in vacuoles, mitochondria, chloroplasts, cytoplasm and peroxisomes (Schenk et al. 2012). The genome-scale models have been recently constructed for Arabidopsis, C4 plants and more than 25 bacterial species based on primary metabolites (Oberhardt et al. 2009; de Oliveira Dal'Molin et al. 2010a, b). In addition the quantitative data obtained from gene expression studies and metabolomics can also be included in both types of model. Although the transcriptomics approaches are useful for understanding the plant-microbe interactions, there is a scope for improvements as in case of transcriptional profiling studies the data generated has not been replicated and thus defy statistical analyses. It is due to the reason that it has high cost and is very complex. The less complex environmental samples from sea water (microbial complexity compared to soil) revealed among the unique reference genes only 17% overlap by repeated pyrosequencing (Stewart et al. 2010) which suggests that current sequencing platforms need to evolve further for gasping the complexities within communities. The sequencing platforms of with better coverage are currently being developed and then are coupled with replicate profiling and statistical analyses and thus these hold a great promise for representation of the expression profiles of interactions accurately. Currently, HiSeq 2000 (maximum 600 Gb, corresponding to 3 billion reads using TreSeq v3 reagent kits; Illumina, Inc.) platform delivers the largest amount of sequence data. New insights will be provided by understanding the detrimental or beneficial plant—*Fusarium* interactions for biotechnological purposes. For the association studies between grasses and endophytes the use of transcriptomics is a tractable system. Similarly, transcriptomics of detrimental association identified candidate genes which play pivotal role in infection processes and in which the interaction switches from being mutalistic to pathogenic between interacting partners (Eaton et al. 2011; Beatty and Good 2011). Thus the knowledge obtained from interactions (beneficial/detrimental) between plants and microbes will provide tremendous opportunities to increase crop productivity. Furthermore, transcriptomics of the environment (soil/water) should be included in systems biology as an essential component and thus integrated along with other omics technologies. The future prospect would be if we as scientists could harness the potential of microbes by engineering such crop plants, which are particularly suited for beneficial interactions with microbes. An example in this direction could be the development of cereals, which can fix nitrogen significantly through their associated N-fixing microbe partners. The cereal crops (such as rice, corn, wheat, barley, sorghum and sugarcane) are consumed worldwide by humans and animals and thus if the nitrogen fixing ability of microbes is controlled among these crops, it will reap enormous financial as well as environmental benefits and the ways to achieve these milestones are being discussed presently the world over (Charpentier and Oldroyd 2010). Despite the recent progresses, we are at the beginning stages and by looking at the current pace of progress the future research upholds great promise and potential and these new insights will be of great value for the development of sustainable production by utilizing new technologies.

Metatranscriptomics Approach

In Metatranscriptomics the environmental samples such as soil or water are used for RNA extraction to analyze the gene expression in microbial communities without their cultivation (Morales and Holben 2011; Simon and Daniel 2011). The study of this multi-species association or interaction with the plant presents with opportunities, which will help in discovering relationships among plant-microbes with potential impact. These metatranscriptomic studies of the whole microbial communities at same time will lead to identification of new plant-microbe interactions either beneficial or detrimental. For example, it is reported that among 150 Pseudomonas sp. from rhizosphere of wheat, 40 % of isolates increased the root growth upon individual inoculation on wheat (Campbell and Greaves 1990). Further, the outcome of the plant-microbe interaction can be determined by the influence of environmental conditions such as abiotic along with biotic stresses on the physiological pathways of plants. Thus, understanding of signaling processes and the cross talk between individual pathways may allow harnessing these pathways to our advantage for solving problems related to pathogenesis. The expression profiling of the genes of signaling pathways have revealed that these mechanisms have protein switches regulated by hormones (e.g. transcription factors, kinases or G-proteins) (Memelink 2009; Yao et al. 2011; Depuyd and Hardtke 2011; Zhao et al. 2010). The signaling responses to abiotic and biotic stresses are complicated such as the abiotic stress-experiencing plants may channel their physiological resources for adaptation towards such factors but in doing so these plants may become susceptible towards attack by pathogen or herbivore (Thaler and Bostock 2004; Trewavas 2009; Hey et al. 2010). The pathways which are activated through the mediation by abscisic acid (ABA) seem to be dominant in providing protection towards abiotic stresses and seem antagonistic to defense pathways controlled by SA (salicylic acid), JA (jasmonic acid) and/or ET (ethylene) (Zabala et al. 2009; Raghavendra et al. 2010; Peleg and Blumwald 2011). In plants under abiotic stresses such as UV radiation, increase in ROS (reactive oxygen species) is triggered and these show resistance against biotrophic pathogens and this response is also achieved when SA is applied to plants (Bechtold et al. 2005; Kunz et al. 2008; Ahmad et al. 2010). In contrast the necrotrophic pathogens are benefitted from any kind of physical damage to cells and JA provides defense against these and further JA is also associated with induced systemic resistance (ISR), which is a priming reaction against subsequent infections (Pieterse et al. 2009; Matilla et al. 2010). Henceforth these recent advanced technical studies of microbial communities on metagenomics and metatranscriptomics (Morales and Holben 2011; Kakirde et al. 2010) could be applied to pathogens and then conjoining with plant transcriptomics, shall provide deeper insights into multiple interactions. In the plant rhizospere 33,000 archeal and bacterial species were found when grown on disease-suppressive soil, by combining a high-density 16S ribosomal DNA oligonucleotide microarray (PhyloChip metagenomics) of the rhizosphere microbiome with culture dependent functional analyses (Mendes et al. 2011). The disease caused by *Rhizoctonia solani* (fungal root-infecting pathogen) was found to be suppressed by the Proteobacteria, Firmicutes and Actinobacteria species, which led to the proposal or conclusion that plants interaction with multiple microbe species can contribute to suppression of the disease (Schenk et al. 2012).

Future Prospects

The prime concerns are related to with the understanding of host-pathogen interactions including plant defense responses. To monitor fungal colonization the systematic re-isolation procedures are required in host-pathogen combinations (compatible and incompatible). This is a challenging issue, which will lead to developing new strategies to control disease. In order to identify the genes required for the hypersensitive response, studies shall focus on performing expression profiling of crop plants. The cDNA-AFLP analysis shall provide expression profiling of both normal and *Fusarium* infected plants, (identification of plant/pathogen genes associated with the infection process). The focus on variability of *Fusarium* sp. infecting Crops is the need of hour and will help to understand the host-pathogen relationship for disease management. The crop colonization by *Fusarium* sp. shall be studied and the samples collected from definite time points irrespective of speed and extent of colonization among the strains. The molecular biology advancements have led to the use of DNA based markers in place of morphological markers. Initially, the RNA samples of (healthy/infected plants) shall be put to cDNA-AFLP analysis for identifying transcripts differentially expressed, which are associated with resistance response and infection process.

Further, the RNA samples from *in vitro* grown fungal strains shall help in identifying *in planta* expressed fungal transcripts and also in identifying *in vitro* differentially expressed fungal genes among the strains. The infected plants samples collected for analysis by cDNA-AFLP at different time intervals will determine early to late stages of infection and will also allow detection of the pathogen transcripts. The expression patterns of the transcripts shall be monitored with several different primer combinations for amplification in a selective manner and for every primer combination; the TDFs (transcript derived fragments) can be viewed as bands on gels. These bands can be visually scored and compared for the differences in intensity for detecting transcripts expressed differentially *in plants* and for comparison these will be eluted and reamplified with appropriate cDNA-AFLP primers. This is followed by sequencing and then the products are screened using public databases for finding homologous sequences with significant alignment. Using the BLAST analysis, sequences identical to crop transcripts, transcripts homologous to known plant sequences in UNIPORT KB Swiss-Prot or TrEMBL http://www.expasy.ch/sprot/databases or NCBI databases (http://blast.ncbi.nlm.nih.gov/Blast.cgi) and transcripts homologous to known *Fusarium* sp. sequences considered as expressing *in planta* shall be identified. The transcripts, which have no matches in any of the databases, shall also be seen; these may represent transcripts that currently lack functional annotations. Further the expression pattern and clustering of crop transcripts shall be done to overview differences between infected and control. The functional annotation of each transcript can be done with help of Gene Ontology Database http://www.geneontology.org and through carefully analyzing the scientific literature. Further for the identification of *Fusarium* genes expressed in crop plant during infection we can include sequences from all *Fusarium* sp available in public databases. Finally, the fungal transcripts differentially expressed among the strains shall be identified when grown *in vitro* searching the *Fusarium* database (Fusarium Comparative Database 2). These studies will provide with information regarding transcriptional changes in crops upon *Fusarium* pathogenesis and also if the wilting symptoms are derived from active plant response besides the infection. The studies shall generate information regarding *In planta* expressed pathogen-derived transcripts during the infection processes which are related potentially to virulence functions as well as the *in vitro* expressed transcripts (expressed differentially between strains). These will further provide the sequences, which will be helpful for distinguishing between races. For the dissipation of information the sequence data generated from these studies shall be deposited in Gene Bank (Gene Expression Omnibus (GEO) database (http://www.ncbi.nlm.nih.gov/geo). The second important concern is related to Identification of the target regions in the fungal genome for probe generation for their use in phylogeny of *Fusarium* sp. Further, the use of a comprehensive EST collection in *Fusarium* for SSR identification shall help to develop EST-SSR markers for genetic mapping which will be extremely useful for diagnostics and research concerned with fungal biology, ecology, and genetics. The analysis will yield the frequency, type and distribution of SSR motifs in ESTs derived from *Fusarium* sp. The Perl script MIcroSAtelitte (MISA) (Thiel et al. 2003) can be used to identify SSRs in the *Fusarium spp.* EST sequences. These data can be used to perform comparative analysis of SSR motif polymorphisms between allelic sequences, and orthologous sequences to conduct and finally to identify functionally associated EST-SSR markers for application in comparative genomics. There is a need to study and focus on the development of EST-SSRs. Due to the lack of sufficient markers for *Fusarium* sp, it becomes necessary to develop enough molecular markers for potential use in *Fusarium* sp. pathogenicity. With the development of *Fusarium* database/projects a vast amount of available EST sequence data has been generated. The screening for repeat motifs (perfect and imperfect) can be done in unigenes within the transcript assembly. The SSRs with a minimum repeat count (n) threshold of $n \geq 5$ can be selected for further analysis and EST-SSR marker development. Flanking forward and reverse primers will be designed for SSRs in unigenes using Primer 3 (http://frodo.wi.mit.edu/cgi-bin/primer3/primer3_www.cgi) and products can be genotyped by sequencing and the allele lengths can be ascertained by gene mapper. The SSR markers can be screened for amplification and length polymorphisms among *Fusarium* strains on agarose. The estimation of EST-SSR individual markers heterozygosities (H) and Genetic distances (G) can be done using the proportion of shared alleles estimator in Microsat, where $G=(1-p)$ and p is the proportion of shared alleles http://hpgl.stanford.edu/projects/microsat/. These data offer an opportunity to identify single sequence repeats (SSRs) in expression sequence tags (ESTs) by data mining. Such kind of studies shall give an insight into the frequency, distribution and type of *Fusarium* EST-SSRs and demonstrate successful development of EST-SSR markers in crop pathogenesis. These EST-SSR markers would be enriching the current resources of molecular markers for the scientific community and would be useful for *Fusarium* identification at species level and breeding programs to develop resistant varieties. Further, the novel EST-SSRs would be useful for comparative genetic mapping which shall give us information about the genetic diversity and polymorphism among the *Fusarium* sp. All these shall provide the novel insights and knowledge about *Fusarium* pathogenesis.

References

Ahmad S, Gordon-Weeks R, Pickett J, Ton J (2010) Natural variation in priming of basal resistance: from evolutionary origin to agricultural exploitation. Mol Plant Pathol 11:817–827

Akhond M, Machray GC (2009) Biotech crops: technologies, achievements and prospects. Euphytica 166:47–59

Alabouvette C, Olivain C, Migheli Q, Steinberg C (2009) Microbiological control of soil-borne phytopathogenic fungi with special emphasis on wilt-inducing *Fusarium oxysporum*. New Phytol 184:529–544

Andersen JR, Lübberstedt T (2003) Functional markers in plants. Trends Plant Sci 8:554–560

Atkinson NJ, Urwin PE (2012) The interaction of plant biotic and abiotic stresses: from genes to the field. J Exp Bot 63(10):3523–3543

Baayen RP, O'Donnell K, Bonants PJM, Cigelnik E, Kroon LPNM, Roebroeck EJA, Waalwijk C (2000) Gene genealogies and AFLP analysis in the *Fusarium oxysporum* complex identify monophyletic and non-monophyletic *formae speciales* causing wilt and rot diseases. Phytopathology 90:891–900

Bacon CW, Hinton DM (1996) Symptomless endophytic colonization of maize by *Fusarium moniliforme*. Can J Bot 74:1195–1202

Bari R, Jones JD (2009) Role of plant hormones in plant defence responses. Plant Mol Biol 69(4):473–488

Beatty PH, Good AG (2011) Future prospects for cereals that fix nitrogen. Science 333:416–417

Bechtold U et al (2005) The influence of the light environment and photosynthesis on oxidative signaling responses in plant–biotrophic pathogen interactions. Plant Cell Environ 28:1046–1055

Beckman CH (1987) The nature of wilt diseases of plants. APS Press, American Phytopathological Society, St. Paul

Berek L, Petri IB, Mesterhazy AA, Teren J, Molnar J (2001) Effects of mycotoxins on human immune functions in vitro. Toxicol In Vitro 15:25–30

Bhattarai KK, Xie QG, Pourshalimi D, Younglove T, Kaloshian I (2007) Coi1-dependent signaling pathway is not required for Mi-1-mediated potato aphid resistance. Mol Plant Microbe Interact 20:276–282

Bhattarai KK, Hagop SA, Kaloshian I, Eulgem T (2010) WRKY72-type transcription factors contribute to basal immunity in tomato and Arabidopsis as well as gene-for-gene resistance mediated by the tomato *R* gene *Mi-1*. Plant J 63:229–240

Boutati EI, Anaissie EJ (1997) *Fusarium*, a significant emerging pathogen in patients with hematologic malignancy: ten years' experience at a cancer center and implications for management. Blood 90:999–1008

Campbell R, Greaves MP (1990) Anatomy and community structure of the rhizosphere. In: Lynch JM (ed) The rhizosphere. Wiley, Chichester, pp 11–34

Chandra NS, Udaya Shankar AC, Niranjana SR, Reddy MS, Prakash HS, Mortensen CN (2008) Control of *Fusarium verticilliodes* causing ear rot of maize by *Pseudomonas fluorescens*. Pest Manag Sci 65:769–775

Chandra SN, Shankar ACU, Niranjana SR, Wulff EG, Mortensen CN, Prakash HS (2010) Detection and quantification of fumonisins from *Fusarium verticillioides* in maize grown in southern India. World J Microbiol Biotechnol 26:71–78

Chang DC, Grant GB, O'Donnell K, Wannemuehler KA, Noble-Wang J, Rao CY et al (2006) Multistate outbreak of Fusarium keratitis associated with use of a contact lens solution. JAMA 296:953–963

Charpentier M, Oldroyd G (2010) How close are we to nitrogen fixing cereals? Curr Opin Plant Biol 13:556–564

Cordeiro GM, Casu R, McIntyre CL, Manners JM, Henry RJ (2001) Microsatellite markers from sugarcane (Saccharum spp.) ESTs cross transferable to erianthus and sorghum. Plant Sci 160:1115–1123

de Oliveira Dal'Molin CG et al (2010a) C4GEM, a genome-scale metabolic model to study C4 plant metabolism. Plant Physiol 154:1871–1885

de Oliveira Dal'Molin CG et al (2010b) AraGEM, a genome-scale reconstruction of the primary metabolic network in Arabidopsis. Plant Physiol 152:579–589

Depuyd S, Hardtke CS (2011) Hormone signaling crosstalk in plant growth regulation. Curr Biol 21:365–373

Desjardins AE, Hohn TM, McCormick SP (1993) Trichothecene biosynthesis in *Fusarium* species: chemistry, genetics, and significance. Microbiol Mol Biol Rev 57:595–604

Desjardines AE, Proctor RH (2007) Molecular biology of Fusarium mycotoxins. Int J Food Microbiol 119:47–50

Di Pietro A, Madrid MP, Caracuel Z, Delgado-Jarana J, Roncero MIG (2003) *Fusarium oxysporum*: exploring the molecular arsenal of a vascular wilt fungus. Mol Plant Path 4:315–325

Diener AC, Ausubel FM (2005) Resistance to *Fusarium oxysporum* 1, a dominant Arabidopsis disease-resistance gene, is not race specific. Genetics 171:305–321

Dilip R, Panthee DR, Chen F (2010) Genomics of fungal disease resistance in tomato. Curr Genomics 11:30–39

Dirlewanger E, Cosson P, Tavaud M, Aranzana MJ, Poizat C, Zanetto A, Arús P, Laigret F (2002) Development of microsatellite markers in peach (*Prunus persica* (L.) Batsch) and their use in genetic diversity analysis in peach and sweet cherry (*Prunus avium* L.). Theor Appl Genet 105:127–138

Doohan FM, Parry DW, Jenkinson P, Nicholson P (1998) The use of species-specific PCR-based assays to analyse Fusarium ear blight of wheat. Plant Pathol 47:197–205

Dubin HJ, Gilchrist L, Reeves L, McNab A (1997) Fusarium head blight: global status and prospects. CIMMYT, Mexico City

Durrant WE, Rowland O, Piedras P, Hammond-Kosack KE, Jones JDG (2000) cDNA AFLP reveals a striking overlap in race-specific resistance and wound response gene expression profiles. Plant Cell 12:963–977

Eaton CJ et al (2011) What triggers grass endophytes to switch from mutualism to pathogenism? Plant Sci 180:190–195

Ellegren H (2000) Microsatellite mutations in the germline: implications for evolutionary inference. Trends Genet 16:551–558

Ellegren H (2004) Microsatellites: simple sequences with complex evolution. Nat Rev Genet 5:435–445

Eujayl I, Sorrells ME, Wolters P, Baum M, Powell W (2002) Isolation of EST-derived microsatellite markers for genotyping the A and B genomes of wheat. Theor Appl Genet 104:399–407

Eujayl I, Sledge MK, Wang L, May GD, Chekhovskiy K, Zwonitzer JC, Mian MA (2004) Medicago truncatula EST-SSRs reveal cross-species genetic markers for Medicago spp. Theor Appl Genet 108:414–422

Fourie G, Steenkamp ET, Ploetz RC, Gordon TR, Viljoen A (2011) Current status of the taxonomic position of *Fusarium oxysporum* formae specialis *cubense* within the *Fusarium oxysporum* complex. Infec Genet Evol 11:533–542

Fracchia S, Garcia-Romera I, Godeas A, Ocampo JA (2000) Effect of the saprophytic fungus *Fusarium oxysporum* on arbuscular mycorrhizal colonization and growth of plants in greenhouse and field trials. Plant Soil 223:175–184

Fravel D, Olivain C, Alabouvette C (2003) *Fusarium oxysporum* and its biocontrol. New Phytol 157:493–502

Fusarium Comparative Database. http://www.broad.mit.edu/annotation/genome/fusarium_group/MultiHome.html

Gaffoor I, Trail F (2006) Characterization of two polyketide synthase genes involved in zearalenone biosynthesis in Gibberella zeae. App Environ Microbiol 72(3):1793–1799

Gaffoor I, Brown DW, Plattner RD, Proctor RH, Qi W, Trail F (2005) Functional analysis of the polyketide synthase genes in the filamentous fungus Gibberella zeae (anamorph Fusarium graminearum). Eukaryotic Cell 4(11):1926–1933

Gaitán-Solís E, Duque MC, Edwards KJ, Tohme J (2002) Microsatellite repeats in common bean (*Phaseolus vulgaris*): isolation, characterization, and cross-species amplification in *Phaseolus* ssp. Crop Sci 42:2128–2136

Gale MD, Devos KM (1998) Comparative genetics in the grasses. Proc Natl Acad Sci U S A 95:1971–1974

Gale LR, Katan T, Kistler HC (2003) The probable center of origin of *Fusarium oxysporum* f. sp. *lycopersici* VCG 0033. Plant Dis 87:1433–1438

Gao LL, Anderson JP, Klingler JP, Nair RM, Edwards OR et al (2007) Involvement of the octadecanoid pathway in bluegreen aphid resistance in *Medicago truncatula*. Mol Plant Microbe Interact 20:82–93

Garvey GS, McCormick SP, Rayment I (2008) Structural and functional characterization of the TRI101 trichothecene 3-O-acetyltransferase from *Fusarium sporotrichioides* and *Fusarium graminearum*. J Biol Chem 283:1660–1669

Genomics and Gene Discovery bEST Resource. http://wheat.pw.usda.gov/bEST/

Glazebrook J (2005) Contrasting mechanisms of defense against biotrophic and necrotrophic pathogens. Annu Rev Phytopathol 43:205–227

Gohre V, Robatzek S (2008) Breaking the barriers: microbial effector molecules subvert plant immunity. Annu Rev Phytopathol 46:189–215

Gonzalez-Ibeas D, Cañizares J, Aranda MA (2012a) Microarray analysis shows that recessive resistance to *Watermelon mosaic virus* in melon is associated with the induction of defense response genes. Mol Plant Microbe Interact 25(1):107–118

Gonzalez-Ibeas D, Blanca J, Cañizares J, Truniger V, Aranda MA (2012b) A cost-effective double-stranded cDNA synthesis for plant microarrays. Plant Mol Biol Rep 30(5):1276–1282

Guo J, Jiang RHY, Kamphuis LG, Govers F (2006) A cDNA-AFLP based strategy to identify transcripts associated with avirulence in *Phytophthora infestans*. Fungal Genet Biol 43(2):111–123

Gupta PK, Varshney RK (2000) The development and use of microsatellite markers for genetic analysis and plant breeding with emphasis on bread wheat. Euphytica 113:163–185

Gupta S, Chakraborti D, Rangi RK, Basu D, Das S (2009) A molecular insight into the early events of Chickpea (*Cicer arietinum*) and *Fusarium oxysporum* f. sp. ciceri (Race 1) interaction through cDNA-AFLP analysis. Phytopathology 99:1245–1257

Gupta S, Chakraborti D, Sengupta A, Basu D, Das S (2010) Primary metabolism of chickpea is the initial target of wound inducing early sensed *Fusarium oxysporum f. sp. ciceri* Race 1. PLoS ONE 5(2):e9030

Gust AA et al (2010) Biotechnological concepts for improving plant innate immunity. Curr Opin Biotechnol 21:204–210

Hackauf B, Wehling P (2002) Identification of microsatellite polymorphisms in an expressed portion of the rye genome. Plant Breed 121:17–25

Han Z, Wang C, Song X, Guo W, Gou J, Li C, Chen X, Zhang T (2006) Characteristics, development and mapping of Gossypium hirsutum derived EST-SSRs in allotetraploid cotton. Theor Appl Genet 112:430–439

Harris LJ, Alexander NJ, Saparno A, Blackwell B, McCormick SP, Desjardins AE, Robert LS, Tinker N, Hattori J, Piche C, Schernthaner J, Watson R, Ouellet T (2007) A novel gene cluster in *Fusarium graminearum* contains a gene that contributes to butenolide biosynthesis. Fungal Genet Biol 44:293–306

Hey SJ et al (2010) The interface between metabolic and stress signaling. Ann Bot Rev 105:197–203

Howard PC, Eppley RM, Stack ME, Warbritton A, Voss KA, Lorentzen RJ, Kovach RM, Bucci TJ (2001) Fumonisin B1 carcinogenicity in a two-year feeding study using F344 rats and B6C3F1 mice. Environ Health Perspect 109(S2):277–282

Howe GA, Jander G (2008) Plant Immunity to Insect Herbivores. Ann Rev Plant Biol 59:41–66

Idnurm A, Howlett BJ (2001) Pathogenicity genes of phytopathogenic fungi. Mol Plant Pathol 2:241–255

Inoue I, Namiki F, Tsuge T (2002) Plant colonization by vascular wilt fungus *Fusarium oxysporum* requires FOW1, a gene encoding a mitochondrial protein. Plant Cell 14:1869–1883

Jonkers W, Estrada AER, Lee K, Breakspear A, May G, Kistler HC (2012) Metabolome and transcriptome of the interaction between Ustilago maydis and *Fusarium verticillioides in vitro*. Appl Environ Microbiol 78(10):3656–3667

Kakirde KS et al (2010) Size does matter: application-driven approaches for soil metagenomics. Soil Biol Biochem 42:1911–1923

Kempema LA, Cui X, Holzer FM, Walling LL (2007) Arabidopsis transcriptome changes in response to phloem-feeding silverleaf whitefly nymphs. Similarities and distinctions in responses to aphids. Plant Physiol 143:849–865

Kim YT, Lee YR, Jin J, Han KH, Kim H, Kim JC, Lee T, Yun SH, Lee YW (2005) Two different polyketide synthase genes are required for synthesis of zearalenone in *Gibberella zeae*. Mol Microbiol 58:1102–1113

Knowledge-based Oryza Molecular Biological Encyclopedia. http://cdna01.dna.affrc.go.jp/cDNA/

Kriek NPJ, Kellerman TS, Marasas WFO (1981) A comparative study of the toxicity of *Fusarium verticillioides* (=*F. moniliforme*) to horses, primates, pigs, sheep and rats. Ondersterpoort J Vet Res 48:129–131

Kunz BA et al (2008) UV-induced DNA damage promotes resistance to the biotrophic pathogen Hyaloperonospora parasitica in Arabidopsis. Plant Physiol 148:1021–1031

Kusnierczyk A, Winge P, Jorstad TS, Troczynska J, Rossiter JT et al (2008) Towards global understanding of plant defence against aphids—timing and dynamics of early Arabidopsis defence responses to cabbage aphid (*Brevicoryne brassicae*) attack. Plant Cell Environ 31:1097–1115

Larsen PE et al (2011) Using next generation transcriptome sequencing to predict an ectomycorrhizal metabolome. BMC Syst Biol 5:70

Lee B, Lee S, Ryu C-M (2012) Foliar aphid feeding recruits rhizosphere bacteria and primes plant immunity against pathogenic and non-pathogenic bacteria in pepper. Ann Bot 110(2):281–290

Leilani AR, James BH, Gary AP (2006) Marker assisted breeding for host resistance to mycotoxin contamination. In: Abbas HK (ed) Aflatoxin and food safety. CRC, Boca Raton, pp 423–426

Leon-Reyes A, Van der Does D, De Lange ES, Delker C, Wasternack C et al (2010) Salicylate-mediated suppression of jasmonate-responsive gene expression in Arabidopsis is targeted downstream of the jasmonate biosynthesis pathway. Planta (Berlin) 232:1423–1432

Levinson G, Gutman GA (1987) Slipped-strand mispairing: a major mechanism for DNA sequence evolution. Mol Biol Evol 4:203–221

Lindsay JA (1997) Chronic sequelae of foodborne disease. Emerg Infect Dis 3:443–452

Liu X, Tang W-H, Zhao X-M, Chen L (2010) A network approach to predict pathogenic genes for *Fusarium graminearum*. PLoS ONE 5(10):e13021 doi:10.1371/journal.pone.0013021

Ma LJ, van der Does HC, Borkovich KA, Coleman JJ, Daboussi MJ, Di Pietro A et al (2010) Comparative genomics reveals mobile pathogenicity chromosomes in *Fusarium*. Nature 464:367–373

Matilla MA et al (2010) Pseudomonas putida KT2440 causes induced systemic resistance and changes in Arabidopsis root exudation. Environ Microbiol Rep 2:381–388

McMullen M, Jones R, Gallenberg D (1997) Scab of wheat and barley: a re-emerging disease of devastating impact. Plant Dis 81:1340–1348

McSteen P, Zhao Y (2008) Plant hormones and signaling: Common themes and new developments. Developmental Cell 14:467–473

Memelink J (2009) Regulation of gene expression by jasmonate hormones. Phytochemistry 70:1560–1570

Mendes R et al (2011) Deciphering the rhizosphere microbiome for disease-suppressive bacteria. Science 332:1097–1100

Michielse CB, Rep M (2009) Pathogen profile update: *Fusarium oxysporum*. Mol Plant Pathol 10:311–324

Michielse CB, van Wijk R, Reijnen L, Cornelissen BJC, Rep M (2009) Insight into the molecular requirements for pathogenicity of Fusarium oxysporum f. sp. lycopersici through large-scale insertional mutagenesis. Genome Biol 10(1):R4. (ISSN 14656906)

Morales SE, Holben WE (2011) Linking bacterial identities and ecosystem processes: can 'omic' analyses be more than the sum of their parts? FEMS Microbiol Ecol 75:2–16

Morgante M, Hanafey M, Powell W (2002) Microsatellites are preferentially associated with nonrepetitive DNA in plant genomes. Nat Genet 30(2):194–200

Mule G, Gonzalez-Jaen MT, Hornok L, Nicholson P, Waalwijk C (2005) Advances in molecular diagnosis of toxigenic *Fusarium* species: a review. Food Addit Contam 22:316–323

Munkvold GP, McGee DC, Carlton WM (1997) Importance of different pathways for maize kernel infection by *Fusarium moniliforme*. Phytopathology 87:209–217

Murphy M, Armstrong D (1995) Fusariosis in patients with neoplastic disease. Infect Med 12:66–67

Mur LAJ, Kenton P, Atzorn R, Miersch O, Wasternack C (2006) The outcomes of concentration-specific interactions between salicylate and jasmonate signaling include synergy, antagonism, and oxidative stress leading to cell death. Plant Physiol 140:249–262

Nayaka SC, Wulff EG, Udayashankar AC, Nandini BP, Niranjana SR, Mortensen CN, Prakash HS (2011) Prospects of molecular markers in *Fusarium* species diversity. Appl Microbiol Biotechnol 90:1625–1639

NCBI Basic Local Alignment and Search Tool. http://www.ncbi.nlm.nih.gov/BLAST/

NCBI Expressed Sequence Tags Database. http://www.ncbi.nlm.nih.gov/dbEST/

Oberhardt MA et al (2009) Applications of genome-scale metabolic reconstructions. Mol Syst Biol 5:320

O'Donnell K, Cigelnik E (1997) Two divergent intragenomic rDNA ITS2 types within a monophyletic lineage of the fungus *Fusarium* are nonorthologous. Mol Phylogenet Evol 7:103–116

O'Donnell K, Cigelnik E, Nirenberg HI (1998) Molecular systematics and phylogeography of the *Gibberella fujikuroi* species complex. Mycologia 90:465–493

O'Donnell K, Kistler HC, Tacke BK, Casper HH (2000) Gene genealogies reveal global phylogeographic structure and reproductive isolation among lineages of *Fusarium graminearum*, the fungus causing wheat scab. Proc Natl Acad Sci U S A 97:7905–7910

O'Donnell K, Sutton DA, Rinaldi MG, Magnon KC, Cox PA, Revankar SG et al (2004) Genetic diversity of human pathogenic members of the *Fusarium oxysporum* complex inferred from multilocus DNA sequence data and amplified fragment length polymorphism analyses: evi-

dence for the recent dispersion of a geographically widespread clonal lineage and nosocomial origin. J Clin Microbiol 42:5109–5120
Parker IM, Gregory SG (2004) The evolutionary ecology of novel plant—pathogen interactions. Annu Rev Ecol Syst 35:675–700
Peakall R, Gilmore S, Keys W, Morgante M, Rafalski A (1998) Cross species amplification of soybean (Glycine max) simple sequence repeat (SSRs) within the genus and other legume genera: implication for transferability of SSRs in plants. Mol Biol Evol 15:1275–1287
Peleg Z, Blumwald E (2011) Hormone balance and abiotic stress tolerance in crop plants. Curr Opin Plant Biol 14:290–295
Pieterse CMJ, Dicke M (2007) Plant interactions with microbes and insects: from molecular mechanisms to ecology. Trends Plant Sci 12:564–569
Pieterse CMJ et al (2009) Networking by small-molecule hormones in plant immunity. Nat Chem Biol 5:308–316
Polesani M, Desario F, Ferrarini A, Zamboni A, Pezzotti M, Kortekamp A, Polverari A (2008) c-DNA-AFLP analysis of plant and pathogen genes expressed in grapevine infected with *Plasmopara viticola*. BMC Genomics 9:142–153
Powell W, Machray GC, Provan J (1996) Polymorphism revealed by simple sequence repeats. Trends Plant Sci 1:215–222
Power W, Machray GC, Povran J (1996) Polymorphism revealed by simple sequence repeats. Trends Plant Sci 1:215–222
Proctor RH, Desjardins AE, Brown DW, McCormick SP, Butchko RAE, Alexander N, Busman M (2006) Biosynthesis of Fusarium mycotoxins and genomics of Fusarium verticillioides. Mycotoxin Res 22(2):75–78
Puthoff DP, Holzer FM, Perring TM, Walling LL (2010) Tomato pathogenesis-related protein genes are expressed in response to *Trialeurodes vaporiorium* and *Bemisia tabaci* type B feeding. J Chem Ecol 36:1271–1285
Raghavendra AS et al (2010) ABA perception and signaling. Trends Plant Sci 15:395–401
Richards RI, Sutherland GR (1992) Heritable unstable DNA sequences. Nat Genet 1:7–9
Roa AC, Chavarriaga-Aguirre P, Duque MC, Maya MM, Bonierbale MW, Iglesias C, Tohme J (2000) Cross-species amplification of cassava (*Manihot esculenta*) (*Euphorbiaceae*) microsatellites: allelic polymorphism and degree of relationship. Am J Bot 87:1647–1655
Rodriguez-Hernandez AM, Gosalvez B, Sempere RN, Burgos L, Arnada MA, Truniger V (2012) Melon RNA interference (RNAi) lines silenced for Cm-eIF4E show broad virus resistance. Mol Plant Pathol 13(7):755–763
Roh SW, Abell GC, Kim KH, Nam YD, Bae JW (2010) Comparing microarrays and next-generation sequencing technologies for microbial ecology research. Trends Biotechnol 28:291–299
Rosewich UL, Pettway RE, Katan T, Kistler HC (1999) Population genetic analysis corroborates dispersal of *Fusarium oxysporum* f. sp *radicis-lycopersici* from Florida to Europe. Phytopathology 89:623–630
Saha MC, Mian MA, Eujayl I, Zwonitzer JC, Wang L, May GD (2004) Tall fescue EST-SSR markers with transferability across several grass species. Theor Appl Genet 109:783–791
Sestili S, Polverari A, Luongo L, Ferrarini A, Scotton M, Hussain J, Delledonne M, Ficcadenti N, Belisario A (2011) Distinct colonization patterns and cDNA-AFLP transcriptome profiles in compatible and incompatible interactions between melon and different races of *Fusarium oxysporum* f. sp. *melonis*. BMC Genomics 12:122. doi:10.1186/1471-2164-12-122
Schenk PM, Carvalhais LC, Kazan K (2012) Unravelling plant—microbe interactions: can multi-species transcriptomics help? Trends Biotechnol 30(3):177–184
Schwessinger B, Ronald PC (2012) Plant innate immunity: perception of conserved microbial signatures. Ann Rev Plant Biol 63:451–482
Scott KD, Eggler P, Seaton G, Rossetto M, Ablett EM, Lee LS, Henry RJ (2000) Analysis of SSRs derived from grape ESTs. Theor Appl Genet 100:723–726
Seefelder W, Humpf H-U, Schwerdt G, Freudinger R, Gekle M (2003) Induction of apoptosis in cultured human proximal tubule cells by fumonisins and fumonisin metabolites. Toxicol Appl Pharmacol 192:146–153

Simon C, Daniel R (2011) Metagenomic analyses: past and future trends. Appl Environ Microbiol 77:1153–1161

Sims JW, Fillmore JP, Warner DD, Schmidt EW (2005) Equisetin biosynthesis in Fusarium heterosporum. Chem Commun 186–188

Stewart F et al (2010) Development and quantitative analyses of a universal rRNA-subtraction protocol for microbial metatranscriptomics. ISME J 4:896–907

Stout MJ, Thaler JS, Thomma B (2006) Plant-mediated interactions between pathogenic microorganisms and herbivorous arthropods. Ann Rev Entomol 51:663–689

Szafranska K, Fusari F, Luongo L, Ferrarini A, Polverari A, Delledonne M, Ficcadenti N, Sestili S, Belisario A (2008) Fusarium wilt infection in melon: a transcriptomic approach to characterize the genetic dialogue between host and pathogen. In: Pitrat M (ed) Cucurbitaceae 2008, Proceedings of the IXth EUCARPIA meeting on genetics and breeding of Cucurbitaceae, Avignon, France, pp 615–619

Takken F, Rep M (2010) The arms between tomato and *Fusarium oxysporum*. Mol Plant Pathol 11:309–314

Temnykh S, DeClerck G, Lukashova A, Lipovich L, Cartinhour S, McCouch S (2001) Computational and experimental analysis of microsatellites in rice (O. sativa L.): frequency, length variation, transposon associations, and genetic marker potential. Genome Res 11:1441–1452

Thaler JS, Bostock RM (2004) Interactions between abscisicacid-mediated responses and plant resistance to pathogens and insects. Ecology 85:48–58

Thaler JS, Agrawal AA, Halitschke R (2010) Salicylate-mediated interactions between pathogens and herbivores. Ecology 91:1075–1082

Thiel T, Michalek W, Varshney RK, Graner A (2003) Exploiting EST databases for the development and characterization of gene derived SSR-markers in barley (Hordeum vulgare L.). Theor Appl Genet 106:411–422

Tian CF, Garnerone A-M, Mathieu-Demazière C, Masson-Boivin C, Batut J (2012) Plant-activated bacterial receptor adenylate cyclases modulate epidermal infection in the *Sinorhizobium meliloti—Medicago* symbiosis. PNAS 109(17):6751–6756

Torben A, Ursula KF, Thomas D, Klaus KN, Thomas L (2007) Frequency, type, and distribution of EST-SSRs from three genotypes of *Lolium perenne*, and their conservation across orthologous sequences of *Festuca arundinacea, Brachypodium distachyon*, and *Oryza sativa*. BMC Plant Biol 7:36

Trail F, Common R (2000) Perithecial development by *Gibberella zeae*: a light microscopy study. Mycologia 92:130–138

Trewavas A (2009) What is plant behavior? Plant Cell Environ 32:606–616

UniProtKB Swiss-Prot and TrEMBL. http://www.expasy.ch/sprot/databases

Varshney RK, Graner A, Sorrells ME (2005a) Genic microsatellite markers in plants: features and applications. Trends Biotechnol 23(1):48–55

Varshney RK, Sigmund R, Boerner A, Korzun V, Stein N, Sorrells ME, Langridge P, Graner A (2005b) Interspecific transferability and comparative mapping of barley EST-SSR markers in wheat, rye, and rice. Plant Sci 168:195–202

Verhage A, van Wees SCM, Pieterse CMJ (2010) Plant immunity: It's the hormones talking, but what do they say? Plant Physiol 154:536–540

Voge C, Marcotte EM (2012) Insights into the regulation of protein abundance from proteomic and transcriptomic analyses. Nat Rev Genet 13:227–232

Vuylsteke M, Peleman JD, van Eijk MJT (2007) AFLP-based transcript profiling (cDNA-AFLP) for genome-wide expression analysis. Nat Protoc 2:1399–1413

Walling LL (2009) Adaptive defense responses to pathogens and pests. Adv Bot Res 51:551–612

Wang X, Tang C, Zhang G, Li Y, Wang C, Liu B, Qu Z, Zhao J, Han Q, Huang L, Chen X, Kang Z (2009) cDNA-AFLP analysis reveals differential gene expression in compatible interaction of wheat challenged with *Puccinia striiformis f. sp. tritici*. BMC Genomics 10:289

Weber JL (1990) Informativeness of human (dC-dA)n, (dG-dT)n polymorphisms. Genomics 7:524–530

Webster RK, Gunnell PS (1992) Compendium of rice diseases. APS Press, The American Phytopathological Society, St. Paul

White DG (1999) Compendium of corn diseases. APS Press, The American Phytopathological Society, St. Paul

White G, Powell W (1997) Isolation and characterization of microsatellite loci in *Swietenia humilis* (*Meliaceae*): an endangered tropical hardwood species. Mol Ecol 6:851–860

Wierdl M, Dominska M, Petes TD (1997) Microsatellite instability in yeast: dependence on the length of the microsatellite. Genetics 146:769–779

Wilson TM, Ross PF, Owens DL, Rice LG, Green SA, Jenkins SJ, Nelson HA (1992) Experimental reproduction of ELEM—a study to determine the minimum toxic dose in ponies. Mycopathologia 117:115–120

Windels, CE (2000) Economic and social impacts of *Fusarium* head blight: changing farms and rural communities in the northern great plains. Phytopathology 90:17–21

Wise RP, Moscou MJ, Bogdanove AJ, Whitham SA (2007) Transcript profiling in host-pathogen interactions. Annu Rev Phytopathol 45:329–369

Wulff EG, Sørensen JS, Lübeck M, Nielsen KF, Thrane U, Torp J (2010) *Fusarium* spp. associated with rice Bakanae: ecology, genetic diversity, pathogenicity and toxigenicity. Environ Microbiol 12:649–657

Wurtzel O, Sesto N, Mellin JR, Karunker I, Edelheit S, Bécavin C, Archambaud C, Cossart P, Sorek R (2012) Comparative transcriptomics of pathogenic and non-pathogenic *Listeria* species. Mol Syst Biol 8:583

Yao DX et al (2011) Transcriptome analysis reveals salt-stress regulated biological processes and key pathways in roots of cotton (*Gossypium hirsutum* L.). Genomics 98:47–55

Yli-Mattila T, Gagkaeva T (2010) Molecular chemotyping of *Fusarium gramineaum, F. culmorum* and *F. cerealis* isolates from Finland and Russia. In: Gherbawy Y, Voigt K (eds) Molecular identification of fungi. Springer, Berlin, pp 159–178

Yli-Mattila T, Paavanen-Huhtala S, Bulat SA, Alekhina IA, Nirenberg HI (2002) Molecular, morphological and phylogenetic analysis of the *Fusarium avenaceum*/*F. arthrosporioides*/*F. tricinctum* species complex a polyphasic approach. Mycol Res 106:655–669

Zabala MD et al (2009) Antagonism between salicylic and abscisic acid reflects early host-pathogen conflict and moulds plant defense responses. Plant J 59:375–386

Zabalgogeazcoa I (2008) Fungal endophytes and their interaction with plant pathogens. Span J Agric Res 6:138–146

Zarate SI, Kempema LA, Walling LL (2007) Silverleaf whitefly induces salicylic acid defenses and suppresses effectual jasmonic acid defenses. Plant Physiol 143:866–875

Zeller KA, Summerall BA, Ballock S, Leslie JF (2003) *Gibberella konza* (*Fusarium konzum*) sp. nov. from prairie grasses, a new species in the *Gibberella fujikuroi* species complex. Mycologia 95:943–954

Zhang PJ, Zheng SJ, van Loon JJA, Boland W, David A et al (2009) Whiteflies interfere with indirect plant defense against spider mites in Lima bean. P Natl Acad Sci USA 106:21202–21207

Zhao XM, Tang W, Chen L (2008a) Motif based identification of pathogenic patterns for filamentous fungi. The second international symposium on optimization and systems biology, pp 115–122

Zhao XM, Wang RS, Chen L, Aihara K (2008b) Uncovering signal transduction networks from high-throughput data by integer linear programming. Nucleic Acids Res 36:e48–e48

Zhao XM, Wang Y, Chen L, Aihara K (2008c) Gene function prediction using labeled and unlabeled data. BMC Bioinformatics 9:57

Zhao XM, Zhang XW, Tang WH, Chen L (2009) FPPI: *Fusarium graminearum* protein—protein interaction database. J Proteome Res 8:4714–4721

Zhao ZX et al (2010) ABA-regulated G protein signaling in Arabidopsis guard cells: a proteomic perspective. J Proteome Res 9:1637–1647

Zvirin T, Herman R, Brotman Y, Denisov Y, Belausov E, Freeman S, Perl- Treves R (2010) Differential colonization and defense responses of resistant and susceptible melon lines infected by *Fusarium oxysporum* race 1–2. Plant Pathol 59:576–585

Chapter 9
Coping Abiotic Stress with Plant Volatile Organic Chemicals (PVOCs): A Promising Approach

Penna Suprasanna and Prasad Shekhar Variyar

Abstract Abiotic stresses including salinity are a major threat to agricultural productivity and hence global food security. Crop plants have adopted specialized strategies to reduce the impact of stress. The biogenic volatile organic compounds (VOCs) emitted from a wide range of plants help enable the buildup defense against biotic and abiotic stresses. Plant VOCs are comprised of different isoprene and monoterpene class of compounds in addition to alkanes, alkenes, carbonyls, alcohols, esters, ethers, and acids which have a demonstrated role against abiotic stress factors. Although it has been shown that several metabolic pathways may be involved in building up the defense, antioxidant route of alleviation is believed to be a common mechanism. The identification of the genes, transcriptomic profiling and proteins of the biosynthetic pathway has enabled ways to manipulate the synthesis of isoprenoid compounds. In recent years, there has been a growing interest in adopting VOC strategy to alleviate abiotic stresses in crop plants.

Introduction

Environmental stress is a major threat to agricultural productivity and plants have adopted specialized strategies to reduce the impact of stress. The abiotic stresses include drought, salinity, cold and high temperature that affect the plant growth, development and yields of crop plants. Plants being sessile, experience multiple stresses in their life cycle and hence the tolerance trait has become complex to be understood and managed. Among the different abiotic stresses, salinity stress is the most severe limiting crop productivity. Salinity interferes with the plant's accessibility to

P. Suprasanna (✉)
Nuclear Agriculture and Biotechnology Division, Bhabha Atomic Research Centre,
Trombay, Mumbai 400085, India
e-mail: prasanna@barc.gov.in

P. S. Variyar
Food Technology Division, Bhabha Atomic Research Centre,
Trombay, Mumbai 400085, India
e-mail: pvariyar@barc.gov.in

K. R. Hakeem et al. (eds.), *Crop Improvement,* DOI 10.1007/978-1-4614-7028-1_9,

nutrients and water. Moreover, it induces osmotic stress; the physiological drought, which typically reduces the growth and photosynthesis in plants (Munns and Tester 2008). Salinity affects plant growth and development in two ways: through osmotic stress by reducing the soil water potential leading to limiting the water uptake and by causing uptake of Na^+ and Cl^- which have an effect on plant metabolism. The mechanism by which plants perceive stress signals and relay their transmission to cellular machinery to trigger adaptive responses is crucial for the improvement of different strategies to impart stress tolerance in crops (Mantri et al. 2012).

The different abiotic stress factors result in the production of reactive oxygen species (ROS) that are extremely reactive and cause damage to biological macromolecules like proteins, lipids, carbohydrates and DNA ultimately leading to oxidative stress. The ROS include, superoxide radicals, hydroxyl radical, perhydroxy radical, alkoxy radicals, hydrogen peroxide and singlet oxygen (Gill and Tuteja 2010). Under normal growth conditions, the ROS molecules are managed by efficient scavenging machinery consisting of various antioxidative defense mechanisms (Foyer and Noctor 2005). The production of ROS and their scavenging needs to be balanced under normal conditions of growth but, however the equilibrium is disturbed by abiotic stress factors including salinity (Tuteja 2007; Mantri et al. 2012).

Volatile Organic Compounds (VOC's) and Their Action

Plants are sessile and have to encounter challenges imposed by other organisms and with the environment mainly by depending on their chemical repertoire. The significance of natural products and their metabolic diversity contribute very much to the survival of the plant kingdom. The biogenic volatile organic compounds (VOCs) released from a wide range of plants help enable the buildup defense against insects, fungi, herbivores and environmental changes (Loreto and Schnitzler 2010; Holopainen and Blande 2012). Plant VOCs are comprised of isoprenoids mainly isoprene and monoterpenes (Variyar et al. 2010). The function of isoprenoid compounds during environmental stress includes protection of the photosynthetic apparatus, detoxification from free radicals and reactive oxygen species (ROS) (Munné-Bosch and Alegre 2000a; Spinelli et al. 2011). Although it has been shown that several metabolic pathways may be involved in building up the defense, antioxidant route is believed to be the common mechanism (Vickers et al. 2009a). The identification of genes in the biosynthetic pathway and transcriptomic profiling has enabled ways to manipulate the synthesis of isoprenoid compounds. Since chloroplasts are the sites of isoprene synthesis a possible relation may occur between isoprene production and environmental stresses affecting the photosynthetic apparatus (Velikova 2008; Loyola et al. 2012). It should thus be of interest to investigate isoprene synthesis in plants in relation to environmental chemistry. The emission of VOCs contributes to an appreciable quantity of photosynthetic carbon fixation under stress conditions, and hence VOCs could also play a significant role in the carbon exchange between the biosphere and the atmosphere (Guenther et al. 2011). Significant research progress has been made in the study of physiological mechanism(s) underlying isoprenoid synthesis under abiotic stress conditions, especially high temperatures and oxidative stress conditions (Fineschi and Loreto 2012).

Isoprenoids protect plants against different abiotic stresses through improving the ability of plants to deal with cellular oxidative modifications, possibly through reaction of isoprenoids with the oxidizing species, or alteration of ROS signaling, or via membrane stabilization. It is postulated that dissolution of VOCs in membranes coupled to interactions with membrane proteins can lead to changes in transmembrane potential and modulation of ion fluxes thereby inducing gene activity and a subsequent cellular response to stress (Vickers et al. 2009a). Plants have developed an efficient antioxidant mechanisms for ROS detoxification (Ahmad et al. 2008; Gill and Tuteja 2010; Ahmad and Umar 2011). Isoprenes can boost plant's defense system not only by keeping the membrane integrity intact and making it less sensitive to denaturation, but also due to the fact that they have the capacity to quench ROS produced under oxidative stress. Vickers et al. (2009a) discussed the possible functions of isoprenes as natural antioxidant machinery in plants.

Plants are endowed with protective mechanisms to cope with a variety of abiotic stresses. When the stress impact goes beyond a certain threshold, plants normally experience stress, resulting in reduced growth and development. Most common and ensuing response, thus, is the production of reactive oxygen spices (ROS). The antioxidant effect of the isoprenoid compounds is mediated by their capacity to swiftly combine with different ROS such as singlet oxygen, superoxide, hydrogen peroxide, hydroxyl radical that are released under stress regime (Holopainen 2004; Fineschi and Loreto 2012). Isoprenes are also known to alleviate visible damage (necrosis) of leaves exposed to ozone through a mechanism involving release of nitric oxide that interacts with increasing levels of ROS especially hydrogen peroxide. The occurrence of conjugated double bonds (delocalized π-electrons) in the isoprene molecule may mediate electron and energy transfers, conferring ROS-scavenging ability (Vickers et al. 2009a). Considering chloroplast as the site of isoprene biosynthesis (Logan et al. 2000), the ROS scavenging ability of isoprene molecule makes it important in plant defense against oxidative stress. Isoprenoids including terpenoids have also been shown to confer a protective effect on photosynthetic process under heat and oxidative stress (Sharkey and Yeh 2001). Isoprenes have also been implicated to protect the photosynthetic system from thermal stress. The mechanism underlying such protective nature is attributed to the stabilization of membrane lipid bilayer by enhancing the hydrophobic (lipid–lipid, lipid–protein and/or protein–protein) interactions (Sharkey et al. 2008). Based on modeling studies with membranes, Siwko et al. (2007) demonstrated that isoprenes are able to partition into the phospholipid membrane enhancing membrane order without major alteration in the dynamic properties of the membrane.

Much less evidence has been accumulated so far on the role of volatile monoterpenes in alleviating oxidative stress. In plants that don't emit monoterpenes, it has been proved that photosynthesis becomes less sensitive to ozone that are externally supplied with volatile monoterpenes (Loreto and Fares 2007). In contrast, when monoterpene synthesis is blocked, ROS rapidly accumulate. The highly volatile

Table 9.1 List of cloned genes involved in the biosynthesis of monoterpenes

Gene	Organism	References
Linalool synthase	*Clarkia breweri*; *Artemisia annua* L.	Dudareva et al. (1996); Cseke et al. (1998); Jia et al. (1999)
(−)-*Limonene synthase*	*Abies grandis*; *Mentha spicata*	Colby et al. (1993); Bohlmann et al. (1997)
(+)-*Limonene synthase*	*Agastache rugoa*	Maruyama et al. (2002)
(−)-*Pinene synthase*	*Abies grandis*	Bohlmann et al. (1997)
Myrcene synthase	*Abies grandis*, *Quercus ilex L.*; *Arabidopsis thaliana*	Bohlmann et al. (1997); Bohlmann et al. (2000); Fischbach et al. (2001)
β-*Ocimene synthase*	*Arabidopsis thaliana*	Bohlmann et al. (2000)
(+)-*Bornyl diphosphate synthase*, *1,8 cineole synthase* and (+)-*Sabinene synthase*	*Salvia offcinalis*	Wise et al. (1998)
(−)-β-*Phellandrene synthase*, (−)-*camphene synthase*, *Terpinolene synthase and* (−)- *limonene*/(−)-α-*pinene synthase*	*Abies grandis*	Bohlmann et al. (1999)
(*E*)-*Beta farnesene synthase*	*Citrus junos*	Maruyama et al. (2001)

monoterpenes exhibit more effectiveness in scavenging antioxidants. On the other hand, the less volatile isoprenes pool up in membrane and intercellular spaces and thus become more effective antioxidants in the aqueous phase. Volatile sesquiterpenes are produced in high levels in ozone-resistant tobacco upon exposure to ozone. It is thus possible that volatile isoprenoids constitute one of the non-enzymatic oxidative defense systems thereby, reducing the oxidative damage caused by abiotic stresses.

Monoterpenes have different effects on plant growth and development, depending on their structure and the quantity. Thus α-pinene exerts protective effect on the photosynthetic apparatus, while α-terpinol shows toxicity. Monoterpenes exogenously applied at levels of 0.5 g/l exhibited toxicity in plant cell cultures (Brown et al. 1987). Monoterpenes such as cineole, thymol, geraniol, menthol and camphor induced oxidative stress and lipid oxidation in maize roots (Zunino and Zygadlo 2004) while, β-myrcene, limonene, β-ocimene and γ-terpinene generated ROS and oxidative damage (Singh et al. 2009). Menthol has shown an increase in cytosolic free calcium ions which can generate signal transduction pathways in cucumber roots (Maffei et al. 2001). The aliphatic monoterpenes (ocimene and myrcene) induced considerable changes in the transcription of several hundred genes in Arabidopsis, many of them are designated as transcription factors, stress and defence genes (Godard et al. 2008). Several genes involved in the biosynthesis of terpenes have now been cloned in different plants (Phillips et al. 2006; Christianson 2006; Degenhardt et al. 2009). Table 9.1 lists some of the genes involved in the synthesis of monoterpenes.

Salt stress is known to mimic water stress limiting CO_2 inflow by lowering conductance of stomata and mesophyll and by impairing carbon metabolism (Delfine et al. 1998, 1999). Loreto and Delfine (2000) tested whether revival from modest salt treatment could result in bursts of isoprene emission and concluded that the progression leading to isoprene release is resistant than photosynthesis to salt stress, and that a secondary source of isoprene, independent of photosynthesis, is induced by salt-stress. In case of short-term drought stress, significant reductions in photosynthesis were observed, whereas isoprene emission was either not repressed or became reduced in *Quercus virginiana* (Tingey et al. 1981) and *Pueraria lobata* (Sharkey and Loreto 1993). On the other hand, there was a good relationship between terpene emission and plant water status. The emission of several monoterpenes and sesquiterpenes was studied in Mediterranean species (*Rosmarinus officinalis, Pinus halepensis, Cistus albidus* and *Quercus coccifera*) upon subjecting them to long term water dehydration stress (Ormeno et al. 2007). There was a slow decrease of emissions in plants exposed to long term water deficit periods in *P. halepensis* and *C. albidus* as compared to decrease in sesquiterpene release of *R. officinali.* Šimpraga et al. (2011) opined that drought stress can affect the VOC emissions in plants. In their experiments with young Common beech, the authors observed sudden burst of non-monoterpene class of VOCs during acute drought stress indicating opportunities for plant sensing using VOCs.

Manipulating the Synthesis of VOCs

Isoprenoids have been demonstrated to confer defense against abiotic stress factors, mainly thermal stress and oxidative stress conditions. A full understanding of the function of terpenes in plant defense process will require experiments at the molecular level, as terpenes may induce the expression of a number of stress-related genes. Studies in this direction by using inhibitors like fossidomycin that can inhibit the MEP pathway, fumigating non-isoprene synthesizing plants with exogenous isoprenoid compounds and transgenic plants either expressing terpene synthesis genes or gene silencing, have yielded results supporting their protection against stresses (Dudareva and Pichersky 2008; Vickers et al. 2009a).

The enzymes leading to the production of monoterpene all appear to be active in the plastids, as all the genes in this pathway possess plastid-targeting signals (Haudenschild and Croteau 1998) and seems to be localized in chloroplasts (Bouvier et al. 2000) and leucoplasts (Turner et al. 1999). The principal functional role of isoprene emission in plants is associated with the protection of leaf physiological processes against oxidative stress induced by heat (Sharkey and Yeh 2001). Behnke et al. (2007) analyzed this 'physiological role' by testing transgenic Grey poplar plants in which expression of *isoprene synthase* (*ISPS*) was either silenced via RNA interference (RNAi) mechanism or upregulated by over-expression of the *ISPS* gene. Despite increased *ISPS* mRNA levels, there was no steady increase in isoprene release in the over-expressing lines, suggesting that *ISPS* could be regulated at the post-transcriptional level while in the RNAi lines, there was no isoprene emission. The researchers also exposed transgenic lines to high temperature with three temporary heat stages (38–40 °C), followed by recovery at 30 °C. During heat stress, the non-isoprene-emitting transgenic poplars exhibited low rates of net assimilation and photosynthetic electron transport, compared to situation where there was no stress. The poplars plants in which isoprene was repressed had an increased zeaxanthin in the absence of stress, suggesting increased non-photochemical quenching or may indicate an increased necessity for antioxidants (Behnke et al. 2007). This study demonstrated that down-regulation of isoprene can influence thermotolerance and induce increased energy dissipation by non-photochemical quenching pathways. Isoprene synthase transcription has been shown to increase as leaves undergo maturity (Wiberley et al. 2005) and is temperature- and light responsive (Sasaki et al. 2005; Cinege et al. 2008). Variation in the accumulation of isoprene synthase protein is also observed under different environmental conditions (Schnitzler et al. 2005; Wiberley et al. 2009; Calfapietra et al. 2007).

Transgenic tobacco (*Nicotiana tabacum* L.) plants transformed with an isoprene synthase gene (from poplar) showed isoprene emission at comparable amounts to a natural situation. These transgenic plants when subjected to heat and combined heat/light exhibited considerable tolerance to stress-induced oxidative stress (Vickers et al. 2009b). Further, Vickers et al. (2011) used transgenic tobacco lines harboring a poplar isoprene synthase gene and then examined control of isoprene emission. In mature transgenic tobacco leaves, it was observed that primary controls on isoprene emission was thought to be via the substrate supply and changes in enzyme kinetics rather than changes in isoprene synthase levels or post-translational regulation of activity. The transgenic tobacco plants also had emission patterns remarkably similar to naturally emitting plants under a wide variety of conditions and the emissions correlated with photosynthetic rates in developing and mature leaves, and with the amount of isoprene synthase protein in mature leaves. Isoprene synthase protein levels did not change under short-term increase in heat/light, despite an increase in emissions under these conditions. In a study with a halophytes (*Kandelia candel*) and *Bruguiera gymnorrhiza*, mRNA expression of four oxidosqualene cyclase (*OSC*) genes namely, *KcMS multifunctional terpenoid synthase* and *KcCAS cyloartenol synthase* (*K. candel*), *BgbAS ß-amyrin synthase* and *BgLUS lupeol synthase* (*B. gymnorrhiza*) in relation to salt concentration was analyzed (Basyunia et al. 2009). The mRNA levels of *KcMS* in both roots and leaves of *K. candel* and *BgLUS* and *BgbAS* in the roots of *B. gymnorrhiza* increased with salt concentration. This result suggested that the function of terpenoids in root is associated with the salt stress.

Attempts have been made to over-accumulate isoprenoids in transgenic plants to study their role in stress alleviation. Over-expression of *Hevea brasiliensis* 3-hydroxy-3-methylglutaryl coenzyme A reductase (HMGR) in transgenic tobacco led to an increase in sterol production (Schaller et al. 1995). Neelakandan et al. (2011) over-expressed Arabidopsis HMGR1 in soybean, resulting in greater seed sterol content. The *Populus alba* isoprene synthase gene was introduced into *Arabidopsis* and has shown to confer elevated heat tolerance in the transgenic lines over wild type (Sasaki et al. 2007). Similarly, the content in some plastidial isoprenoids has also been successfully enhanced in plants through genetic engineering. Transgenic mint over-expressing 1-deoxy-D-xylulose-5-phosphate synthase, one of the entry enzymes into the MEP pathway (DXS), showed increased essential oil content (Mahmoud and Croteau 2001). Arabidopsis plants over-expressing *Brassica juncea* 3-hydroxy-3-methylglutaryl-CoA synthase gene (*BjHMGS*), coding for the second enzyme in the cytosolic isoprenoid biosynthesis pathway, have been shown to provide enhanced fungal and hydrogen peroxide-tolerance (Wang et al. 2011). The Brassica gene was found to be down-regulated by abscisic acid, mannitol, and water stress, but up-regulated by growth regulators like salicylic acid, methyl jasmonate, and wounding, suggesting that it could have a role in plant stress resistance.

The genetic engineering of volatile compounds have also brought to light some genetic changes on plant growth and development, and challenges to accomplish efficient production of the suitable volatile terpenoid compounds in a spatial and temporal mode (Dudareva and Pichersky 2008). For example in Arabidopsis, over-expression of *FaNES1* resulted in the diversion of carbon to linalool production, without affecting the levels of chlorophylls, lutein and bcarotene, and resulting in a growth-retardation phenotype that was stable through several generations (Aharoni et al. 2003). Transgenic potato engineered for linalool production resulted in growth retardation and leaf bleaching of plants when grown in the greenhouse (Aharoni et al. 2006). Transgenic tobacco containing high levels of patchoulol as a result of the expression of *PTS* coupled with *FPP synthase*, both targeted to the plastids, led to plants with growth disturbances like leaf chlorosis, vein clearing, and reduced stature (Wu et al. 2006). Such growth abnormalities are attributed to the consequences of the reduction of isoprenoid precursors for other metabolites which are otherwise are essential for plant growth and development, or that the newly introduced terpenoids could become toxic to plant cells.

A number of plant species synthesize myriad of isoprenoid for plant growth, development and for adaptation to environment (Leivara et al. 2011). The enzyme 3-hydroxy-3-methylglutaryl-CoA reductase (HMGR) in the mevalonate pathway is modulated by many endogenous and external stimuli. Two B″ regulatory subunits (B″α and B″β) of protein phosphatase 2A (PP2A) interact with HMGR1S and HMGR1L, the two major isoforms of *Arabidopsis thaliana* HMGR (Leivara et al. 2011). Since B″α and B″β are Ca^{2+} binding proteins of the EF-hand type, it was found that PP2A modulates HMGR transcript. Under salt stress conditions, the B″α and PP2A mediated the decrease and subsequent increase of HMGR activity in Arabidopsis seedlings, resulting from a steady rise of HMGR1-encoding transcript level and an early sharper reduction of HMGR protein level. In the non-stress conditions, the PP2A operates as a posttranslational negative regulator of HMGR activity with the involvement of B″β. The authors suggested that PP2A can exert multilevel regulation on HMGR through the five-member B″ protein family in response to stress conditions (Leivara et al. 2011).

The mevalonate pathway that mediates the production of isoprenoids has been operative in higher eukaryotes. Brodersen et al. (2012) studied the necessity of isoprenoid biosynthesis for plant miRNA activity in Arabidopsis. In plants

ARGONAUTE (AGO) protein complexes are guided by microRNAs (miRNAs) to regulate expression of complementary RNAs. Brodersen et al. (2012) used *mad3* and *mad4*, the miRNA action deficient (*mad*) mutants, for the isolation of genes involved in isoprenoid biosynthesis. The 3-hydroxy-3-methylglutaryl CoA reductase (HMG1), acting in the initial C5 building block biogenesis that precedes isoprenoid metabolism and acts as a key regulatory enzyme controlling the amounts of isoprenoid end products is encoded by *MAD3* while, the sterol C-8 isomerase that acts downstream in dedicated sterol biosynthesis is encoded by *MAD4*. Complementation studies using yeast system and treatment *in planta* with an inhibitor of HMG1 (lovastatin), indicated that lack of catalytic activity in HMG1 is adequate to inhibit miRNA activity. Further knockdown of HMG1/MAD3 reduced AGO1-membrane interaction and specific hypomorphic mutant alleles of AGO1 displayed compromised membrane association. The study has shown an interesting possibility that for the activity of plant miRNAs, isoprenoid synthesis could be required and this could unravel underlying mechanisms of microRNA function and regulation.

Conclusions and Future Perspectives

Abiotic stresses including salinity, drought and high temperature limit crop productivity. In this regard, PVOCs either emitted or induced from different plant species can be applied to confer better defense. Understanding of the biosynthesis of volatile compounds and the genetic machinery involved has greatly contributed to use this chemical repertoire for integrating biochemical, molecular and functional data into stress alleviation. A complete picture of metabolic network of PVOC synthesis and information on their regulation will necessitate further investigation. In addition, screening and use of suitable compounds involved in the biosynthesis of volatile-induced plant defenses will greatly facilitate fine tuning of plant responses to stress factors. In the past decade, considerable progress has been made in the metabolic engineering of the isoprenoid biosynthetic pathway in plants (Mahmoud and Croteau 2001; Lucker et al. 2001; Nagegowda 2010). An increasing number of successful attempts have raised hopes that their manipulation could offer a promising tool for increasing isoprenoid content for varied applications in stress tolerance and protection from environmental damage.

Another direction in PVOCs is by using priming approach by which planting a few transgenic plants that release defense volatiles in the field may contribute to plant protection and provide an advantage to non-transgenic plants (Dudareva and Pichersky 2008). In order to derive such benefits, it is imperative that we need to investigate the molecular mechanisms underlying priming induced capacitance, the detection of volatile signal components that activate the capacitance, species specific responses and molecular markers for the primed state in crop plants. It has also been suggested that histone modifications that are operative during a primary event might create memory associated reaction to a second stress exposure (Jaskiewicz et al. 2011).

Plants produce a plethora of volatile compounds for both general and specialized functions (Ueda et al. 2012). The plant volatilome is defined as the complex consortium of volatile organic compounds through different biosynthetic pathways and produced by plants, constitutively and/or after induction, as a defense strategy against biotic and abiotic stress (Maffei et al. 2007). An integrated approach will greatly help our understanding about the metabolism, genomics and interactome of the VOCs in plant's adaptation to environmental stresses.

References

Aharoni A, Giri AP, Deuerlein S, Griepink F, de Kogel WJ, Verstappen FWA, Verhoeven HA, Jongsma MA, Schwab W, Bouwmeester HJ (2003) Terpenoid metabolism in wild-type and transgenic Arabidopsis plants. Plant Cell 15:2866–2884

Aharoni A, Jongsma MA, Kim TY, Ri MB, Giri AP, Verstappen FWA, Schwab W, Bouwmeester HJ (2006) Metabolic engineering of terpenoid biosynthesis in plants. Phytochem Rev 5:49–58

Ahmad P, Sarwat M, Sharma S (2008) Reactive oxygen species, antioxidants and signaling in plants. J Plant Biol 51(3):167–173

Ahmad P, Umar S (2011) Oxidative stress: role of antioxidants in plants. Studium Press Pvt Ltd, New Delhi

Basyunia M, Babab S, Inafukua M, Iwasakia H, Kinjoc K, Okua H (2009) Expression of terpenoid synthase mRNA and terpenoid content in salt stressed mangrove. J Plant Physiol 166:1786–1800

Behnke K, Ehlting B, Teuber M, Bauerfeind M, Louis S, Hänsch R, Polle A, Bohlmann J, Schnitzler JP (2007) Transgenic, non-isoprene emitting poplars don't like it hot. Plant J 51:485–499

Bohlmann J, Martin D, Oldham NJ, Gershenzon J (2000) Terpenoid secondary metabolism in *Arabidopsis thaliana*: cDNA cloning, characterization, and functional expression of a *myrcene/(E)-beta-ocimene synthase*. Arch Biochem Biophys 375:261–269

Bohlmann J, Phillips M, Ramachandiran V, Katoh S, Croteau R (1999) cDNA cloning, characterization, and functional expression of four new monoterpene synthase members of the *Tpsd* gene family from grand fir (*Abies grandis*). Arch Biochem Biophys 368:232–243

Bohlmann J, Steele CL, Croteau R (1997) Monoterpene synthases from grand fir (*Abies grandis*). cDNA isolation, characterization, and functional expression of myrcene synthase, (−)-(4S)-limonene synthase, and (−)-(1S,5S)-pinene synthase. J Biol Chem 272:21784–21792

Bouvier F, Suire C, d' Harlingue A, Backhaus RA, Camara B (2000) Molecular cloning of geranyl diphosphate synthase and compartmentation of monoterpene synthesis in plant cells. Plant J 24:241–252

Brodersen P, Sakvarelidze-Achard L, Schaller H, Khafif M, Schott G, Bendahmane A, Voinnet O (2012) Isoprenoid biosynthesis is required for miRNA function and affects membrane association of ARGONAUTE 1 in Arabidopsis. Proc Natl Acad Sci USA 109:1778–1783

Brown JT, Hegarty PK, Charlwood BV (1987) The toxicity of monoterpenes to plant cell cultures. Plant Sci 48:195–201

Calfapietra C, Wiberley AE, Falbel TG, Linskey AR, Mugnozza GS, Karnosky DF, Loreto F, Sharkey TD (2007) Isoprene synthase expression and protein levels are reduced under elevated O3 but not under elevated CO_2 (FACE) in field-grown aspen trees. Plant Cell Environ 30:654–661

Christianson DW (2006) Structural biology and chemistry of the terpenoid cyclases. Chem Rev 106:3412–3442

Cinege G, Louis S, Hänsch R, Schnitzler JP (2008) Regulation of isoprene synthase promoter by environmental and internal factors. Plant Mol Biol 69:593–604

Colby SM, Alonso WR, Katahira EJ, McGarvey DJ, Croteau R (1993) *4S-limonene synthase* from the oil glands of spearmint (*Mentha spicata*): cDNA isolation, characterization, and bacterial expression of the catalytically active monoterpene cyclase. J Biol Chem 268:23016–23024

Cseke L, Dudareva N, Pichersky E (1998) Structure and evolution of linalool synthase. Mol Biol Evol 15:1491–1498

Degenhardt J, Köllner TG, Gershenzon J (2009) Monoterpene and sesquiterpene synthases and the origin of terpene skeletal diversity in plants. Phytochemistry 70:1621–1637

Delfine S, Alvino A, Zacchini M, Loreto F (1998) Consequences of salt stress on diffusive conductances, Rubisco characteristics and anatomy of spinach leaves. Aust J Plant Physiol 25:395–402

Delfine S, Alvino A, Villani MC, Loreto F (1999) Restrictions to CO_2 conductance and photosynthesis in spinach leaves recovering from salt stress. Plant Physiol 119:1101–1106

Dudareva N, Cseke L, Blanc VM, Pichersky E (1996) Evolution of foral scent in Clarkia: novel patterns of S-linalool synthase gene expression in the *C. breweri* flower. Plant Cell 8:1137–1148

Dudareva N, Pichersky E (2008) Metabolic engineering of plant volatiles. Curr Opin Biotechnol 19:1–9

Fineschi S, Loreto F (2012) Leaf volatile isoprenoids: an important defensive armament in forest tree species. iForest 5:13–17

Fischbach RJ, Zimmer W, Schnitzler JP (2001) Isolation and functional analysis of a cDNA encoding a myrcene synthase from holm oak (*Quercus ilex* L.). Eur J Biochem 268:5633–5638

Foyer CH, Noctor G (2005) Redox homeostis and antioxidant signaling: a metabolic interface between stress perception and physiological responses. Plant Cell 17 1866–1875

Gill SS, Tuteja N (2010) Reactive oxygen species and antioxidant machinery in abiotic stress tolerance in crop plants. Plant Physiol Biochem 48:909–930

Godard KA, White R, Bohlmann J (2008) Monoterpene induced molecular responses in *Arabidopsis thaliana*. Phytochemistry 69:1838–1849

Guenther A, Kulmala M, Turnipseed A, Rinne J, Suni T, Reissell A (2011) Land ecosystem—atmosphere observational networks. iLeaps Newsletter 11:5–13

Haudenschild CD, Croteau RB (1998) Molecular engineering of monoterpene production. Genet Eng 20:267–280

Holopainen JK (2004) Multiple functions of inducible plant volatiles. Trends Plant Sci 9(11): 529–533

Holopainen JK, Blande JD (2012) Molecular plant volatile communication. Adv Exp Med Biol 739:17–31

Jaskiewicz M, Conrath U, Peterhänsel C (2011) Chromatin modification acts as a memory for systemic acquired resistance in the plant stress response. EMBO Rep 12:50–55

Jia JW, Crock J, Lu S, Croteau R, Chen XY (1999) (3R)-*linalool synthase* from *Artemisia annua* L. cDNA isolation, characterization, and wound induction. Arch Biochem Biophys 372:143–149

Leivara P, Antolín-Llovera M, Ferrero S, Closa M, Arró M, Ferrer A, Boronat A, Campos N (2011) Multilevel control of Arabidopsis 3-hydroxy-3-methylglutaryl coenzyme A reductase by protein phosphatase 2A. Plant Cell 23:1494–511

Logan BA, Monson RK, Potosnak MJ (2000) Biochemistry and physiology of foliar isoprene production. Trends Plant Sci 5:477–481

Loreto F, Delfine S (2000) Emission of Isoprene from Salt-Stressed *Eucalyptus globulus* leaves. Plant Physiol 123:1605–1610

Loreto F, Fares S (2007) Is ozone flux Inside leaves only a damage indicator? Clues from volatile isoprenoid studies. Plant Physiol 143:1096–1100

Loreto F, Schnitzler JP (2010) Abiotic stress and induced BVOCs. Trends Plant Sci 15:154–166

Loyola J, Verdugo E, Gonzalez I, Casaretto JA, Ruiz-Lara S (2012) Plastidic isoprenoid biosynthesis in tomato: physiological and molecular analysis in genotypes resistant and sensitive to drought stress. Plant Biol 14:149–156

Lucker J, Bouwmeester HJ, Schwab W, Jan B, Linus H, van der Plas W, Verhoeven HA (2001) Expression of Clarkia S-linalool synthase in transgenic petunia plants results in the accumulation of S-linalyl-b-Dglucopyranoside. Plant J 27:315–324

Maffei M, Camusso W, Sacco S (2001) Effect of *Mentha x piperita* essential oil and monoterpenes on cucumber root membrane potential. Phytochemistry 8:703–707

Maffei ME, Mithofer A, Boland W (2007) Insects feeding on plants: rapid signals and responses preceding the induction of phytochemical release. Phytochemistry 68:2946–2959

Mahmoud SS, Croteau R (2001) Metabolic engineering of essential oil yield and composition in mint by altering expression of *deoxyxylulose phosphate reductoisomerase* and *menthofuran synthase*. Proc Natl Acad Sci USA 98:8915–8920

Mantri N, Patade V, Suprasanna P, Rebecca Ford, Edwin Pang (2012) Abiotic stress responses in plants—present and future. In: Parvaiz A, Prasad MNV (eds) Environmental adaptations to changing climate: metabolism, productivity and sustainability. Springer, pp 1–20

Maruyama T, Ito M, Honda G (2001) Molecular cloning, functional expression and characterization of (E)-beta farnesene synthase from *Citrus junos*. Biol Pharm Bull 24:1171–1175

Maruyama T, Saeki D, Ito M, Honda G (2002) Molecular cloning, functional expression and characterization of d-limonene synthase from *Agastache rugoa*. Biol Pharm Bull 25:661–665

Munné-Bosch S, Alegre L (2000a) Changes in carotenoids, tocopherols and diterpenes during drought and recovery, and the biological significance of chlorophyll loss in *Rosmarinus officinalis* plants. Planta 210:139–146

Munns R, Tester M (2008). Mechanisms of salinity tolerance. Annu Rev Plant Biol 59:651–681

Nagegowda DA (2010) Plant volatile terpenoid metabolism: biosynthetic genes, transcriptional regulation and subcellular compartmentation. FEBS Lett 584:2965–2973

Neelakandan AK, Chamala S, Valliyodan B, Nes WD, Nguyen HT (2011) Metabolic engineering of soybean affords improved phytosterol seed traits. Plant Biotechnol J 10:12–19

Ormeno E, Mévy JP, Vila B, Bousquet-Mélou A, Greff S, Bonin G, Fernandez C (2007) Water deficit stress induces different monoterpene and sesquiterpene emission changes in Mediterranean species. Relationship between terpene emissions and plant water potential. Chemosphere 67:276–284

Phillips DR, Rasbery JM, Bartel B, Matsuda SPT (2006) Biosynthetic diversity in plant triterpene cyclization. Curr Opin Plant Biol. 9:305–314

Sasaki K, Ohara K, Yazaki K (2005) Gene expression and characterization of isoprene synthase from *Populus alba*. FEBS Lett 579:2514–2518

Sasaki K, Saito T, Lämsä M, Oksman-Caldentey KM, Suzuki M, Ohyama K, Muranaka T, Ohara K, Yazaki K (2007) Plants utilize isoprene emission as a thermotolerance mechanism. Plant Cell Physiol 48:1254–1262

Schaller H, Grausem B, Benveniste P, Chye ML, Tan CT, Song YH, Chua NH (1995) Expression of the *Hevea brasiliensis* (H.B.K.) Müll. Arg. 3-hydroxy-3-methylglutaryl-coenzyme A reductase 1 in tobacco results in sterol overproduction. Plant Physiol 109:761–770

Schnitzler JP, Zimmer I, Bachl A, Arend M, Fromm J, Fischbach RJ (2005) Biochemical properties of isoprene synthase in poplar (*Populus × canescens*). Planta 222:777–786

Sharkey TD, Loreto F (1993) Water stress, temperature, and light effects on the capacity for isoprene emission and photosynthesis of kudzu leaves. Oecologia 95:328–333

Sharkey TD, Yeh S (2001) Isoprene emission from plants. Annu Rev Plant Physiol Plant Mol Biol 52:407–436

Sharkey TD, Wiberley AE, Donohue AR (2008) Isoprene emission from plants: why and how. Ann Bot 101:5–18

Šimpraga M, Verbeeck H, Demarcke M, Joó É, Okorska O, Amelynck C, Schoon N, Dewulf J, Van Langenhove H, Heinesch B, Aubinet M, Laffineur Q, Müller J-F, Steppe K (2011) Clear link between drought stress, photosynthesis and biogenic volatile organic compounds in *Fagus sylvatica* L. Atmos Environ 45(30):5254–5259

Singh HP, Kaur S, Mittal S, Batish DR, Kohli RK (2009) Essential oil of *Artemisia scoparia* inhibits plant growth by generating reactive oxygen species and causing oxidative damage. J Chem Ecol 35:154–62

Siwko ME, Marrink SJ, de Vries AH, Kozubek A, Schoot Uiterkamp A JM, Mark AE (2007) Does isoprene protect plant membranes from thermal shock? A molecular dynamics study. Biochim Biophys Acta 1768:198–206

Spinelli F, Cellini A, Marchetti L, Mudigere NK, Piovene C (2011) Emission and function of volatile organic compounds in response to abiotic stress. In: Shanker A, Venkateswarlu B (eds) Abiotic stress in plants-mechanisms and adaptations. InTech Publ. ISBN: 978-953-307-394-1
Tingey DT, Evans R, Gumpertz M (1981) Effects of environmental conditions on isoprene emission from live oak. Planta 152:565–570
Turner G, Gershenzon J, Nielson EE, Froehlich JE, Croteau R (1999) Limonene synthase, the enzyme responsible for monoterpene biosynthesis in peppermint, is localized to leucoplasts of oil gland secretory cells. Plant Physiol 120:879–886
Tuteja N (2007) Mechanisms of high salinity tolerance in plants. Meth Enzymol 428:419–438
Ueda H, Kikuta Y, Matsuda K (2012) Plant communication—mediated by individual or blended VOCs? Plant Signal Behav 7(2):222–226
Variyar PS, Chatterjee S, Sharma A (2010) Chemistry and biochemistry of odors. In: Logan EW, Jason MA (eds) The biology of odors: sources, olfaction and response. Nova Science Publisher, USA, pp 263–294
Velikova VB (2008) Isoprene as a tool for plant protection against abiotic stresses. J Plant Interact 3:1–15
Vickers CE, Gershenzon J, Lerdau MT, Loreto F (2009a) A unified mechanism of action for volatile isoprenoids in plant abiotic stress. Nat Chem Biol 5:283–291
Vickers CE, Possell M, Cojocariu CI, Velikova VB, Laothawornkitkul J, Ryan A, Mullineaux PM, Nicholas Hewitt C (2009b) Isoprene synthesis protects transgenic tobacco plants from oxidative stress. Plant Cell Environ 32:520–531
Vickers CE, Possell M, Laothawornkitkul J, Ryan AC, Hewitt CN, Mullineaux PM (2011) Isoprene synthesis in plants: lessons from a transgenic tobacco model. Plant Cell Environ 34:1043–1053
Wang H, Nagegowda DA, Rawat R, Bouvier-Navé P, Guo DJ, Bach TJ, Chye ML (2011) Overexpression of *Brassica juncea* wild-type and mutant HMG-CoA synthase 1 in Arabidopsis up-regulates genes in sterol biosynthesis and enhances sterol production and stress tolerance. Plant Biotechnol J 10:31–42
Wiberley AE, Linskey AR, Falbel TG, Sharkey TD (2005) Development of the capacity for isoprene emission in kudzu. Plant Cell Environ 28:898–905
Wiberley AE, Donohue AR, Westpha, MM, Sharkey TD (2009) Regulation of isoprene emission from poplar leaves throughout a day. Plant Cell Environ 32:939–947
Wise ML, Savage TJ, Katahira E, Croteau R (1998) Monoterpene synthases from common sage (*Salvia offcinalis*): cDNA isolation, characterization, and functional expression of (+)-sabinene synthase, 1,8-cineole synthase, and (+)-bornyl diphosphate synthase. J Biol Chem 273:14891–14899
Wu SQ, Schalk M, Clark A, Miles RB, Coates R, Chappell J (2006) Redirection of cytosolic or plastidic isoprenoid precursors elevates terpene production in plants. Nat Biotechnol 24:1441–1447
Zunino MP, Zygadlo JA (2004) Effects of monoterpenes on lipid oxidation in maize. Planta 219:303–309

Chapter 10
An Overview of Omics for Wheat Grain Quality Improvement

Awais Rasheed, Tariq Mahmood, Alvina Gul-Kazi and Abdul Mujeeb-Kazi

Abstract Cereal grain quality aspects are integral aspects of a complex food chain, which assimilate outputs achievable by breeding, production and processing. In order to get better economic gains and be internationally competitive in diverse market scenarios, it is paramount to breed wheat cultivars with better grain quality. Higher grain quality demands are exponentially increasing due to novel processing technologies, environmental changes and change in consumer preferences due to striking demographic shifts. Advances in the genomic arena of grain quality are considered crucial for defining genes and their networks underpinning functional flour qualities. The complexities associated with the genes underlying these traits can be resolved by elucidating functional and comparative genomics information of relevant genes and the efficient transfer of such information across cultivars. Wheat, due to wider consumption as a staple food, has been a subject of intensive cytogenetic investigations which are now extended further in the genomics era using powerful tools of molecular biology and new genetic stocks. The recent progress in wheat genomics research particularly the use of molecular markers for a variety of purposes and advances in map based positional cloning of several genes has been remarkable. As a result we have been able to better understand the wheat genome and the mechanisms involved in the function of different quality encoding genes. Additionally, we have also utilized information generated from genomics research in producing better quality grains. The advances in the genomics of

A. Rasheed (✉) · T. Mahmood
Department of Plant Sciences, Quaid-i-Azam University, Islamabad, Pakistan
e-mail: awaispbg@gmail.com

T. Mahmood
e-mail: tmahmood@qau.edu.pk

A. Gul-Kazi
Atta-ur-Rahman School of Applied Biosciences (ASAB), National University of Sciences and Technology (NUST), Islamabad, Pakistan
e-mail: alvina_gul@yahoo.com

A. Mujeeb-Kazi
National Institute for Biotechnology and Genetic Engineering (NIBGE), Faisalabad, Pakistan
e-mail: m.kazi@cgiar.org

K. R. Hakeem et al. (eds.), *Crop Improvement,* DOI 10.1007/978-1-4614-7028-1_10,

quality presented in this chapter provide ample information to the underlying gene networks controlling quality traits thereby addressing the challenges of the brisk changes prevalent within the wheat based food systems. Aiding the exploitation of novel genome diversity for quality value addition, research has benefitted from the unique germplasm resource generated by synthesizing wheat from genomic/allelic variability residing in the wheat progenitor accessional resource. These under-utilized diploid wheat progenitor accessions are a promising conduit to wheat productivity enhancement and the novel genomic resource contributing to wheat quality as elucidated here.

Introduction

Bread wheat (*Triticum aestivum*) is one of the most important crop species, with global annual production currently over 600 million tonnes providing approximately one fifth of the world's total calorific input (FAO 2009). Continually raising the yield potential of wheat to match human population growth and stabilizing yield against the damaging effects of climate change is a top priority for agricultural science (Reynolds et al. 2009). The multitude demands of variable wheat products are challenging to fulfill in a scenario of maintaining competitiveness in international marketplace. Especially, in the most rapidly growing markets of South-Asia and China where the grain quality improvement has a critical role to play in establishing the linkages with customers. The traditional quality aspects of wheat need to be evolved due to the advent of new processing technologies and changes in the market place resulting from the striking demographic changes in the region. The integration of several disciplines like functional genomics, biotechnology and exploitation of the genetic resources is stimulating the identification of genetic, biochemical and physiological basis of quality encoding traits in wheat. The ongoing activities for wheat quality improvement aim to address the major challenge of capturing the information from both wheat and model organisms, such as rice and *Arabidopsis*, in order to define genes that underpin the unique quality attributes of wheat. The resources being developed using biotechnology, comparative and functional genomics include comprehensive mapping initiatives, genome-wide expression studies and exploring the molecular basis of quality characteristics. The linkage of large information generated from these tools need to be incorporated in wheat-breeding programs in conjunction with high-throughput screening in order to provide the solution to efficiently develop new, improved quality wheat varieties.

Grain yield and quality, both are determined by the size and composition of wheat endosperm. Biochemical and genetic studies in the past three decades have considerably increased the understanding of genetics, structure and composition of different proteins stored in endosperm which highly influence end-use quality traits (Ma et al. 2007). Wheat storage proteins include glutenins, gliadins, secalins and puroindolines within endosperm which largely determine the rheological properties of wheat flour, the most important quality attribute. Additionally the mineral and phytate competition is the major determinant of bioavailability of essential minerals which is an important component of wheat grain quality. The recent discovery of NAC gene transcription factor, *TtNAM-B1*, role in enhancing grain mineral concentration (Uauy et al. 2006b) has opened new ways to efficiently utilize the genomic approaches to harness quality related genes from wild relatives in order to get desirable products.

Some of the quality-encoding traits are polygenic, while the others are simply inherited. The development and utilization of the functional markers for monogenic quality traits like high molecular weight glutenins (HMW-GS), Low molecular weight glutenins (LMW-GS), grain hardiness genes (*PINA* and *PINB*) and waxy alleles have dramatically changed the selection of appropriate breeding material having desirable genes. Similarly, efforts are underway to identify gene networks underlying the quality traits through bi-parental quantitative trait and analysis (QTL) and genome wide association studies (GWAS).

Genomics of Wheat Storage Proteins

The composition and amount of seed storage proteins play an important role in determining wheat quality (Payne 1987). Beccari in 1745, first isolated the gluten proteins and until now, the gluten proteins have been active area of investigation at genomic and proteomic level over a period of 250 years, in order to determine their structure and properties and to provide basis for manipulating and improving end use quality (Shewry et al. 2002). At the beginning of the 20th century, Osborne (1907) developed a systematic way to classify wheat storage proteins based on their graded extraction and differences in solubility. According to Osborne (1907) four different protein groups can be recognized in wheat flour. These groups include albumins (water soluble), globulins (water insoluble and soluble in saline solutions), prolamins (soluble in 70–90 % ethanol) and glutens (soluble in dilute acid or alkali). The most important protein, gluten, gives rise to two distinct groups based on their solubility in 70 % ethanol, known as glutenins and gliadins (Wrigley et al. 1996).

Glutenins

High Molecular Weight Glutenin Subunits (HMW-GS)

HMW-GS reportedly account for 12 % of the total seed storage protein which correspond to about 1–1.7 % of the flour dry weight (Halford et al. 1992). However much of the work has been focused on functional and structural aspects of HMW-GS due to their largest contribution in wheat end-use quality. HMW-GS are encoded at *Glu-1* loci present on the long arm of homeologous group 1 chromosomes (1AL, 1BL and 1DL), each locus contributing two gene subunits that differ in their properties and are called x-type and y-type subunits (Payne et al. 1980). These loci are named *Glu-A1*, *Glu-B1* and *Glu-D1*, respectively. The x- and y-type subunits have a comparatively high and low molecular weight, respectively. In earlier studies on the allelic variation of *Glu-1* loci, the number of alleles at all three loci differed greatly (Payne and Lawrence 1983). *Glu-A1* had three allelic forms, eleven alleles for *Glu-1B*, and six alleles for *Glu-1D* were found. However, in the subsequent studies several alleles were found and the latest information has been documented by McIntosh et al. (2010).

Gene Expression

All bread wheat cultivars express 1Bx, 1Dx, and 1Dy subunits while some cultivars also express *1By* and *1Ax* subunit as well. The gene encoding the *1Ay* subunit usually remains silent. Nevertheless, many accessions of A-genome related species *T. monococcum* and *T. urartu* express *1Ay* subunits (Waines and Payne 1987; Rasheed et al. Unpublished; Alvarez et al. 2009; Caballero et al. 2008; Gutierrez et al. 2011). Some bread wheat with six HMW-GS have also been reported (Margiotta et al. 1996). The extensive studies on the electrophoretic mobility of glutenin subunits revealed that durm and bread wheat genotypes lack certain subunits (Lafiandra et al. 1988). Beitz et al. (1975) reported some mutants lack 1D encoded HMW-GS in landraces from Nepal. Contrastingly, some tetraploid and hexaploid genotypes with four and six subunits, respectively, were developed by replacing the silenced subunit of *Glu-A1* by the expressed ones. These genotypes showed an increment in polymeric glutenin quantity, hence better flour characteristics (Lafiandra et al. 1998). Allelic variation at *Glu-D1* has most profound effect on bread-making quality, although limited numbers of alleles have been reported at this locus. In addition to hexaploid wheat, *Aegilops tauschii* Cosson, the diploid ancestor of the D-genome, conserves many unique *Glu-D1* alleles. So far, 14 x-types and 10 y-type subunits in *Ae. tauschii* have been identified resulting into combination of 85 different *Glu-D1* alleles (Rehman et al. 2008).

Amino Acid Composition and Structure

HMW-GS is a highly complex mixture of proteins and high level of polymorphism is the main limiting factor to study the structure of gluten proteins. However in the last 30 years, much of the work has focused on HMW-GSs of wheat (Shewry et al. 2002). There is a high resemblance in primary structure of x- and y-type subunits. Both subunits consist of a signal peptide, N- and C-terminal domains and a repetitive central domain (Shewry and Halford 2002). The significant difference is in the number of conserved cysteine residues which are four in majority of x-type subunits and usually seven in the y-type subunits. These cysteine residues had a major role in the formation of disulphide bonds within and between subunits. Therefore these are primarily important in structure and function of the elastic properties of gluten protein result in risen loaf of wheat dough (Shewry and Tatham 1997). The repetitive domains are consisted of short and repeated peptide motifs in both x- and y-type subunits. These peptide motifs may be tripeptide, hexapeptide and nanopeptide, while the presence of the tripeptide motif is unique to the repetitive domain of x-type subunits (Shewry and Tatham 1997). Because the HMW-GS have very high glutamic acid contents, therefore proline and glycine contents are also very high, while lysine contents are very low. It is also evident from amino acid composition of HMW-GS that the central repetitive domain has hydrophilic nature and N- and C-terminal domains have hydrophobic characteristics (Shewry et al. 1989). The polypeptide motifs determine the proportion of the different amino acids in HMW-GS. Shewry and Tatham (1997) revealed that >90 % of the repetitive domains are formed from the variations in the consensus repeat sequences (PGQGQQ and GYYPTSPQQ). Moreover, x-type subunits are characterized by the presence of unique tri-peptide motif (GQQ). While in y-type subunits, the second proline is replaced by a leucine in the GYYPTSPQQ repeat motif.

Alleleic identification

The invaluable platform for HMW-GS diagnosis is, no doubt, sodium dodecyl sulphate poly-acrylamide gel electrophoresis (SDS-PAGE). However some limitations, like co-migration of some subunits, difficulty in detecting differences in expression levels, results in inaccurate identification of alleles differing in functional properties. Moreover, this technique is only possible from the flour of mature grains. The advancements in molecular biology has enabled us to overcome these limitations by using allele specific PCR markers. These markers are developed based on DNA polymorphism present among the glutenin subunit genes are considered perfect to study allelic variations for HMW-GS. The major advantage is the high-throughput analysis of different alleles in breeding materials which is also possible during the vegetative growth stages (Liu et al. 2008a). We have discussed in detail the molecular diagnosis approaches for HMW-GS identification in "Functional markers" section of this chapter.

Among the other proteomics based technologies used to detect HMW-GS include reversed-phase high-performance liquid chromatography (RP-HPLC) and the most recent matrix assisted laser desorption time-of-flight mass spectrometry (MALDI-TOF). Gao et al. (2010) analyzed HMW-GS separation and characterization of bread wheat and wild accessions on MALDI-TOF, SDS-PAGE and RP-HPLC. Comparative analysis demonstrated merits and demerits of each methodology. Incorrect identification due to low resolution and overestimation has been the main drawback of SDS-PAGE. Irrespective of its disadvantages, SDS-PAGE is the simplest and cheaper technique, therefore, suitable for large-scale and high-throughput HMW-GS screening for wheat genotypes especially when the glutenin composition is clear in the breeding material. The most recent mass spectroscopy MALDI-TOF had several technical advantages including high throughput, high resolution, and accuracy. However, high equipment cost is the main hindrance to access this technology for many breeding programs.

Low Molecular Weight Glutenin Subunits

Low molecular weight glutenin subunits (LMW-GS) are also major fraction of glutenins and are also known as prolamine due to high amino acid, glutamines and pralines. LMW-GS represent about one third of the storage proteins and about 60 % of total glutenins (Beitz and Wall 1973). The genes controlling LMW-GS are present at short arm of group 1 chromosomes, *Glu-3 loci* (*Glu-A3, Glu-B3 and Glu-D3*), which are tightly linked to the *Gli-1* loci. Additionally, three new loci *Glu-2, Glu-4* (Jackson et al. 1985; Liu and Shepherd 1995) and *Glu-5* (Sreeramulu and Singh 1997) located on chromosomes 1B, 1D and 7D respectively, with molecular weights ranging from 30-31,000 Da, have also been reported to encode LMW-GS. The LMW-GS are more difficult to characterize and study as compared to HMW-GS, due to their heterogeneity. However, different analytical tools become available with the advancement of technology which made easier their characterization. Biochemical classification revealed three classes of LMW-GS i.e. B, C and D types (Jackson et al. 1983). Additionally, B type LMW-GS are further classified into three classes, LMW-m, LMW-s and LMW-i, based on the first amino acid residue which may be Methionine, Serine and Ile, respectively. Genes present at *Glu-A3* locus manly encode LMWi type subunits, which is the most recently identified class of LMW-GS (Zhang et al. 2004). Additionally, LMW-i had significant structural differences from LMW-m and LMW-s groups due to the lack of an N-terminal region and localization of cysteine residues in C-terminal region. However, eight number of cysteine residues are common in all groups. This structural difference is proposed to encode quality differences by glutenin polymer formation and gluten interaction. The C and D type subunits are composed mainly of proteins related to α/β-, γ- and ώ-gliadins which have variable numbers of cysteine residues. D'Ovidio and Masci (2004) proposed these subunits are incorporated into the polymeric network by virtue of unpaired cysteines. The dough quality is known to be influenced greatly by allelic variations at the *Glu-3* loci and ranking of different alleles due to their functional properties have been reported (Juhász and Gianibelli 2006).

Gene Expression and Polymorphism

In MacGene (2010) 17, 26 and 11 alleles differentiated by different diagnostic techniques are documented at all *Glu-3* loci wheat. However, this does not include the allelic variants encoded and confirmed in different wild relatives of Triticeae. In earlier studies, 20 different LMW-GSs were identified in 222 common wheat varieties from 32 countries (Gupta and Shepherd 1990). It was revealed that six alleles were encoded at *Glu-A3*, nine at *Glu-B3* and five at *Glu-D3* locus. Allelic richness of chromosome 1A encoded subunits was relatively low and even some cultivars did not express any LMW-GS. While the B-genome encoded LMW-GS, *Glu-B3*, showed highest polymorphism. Gupta and Shepherd (1993) provided evidence for the presence of LMW-GS genes on group 6 chromosomes. There is a close linkage between gliadins encoded by *Gli-1* loci and LMW-GS encoded by *Glu-3* loci

(Gianibelli et al. 2001). This linkage between alleles of two loci helped in diagnosis of several *Glu-B3* and *Glu-D3* alleles in wheat genotypes. Several glaidins were found as reliable markers LMW-GS allele diagnosis, due to their easy detection (Jackson et al. 1996).

Amino Acid Composition and Structure

The N-termninal sequence is most important for identification of LMW-GS, therefore seven main types of LMW-GS have been identified on the basis of first amino acid in the N-terminal sequences of the proteins. These include, seven LMW-s with starting with sequence SHIPGL-, three LMW-m with N-terminal sequences of METSHIPGL-, METSRIPGL and METSCIPGL respectively. In three LMW-GS, the N-terminal sequences resembles to the α-, β- and γ-type gliadins (Cloutier et al. 2001). The further classification of LMW-GS is based on deduced amino acid sequences and cysteine residue position facilitating inter-molecule disulfide linkage (Ikeda et al. 2002). Twelve such LMW-GS groups have been identified in wheat. Collectively, more than 100 genes of LMW-GS have been characterized and sequenced from common wheat including several partial and pseudogenes (Cloutier et al. 2001; Zhang et al. 2004). There was an effort by Long et al. (2005) to develop LMW-GS group specific primers, which they developed from the analysis of 69 known gene sequences from GenBank and classified them into nine groups by the deduced amino acid sequence of the highly conserved N-terminal domain. Later on, Ikeda et al. (2006) also developed 10 primers, based on the available sequences in the GenBank, which were group specific. In wheat varieties from Australia, Zhao et al. (2006, 2007) identified 6 different gene sequences and 12 gene haplotypes at the *Glu-D3* locus.

There is high similarity between the secondary structures of LMW-GS and structure of the S rich gliadins and the only exception of D type LMW-GS (D'Ovidio et al. 1995). There are about 250–300 residues are reported in polypeptides. Several workers have reported the further modification in the two domian structure of LMW-GS (Kasarda et al. 1984; Wieser 1995). In both domains, the N-terminal repetitive domains is rich in β-turns while short nonrepetitive domain is rich in α-helix and is more compact (Thomson et al. 1992).

Allelic identification

SDS-PAGE is considered one of the simplest techniques to identify LMW-GS with some restrictions. At *Glu-A3* locus, SDS-PAGE could identify five out of seven alleles, while 2-D gel electrophoresis and PCR based markers identified all the allelic variation at this locus in bread wheat (Liu et al. 2010). The *Glu-B3* alleles are easier to be identified by SDS-PAGE, MALDI-TOF and PCR based markers but some additional validation is more reliable by 2-DE method. Liu et al. (2010) compared four techniques (SDS-PAGE, 2-DE, MALDI-TOF, PCR based markers) as a conduit to test their suitability to be integrated in breeding programs. They established that the PCR based markers are the simplest, most accurate, lowest cost technique and therefore recommended this method for the identification of *Glu-A3* and *Glu-B3* alleles in breeding programs. However, the combination of different techniques was required to identify certain alleles, and would be especially useful when characterizing new alleles from new genetic resources. They also recommended a standard set of 30 cultivars for use in future studies to represent all LMW-GS allelic variants in the collection.

Gliadins

In wheat storage proteins, gliadin is an important fraction that accounts for about 40–50 % of the total proteins. It has great impact on processing and nutritional quality of flour, followed by HMW-GS and LMW-GS. Gliadins are soluble in 70 % ethanol and are heterogeneous mixtures of single-chained polypeptides. Gliadins can be seperated in A-PAGE (acid-PAGE) based on the differences in their mobility. The four different groups identified include α-, β-, γ-, and ω-gliadins, of which α- gliadins has fastest while ω-gliadins has slowest mobility. Gliadins are controlled by *Gli-1* loci, which are complex and comprise the ω-gliadin and γ-gliadin (Mecham et al. 1978) multi-gene families (Harberd et al. 1985), which in some circumstances may be divided into *Gli-1-1* and *Gli-1-2*, respectively. The LMW glutenin multigene families, which are closely linked to the *Gli-1* loci (Jackson et al. 1983), are listed separately as the *Glu-3* set (Singh and Shepherd 1985); information on map distance and gene order in relation to *Glu-3* and the centromere is given in the preamble for the *Glu-3* loci.

Gene Expression and Polymorphism

It was identified that short arms of group 1 and group 6 chromosomes encode gliadin genes. The genes present on *Gli-1* loci are controlled by *Gli-A1*, *Gli-B1*, and *Gli-D1* loci on 1AS,1BS and 1DS, respectively. Similarly, *Gli-2* genes are controlled by *Gli-A2*, *Gli-B2*, and *Gli-D2* loci on chromosomes 6AS, 6BS and 6DS, respectively. The genetic analysis revealed that all ω- and many γ-gliadins are encoded by *Gli-1* loci and all α-, many of β-, and some of γ-gliadins are encoded by *Gli-2* loci. The gene clusters of gliadins encode polypeptides as Mendalian factor and multiple allelism phenomenons have been observed at both loci (Metakovsky 1991). In MacGene (2010), 23 alleles are listed for *Gli-A1*, 24 for *Gli-B1* and 15 alleles for *Gli-D1* in bread and durum wheat. Similarly, among *Gli-2* loci, *Gli-A2* encodes 36 alleles, *Gli-B2* encodes 47 and *Gli-D1* encodes 31 alleles. Apart from these two loci, several workers reported many other loci. These include *Gli-3* (4 alleles at *Gli-A3* and 3 alleles at *Gli-B3*) coding for ω-gliadins on short arms of group 1 chromosomes (Galili et al. 1984; Sobko 1984), *Gli-5* (2 alleles at both *Gli-A5* and *Gli-B5* loci) coding for ω-gliadins on short arm of chromosomes 1A and 1B distal to *Gli-1* (Pogna et al. 1993), *Gli-6* and *Gli-7* on short arm of chromosome 1A and 1D (Metacovsky et al. 1996; Hassani et al. 2006).

Amino Acid Composition and Structure

Among the four gliadin groups, ω-gliagins have high level of glutamione and proline while low level of sulfurous amino acids (Gianibelli et al. 2002). Comparatively, they had few amino acids and high phenylalanine levels as compared to other gliadin groups (Kasarda et al. 1983; Tatham and Shewry 1995). Difference does exist among gliadin groups for surface hydrophobicity and ω-gliagins are lower hydrophobic than that of the α- and γ-type gliadins. Popineau and Pineau (1987) identified the gliadins as first peptides elute from the reverse phase-HPLC column. Among all the gluten protein fractions, gliadins are highest hydrophilic with reference to amino acid composition with only a few residues with charged side chains (Dupont et al. 2000). Three different types of ω-gliagins have been observed on the basis of the N-terminal sequences. The nomenclature of ω-gliadins is followed from the first three amino acids in their N-terminal sequences (Kasarda et al. 1983). Therefore, these three groups are called ARQ-, KEL-, and SRL-types based on the aforesaid nomenclature system (Tatham and Shewry 1995).

Similar to the ω-gliadins, α-, β- and γ-gliadins are also rich in glutamine and proline. In these groups, about 90 % of the glutamic and aspartic acid residues are amidated (Bietz et al. 1977; Kasarda et al. 1983). They are also characterized by low levels of basic amino acids and high leucine. In α/β- and γ-gliadins cysteine residues are 6 and 8, respectively and both all types are rich in sulfur. Müller and Wieser (1995, 1997) confirmed that 3–4 disulfide bonds between molecules are formed. The α/β-gliadins are represented by a very small sequence of five amino acid residues (VRVPV) on the basis of N-terminal sequences (Bietz et al. 1977). In α/β-gliadin, the pentapeptide motifs (PQQQP and PQQPY) are always present in a repetitive region that follows the N-terminal region of the proteins (Shewry et al. 1986). Contrastingly, in γ-gliadins the N-terminal region is consisted twelve amino acid residues (NMQVDPSGQVQW) followed by several repeats of consensus motif PQQPFPQ (Kasarda et al. 1983; Shewry and Tatham 1990).

Recently, Qi et al. (2009) analyzed 170 γ-gliadin genes isolated from common wheat and its closely related species, among which 138 sequences are putatively functional. The ORF lengths of these sequences range from 678 to 1089 bp, and the repetitive region is mainly responsible for the size heterogeneity of γ-gliadins. The repeat motif

$$\mathbf{P}(Q/L/S/T/I/V/R/A)\mathbf{F}(S/Y/V/Q/I/C/L)\mathbf{P}(R/L/S/T/H/C/Y)\mathbf{Q}_{1-2}(\mathbf{P}(S/L/T/A/F/H)\mathbf{QQ})_{1-2}$$

is repeated from 7 to 22 times. They found a wide range of amino acid composition in γ-gliadins, and those γ-gliadins from subgroup SG-10 and SG-12 and γ-gliadins with a short repetitive domain are more nutritional.

Kernel Texture (*Ha Locus*)

In wheat, the grain texture is encoded by the friabilin group of proteins having Mr~13 k Da. These proteins have strong association with starch. In soft grains, these are frequently associated with water-washed starch. While in hard grains and durum they show limited and no association with starch, respectively. It was generally suggested that friabilins are non-sticky proteins minimizing sticking of starch granules and protein matrix, thus allowing their easier separation. Friabilin give rise to two major polypeptides upon electrophoresis separation and amino acid sequencing. There are three polypeptides viz. puroindoline a, puroindoline b and grain softness protein-1 which are designated as *PINA*, *PINB* and *GSP-1*, respectively. Morris (2002) reviewed the properties, purification methods and discovery of friabilin. Hard and soft kernel textures are not due to difference in amount of friabilin because both classes have similar amount of friabilin. Rather difference is due to association of strach granules with friabilin during aqueous isolation. The major locus, *Ha*, is responsible for textural properties (Symes 1965), and has been identified on 5DS chromosome (Mattern et al. 1973; Sourdill et al. 1996). Puroindoline proteins encoded by two strongly linked genes at this locus were identified that is associated with variation for grain hardiness (Gautier et al. 1994). A mutation in *Pinb* gene results in change in amino acid giving rise to altered protein structure to bind with membrane polar lipids (Giroux and Morris 1998). This results in alteration of binding strength between protein matrix and starch granules. Apart from this mutation, Giroux and Morris (1998) also identified a null allele, *Pina-Da*, on other puroindoline gene. It was concluded that a variety will have the hard texture having mutant alleles (*Pina-D1b* or *Pinb-D1b*) at both puroindoline genes. Although, higher allelic variation observed at these loci (Morris 2002) but genotypes having alleles *Pina-D1a/Pinb-D1a* (soft), *Pina-D1a/Pinb-D1b* (hard) and *Pina-D1b/Pinb-D1a* (extra hard) are predominant (Cane et al. 2004). A positive correlation of these genotypic classes with water absorption was observed. Genotypes with extra hard texture absorbed 3.5 % more water than varieties with the hard texture and 8.3 % more than those with the soft texture. Contrastingly, they, did not find any difference for water absorption in the "extra hard" and "hard" classes. However a drop in milling yield in extra hard genotypes was observed by both. Apart from *Ha* locus, several QTLs underlying grain hardness characteristics have been discovered.

Genes and Polymorphism

Bhave and Morris (2008) reviewed the molecular genetics, gene regulation and structure of puroindolines. MacGene (McIntosh et al. 2010) listed 17 alleles at *Pina-D1*, 29 alleles at *Pinb-D1* and 9 alleles at *Gsp-D1* locus. Recently, Chen et al. (2009) identified 9 haplotypes in 56 sequences from einkorn wheat. Guzman et al. (2012)

also observed genetic polymorphism and nucleotide diversity of puroindoline genes in einkorn wheat and *T. urartu* accessions. A new translocation lines ($5A^{m}S.5AS$) from CS background carries puroindolines and *GSP-1* alleles that confer softer kernel texture. Massa et al. (2004) analyzed 50 accessions of *Ae. tauschii* which held enormous diversity at *Ha* locus. Simeone et al. (2006) identified putative puroindoline proteins with 22 and 28 amino acid replacements while working on diploid species. In some accessions of *Ae. speltoides* and *Ae. searsii*, a stop codon at final Tryptophane of *PINA* was observed while in other accessions one-residue was deleted inin *PINB* (Lillemo et al. 2002).

Structure of Genes at Ha Locus

The intronless coding regions of the *Pina-D1* and *Pinb-D1* genes of bread wheat are 447 bp long. The genes are 70.2 % identical in the coding regions but only ~53 % identical in the 30 un-translated region (Gautier et al. 1994). The physical order of these genes within BAC clones was identified as *Pinb-Pina-Gsp-1* in *T. monococcum* (Tranquilli et al. 1999) *and Ae. tauschii* (Turnbull et al. 2003). There are several reports where additional partial copy of *Pina-D1* has been found on 5DL. Bahve and Morris (2008) discussed in detail the gene expression their regulation and promoter sequences of puroindolines. The characterization of nucleotide sequence in A and B genome of tetraploid wheat, the 50 boundary of the *Ha* locus was defined by *Gsp* gene. A gene cluster known as Gene7 and Gene8 was also validated by the presence of the 30 boundary. Therefore, a ca 55 kb gDNA segment is defined as the *Ha* locus having the *Pina*, *Pin*b, two *Pinb* degenerated copies, Gene3 (only in D-genome) and Gene5. Chantret et al. (2005) identified a deletion of 38 kb in common wheat instead of Gene3 and Gene5. Instead these genes were annotated in diploid progenitors of hexaploid wheat. In barley, analysis of *Ha* locus identified some genes clusters were found conserved between wheat and barley (Caldwell et al. 2004). Some rearrangements were observed in barley hordoindolines (equivalents of puroindilines) like upstream position of gene from GC2 instead of downstream (as in wheat) and they are also in the opposite orientation. However, in barley grain hardiness is focused to their resistance to pest and disease invasion rather than milling attributes.

The other important part of *Ha* locus is *Gsp-1* gene which is tightly linked to *Pina* and *Pinb* on chromosome 5DS. Unlike Pin genes, *GSP-1* genes are present on all three group 5 chromosomes in wheat (Chantret et al. 2005). The nucleotide coding region of these genes consists of 495 nucleotides without introns and resembles 90–100 % with one another and ~42 % with Pin genes (Bhave and Morris 2008). Some studies suggest multiple *Gsp-1* copies per genome in at least some genomes/accessions (Gollan et al. 2007) whereas mapping studies show only one gene at the *Ha* locus (Chantret et al. 2005).

Grain Protein Content (*GPC*) and Nutritional Aspects

Improving the grain protein content has been area of main focus for wheat breeders due to its major contribution in bread and pasta-making quality and has a major contribution in improving nutritional status of masses. Despite of its importance, constrains prevail to increase protein contents due to its quantitative inheritence and high influence of the environment (Simmonds, 1995). Several reports are available on QTLs controlling GPC and linked markers are available for MAS. An authentic source of high protein content has been identified in accession of *Triticum turgidum* L. ssp. dicoccoides (referred to as DIC) during a survey (Avivi (1978). Cantrell and Joppa (1991) substituted each chromosome of DIC with the durum cultivar 'Langdon (LDN)' and later it was found that 6B substitution line of DIC into LDN (DIC-6B) had highest protein contents (Joppa et al. 1997). The DIC-6B substitution line and LDN were used a parents and a RILs mapping population was developed to map QTL conferring high GPC, which was found on the chromosome 6BS. Another secondary mapping population (RILs) was developed to further shorten this QTL and it was mapped between RFLP probes *Xcdo365* and *Xucw65* as a single Mendelian locus (*Gpc-B1*) within a 2.7 cm region (Olmos et al. 2003). Some new markers were developed in this region for high density mapping through Rice-Wheat micro-colinearity studies. Some additional recombination was initiated by developing more RILs and *Gpc-B1* locus was reduced with a 0.3 cm segment using newly developed markers (Distelfeld et al. 2004). The *Gpc-B1* gene within the 0.3 cm segment was physically mapped which spanned about 250 kb region (Distelfeld et al. 2006). The Gpc-B1 allele in DIC accelerates leaf senescence and Uauy et al. (2006a) suggested the differences in GPC are actually pelotropic effects of the in senescence. Kade et al. (2005) discovered the effect of DIC *Gpc-B1* allele during senescence explained the higher levels of soluble proteins and amino acids in flag leaves after anthesis relative to those with the LDN allele. Higher mineral contents in DIC were found to be associated with chromosome 6B (Cakmak et al. 2004), but its association with 250 kb region including *Gpc-B1* was validated later (Distelfeld et al. 2007). The major discovery was reported when map based cloning identified the *Gpc-B1* as a NAC transcription factor (*TtNAM-B1*) and it was established that wild emmer wheat has a functional allele whereas modern wheat varieties carry a nonfunctional allele originated by a frame shift mutation (Uauy et al. 2006b). The functional NAM-B1 orthologous has been found on chromosome 6A and 6D (*TtNAM-A1* and *TaNAM-A1*) and 6D (*TaNAM-D1*), and closely related paralogues on chromosomes 2B (*TtNAM-B2* and *TaNAM-B2*) and 2D (*TaNAM-D2*). In *RNAi* studies, RNA levels of these NAM homologs was reduced which in turn delayed senescence for more than 3 months and reduced grain protein and mineral contents by more than 30 % as compared to control lines (Uauy et al. 2006b). Most Recently, Cantu et al. (2011) employed mRNA-seq approach to detect small differences in transcript levels and identified the monocarpic senescence as an active process leading to large-scale changes in gene expression which begins considerably before the appearance of visual symptoms of senescence. As a result several GPC-regulated genes including transporters, hormone regulated genes, and transcription factors are activated. These GPC-regulated genes, particularly those up-regulated during senescence, provide valuable entry points to dissect the early stages of monocarpic senescence and nutrient remobilization in wheat.

Another main miner bioavailability limiting factor is the presence of phytic acid (PA). PA is stored in the aleurone layer and hampers the intestinal absorption of mineral cations by making insouluble complexes (Cheryan, 1980). Phytase activity of the flour strongly reduces the PA breakdown. Therefore, the mineral bio-availability depends on, both, mineral and phytase concentrations and these should be taken into account in wheat improvement for biofortification. Recently, Ram et al. (2011) indicated the presence of higher genetic variability of phytase in synthetic hexaploids as compared to Indian cultivars. There is a greater scope for manipulating phytase levels as compared to phytate in wheat breeding, due to the larger genetic effects and greater genetic variability of the phytase in wheat. Thus, D-genome synthetics hold significance to be used as source for increasing phytase levels. The release of cultivars with high mineral concentrations complemented with high intrinsic phytasic activity could greatly improve the nutritional value of bread, provided that less refined flour is utilized to preserve the source of the minerals. CIMMYT nearly a decade ago screened some wheat progenitor resources and identified accessions of *T. dicoccon* with elevated levels of iron and zinc. On these tetraploids, synthetic hexaploids were developed by the wide crossing unit and produced stocks (*T. dicoccon/Ae. tauschii)* for wheat breeding program. A nursery set has been deployed in India and Pakistan from which promise has been observed but impacting findings are still awaited.

QTLS for Grain Quality Traits

Understanding the genetic architecture underlying quality traits is essential to isolate and characterize the desirable genes. Extensive QTL mapping studies have been performed to study the genetic control of quality traits. The recent trend shifted to association mapping is the further extension to bi-parental mapping to study the QTLs with accuracy and precision (Mascri et al. 2012). Earlier, Campbell et al. (1999, 2001) identified QTL for kernel, milling, and baking traits. QTL for kernel traits are located on chromosomes 1A, 2B, 2D, 3B, 7A, and 7B. Earlier to this, Parker et al. (1998) identified two major loci for flour color on chromosomes 3A and 7A, using RFLP marker in 150 RILs. In quantitative terms, the most important trait is flour yield and several QTLs have been identified for this trait on chromosomes 4A, 4D, 5D (Nelson et al. 2006), 5D (Campbell et al. 2001) and 7D (McCartney et al. 2006). Recently, Carter et al. (2012) identified two QTLs on chromosome 7B explaining the 17 and 19 % variability in 188 RILs population. Similarly, flour and grain protein identified as a key quality trait largely influencing the quality attributes of dough had several QTLs identified in RILs (Nelson et al. 2006; Carter et al. 2012) and double haploid (McCartney et al. 2006; Huang et al. 2006) mapping populations. Two QTLs on chromosomes 2D and 4D explained about 30 % of the phenotypic variability and were also validated by other researchers.

The genetic control of milling and baking qualities is of paramount interest for industry. QTL mapping population from a soft x hard wheat was used by Breseghello et al. (2005) and about 15 QTL conferred control of milling traits, protein content and baking assay, were detected on group 1 and group 2 chromosomes with some QTLs on 3A/B, 4B, 5B, and 6B. In another mapping population from RL4452 x 'AC Domain' hard wheat cross, about 99 QTLs were found on 18 chromosomes for 41 quality traits (McCartney et al. 2006). A major QTL on 4D controlling plant height (Rht-D1b) flanked about 20 QTLs while a crop maturity controlling locus adhered about 10 QTLs controlling grain quality characteristics like starch contents, mixograph, farinograph and baking performance.

Apart from the bi-parental QTL studies, recently few association-mapping studies have been attempted to detect QTL for quality traits in sets of soft wheat germplasm. Several QTLs conferring control of kernel morphology were detected on chromosomes 2D, 5A, and 5B. Similarly, several quality-encoding QTLs were detected on 15 different chromosomes in an association mapping population by Reif et al. (2011). The majority of QTLs for flour characteristics, retention capacity of solvent and softness equivalent were found to be located on chromosome 1B and 2B in soft wheat bi-parental mapping population. We have presented an overview of QTL detection efforts for grain quality traits in wheat in Table 10.1.

Functional Markers for Wheat Grain Quality Traits

During the past two decades, there are extensive studies on the molecular mapping of the genes underlying the grain quality traits and a brief overview has been presented in earlier heading. These QTL analyses identified linked molecular markers such as SSRs, RAPDs, AFLPs, RFLPs and DArTs with the key quality traits. The low detection power, distance from the genes and allele specificity to the population and parents are the key characteristics of the neutral markers which effect on their predictive value in the diverse populations. Therefore, the diagnostics by the linked markers (MAS) is questioned in breeding programs and their use is restricted with some exceptional cases. Due to recent developments in molecular biology, there is overwhelming response for the use of functional markers as a selection tool due to their apparent advantage over the linked molecular markers. These are developed from the nucleotide sequence of the functional gene and it has powerful tendency to distinguish allelic variation on a single locus, thus are considered perfect markers for MAB (Varshney et al. 2005). Nevertheless, with the progress in gene cloning during the last years, the development of corresponding functional markers is getting fast track. In quantitative terms, 97 markers have been developed as a result of cloning of 30 genes. These markers had ability to identify 93 disease resistance alleles, agronomic and grain quality traits. In wheat, the grain processing and baking quality is controlled by high- and low-molecular weight glutenins, grain hardiness, and starch contents, polyphenol oxidase (PPO) activity, lipoxynase (LOX) activity and yellow pigment content (YPC). In total, 56 functional markers have been

Table 10.1 An overview of studies on QTLs for grain quality traits in wheat

Trait	MP*	No of lines	QTL/Marker Interval	Position	R^2**	Reference
Flour yield	RILs	114	DR Oxo1-Xcdo949	4D	0.16	Nelson et al. ((2006)
	RILs	114	XksuD9-Xcdo475	4A	0.17	Nelson et al. (2006)
	RILs	114	Xgwm190-Xmta9	5D	0.46	Nelson et al. (2006)
	RILs	78	Pinb	5D	0.53	Campbell et al. (1999)
	DH	182	QFyd.crc-7D	7D	0.18	McCartney et al. (2006)
	RILs	150	Xbcd115-3A-Xpsr754-3A	3A	0.22	Chalmers et al. (1999)
	RILs	150	–	7D	0.19	Chalmers et al. (1999)
	RILs	188	QFyeld.wak-3B	3B	0.19	Carter et al. (2012)
	RILs	188	QBkyeld.wak-3B	3B	0.17	Carter et al. (2012)
Grain protein	RILs	114	Xbcd152-Xfbb329	2A	0.15	Nelson et al. (2006)
	RILs	114	Xbcd102-Xbcd18	2D	0.32	Nelson et al. (2006)
	RILs	114	Xcdo1312-Xabg391	5A	0.19	Nelson et al. (2006)
	DH	182	QGpc.crc-4D	4D	0.29	McCartney et al. (2006)
	DH	185	QGpc.crc-4D	4D	0.32	Huang et al. (2006)
	DH	185	QGpc.crc-7B	7B	0.12	Huang et al. (2006)
	DH	–	QGpc2B	2B	0.14	Zhao et al. (2010)
	RILs	188	QPro.wak-3B	3B	0.07	Carter et al. (2012)
Flour protein	DH	182	QFpc.crc-4D	4D	0.28	McCartney et al. (2006)
	DH	182	QFpc.crc-2B	2B	0.16	McCartney et al. (2006)
	RILs	114	Xabc158-Xgwm60	7A	0.16	Nelson et al. (2006)
	DH	185	QFpc.crc-7B	7B	0.16	Huang et al. (2006)
	DH	185	QFpc.crc-4D	4D	0.28	Huang et al. (2006)
SDS sedimentation	RILs	114	Xwg177-DRCht1	3A	0.11	Nelson et al. (2006)
	RILs	114	Xfbb9-Xfba62	2D	0.21	Nelson et al. (2006)
	DH	182	QSsd.crc-1B	1B	0.18	McCartney et al. (2006)
	DH	185	QSv.crc-1B	1B	0.15	Huang et al. (2006)

Table 10.1 (continued)

Trait	MP*	No of lines	QTL/Marker Interval	Position	R^2**	Reference
	DH	185	QSv.crc-2D	2D	0.14	Huang et al. (2006)
	SWW	207	wmc419	1B	0.23	Reif et al. (2011)
	RILs	188	QSev.wak-3B	3B	0.1	Carter et al. (2012)
Zeleny score	RILs	114	Xbcd102-Xbcd18	5D	0.18	Nelson et al. (2006)
	RILs	114	Xcdo426-XksuD14.1	1A	0.14	Nelson et al. (2006)
Loaf volume	RILs	114	Xabc174-Xgwm108	3B	0.18	Nelson et al. (2006)
	RILs	78	Pkaba1c	2B	0.12	Campbell et al. (2001)
	DH	182	Xgwm666	3A	0.11	Kuchel et al. (2006)
	RILs	105	QGpc2B	2B	0.29	Elangovan et al. (2008)
	RILs	105	QLv.ncl-6D.3	6D	0.45	Elangovan et al. (2008)
	DH	182	QBlvl.crc-4D	4D	0.2	McCartney et al. (2006)
Alveogram (L)	RILs	114	Xcdo426-XksuD14.1	1A	0.3	Nelson et al. (2006)
	RILs	114	Xmwg706-Xmwg733	1A	0.19	Nelson et al. (2006)
P/L	RILs	114	Xcdo426-XksuD14.1	1A	0.31	Nelson et al. (2006)
	RILs	114	Xgwm550-XksuD14.1	1B	0.38	Nelson et al. (2006)
	RILs	78	QPlr.ipk-1A	1A		Campbell et al. (2001)
Alveogram (W)	RILs	114	DR Fmt.1-Xgwm550	1B	0.28	Nelson et al. (2006)
	RILs	114	Xwg686-Xcdo686	7B	0.14	Nelson et al. (2006)
Alveogram (P)	RILs	114	XksuH7-XgbxR080	3B	0.26	Nelson et al. (2006)
	RILs	114	Xgwm550-XksuD14	1B	0.24	Nelson et al. (2006)
Pelshenke score	RILs	114	DR Fmt.1-Xgwm550	1B	0.32	Nelson et al. (2006)
Yellow pigment	RILs	240	QYpc-1B	1B	0.32	Zhang et al. (2009)
	RILs	240	QYpc-7A	7A	0.34	Zhang et al. (2009)
Flour color	RILs	150	Xbcd828-3A	3A	0.13	Parker et al. (1998)
	RILs	150	Xcdo347-Xwg232	7A	0.6	Parker et al. (1998)

Table 10.1 (continued)

Trait	MP*	No of lines	QTL/Marker Interval	Position	R^2**	Reference
	DH	155	Xgwm193	6B	0.21	Pozniak et al. (2007)
	DH	155	Psy1-1	7B	0.23	Pozniak et al. (2007)
Amylose content	SCRILS	98	QAmc.ocs-4A.1	4A	0.17	Araki et al. (1999)
Particle size index	DH	182	QPsi.crc-7D	7D	0.28	McCartney et al. (2006)
Test weight	DH	185	QTw.crc-4D	4D	0.13	Huang et al. (2006)
	Soft winter wheat	207	gwm219	6B	0.34	Reif et al. (2011)

**MP* Mapping population, *RILs* Recombinant inbred lines, *DH* Double haploids

**In case of several QTLs identified in a study for a single trait, only QTLs with highest R^2 values were selected developed for quality traits by cloning of 62 alleles at 16 loci. Functional markers that have application for wheat quality improvement are presented in Table 10.2.

Isolation and characterization of functional motifs within genes controlling phenotypic variability is critical to develop allele specific markers. These motifs are usually characterized by single nucleotide polymorphisms (SNPs) or insertions/deletions (InDels) within the nucleotide sequences of different alleles. Using the map-based cloning approach, several genes have been isolated in plants; however, the very large genome is the main problem in common wheat which makes map-based cloning difficult as compared to rice and maize. Rice-wheat micro-colinerarity and different comparative genomics tools provides an alternate and efficient way to dissect target genes in wheat. This is due to the fact that rthologs descended from a common ancestor often have conserved functions and are expected to produce similar phenotypes across species (Devos 2005). The whole genome sequence is available for several grasses including rice, maize and Brachypodium which provided powerful tools for gene discovery in wheat (Vogel et al. 2010). The in silico technology is now widely used for discovery of genes of interest in wheat (Ma et al. 2012). A major breakthrough is the availability of expressed sequence tags (EST) database, from where the sequences of putative wheat genes can be obtained by aligning and joining of orthologous genes with the same function in the grass.

Description of Quality Traits and Their Functional Markers

Significance of HMW-GS and LMW-GS ingrain quality has been described earlier. There are several reports on the nucleotide sequences of the cloned genes for HMW-GS and LMW-GS. The nucleotide sequence of these cloned genes provided the basis for marker development for their further use in breeding. Zhang et al. (2004) developed markers for *Glu-A3* alleles based on DNA polymorphisms identified between the LMW glutenin genes. However markers developed by Wang et al. (2009b, 2010) for *Glu-A3* and *Glu-B3* are more efficient and easier to use. However, due to limited variation among *Glu-D3* haplotypes, no allele specific marker was developed (Liu et al. 2010), and comparatively their impact is also very small compared to *Glu-A3* and *Glu-B3* loci (Gupta et al. 1989). Zhao et al. (2007b) attempted to develop markers for *Glu-D3* haplotypes and later Appelbee et al. (2009) tried to use those haplotype specific marker combinations for diagnosis of some specific alleles like *Glu-D3a*, *b*, c and *f*. A total of seven allele specific markers for *Glu-A3* and ten markers for *Glu-B3* loci have been reported. Additionally, Wang et al. (2009b, 2010) also established multiplex PCR strategies to reduce the cost of technique in practical breeding programs. The practical usage of functional markers for HMW-GS and LMW-GS to test wheat cultivars and lines has been established and reported (Liang et al. 2010; Jin et al. 2011; Ram et al. 2011; Khalid et al. 2013).

The GBSS I (granule-bound starch synthase) is defined as Waxy (*Wx*) protein, and it is a primary enzyme involved in synthesis of amylose in wheat endosperm. The amylose contents play an important role in determining noodle quality. The

Table 10.2 An overview of functional (allele specific) markers for grain quality traits in wheat

Trait	Locus	Primer	Allele	Size (bp)	PCR conditions	Reference
HMW-GS	Glu-A1	F: CGAGACAATATGAGCAGCAAG	Ax2*	344	94 °C/60 s–60 °C/60 s–72 °C/2 m	Liu et al. 2008
		R: CTGCCATGGAGAAGTTGGA	Ax1, AxNull	362		
		F: ATGACTAAGCGGTTGGTTCTT	Ax2*	1319	94 °C/30 s–58 °C/30 s–72 °C/2 m	Ma et al. 2003
		R: ACCTTGCTCCCCTTGTCTTT				
	Glu-B1	F: CCTCAGCATGCAAACATGCAGC	Bx7^{OE}	563	95 °C/30 s–58 °C/30 s–72 °C/1 m	Butow et al. 2004
		R: CTGAAACCTTTGGCCAGTCATGTC				
		F:CACTGAGATGGCTAAGCGCC	*Bx6*	321	95 °C/30 s–50 °C/30 s–72 °C/1m	Schwarz et al. 2004
		R: GCCTTGGACGGCACCACAGG				
		F: ACGTGTCCAAGCTTTGGTTC	*Bx7*^{OE}	447	94 °C/35 s–63 °C/30 s–72 °C/60 s	Ragupathy et al. 2008
		R: GATTGGTGGGTGGATACAGG				
		F: CCACTTCCAAGGTGGGACTA	*Bx7*^{OE}	844	94 °C/35 s–63 °C/30 s–72 °C/60 s	
		R: TGCCAACACAAAAGAAGCTG				
		F: CGCAACAGCCAGGACAATT	*Bx17*	669	94 °C/30 s–58 °C/30 s–72 °C/2m	Ma et al. 2003
		R: AGAGTTCTATCACTGCCTGGT				
		F: TTAGCGCTAAGTGCCGTCT	*By8*	527	94 °C/30 s–64 °C/30 s–72 °C/90 s	Lei et al. 2006
		R: TTGTCCTATTTGCTGCCCTT				
		F: TTCTCTGCATCAGTCAGGA	*By9*	662	94 °C/30 s–59 °C/30 s–72 °C/90 s	
		R: AGAGAAGCTGTGTAATGCC	*nonBy9*	707		
		F: GCAGTACCCAGCTTCTCAA	*By16, ByNull*		94 °C/30 s–62 °C/30 s–72 °C/90 s	
		R: CCTTGTCTTGTTTGTTGCC	*Bx20*			
	Glu-D1	UMN25F: GGGACAATACGAGCAGCAAA	*Dx2*	299	94 °C/30 s–60 °C/30 s–72 °C/2 m	Liu et al. 2008b
		UMN25R: CTTGTTCCGGTTGTTGCCA	*Dx5*	281		
		Forward: CGTCCCTATAAAAGCCTAGC	*Dx5*	478	94 °C/30 s–58 °C/30 s–72 °C/2 m	Ma et al. 2003
		Reverse: AGTATGAAACCTGCTGCGGAC				

Table 10.2 (continued)

Trait	Locus	Primer	Allele	Size (bp)	PCR conditions	Reference
		UMN26F: CGCAAGACAATATGAGCAAACT	*Dy10*	397	94 °C/30 s–60 °C/30 s–72 °C/2 m	Liu et al. 2008b
		UMN26R: TTGCCTTTGTCCTGTGTGC	*Dy12*	415		
LMW-GS	Glu-A3	F: AAACAGAATTATTAAAGCCGG R: GGTTGTTGTTGTTGCAGCA	Glu-A3a	529	94 °C/35 s–60 °C/45 s–72 °C/90 s	Wang et al. 2010
		F: TTCAGATGCAGCCAAACAA R: GCTGTGCTTGGATGATACTCTA	Glu-A3b	894	94 °C/35 s–60 °C/45 s–72 °C/90 s	
		F: AAACAGAATTATTAAAGCCGG R: GTGGCTGTTGTGAAAACGA	Glu-A3ac	573	94 °C/35 s–60 °C/45 s–72 °C/90 s	
		F: TTCAGATGCAGCCAAACAA R: TGGGGTTGGGAGACACATA	Glu-A3d	967	94 °C/35 s–60 °C/45 s–72 °C/90 s	
		F: AAACAGAATTATTAAAGCCGG R: GGCACAGACGAGGAAGGTT	Glu-A3e	158	94 °C/35 s–60 °C/45 s–72 °C/90 s	
		F: AAACAGAATTATTAAAGCCGG R: GCTGCTGCTGCTGTGTAAA	Glu-A3f	552	94 °C/35 s–60 °C/45 s–72 °C/90 s	
		F: AAACAGAATTATTAAAGCCGG R: AAACAACGGTGATCCAACTAA	Glu-A3g	1345	94 °C/35 s–60 °C/45 s–72 °C/90 s	
	Glu-B3	F: CACAAGCATCAAAACCAAGA R: TGGCACACTAGTGGTGGTC	Glu-B3a*	1095	94 °C/35 s–56 °C/35 s–72 °C/90 s	Wang et al. 2009b
		F: ATCAGGTGTAAAAGTGATAG R: TGCTACATCGACATATCCA	Glu-B3b	1570	94 °C/35 s–56 °C/35 s–72 °C/90 s	
		F: CAAATGTTGCAGCAGAGA R: CATATCCATCGACTAAACAAA	Glu-B3c	472	94 °C/35 s–56 °C/35 s–72 °C/90 s	
		F: CACCATGAAGACCTTCCTCA R: GTTGTTGCAGTAGAACTGGA	Glu-B3d	662	94 °C/35 s–58 °C/35 s–72 °C/90 s	
		F: GACCTTCCTCATCTTCGCA R: GCAAGACTTTGTGGCATT	Glu-B3e	669	94 °C/35 s–58 °C/35 s–72 °C/90 s	
		F: TATAGCTAGTGCAACCTACCAT	Glu-B3fg*	812	94 °C/35 s–63 °C/35 s–72 °C/90 s	

Table 10.2 (continued)

Trait	Locus	Primer	Allele	Size (bp)	PCR conditions	Reference
		R: CAACTACTCTGCCACAACG				
		F: CCAAGAAATACTAGTTAACAC-TAGTC	Glu-B3g*	853	94 °C/35 s–61 °C/35 s–72 °C/90 s	
		R: GTTGGGGTTGGGAAACA				
		F: CCACCACAACAAACATTAA	Glu-=B3h	1022	94 °C/35 s–60 °C/35 s–72 °C/90 s	
		R: GTGGTGGTTCTATACAACGA				
		F: TATAGCTAGTGCAACCTACCAT	Glu-B3i	621	94 °C/35 s–58 °C/35 s–72 °C/90 s	
		R: TGGTTGTTGCGGTATAATTT				
		F: GCATCAACAACAAATAGTAC-TAGAA	Glu-B3bef	750	94 °C/35 s–60 °C/35 s–72 °C/90 s	
		R: GGCGGGTCACACATGACA				
Grain Texture	Pin-a	F: TCAACATTCGTGCATCATCA	Pina-D1r	436	94 °C/50 s–53 °C/1 m–72 °C/1 m	Chen et al. 2012
		R: CTTCATTCGTCAGAGTTCCAT				
	Pin-b	F: ATGAAGACCTTATTCCTCCTA	Pinb-D1a	250	94 °C/45 s–55 °C/45 s–72 °C/1 m	Giroux and Morris 1997
		R: CTCATGCTCACAGCCGCC				
		F: ATGAAGACCTTATTCCTCCTA	Pinb-D1b	250	94 °C/45 s–55 °C/45s–72 °C/1 m	
		R: CTCATGCTCACAGCCGCT				
Starch property	Wx-B1	F: CTGGCCTGCTACCTCAAGAG-CAACT	Wx-B1a	425	94 °C/30 s–65 °C/30 s–72 °C/2 m	Nakamura et al. 2002
		R: CTGACGTCCATGCCGTTGACGA				
		F: CTGGCCTGCTACCTCAAGAG-CAACT	Wx-B1	778	94 °C/30 s–65 °C/30 s–72 °C/2 m	Saito et al. 2009
		R: GGTTGCGGTTGGGGTCGATGAC				
		F: CGTAGTAAGGTGCAAAAAAGT-GCCACG	Wx-B1 (Null)	668		
		R: ACAGCCTTATTGTAC-CAAGACCCATGTGTG			94 °C/30 s–65 °C/30 s–72 °C/2 m	

Table 10.2 (continued)

Trait	Locus	Primer	Allele	Size (bp)	PCR conditions	Reference
GPC	GPC-B1	F: AGCCAGGGATAGAGGAGGAA	Gpc-B1	217	94 °C/30 s–58 °C/45 s–72 °C/40 s	Uauy et al. 2006
		R: AGCTGTGAGCTGGTGTCCTT				
		F: TCTCCAAGAGGGGAGAGACA	Gpc-B1	126	94 °C/30 s–58 °C/45 s–72 °C/40 s	Distelfeld et al. 2006
		R: TTCCTCTACCCATGAATCTAGCA				
PPO activity	*Ppo-A1*	F: AACTGCTGGCTCTTCTTCCCA	*Ppo-A1a*	685	94 °C/60 s–65 °C/60 s–72 °C/60 s	Sun et al. 2005
		R: AAGAAGTTGCCCATGTCCGC	*Ppo-A1b*	876		
		F: CCAGATACACAACTGCTGGC	*Ppo-A1a*	290	95 °C/30 s–66–54 °C/30 s–72 °C/60 s*	He et al. 2007
		R: TGATCTTGAGGTTCTCGTCG	*Ppo-A1b*	481		
	Ppo-D1	F: TGCTGACCGACCTTGACTCC	*Ppo-D1a*	713		
		R: CTCGTCACCGTCACCCGTAT				
		F: TGAAGCTGCCGGTCATCTAC	*Ppo-D1b*	490		
		R: AAGTTGCCCATGTCCTCGCC				
Lipoxygenase activity	*TaLox-B1*	F: CCATGACCTGATCCTTCCCTT	*TaLox-B1a*	489		Geng et al. 2012
		R: GCGCGGATAGGGGTGGT				
		F: ACGATGTGAGTTGTGACTTGTGA	*TaLox-B1b*	791		
		R: GCGCGGATAGGGGTGC				
Yellow pigment content	*Psy-A1*	F: GGACCTTGCTGATGACCGAG	*PsyA1a*	194	95 °C/30 s–65 °C/30 s–72 °C/30 s	He et al. 2008
		R: TGACGGTCTGAAGTGAGAATGA	*PsyA1b*	231		
		F: GCCAGCCCTTCAAGGACATG	*PsyA1a*	1686	95 °C/60 s–60 °C/2 m–72 °C/30 s	He et al. 2009a
		R: CAGATGTCGCCACACTGCCA	*PsyA1c*	1001		
	Psy-B1	F: GCCACAACTTGAATGTGAAAC	*Psy-B1a*	151	95 °C/60 s–60 °C/2 m–72 °C/30 s	He et al. 2009a
		R: ACTTCTTCCATTTGAACCCC	*Psy-B1b*	156		
		F: GCCACCCACTGATTACCACTA	*Psy-B1c*	428		
		R: CCAAGGTGAGGGTCTTCAAC				

Table 10.2 (continued)

Trait	Locus	Primer	Allele	Size (bp)	PCR conditions	Reference
		F: GAGTAAGCCACCCACTGATT	*Psy-B1d*	884		
		R: TCGCTGAGGAATGTACTGAC				
		F: AGGTACCAGCCAGCCCATA	*Psy-B1e*	716		
		R: CTCGTCAAAGTTCGTGTACC				
	Psy1-D1	F: TCCGACACCATCACCAAGTTCC	*Psy1-D1a*	1074	95 °C/30 s–58 °C/30 s–72 °C/60 s	Wang et al. 2009a
		R: CGTTGTAGGTTTGTGGGAGT	*Psy1-D1g*	1093		
		F: ACTCCCACAAACCTACAACG	*Psy1-D1a*	967	95 °C/30 s–58 °C/30 s–72 °C/60 s	
		R: ACGCTCATCAACCCCACG	*Psy1-D1g*	1046		
	TaZds-A1	F: CCCTAAGGAAGCCGAGCAAAT	*TaZds-A1a*	183		Dong et al. 2012
		R: GTGAGAGTACTAATGTTATGACCG	*TaZds-A1b*	179		
	TaZds-D1	F: GTGGGATCCTGTTGCTTATGC	*TaZds-D1a*	Null		Zhang et al. 2011
		R: GTAGATTATCCAAGCCAACTGCC	*TaZds-D1b*	981		

Wx-B1 null allele in a wheat cultivar is characterized by low amylose level that is associated with high noodle making quality. Several allele specific markers have been developed to facilitate marker assisted introgression of desired waxy allele (Saito et al. 2009; Nakamura et al. 2002).

The description of puroindoline genes determining the kernel hardness has been described earlier. Two genes designated as *Pina-D1* and *Pinb-D1* encoding puroindoline a and puroindoline b proteins, respectively, has a profound effect on wheat milling quality. Molecular markers for identifying alleles at both loci have been developed by many workers (Giroux and Morris 1997; Gazza et al. 2005; Chen et al. 2012). The allele specific marker for *Pinb-D1b* is diagnostic for superior milling and processing qualities and is in extensive use in wheat breeding (Chen et al. 2012).

Polyphenol oxidase (PPO) activity is responsible for brown discoloration of the wheat products especially Asian noodles, which is an undesirable character. It is important to screen cultivars for low PPO activity and several markers have been developed to serve the purpose. The PPO genes are present on chrosomosomes 2A and 2D (Sun et al. 2005; He et al. 2007). Low PPO activity encoded by the *Ppo-A1* gene is characterized by the presence of 87 and 481 PCR fragments amplified by PPO18 and PPO33 markers, respectively. Similarly, two other allele specific markers PPO16 and PPO29 can distinguish alleles *Ppo-D1a* and *Ppo-D1b* which are associated with lower and higher PPO activity, respectively. Practical usage of these markers in wheat breeding for identification of genotypes with lower PPO activity is scientifically valid (Liang et al. 2010). Nevertheless, PPO gene located on 2B chromosome had limited polymorphism in Chinese wheat to develop a functional marker. Lipoxygenase activity is also major determinant of color and processing quality of wheat products (Geng et al. 2012). A lipoxygenase (LOX) gene has been localized on chromosome 4BS (*TaLox-B1*) and two allele specific markers LOX16 and LOX18 amplify 48 and 791 bp PCR fragments in cultivars with higher and lower LOX activities, respectively (Geng et al. 2012). The gene, TaLox-B1 was sequenced and a SNP was identified in the third exon which was helped in development of two markers for identifying alleles *TaLox-B1a* and *TaLox-B1b*.The color of wheat derived products is due to the yellow pigment content. Regional preference for color does exist, like bright white color is preferred for Chinese white salted noodles, whereas yellow alkaline noodles with bright yellow color are widely preferred in southeastern Asia and Japan (Parker et al. 1998). Carotenoids are responsible for yellow pigment (He et al. 2008) while phytoene synthase (PSY) and zeta-carotene desaturase (ZDS) are important enzymes in the biosynthetic pathway for carotenoid synthesis in wheat (Zhang et al. 2011; Dong et al. 2012). PSY genes are present chromosomes 7AL, 7BL and 7DL and several allele specific markers for PSY genes have been developed (He et al. 2008, 2009a; Wang et al. 2009a). Similarly, markers for ZDS genes on chromosomes 2A and 2D can discriminate alleleic difference in wheat (Zhang et al. 2011; Dong et al. 2012).

Genetic Resources for Quality Improvement

The tribe *Triticeae* had almost 350 species, of which wheat and barley are the members. This natural diversity in the *Triticeae* gene pool can be incorporated in controlled and well directed manner for which the priority is given to annual and perennial *Triticeae* species. The species resources are distributed within gene pools and wheat improvement for environmental stresses can be realized by genetic transfer from these gene pools over short- and long term time frames. The allelic diversity within *Triticeae* is crucial to harness for meeting the projected global demand of wheat. The key resources of variability at priority are the primary gene pool diploid D genome donor accessions of wheat; *Aegilops tauschii*, and some sources from the tertiary gene pool possessing high potential values. Utilization of these genetic resources to develop genetically compatible germplasm readily available for wheat improvement needs integrated breeding approach with association of emerging technologies and multidisciplinary specialties facilitating exploitation of molecular tools of MAS, SMART (Selectison with marker and advanced reproductive technologies) breeding and QTLs hopefully to add to breeding efficiency.

The genes coding for high molecular weight glutenins has been dissected from several species of Triticeae including *Hordeum, Secale, Taeniatherum, Thinopyrum, Aegilops, Crithopsis, Dasypyrum* and their different ploidy members (Wan et al. 2002; Yan et al. 2002; De Bustos and Jouve 2003; Liu et al. 2003; Sun et al. 2004; Wang et al. 2006; Cao et al. 2007; Liu et al. 2007). Due to wheat domestication syndrome, *Glu-Ay* always remains silent in durum and bread wheat however several A-genone wild species and wild tetraplopid species (*T. dicoccoides and T. dicoccon*) express this gene (Waines and Payne 1987). The presence of active *Ay* genes had significant positive effect on the bread-making quality (Ciaffi et al. 1995). The narrow allelic diversity for *Glu-A1* locus in bread and durum wheat which encodes limited number of x-type subunits and does not express an active y-type subunit require attention to expand it by using novel allelic variants reported by several workers in *T. urartu* and *T. monococcum* (Waines and Payne 1987; Ciaffi et al. 1998; Alvarez et al. 2009; Caballero et al. 2008; Gutierrez et al. 2011). There are extensive studies on identification and characterization of allelic variation for *Glu-D^{t}1* loci from *Ae. tauschii* and D-genome synthetic hexaploids (An et al. 2009; Yan et al. 2003; Gianibelli et al. 2001; Rehman et al. 2008; Xu et al. 2010; Bibi et al. 2012; Rasheed et al. 2012). A higher variability of HMW-GS due to their electrophoretic mobility has been observed in A-genome species (*T. monococcum* and *T. urartu*) Lee et al. 1999b; Caballero et al. 2008; Gutierrez et al. 2011), AB genome species (*T. dicoccoides*) (Ciaffi et al. 1993), and D-genome species (*T. tauschii*) (Rehman et al. 2008). More recently, Niu et al. (2011) analyzed HMW-GS in *Th. bessarabicum, Th. intermedium, Lophopyrum elongatum, Ae. markgrafii* and their addition lines. The information provided is useful for the development of molecular markers that will facilitate the introgression of desirable genes from the alien chromosomes into wheat genomes. The identified novel HMWGS alleles may serve as new genetic resources for wheat quality improvement.

Likewise HMW-GSs, many genes encoding LMW-GS have been isolated and analyzed in cultivated and wild species of the family *Triticeae*. The genes coding LMW-GS have been studied in the genera *Elytrigia* (Gupta and Shephard, 1990) *Elymus* (Obukhova et al. 1997), *Dasypyrum* (Blanco et al. 1991) and *Hordeum* (Atienza et al. 2002). *Ae. tauschii* (DD) has been an important source for genetic studies of LMW-GS (Gianibelli et al. 2000, 2002a; Hsam et al. 2001; Pfluger et al. 2001; Vensel et al. 1997; Zhao et al. 2008) and exhibited greater variation in the coding sequence of LMW-GS (Masci et al. 1991; Lafiandra et al. 2000). Similarly, the other speices that have been analyzed for LMW-GS include *T. monococcum, T. urartu* (Tranquilli et al. 2002; Lee et al. 1999), *T. turgidum* var. *dicoccoides* (AABB) (Ciaffi et al. 1993), *T. dicoccum* (AABB) (Galterio et al. 2001), *T. polonicum* (AABB) (Liu and Shepherd, 1996), *T. macha* (Xiong et al. 2010), *Ae. bellulata, Ae. comosa, Ae. markgrafii* and *Ae. speltoides* (Li et al. 2010), hexaploid obsolete cultivars and landraces (Ovesna et al. 2001). The variability found for LMW-GS in wheat wild relatives indicates the valuable potential is available to improve the properties demanded to make variable products. The advancements have been reported on the molecular characterization of *Glu-3* genes from different *Triticeae* species. For example nucleotide sequences are available from several species of *Aegilops* spp. (Jiang et al. 2008; Li et al. 2008), *Agropyron elongatum* (Luo et al. 2005), *Secale sylvestre* (Shang et al. 2005), *Crithopsis delileana* (Guo et al. 2008), *Hordeum chilense*, and *H. brevisubulatum* (Piston et al. 2005). The nucleotide diversity of LMW-GS in these wild species indicated the allelic rich of *Glu-3* loci in *Triticeae*. The comparative analysis of nucleotide sequences of LMW-GS revealed some important differences among species. For example, *Hordeum chilense* and *A. elongatum* lacks the N-terminal regions in the predicted mature proteins (Piston et al. 2005). However, further efforts need to be continued to study the evolutionary pattern and structure of LMW-GS gene in *Triticeae* which will further facilitate their utilization for wheat quality improvement.

A wide survey to isolate hundreds of *Pina, Pinb* and *GSP* genes from wild accessions of *T. aestivum, T. turgidum, T. urartu, T. monococcum, T. timopheevii, T. zhukovskyi, Ae. tauschii, Ae. speltoides, Secale* and *Hordeum* have been conducted (Morris 2002). The wild ancestors are known to have very soft texture as compared to domesticated derivatives (Morris 2002) however the exact variability for texture is not well established in diploid species. Diploid and hexaploid accessions of wild species had starch-associated friabilin which are generally absent in tetraploid species. However puroindoline genes are present in accessions of diploid *T. urartu, T. monococcum, Ae. tauschii* and *Ae. speltoides* (Lillemo et al. 2002). SKCS based characterization of 67 accessions of *T. monococcum* revealed the soft texture (Pogna et al. 2002). Similarly, scanning electron microscopy based characterization of texture revealed that *Aegilops* accessions of different genomes and ploidy were usually soft (Chen et al. 2005) with exception of a single *Ae. Sharonensis* accession. The species which lack *Pina* sequences include S-genome species *Ae. bicornis* and *Ae. longissima* which was contradictory to the findings of Simeone et al. (2006). They analyzed many combinations of 13 and 24 variable amino acids in the seven new haplotypes of *Pina* and *Pinb*, respectively. A null allele at PINA locus was found, which carries a premature stop codon, in two *Ae. kotschyi* (UUSS) accessions. In *Ae. sharonensis* (S^shS^sh) novel haplotypes of *Pina* and *Pinb* were observed with their possible pseudogenes. The credibility of cDNA of intron-less genes is questioned due to the lack of cDNA equivalents for some genomic copies. Recently, Chen et al. (2009) studied several accessions of einkorn wheat and identified 56 sequences encoding the *pina* protein. All the gene sequences from *T. urartu* grouped together, whereas some sharing by three and two clusters was observed for *T. monococcum* ssp. aegilopoides and *T. monococcum* ssp. monococcum, respectively. Guzman et al. (2012) also identified various alleles for *Pina* and *Pinb* genes including three novel alleles for the *Pinb* locus, *Pinb-A^{m}1i, Pinb-A^{m}1j* and *Pinb-A^{m}1k*, from *T. monococcum*.

The breeding of food crops for biofortification with high iron and zinc contents is primarily important component within the food security nexus, especially in developing countries. There is need to develop special conventional and molecular breeding approaches for cost effective nutritional improvement in cereal crops (Bouis and Welch 2010). Currently, the cultivated durum and bread wheat varieties are low in grain iron and zinc contents than the related wild *Triticum* and *Aegilops* species (Chhuneja et al. 2006). Therefore the wild relatives should be emphasized for screening for the targeted biofortification traits. Due to ease of genetic transfer, preference should be given to the *T. monococcum* L., *Triticum turgidum* L. ssp. dicoccoides (Korn. ex Asch. et Graebn.) Thell, *Triticum turgidum* L. ssp. dicoccon (Schrank) Thell., and *Ae. tauschii* accessions. Several QTLs have been identified for higher grain iron and zinc contents in a *Triticum monococcum* x *T. boeoticum* mapping population consisting of RILs Tiwari et al. (2009). Two chromosomes 2A and 7A were found important for the presence of QTLs controlling iron zinc concentrations. Several *Aegilops* species have been identified as potential donors of useful variability for high iron and zinc concentration Rawat et al. (2009). These species include *Ae. kotschyi* Boiss., *Ae. peregrina* (Hack.) Maire et Weill., *Ae. geniculata* Roth, *Ae. ventricosa* Tausch, and *Ae. cylindrica* Host. Recently, Rawat et al. (2011) characterized addition and substitution lines chromosome 1, 2 and 7 from *Ae. kotschyi* which possess genes for high grain micronutrients. Similarly, Neelam et al. (2011) also identified the introgression of group 4 and 7 chromosomes from *Ae. peregrine* enhances 100–200 % grain iron and zinc density. A series of wheat–*Ae. longissima* amphiploids were also reported to have high grain iron and zinc concentrations (Tiwari et al. 2008) and could be used as immortal sources of variability for biofortification of wheat for high grain micronutrient concentrations.

Conclusions and Future Perspectives

The implementation of marker-trait combination is pre-requisite for genomics based wheat improvement. There is rapid advancement in high-throughput protein and gene analyses techniques offering large scale comparative analysis of genes from wild and domestication sources. Advances in developing functional markers for quality traits is aiding in diagnostics and introgression of favorable alleles from different genetic resources of higher and lower ploidy including land races and wild species. Harnessing the new allelic variability from wild sources using molecular markers will catalyze the genetic improvement by broadening the gene pool for maximizing the genetic gain of desirable alleles. The information discussed in this chapter ensures that the advances in the molecular diagnostics and cytological introgression approaches would resolve the complexities of the gene networks underpinning quality attributes that would help to meet the challenges presented by the swift changes occurring within the food chain.

Large scale genome sequencing and integration of bioinformatics will accelerate the analysis structure and function of quality genes. Analysis of huge databases generated from genome sequencing and high-throughput marker analyses (SNPs and microarray) of the expressed genes in developing grain and their integration with web-based comparative genomic tools are formulating the strategies leading towards stringent objectivity. Another avenue is the use of TILLING and small RNAs where the specific functions are assigned to quality encoding genes by identifying mutants and deleting the mRNA, respectively. The recent advent of Multi-parent Advanced Generation Inter-Cross (MAGIC) approach will identify more precisely the quality encoding genes and resolve the complexities of gene networks underpinning the quality attributes to meet the upcoming challenges in grain quality.

References

Alvarez JB, Cabalero L, Nadal S, Ramirez MC, Martin A (2009) Development and gluten strength evaluation of introgression lines of *Triticum urartu* in durum wheat. Cereal Res Commun 37:243–248

An XL, Li XH, Xiong XJ, Yan YM, Zhang YZ, Gao LY, Wang AL, Wang K, Zeller FJ, Hsam, SLK (2009) Identification and isolation of a new x-type HMW glutenin subunit 1Dx1.6t gene from Aegilops tauschii. Plant Breed 128:41–45

Appelbee MJ, Mekuria GT, Nagasandra V, Bonneau JP, Eagles HA, Eastwood RF, Mather DE (2009) Novel allelic variants encoded at the Glu-D3 locus in bread wheat. J Cereal Sci 49:254–261

Araki E, Miura, H, Sawada S (1999) Identification of genetic loci affecting amylose content and agronomic traits on chromosome 4A of wheat. Theor Appl Genet 98:977–984

Atienza SG, Alvarez JB, Villegas AM, Giménez MJ, Ramírez MC, Martín A, Martín LM (2002) Variation for the low-molecular-weight glutenin subunits in a collection of Hordeum chilense. Euphytica 128:269–277

Avivi L (1978) High protein content in wild tetraploid *Triticum dicoccoides* Korn. In: Ramanujam S (ed) Proceedings of the 5th international wheat genetics symposium. Indian Society of Genetics and Plant Breeding (ISGPB), New Delhi pp 372–380

Bhave M, Morris CF (2008) Molecular genetics of puroindolines and related genes: allelic diversity in wheat and other grasses. Plant Mol Biol 66:205–219

Bibi A, Rasheed A, Kazi AG, Mahmoood T, Ajmal S, Mujeeb-Kazi A (2012) High-molecular-weight (HMW) glutenin subunit composition of the Elite-II synthetic hexaploid wheat subset (*Triticum turgidum* x *Aegilops* tauschii; 2n=6x=42; AABBDD). Plant Genet Resour 10:1–4

Bietz JA, Huebner FR, Sanderson JE, Wall JS (1977) Wheat gliadin homology revealed through N-terminal amino acid sequence analysis. Cereal Chem 54:1070–1083

Bietz JA, Shepherd KW, Wall JS (1975) Single-kernel analysis of glutenin: use in wheat genetics and breeding. Cereal Chem 52:513–532

Bietz JA, Wall JS (1973) Isolation and characterization of gliadin-like subunits from glutenins. Cereal Chem. 50:537–547

Blanco A, Resta P, Simeone R, Parmar S, Shewry PR, Sabelli P, Lafiandra D (1991) Chromosomal location of seed storage protein genes in the genome of Dasypyrum villosum. Theor Appl Genet 82:358–362

Bouis HE, Welch RM (2010) Biofortification—a sustainable agricultural strategy for reducing micronutrient malnutrition in the global south. Crop Sci 50:20–32

Breseghello F, Finney PL, Gaines C, Andrews L, Tanaka J, Penner G, Sorrells ME (2005) Genetic loci related to kernel quality differences between a soft and a hard wheat cultivar. Crop Sci 45:1685–1695

Butow BJ, Gale KR, Ikea J, Juhasz A, Bedo Z, Tamas L, Gianibelli MC (2004) Dissemination of the highly expressed Bx7 glutenin subunit (Glu-B1al allele) in wheat as revealed by novel PCR markers and RP-HPLC. Theor Appl Genet 109:1525–1535

Caballero L, Martin MA, Alvarez JB (2008) Allelic variation for the high- and low-molecular-weight glutenin subunits in wild diploid wheat (Triticum urartu) and its comparison with durum wheat. Aus J Agric Res 59:906–910

Cakmak I, Torun A, Millet E, Feldman M, Fahima T, Korol AB, Nevo E, Braun HJ, Ozkan H (2004) Triticum dicoccoides: an important genetic resource for increasing zinc and iron concentration in modern cultivated wheat. Soil Sci Plant Nutr 50:1047–1054

Caldwell DG, McCallum N, Shaw P, Muehlbauer GJ, Marshall DF, Waugh R (2004) A structured mutant population for forward and reverse genetics in barley (Hordeum vulgare L.). Plant J 4:143–150

Campbell KG, Bergman CJ, Gualberto DG, Anderson JA, Giroux MJ, Hareland G, Flucher GR, Sorrells ME, Finney PL (1999) Quantitative trait loci associated with kernel traits in a soft× hard wheat cross. Crop Sci 39:1184–1195

Campbell KG, Finney PL, Bergman CJ, Gualberto DG, Anderson JA, Giroux MJ, Siritunga D, Zhu J, Gendre F, Roue′ C, Ve′rel A, Sorrells ME(2001) Quantitative trait loci associated with milling and baking quality in a soft by hard wheat cross. Crop Sci 41:1275–1285

Cane K, Spackman M, Eagles HA (2004) Puroindoline genes and their effects on grains quality traits in southern Australian wheat cultivars. Aus J Agric Res 55:89–95

Cantrell RG, Joppa LR (1991) Genetic analysis of quantitative traits in wild emmer (*Triticumturgidum* L. var. dicoccoides). Crop Sci 31:645–649

Cantu D, Pearce SP, Distelfeld A, Christiansen MW, Uauy C, Akhunov E, Fahima T, Dubcovsky J (2011) Effect of the down-regulation of the high grain protein content (GPC) genes on the wheat transcriptome during monocarpic senescence. BMC Genomics 12:1–12

Cao S, Xu H, Li Z, Wang X, Wang D, Zhang A (2007) Identification and characterization of a novel Ag. intermedium HMW-GS gene from T. aestivum-Ag. intermedium addition lines TAI-I series. J Cereal Sci 45:293–301

Carter AH, Garland-Campbell K, Morris FC, Kidwell KK (2012) Chromosomes 3B and 4D are associated with several milling and baking quality traits in a soft white spring wheat (*Triticum aestivum* L.) population. Theor Appl Genet 124:1079–1096

Chalmers KJ, Rathjen AJ, Langridge P (1999) Mapping loci associated with milling yield in wheat (*Triticum aestivum* L.). Mol Breed 5:561–568.

Chantret N, Salse J, Sabot F, Rahman S, Bellec A, Laubin B, Dubois I, Dossat C, Sourdille P, Joudrier P, Gautier MF, Cattolico L, Beckert M, Aubourg S, Weissenbach J, Caboche M, Bernard M, Leroy P, Chalhoub B (2005) Molecular basis of evolutionary events that shaped the hardness locus in diploid and polyploid wheat species (Triticum and aegilops). Plant Cell 17:1033–1045

Chen F, Zhang FY, Xia XC, Dong ZD, Cui DQ (2012) Distribution of puroindoline alleles in bread wheat cultivars of the Yellow and Huai valley of China and discovery of a novel puroindoline a allele without PINA protein. Mol Breed. doi:10.1007/s11032-011-9553-2

Chen M, Wilkinson M, Tosi P (2005) Novel puroindoline and grain softness protein alleles in Aegilops species with the C, D, S, M and U genomes. Theor Appl Genet 111:1159–1166
Chen Q, Qi P, Wei Y, Wang J, Zheng Y (2009) Molecular Characterization of the pina Gene in Einkorn Wheat. Biochem Genet 47:384–396
Cheryan M (1980) Phytic acid interactions in food systems. CRC Crit Rev Food Sci Nutr 13:297–335
Chhuneja P, Dhaliwal HS, Bains NS, Singh K (2006) *Aegilops kotschyi* and *Aegilops tauschii* as sources for higher levels of grain iron and zinc. Plant Breed 125:529–531
Ciaffi M, Dominici L, Lafiandra D (1998) High molecular weight glutenin subunit variation in wild and cultivated einkorn wheat. Plant Syst Evol. 209:123–137
Ciaffi M, Lafiandra D, Porceddu E, Benedettelli S (1993) Storageprotein variation in wild emmer wheat (Triticum turgidum ssp. dicoccoides) from Jordan and Turkey. I. Electrophoretic characterization of genotypes. Theor Appl Genet 86:474–480
Ciaffi M, Lafiandra D, Turchetta T, Ravaglia S, Bariana H, Gupta R, MacRitchie F (1995) Bread making potential of durum wheat lines expressing both x- and y-type subunits at the Glu-A1 locus. Cereal Chem 72:465–469
Cloutier S, Rampitsch C, Penner GA (2001) Cloning and expression of a LMW-i glutenin gene. J Cereal Sci 33:143–154
D'Ovidio R, Masci S (2004) The low-molecular-weight glutenin subunits of wheat gluten. J Cereal Sci 39:321–339
D'Ovidio R, Masci S, Porceddu E (1995) Development of a set of oligonucleotide primers specificc for genes at the Glu-1 complex loci of wheat. Theor Appl Genet 91:189–194
De Bustos A, Jouve N (2003) Characterisation and analysis of new HMW-glutenin alleles encoded by the Glu-R1 locus of Secale cereale. Theor Appl Genet 107:74–83
Devos KM (2005) Updating the 'crop circle'. Current Opin Plant Biol 8:155–162
Distelfeld A, Cakmak I, Peleg Z, Ozturk L, Yazici AM, Budak H, Saranga Y, Fahima T (2007) Multiple QTL-effects of wheat Gpc-B1 locus on grain protein and micronutrient concentrations. Physiol Plantarum 127:635–644
Distelfeld A, Uauy C, Fahima T, Dubcovsky J (2006) Physical map of the wheat high-grain protein content gene Gpc-B1 and development of a high-throughput molecular marker. New Phytologist 169:753–763
Distelfeld A, Uauy C, Olmos S, Schlatter AR, Dubcovsky J, Fahima T (2004) Microcolinearity between a 2 cm region encompassing the grain protein content locus Gpc-6B1 on wheat chromosome 6 and a 350-kb region on rice chromosome 2. Func Integrat Genom 4:59–66
Dong CH, Ma ZY, Xia XC, Zhang LP, He ZH (2012) Allelic variation at the TaZds-A1 locus on wheat chromosome 2A and development of a functional marker in common wheat. Agric Sci China (in press)
DuPont FM, Vensel WH, Chan R, Kasarda DD (2000) Characterization of the 1Btype ω-gliadins from *Triticum aestivum* cultivar Butte. Cereal Chem 77:607–614
Elangovan M, Rai R, Dholakia BB, Lagu MD, Tiwari R, Gupta RK, Rao VS, Roder MS, Gupta VS (2008) Molecular genetic mapping of quantitative trait loci associated with loaf volume in hexaploid wheat *(Triticum aestivum)*. J Cereal Sci 47:587–598
FAO (2009) World wheat statistics. Food and Agriculture Organization, Rome
Galili G, Feldman M (1984) Mapping of glutenin and gliadin genes located on chromosome 1B of common wheat. Mol Gen Genet 193:293–298
Galterio G, Cardarilli D, Codianni P, Acquistucci R (2001) Evaluation of chemical and technological characteristics of new lines of Triticum turgidum ssp dicoccum. Nahrung 45:263–266
Gao L, Ma W, Chen L, Wang K, Ki J, Wang S, Békés F, Appels R, Yan Y (2010) Characterisation and comparative analysis of wheat high molecular weight glutenin subunits by SDS-PAGE, RP-HPLC, HPCE, and MALDI-TOF-MS. J Agric Food Chem 58:2777–2786
Gautier MF, Aleman ME, Guirao A, Marion D, Joudrier P (1994) *Triticum aestivum* puroindolines, two basic cystine-rich seed proteins: CDNA sequence analysis and developmental gene expression. Plant Mol Biol 25:43–57

Geng HW, He ZH, Zhang LP, Qu YY, Xia XC (2012) Cloning the lipoxygenase gene on chromosome 4BS and development of functional markers in common wheat. Crop Sci 52. doi:10.2135/cropsci2011.07.0365

Gianibelli MC, Gupta RB, Lafiandra D, Margiotta B, MacRitchie F (2001) Polymorphism of high Mr glutenin subunits in Triticum tauschii: characterization by chromatography and electrophoresis methods. J Cereal Sci 33:39–52

Gianibelli MC, Gupta RB, MacRitchie F (2000) HMW and LMW subunits of glutenin of Triticum tauschii, the D genome donor to hexaploid wheat. In: Schofield JD (ed.), Wheat Structure, Biochemistry and Functionality. The Royal Society of Chemistry, UK pp 139–145

Gianibelli MC, Larroque OR, MacRichie F, Wrigley CW (2001) Biochemical, genetic and molecular characterization of wheat glutenin and its component subunits. Cereal Chem 78:635–646

Gianibelli MC, Masci S, Larroque OR, Lafiandra D, MacRitchie F (2002) Biochemical characterization of a novel polymeric protein subunit from bread wheat (*Triticum aestivum* L.). J Cereal Sci 35:265–276

Giroux MJ, Morris CF (1997) A glycine to serine change in puroindoline b is associated with wheat grain hardness and low levels of starch-surface friabilin. Theor Appl Genet 95:857–864

Giroux MJ, Morris CF (1998) Wheat grain hardness results from highly conserved mutations in the friabilin components Puroinodoline a and b. Proceedings Nat Acad Sci U S A 95:6262–6266

Gollan P, Smith K, Bhave M (2007) Gsp-1 genes comprise a multigene family in wheat that exhibits a unique combination of sequence diversity yet conservation. J Cereal Sci 45:184–198

Guo ZF, Dong P, Long XY, Wei YM, Zhang LJ, Zheng YL (2008) Molecular characterization of LMW prolamines from Crithopsis delileana and the comparative analysis with those from Triticeae. Hereditas 145:204–211

Gupta RB, Shephard KW (1990) Two-step one-dimensional SDSPAGE analysis of LMW subunits of glutelin. 2. Genetic control of the subunits in species related to wheat. Theor Appl Genet 80:183–187

Gupta RB, Shepherd KW (1993) Production of multiple wheat rye 1RS translocation stocks and genetic analysis of LMW subunits of glutenin and gliadins in wheat using these stocks. Theor Appl Genet 85:719–728

Gupta RB, Singh NK, Shepherd KW (1989) The cumulative effects of allelic variation in LMW and HMW glutenin subunits on physical dough properties in the progeny of two bread wheat. Theor Appl Genet 77:57–62

Gutierrez MV, Guzman C, Martin LM, Alvarez JB (2011) Molecular characterization of the Glu-Ay gene from *Triticum urartu* for its potential use in quality wheat breeding. Plant Genet Resour 9:334–337

Guzman C, Caballero L, Martın MA, Alvarez JB (2012) Molecular characterization and diversity of the Pina and Pinb genes in cultivated and wild diploid wheat. Mol Breed doi:10.1007/s11032-011-9599-1

Halford NG, Field JM, Blair H, Urwin P, Moore K, Robert L, Thompson R, Flavell RB, Tatham AS, Shewry PR (1992) Analysis of HMW glutenin subunits encoded by chromosome 1A of bread wheat (*Triticum aestivum* L.) indicates quantitative effects on grain quality. Theor Appl Genet 83:373–378

Harberd NP, Bartels D, Thompsom RD (1985) Analysis of the gliadin multigene loci in bread wheat using nullisomic-tetrasomic lines. Mol Gen Genet 198:234–242

Hassani ME, Shariflou MR, Gianibelli MC, Sharp PJ (2006) Gli -Dtt1 and a novel γ-gliadin gene in *Aegilops tauschii*. Plant Breed 125:27–31

He XY, He ZH, Ma W, Appels R, Xia XC (2009a) Allelic variants of phytoene synthase 1 (Psy1) genes in Chinese and CIMMYT wheat cultivars and development of functional markers for flour colour. Mol Breed 23:553–563

He XY, He ZH, Zhang LP, Sun DJ, Morris CF, Fuerst EP, Xia XC (2007) Allelic variation of polyphenol oxidase (PPO) genes located on chromosomes 2A and 2D and development of functional markers for the PPO genes in common wheat. Theor Appl Genet 115:47–58

He XY, Wang JW, Ammar K, Pena RJ, Xia XC, He ZH (2009b) Allelic variants at the Psy-A1 and Psy-B1 loci in durum wheat and their associations with grain yellowness. Crop Sci 49:2058–2064

He XY, Zhang YL, He ZH, Wu YP, Xiao YG, Ma CX, Xia XC (2008) Characterization of phytoene synthase 1 gene (Psy1) located on common wheat chromosome 7A and development of a functional marker. Theor Appl Genet 116:213–221

Hsam SLK, Kieffer R, Zeller FJ (2001) Significance of *Aegilops tauschii* glutenin genes on breadmaking properties of wheat. Cereal Chem 78:521–525

Huang XQ, Cloutier S, Lycar L, Radovanovic N, Humphreys DG, Noll JS, Somers DJ, Brown PD (2006) Molecular detection of QTL for agronomic and quality traits in a doubled haploid population derived from two Canadian wheats (*Triticum aestivum* L.). Theor Appl Genet 113:753–766

Ikeda TM, Araki E, Fujita Y, Yano H (2006) Characterization of low-molecular-weight glutenin subunit genes and their protein products in common wheat. Theor Appl Genet 112:327–334

Ikeda TM, Nagamine T, Fukuoka H, Yano H (2002) Identification of new low-molecular weight glutenin subunit genes in wheat. Theor Appl Genet 104:680–687

Jackson EA, Holt LM, Payne PI (1983) Characterisation of high-molecular-weight gliadin and low-molecular-weight glutenin subunits of wheat endosperm by two dimensional electrophoresis and the chromosomal localisation of their controlling genes. Theor Appl Genet 66:29–37

Jackson EA, Holt LM, Payne PI (1985) Glu-B2, a storage protein locus controlling the D group of LMW glutenin subunits in bread wheat (*Triticum aestivum*). Genet Res 47:11–17

Jackson EA, Morel MH, Sontag-Strohm T, Branlard G, Metakovsky EV, Redaelli R (1996) Proposal for combining the classification systems of alleles of Gli-1 and Glu-3 loci in bread wheat (*Triticum aestivum* L.) J Genet Breed 50:321–336

Jiang CX, Pei YH, Zhang YZ, Li XH, Yao DN, Yan YM, Hsam SLK, Zeller FJ (2008) Molecular cloning and characterization of four novel LMW glutenin subunit genes from *Aegilops longissima, Triticum dicoccoides* and *T. zhukovskyi*. Hereditas 145:92–98

Jin H, Yan J, Peña RJ, Xia XC, Morgounov A, Han LM, Zhang Y, He ZH (2011) Molecular detection of high- and low-molecular weight glutenin subunit genes in common wheat cultivars from 20 countries using allele-specific markers. Crop Pasture Sci 62:746–754

Joppa LR, Hart GE, Hareland GA (1997) Mapping a QTL for grain protein in tetraploid wheat (*Triticum turgidum* L.) using a population of recombinant inbred chromosome lines. Crop Sci 37:1586–1589

Juhász A, Gianibelli MC (2006) Low-molecular-weight glutenin subunits: insight into this abundant subunit group present in glutenin polymers. In: Wrigley CW, Békés F, Bushuk W (eds) Gliadin and glutenin: the unique balance of wheat quality. AACC International Press, St Paul, pp 171–212

Kade M, Barneix AJ, Olmos S, Dubcovsky J (2005) Nitrogen uptake and remobilization in tetraploid 'Langdon' durum wheat and a recombinant substitution line with the high grain protein gene Gpc-B1. Plant Breed 124:343–349

Kasarda DD, Autran JC, Lew EJL, Nimmo CC, Shewry PR (1983) N-terminal amino acid sequences of ω-gliadins and ω-secalins; Implications for the evolution of prolamin genes. Biochim Biophy Acta 747:138–150

Kasarda DD, Okita TW, Bernardin JE, Baecker PA, Nimmo CC, Lew EJL, Dietler MS, Green FC (1984) Nucleic (cDNA) and amino acid sequences of α-type gliadins from wheat (*Triticum aestivum*). Proceedings Nat Acad Sci U S A 81:4712–4716

Khalid M, Mahmood T, Rasheed A, Kazi AG, Ali A, Mujeeb-Kazi A (2013) *GluDt1* allelic variation in synthetic hexaploid wheats derived from durum cultivar 'Decoy' x *Aegilops tauschii* accessional crosses. Pak J Bot 45:409–414

Kuchel H, Langridge P, Mosionek L, Williams K, Jefferies SP (2006) The genetic control of milling yield, dough rheology and baking quality of wheat. Theor Appl Genet 112:1487–1495

Lafiandra D, Benedettelli S, Porceddu E (1988) Null forms for storage proteins in bread wheat and durum. pp 963–967 In: Miller TE, Koebner RMD (eds). proceeding the 7th International Wheat Genetics Symposium. Inst. Plant Sci Res, Cambridge

Lafiandra D, Margiotta B, Colaprico G, Masci S, Roth MR, MacRitchie F (2000) Introduction of the D-genome related high- and low-Mr glutenin subunits into durum wheat and their effect on technological properties. In: Shewry PR., Thatam AS (eds.), Wheat Gluten, Royal Society of Chemistry, UK, pp 51–54

Lafiandra D, Masci S, Margiotta B, Ambrogio ED (1998) Development of durum and bread wheat with increased number of high molecular weight glutenin subunits. pp 261–264, 146 In: Proceeding 9th International Wheat Genetics Symposium, Vol. 4. Grain Quality. Slinkard AE (ed). University of Saskatchewan, Canada

Lee YK, Bekes F, Gras P, Ciaffi M, Morell MK, Appels R (1999) The low-molecular-weight glutenin subunit proteins of primitive wheat. IV. Functional properties of products from individual genes. Theor Appl Genet 98:149–155

Lei ZS, Gale KR, He ZH, Gianibelli C, Larroque O, Xia SX, Butow BJ, Ma W (2006) Y-type gene specific markers for enhanced discrimination of high-molecular weight glutenin alleles at the Glu-B1 locus in hexaploid wheat. J Cereal Sci 43:94–101

Li XH, Ma WJ, Gao LY, Zhang YZ, Wang AL, Ji KM, Wang K, Appels R, Yan YM (2008) A novel chimeric low molecular-weight glutenin subunit gene from the wild relatives of wheat *Aegilops kotschyi* and *Ae. juvenalis*: evolution at the Glu-3 loci. Genetics 180:93–101

Li XH, Wang K, Wang SL, Gao LY, Xie XX, Hsam SLK, Zeller FJ, Yan YM (2010) Molecular characterization and comparative transcriptional analysis of LMW-m-type genes from wheat (*Triticum aestivum* L.) and *Aegilops* species. Theor Appl Genet 121:845–856

Liang D, Tang JW, Pen˜a RJ, Singh R, He XY, Shen XY, Yao DN, Xia XC, He ZH (2010) Characterization of CIMMYT bread wheat for high- and low-molecular weight glutenin subunits and other quality-related genes with SDS-PAGE, RP-HPLC and molecular markers. Euphytica 172:235–250

Lillemo M, Simeone MC, Morris CF (2002) Analysis of puroindoline a and b sequences from *Triticum aestivum* cv. 'Penawawa' and related taxa. Euphytica 126:321–331

Liu CY, Shepherd KW (1996) Variation of B subunits of glutenin in durum, wild and less-widely cultivated tetraploid wheat. Plant Breed 115:172–178

Liu L, Ikeda TM, Branlard G, Peña RJ, Rogers WJ, Lerner SE, Kolman MA, Xia XC, Wang LH, Ma WJ, Appels R, Yoshida H, Wang AL, Yan YM, He ZH (2010) Comparison of low molecular weight glutenin subunits identified by SDS-PAGE, 2-DE, MALDI-TOF-MS and PCR in common wheat. BMC Plant Biol 10:124

Liu SW, Gao X, Xia GM (2008b) Characterizing HMW-GS alleles of decaploid Agropyron elongatum in relation to evolution and wheat breeding. Theor Appl Genet 116:325–334

Liu SW, Zhao SG, Chen FG, Xia GM (2007) Generation of novel high quality HMW-GS genes in two introgression lines of *Triticum aestivum*/Agropyron elongatum. BMC Evol Biol 7:76

Liu SX, Chao SM, Anderson JA (2008a) New DNA markers for high molecular weight glutenin subunits in wheat. Theor Appl Genet 118:177–183

Liu Z, Yan Z, Wan Y, Liu K, Zheng Y, Wang D (2003) Analysis of HMW glutenin subunits and their coding sequences in two diploid Aegilops species. Theor Appl Genet 106:1368–1378

Long H, Wei YM, Yan ZH, Baum B, Nevo E, Zheng YL (2005) Classification of wheat low-molecular-weight glutenin subunit genes and its chromosome assignment by developing LMW-GS group-specific primers. Theor Appl Genet 111:1251–1259

Luo Z, Chen FG, Feng DS, Xia GM (2005) LMW glutenin genes in Agropyron elongatum and their potential value in wheat breeding. Theor Appl Genet 111:272–280

Ma DY, Yan J, He ZH, Wu L, Xia XC (2012) Characterization of a cell wall invertase gene TaCwi-A1 on common wheat chromosome 2A and development of functional markers. Mol Breed 29:43–52

Ma W, Anderson O, Kuchel H, Bonnardeaux Y, Collins H, Morell MK, Langridge P, Appels R (2009) Genomics of quality traits. In: Feuillet C, Muehlbauer GJ (eds.) Genetics and Genomics of the Triticeae. Springer, New York, pp 611–652

Ma W, Zhang W, Gale KR (2003) Multiplex-PCR typing of high molecular weight glutenin alleles in wheat. Euphytica 134:51–60

Margiotta B, Urbano M, Colaprico G, Johansson E, Buonocore F, D'Ovidio R, Lafiandra D (1996) Detection of y-type subunit at the Glu-A1 locus in some Swedish bread wheat lines. J Cereal Sci 23:203–211

Masci SM, Porceddu E, Colaprico G, Lafiandra D (1991) Comparison of the B and D subunits of glutenin encoded at the Glu-D3 locus in two biotypes of the common wheat cultivar Newton with different technological characteristics. J Cereal Sci 14:35–46

Massa AN, Morris CF, Gill BS (2004) Sequence diversity of puroindoline-a, puroindoline-b, and the grain softness protein genes in Aegilops tauschii coss. Crop Sci 44:1808–1816

Mattern PJ, Morris R, Schmidt JW, Johnson VA (1973) Location of genes for kernel properties in the wheat cultivar 'Cheyenne' using chromosome substitution lines. In: Sears ER, Sears LMS (eds.) Proceeding 4th Int. Wheat Genet. Symp. University of Missouri, Columbia pp 703–707

McCartney CA, Somers DJ, Lukow O, Ames N, Noll J, Cloutier S, Humphreys DG, McCallum BD (2006) QTL analysis of quality traits in the spring wheat cross RL4452 by 'AC Domain'. Plant Breed 125:565–575

McIntosh R, Dubcovsky J, Rogers J, Morris C, Appels R, Xia X (2010) Catalogue of gene symbols for wheat 2010 supplement. Annual Wheat Newsletter 56:273–282

Mecham DK, Kasarda DD, Qualset CO (1978) Genetic aspects of wheat gliadin proteins. Biochem Genet 16:831–853

Metakovsky EV (1991) Gliadin allele identification in common wheat II. Catalogue of gliadin alleles in common wheat. J Genet Breed 45:325–344

Metakovsky EV, Chernakov VM, Upelniek VP, Redaelli R, Dardevet M, Branlard G, Pogna NG (1996) Recombination mapping of minor omega-gliadin-coding loci on chromosome 1A of common wheat: a revision. J Genet Breed 50:277–286

Metakovsky EV, Novoselskaya AY, Sozinov AA (1984) Genetic analysis of gliadin components in winter wheat using two-dimensional polyacrylamide gel electrophoresis. Theor Appl Genet 69:31–37

Morris CF (2002) Puroindolines: the molecular genetic basis of wheat grain hardness. Plant Mol Biol 48:633–647

Morris CF, Bhave M (2008) Reconciliation of D-genome puroindoline allele designations with current DNA sequence data. J Cereal Sci 47:277–287

Müller S, Wieser H (1995) Disulphide bonds of α-type gliadins. J Cere Sci 22: 21–27

Müller S, Wieser H (1997) The location of disulphide bonds in monomeric γ-type gliadins. J Cereal Sci 26:169–176

Nakamura T, Vrinten P, Saito M, Konda M (2002) Rapid classification of partial waxy wheat using PCR-based markers. Genome 45:1150–1156

Neelam K, Rawat N, Tiwari VK, Kumar S, Chhuneja P, Singh K, Randhawa GS, Dhaliwal HS (2011) Introgression of group 4 and 7 chromosomes of Ae. peregrina in wheat enhances grain iron and zinc density. Mol Breed 28:623–634

Nelson JC, Andreescu C, Breseghello F, Finney PL, Daisy G, Gualberto DG, Bergman CJ, Pena RJ, Perretant MR, Leroy P, Qualset CO, Sorrells ME (2006) Quantitative trait locus analysis of wheat quality traits. Euphytica 149:145–159

Niu ZX, Klindworth R, Wang R R-C, Jauhar PP, Larkin PJ, Xu SS (2011) Characterization of HMW glutenin subunits in Thinopyrum intermedium, Th. bessarabicum, Lophopyrum elongatum, Aegilops markgrafii, and their addition lines in wheat. Crop Sci 51:667–677

Obukhova LV, Generalova GV, Agafonov AV, Kumarev VP, Popova NA, Gulevich VV (1997) A comparative molecular genetic study of glutelins in wheat and Elymus. Genetika 33:1001–1004

Obukhova LV, Maystrenko OI, Generalova GV, Ermakova MF, Popova RK (1997) Composition of high molecular weight glutenin subunits in common wheat substitution lines created from cultivars with contrasting bread-making qualities. Russian J Genetics 33:5–100

Olmos S, Distelfeld A, Chicaiza O, Schlatter AR, Fahima T, Echenique V, Dubcovsky J (2003) Precise mapping of a locus affecting grain protein content in durum wheat. Theor Appl Genet 107:1243–1251

Osborne TB (1907) The protein of the wheat kernel. Publication No. 84. Carnegie Institute: Washington, DC

Ovesna J, Novakova I, Kucera L, Dotlacil L (2001) Characterisation of variability at Glu-3 loci in some European wheat obsolete cultivars and landraces using PCR. Cereal Res Commun 29:205–213

Parker GD, Chalmers KJ, Rathgen AJ, Langridge P (1998) Mapping loci associated with flour colour in wheat (*Triticum aestivum* L.). Theor Appl Genet 97:238–245

Payne PI (1987) Genetic of wheat storage proteins and the effect of allelic variation on bread making quality. Ann Rev Plant Physiol 38:141–153

Payne PI, Law CN, Mudd EE (1980) Control by homoeologous group 1 chromosomes of the high-molecular-weight subunits of glutenin, a major protein of wheat endosperm. Theor Appl Genet 58:113–120

Payne PI, Lawrence GJ (1983) Catalogue or alleles for the complex gene loci, Glu-A1, Glu-B1 and Glu-D1 which code for the high-molecular weight subunit of glutenin whose in hexaploid wheat. Cereal Res Commun 11:29–35

Payne PI, Nightingale MA, Krattinger AF, Holt LM (1987) The relationship between HMW glutenin subunit composition and bread making quality of British grown wheat varieties. J Sci Food Agric 40:51–65

Pfluger LA, D'Ovidio R, Margiotta B, Pena R, Mujeeb-Kazi A, Lafiandra D (2001) Characterisation of high-and low-molecular weight glutenin subunits associated to the D genome of *Aegilops tauschii* in a collection of synthetic hexaploid wheat. Theor Appl Genet 103:1293–1301

Piston F, Martin A, Dorado G, Barro F (2005) Cloning and molecular characterization of B-hordeins from Hordeum chilense (Roem. et Schult.). Theor Appl Genet 111:551–560

Pogna NE, Gazza L, Boggini G (2002) Puroindoline and kernel hardness in *Triticum aestivum* and *Triticum monococcum*. Annual Wheat Newsletter. Experimental Institute for Cereal Research, Italy p 60

Pogna NE, Metakovsky EV, Redaelli R, Raineri F, Dachkevitch T (1993) Recombination mapping of Gli-5, a new gliadin-coding locus on chromosomes 1A and 1B in common wheat. Theor Appl Genet 87:113–121

Popineau Y, Pineau F (1987) Investigation of surface hydrophobicities of purified gliadins by hydrophobic interaction chromatography, reversed-phase high performance chromatography and a polar ligand binding. J Cereal Sci 5:215–231

Pozniak CJ, Knox RE, Clarke FR, Clarke JM (2007) Identification of QTL and association of a phytoene synthase gene with endosperm colour in durum wheat. Theor Appl Genet 114:525

Qi PF, Wei YM, Ouellet T, Chen Q, Tan Z, Zheng YL (2009) The gamma-gliadin multigene family in common wheat (*Triticum aestivum*) and its closely related species. BMC Genomics 10:168

Ragupathy R, Naeem HA, Reimer E, Lukow OM, Sapirstein HD, Cloutier S (2008) Evolutionary origin of the segmental duplication encompassing the wheat Glu-B1 locus encoding the over expressed Bx7 (Bx7OE) high molecular weight glutenin subunit. Theor Appl Genet 116:283–296

Ram S, Sharma S, Verma A, Tyagi BS, Pena RJ (2011) Comparative analyses of LMW glutenin alleles in bread wheat using allele specific PCR and SDS-PAGE. J Cereal Sci 54:488–493

Ram S, Verma A, Sharma S (2010) Large variability exist in phytase levels among indian wheat varieties and synthetic hexaploids. J Cereal Sci 52:486–490

Rasheed A, Mahmood T, Kazi AG, Ghafoor A, Mujeeb-Kazi A (2012) Allelic variation and composition of HMW-GS in advanced lines derived from D-genome synthetic hexaploid/bread wheat (*Triticum aestivum* L.). J Crop Sci Biotechnol 15:1–7

Rawat N, Neelam K, Tiwari VK, Randhawa GS, Friebe B, Gill BS, Dhaliwal HS (2011) Development and molecular characterization of wheat—*Aegilops kotschyi* addition and substitution lines with high grain protein, iron, and zinc. Genome 54:943–953

Rawat N, Tiwari VK, Neelam K, Randhawa GS, Singh K, Chhuneja P, Dhaliwal HS (2009) Development and characterization of wheat-Aegilops kotschyi amphiploids with high grain iron and zinc. Plant Genet Resour 7:271–280

Rehman A, Evans N, Gianibelli MC, Rose RJ (2008) Allelic variation in high and low molecular weight glutenins at the Glu-Dt locus of *Aegilops tauschii* as a potential source for improving bread wheat quality. Aus J Agric Res 59:399–405

Reif JC, Gowda M, Maurer HP, Longin CFH, Korzun V, Ebmeyer E, Bothe R, Pietsch C, Wurschum T (2011) Association mapping for quality traits in soft winter wheat. Theor Appl Genet 122:961–970

Reynolds MP, Foulkes JM, Slafer GA, Berry P, Parry MAJ, Snape J, Angus WJ (2009) Raising yield potential in wheat. J Exp Bot 60:1899–1918

Saito M, Vrinten P, Ishikawa G, Graybosch R, Nakamura T (2009) A novel codominant marker for selection of the null Wx-B1 allele in wheat breeding programs. Mol Breed 23:209–217

Schwarz G, Felsenstein FG, Wenzel G (2004) Development and validation of a PCR-based marker assay for negative selection of the HMW glutenin allele Glu-B1–1d (Bx-6) in wheat. Theor Appl Genet 109:1064–1069

Shang HY, Wei YM, Long H, Yan ZH, Zheng YL (2005) Identification of LMW glutenin-like genes from Secale sylvestre Host. Russian J Genet 41:1372–1380

Shewry PR, Halford NG (2002) Cereal seed storage proteins: structures, properties and role in grain utilization. J Exp Bot 53:947–958

Shewry PR, Halford NG, Belton PS, Tatham AS (2002) The structure and properties of gluten: an elastic protein from wheat grain. Phil Trans R Soc Lond B 357:133–142

Shewry PR, Halford NG, Tatham AS (1989) The high molecular weight subunits of wheat, barley and rye: genetics, molecular biology, chemistry and role in wheat gluten structure and functionality. Oxford Surv. Plant Mol Cell Biol 6:163–219

Shewry PR, Tatham AS (1990) The prolamin storage proteins of cereal seeds: structure and evolution. Biochem J 267:1–12

Shewry PR, Tatham AS (1997) Disulphide bonds in wheat gluten proteins. J Cereal Sci 25:207–227

Shewry PR, Tatham AS, Forde J, Kreis M, Miflin BJ (1986) The classification and nomenclature of wheat gluten proteins: a reassessment. J Cereal Sci 4:97–106

Shewry, PR, Halford NG, Tatham AS (1992) High molecular weight subunits of wheat glutenin. J Cereal Sci 15:105–120

Simeone MC, Gedye KR, Mason-Gammer R (2006) Conserved regulatory elements identified from a comparative puroindoline gene sequence survey of *Triticum* and Aegilops diploid taxa. J Cereal Sci 44:21–33

Simmonds N (1995) The relation between yield and protein in cereal grain. J Sci Food Agric 67:309–315

Singh NK, Shepherd KW (1985) The structure and genetic control of a new class of disulphide-linked proteins in wheat endosperm. Theor Appl Genet 7:79–92

Singh NK, Shepherd KW (1988) Linkage mapping of genes controlling endosperm storage proteins in wheat. 1. Genes on the short arms of group 1 chromosomes. Theor Appl Genet 75:628–641

Sobko TA (1984) Identification of a new locus which controls the synthesis of alcohol-soluble protein in the endosperm of winter common wheat. J Agric Sci 7:78–80

Sourdille P, Perretant MR, Charmet G, Leroy P, Gautier M.-F, Joudrier P, Nelson JC, Sorrells ME, Bernard M (1996) Linkage between RFLP markers and genes affecting kernel hardness in wheat. Theor Appl Genet 93:580–586

Sreeramulu G, Singh NK (1997) Genetic and biochemical characterization of novel low molecular weight glutenin subunits in wheat (*Triticum aestivum* L.) Genome 40:41–48

Sun DJ, He ZH, Xia XC, Zhang LP, Morris CF, Appels R, Ma WJ, Wang H (2005) A novel STS marker for polyphenol oxidase activity in bread wheat. Mol Breed 16:209–218

Sun M, Yan Y, Jiang Y, Xiao Y, Hu Y, Cai M, Li Y (2004) Molecular cloning and comparative analysis of a y-type inactive HMW glutenin subunit gene from cultivated emmer wheat (*Triticum dicoccum* L.). Hereditas 141:46–54

Symes KJ (1965) The inheritance of grain hardness in wheat as measured by the particle size index. Australian J Agric Res 16:113–123

Tatham AS, Shewry PR (1985) The conformation of wheat gluten proteins. The secondary structures and thermal stabilities of alpha-, beta-, gamma- and omega- gliadins. J Cereal Sci 3:103–113

Tatham AS, Shewry PR (1995) The S-poor prolamins of wheat, barley, and rye. J Cereal Sci 22:1–16

Thomson NH, Miles MJ, Tatham AS, Shewry PR (1992) Molecular images of cereal proteins by STM. Ultramicroscopy 42–44:1118–1122

Tiwari VK, Rawat N, Neelam K, Randhawa GS, Singh K, Chhuneja P, Dhaliwal HS (2008) Development of Triticum turgidum subsp. durum-Aegilops longissima amphiploids with high iron and zinc content through unreduced gamete formation in F1 hybrids. Genome 51:757–766

Tiwari VK, Rawat N, Chhuneja P, Neelam K, Aggarwal R, Randhawa GS, Dhaliwal HS, Keller B, Singh K (2009) Mapping of quantitative trait loci for grain iron and zinc concentration in A genome diploid wheat. J Hered 100:771–776

Tranquilli G, Cuniberti M, Gianibelli MC, Bullrich L, Larroque OR, MacRitchie F, Dubcovsky J (2002) Effect of Triticum monococcum glutenin loci on cookie making quality and on predictive tests for bread making quality. J Cereal Sci 36:9–18

Tranquilli G, Lijavetzky D, Muzzi G, Dubcovsky J (1999) Genetic and physical characterization of grain texture-related loci in diploid wheat. Mol Gen Genet 262:846–850

Turnbull K-M, Turner M, Mukai Y, Yamamoto M, Morell MK, Appels R, Rahman S (2003) The organization of genes tightly linked to the *Ha* locus in *Aegilops tauschii,* the D-genome donor to wheat. Genome 46:330–338

Uauy C, Brevis JC, Dubcovsky J (2006a) The high grain protein content gene Gpc-B1 accelerates senescence and has pleiotropic effects on protein content in wheat. J Exp Bot 57:2785–2794

Uauy C, Distelfeld A, Fahima T, Blechl A, Dubcovsky J (2006b) A NAC gene regulating senescence improves grain Protein, zinc, and iron content in wheat. Science 314:1298–1301

Varshney RK, Graner A, Sorrells ME (2005) Genomics-assisted breeding for crop improvement. Trends Plant Sci 10:621–630

Vensel WH, Adalsteins AE, Kasarda DD (1997) Purification and characterization of the glutenin subunits of *Triticum tauschii,* progenitor of the D genome in hexaploid bread wheat. Cereal Chem 74:108–114

Vogel JP, Garvin DF, Mockler TC. 2010. Genome sequencing and analysis of the model grass Brachypodium distachyon. Nature 463:763–768

Waines JG. Payne PI (1987) Electrophoretic analysis of the high molecular-weight glutenin subunits of *Triticum monococcum, T. urartu,* and the *A genome* of bread wheat (*T. aestivum*). Theor Appl Genet 74:71–76

Wan Y, Wang D, Shewry PR, Halford NG (2002) Isolation and characterization of five novel high molecular weight subunit of glutenin genes from Triticum timopheevi and Aegilops cylindrical. Theor Appl Genet 104:828–839

Wang JR, Yan Z, Wei YM, Zheng YL (2006) Characterization of high-molecular-weight glutenin subunits genes from Elytrigia elongata (Host) Nevski. Plant Breed 125:89–95

Wang JW, He XY, He ZH, Wang H, Xia XC (2009a) Cloning and phylogenetic analysis of phytoene synthase 1 (Psy1) genes in common wheat and related species. Hereditas 146:208–256

Wang LH, Zhao XL, He ZH, Ma W, Appels R, Pena RJ, Xia XC (2009b) Characterization of low-molecular-weight glutenin subunit Glu-B3 genes and development of STS markers in common wheat (*Triticum aestivum* L.). Theor Appl Genet 118:525–539

Wang LH, Li GY, Pena RJ, Xia XC, He ZH (2010) Development of STS markers and establishment of multiplex PCR for Glu-A3 alleles in common wheat (*Triticum aestivum* L.). J Cereal Sci 51:305–312

Wieser H (1995) The precipitating factor in coeliac disease. In: Howdle PD (ed.) pp 191–207 in: clinical Gastroenterology, Int Pract Res. Bailliére Tindall, London

Wrigley CW, Bushuk W, Gupta R (1996) Nomenclature: establishing a common gluten language. In: Gluten 96. Wrigley CW (ed.) pp 403–407 RACI, Melbourne

Wrigley CW, Shepherd KW (1973) Electrofocusing of grain proteins from wheat genotypes. Annals NY Acad Sci 209:154–162

Xiong L, Wang J, Zheng Y (2010) Analysis of LMW-GS, α- and β-Gliadin gene coding sequences from *Triticum macha*. Agric Sci China 9:163–169

Xu SS, Khan K, Klindworth DL, Nygard G (2010) Evaluation and characterization of high-molecular weight 1D glutenin subunits from *Aegilops tauschii* in synthetic hexaploid wheat. J Cereal Sci 52:333–336

Yan LL, Loukoianov A, Blechl A, Tranquilli G, Ramakrishna W, SanMiguel P, Bennetzen JL, Echenique V, Dubcovsky J (2004b) The wheat VRN2 gene is a flowering repressor down-regulated by vernalization. Science 303:1640–1643

Yan Y, Hsam SLK, Yu JZ, Jiang Y, Ohtsuka I, Zeller FJ (2003a) HMW and LMW glutenin alleles among putative tetraploid and hexaploid European spelt wheat (*Triticum spelta* L.) progenitors. Theor Appl Genet 107:1321–1330

Yan Y, Hsam SLK, Yu JZ, Jiang Y, Zeller FJ (2003b) Allelic variation of the HMW glutenin subunits in Aegilops tauschii accessions detected by sodium dodecyl sulphate (SDS-PAGE) acid polyacrylamide gel (A-PAGE) and capillary electrophoresis. Euphytica 130:377–385

Yan Y, Zheng J, Xiao Y, Yu J, Hu Y, Cai M, Li Y, Hsam SLK, Zeller FJ (2004a) Identification and molecular characterization of a novel y-type Glu-Dt1 glutenin gene of Aegilops tauschii. Theor Appl Genet 108:1349–1358

Yan ZH, Wan YF, Liu KF, Zheng YL, Wang DW (2002) Identification of a novel HMW glutenin subunit and comparison of its amino acid sequence with those of homologous subunits. Chin Sci Bull 47:220–225

Zhang CY, Dong CH, He XY, Zhang LP, Xia XC, He ZH (2011) Allelic variation at the TaZds-D1 locus on wheat chromosome 2DL and their association with yellow pigment content. Crop Sci 51:1580–1590

Zhang W, Gianibelli MC, Rampling LR, Gale KR (2004) Characterisation and marker development for low molecular weight glutenin genes from Glu-A3 alleles of bread wheat (*Triticum aestivum* L). Theor Appl Genet 108:1409–1419

Zhang Y, Wu Y, Xiao Y, Yan J, Zhang Y, Zhang Y, Ma C, Xia X, He Z (2009) QTL mapping for milling, gluten quality, and flour pasting properties in a recombinant inbred line population derived from a Chinese soft x hard wheat cross. Crop Pasture Sci 60:587–597

Zhao XL, Xia XC, He ZH, Gale KR, Lei ZS, Appels R, Ma W (2006) Characterization of three low-molecular-weight Glu-D3 subunit genes in common wheat. Theor Appl Genet 113:1247–1259

Zhao XL, Xia XC, He ZH, Lei ZS, Appels R, Yang Y, Sun QX, Ma W (2007a) Novel DNA variations to characterize low molecular weight glutenin Glu-D3 genes and develop STS markers in common wheat. Theor Appl Genet 114:451–460

Zhao XL, Ma W, Gale KR, Lei ZS, He ZH, Sun QX, Xia XC (2007b) Identification of SNPs and development functional markers for LMW-GS genes at Glu-D3 and Glu-B3 loci in bread wheat (*Triticum aestivum* L.). Mol Breed 20:223–231

Zhao L, Zhang K-P, Liu B, Deng Z, Qu H-L, Tian J-C (2010) A comparison of grain protein content QTLs and flour protein content QTLs across environments in cultivated wheat. Euphytica 174:325–335

Chapter 11
From Agronomy to Molecular Genetics and Proteomics in an Effort to Improve Nitrogen Use Efficiency in Crops

Ruby Chandna and Khalid Rehman Hakeem

Abstract Agriculture plays significant role in the sustaining human society among most of the developing countries. The agricultural practices are dependent on the application of the nitrogenous fertilizers. The excessive application of nitrogenous fertilizer contributes enormously to the environmental pollution. So, in today's scenario there is growing need to reduce N fertilizer applications thereby improving plant's N-use efficiency (NUE). Initially, various studies have been carried out to improve inputs of N fertilizers interaction with soil, water and air but low efficiency of the plant to make use of available N has initiated biological interferences. In this article, we will be discussing the possible technologies applied towards understanding the genetic control of nitrogen use efficiency and its improvement in crops. The classification/identification of suitable target candidates like phenotypes, genotypes or molecular markers, for the upgrading of NUE poses big confront. Therefore, it is necessary to understand NUE and its importance with respect to economy and environment. Also, to figure out the diverse approaches for progress towards NUE enhancement and possibilities for future development.

Introduction

The rate at which India's population is growing it is expected to reach up to a total of 1.5 billion by the year 2030 (FAO 2008).Over the next few decades food security will be a major concern in India. However, there are limited options to meet

R. Chandna (✉)
Department of Botany, Jamia Hamdard New Delhi,
New Delhi 110062, India
e-mail: rubychandna@gmail.com

K. R. Hakeem
Faculty of Forestry, Universiti Putra Malaysia,
43400 Serdang, Selangor, Malaysia
e-mail: hakeem.bot@gmail.com

K. R. Hakeem et al. (eds.), *Crop Improvement,* DOI 10.1007/978-1-4614-7028-1_11,

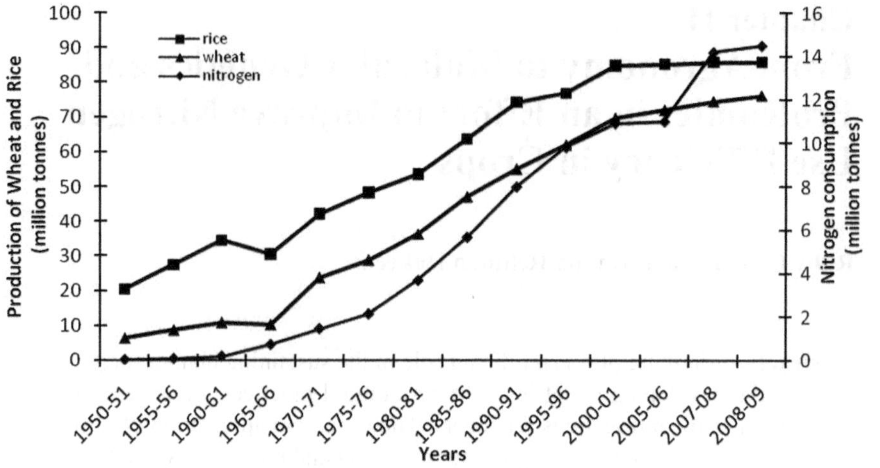

Fig. 11.1 Crop production of India in the past 50 years

food production for increasing population, since most of the available is already under cultivation, and in many areas land cannot be used for intensified productivity because of rapid urbanization and increasing environmental pollution. Together, all of these issues have forced a great challenge on Indian agriculture for developing novel approaches that can increase crop productivity on the cultivable land. Development of high-yielding wheat and rice varieties and use of chemical fertilizers have made enormous contribution in doubling crop production of India in the past 50 years. The production of wheat and rice in 1950–1951 was 6.46 and 20.57 million tonnes, respectively, which has increased to 76.2 and 85.9 million tonnes in 2008–2009 (Directorate of Economics and Statistics 2008) (Fig. 11.1). In order to increase crop production, the consumption of N fertilizer has increased up to 10-folds in last 50 years globally (Lian et al. 2005), because high-yielding present crop varieties have high demands for N. The problem lies in the, nitrogen use efficiency (NUE) which is as low as 33 % for cereals on global basis (Raun and Johnson 1999). The unutilized 60–70 % caused severe environmental hazards. It is, therefore, necessary to control too much input of chemical N fertilizers into the field by enhancing the NUE of the crops.

N-assimilation Processes in Plants

Nitrate (NO_3-) is the major resource of nitrogen for most of the cultivated crops. The first step in acquisition of nitrate is its uptake by root cells. NO_3- reaches the aerial organs by moving out in the external medium from root cell or by unloading in the xylem vessel (Kant et al. 2011). NO_3- assimilation takes place in leaves and root; it is first reduced to nitrite in the cytosol by nitrate reductase (NR). Nitrite gets translocated to the chloroplast and reduced into ammonium by nitrite reductase (NiR), NH_{4^+} the end product is finally integrated into the amino acids via the GS/GOGAT pathway (Mokhele et al. 2012)

Nitrate Reductase (EC 1:6:6:1)

Nitrate reductase (NR) in higher plants is thought to be a homodimer. The estimated subunit size of the monomers is between 100–120 kDa (Caboche and Rouze 1990). The monomer size is about 100 kDa. In higher plants each monomer constitutes three functional domains, each of which has three redox centers formed by FAD, heme and a molybdenum Co-factor (MoCo). These are present in the steochiometry of 1:1:1 (Caboche and Rouze 1990). These redox centres catalytically transfer two electrons from NAD(P)H to NO_{3}–. According to Lillo and Ruoff (1992), there is a second site for NADH, which is occupied in the allosteric, regulating the NR activity. Thus, electrons are able to move from redox centres in one subunit to the redox centres in second subunit, hence enabling the total electron transfer from NADPH to NO_{3}– in heterodimers. NR is found in cytoplasm of shoot and root cells identified using either cell-fractionation and biochemical techniques or immune cytochemistry (Fedorova et al. 1994). In the cytoplasm of the leaf cells, the NADH required for the functioning of NR is supplied by either of two shatter mechanisms, one involving the phosphate translocator located in the chloroplast envelope and the other involving the malate oxaloacetate translocator located in the envelopes of mitochondria. In root cells where NR can utilize both NADH and NADPH as reductants, either the glucose 6-phosphate dehydrogenase or 6-phospogluconate dehydrogenase, present in the cytoplasm or plastid, can supply NADPH (Bowsher et al. 1993).

Nitrite Reductase (EC 1. 7.7 .1)

Nitrite reductase (NiR) catalyses the 6 electron transfer reaction from reduced ferredoxin to NO_{2}–, leading to the synthesis of NH_{4^+}. It is localized inside chloroplasts of the leaf, also in plastids of the root tissues (Sechley et al. 1992). In both roots and leaves, the reduced ferredoxinacts as an electron donor. The NiR enzymes are monomeric proteins of about 63 kDa containing siroheem and a 4Fe-4S centre as prosthetic groups (Seigel and Wilkerson 1989). Wray 1993 confirmed that the NiR apoprotein is synthesised as a precursor of transit peptide carrying an N-terminal extension which helps in recognition of the protein in chloroplast/plastid. This NiR apoprotein is encoded by nuclear DNA.

Glutamine Synthetase (EC 6.3.1.2)

Glutamine synthetase (GS) catalyzes the critical inclusion of inorganic ammonium into glutamine. GS catalyzes the ATP-dependent condensation of NH_{4+} with glutamate to yield glutamine. The native GS protein weighs 350 kDa and is composed of 8 almost identical subunits (Sechley et al. 1992; Nogueira et al. 2005). In leaves, it is present in both chloroplasts (GS2) and cytoplasm (GS1) (Scarpeci et al. 2007). The chloroplast form appears to play role in the assimilation of photorespiratory NH_{4+} (Freeman et al. 1990). The root enzyme too, is found in both cytoplasm and plastids. In pea, the dominant form is plastidic, whereas in maize it is cystosolic (Sakakibara et al. 1992). Studies carried out to encode the genes for GS has helped to elucidate the function of each isoform are involved. Chloroplastic GS (GS2) is believed to be having function in the re-assimilation of photorespiratory conditions (Freeman et al. 1990). The gene for the cystosolic GS (GS3A in pea) is found to be active in the phloem of the transgenic tobacco and alfalfa, indicating that it functions primarily to produce glutamine for intercellular transfer (Fei et al. 2003). In rice plants, cytosolic GS has been reported to be present in vascular bundle. In leaf tissue, it is active in exporting nitrogen to 'sink' tissues (Gallais et al. 2006).

Glutamate Synthase (EC 1.4.7.1 and 1.4.1.14)

Glutamate synthase (glutamine-2-oxoglutarate aminotransferase, GOGAT) is involved in the reductive relocation of the GS to 2-oxoglutarate to create two molecules of glutamate. One of the glutamate molecules can then be cycled back as the substrate for the GS reaction. This is GS-GOGAT cycle was defined by Lea and Miflin (1974). Based on the nature of electron donor, two forms of GOGAT exists, the ferredoxin-GOGAT and the NAD(P)H-GOGAT (Suzuki and Knaff 2005). In rice leaves Fd-GOGAT is known to be present in mesophyll cells, consistent with a job in photorespiratory nitrogen metabolism (Hayakawa et al. 1994). The NAD(P)H-GOGAT occurs in vascular bundles of developing leaf blades, indicating a role in the synthesis of glutamate from glutamine that is imported to the vascular bundle from roots and senescing tissues (Tabuchi et al. 2007).

Glutamate Dehydrogenase (EC 1.4.1.2)

Glutamate dehydrogenase (GDH) is capable either of synthesizing or de-aminating glutamate (Pahlich 1996). One isoform of the enzyme is localized in the mitochondria. It uses NADH as the electron donor (Sechley et al. 1992). Another form that has a specific requirement for NAD(P)H is present in the chloroplasts of photosynthetic tissues. Levels of the NADH form of GDH increase with senescence or after adding NH_4+ to the medium (Pahlich 1996). These factors also lead to higher levels of GDH protein in maize root tissues (Oaks 1994). A primary role of GDH is known to replenish TCA cycle intermediates through their oxidation to 2-oxoglutarate. However, in vivo, aspartate amino transferase and glutamate decarboxylase also maintain the supply of carbon to the citric acid cycle (Kisaka et al. 2007). Glutamate is deaminated to 2-oxoglutarate in isolated mitochondria; however in the presence of amino-oxy acetate, glutamate no longer contributes to mitochondrial respiration (Sechley et al. 1992). This observation indicates that GDH does not oxidize glutamate. Thus, the correct in vivo role of GDH in nitrogen metabolism of higher plants remains to be defined (Pahlich 1996).

Acquisition of N by Roots and Its Regulation

The nitrate uptake is the foremost step that adds to N use of any plant. The identification of genes and the proteins responsible for NO_3- transport and distribution is required for enhanced understanding of the mechanisms that takes place within the plant. Nitrate transport is a proton-symport type of transporter system (Crawford 1995). Nitrate uptake and its release into cells are mediated by nitrate transporter(s) located into the plasma membrane of the root. Three transporters have been identified by kinetic measurements in plant roots (Remans et al. 2006). These are constitutive High Affinity Nitrate Transporter (cHATS), inducible High Affinity Nitrate Transporter (iHATS) and Low Affinity Nitrate Transporter (LATS) (Okamoto et al. 2006; Chandna et al. 2011). Two of these display saturable kinetics; a low capacity constitutive system and a high capacity inducible system (Okamoto et al. 2006). In addition, a non-saturating low affinity, high capacity system becomes apparent only at higher external NO_3- concentration (Kronzucker et al. 1995; Cerezo et al. 2000, 2001). The high affinity transport system (HATS) works at low concentrations (1 μM–1 mM). HATS transporters are constitutively expressed (cHATS) as well as nitrate-inducible (iHATS), and are subjected to negative feedback regulation by the products of nitrate assimilation. Both types of HATS happen to active during low N (<1 mM) concentration in the medium, they show up-regulation on the availability of nitrate. Thus, constitutive HATS seems to offer a high affinity and low capacity passage for nitrate entry in un-induced plants, a 3-fold increase in their expression is observed on introduction of nitrate (Crawford and Glass 1998). Inducible HATS are known to be induced on presence of nitrate or nitrite (Orsel et al. 2006). The low affinity transport system (LATS) is known to work under high external nitrate concentrations i.e: >1 mM (Glass 2003). NO_3- uptake appears to increase linearly with increasing NO_3- concentration with no indication of saturation even at 100 mM (Omata et al. 1989). The linear concentration dependence of the LATS has been observed in a wide variety of organisms. LATS for NO_3- in barley are referred to as constitutive (Kronzucker et al. 1995). LATS which are constitutive, perhaps also have a signalling function to play in induction of HATS and other nitrate assimilatory genes, which might plays a nutritional role when above a certain threshold.

C and N Interactions in Plants

Carbon (C) and Nitrogen (N) both are important component that play crucial role in carrying out basic cellular activities of plants. C and its compounds is part of various carbohydrates like sucrose and glucose. Both the C-skeletons and energy are provided by carbohydrates during amino acid biosynthesis. N nutrients from the part of inorganic compounds such as ammonium and nitrate, also participates in building organic compounds like amino acids. Research has disclosed the tight coordination between cellular C and N metabolism, suggesting their importance for plant growth (Xu et al. 2012). CO_2 is assimilated through photosynthesis that results in formation of sucrose and glucose via glycolysis and tricarboxylic acid cycle to 2-oxoglutarate (2OG) or α-ketoglutarate. Nitrate reductase reduces nitrate (NO_3-) to nitrite (NO_2-). NO_2 by nitrite reductase finally converting it to ammonium (NH_{4^+}). 2-oxoglutarate provide C skeleton for the synthesis of glutamate (Glu) by incorporating NH_{4^+} formed by photorespiration. NH_{4^+} that is gained from the assimilation of N is induced in Glu, that results in the formation of glutamine (Gln). Glu and GlncontributeNH$_{4^+}$ that is finally used for the synthesis of all other amino acids, that also includes aspartate (Asp) or asparagine (Asn), which in-turn serves as NH_4+ donor. Proteins particularly enzymes are essential for all cellular functions, like metabolic reactions, and are involved in C and N metabolism. Therefore, it is necessary to maintain an appropriate proportion of C and N nutrients (Zheng 2009).

Concepts of Nitrogen Use Efficiency

The term NUE has two basic components: (1) Nitrogen uptake, recovery or acquisition efficiency (2) Nitrogen use, physiological N use, or internal N use efficiency The terms NUE has been used as a ratio that considers an output (i.e., grain yield, total plant dry matter yield, N accumulation in grain, or N accumulation in total plant dry matter) as a numerator and input (i.e., total N supply, soil N supply or fertilizer N supply) as a denominator. NUE is based on different parameters of efficiency, including N uptake, N utilization efficiency, and N-use efficiency, it is expressed as a ratio of output (biomass produced) and input (total N supplied) (Xu et al. 2012). Agronomic, Recovery and Physiological efficiency ratios have been widely used to quantify NUE. Agronomic efficiency of nitrogen (AEN) or partial factor productivity (PFPN), relates integrative index of total economic outputs relative to the use of all sources of N. Nitrogen recovery efficiency (REN) measures the efficiency of the plant to assimilate N provided. Physiological N use efficiency (PEN) defines the rate at which plant uses N from available N to produce grain (Table 11.1). The average AEN has a narrow range of 16–22 kg grain increase per kg N applied, studies have shown it to be smallest in maize and largest in rice. Whereas PFPN has large differences as observed by Ladha 2005 in Maize and rice having similar PFPN values of 65–70, whereas wheat had 44 (Ladha 2005). These large differences in

Table 11.1 The key parameters involved in uptake and utilization efficiencies of nitrogen in plants

S.N.	Components of NUE	Formula
1	NUE	NUE = *Sw*/*N*
2	Usage index	UI = Sw*(*Sw*/*N*)
3	Utilization efficiency	UtE = *Gw*/*N*t
4	Agronomic efficiecy	AE = (*G* − *G*)/*N*f
5	Physiological efficiency uptake	PE = (*G*wf–*G*wc)/*N*f − *N*c uptake
6	Uptake efficiency	UpE = *N*t/*N*s
7	Apparent plant recovery	AR = (*N*t uptake − *N*c N nitrogen uptake)/*N*t*100

Sw shoot weight, *N* total nitrogen content of shoots, *Gw* grain weight, *Ns* nitrogen supplied in gram per plant, *Nt* total nitrogen in plant, *Gwf* grain weight with fertilizer, *Gwc* grain weight without fertilizer (control), *Nf* nitrogen fertilizer applied, *Nf uptake* plant nitrogen with fertilizer, *Nc uptake* plant nitrogen unfertilized control, *PE* physiological N-use efficiency, *NUE* N-use efficiency

PFPN, indicates that maize and rice are able to produce large economic outputs with respect to applied N fertilizer. These differences may be due to differences in (1) internal N requirement for plant growth, (2) ability of the plant to translocation and distribution of N, (3) flag leaf N import/export and leaf senescence pattern and (4) plant's efficiency in converting CO_2 to carbohydrate (Ladha et al. 1998). Proper N-application rates and timing of application are very important to meet plant N demand and improve NUE. Studies done by Abdin et al. (2005) states that timing of N fertilizer applications does show noticeable results in plant growth and N uptake with respect to its application. In addition, the application of the fertilizer at different growth stages of plants determines NUE, which also showed genotypic variation (Hirel et al. 2007).The amount of N that is finally available to the plant can be improved by using various simple techniques like sustained-release fertilizers, split applications and other nutrient and crop management strategies (Abdin et al. 2005). Nitrogen applications in split doses have shown to enhance the yield, NUE, and N uptake efficiency in hard red winter wheat under temperate conditions when compared with fall N applications (Sowers et al. 1994). Nitrogen use efficiency (NUE) in the perspective of photosynthesis is called as photosynthetic nitrogen use efficiency (PNUE), which is measured as the rate of carbon assimilation per unit leaf nitrogen (Kumar et al. 2002).

Analysis of Variation Nitrogen Use Efficiency

Genotypic differences in the NR levels have been studied and reported by Abdin et al. (1992) in corn, wheat, barley and sorghum. Hakeem et al. (2011, 2012) also observed the genotypic differences in the NR levels in rice. Further, Bhatt et al. (1979) showed in sorghum, a decrease in the height of the plant with the enhancement of NR activity while such relationship with tall and dwarf cultivars of wheat were not observed (Abdin and Abrol 1997). Abdin and Abrol 1997 revealed in Wheat genotypes more than two-fold variation in NR activity was observed, that might be due to genetic levels of NR enzyme which is heritable. These genetic differences in the NR activity are also reflected in N harvest and it may be associated with improvement in growth and yield in some genotypes. In some of the high NR genotypes, the grain N concentration was significantly higher. It was observed that high NR (HNR) genotypes maintained higher levels of NR activity even under low N levels (Abdin et al. 1992, 1996) this may be because of high levels of NADH that might enhance NR activity in high NR genotypes (Bauwe and Kolukisaoglu 2003). The activity was especially maintained at the later stages of growth i.e. at the time of flag leaf emergence and anthesis (Jain and Abrol 2005). Studies also indicated that the activity of the NR was regulated at the level of gene expression (Jain and Abrol 2005; Skiba et al. 2011). Recent studies of the genotypes that differed in the levels of their NR activity have revealed that not only the single enzyme NR but the whole N metabolism pathway operates at the elevated level viz. all the enzymes of the pathway nitrite reductase, glutamine synthetase and glutamate synthase function at significantly higher levels in the high NR genotypes as compared with the LNR genotypes, leading to higher accumulation of grain N (Xu et al. 2012). The selection of genotypes with a more efficient mechanism of N uptake and metabolism is a strategy aimed at increasing N utilization efficiency of the maize crop. Several trials for efficient use of N under conditions of low N availability have been carried out with maize (Machado et al. 1992). In order to characterize and select genotypes for efficient use of N, several authors have used physiological and biochemical parameters, such as high nitrate in leaves (Mollaretti et al. 1987), enlarged nitrate reductase activity (Feil et al. 1993), glutamine synthetase activities (Machado and Magalhães 1995), or increased enlistment of N from leaves and stems to the kernels (Machado et al. 1992). Genetic diversity has several ‘indicators’, which are measured using various tools such as Mendelian genetic analysis that are employed to assess disparity in single known gene (qualitative traits), such as resistance to disease (Smale and McBride 1996) or multivariate traits/quantitative traits. Also, pair-wise coefficients of parentage are calculated from pedigree information that serves as genetic diversity indicators of (Cox et al. 1996). Nitrogen use efficiency is a complex quantitative trait which is governed by many genes depending on the number of internal and external factors like nutritional and environmental that limits the nitrogen availability.Quantitative genetic studies are associated with molecular markers provides insight to the identification of Quantitative Trait Loci (QTL) that is known to be part of the genetic variation of a complex character such as NUE (Harrison et al. 2004; Hirel et al. 2007) and gives a new turning point in identification of agronomic traits.

Crop Management Practices and Source-sink Relationships for Improving NUE

The strategies developed to work in favour of nitrogen utilization efficiency included use of variety of fertilizers and their manner of application also avoiding runoff and limiting the fertilizer loss from soil. Use of slow-release fertilizers and of organic manures also minimizes N fertilizer use and their loss. The legumes cropping systems have added advantage of correcting the imbalanced use and nutrient management (Wang et al. 2012a). Strategies that increase fertilizer N use by crop-can also be the part of focus by increasing the fertilizer N use during the growing season since that will decrease the N loss thus, higher NUE (Balasubramanian et al. 2004). Other management practices like soybean-corn rotations, forage-only production systems, conservation tillage systems have low N losses and improved NUE (Wen-Yuan et al. 1996).When differences were studied by Moll et al. 1982 among corn hybrids after application of N before anthesis, under low N levels, observed improved NUE. Machado et al (1992) observed that NUE status is parallel that off water use efficiency (WUE) in corn. Another way is to distribute more of N resources to the organ of interest, such as grains. Since, the efficiency of protein synthesis is known to be dependent on the light and dark regulation of aspargine synthase (AS) of leaf that leads to the elevationsasparagine levels that is an important parameter used to screen for high grain-protein cultivars of maize and rye (Dembinski et al. 1995). Also, by controlling the expression of the ASN 1 gene that control the ASN levels might lead to manipulation of the relationship between Asn and seed N status and enable another way to enhance nutritional quality which needs to be tested.

Biotechnological Inventions for Improving Nitrogen Use Efficiency

NUE is governed by multigenes that involve huge number of genes and expand beyond the key steps of nitrate metabolism and incorporation. The efforts of creating transgenic that can targets those genes which participating in N uptake, transport, assimilation, and carbon metabolism. Manipulation of signalling and metabolic pathway regulatory elements is emerging as an important target for biotechnological advancements. Different transgenic experiments that were attempted in a hope to improve NUE are summarized below and some of the significant work is listed in Table 11.2.

Nitrate reductase enzyme considered being the rate-limiting step of nitrate assimilation, but its genetic manipulation in *Nicotiana* spp. indicated its importance in other steps as well (Ali et al. 2007). The constitutive NR expression in tobacco showed 2-fold increase in its activity while a 20 % decrease in foliar nitrate content was observed. Also an increase in total amino acids contents, but simultaneously

Table 11.2 List of transgenic effort towards improving N-use efficiency (NUE)

Gene	Target plant	Phenotype observed
Nrt1.1—high affinity nitrate transporter	*Arabidopsis*	Increase in constitutive nitrate uptake but not induced
NR—nitrate reductase	*N. tabacum*	Three to four fold drop in NR protein and activity, no change in NR transcript
NiR—Nitrite reductase	*N. plumbaginifolia, Arabidopsis*	NiR activity, no phenotypic difference
GS2—chloroplastic glutamine synthetase	*N. tabacum*	Improved photorespiration capacity, and increased resistance to photo-oxidation
Fd-GOGAT—Fd dependent glutamate synthase	*N. tabacum*	Diurnal changes in NH3 assimilation
GS1—cytosolic glutamine synthetase	*N. tabacum, P. sativum*	Enhanced capacity to accumulate nitrogen and enhanced growth under N starvation higher biomass and leaf proreins
NADH-GOGAT—NADH dependent glutamate synthase	*O. sativa, N. tabacum*	Enhanced grain filling, increased grain weight and higher total C and N content, increased dry wt
GDH—glutamate dehydrogenase	*N. tabacum*	Increased biomass and dry weight
Dof1—transcription factor	*Arabidopsis*	Enhanced growth rate under N-limited conditions, increase in amino acid content
ANR1—MADS transcription factor	*Arabidopsis*	Lateral root induction and elongation

no changes in total N, starch and productivity parameters were observed (Quillere et al. 1994). The transgenics with NR double mutant Nia30 were not able to show any detectable NR activity. While the when plant was transformed with the Nia2-cDNA, a decreased in NR activity with enhanced levels of nitrate accumulation was observed (Hansch et al. 2001). Transformed *Nicotiana plumbaginifolia* plants that were constitutively expressing nitrate reductase (NR) showed a momentarily delayed in drought-induced loss of NR activity, hence permitting speedy recovery of N assimilation. Since NR enzyme is post-translationally controlled by phosphorylation and also with binding of 14-3-3 proteins, several attempts to reduce the inhibitory effect on NR regulation have been made. 56 amino acids were deleted in the amino terminal domain of NR that was known to impair this type of regulation in *Nicotiana plumbaginifolia* (Provan et al. 2000). Over-expression of NR genes from various plants have been worked on since past ten years (Lea et al. 2006), but without any important outcome for the improvement in NUE.

In an effort for improving NUE, over expression of NiR genes were studied in *Arabidopsis* and tobacco that though increased the NiR transcript levels but showed decrease in enzyme activity levels. This may be because of post-translational modifications (Shigeto et al. 2006). Till now there is no confirmations fromNiR overexpression in terms of improvement in NUE.

Hirel et al. (2005) signified glutamine synthetase (GS) involvement in kernel production of maize, through QTLs for studying the GS activity in leaf showed that the GS activity to coincide with QTLs for yield. One QTL was concurrent for thousand kernels weight with a GS (Gln1-4) locus, and two QTLs conceded for GS (Gln1-3) locus of thousand kernel weight and yield. This shows the positive association between kernel yield and GS activity.In another two experiments by Li et al. 1993 and Martin et al. 2006 identified the two cytosolic GS isoenzymes (GS1) in maize, and their molecular and physiological properties were examined using knockout mutation on kernel yield, thereby examining the plants grown under N deficient conditions.Martin et al. 2006 observed the over expressing transgenic lines of Gln1-3 in leaves showed rise in kernel number, proving that the GS1-3 isoenzyme plays an important role in regulating kernel yield under optimal N-fertilization conditions. Yanagisawa et al. 2004; Coque and Gallais 2006; Hirel et al. 2007 also observed the GS mutants and the GS-overexpressing lines that were grown under N-limiting conditions and observed the reduction inkernel number when compared to wild type. Huang et al. (2005) carried out experiments in wheat by adding an extra GS gene and observed no overall increase in the amount of GS. Many other studies in various crop plants like tobacco and rice employed the role of GS and enhanced N-assimilation efficiency (Oliveria et al. 2002; Man 2005; Sun et al. 2005; Cai et al. 2009)

Transgenic over-expression and antisense technologies in alfalfa and rice plants were engaged to alter the expression of NADH-GOGAT (Yamaya et al. 2002). The studies highlighting the antisense RNA for NADH-GOGAT carried out in transgenic rice plants throws some light towards the possible improvement of the transport of N via phloem in senescing leaves. Andrews et al. 2004, studied the expression of NADH-GOGAT in initial growing leaf blades and spikelets, also showed the of glutamine transported from senescing organs. Tabuchi et al. 2007 showed changes in various nitrogenous metabolites and decreased leaf protein, rubisco activity and nitrate contents in Barley mutants having reduced Fd-GOGAT. Genes does emerge to be good candidates that can be employed for improving NUE, but the use depends on crop and cropping conditions (Shrawat and Good 2008).

Improving NUE Through Manipulation of Signalling Targets

The failure to improve NUE through over expression of candidate genes in transgenic plants involved in nitrate and ammonia assimilation concluded that that metabolic flux of these pathways are being controlled by regulatory switches outside these pathways. Yet, the exact mechanism involved in nitrate signalling still needs to be understood. Studies have revealed that nitrate signalling is affected by light, though there are researches stating the involvement of 14-3-3 proteins also.Ca_{2+} and protein kinases/ phosphatise are also known to be associated with nitrate signalling. They have known to be involved in mediating the expression of NR, NiR and GS2 m RNAs through nitrate signal (Wang et al. 2012b). Krapp et al. (2002) highlighted the role played by Ca_{2+} dependent kinases in implicating nitrate and the other involved signals. The role of light as an additional signal in regulation of NR gene expression has been studied in number of previous studies. Lillo and Appenroth 2001 expressed the role of light in signalling and N-use efficiency, by revealing the signals being transmitted through photosynthesis and sugars.Castaings et al. 2011 found ANR1 in *Arabidopsis thaliana*, an N-use efficiency transcription factor homologous to the MADS box family and also known to play role in signalling. He showed the nitrate-dependent stimulation leading to lateral-root proliferation in transgenic plants. However, it is reported that ANR1 does not bring about transcription of all the known nitrate responses even in the root.Transgenic Arabidopsis lines over expressing, a maize protein Dofl was created by Yanagisawa et al. (2004). Dofl belonged to Dof family of plant-specific transcription factors that are known to stimulate C-metabolizing genes. This research revealed the fact that Dofl transcription factor can carry coordinated expression of nitrate-responsive genes involved in N and C metabolites. This approach could be one of the new targets for future metabolic-engineering efforts (Lochab et al. 2007).

Understanding NUE Through Proteomics

Uptake of nutrients from the surroundings in order to maintain energy, metabolism and growth is the main concern all living organisms. Organism therefore, enfolds and evolves numerous programs so that they are able to adapt to changing environment. Such processes are involved in immediate responses that include changes in metabolites, activation or inhibition of enzymes, slower processes are also involved like changes in the levels of macromolecules. The online availability of genome sequences of model plants has assisted technologies that permit inclusive analysis of global mRNA profiles, expanding the horizon to screen the transcriptional programming within cells in response to change in their environment (Daran-Lapujade et al. 2004). The use of methods for protein identification has brought about advancement in studying descriptive analysis of protein patterns. Two-dimensional gel electrophoresis (2-DE), has been a turnover in this field that brought about transformation. Thus, combining metablomics, transcriptomics and proteomics techniques, together forms a powerful tool for functional genomics analysis, these days effectively used in plant studies (Kusano et al. 2011; Amiour et al. 2012). Proteomics has proved itself as a vital tool for analysing the differences occurring in the protein profile caused due to environmental conditions, gene mutations, introduction or silencing of genes, fairly being fast, sensitive and productive. Proteomics science has become an important source for generating information on physiological, biochemical, genetic and architectural aspects. This approach has gained recognition in revealing/characterising individuals or mutants or lines, estimation of genetic variability, establishment of genetic distances to be used in phylogentic studies (Thiellement et al. 1999). It has turned out to be most promising technique that is able to characterize proteins showing differential expression or post-transcriptionally modified during a complex developmental process like senescence. Application of proteomics has brought about a great deal of improvement in agricultural production (Xu et al. 2006). It has been revealed by Salon et al. 2001, using proteomics that during seed filling, supply of N from the mother plant helps in assimilation of proteins in the seeds. In legumes, this N is accumulated either from exogenous nitrogen supplied in soil/atmosphere assimilated by the symbiotic fixation or from the nitrogen that is translocated from vegetative parts. In monocarpic species, nitrogen mobilization such as in pea, for seed filling is tightly linked to the senescence of vegetative parts, which is brings about decrease in protein and chlorophyll content, followed by leaf yellowing.Nitrogen availability and type of nitrogen source also initiate a complex and still not fully understood metabolic rearrangement (Wek et al. 2004; Wang et al. 2012b). Kolkman et al. 2006, showed that limitations and availability of N results in the induced proteins belonging to the category 'metabolism' reflecting a significant metabolic rearrangement in yeast enabling it to adapt to the nutrient availability. Lin et al. 2005 in an effort to understand N metabolism and its regulation studied the response of nitrogen limitation in cultured Monascus cells and identified proteins playing role during the nitrogen-limited media and C/N ratio. These researches demonstrates that proteomics provide means of exploring biological processes by methodical examination of a large number of expressed proteins that bridges sequence information and functional genomics.

Conclusions and Future Perspectives

The developed countries have contributed enormously in research in crop improvements, development of new fertilizers and adapted better management practices. This have attributed to improved NUE that greatly exceeds from that in developing countries. Promotion of improved N management practices and technologies might reduce N losses in the environment. However, efforts to increase NUE at farm level needs a combination of improved technologies/knowledge and carefully crafted local policies that will help in sustaining yield increases. In this Omics era, the trend is now to search the mechanism/s in detail to understand the nutrient use efficiencies by the crop plants. Though some basic steps in this path has been taken, there is a lot of scope to explore the hidden mechanism fully by omics tools.

References

Abdin MZ, Abrol YP (1997) Effect of split nitrogen application on grain nitrogen and nitrogen harvest of wheat (*Triticum aestivum* L.) genotypes. In: T Ando, H Matsumoto, S Moni and S Seksya (eds) for sustainable food production and environment. Kluwer Academic Publishers, Dordecht, pp 635–638

Abdin MZ, Kumar PA, Abrol YP (1992) Biochemical basis of variability in nitrate reductase activity in wheat (*Triticum aestivum* L.). Plant Cell Physiol 33:951–956

Abdin MA, Dwivedi RK, Abrol YP (2005) Nitrogen in agriculture. In: Singh RP, Shankar N and Jaiwal PK (eds) Nitrogen nutrition and plant productivity. Studium Press LLC, Houston, pp 1–41

Abdin MZ, Bansal KC, Kumar PA, Abrol YP (1996) Effect of split nitrogen application on growth and yield of wheat (*Triticum aestivum* L.) genotypes with different N assimilation potential. J Agron Crop Sci 176:88–92

Ali A, Sivakami S, Raghuram N (2007) Effect of nitrate, nitrite, ammonium, glutamate, glutamine and 2-oxoglutarate on the RNA levels and enzyme activities of nitrate reductase and nitrite reductase in rice. Physiol Mol Biol Plants 13:17–25

Amiour N, Imbaud S, Clément G, Agier N, Zivy M, Valot B, Balliau T, Armengaud P, Quilleré I, Cañas R, Tercet-Laforgue T and Hirel B (2012) The use of metabolomics integrated with transcriptomic and proteomic studies for identifying key steps involved in the control of nitrogen metabolism in crops such as maize. J Exp Bot 63:5017–5033

Andrews M, Lea PJ, Raven JA, Lindsey K (2004) Can genetic manipulation of plant nitrogen assimilation enzymes result in increased crop yield and greater N-use efficiency? An assessment. Ann App Biol 145:25–40

Balasubramanian V, Alves B, Aulakh MS, Bekunda M, Cai ZC, Drinkwater L, Mugendi D, Van Kessel C, Oenama O (2004) Crop, environmental and management factors affecting N use efficiency. In: Mosier AR, Syers JK and Freney JR (eds) Agricultural and the N cycle: assessing the impacts of fertilizer use on food production and the environment. SCOPE 65, Paris, pp 19–33

Bauwe H, ner Kolukisaoglu UÈ (2003) Genetic manipulation of glycine decarboxylation. J Exp Bot 54:1523–1535

Bhatt TN, Sasisekharan V, Vijayan M (1979) An analysis of side-chain conformation in proteins. Int J Peptide Protein Res 13:170

Bowsher CG, Hucklesby DP, Emes MJ (1993) Induction of ferrodoxin-$NADP^+$ oxidoreductase and ferrodoxin synthesis in pea root plastids during nitrate assimilation. Plant J 3:463–467

Caboche M, Rouzé P (1990) Nitrate reductase: a target for molecular and cellular studies in higher plants. Trends Genet 6:187–192

Cai H, Zhou Y, Xiao J, Li X, Zhang Q, Lian X (2009) Overexpressed glutamine synthetase gene modifies nitrogen metabolism and abiotic stress responses in rice. Plant Cell Rep 28:527–537

Castaings L, Marchive C, Meyer C, Krapp A (2011) Nitrogen signalling in Arabidopsis: how to obtain insights into a complex signalling network. J Exp Bot 62:1391–1397

Cerezo M, Flors V, Legaz F, Garcia-Agustin P (2000) Characterization of the low affinity transport system for NO_3– uptake by Citrus roots. Plant Sci 160:95–104.

Cerezo M, Tillard P, Filleur S, Munos S, Daniel-Vedele F, Gojon A (2001) Major alterations of the regulation of root NO_3– uptake are associated with the mutation of NRT2.1 and NRT2.2 genes in Arabidopsis. Plant Physiol 127:262–271

Chandna R, Kaur G, Iqbal M, Khan I, Ahmad A (2011) Differential response of wheat genotypes to applied nitrogen: biochemical and molecular analysis. Arch Agro Soil Sci. doi:10.1080/03650340.2011.555761

Coque M, Gallais A (2006) Genomic regions involved in response to grain yield selection at high and low nitrogen fertilization in maize. Theo App Genet 112:1205–1220

Cox T, Murphy JP, Rodgers DM (1996) Changes in genetic diversity in the red winter wheat regions of the United States. Proc Natl Acad Sci 83:5583–5586
Crawford NM (1995) Nitrate: nutrient and signal for plant growth. Plant Cell 7:859–868
Crawford NM, Glass ADM (1998) Molecular and physiological aspects of nitrate uptake in plants. Trends in Plant Sci 3:389–395
Daran-Lapujade P, Jansen ML, Daran JM, van Gulik W, de Winde JH, Pronk JT (2004) Role of transcriptional regulation in controlling fluxes in central carbon metabolism of Saccharomyces cerevisiae. A chemostat culture study. J Biol Chem 279:9125–9138
Dembinski E, Bany S, Raczynska-Bojanowska K (1995) Asparagine and glutamine in the leaves of high and low protein maize. Acta Physiol Plant 17:361–365
Directorate of Economics and Statistics (2008) The statistical hand book. New Delhi
FAO (2008) Yearbook. Food and Agriculture Organization, India
Fedorova E, Greenwood JS, Oaks A (1994) In situ localization of nitrate reductase in maize roots. Planta 194:279–286
Fei H, Chaillou S, Hirel B, Mahon JD, Vessey JK (2003) Overexpression of a soybean cytosolic glutamine synthetase gene linked to organ-specific promoters in pea plants grown in different concentrations of nitrate. Planta 216:467–474
Feil B, Thiraporn R, Stamp P (1993) In vitro nitrate reductase activity of laboratory-grown seedlings as an indirect selection criterion for maize. Crop Sci Madison 33:1280–1286
Freeman J, Marquez AJ, Wallsgrove RM, Sarrelainen R, Forde BG (1990) Molecular analysis of barley mutants efficient in chloroplast glutamine synthetase. Plant Mol Biol 14:297–311
Gallais A, Coque M, Quilleré I, Prioul JL, Hirel B (2006) Modelling post-silking N-fluxes in maize using 15 N-labelling field experiments. New Phytol 172:696–707
Glass ADM (2003) Nitrogen use efficiency of crop plants: physiological constraints upon nitrogen absorption. Crit Rev Plant Sci 22:453–470
Hakeem KR, Ahmad A, Iqbal M, Gucel S, Ozturk M (2011) Nitrogen-efficient rice cultivars can reduce nitrate pollution. Env Sci Poll Res 18:1184–1193
Hakeem KR, Chandna R, Ahmad A, Iqbal M (2012). Physiological and molecular analysis of applied nitrogen in rice (*Oryza sativa* L.) genotypes. Rice Sci 19(1)
Hansch R, Fessel DG, Witt C, Hesberg C, Hoffman C, Walch-Liu P, Engels C, Kruse J, Rennenberg H, Kaiser W, Mendel RR (2001) Tobacco plants that lack expression of functional nitrate reductase in roots show changes in growth rates and metabolites accumulation. J Exp Bot 52:1251–1258
Harrison J, Hirel B, Limami A (2004) Variation in nitrate uptake and assimilation between two ecotypes of *Lotus japonicus* L. and their recombinant inbred lines. Physiol Planta 120:124–131
Hayakawa T, Nakamura T, Hattori F, Mae T, Ojima K, Yamaya T (1994) Cellular localization of NADH-dependent glutamate synthase protein in vascular bundles of unexpanded leaf blades and young grains of rice plants. Planta 193:455–460
Hirel B, Le Gouis J, Ney B, Gallais A (2007) The challenge of improving nitrogen use efficiency in crop plants: towards a more central role for genetic variability and quantitative genetics within integrated approaches. J Exp Bot 58:2369–2387
Hirel B, Martin Tercé-Laforgue T, Gonzalez-Moro MB, Estavillo JM (2005) Physiology of maize. I. A comprehensive and integrated view of nitrogen metabolism in a C4 plant. Physiol Planta 124:167–177
Huang QM, Liu WH, Sun H, Deng X, Su J (2005) *Agrobacterium tumefaciens* mediated transgenic wheat plants with glutamine synthetases confer tolerance to herbicide. J Plant Ecol 29:338–344
Jain V, Abrol YP (2005) Regulation of nitrate reductase activity by phytohormones in wheat. J Plant Biol 32:43–48
Kant S, Bi Y M, Rothstein SJ (2011) Understanding plant response to nitrogen limitation for the improvement of crop nitrogen use efficiency. J Exp Bot 62:1499–1509
Kisaka H, Kida, T, Miwa T (2007) Transgenic tomato plants that overexpress a gene for NADH-dependent glutamate dehydrogenase (legdh1). Breed Sci 57:101–106

Kolkman A, Daran-Lapujade P, Fullaondo A, Olsthoorn MMA, Pronk JT, Monique S, Heck AJR (2006) Proteome analysis of yeast response to various nutrient limitations. Mol Syst Biol 2:1–26

Krapp A, Ferrario-Mery S, Touraine B (2002) Nitrogen and signalling. In: Foyer CH and Noctor G (eds) Photosynthetic nitrogen assimilation and associated carbon and respiratory metabolism. Kluwer Academic Publishers, Netherlands, pp 205–225

Kronzucker HJ, Siddiqi MY, Glass ADM (1995) Kinetics of NO_3– influx in spruce. Plant Physiol 109:319–326

Kumar PA, Parry MAJ, Mitchell RAC, Ahmad A, Abrol YP (2002) Photosynthesis and nitrogen use-efficiency. In: Foyer CH and Noctor G (eds) Photosynthetic nitrogen assimilation and associated carbon and respiratory metabolism. Kluwer Academic Publishers, Dordecht, pp 23–34.

Kusano M, Fukushima A, Redestig H, Saito K (2011) Metabolomic approaches toward understanding nitrogen metabolism in plants. J Exp Bot 62:1439−1453

Ladha JK (2005) Improving the recovery efficiency of fertilizer nitrogen in cereals. J India Soc Soil Sci 53:472–483

Ladha JK, Kirk GJD, Bennett J, Peng S, Reddy CK, Reddy PM, Singh U (1998) Opportunities for increased nitrogen use efficiency from improved lowland rice germplasm. Field Crops Res 56:41–71

Lea PJ, Miflin BJ (1974) An alternative route for nitrogen assimilation in plants. Nature 251:680–685

Lea US, Leydecker MT, Quilleré I, Meyer C, Lillo C (2006) Posttranslational regulation of nitrate reductase strongly affects the levels of free amino acids and nitrate, whereas transcriptional regulation has only minor Influence. Plant Physiol 140:1085–1094

Li MG, Villemur R, Hussey PJ, Silflow CD, Gantt JS, Snusta DP (1993) Differential expression of six glutamine synthetase genes in Zea mays. Plant Mol Biol 23:401–407

Lian X, Xing Y, Yan H, Xu C, Li X, Zhang Q (2005) QTLs for low nitrogen tolerance at seedling stage identified using a recombinant inbred line population derived from an elite rice hybrid. Theor Appl Genet 112:85–96

Lillo C, Ruoff P (1992) Hysteretic behavior of nitrate reductase: evidence of an allosteric site for reduced pyridine nucleotides. J Biol Chem 267:13456–13459

Lillo C, Appenroth KJ (2001) Light regulation of nitrate reductase in higher plants: which photoreceptors are involved? Plant Biol 3:455–465

Lin B, White JT, Lu W, Xie T, Utleg AG, Yan X, Yi EC, Shannon P, Khrebtukova I, Lange PH, Goodlett DR, Zhou D, Vasicek TJ, Hood L (2005) Evidence for the presence of disease-perturbed networks in prostate cancer cells by genomic and proteomic analyses: a systems approach to disease. Cancer Res 65:3081–91.

Lochab S, Pathak RR, Raghuram N (2007) Molecular approaches for enhancement of nitrogen use efficiency in plants. In: Abrol YP, Raghuram N and Sachdev MS (eds) Agricultural nitrogen use and its environmental implications. IK International, New Delhi, pp 327–350

Machodo AT, Magalhães JR (1995) Melhoramento de milho para uso eficiente de N sob condições de estresse. In: SIMPÓSIO INTERNACIONAL SOBRE ESTRESSE AMBIENTAL: O MILHO EM PERSPECTIVA, 1992, Belo Horizonte. Anaislysis Sete Lagoas, Embrapa-CNPMS, pp 321–342

Machado AT, Magalhães JR, Magnavaca R, Silva MR (1992) Determinação da atividade de enzimas envolvidas no metabolismo do N em diferentes genótipos de milho. Revista Brasileira de Fisiologia Vegetal, Brasília, v. 4, n. 1, pp 45–47

Man HM, Boriel R, El-Khatib R, Kirby EG (2005) Characterization of transgenic poplar with ectopic expression of pine cytosolic glutamine synthetase under conditions of varying nitrogen availability. New Phytol 167:31–41

Martin A, Lee J, Kichey T, Gerentes D, et al (2006) Two cytosolic glutamine synthetase isoforms of maize (*Zea mays* L.) are specifically involved in the control of grain production. The Plant Cell 18:3252–3274

Mokhele B, Zhan X, Yang G, Zhang X (2012) Review: nitrogen assimilation in crop plants and its affecting factors. Can J Plant Sci 92:399–405

Moll RH, Kamprath EJ, Jackson WA (1982) Analysis and interpretation of factors which contribute to efficiency of nitrogen utilization. Agro J 74:562–564

Mollaretti G, Bosio M, Gentinetta E, Motto M (1987) Genotypic variability for N related traits in maize. Identification of inbread lines with high or low levels on NO_{3}– N in the stalks. Maydica 32:309–323

Nogueira EDM, Olivares FL, Japiassu JC, Vilar CV, Vinagre F, Baldini JI, Hemerly AS (2005) Characterisation of glutamine synthetase genes in sugarcane genotypes with different rates of biological nitrogen fixation. Plant Sci 169:819–832

Oaks A (1994) Primary nitrogen assimilation in higher plants and its regulation. Can J Bot 72:739–750

Okamoto M, Kumar A, Li W, Wang Y, Siddiqi MY, Crawford NM, Glass ADM (2006) High-affinity nitrate transport in roots of Arabidopsis depends on expression of the NAR2-like gene AtNRT3.1. Plant Physiol 140:1036–1046

Oliveria IC, Brears T, Knight TJ, Clark A, Coruzzi GM (2002) Overexpression of cytosolic glutamate synthethase in Arabidopsis thaliana. Plant Physiol 121:301–309

Omata T, Ohmori M, Arai N, Ogawa T (1989) Genetically engineered mutant of the cyanobacterium Synechococcus PCC 7942 defective in nitrate transport. Proc Nat Acad Sci USA 86:6612–6616

Orsel M, Chopin F, Leleu O, Smith SJ, Krapp A, Daniel-Vedele F, Miller AJ (2006) Characterization of a two-component high-affinity nitrate uptake system in Arabidopsis. Physiology and protein-protein interaction. Plant Physiol 142:1304–1317

Pahlich E (1996). Remarks concerning the dispute related to the function of plant glutamate dehydrogenase: commentary. Can J Bot 74:512–515

Provan F, Aksland LM, Meyer C, Lillo C (2000). Deletion of the Nitrate Reductase N-terminal domain still allows binding of 14-3-3 proteins but affects their inhibitory properties. Plant Physiol 123:757–764

Quillere I, Dufosse C, Roux Y, Foyer CH, Caboche M, Morot-Gaudry JF (1994) The effects of deregulation of NR gene expression on growth and nitrogen metabolism of *Nicotiana plumbaginifolia* plants. J Exp Bot 45:1205–1211

Raun WR, Johnson GV (1999) Improving nitrogen use efficiency for cereal production. Agron J 91:357–363.

Remans T, Nacry P, Pervent M, Filleur S, Diatoff E, Mounier E, Tillard P, Forde BG, Gojon A (2006) The Arabidopsis NRT1.1 transporter participates in the signalling pathway triggering root colonization of nitrate rich patches. Proc Natl Acad Sci U S A 103:19206–19211

Sakakibara H, Kawabata S, Takahashi H, HaseT, Sugiyama T (1992) Molecular cloning of the family of glutamine synthetase genes from maize: expression of genes for glutamine synthetase and ferredoxin-dependent glutamate synthase in photosynthetic and non-photosynthetic tissues. Plant Cell Physiol 33:49–58

Salon C, Munier-Jolain NG, Duc G, Voisin AS, Grangirard D, Larmure A, Emery RJN, Ney B (2001) Grain legume seed filling in relation to nitrogen acquisition: a review and prospects with particular reference to pea. Agronomie 21:539–552

Scarpeci TE, Marro ML, Bortolotti S, Boggio SB, Valle EM (2007) Plant nutritional status modulates glutamine synthetase levels in ripe tomatoes (*Solanum lycopersicum* cv. Micro-Tom). J Plant Physiol 164:137–145

Sechley KA, Yamaya T, Oaks A (1992) Compartmentation of nitrogen assimilation in higher plants. Int Rev Cytol 134:85–163.

Seigel LM, Wilkerson JO (1989) In: Wray JL and Kinghorn JR (eds) Molecular and Genetic aspects of nitrate assimilation. Oxford Science Publications, Oxford, pp 263–283

Shigeto J, Yoshihara S, Adam Suaad EH, Sueyoshi K, Sakamoto A, Morikawa H, Takahashi M (2006) Genetic engineering of nitrite reductase gene improves uptake and assimilation of nitrogen dioxide by *Rhaphiolepis umbellate* (Thunb.) Makino. Plant Biotech 23:111–116

Shrawat AK, Good AG (2008) Genetic engineering approaches to improving nitrogen use efficiency. ISB News Report, May 2008

Skiba MW, George TS, Baggs EM, Daniell TJ (2011) Plant influence on nitrification. Biochem Soc Trans 39:275–278

Smale M, McBride T (1996) Understanding global trends in the use of wheat diversity and international flows of wheat genetic resources. Part 1 of CIMMYT 1995/96 world wheat facts and trends: understanding global trends in the use of wheat diversity and international flows of wheat genetic resources. Mexico, D.F.: CIMMYT (Centro Internacional de Mejoramiento de Maiz y Trigo—International Center for the Improvement of Wheat and Maize)

Sowers KE, Pan WL, Miller BC, Smith JL (1994) Nitrogen use efficiency of split nitrogen applications in soft white winter wheat. Agron J 86:942–948

Sun H, Huang QM, Su J (2005) Overexpression of glutamine synthetase confers transgenic rice herbicide resistance. High Technol Lett 11:75–79

Suzuki A, Knaff DB (2005) Glutamate synthase: structural, mechanistic and regulatory properties, and role in the amino acid metabolism. Photosyn Res 83:191–217

Tabuchi M, Abiko T, Yamaya T (2007) Assimilation of ammonium ions and reutilization of nitrogen in rice (*Oryza sativa* L.). J Exp Bot 58:2319–2327

Thiellement H, Bharmann N, Damerval C (1999) Proteomics for genetic and physiological studies in plants. Electropho 20:2013–2026

Wang X, Bian Y, Cheng K, Zou H, Sun SS-M, He J-X (2012a) A comprehensive 1 differential proteomic study of nitrate deprivation in *Arabidopsis* reveals complex regulatory networks of plant nitrogen responses. J Proteome Res 11:2301−2315

Wang YY, Hsu PK, Tsay F-Y (2012b) Uptake, allocation and signalling of nitrate. Trend Plant Sci 17:458–467

Wek RC, Staschke KA, Narasimhan J (2004) Regulation of the yeast general amino acid control pathway in response to nutrient stress. In: Winderickx JG (ed) Nutrient-induced responses in eukaryotic cells. Springer-Verlag, Berlin, pp 171–199

Wen-Yuan H, Shank D, Herwitt TI (1996) On-farm costs of reducing residual Nitrogen on cropland vulnerable to nitrate leaching. Rev Agri Eco 18:325–339

Wray JL (1993) Molecular biology, genetics and regulation of nitrite reductase in higher plants. Plant Physiol 89:607–612

Xu G, Fan X, Miller AJ (2012) Plant nitrogen assimilation and use efficiency. Ann Rev Plant Biol 63:153–182

Xu K, Xu X, Fukao T, Canlas P, Maghirang-Rodriguez R, Heuer S, Ismail AM, Bailey-Serres J, Ronald PC, Mackill DJ (2006).Sub1A is an ethylene-response-factor-like gene that confers submergence tolerance to rice. Nature 442:705–708

Yamaya T, Obara M, Nakajima H, Sasaki S, Hayakawa T, Sato T (2002) Genetic manipulation and quantitative-trait loci mapping for nitrogen recycling in rice. J Exp Bot 53:917–925

Yanagisawa S, Akiyama A, Kisaka H, Uchimiya H, Miwa T (2004) Metabolic engineering with Dof1 transcription factor in plants: improved nitrogen assimilation and growth under low nitrogen conditions. Proc Natl Aca Sci U S A 101:7833–7838

Zheng Z-L (2009) Carbon and nitrogen nutrient balance signalling in plants. Plant Signal Behav 4:584–591

Chapter 12
Arsenic Toxicity and Tolerance Mechanisms in Plants: An Overview

Saiema Rasool, Muneeb U. Rehman, Mohamed Mahgoub Azooz, Muhammad Iqbal, Tariq Omar Siddiqi and Parvaiz Ahmad

Abstract Heavy metal stress is increasing at an alarming rate in agricultural soils through out the world. Heavy metal (Cd, Cu, Zn, Ni, Co, Cr, Pb and As) toxicity have been reported to be responsible for the reduced crop production. Among the heavy metals arsenic (As) is non-essential and toxic to both plants and animals. As can exist in environment in the form of oxidized arsenate (AsV) and reduced arsenite (AsIII). As stress has become a global concern, the uptake of As in the plants through contaminated soil will make its entry into the human food chain. As toxicity can lead to skin, bladder, lung and prostate cancer. Soil contaminated with As is the main source of arsenic in drinking water. Uptake of As by plants is very important in understanding its physiological effects and its metabolism within plants. How plants respond to the arsenic stress in plants is a major concern to biologists. As per the published literature numerous physiological processes are affected by the As

P. Ahmad (✉)
Department of Botany, A. S. College, University of Kashmir, Srinagar 190008, India
e-mail: parvaizbot@yahoo.com

S. Rasool · M. Iqbal · T. O. Siddiqi
Department of Botany, Faculty of Science, Jamia Hamdard, New Delhi 110062, India
e-mail: saimu083@yahoo.co.in

M. Iqbal
e-mail: miqbal@jamiahamdard.ac.in

T. O. Siddiqi
e-mail: tosiddiqui@jamiahamdard.ac.in

M. U. Rehman
Department of Medical Elementology and Toxicology,
Faculty of Science, Jamia Hamdard, New Delhi 110062, India
e-mail: reiazrehman@yahoo.co.in

M. M. Azooz
Department of Biological Sciences, Faculty of Science,
King Faisal University, Ahsaa, Saudi Arabia
e-mail: azzozm@yahoo.com

K. R. Hakeem et al. (eds.), *Crop Improvement,* DOI 10.1007/978-1-4614-7028-1_12,

toxicity. As is also responsible for oxidative stress in plants through the generation of reactive oxygen species (ROS) which attack the biomolecules like, membranes proteins, carbohydrates, nucleic acids etc. At the same time activity of enzymatic and non-enzymatic antioxidants are increased which helped the plant to withstand the toxicity of As. The present chapter throws light on the arsenic toxicity in plants and their tolerance mechanism in plants. The chapter also highlights the generation of reactive oxygen species and antioxidants during As stress.

Introduction

In human history, metal pollutants have been their part. However, due to the onset of industrial revolution, the biosphere has been intensely polluted by toxic metals leading to the major environmental and health problems (Bhattacharya et al. 2012; Lui et al. 2012). Arsenic (As) is of major concern among all the metal pollutants because it is a persistent bioaccumulative carcinogen (Kaur et al. 2011). The anthropogenic activities have increased many folds the level of arsenic as humans constantly endeavored to improve quality of life when compared to the naturally occurring elements occurring in the earth's crust (Moreno-Jimenez et al. 2012). The global input of arsenic to soil by human activities was estimated between 52,000 and 1,12,000 t per year (Chandra and Srivastava 2003). Severe problems like vegetation loss, contamination of ground water, and toxicity of arsenic in animals, plants and have been caused by arsenic contamination (Fowler 1983; Zhao et al 2010). Although several studies have reported the detoxification of arsenic ions through metal binding peptides (Schmoger et al. 2000; Bondada and Ma 2002), due to the lack of literature regarding to the role of ascorbate -glutathione pathway in cellular defense against arsenic in plants.

Heavy metals produce oxidative-stress possibly by free-radical generation and active oxygen species (Hall 2002). The reaction of these oxyradicals with proteins, lipids, pigments and nucleic acids causes lipid peroxidation, membrane damage and inactivation of enzymes, thus affecting the cell viability. Two major roles are played by active oxygen species: (1) exacerbating damage and (2) signaling the activation of defence responses. Recently these two functions have been verified by several abiotic stress responses (Dat et al. 2000). In higher vascular plants, change in cellular metabolism is required to counteract with heavy metal stresses. Various antioxidant enzymes like superoxide dismutase (SOD), catalase (CAT), ascorbate peroxidase (APX) and glutathione reductase (GR) of the ascorbate glutathione pathway is the main mechanism for ROS quenching under heavy metal stress. (Clijsters et al. 1999). Apart from these enzymes, low molecular weight antioxidants metabolites like ascorbate and reduced glutathione (GSH) also play an important role in plants by protecting them from oxidative. Antioxidant responses to arsenic in higher plants have not been studied so far.

Occurrence and Distribution of Arsenic

In relation to the other element, arsenic ranks twentieth in abundance and is natural constituent on earth's crust. In continental crust, the average As content varies between 1 and 2 mg As/kg. It is distributed in a variety of minerals but occurs commonly as arsenides of copper, iron, lead, silver and gold or as sulfides. Realgar (As_4S_4) and orpiment (As_2S_3) are the two common As sulphides where As occurs in oxidized form in the mineral arsenolite (As_2O_3). Loellingite ($FeAs_2$), safforlite (CoAs), niccolite (NiAs), rammelesbergite ($NiAs_2$), arsenopyrite (FeAsS), Coballite (CoAsS), enargite (CU_3), (As_2S_4), gerdsorfite (NiAsS), glucodal [(CO, Fe)AsS] and elemental As are other naturally occurring As-bearing minerals.

Sources

From its origin in the earth's crust, As can enter the environment through natural and anthropogenic processes. Two principal pathways of As emission in the environment, are (a) natural processes and (b) industrial activities. In natural environment, arsenic is released through natural processes such as weathering and volcanic eruptions and as a suspended particulate arsenic may be transported over long distances through air or water. However, the most important source of As emission is industry and accounts for widespread section, we discuss these two principal modes of As emissions and their comparison among these two courses.

Natural Sources

Mean global atmospheric emission of As from natural sources is about 12.2 Gg. These sources include wind bloom dust from weathered continental crust, forest fires, volcanoes, sea spray, hot springs, and geysers. Emissions of As from volcanic eruptions very considerably as high as 8.9 Gg 1 year from Mount Saint Helens in the united states to about 0.04 Gg 1 year from poas in Costa Rica. Arsenic emission through volcanic ca.2.3 Gg/year compared to nearly 0.01 Gg/year as volatile forms.

Anthropogenic Sources

The major producers of As_2O_3 ("white arsenic") are the United States, Sweden, France, the former USSR, Mexico and South West Africa. The As compounds such as monosodium methylarsonate ($NaCH_3HAsO_3$), disodium methyle arsonate

(Na_2CH_3, AsO_3), and diethylarsenic acid [$(CH_3)_2AsO(OH)$] are widely used as agricultural insecticides, larvicides, and herbicides. Sodium arsenite (NaH_2AsO_4) is used for aquatic weed control and for sheep and cattle dips. Arsenic acid (H_3AsO_4) is used to defoliate cotton balls prior to harvesting and as a wood preservative. As_2O_3 is used in the manufacture of pharmaceuticals. Elemental As is mainly used in Pb, Cu, Sb, Sn, Al, and Ga alloys.

Mining, smelting and are beneficiation, pesticides, fertilizers and chemical industries, thermal power plants using coal, wood preservation industries using CCA and incineration of preserved wood wastes contribute to significant influx of As to the environment. Global emission of As in the atmosphere has been estimated to be 0.019 Gg (0.012–0.026 Gg), but in soil and aquatic environment, the estimated figures are 0.082 and 0.042 Gg, respectively. However, there has been a substantial decrease in the atmospheric emission of As in Europe, from Circa 0.005 Gg in 1986 to 0.00031 Gg in 1995.

Mining and Ore Beneficiation

Elevated concentration of As, as well as other metals such as cadmium copper, nickel and zinc are commonly encountered in the acid mine effluents. The principal source of As in mine tailings is the oxidation of arsenopyrite (FeAsS) following the reaction.

$$FeAsS(s) + 13\ Fe^{3+} + 8H_2O = 14\ Fe^{2+} + SO^2{}_4 + 13H^+ + H_3AsO_4$$

Arsenopyrite can be oxidized by both O_2 and Fe, but the rate of oxidation by Fe^{III} is faster than pyrite. The rate of this reaction was reported as 1.7 μ mal/m^2, a reaction faster than a similar oxidation reaction for pyrite. Under extremely acidic environment, with a pH of about 1.5 and an aqueous As concentration at >10 m mol/l, As precipitates as scordite ($FeAsO_4.2H_2O$). Under acidic conditions (pH<3), As^V may substitute SO_4 in the structure of Jarosite [$KFe_3(SO_4)_2(OH)_6$] in different mine wastes. Adsorption of As on $Fe(OH)_3$ surfaces was found to be the principal sink for As in studies of acid mine drainage. However, the adsorption of As by $Fe(OH)_3$ may be only transient as changes in redone conditions (Eh) and pH may result in dissolution of $Fe(OH)_3$ with consequent mobilization of As. Effluents and water in tailings ponds are often treated with lime to increase pH levels to stabilize the dissolved As and other metals as precipitates.

Agriculture

Over hundreds of years, inorganic arsenicals (arsenic trioxide, arsenic-acid, arsenates of calcium, Cu, Pb, and Na and arsenites of Na and K) have been widely used in pigments, pesticides, insecticides, herbicides and fungicides. At present, As is no longer used in agriculture in the west, but persistence of the residue of the inorganic arsenicals in soil is an issue of environmental concerns. Studies by Kenyon et al. 1979 and Aten et al. 1980 have indicated elevated concentrations of As in vegetables grown in soils contaminated by lead arsenate used as an insecticide in apple orchards. The recalcitrant nature of arsenical herbicides has, however, been observed in agricultural soils particularly around old orchards. Biomethylation of As is a mechanism through which a significant quantity of methylarsines may be released into the atmosphere following the application of As compounds to the soil. A relatively faster production of dimethyl and trimethylarsinics has been reported from grasslands treated with methylarsenic compounds while grass treated with sodium arsenite indicated slow release of methylorsene into the atmosphere.

Wood Preservation

The extensive contamination of soils and aquatic environments has been due to the use of CCA and other As-based chemicals in wood preservation industries. The wood preservative chemical like CCA has attained wide scale industrial application due to biocidic characteristics of Cu^{II} and As^{v}. The preservative chemical used for pressure impregnation comprises a water-based mineral of diachronic acid ($H_2Cr_2O_7$) arsenic acid (H_3AsO_4) and Cu^{II} as divalent cation at variable proportions. Cr is used to bind As and Cu into the cellular structure of the wood. Fixation of CCA is dependent on the transformation of Cr^{III} to Cr^{III}, a reaction that is dependent on the temperature and water content of the wood. Cr^{III} forms insoluble complexes with both As and Cu. Further stabilization of these complexes takes place after complete fixation of the As and Cu in the wood tissues and minimizes the risk of leaching of the CCA components from processed wood. Among the active ingredients of CCA wood preservatives. As is most mobile and toxic to a broad range of organisms including human beings.

Chemistry of Arsenic in Soil

The natural content of As in soils varies significantly but is mostly in a range below 10 mg/kg. The background concentration of As in soils is governed by the lithalogy of the parent rocks. As concentration in Swedish tills (<0.06 mm) range between <5 and 175 mg/kg with a medium value of 8 mg 1 kg (Selenics, Personal Communication 2000). Availability and dispersal of As in the soil environment are influenced by several factors. Rainfall, surface runoff, rate of infiltration, and the groundwater level like climatic and geographic characteristics and their fluctuations affect the mobility and distribution of As. The speciation and mobility of As in soils also depends upon the soil physical characteristics like grain size and mineralogy and chemical characteristics such as redox potential (Eh) and pH conditions of the soils.

Physical Characteristics of Arsenic

- Arsenic is tin white tarnishes to dark grey or black in colour.
- It is metallic but tarnishes to dull in luster.
- Its crystals are opaque.
- Crystals are rare pseudocubic rhombohediral and acircular radial aggregates. It is usually found in fine-grained masses with concentric bands or botryoidal crusts. Allied minerals of arsenic are barite, cinnabar and neckline. Others are poisonous.
- Arsenic is found to occur in France, Kangsberg, Norway, Somany and Harz Mountains, Germany, Honshu, Japah, England, Italy, and Santa Cruz, Arizona and New Jersey, USA.
- Best field indicators are tarnish, density, softness, crystal habits, color, garlic, small and association.

Behaviour of Arsenic

Both arsenic and phosphorus have similar chemical properties, therefore they act similarly in soil. Phosphorous and arsenic competes for soil fixation sites and for plant uptake (Adriano 1986). By decreasing soil phosphorus level the phytotoxicity of arsenic increases (Rumberg et al. 1960). Other experiments showed that additional phosphorus may increase arsenic phytotoxicity by releasing more arsenic into solution (Jacobs and Keeney 1970).

Availability of Arsenic

Arsenic availability cannot be determined by the total arsenic concentration in soils (Adriano 1986). Even though a limited quantity of arsenic in soil is readily mobile and the rest is not available to plants because it is associated with iron (Fe) and almunium (Al).

In soil the solubility of reduced form of arsenic (arsenite) is more oxidized form (arsenate). Soluble arsenic concentration is directly proportional to the plant arsenic toxicity, (Kabata-Pendias and Pendias 2001; Sturchio et al. 2011).

Many soil factors like soil pH affect the arsenic availability (Adriano 2001). Therefore, soil pH is important because it plays a vital role in arsenic speciation and leachability. The optimum adsorption for arsenite is approximately at pH 7.0 while as arsenate adsorbs at pH 4.0 (Pierce and Moore 1982). On the whole, low soil pH, clays makes the amorphous silicates and metal oxides protonated and is then able to absorb arsenic anions present in the soil. Arsenic is less mobile at low pH as most arsenic is present as arsenate in soils and at this pH there are high concentrations of arsenic- binding species such as iron and aluminum. With the increase in pH protonated sites allow the arsenic to become more mobile.

Arsenic does not have the capability to form a strong association with calcium (calcite) at higher pH. According to Woolson (1983) under high arsenic concentrations this association may be found where calcium is the secondary preference over aluminum. Acidic conditions i.e. lower pH allows the calcite to be dissolved and the arsenic is released.

The other aspect that affects the arsenic availability is soil texture (Adriano 2001). Soil surface area is affected by soil texture e.g. textured soils (silt and clays) have much more surface area than sandy soils therefore they are more reactive. Higher amounts of trace elements are likely to be retained by fine textured soils as compared to sandy soils (Chen et al. 1999). Fine textured soils have higher cation enhance capacity (CEC) apart from increased surface area. Higher retention of cationic species like copper is caused due to higher cationic enhancement capacity (Chen et al. 1999).

High organic matter (OM) is found in finer textured soils with a higher CEC, compared to sandy soils with low CEC. Often, higher organic matter leads to higher cationic enhance capacity. In fine textured soils the pH dependent charge conditions also favours the organic matter accumulation and retention. Retention of both anionic and cationic species increases in organic matter because cations bridge with the iron and aluminum, resulting in anion retention and the dissociation of organic matter complexes in response to change in pH.

Sandy textured soils increases the arsenic toxicity in plants and arsenic mobility compared to soils with the clayed textures (Jacobs and Keeney 1970; Adriano 1986). Presence of aluminium and iron oxides play a vital role in the ability of soil to retain arsenic (Adriano 2001; Jacobs and Keeney 1970). Moreover, concentrations of iron and phosphorus influences arsenic concentrations in florida soils (Chen et al. 2002). In sand grains the phophorus with clay coatings increases the retention of elements as compared to bare quartz grains (Harris et al. 1987a, b). Metal oxides and alumino-silicates are coating compounds that have higher affinity for trace elements such as arsenic soil horizons (i.e, albic horizons in spodosols) that have been exposed to extreme weathering and leaching. And the weathering results in sand grains that are exposed to clay coatings (Harris et al. 1987a, b). Rhue et al. (1994) found that these horizons are able to retain these clay coatings and they exhibit higher retention as compared to those that did not retain their coatings.

Arsenic Uptake

Plants metabolize the elements through the plasma membrane of the roots. Electrochemical potential is created by H^+-ATPases across the membrane (Kennedy and Gonsalves 1987; Palmgren 2001). Membrane potential is depolarized by plasma membrane and acidifies the cytoplasm by the excess of positively charged ions passing through it (Cumming and Taylor 1990: Axelsen and Palmgren 2001). Membrane potential is disturbed by the arsenates of Pb and Zn (Barlian-Aidid and Okamoto 1992) e.g. low Zn concentration in the plasma membrane of roots of *Zea mays* and enhances the H^{+}ATPase activity by competing with Mg (Axelsen and Palmgren 2001), but 3 mM of Zn is inhibitory. Whereas lead does not activate the ATPase but depolarizes the membrane potential slowly (Kennedy and Gonsalves 1987). In *Lemma gibba* arsenate depolarization is dependent on the phosphate level (Ulrich-Eberius et al. 1989), and in *Impatients balbaniana* stem sections Pb (0.5 mM) xylem parenchyma membrane potential is depolarized (Barlian-Aidid and Okamoto 1992). Recent studies have confirmed the phosphate and arsenate competition at the uptake level (Clements and Munson 1947) e.g. in *Oryza sativa* (Abedin et al. 2002), *Halcustantus* (Hartley-Whitaker et al. 2001a, b), *L. gibba* (Ulrich-Eberius et al. 1989), *Brassica juncea* (Pickering et al. 2000) and pteris ferns (Zhao et al. 2002). Uptake metabolism of As in plants has been reviewed by Meharg and Hartley-Whitaker (2002). Negative charged root cells absorb anions instead of moving them in to apoplastic space of the root cortex (Clarckson 1996). Uptake of arsenic is characterized as proton anion co-transport in *Lemna gibba* (Ulrich-Eberius et al. 1989). Uptake system is shared by arsenate and phosphate in higher plants (Abedin et al. 2002), mycorrhizae (Sharples et al. 2000) and bacteria (Bruins et al. 2000), and the further details are being investigated (Meharg and Hartley-Whitaker 2002). Accumulation of arsenate follows the Michaelis-Menten kinetics in which concentration range coincides with the level of activity of the high affinity phosphate uptake (Sharples et al. 2000; Abedin et al. 2002). In micro-organisms, two types of arsenate transporters have been recognized operating in the pumping arsenite either into the vacuole or in efflux from cells (Rosen 1999; Ali et al. 2009). It means that accumulation, uptake and toxicity, varies within and between plant species and in general more the As in soil higher will be the concentration in plants (Banejad and Olyaie 2011).

Arsenic Tolerance and Toxicity

Contaminated and naturally enriched soils will be used for agriculture in future with higher concentrations of one or more elements (Abedin et al. 2002). Metal resistance enhancement in crop varieties is important only as long as the food plants with metal concentration do not exceed health levels. The variable response of crop plants to soil toxicants extends itself to their nutrient efficiency (Aniol and Gustafson 1989). Tolerance "represents a genotype environment interaction" (Macnair 1993) and the plants have been divided into two groups such as accumulators or excluders (Baker 1987; Tangahu et al. 2011). According to Aniol and Gustafson 1989 many crop plants are accumulators. Excluder plant reduces the elements uptake, Baker (1987). Exclusion capacity in higher plants is poor or absent (Ernst, 1976), while as in bacteria (Nies and Silver 1995) and some mycorrhiza (Sharples et al. 2000) are able to efflux toxicants. Tolerance is under genetic control, although the number of genes from one to smaller or larger number of genes varies and the action may be influenced by modified genes (Macnair 1993; Schat et al. 1996). Tolerance may be constitutive and adaptive i.e depending upon external factor and both types are interlinked (Macnair 1993). For example, an increase in soil toxicity leads to selection pressure, which plays an important role in tolerance (Schat et al. 1996). Metal tolerance can be separate i.e regulated by separated genes for each metal tolerance or co-tolerance (Pleiotropy), although the multiple tolerance developed by plant is by growing them on soils with more than one metal (Macnair 1993) e.g., *Silene vulgar* population originating from metalliferous sites in Ireland and Germany and from non-metalliferous site in Netherlands, two main co-additive gene control of Zn tolerance in this species (Schat et al. 1996). The understanding of cellular level processes has been progressed by metal trafficking proteins, but the relationship between tolerance to toxic metals or metalloids and element homeostasis of the entire organism is less known (Clemens 2001). Parameters like yield reduction, shoot and root length or fresh and dry matter describes the toxicity (Berry and Wallace 1981; Odjegba 2012), but the reversibility of plasmolysis enzyme activities, chlorophyll contents and other physiological parameters are also employed (Baker and Walker 1989). Several indices have been developed for the measurement of tolerance and toxicity. The relationship between root growth with and without a toxic element is expressed by tolerance index (TI). The dose CD50 toxicant causes the reduction, it may be expressed as EC10 to EC50 i.e. 'effective concentrations (EC) to lower the yield by 10–50 % (Ernst 1997a; Kooijman 1997). Macnicol and Beckett (1985) found that critical tissue toxicant concentration can also be used e.g. in soybean and cabbage soil culture, 10 % toxicity occurs at the upper critical tissue by the As level between 1 and 4 mgAs kg^{-1} plant shoot and leaf dry weight. For soybean, bush beans and pea, the values for the essential micronutrient like Zn are higher and the values are 450, 250, and 380–500 $mgZnkg^{-1}$ for shoot or leaf dry weight (Macnicol and Beckett 1985).

Physiological Response Mechanisms

Modes of action of plants under exposure instead of the term tolerance mechanism are used in the meaning of 'response mechanism'. Tolerance or toxicity mechanisms are not fully defined as yet (Schat et al. 1996), and tolerance mechanism also includes responses like altered permeability, enhanced metal binding capacity of the root apoplasm and root exudates. Cellular mechanism includes the synthesis of phytochelatins, organic acids, proteins and membrane adjusting functions together with the synthesis of specific transporters (Hall 2002). Still the question arises as which mechanism provides contribution to the primary and secondary responses. These mechanisms shows the element and plant species dependency and more than one mechanism is active simultaneously in a species (Macnair 1993). Recently Fodor (2002) has reviewed heavy metal responses of higher plants.

Phytochelatins

Synthesis of phytochelatins (PCs) and metallothioniens (MTs) is the response of plants applied with the high concentration of metals or metalloids. Due to their similarity with metallothioniens, phytochelatins have been called as class III metallothioniens (Cobbett and Goldsbrough 2002). The distribution task between PCs and MTs is repeatedly reviewed (Cobbett 2000; Cobbett and Goldsbrough 2002). At present, PCs primarily functions in detoxification while as MTs have been given other roles, e.g. in chaperoning the metallic element translocation. MTs belong to the gene family and PCs are enzymatically produced (Cobbett and Goldsbrough 2002). The small molecular weight phytochelatins are cysteine rich polypeptides with $n=2$–11 (Reddy and Prasad 1990) or $n=2$–5 (Cobbett and Goldsbrough 2002). They were first detected in cell suspension culture of *Rauwalfia serpentina* exposed to 0.2 mM $CdSO_4$ (Grill et al. 1985). Phytochelatins are synthesized in response to Cd, Au, Cu, Ag, Ni, Pb, Sb, Sn, Hg, Te and Zn (Grill et al. 1987) and selenate and arsenate (Grill et al. 1986). The proposed sequence to the intensity of induction in metal specific is: Hg > Cd > As > Te > Ag > Cu > Ni > Sb > Au > Sn > Se > Bi > Pb > W > Zn (Ernst 1997b). Many exceptions to this sequence raised, e.g. Pb is strong inducer of phytochelatins in some legumes (Piechalak et al. 2002). To confirm this the sequence of the root culture of *Rubia tinctorium*; Ag > Cd > Pb > Hg > As (III) > Cu > As (V) > Zn > Pd > In > Ga > Se > Ni has been suggested (Maitani et al. 1996). Phytochelatins production leads to toxicity rather than tolerance (Ebbs et al. 2002). Current studies revealed that phtochelatins do not contribute to Zn, Cd (Ebbs et al. 2002; Schat et al. 2002) or Cu tolerance (Schat et al. 2002). In the root cells of *Silene cucubalus* the Cu tolerance is associated with the glutathione level, i.e by restricting the influx of Cu to these cells and PC synthesis reduction (De Vos et al. 1992). Apart from toxic element inactivation, PCs perform other functions as well like micronutrient homeostasis (Schat et al. 2002), sulphur metabolism (Tomaszewska et al. 1996) maintainance of enzyme activity (Kneer and Zenk 1992) translocation of metals (Cobbett and Goldsbrough 2002) and transport and storage of As (Hartley-Whitaker et al. 2001b). It has been anticipated that PCs primary function is homeostasis and inactivation is secondary (Steffens 1990). In higher plants and microorganisms, the trafficking of PC-complex metals in energy consuming across tonoplast, the transport is mediated by ABC-type cassette binding ATPase (Nies and Silver 1995). *Avena sativa* roots have shown that these ATPases transport Cd-PCs in tonoplast vesicles in the presence of Mg^{2+} (Salt and Rauser 1995). Most likely, higher plants have a gene homologue to the hunt gene that regulates the production of these transporter proteins in *Schizosaccharomyces pombe* (Ortiz et al. 1992). Five types of PCs have been known on the basis of C-terminal amino-acid and the length of the chain (PC0-PC4) (Rauser 1995). In addition to PCs, i.e. the polymer of glutamyl-cysteinyl glycine, homo phytochelatins (h-PCs), polymer of glutamyl-cysteinyl alanine occurs in legumes (Piechalak et al. 2002). Evidences point towards glutathione and homoglutathione acts as precursor of PCs and h-PCs (Cobbett and Goldsbrough 2002). The polymerization of glutathione and PC synthase the metal of PC reaction is catalysed by glutamyl-cysteine synthetase (Piechalak et al. 2002). PC-metal complex regulates the PC synthase (Cobbett 2000). In many plants PC synthase is present in the cytoplasm and also in the roots of *Pisum sativum* (Klapheck et al. 1995) metal ion is most likely to be activated although de novo synthesis may occur (Cobbett 2000).

Oxidative Stress and Antioxidative System in Plants

Oxidative stress is an essential regulated process, as the equilibrium between the oxidative and antioxidative capacities determines the fate of the plant. The antioxidant defence system provides sufficient protection against active oxygen and free radicals under non-stressful conditions. Natural and anthropogenic both stresses provoke the high production of toxic oxygen derivatives. In this condition, the response of the capacity of the antioxidative defence system is increased, but in many situations the response is moderate. In addition, important sites like reaction center and apoplastic space have very little protection against this oxidative damage.

About two billion years ago, ROS have been the unwanted companions of aerobic metabolism. Apart from molecular oxygen (O_2), partially reduced or activated derivatives of oxygen (O_2^{-1}, H_2O_2 and HO) are highly reactive and toxic and cause the oxidative destruction of cells. This results that evolution of all aerobic organism has been dependent on the development of efficient ROS scavenging mechanisms. Recently new rate for ROS was known like the control and regulation of biological processes such as cell death, stress responses, hormonal signaling and development. Therefore the understandings of ROS suggest its dual role in plant biology:

- Toxic products of aerobic metabolism
- Key regulators of metabolic and defense

The ROS Cycle

Rate of ROS in different cellular compartments is determined by the relationship between multiple ROS-producing pathways and ROS scavenging mechanisms. ROS signal transduction pathways control these mechanisms and form the 'basic ROS cycle'. This pathway monitors the level of ROS, produced by aerobic metabolism, and controls the expression and ROS-scavenging pathways during normal growth and development. The ROS cycle also performs good metabolic timing e.g. Photosynthesis control to reduce the production rate of ROI. The ROS source in plants belongs to the aerobic metabolic reactions, such as photosynthesis and respiration and others belong to pathways enhanced during abiotic stresses like photorespiration. Recently, NADPH oxidase, amino-oxidases and cell wall bound peroxidases were identified to be the new sources of ROS in plants. They participate and are tightly regulated in control processes like stress responses, programmed cell death and pathogen defense pathways. The estimated constant rate of 240 μMO_2 and steady level of 0.5 μH_2O_2 for ROS production in cells are optimal growth conditions. On the other hand, stresses that destroy the cellular homeostasis of cells result in the enhancement production of ROS (i.e. up to 720 μMO_2 and steady state level of 5–15 μMH_2O_2). These stresses consist of salt, drought, chilling, heavy metal, heat shock, UV radiations, desiccation and air pollutants such as ozone and SO_2, nutrient deprivation, mechanical stress, pathogen attack and high light. Stress enhanced the production of ROS that creates threat to cells and these conditions also enhance the expression of ROS scavenging enzymes.

Under stress conditions, ROS are produced by cells (e.g. by NADPH oxidase) and the signals for the defence pathways are also produced. Therefore, ROS may be considered as the cellular byproduct of stress metabolism as well as secondary messenger for signal transduction pathway in stress response.

Plant cells require different mechanisms to regulate their intracellular ROS concentrations by scavenging ROS because ROS are toxic and participate in key signal events as well. These include superoxide dismutase (SOD), ascorbate peroxidase (APX) and catalase (CAT). The equilibrium between SOD and APX activity in cells is considered to be crucial for estimating the steady state level of O_2^{-1} and H_2O_2. And the balance with metal ions like Fe and Cu by ferritin and copper binding proteins is also important to prevent formation of highly toxic OH by metal-dependent Haber-Wiess or Fenton reaction. Other antioxidants important for the defence of plants against oxidative stress are ascorbic acid and glutathione that are found at high concentrations in chloroplasts and other cellular compartments. Though the ROS scavenging enzymes expression increases the tolerance of plants under abiotic stress. However, in chloroplasts and mitochondria a group of enzymes called alternative oxides also decreases the ROS production in cells by alternative channelling of electrons in electron transport chain.

Conclusion and Future Perspectives

In this review article, arsenic occurrence, distribution, sources, chemistry, physiological response mechanism and oxidative stress were discussed. We found that arsenic from both natural and anthropogenic sources have been considered as one of the most toxic element affecting millions of people in the world. And several problems like vegetation loss, contamination of ground water and toxicity in animals, plants has been due to arsenic contaminations. From many published reports, it is now clear that arsenic induces cellular toxicity by damaging the oxidative defense mechanism that can be prevented by the phytochelation method. But we need more sound information related to arsenic as treatments of residues from smelting or mining, preventing the use of agrochemicals containing As or simple methods for soil/water testing in field or laboratory which will allow us in making decision for remediation and an adequate disposal plan.

It is clear now that still we are far away from having a safe, specific and effective chelating agent for the treatment of metal poisoning. Several factors must be considered in order to accomplish a high performance of remediation result. Phytoremediation is the most thriving way to remediate arsenic contaminated environment, as it has many advantages as compared to other conventional technologies.

References

Abedin MJ, Feldmann J, Meharg AA (2002) Uptake kinetics of arsenic species in rice plants. Plant Physiol 128:1120–1128

Adriano DC (2001) Trace elements in the terrestrial environment. Springer, New York,

Adriano DC (1986) Trace elements in the terrestrial environment. Springer, New York,

Ali W, Isayenkov SV, Zhao FJ, Maathuis F (2009) Arsenite transport in plants. Cell Mol Life Sci 66:2329–2339

Aniol A, Gustafson JP (1989) Genetics of tolerance in agronomic plants. In: Shaw AJ (ed) Heavy metal tolerance in plants: evolutionary aspects. CRC Press, Florida, pp 255–267

Aten CF, Bourke JB, Martini JH et al (1980) Arsenic and lead in an orchard environment. Environ Toxicol 24:108−115

Axelsen K, Palmgren M (2001). Inventory of the superfamily of P-type ion pumps in Arabidopsis. Plant Physiol 126:696–706

Baker AJM (1987) Metal tolerance. New Phytol 106(Suppl):93–111

Baker AJM, Walker PL (1989) Physiological responses of plants to heavy metals and the quantification of tolerance and toxicity. Chem Speciat Bioavailab 1:7–17

Banejad H, Olyaie E (2011) Arsenic toxicity in the irrigation water-soil-plant system: a significant environmental problem. J Am Sci 7(1):125–131

Barlian-Aidid S, Okamoto H (1992) Effects of lead, cadmium and zinc on the electric membrane potential at the xylem/symplast interface and cell elongation of *Impatiens balsamina*. Environ Exp Bot 32:439–448

Berry WL, Wallace A (1981) Toxicity: the concept and relationship to the dose response curve. J Plant Nutr 3:13–19

Bhattacharya S, Goswami MR, Dhar A, Chattopadhyay D, Mukhopadhyay A, Ghosh UC (2012) Effects of arsenic stress on germination, seedling growth and peroxidase activity in pea seeds and kidney bean seeds. J Universal Pharm Life Sci 2(3):760–770

Bondada B, Ma LQ (2002) Tolerance of heavy metals in vascular plants: arsenic hyperaccumulation by Chinese brake fern (*Pteris vittata* L.), In: Nayar BK, Chandra S, Srivastava M (eds) Pteridology in New Millennium, Kluwer Academy Publishers, 397–420

Bruins MR, Kapil S, Oehme FW (2000) Microbial resistance to metals in the environment. Ecotoxicol Environ Safety 45B:198–207

Chandra S, Srivastava M (2003) Pteridology in the new millennium. Kluwer Academic Publishers, Netherlands

Chen M, Ma LQ, Harris WG (1999) Baseline concentrations of 15 trace elements in Florida surface soils. J Environ Qual 28:1173–1181

Chen T, Wei C, Huang Z, Huang Q, Lu Q, Fan Z (2002) Arsenic hyper-accumulator *Pteris vittata* L. and its arsenic accumulation. Chinese Sci Bull 47:902–905

Clarkson DT (1996) Root structure and sites of ion uptake. In: Waisel Y, Eshel A, Kafkafi U (eds) Plant roots. The hidden half, Marcel Dekker, New York, pp 483–510

Clemens S (2001) Molecular mechanisms of plant metal tolerance and homeostasis. Planta 212:475–486

Clements HF, Munson J (1947) Arsenic toxicity studies in soil and in culture solution. Pacific Sci 1:151–171

Clijsters H, Cuypers A, Vangronsveld J (1999) Physiological responses to heavy metals in higher plants; defence against oxidative stress. Z Naturforsch 54c:730–734
Cobbett C, Goldsbrough P (2002) Phytochelatins and metallothioneins: roles in heavy metal detoxification and homeostasis. Annu Rev Plant Biol 53:159–182
Cobbett CS (2000) Phytochelatins and their roles in heavy metal detoxification. Plant Physiol 123:825–832
Cumming JR, Taylor GJ (1990) Mechanisms of metal tolerance in plants: physiological adaptations for exclusion of metal ions from the cytoplasm. In: Alscher RG, Cumming JR (eds) Stress responses in plants: adaptation and acclimation mechanisms. Wiley, New York, pp 329–356
Dat J, Vandenabeele S, Vranová E, Van Montagu M, Inzé D, Van Breusegem F (2000) Dual action of the active oxygen species during plant stress responses. Cell Mol Life Sci 57:779–795
De Vos CHR, Vonk MJ, Vooijs R, Schat H (1992) Glutathione depletion due to copper-induced phytochelatin synthesis causes oxidative stress in *Silene cucubalus*. Plant Physiol 98:853–858
Ebbs S, Lau I, Ahner B, Kochian L (2002) Phytochelatin synthesis is not responsible for Cd tolerance in the Zn/Cd hyperaccumulator *Thlaspi caerulescens* (J. & C. Presl). Planta 214:635–640
Ernst W (1976) Physiological and biochemical aspects of metal tolerance. In: Mansfield TA (ed) Effects of air pollutants on plants. Cambridge University Press, Cambridge, pp 115–133
Ernst WHO (1997a) Population dynamics of plants under exposure and the selection of resistance. In: Schüürmann G, Markert B (eds) Ecotoxicology. Ecological fundamentals, chemical exposure, and biological effects. Wiley, New York, pp 117–132
Ernst WHO (1997b) Effects of heavy metals in plants at the cellular and organismic level. In: Schüürmann G, Markert B (eds) Ecotoxicology. Ecological fundamentals, chemical exposure, and biological effects. Wiley, New York, pp 587–620
Fodor F (2002) Physiological responses of vascular plant to heavy metals. In: Prasad MNV, Strzalka K (eds) Physiology and biochemistry of metal toxicities and tolerance in plants. Kluwer Academic Publishers, Dordrecht, pp 149–177
Fowler BA (1983) Biological and environmental effects of arsenic. Elsevier Sci. Publ., Amsterdam
Grill E, Gekeler W, Winnacker E-L, Zenk HH (1986) Homo-phytochelatins are heavy metal-binding peptides of homo-glutathione containing Fabales. FEBS Lett 205:47–50
Grill E, Winnacker E-L, Zenk MH (1985) Phytochelatins: the principal heavy-metal complexing peptides of higher plants. Science 230:674–676
Grill E, Winnacker E-L, Zenk MH (1987) Phytochelatins, a class of heavy-metal-binding peptides from plants, are functionally analogous to metallothioneins. Proc Natl Acad Sci USA 84:439–443
Hall JL (2002) Cellular mechanisms for heavy metal detoxification and tolerance. J Exp Bot 53:1–11
Harris WG, Carlisle VW, Chesser SL (1987a) Clay mineralogy as related to morphology of Florida soils with sandy epipedons. Soil Sci Soc Am J 51:1673–1677
Harris WG, Carlisle VW, van Rees KCJ (1987b) Pedon zonation of hydroxy-interlayered minerals in Ultic Haplaquods. Soil Sci Soc Am J 51:1367–1372
Hartley-Whitaker J, Ainsworth G, Meharg AA (2001a) Copper and arsenate induced oxidative stress in *Holcus lanatus* L. clones with differential sensitivity. Plant Cell Environ 24:713–722
Hartley-Whitaker J, Ainsworth G, Vooijs R, Ten Bookum W, Schat H, Meharg AA (2001b) Phytochelatins are involved in differential arsenate tolerance in *Holcus lanatus*. Plant Physiol 126:299–306
Jacobs LW, Keeney DR (1970) Arsenic-phosphorus interactions in corn. Commu Soil Sci Plant Analy 1:85–93
Kabata-Pendias A, Pendias H (2001) Trace elements in soils and plants. CRC Press, Boca Raton, p 413
Kaur S, Kamli MR, Ali A (2011) Role of arsenic and its resistance in nature. Can J Microbiol 57(10):769–774
Kennedy CD, Gonsalves FAN (1987) The action of divalent zinc, cadmium, mercury, copper and lead on the trans-root potential and H^+efflux of excised roots. J Exp Bot 38:800–817

Kenyon DJ, Elfving DC, Pakkala IS, Bache CA, Lisk DJ (1979) Residues of Lead and Arsenic in crops cultured on old orchard soils. Bul Environ Contam Toxicol 22:221–223

Klapheck S, Schlunz S, Bergmann L (1995) Synthesis of phytochelatins and homophytochelatins in *Pisum sativum* L. Plant Physiol 107:515–521

Kneer R, Zenk MH (1992) Phytochelatins protect plant enzymes from heavy metal poisoning. Phytochemistry 31:2663–2667

Kooijman SALM (1997) Process-oriented descriptions of toxic effects. In: Schüürmann G, Markert B (eds) Ecotoxicology. Ecological fundamentals, chemical exposure, and biological effects. Wiley, New York, pp 483–520

Lui Yl, Xu Y, Ke-Bing D, Tu BK (2012) Absorption and metabolism mechanisms of inorganic arsenic in plants: a review. Ying Yong Sheng Tai Xue Bao 23(3):842–848

Macnair MR (1993) The genetics of metal tolerance in vascular plants. New Phytol 124:541–559

Macnicol RD, Beckett PHT (1985) Critical tissue concentrations of potentially toxic elements. Plant Soil 85:107–129

Maitani T, Kubota H, Sato K, Yamada T (1996) The composition of metals bound to Class III metallothionein (phytochelatin and its desglycyl peptide) induced by various metals in root cultures of *Rubia tinctorum*. Plant Physiol 110:1145–1150

Meharg AA, Hartley-Whitaker J (2002) Arsenic uptake and metabolism in arsenic resistant and non-resistant plant species. New Phytol 154:29–43

Moreno-Jimenez E, Esteban E, Penalosa JM (2012) The fate of arsenic in soil-plant systems. Rev Environ Contam Toxicol 215:1–37

Nies DH, Silver S (1995) Ion efflux systems involved in bacterial metal resistances. J Indust Micro biol 14:186–199

Odjegba VJ (2012) Exogenous salicylic acid alleviates arsenic toxicity in *Arabidopsis thaliana*. Indian J Innovations Dev 1(7):516–522

Ortiz DF, Kreppel L, Speiser DM, Scheel G, McDonald G, Ow DW (1992) Heavy metal tolerance in the fission yeast requires an ATP-binding cassette-type vacuolar membrane transporter. EMBO J 11:3491–3499

Palmgren M (2001) Plant plasma membrane H^+-ATPase: powerhous for nutrient uptake. Ann Rev Plant Physiol Mol Biol 52:817–845

Pickering IJ, Prince RC, George MJ, Smith RD, George GN, Salt DE (2000) Reduction and coordination of arsenic in Indian mustard. Plant Physiol 122:1171–1178

Piechalak A, Tomaszewska B, Baralkiewicz D, Malecka A (2002) Accumulation and detoxification of lead ions in legumes. Phytochemistry 60:153–162

Pierce ML, Moore CB (1982) Adsorption of arsenite and arsenate on amorphous iron hydroxide. Water Residue 16:1247–1253

Rauser WE (1995) Phytochelatins and related peptides. structure, biosynthesis, and function. Plant Physiol 109:1141–1149

Reddy GN, Prasad NMV (1990) Heavy metal-binding proteins/peptides: occurrence, structure, synthesis and functions. A review. Environ Exp Bot 30:251–264

Rhue RD, Harris WG, Kidder G, Brown RB, Littell RC (1994) A soil based phosphorous retention index for animal waste disposal on sandy soil. Final project report. Florida Depart Environ Protec EPA grant no.9004984910

Rosen BP (1999) Families of arsenic transporters. Trends Microb 7:207–212

Rumberg CB, Engel RE, Meggitt WF (1960) Effect of phosphorus concentration on the absorption of arsenate by oats from nutrient solution. Agronomy J 52:452–453

Salt DE, Rauser WE (1995) MgATP-dependent transport of phytochelatins across the tonoplast of oat roots. Plant Physiol 107:1293–1301

Schat H, Llugany M, Vooijs R, Hartley-Whitaker J, Bleeker PM (2002) The role of phytochelatins in constitutive and adaptive heavy metal tolerances in hyperaccumulator and non-hyperaccumulator metallophytes. J Exp Bot 53:2381–2392

Schat H, Vooijs R, Kuiper E (1996) Identical major gene loci for heavy metal tolerances that have independently evolved in different local populations and subspecies of *Silene vulgaris*. Evolution 50:1888–1895

Schmoger ME, Oven M, Grill E (2000) Detoxification of arsenic by phytochelatins in plants. Plant Physiol 122:793–801

Sharples JM, Meharg AA, Chambers SM, Cairney JWG (2000) Mechanism of arsenate resistance in the ericoid mycorrhizal fungus *Hymenoscyphus ericae*. Plant Physiol 124:1327–1334

Steffens JC (1990) The heavy metal-binding peptides of plants. Annu Rev Plant Physiol Plant Mol Biol 41:553–575

Sturchio E, Boccia P, Meconi C, Zanellato M, Marconi S, Beni C, Aromolo R, Ciampa A, Diana G, Valentini M (2011) Effects of arsenic on soil-plant systems. Chem Ecol 27(1):67–78

Tangahu BV, Sheikh Abdullah SR, Basri H, Idris M, Anuar N, Mukhlisin M (2011) A Review on Heavy metals (As, Pb, and Hg) uptake by plants through phytoremediation. Int J Chem Engineer 2011: doi:10.1155/2011/939161

Tomaszewska B, Tukendorf A, Baralkiewicz D (1996) The synthesis of phytochelatins in lupin roots treated with lead ions. Sci Legume 3:206–217

Ullrich-Eberius CI, Sanz A, Novacky AJ (1989) Evaluation of arsenate and vanadate associated changes of electrical membrane potential and phosphate transport in *Lemna gibba* G1. J Exp Bot 40:119–128

Woolson EA (1983) Emissions, cycling, and effects of arsenic in soil ecosystems. Elsevier, Amsterdam, pp 52–125

Zhao F, MecGrath SP, Meharg AA (2010) Arsenic as a food chain contaminant: mechanisms of uptake and metabolism and mitigation strategies. Ann Rev Plant Biol 61:535–559

Zhao FJ, Dunham SJ, McGrath SP (2002) Arsenic hyperaccumulation by different fern species. New Phytol 156:27–31

Chapter 13
Arsenic Stress in Plants: An Inside Story

Iti Sharma

Abstract Arsenic (As) toxicity is a global concern due to increasing contamination of metalloid in water, soil and crops especially in South East Asia. Arsenic poses a serious threat of food chain contamination by accumulating in various crops through the phosphate transporters as a phosphate analogue. After accumulating in plant tissues arsenic interferes with various metabolic processes and thereby adversely affects the plant metabolism, and ultimately leads to reduced plant productivity. Alteration of phosphate, nitrogen, sulfur metabolism and disorder in major physiological reactions like respiration, photosynthesis and transpiration are responsible for metabolic dysfunction of plants exposed to arsenic. This chapter discusses recent advances in plant arsenic interaction at molecular, biochemical and physiological levels. It is necessary to develop a detailed biochemical understanding about interaction of arsenic with plants to limit detrimental effects of arsenic on crops and also for better agronomic production.

Introduction

Arsenic (As) is an environmental and food chain contaminant. The toxicity of arsenic was known as early as in 300 B.C. It has been used at least from the 12th century A.D in metallic form. It bears a sinister connotation linked to suicides, witchcraft and murder and it was a source of despair and inspiration to alchemists wishing to transform metals into gold (Azcue and Nriagu 1994). Arsenic is a metalloid of group VA in the periodic system. It is a natural constituent of the lithosphere and occurs in some 200 minerals, frequently as mixed sulphides. Metal oars such as lead, gold, zinc and copper, volcanic eruptions and sea spray liberate arsenic naturally, and it can be methylated, demethylated and volatilized by soil bacteria (Matscullat 2000). Anthropogenic sources of arsenic pollution are spreading of herbicides or pesticides, coal combustion and timber preservatives. High-temperature urban

I. Sharma (✉)
Department of Bioscience and Biotechnology,
Banasthali University,
Banasthali 304022, India
e-mail: itisharma@hotmail.com

K. R. Hakeem et al. (eds.), *Crop Improvement,* DOI 10.1007/978-1-4614-7028-1_13,

waste combustion generates particulate emissions of arsenic, that contribute to dry (particles) and wet deposits (rain) (Melanen et al. 1999). Inorganic arsenic is a potent human carcinogen, associated with increased risk for cancer of the skin, lungs, urinary bladder and kidneys, as well as hyperkeratosis, pigmentation changes, and effects on the circulatory and nervous systems. It is also known to generate oxidative stress in humans (Benton et al. 2011).

Distribution of Arsenic in World

Arsenic contamination has become a problem in many parts of the world. The World Health Organization (WHO) has set a guideline of 10 μg As L^{-1} as the drinking water standard (WHO 2008). However, As concentrations in the range of > 1,000 μg L^{-1} have been reported from various places throughout the world (Tuli et al. 2010). Especially Australia, Canada, Japan, Mexico, Thailand, United Kingdom, United States, Argentina, Bangladesh, Cambodia, Chile, China, Ghana, Hungary, Inner Mongolia, Mexico, Nepal, New Zealand, Philippines, Taiwan, the United States and Vietnam are reported as highly contaminated countries (WHO 2002).

In India sever contamination of arsenic in ground water has been reported in West Bangal (69 times more arsenic than WHO standard) whereas Jharkhand, Assam, Bihar, Utter Pradesh, Manipur and Chhattisgarh are some other states well effected by arsenic toxicity. In Rajasthan also, arsenic bearing groundwater has been reported in the vicinity of Cu (Khetri) and Zn mines (Zawar). It has been reported that agricultural land contains ≤ 10 mg kg^{-1} in non-contaminated soil whereas it increases up to 1,000–3,000 mg kg^{-1} in contaminated soil (Panda et al. 2009). Growing crops in arsenic contaminated soil and irrigation with arsenic contaminated water effects the crops in terms of growth, yield, biomass production and arsenic accumulation which can directly leads to food chain contamination.

Arsenic Species Found in the Environment

The oxidation states and electron orbital of arsenic is similar to those of phosphorus. Arsenic forms alloys with various metals and covalent bonds with carbon, hydrogen, oxygen, and sulfur. In nature, arsenic exists predominantly in inorganic form as trivalent arsenite (III) or pentavalent arsenate (V). The major species found in the environment are arsenite As(III), arsenate As(V), mono methylarsonic acid (MMA), dimethyal arsenic acid (DMA), arseno betain and arseno choline (Tangahu et al. 2011). In the environment arsenic is present in both organic and inorganic forms. The inorganic species arsenate As(V) and arsenite As(III) are more abundant in soil as compare to organic species, monomethyarsonic (MMA), Dimethylarsenic (DMM) etc. (Takamatsu et al. 1982). These forms have different physical and chemical characteristics resulting in various degrees of mobility, bioavailability and toxicity. In general inorganic arsenic species are more toxic than organic arsenic species (Adriano 2001).

Arsenic accumulation and speciation in plants are affected by root aeration and porosity (Wu et al. 2011). Behavior of arsenic in soil and also in plants is different due to its dynamic and complex chemical species with inter conversable forms regulated by biotic and abiotic process (Fig. 13.1).

This chapter evaluates all the changes in plants at physiological, biochemical and molecular level to better understand toxicity and resistance mechanisms in plants.

Interaction of Arsenic with Plants

Uptake of Arsenic in Plants

Different soil parameters like pH and redox state has a major influence on the toxicity of arsenic species due to altered availability (solubility and mobility). The uptake of arsenic affected by some factors such as soil type, nutrient supply, pH of the medium and mugineic acid which is excreted by some graminaceous (grassy) plants (Sultana and Katsuichiro 2011). Among all factors phosphorus and pH are the most important ones influencing plant growth and As uptake. As(V) has been shown to be taken up by the high affinity phosphate uptake system (Ullrich-Eberius et al. 1989) while As(III) uptake is thought to be through aquaporins in the roots (Meharg and Jardine 2003) while low level uptake of organic As species, such as MMA and DMA has also been observed in rice but the underlying transport pathways are unknown (Meharg 2004). Recently, Arsenic transformation and volatilization is also investigated. The uptake of As species into rice roots is in the order of arsenate [As(V)] > monomethylarsonic acid [MMAs(V)] > dimethylarsinic acid [DMAs(V)] > trimethylarsine oxide (TMAs(V)O), but the order of the root-to-shoot transport index (Ti) is reverse. The volatilization of trimethylarsine (TMAs) from rice plants is also observed when plants were treated with TMAs(V)O but not with As(V), DMAs(V) and MMAs(V) (Jia et al. 2012). Lomax et al. (2012) has reported a latest finding on the basis of GeoChip analysis of microbial genes in a Bangladeshi paddy soil that plants are unable to methylate inorganic As, and instead take up methylated As produced by microorganisms. Recently, specific accumulation patterns were observed among growth habitat and plant groups, it was found that submerged plants have a higher accumulation than emergent and terrestrial plants. Whereas (*Oryza sativa*) grown in multiple sites (Norton et al. 2012).

Once arsenic enters inside root cells, As(V) is quickly reduced to As(III) and become complexed with phytochelatines. Phosphorus nutrition influences As(V) uptake and toxicity in Gymnosperms had a high $[As]_{(shoot)}$: $[As]_{(root)}$ ratio (Bergqvist 2012). Statistically significant effect of year, location and flooding management are also reported as an important factors to develop variation in grain arsenic evaluated in a varied section of rice plants, whilst silicon has similar influences

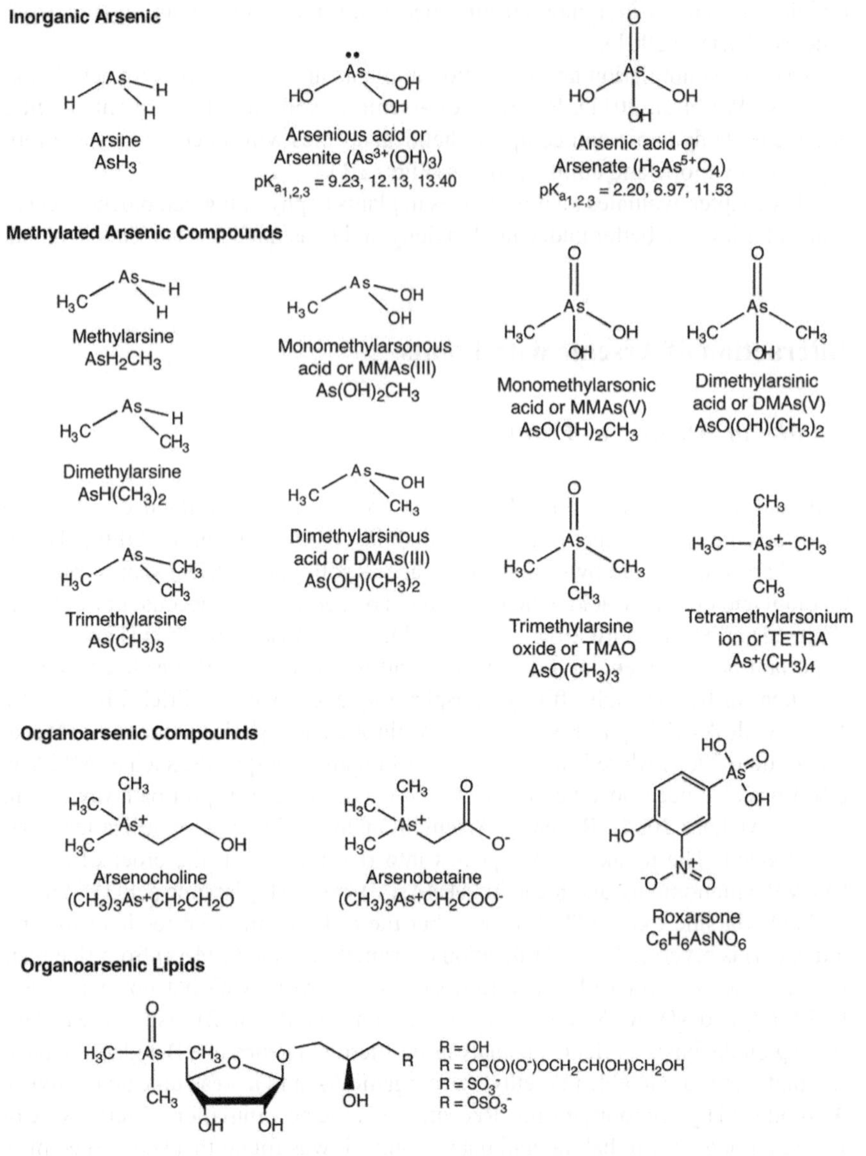

Fig. 13.1 Various forms of arsenic present in the environment. (Adapted from http: elements. geoscienceworld.org)

on As(III). Phytotoxic effects commonly observed from As exposure includes growth inhibition, chlorophyll degradation, nutrient depletion and oxidative stress (Moreno-Jiménez et al. 2012). As(III) can react with sulphahydryl groups of enzymes and proteins, which lead to loss of function and can cause cell death. As(V) can compete with phosphate, replacing it in key molecules, including ATP (Scott et al. 1993; Meharg and Hartley-Whitaker 2002; Quaghebeur and Rengel 2003). Exposure of plants to inorganic As leads to the synthesis of phytochelatins that complex with As(III), those complexes or as alone, being transported across the tonoplast by ABC-type transporters or can be efflux from the cytoplasm by As(III) efflux transporters. Once arsenic enters in plant cell it causes various changes in normal metabolic activities. The changes occur due to arsenic toxicity and also in process to cope up from arsenic toxicity to some extent. But at higher concentration of arsenic toxicity plants fails to counterbalance between toxicity and resistance.

Anatomical Changes in Plants Exposed to Arsenic

Arsenic causes many physiological changes and damages in plants (Wells and Gilmor 1997). There are several anatomical parameters, in which reduction in growth is the earliest As toxicity response. Arsenic affects root growth more severely than shoot growth possibly due to the retention of As in the roots in higher amount than in the stem. Stoeva et al. (2005) also reported that arsenic accumulated mainly in the root system and to a lesser extent in the overgrown organs. This also confirmed from the study of mung bean and *Anadenanthera Peregrina* (Pal et al. 2006; Gomes et al. 2012). There is however, contrasting reports showing that the effect of arsenic on stem and root growth varies depending on the plant species, level of contamination and plant tissue ability to As. Arsenic also inhibits fresh and dry biomass accumulation (Wells and Gilmor 1997) which may be possibly due to plant growth inhibition. From the study of mung bean it is also observed that reduction in root elongation is accompanied by the anatomical changes, which occurs on exposure to arsenic toxicity (Pal et al. 2006). The anatomical changes are severe decrease or completely loss of root hairs, damage to epidermal cells and the cortex, with those cells losing their shape and showing signs of shriveling and disintegration while the untreated epidermal root cells are intact and the root hairs are turgid. Further compared to the control roots, where the stele is in a tetrarch condition, there is a lack of complete differentiation and pith formation in arsenic treated root cells. Also, arsenic cause necrosis and reduction of the number of raminification in root system (Stoeva et al. 2005). According to a study by Kopittke et al. (2012) the accumulation of As causes permanent damage to the meristem but root border cells accumulates high levels of As and limiting its movement into the root. When roots are counteract with arsenic in soil environment a greater diffusion of oxygen from the roots indicates increased root oxidizability (RO). This is considering as avoidance from the toxicity. The TIC, salt used to measure RO, which absorbs electrons from the mitochondrial transport chain and correlates positively with the respiratory activity, in term associated with enhanced RO is also an indicator of higher reactive oxygen species generation (Singh et al. 2007).

Besides roots, above ground part of plant is also affected by arsenic toxicity. There is reduction in leaf area, necrosis and chlorosis of leaf tips are occurred on arsenic toxicity (Stoeva et al. 2005). Arsenic also changes the osmotic adjustment of plant. The leaf water potential decreases while the relative water content (RWC) slightly decreases in toxic condition. All these changes, collectively lead to the changes in important physiological activities of plant like photosynthesis, transpiration, stomatal conductance etc.

Arsenic causes a reduction in photosynthetic rate (Miteva and Merakchiyska 2002). The reduced photosynthetic rate can be due to the many factors like result of disturbance in assembling of the pigment protein complex and thylakoid membrane. Arsenic can release protein; lipid and element components of thylakoid membrane indispensable for photosynthesis activity of proteins especially connected with the water splitting system and of lipid, glactolipid probably connected with PSII are liberated. Another reason for decreasing the photosynthesis rate could be disturbance in the pigment apparatus. The photosynthesis pigments are some of the most important internal factor which in certain cases is able to limit the photosynthesis rate. It is believed that they are target of toxic As effect (Miteva and Merakchiyska 2002). There is considerable decrease of chlorophyll and carotenoid contents on exposure to arsenic (Stoeva et al. 2005). Also photosynthetic reactions are closely related with stomata behavior, due to closure of stomata in stress condition, diminish or cessation of CO_2 uptake, by which photosynthetic rate also reduced (Biehler et al. 1996). The insufficient water supply in tissue may also induce photo inhibition, which influence rate of electron transport. The transpiration processes also reduced on exposure to arsenic probably is a result of the disturbed uptake and transport of water, caused by the changes in root system. (Stoeva et al. 2005). As(V) affect cell wall, primary and secondary metabolism, abscisic acid metabolism and germination of the seedlings whereas, As(III) mainly affects hormonal and signaling processes (Chakrabarty et al. 2009). Anarchy in mitotic and labeling index, mitotic arrays of microtubules, increased percentage of metaphase and DNA fragmentation are also observed in roots of *Pisum sativum* exposed arsenic (Faria et al. 2010).

Biochemical Changes in Plants Exposed to Arsenic

At biochemical level plants shows many changes during exposure to arsenic. Some of the constituents are decreased due to toxicity of arsenic and some are increased to combat from detrimental effects of toxicant. Arsenate acts as a phosphate analogue and is transported across the plasma membrane via phosphate co-transport system (Ullrich-Eberius et al. 1989). Once inside the cytoplasm it competes with phosphate from several vital reactions like in the glycolysis (metabolism of glucose) the conversion of 1, 3 bisphosphoglycerate to 3 phosphoglycerate, ATP molecule is formed by using inorganic phosphate but in the toxicity of arsenic, arsenate competes with phosphate and give l-arseno-3 phosphoglycerate which hydrolyses spontaneously to 3 phosphoglycerate without forming ATP so deprives the cell from the energy sources. Also arsenate can replace phosphate group from DNA causing disorganize the structure of nucleic acid which effects directly to gene function. Similarly arsenite is a well known thiol reagent that combines rapidly with dithiol groups on pro-

Table 13.1 Enzymes affected by arsenic toxicity in plants

Mode of as toxicity	Target enzyme	References
Phosphate replacement with arsenic	F1-F0 ATP synthases	Gresser (1981)
	GDPH (glycolytic enzyme)	Orsit and Cleland (1972)
	Aspartate-β-semialdehyde dehydrogenase	Kish and Viola (1999)
	Purine Nucleoside Phosphorylase (PNP)	Park and Agrawal (1972)
Binding with thiol groups	Dihydrolipomide (Co-factor of pyruvate dehydrogenase complex (mt PDC, pt PDC)	Bergquist et al. (2009)
	Gly decarboxylase complex (GDC)	Peters et al. (1946)
	Branched Chain 2-oxoaciddecarboxylase complex (BCOADC)	Bergquist et al. (2009)

teins and act as an effective inhibitor of enzymes requiring free sulfahydryl groups (Webb et al. 2003). Thus phosphate replacement and inactivation of enzymes by binding with their thiol groups are the main modes of arsenic toxicity (Table 13.1).

Arsenic interferes with various events of respiratory cycle. Arsenite act as an inhibitor of α-ketoglutrate dehydrogenase enzyme in TCA cycle, causing the accumulation of substrate α-ketoglutrate and no product formation (succinyl Co-A). The nutrient uptake is also affected due to arsenic toxicity. Phosphorus uptake decreases with increasing arsenic concentration due to similarity with arsenate. Also uptake of nitrate is reduced on exposure to arsenic. The uptake of nitrate and further assimilation to ammonium is also altered, possibly due to interference of arsenic with the involving enzymes, nitrate reductase and nitrite reductase (Table 13.2).

Arsenic Induced Oxidative Stress in Plants

There is significant evidence that exposure to inorganic arsenic species (ROS) results in the generation of reactive oxygen species caused oxidative damages to plants (Sharma 2012). This probably occurs through the conversion of arsenate to arsenite, a process which readily occurs in plants. After this reduction, arsenic may be potentially further metabolized to methylated species leading to further oxidative stress (Zaman and Pardini 1996) because methylation is thought to be redox driven and such reactions could give rise to reactive oxygen species (ROS). However there is no sufficient evidence for methylation in higher plants but it can be seen in the cell suspension of *Cathranthus roseus* (Cullen and Hettipathirana 1994) and in phosphate starved tomato plants (Nissen and Benson 1982). Also Wu et al. (2002) has been shown in vitro methylation of arsenic in cell extracts from bent grass (*Agrostis tenuis*). These reactive oxygen species causes peroxidation of lipid by reacting with lipid bilayer and thereby leading to membrane leakage. Lipid peroxidation can be measured in terms of MDA content. An increase in MDA con-

Table 13.2 Alterations in some vital biochemical processes of the plants exposed to arsenic

Biochemical reaction altered by arsenic	Steps of the reaction	Plant species	References
Photosynthesis			
Light reaction	Decrease in chlorophyll content	Alfalfa	Porter and Sheridan (1981)
		Oat	Stoeva and Bineva (2003)
		Pteris vittata	Singh et al. (2006)
		Rice	Rahman et al. (2007)
		Duck weed	Duman et al. (2010)
	Decrease in PS II activity	Oat	Stoeva and Bineva (2003)
	By Decrease in electron flow from thylakoid membrane		
Photophosphorylation	Replacement of Pi with As in ATP	Spinach	Avron and Jagendorf (1959)
		Pea	Watling-Payne and Selwyn (1974)
Dark reaction	Decrease the amount of Rubisco large subunit content	Rice	Ahsan et al. (2010)
	AsIII inhibits the light activation of photosynthetic CO_2 fixation	Pea	Marques and Anderson (1986)
	Inhibition of maltose libration from starch by inhibiting amylolytic activity	Wheat	Liu et al. (2005)
	Substitution of Pi in glucose 1-Po_4 by glucose 1-As(V)	Pea	Levi and Preiss (1978)
	Decrease in glceraldehyde -3-Phosphate dehydrogenase activity	Rice	Ahsan et al. (2010)
	Decrease in seduheptulose-1,7-bisphosphatase, malate dehydrogenase triose-phosphate isomerase, and a subunit of pyruvate dehydrogenase	*Pteris vittata*	Bona et al. (2010)
Photorespiration	Inactivation of GDP, BCOADC, mt PDC, Pt PDC, OGDC complex	Arabidopsis	Chen et al. (2010)
	Activation of alternative photorespiration in chloroplast and mitochondria	Barley *Amaranthus edulis*	Wingler et al. (1999)
Respiration			
Glycolysis	Decrease in glyceraldehyde -3-phosphate dehydrogenase activity	Rice	Ahsan et al. (2010)

Table 13.2 (*Continued*)

Biochemical reaction altered by arsenic	Steps of the reaction	Plant species	References
	Replacement of 1,3 diphoshogyceric acid with 3-arseno glcerate (uncoupling of ATP synthesis)	*Arabidopsis thaliana*	Abercrombie et al. (2008)
		Rice	Norton et al. (2008) and Chakrabarty et al. (2009)
Citric acid cycle	As III blocks the enzymes mt PDC, OGDC, GDC,PDC and block the respiration and its products	Potato	Araújo et al. (2008)
	Disturbance in glycolytic and citric acid cycle metabolite pool by Increase in abundance of the succinyl-CoA synthetase α subunit	Maize	Requejo and Tena (2005)
	Decrease activity of mt PDC resulting in 2-oxoglutarate and pyruvate metabolism	*Arabidopsis thaliana*	Chen et al. (2010)
	Loss of melate dehydrogenase and ATP synthase FAd subunit	*maize*	Requejo and Tena (2005)
	Inhibition of dark respiration	alfalfa	Porter and Sheridan (1981)
	Respiratory O_2 consumption was more resistant to As V supply, than photosynthetic O_2 evolution	Rice	Marin et al. (1993)
	AsV-dependent uncoupling of ATP synthesis and formation of ADP-As V from electron transport chain	*Brassica oleracea L.* *Beta vulgaris L.*	Wickes and Wiskich (1975)
	Failed to induce AOX transcripts during microarray studies	Arabidopsis	Abercrombie et al. (2008)
		Rice	Norton et al. (2008) and Chakrabarty et al. (2009)
	A homolog to the *Arabidopsis* NDA1 alternative NAD (P) H dehydrogenase induced at the transcript level by both AsV and AsIII	Rice	Chakrabarty et al. (2009)
	Mitochondrial substrate carrier protein dicarboxylate carrier2(DIC2) get repressed	Arabidopsis	Abercrombie et al. (2008)
	Decreased amount of mitochondrial ATP synthatase subunit	*Agrostis tenuis*	Duquesnoy et al. (2009)

Table 13.2 (*Continued*)

Biochemical reaction altered by arsenic	Steps of the reaction	Plant species	References
Nitrogen metabolism	Symbiotic N_2 fixation in legumes by Rhizobium	Alfalfa	Porter and Sheridan (1981)
	Decrease the number of root nodules, root necrosis, root hair damage, short length of root zone	*Medicago sativa*	Pajuelo et al. (2008)
	Decreased amount of transcript for NO_3 and NH_4+ transporters	Rice	Norton et al. (2008)
	Decrease in protein content	Rice	Dwivedi et al. (2010)
		P. ensiformis	Singh et al. (2006)
		P. vittata	Mascher et al. (2002)
		Red clover maize	Stoeva et al. (2005)
Sulfur metabolism	Induction of GSH and PC biosynthesis	Arabidopsis	Muñoz-Bertomeu et al. (2009)
	Decrease in cellular Cys pools	*Arabidopsis thaliana*	Sung et al. (2009)

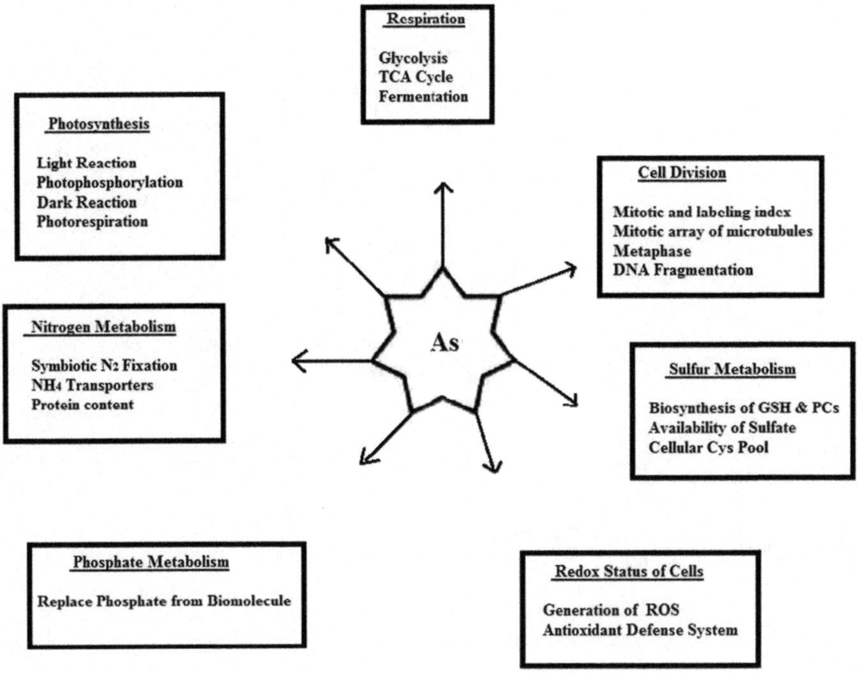

Fig. 13.2 Major plant physiological reactions severely affected by exposure to arsenic

tent indicates the occurrence of membrane damage due to peroxidation of polyunsaturated fatty acids, resulting in the generation of ROS and subsequent oxidative stress. There are several reports that confirm that MDA content increase on exposure to arsenic (Stoeva et al. 2005; Srivastava et al. 2007).

The amount of soluble protein content is reduced in the arsenic toxicity. In the study of Stoeva et al. (2005) oat root has been shown that protein content decreased linearly on increasing the arsenic concentration which may be most probably a result of the reduced anabolic or the accelerated catabolic processes. The protein degradation to amino acids is in fact an adaptation of the cells to the carbohydrate deficiency. On the other hand, the accelerated catabolism is probably due to the considerable disturbance in the membrane systems, in response to metal phytotoxicity (Fig. 13.2).

As well as different mechanisms of heavy metal toxicity, several defense strategies related to metal tolerance in plants have also been well characterized. Similarly plants have also several detoxification mechanisms related to arsenic toxicity, these are suppression of high affinity arsenate/phosphate uptake systems, reduction of arsenate to arsenite, chelation of arsenite to metal binding peptides, induction of different antioxidant enzymes such as superoxide dismutase (SOD), ascorbate peroxidase (APX), catalase (CAT), and nonenzymatic antioxidants like glutathione, ascorbate, tochopherol etc., effluxation of arsenic from cytoplasm, and

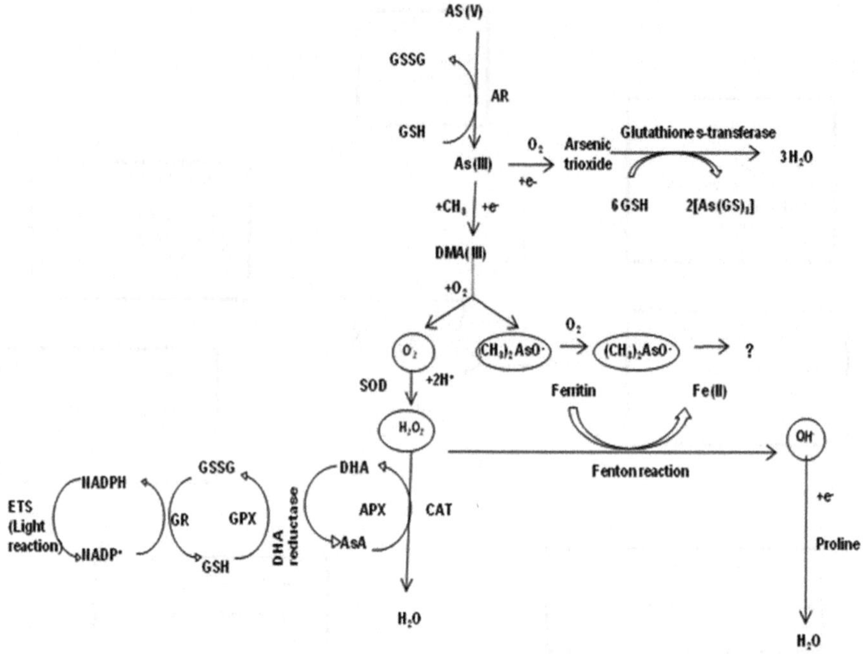

Fig. 13.3 Possible mechanism of arsenic induced oxidative stress and antioxidant defense system. (Adopted from Sharma 2012)

finally sequestration of arsenic in vacuoles (Sharma et al. 2007). Increasing activity of Guaiacol peroxidase, Catalase, Ascorbate peroxidase contributes to high accumulation of the arsenic species by the plants (Srivastava et al. 2011). Similarly, another non enzymatic antioxidant of glutathione-ascorbate cycle like ascorbate (AsA) and dehydroascorbate (DAsA), were also analyzed in some plants during As(V) exposure. As(V) treatment caused an increase in the ratio of AsA/DAsA in *P. vitatta, P. ensiformis, H. verticillata, and O. sativa* (Singh et al. 2006; Srivastava et al. 2011; Tripathi et al. 2012b) indicating the significant role of ascorbate in As induced stress tolerance. Also Requejo and Tena (2005) confirmed that the level of these enzyme increased because these are involved in cellular homeostasis for redox perturbation by the study of proteome analysis in maize roots (Fig. 13.3).

Reduction of arsenate to arsenite is catalyzed by enzyme arsenate reductase, it is also considered as a mechanism involved in detoxification because arsenite can bind with phytochelatins. Arsenate reduction is coupled to NADP (NADPH) oxidation via the reduction of oxidized glutathione by glutathione reductase (GR) and with the resulting glutathione (GSH) serving as the electron donor for arsenate reductase (Ellis et al. 2006).

The activity of arsenate reductase (AR) is well studied in yeast (Rosen 2002). Root extracts from the arsenic hyperaccumulator fern *Pteris vittata* also show the ability to reduce arsenate to arsenite by same enzyme; this reaction is similar to yeast in terms of arsenate reduction, substrate specificity and sensitivity towards inhibitors.

In the resistance mechanism of arsenic the level of phytochelatins is increased because arsenite binds with phytochelatins. Phytochelatins are heavy metal binding peptides derived from glutathione (GSH) with the general structure (γ-glu-cys)nGly. The biosynthesis of phytochelatins involves the transpeptidation of γ-glutamyl-cysteinyl dipeptides from GSH by the action of constitutively expressed phytochelatins synthetase (Grill et al. 1987). Recent studies have confirmed that As(III) is complexed with phytochelatins in a range of terrestrial plant species, suggesting that phytochelatins play an important role in decreasing the toxicity of arsenic in crops (Hartley-Whitaker et al. 2001; Zhang et al. 2012). Recently, Duana et al. (2011) has suggest that PC complexation of arsenite in rice leaves reduces As translocation from leaves to grains, and implicate that manipulation of PC synthesis might mitigate As accumulation in rice grain. Similarly induced levels of PCs were also observed in *O. sativa* (Tripathi et al. 2012a) under As stress.

Molecular Changes in Plants Exposed to Arsenic

There are several molecular responses by plants towards exposure to arsenic, due to both arsenic toxicity and also due to arsenic tolerance. Arsenic affects the expression of many genes that are involved in various essential cellular processes in plants. A number of genes involved in cell growth, cellular morphogenesis and cell cycle are down regulated on exposure to arsenic. From the study of the rice genome, two expansion genes (OsOlg14660 and Os04g46650), two tubulin genes (Os03g45920 and Os03g56810), an actin gene (OsOlg64630) and two microtubule genes (Os03g13460 and Os09g27700) which are involved in cell cycle and cell growth are less expressed when exposed to arsenic at low concentration for long term exposure (Norton et al. 2008). In As(V)-treated rice seedling, a triose-phosphate/Pi translocator gene was transcriptionally up-regulated (Chakrabarty et al. 2009). This protein would be expected to transport Pi and As(V) across the plastid inner membrane in exchange for triose-phosphate. The effect of As exposure on genome-wide expression was also examined in rice (Yu et al. 2012). An As tolerance gene has been identified and mapped to chromosome 6 in rice (Tripathi et al. 2012a). A signaling molecule Nitric oxide, was also found to be induced during As(V) stress condition in *A. thaliana* (Leterrier et al. 2012).

The genes responsible for the expression of nutrient uptake transporters are responding differentially in arsenic toxicity. Norton et al. (2008) has also revealed that genes for transporters of various nutrients differentially expressed in rice genome. For phosphate, chloride, ammonium and nitrate transporters are down regulated which possibly cause nutrient deficiency in plants while sulphate transporters are up regulated, this is may be due to increased uptake of sulphur for the synthesis of glutathione which is precursor of phytochelatins because chelation capacity of a plant is increased under metal toxicity (Cherian and Oliveira 2005). As previously describes that As(III) is taken up by aquaporins, two genes for these aquaporins in rice (Os05g14240 and Os12g10280) are also down regulated in the study of Norton et al. (2008). It is also reported that lower expression of genes expression of PHT1 encoding phosphate transporters contributes to arsenic tolerance and accumulation in *shrub willow* (Puckett et al. 2011). The differential expression pattern of sulphate transporters were observed after As(V) exposure (Kumar et al. 2011). Three members of rice PIP subfamily of aquaporins have been recently reported to mediate As(III) transport (Mosa et al. 2012). Overexpression of OsPIP2;4, OsPIP2;6, and OsPIP2;7 proteins in *Arabidopsis* resulted in increased As(III) resistance (Maciaszczyk-Dziubinska et al. 2012).

By analyzing maize root proteome Requejo and Tena (2005) revealed that oxidative stress is the main contributing factor to plant arsenic toxicity. They reported that three superoxide dismutase, two glutathione peroxidase, one peroxidation and one p-benzoquinone are up regulated and these are involved in cellular homeostasis for redox perturbation. Conversely, recently by the study of rice genome very surprising results came in which enzymes involved in detoxifying various reactive oxygen species and free radicals, gives no response to low concentration of arsenic for long term exposure (Norton et al. 2008). Only some of the Tau classes of GSTs (Glutathione-S-transferase) showed remarkable changes in expression which was only agreement with Mylona et al. (1998).

As part of arsenic detoxification, the majority of arsenate is reduced to arsenite by the enzyme arsenate reductase (AR). Recently, AR genes have been identified in plants, including *Arabidopsis thaliana* (AtAsr/AtACR2), *Holcus lanatus* (Hl-Asr) and *Pteris vittata* (PvACR2) (Dhankher et al. 2006; Bleeker et al. 2006; Ellis et al. 2006). The Arabidopsis, fungal, protist ACR2 sequences show homology a region within the CDC25 super family of protein tyrosine phosphatases (PTPase) that also contains the conserved HCX_s-R motif and also has similar catalytic activity like arsenate reductase. The arsenic detoxification gene is well studied in yeast in which three genes in cluster ACR1, ACR2, ACR3 are present for As tolerance. ACRI encodes a transcription factor, ACR2 encodes an arsenate reductase and ACR3 encodes a plasma membrane As(III) efflux transporter (Ghosh et al. 1999; Rosen 2002).

Many studies showed that alteration in the expression of arsenate reductase genes leads to the more arsenic tolerance in plants. Dhankher et al. (2006) cloned an Arsenate reductase gene from *Arabidopsis thaliana* (At ACR2) and silenced its expression in root (because of it appears that in most plants, arsenite is sequestered in roots, preventing it from moving up into stems, leaves and reproductive organs but to enhance phytoremediation it should be stored in aboveground tissues) and obtaining RNAi transgenics that accumulated 10–16 fold more arsenic in the shoots and retained less root As compared with WT plants when grown in the presence of low levels of As(V). However transgenic lines are more sensitive than WT plants when exposed to high concentrations of As(V), probably because of the negative effects of As(V) on phosphate metabolism. By contrast, Bleeker et al. (2006) found that in Arabidopsis (equivalent to AtACR2) loss of function mutants are more As sensitive than WT plants, even to low levels of As(V) and when they over expressed AtASR the plant showed enhance tolerance to mildly toxic levels of As(V) to more toxic As(III). Recently two arsenate reductase genes have been identified in rice (OsArs 1, Oslog39860 and OsArs2, Os03g0 1770) (Duan et al. 2007). But interestingly, besides their major role in arsenic detoxification these genes are not differentially regulated on exposure to arsenic in the whole rice genome study (Norton et al. 2008).

Further arsenite binds with phytochelatins as described earlier in respect to reducing arsenic toxicity. Overexpression of genes involving in phytochelatins synthesis, like phytochelatins synthatase, γ-glutamylcystein synthatase and glutathione synthetase provides tolerance to arsenic toxicity. Li et al. (2004) overexpressed the AtPCSI resulted in a substantial increase in arsenic resistance, with a 20–100 times greater biomass in transgenic plants after exposure to arsenic, but led to Cd hypersensitivity. In contrasts in the study of Picault et al. (2006) overexpression of cytoplasmic AtPCS 1 markedly increased tolerance in transgenic plants to arsenic, whereas chloroplast-targeted overexpression of the same gene resulted in decreased tolerance of transgenic plants to arsenic. This is may be due to the limiting supply of essential metabolites such as cystein, γ-glutamylcystein and glutathione, which are needed for the production of phytochelatins. Because of phytochelatins production on exposure to arsenic leads to depletion of GSH, overexpression of components involved in GSH biosynthesis, such as γ-glutamylcystein synthetase (γ ECS) and glutathione synthetase (GS), will lead to increased tolerance to arsenic. Li et al. (2005) overexpressed γ ECS in *Arabidopsis thaliana* and found a 3–20 fold greater production of γ-glutamylcystein, glutathione and phytochelatins in plants exposed to arsenic. These studies shows that the expression of all these genes play important role in balance between toxicity and resistance to arsenic.

As part of the detoxification mechanism, Arsenic can be effluxed from the cytoplasm through As(III)-efflux transporters and can be sequestered in vacuoles by ABC type transporters. The genes for these transporters are not well identified in plants till now but it has homology to yeast. In yeast in ACR gene cluster, ACR3 gene encoded a plasma membrane As(III)-efflux transporter, for the removal of cytosolic arsenic and an ABC type transporter, yeast cadmium factor (YCF I) is located at the vacuolar membrane by which arsenic sequester in vacuoles (Rosen 2002). Ali et al. (2012) reported that heterologous expression of the yeast arsenite efflux system ACR3 improves *Arabidopsis thaliana* tolerance to arsenic stress.

In general, the expression of genes that encode various transporters is higher in hyperaccumulater plants than non-accumulators (Krammer 2005). By some studies it is confirm that by increase in the over expression of these transporters can increase the tolerance of arsenic. Ellis et al. (2006) isolate and characterized arsenate reductase (PvACR2) and arsenite transporter (PvACR3) from *pteris vittata* that can suppress the arsenic sensitivity. The expression of PvACR3 in *Pteris vittata* is rapidly increased within 24 h after exposure to arsenate. But to date, there is lack of information regarding these type of transporters mediated arsenic detoxification in higher plants and effect on their expression in arsenic toxicity. Little information are present upon the expression of these transporter from the study of whole genome analysis in rice that three genes with the predicted putative function of an ABC transporter family protein (Os 04g49890, Os04g52900, and Os11g05700) is up regulated in the arsenic toxicity. Phylogenetic analysis of these ABC transporters gene sequences revealed that they are also MRP transporter. MRPs are a subclass of that ATP-binding cassette (ABC) transporter, which are involved in the transport of glutathione conjugated compounds into the vacuoles of plants (Rea et al. 1998). And also a single gene with an annotated putative function of a glutathione conjugate transporter (Os04g 1321 0) is also upregulated on exposure to arsenic (Norton et al. 2008).

A number of genes involved in N transport also appear to alter expression in response to As. Amino acid transporters are down- regulated in response to As(V) in roots and seedlings of rice (Norton et al. 2008; Chakrabarty et al. 2009). However, amino acid transporter gene transcript levels were not influenced by As(III) (Chakrabarty et al. 2009). Peptide and oligopeptide transporters have also been reported to be As(V) responsive in rice, but reports disagree on the direction (Norton et al. 2008; Chakrabarty et al. 2009).

Arsenic is also known to interfere with sulfur metabolism. It is reported at molecular level that in As(V)-treated rice, up to five sulfate transporter genes are upregulated in roots (Norton et al. 2008), and at least one sulfate transporter is up-regulated in *Arabidopsis* (Sung et al. 2009). As(III) also induces a sulfate transporter gene in rice and *B. juncea* seedlings (Chakrabarty et al. 2009). Although it is not yet clear whether As(V) and As(III) influence the expression of these transporters in the same way, but at least one of the transporter genes is induced by both forms of As (Chakrabarty et al. 2009). During transfer of sulfate from soil to plant once sulfate is reduced to sulfide, the sulfide is combined with *O*-acetylserine to form Cys in a reaction catalyzed by *O*-acetylserine(thiol)-lyase also known as Cys synthase. It appears that As(V) and As(III) exposure may cause a down-regulation of (OAS-TL), (OAS-TL)in As-sensitive plants. OAS-TL protein disappeared from maize shoots exposed to As (Requejo and Tena 2006), while OAS-Tl activity was repressed in an As-sensitive line of *B. juncea* (Srivastava et al. 2009).

Moreover, in rice, several methyl transferase genes are induced by As(V)-treatment (Norton et al. 2008). Two of these are homocysteine S-methyltransferases, which catalyze the formation of *S*-adenosyl-l-homocysteine and Met from *S*-adenosylmethionine and l-homocysteine.The enzyme is involved in the synthesis of *S*-methylmethionine (Ranocha et al. 2001), and may play a key role in maintaining a pool of soluble Met, in the cycling of methyl groups within cells, or as a phloem-mobile form of Met that can be used to translocate sulfur derived from protein degradation (Bürstenbinder and Sauter 2012).

Conclusion and Future Perspectives

Arsenic contamination in food chain is challenging issue for researchers. Various aspects of the study include biochemical, molecular, physiological and anatomical changes appeared in the plants as a result of arsenic exposure. Moreover, there are new findings regarding methylation of inorganic arsenic, uptake mechanism and accumulation of arsenic in plant. The role of transporters in uptake and vacuolar localization reveal that these processes are the key to understand As-resistance. There is considerable knowledge gap regarding arsenic mediated hindrance in plant physiological reactions, nutrient uptake, pathway of oxidative stress and effect on crop yield. This basic information will be significant in phytoremediation studies to eradicate the arsenic contamination from agricultural soil. We can acquire new insight about arsenic and plant interaction by using molecular advances and sophisticated analytical technologies. So, research should focus on combing physiology and genetics to breed plants with low arsenic in edible plant parts and productivity.

Acknowledgements I thank to Professor Aditya Shastri, Director, Banasthali University for his kind support and necessary facilities for carrying out the present study. The sincere cooperation and support of my mentor Dr. Bhumi Nath Tripathi is also gratefully acknowledged. The work was financially supported by Department of Science and Technology (DST), Govt. of India, New Delhi, in the form of Women Scientist-A Scheme (WOS-A).

References

Abercrombie JM, Halfhill MD, Ranjan P, Rao MR, Saxton AM, Yuan JS, Stewart CNJr (2008) Transcriptional responses of *Arabidopsis thaliana* plants to As(V) stress. BMC Plant Biol 8:87

Adriano DC (2001) Trace elements in the terrestrial environment. Springer, New York

Ahsan N, Lee DG, Kim KH, Alam I, Lee SH, Lee KW, Lee H, Lee BH (2010) Analysis of arsenic stress-induced differentially expressed proteins in rice leaves by two-dimensional gel electrophoresis coupled with mass spectrometry. Chemosphere 78:224–231

Ali W, Isner JC, Isayenkov SV, Liu W, Zhao FJ, Maathuis FJM (2012) Heterologous expression of the yeast arsenite efflux system ACR3 improves *Arabidopsis thaliana* tolerance to arsenic stress. New Phytol 194:716–723

Araújo WL, Nunes-Nesi A, Trenkamp S, Bunik VI, Fernie AR (2008) Inhibition of 2-oxoglutarate dehydrogenase in potato tuber suggests the enzyme is limiting for respiration and confirms its importance in nitrogen assimilation. Plant Physiol 148:1782–1796

Avron M, Jagendorf A (1959) Evidence concerning the mechanism of adenosine triphosphate formation by spinach chloroplasts. J Biol Chem 234:967–972

Azcue JM, Nriagu JO (1994) Arsenic: historical perspectives. In: Nriagu JO (ed) Arsenic in the environment. Part I: cycling and characterization. John Wiley & Sons, New York, pp 1–15

Benton MA, Rager JE, Smeester L, Fry RC (2011) Comparative genomic analyses identify common molecular pathways modulated upon exposure to low doses of arsenic and cadmium. BMC Genomics 12:173

Bergqvist C (2012) Arsenic accumulation and speciation in plants from different habitats. Appl Geochem 27(3):615–622

Bergquist ER, Fischer RJ, Sugden KD, Martin BD (2009) Inhibition by methylated organo arsenicals of the respiratory 2-oxo-acid dehydrogenases. J Organomet Chem 694:973–980

Biehler K, Migge A, Fock H (1996) The role the malate dehydrogenase in dissipating excess energy under water stress in two wheat species. Photosynthetica 32:431–438

Bleeker PM, Hakvoort HWJ, Bliek M, Souer E, Schat H (2006) Enhanced arsenate reduction by a CDC25-like tyrosine phosphatase explains increased phytochelatin accumulation in arsenate-tolerant *Holcus lanatus*. Plant J 45:917–929

Bona E, Cattaneo C, Cesaro P, Marsano F, Lingua G, Cavaletto M, Berta G (2010) Proteomic analysis of *Pteris vittata* fronds: two arbuscular mycorrhizal fungi differentially modulate protein expression under arsenic contamination. Proteomics 10:3811–3834

Bürstenbinder K, Sauter M (2012) Early events in the ethylene biosynthetic pathway—regulation of the pools of methionine and S-adenosylmethionine. In: McManus MT (ed) Annual plant reviews, vol. 44. Wiley, Hoboken pp 19–52

Chakrabarty D, Trivedi PK, Misra P, Tiwari M, Shri M, Shukla D, Kumar S, Rai A, Pandey A, Nigam D, Tripathi RD, Tuli R (2009) Comparative transcriptome analysis of arsenate and arsenite stresses in rice seedlings. Chemosphere 74:688–702

Chen W, Chi Y, Taylor, NL, Lambers H, Finnegan PM (2010) Disruption of ptLPD1or ptLPD2, genes that encode isoforms of the plastidial lipoamide dehydrogenase, confers arsenate hypersensitivity in Arabidopsis. Plant Physiol 153:1385–1397

Cherian S, Oliveira MM (2005) Transgenic plants in phytoremediation: recent advances and new possibilities. Environ Sci Technol 39:9377–9390

Cullen WR, Hettipathirana DI (1994) Application of whole cell NMR techniques to study the interaction of arsenic compounds with *Catharanthus roseus* cell suspension cultures. Appl Organomet Chem 8:463–471

Dhankher OP, Rosen BP, McKinney EC, Meagher RB (2006) Hyperaccumulation of arsenic in the shoots of Arabidopsis silenced for arsenate reductase (ACR2). Proc Natl Acad Sci USA 103:5413–5418

Duan G, Zhou Y, Tong Y, Mukhopadhyay R, Rosen BP, Zhu Y (2007) A CDC25 homologue from rice functions as an arsenate reductase. New Phytol 174:311–321

Duana GL, Hu Y, Liu Wen-Ju, Kneer R, Zhao F-J, Zhu Y-G (2011) Evidence for a role of phytochelatins in regulating arsenic accumulation in rice grain. Environ Exp Bot 71:416–421

Duman F, Ozturk F, Aydin Z (2010) Biological responses of duck weed (*Lemna minor* L.) exposed to the inorganic arsenic species As(III) and As(V): effects of concentration and duration of exposure. Ecotoxicology 19:983–993

Duquesnoy I, Goupil P, Nadaud I, Branlard G, Piquet-Pissaloux A, Ledoigt G (2009) Identification of *Agrostis tenuis* leaf proteins in response to As(V) and As(III) induced stress using a proteomics approach. Plant Sci 176:206–213

Dwivedi S, Tripathi RD, Tripathi P, Kumar A, Dave R, Mishra S, Singh R, Sharma D, Rai UN, Chakrabarty D, Trivedi PK, Adhikari B, Bag MK, Dhankher OP, Tuli R (2010) Arsenate exposure affects amino acids, mineral nutrient status and antioxidants in rice(*Oryza sativa* L.) genotypes. Environ Sci Technol 44:9542–9549

Ellis DR, Gumaelius L, Indriolo E, Pickering I, Banks JA, Salt DE (2006) A novel arsenate reductase from the arsenic hyperaccumulating *Pteris vittata*. Plant Physiol 141:1544–1554

Faria D, Wanda C, Mucciarelli M, Fusconi A, (2010) Arsenate toxicity on the apices of *Pisum sativum* L. seedling root: effect on mitotic activity, chromatin integrity and microtubules. Environ Exp Bot 69(1):17–23

Ghosh M, Shen J, Rosen BP (1999) Pathways of As(III) detoxification in *Saccharomyces cerevisiae*. Proc Natl Acad Sci USA 96:5001–5006

Gomes MP, Carvalho M, Marques TCLSáe Melo, Duarte DM, Nogueira COG, Soares AM, Garcia QS (2012) Arsenic-sensitivity in *Anadenanthera Peregrina* eue to arsenic-induced lipid peroxidation. Int J Appl Sci Technol 2(2):55–63

Gresser MJ (1981) ADP-arsenate formation by submitochondrial particles under phosphorylating conditions. J Biol Chem 256:5981–5983

Grill E, Winnacker EL, Zenk MH (1987) Phytochelatins a class of heavy-metal-binding peptides from plants are functionally analogous to metallothioneins. Proc Natl Acad Sci USA 84:439–443

Hartley-Whitaker J, Ainsworth G, Meharg AA (2001) Copper and arsenate-induced oxidative stress in *Holcus lanatus* L. clones with differential sensitivity. Plant Cell Environ 24:713–722

Jia Y, Huang H, Sun GX, Zhao FJ, Zhao FJ, Zhu YG (2012) Pathways and relative contributions to arsenic volatilization from rice plants and paddy soil. Environ Sci Technol 46(15):8090–8096

Kish MM, Viola RE (1999) Oxy anion specificity of aspartate beta-semialdehyde dehydrogenase. Inorg Chem 38:818–820

Kopittke PM, Jonge MD, Menzies NW, Peng W, Donner E, McKenna BA, Paterson D, Howard DL, Lombi E (2012) Examination of the distribution of arsenic in hydrated and fresh cowpea roots using two- and three-dimensional techniques. Plant Physiol 159(3):1149–1158

Krammer U (2005) Phytoremediation: novel approaches to cleaning up polluted soils. Curr Opin Biotechnol 16:133–141

Kumar S, Asif MH, Chakrabarty D, Tripathi RD, Trivedi PK (2011) Differential expression and alternative splicing of rice sulphate transporter family members regulate sulphur status during plant growth, development and stress conditions. Funct Integr Genomics 11:259–273

Leterrier M, Airaki M, Palma JM, Chaki M, Barroso JB, Corpas FJ (2012) Arsenic triggers the nitric oxide (NO) and S-nitrosoglutathione (GSNO) metabolism in *Arabidopsis*. Environ Poll 166:136–143

Levi C, Preiss J (1978) Amylo pectin degradation in pea chloroplast extracts. Plant Physiol 61:218–220

Li Y, Dhankher OP, Carreira L, Lee D, Chen A, Schroeder JI, Balish RS, Meagher RB (2004) Overexpression of phytochelatin synthase in Arabidopsis leads to enhanced arsenic tolerance and cadmium hypersensitivity. Plant Cell Physiol 45:1787–1797

Li Y, Dhankher OP, Carreira L, Balish R, Meagher R (2005) Engineered overexpression of g-glutamylcysteine synthetase in plants confers high level arsenic and mercury tolerance. Environ Toxicol Chem 24:1376–1386

Liu X, Zhang S, Shan X, Zhu YG (2005) Toxicity of arsenate and arsenite on germination, seedling growth and amylolytic activity of wheat. Chemosphere 61:293–301

Lomax C, Liu WJ, Wu L, Xue K, Xiong J, Zhou J, McGrath SP, Meharg AA, Miller AJ, Zho FJ (2012) Methylated arsenic species in plants originate from soil microorganisms. New Phytol 193(3):665–672

Maciaszczyk-Dziubinska E, Wawrzycka D, Wysocki R (2012) Arsenic and antimony transporters in eukaryotes. Int J Mol Sci 13(3):3527–3548

Marin AR, Masscheleyn PH, Patrick WH (1993) Soil redox-pH stability of arsenic species and its influence on arsenic uptake by rice. Plant Soil 152:245–253

Marques IA, Anderson LE (1986) Effects of arsenite, sulfite, and sulfate on photosynthetic carbon metabolism in isolate pea (*Pisum sativum* L.,cv Little Marvel) chloroplasts. Plant Physiol 82:488–493

Mascher R, Lippmann B, Holzinger S, Bergmann H (2002) Arsenate toxicity:effects on oxidative stress response molecules and enzymes in red clover plants. Plant Sci 163:961–969

Matscullat J (2000) Arsenic in the geosphere-a review. Sci Total Environ 249:297–312

Meharg AA (2004) Arsenic in rice—understanding a new disaster for SouthEast Asia. Trends Plant Sci 9:415–417

Meharg AA, Hartley-Whitaker J (2002) Arsenic uptake and metabolism in arsenic resistant and non resistant plant species. New Phytol 154:29–43

Meharg AA, Jardine L (2003) Arsenite transport into paddy rice (*Oryza sativa*) roots. New Phytol 157:39–44

Melanen M, Ekqvist M, Mukherjee AB, Aune la-Tapola L, Verta M, Salmikan gas T (1999) Raskasmetallien päästöt ilmaan Suomes sa 1990)-luvulla. Suomen Ym päristö Suomen Ym päristökeskus, Edita (distributor), Hel sin ki 329, 92 pp

Miteva E, Merakchiyska M (2002) Response of chloroplasts and photosynthetic mechanism of bean plants to excess arsenic in soil. Bulg J Agric Sci 8:151–156

Moreno-Jiménez E, Esteban E, Penalosa JM (2012) The fate of arsenic in soil-plant systems. Rev Environ Contam Toxicol 215:1–37

Mosa KA, Kumar K, Chhikara S, McDermott J, Liu Z, Musante C, White JC, Dhankher OP (2012) Members of rice plasma membrane intrinsic proteins subfamily are involved in arsenite permeability and tolerance in plants. Transgenic Res. doi:.1007/s11248-012-9600-8.

Muñoz-Bertomeu J, Cascales-Miñana B, Mulet JM, Baroja-Fernández E, Pozueta-Romero J, Kuhn JM, Segura J, Ros R (2009) Plastidial glyceraldehyde-3-phosphate dehydrogenase deficiency leads to altered root development and affects the sugar and amino acid balance in Arabidopsis. Plant Physiol 151:541–558

Mylona PV, Polidoros AN, Scandalios JG (1998) Modulation of antioxidant responses by arsenic in maize. Free Radical Biol Med 25:576–585

Nissen P, Benson AA (1982) Arsenic metabolism in freshwater and terrestrial plants. Physiol Plantarum 54:446–450

Norton GJ, Lou-Hing Daniel E, Meharg AA, Adam H (2008) Rice-arsenate interactions in hydroponics: whole genome transcriptional analysis. J Exp Bot 59:2267–2276

Norton GJ, Pinson SR, Alexander J, McKay S, Hansen H, Duan GL, Rafiqul Islam M, Islam S, Stroud JL, Zhao FJ, McGrath SP, Zhu YG, Lahner B, Yakubova E, Guerinot ML, Tarpley L, Eizenga GC, Salt DE, Meharg AA, Price AH (2012) Variation in grain arsenic assessed in a diverse panel of rice (*Oryza sativa*) grown in multiple sites. New Phytol 193(3):650–664

Orsit BA, Cleland WW (1972) Inhibition and kinetic mechanism of rabbit muscle glyceraldehyde-3-phosphate dehydrogenase. Biochemistry 11:102–109

Pajuelo E, Rodríguez-Llorente ID, Dary M, Palomares AJ (2008) Toxic effects of arsenic on Sino rhizobiume-*Medicago sativa* symbiotic interaction. Environ Pollut 154:203–211

Pal M, Horvath E, Janda T, Paldi E, Szalai G (2006) Physiological changes and defence mechanisms induce by cadmium stress in maize. J Plant Nutr Soil Sci 169:239–246

Panda SK, Upadhyay RK, Nath S (2009) Arsenic stress in plants. J Agron Crop Sci. doi: 10.1111/j1439-037X.2009.00407.x

Park RE Jr, Agrawal RP (1972) Purine nucleoside phosphorylase. In: Boyer PD (ed) The enzymes. Academic Press, New York, pp 483–514

Peters RA, Sinclair HM, Thompson RHS (1946) An analysis of the inhibition of pyruvate oxidation by arsenicals in relation to the enzyme theory of vesication. Biochem J 40:516–524

Picault N et al (2006) Chloroplast targeting of phytochelatin synthase in Arabidopsis: effects on heavy metal tolerance and accumulation. Biochimie 88:1743–1750

Porter JR, Sheridan RP (1981) Inhibition of nitrogen fixation in alfalfa by arsenate, heavy metals, fluoride, and simulated acid rain. Plant Physiol 68:143–148

Puckett EE, Serapiglia MJ, DeLeon AM, Long S, Minochac R, Smart LB (2011) Differential expression of genes encoding phosphate transporters contributes to arsenic tolerance and accumulation in *shrub willow* (Salix spp.). Environ Exp Bot. doi:10.1016/j.envexpbot.2011.07.008

Quaghebeur M, Rengel Z (2003) The distribution of arsenate and arsenite in shoots and roots of *Holcus lanatus* is influenced by arsenic tolerance and arsenate and phosphate supply. Plant Physiol 132:1600–1609

Rahman MA, Hasegawa H, Rahman MM, Islam MN, Miah M AM, Tasmen A (2007) Effect of arsenic on photosynthesis, growth and yield of five widely cultivated rice (*Oryza sativa* L.) varieties in Bangladesh. Chemosphere 67:1072–1079

Ranocha P, McNeil SD, Ziemak MJr, Li C, Tarczynski MC, Hanson Adb (2001) The S-methylmethionine cycle in angiosperms:ubiquity, antiquity and activity. Plant J 25:575–584

Rea PA, Li Z, Lu Y, Drozdowicz YM, Martinoia E (1998) From vacuolar GS-X pumps to multi specific ABC transporters. Ann Rev Plant Biol 49:727–760

Requejo R, Tena M (2005) Proteome analysis of maize roots reveals that oxidative stress is a main contributing factor to plant arsenic toxicity. Phytochemistry 66:1519–1528

Requejo R, Tena M (2006) Maize response to acute arsenic toxicity as revealed by proteome analysis of plant shoots. Proteomics 6:156–162

Rosen, BP (2002) Biochemistry of arsenic detoxification. FEBS Lett 529:86–92

Scott N, Hatlelid KM, MacKenzie NE, Carter DE (1993) Reactions of arsenic(III) and arsenic(V) species with glutathione. Chem Res Toxicol 6:102–106

Sharma I, Singh R, Tripathi BN (2007) Biochemistry of arsenic toxicity and tolerance in plants. Biochem Cell Arch 7:165–170

Sharma I (2012) Arsenic induced oxidative stress in plants. Biologia 67(3):447–453

Singh HP, Batish DR, Kohali RK, Arora K (2007) Arsenic- induced root growth inhibition in mung bean (*Phaseolus aureus* Roxb.) is due to oxidative stress resulting from enhanced lipid peroxidation. Plant Growth Regul 53:65–73

Singh N, Ma LQ, Srivastava M, Rathinasabapathi B (2006) Metabolic adaptations to arsenic induced oxidative stress in *Pteris vittata* L. and *Pteris ensiformis* L. Plant Sci 170:274–282

Srivastava S, Mishra S, Trtpathi RD, Dwivedi S, Trivedi PK, Tandon PK (2007) Phytochelatins and antioxidant systems respond differentially during arsenite and arsenate stress in *Hydrilla verticillata* (L.f.) Royle. Environ Sci Technol 41:2930–2936

Srivastava S, Srivastava AK, Suprasanna P, D'Souza SF (2009) Comparative biochemical and transcriptional profiling of two contrasting varieties of *Brassica juncea* L. in response to arsenic exposure reveals mechanisms of stress perception and tolerance. J Exp Bot 60:3419–3431

Srivastava S, Shrivastava M, Suprasanna P, Souza SFD (2011) Phytofiltration of arsenic from simulated contaminated water using *Hydrilla verticillata* in field conditions. Ecol Eng 37:1937–1941

Stoeva N, Berova M, Zlaten Z (2005) Effect of arsenic on some physiological parameters in bean plants. Biologia Plantarum 49:293–296

Stoeva N, Bineva T (2003) Oxidative changes and photosynthesis in oat plants grown in As-contaminated soil. Bulg J Plant Physiol 29:87–95

Sultana R, Katsuichiro K (2011) Potential of barnyard grass to remediate arsenic-contaminated soil. Weed Biol Manag 11:12–17

Sung DY, Kim TH, Komives EA, Mendoza-Cózatl DG, Schroeder JI (2009) ARS5 is a component of the 26S proteasome complex, and negatively regulates thiol biosynthesis and arsenic tolerance in Arabidopsis. Plant J 59:802–812

Takamatsu T, Aoki H, Yoshida T (1982) Determination of arsenate, arsenite, monomethylarsonate, and dimethylarsinate in soil polluted with arsenic. Soil Sci 133:239–246

Tangahu BV, Abdullah SRS, Basri H, Idris M, Anuar N, Mukhlisin M (2011) A review on heavy metals (As, Pb, and Hg) uptake by plants through phytoremediation. Int J Chem Eng 939161:1–31

Tripathi RD, Tripathi P, Dwivedi S, Dubey S, Chatterjee S, Chakrabarty D, Trivedi PK (2012a) Arsenomics: omics of arsenic metabolism in plants. Front Physiol 3(275):1

Tripathi P, Mishra A, Dwivedi S, Chakrabarty D, Trivedi PK, Singh RP, Tripathi RD (2012b) Differential response of oxidative stress and thiol metabolism in contrasting rice genotypes for arsenic tolerance. Ecotoxicol Environ Saf 79:189–198

Tuli R, Chakrabarty D, Trivedi PK and Tripathi RD (2010) Recent advances in arsenic accumulation and metabolism in rice. Mol Breed 26:307–323

Ullrich-Eberius CI, Sanz A, Novacky AJ (1989) Evaluation of arsenate and vanadate-associated changes of electrical membrane potential and phosphate transport in *Lemna gibba* G I. J Exp Bot 40:119–128

Watling-Payne AS, Selwyn MJ (1974) Inhibition and uncoupling of photophosphorylation in isolated chloroplasts by organotin, organomercury and diphenyleneiodonium compounds. Biochem J 142:65–74

Webb S.M, Gaillard JF, Ma LQ, Tu C (2003) XAS speciation of arsenic in a hyperaccumulating fern. Environ Sci Technol 37:754–760

Wells BR, Gilmor J (1997) Sterility in rice cultivars as influenced by MSMA rate and water management. Agron J 69:451–454

WHO (2002) United Nations synthesis report on arsenic in drinking water. http://www.who.int/water_sanitation_health/Arsenic/ArsenicUNRep6. Accessed 01 July 2002

WHO (2008) Guidelines for drinking-water quality, 3rd edition incorporating 1st and 2nd addenda,vol. 1. Recommendations. World Health Organization, Geneva, pp 306–308b. http:// www.who.int/water_sanitation_health/dwq/GDW12rev1and2.pdf. Accessed 10 Oct 2012

Wickes WA, Wiskich JT (1975) Arsenate uncoupling of oxidative phosphorylation in isolated plant mitochondria. Aust J Plant Physiol 3:153–162

Wingler A, Lea PJ, Leegood RC (1999) Photorespiratory metabolism of glyoxylate and formate in glycine-accumulating mutants of barley and *Amaranthus edulis*. Planta 207:518–526.

Wu C, Ye Z, Shu W, Zhu Y, Wong M (2011) Arsenic accumulation and speciation in rice are affected by root aeration and variation of genotypes. J Exp Bot 1–10. doi:10.1093/jxb/erq462

Wu JH, Zhang R, Lilley RM (2002) Methylation of arsenic in vitro by cell extract from bentgrass (*Agrostis tenuis*): effect of acute exposure of plants to arsenate. Funct Plant Biol 29:73–80

Yu LJ, LuoYF, Liao B, Xie LJ, Chen L, Xiao S, Li JT, Hu SN, Shu WS (2012) Comparative transcriptome analysis of transporters, phytohormone and lipid metabolism path ways in response to arsenic stress in rice (*Oryzasativa*). New Phytol 195:97–112

Zaman K, Pardini RS (1996) An overview of the relationship between oxidative stress and mercury and arsenic. Toxic Subst Mech 15:151–181

Zhang X, Uroic MK, Xie WY, Zhu YG, Chen BD, McGrath SP, Feldmann J, Zhao FJ (2012) Phytochelatins play a key role in arsenic accumulation and tolerance in the aquatic macrophyte *Wolffia globosa*. Environ Pollut 165:18–24

Chapter 14
In vitro Production of Secondary Metabolites Using Elicitor in *Catharanthus roseus*: A Case Study

Zahid Hameed Siddiqui, Abdul Mujib, Mahmooduzzafar, Junaid Aslam, Khalid Rehman Hakeem and Talat Parween

Abstract Secondary metabolites are mainly derived from plants and are used by humans from time immemorial. A plant cell, tissue, and organ culture has an inherent capacity to manufacture valuable chemical compounds as the parent plant does in nature. *In vitro* plant materials are one of the good sources for the production of secondary metabolite and elicitation can be used as one of the important tool in order to improve the synthesis of these compounds. In a variety of plant cell cultures, elicitors have increased production of terpenoid indole alkaloids, isoflavonoid phytoalexins, serquiterpenoid phytoalexin, coumarins etc. Although elicitation has been carried out in large number of medicinal plants, we extensively studied it in *Catharanthus roseus*, because it is an important source of anticancer compounds Vinblastine (VLB) and Vincristine (VCR). The use of elicitor is also important in order to meet the market demands, for reducing production costs and for in-depth investigation of biochemical and metabolic pathways. This information helps us in manipulation of biosynthetic pathways which can be used as a powerful tool to make natural product-like compounds.

Z. H. Siddiqui (✉)
Department of Botany, Zakir Husain Delhi College, University of Delhi,
Jawahar Lal Nehru Marg, New Delhi 110002, India
e-mail: siddiquizh84@gmail.com

A. Mujib · Mahmooduzzafar
Department of Botany, Hamdard University, New Delhi 110062, India
e-mail: amujib@jamiahamdard.ac.in

Mahmooduzzafar
e-mail: mahmooduzzafar@jamiahamdard.ac.in

J. Aslam
Department of Biotechnology, Hamdard University, New Delhi 110062, India
e-mail: Junaidg1@gmail.com

K. R. Hakeem
Faculty of Forestry, Universiti Putra Malaysia, 43400 Serdang, Selangor, Malaysia
e-mail: hakeem.bot@gmail.com

T. Parween
Department of Bioscience, Jamia Millia Islamia, New Delhi 110025, India
e-mail: talat.jmi@gmail.com

K. R. Hakeem et al. (eds.), *Crop Improvement,* DOI 10.1007/978-1-4614-7028-1_14,

Introduction

Plants with medicinal properties (secondary metabolites) have been used by humans to treat infections, health disorders, and illness since the early days of mankind (Lovkova et al. 2001; Wyk and Wink 2004; Toso 2010). Plants and plant cell cultures have served as resources for flavours (food additives), aromas and fragrances, cosmetics (cosmeceuticals), biobased fuels, insecticides, perfumes, fine chemicals and bioactive compounds (Balandrin and Klocke 1988) and are collectively known as secondary metabolites. These compounds are also well known to play a key role in the adjustment of plants to their surroundings (Rao and Ravishankar 2002). During the last 50 years research work on plant secondary metabolites has been increasing because the daily lives including health care are essentially depends on these plant products (Mulabagal and Tsay 2004). Therefore, in order to achieve the market demand, cultivation of medicinal plants and *in vitro* production of plant secondary metabolites are the only sustainable ways.

Plant cell, tissue, and organ cultures has an inherent capacity to manufacture valuable chemical compounds as the parent plant does in nature which has been recognized since the commencement of *in vitro* technology. *In vitro* plant materials are one of the good sources for the production of secondary metabolite and also provide an excellent environment for in-depth investigation of biochemical and metabolic pathways (Mulabagal and Tsay 2004; Karuppusamy 2009). *In vitro* studies including plant tissues and suspension cultures are continued in diverse directions for the commercial production of secondary metabolites (Ramawat and Merillon 2007; Ghorpade et al. 2011). The accumulation of secondary compounds during plant cell cultures varies significantly due to the elements of the culture medium and environmental conditions (Stafford et al. 1986). Various efforts have been made to circumvent these biological and technological limitations (Lee and Shuler 1991). Robins (1994) reported different strategies in order to improve the synthesis of secondary metabolites in suspension cultures. Different media have been employed for different species and the use of biotic and abiotic elicitors has also been engaged because of their strong and rapid improving effects on indole alkaloid synthesis (Moreno et al 1995). There are many reports of cell culture in which secondary metabolites has been produced, such as solasodine production from calli of *Solanum eleagnifolium* and from root cultures of *Senecio* production of pyrrolizidine alkaloids (Nigra et al. 1987; Toppel et al. 1987) and production of anthraquinones in cell cultures of *Rubia tinctorum*(Abd El-Mawla 2012). Jha et al. (1988) separated cephaelin and emetine from callus cultures of *Cephaelis ipecacuanha* and quinoline alkaloids in considerable amount was separated from *Cinchona ledgeriana* cell suspension cultures (Scragg et al. 1992). In *C. roseus*, Zhao et al. (2001e) reported enhanced alkaloid biosynthesis in suspension culture, Ravishankar and Grewal (1991) reported production of diosgenin in callus culture of *Dioscorea deltoidea* by assessing the effect of media constituents and nutrient stress. Cardenolides biosynthesis was noted to be maximum in the hairy root culture of *Digitalis lanata* as compared to leaf (Pradel et al. 1997) and *in vitro* synthesis of azadirachtin and nimbin was less in field grown plant as compared to cultured shoots and roots of *Azadirachta indica* (Srividya and Devi 1998). It has also been found that because of the source and the type of explants lepidine content in *Lepidium sativum* vary considerably (Pande et al. 2002). It shows that biosynthetic efficiency of populations varies; therefore we should select a high yielding variety as a starting material (Tripathi and Tripathi 2003). The basic prerequisite in all this is a good yield of the compound, and cheap cost compared to the natural synthesis by the plants.

Elicitation can be used as one of the important strategy in order to get better productivity of the bioactive secondary products (Chong et al. 2005; Smetanska 2008; Sharma et al. 2011; Hussain et al. 2012) and reducing production costs (Miao et al. 2000; Zhang et al. 2003). There are reports where elicitors have increased production of isoflavonoid phytoalexins (Smith and Banks 1986), serquiterpenoid phytoalexin (Threlfall and Whitehead 1988), coumarins (Hamerski et al. 1990), podophyllotoxin (Muranaka et al. 1992), azadirachtin (Prakash and Srivastava 2008), terpenoids (Gao et al. 2011), hypericins (hypericin and pseudohypericin) and hyperforin (Coste et al. 2011), capsaicinoid (Gururaj et al. 2012), tanshinone (Kai et al. 2012), vascicine (Bhambhani et al. 2012) and resistance to pathogens (Benhamou et al. 2001). Mihai et al. (2011) reported that biotic and abiotic elicitors stimulate biosynthesis and increase of resveratrol in *Vitis vinifera* callus cultures. Ahmed et al. (2012) reported that different concentration fungal extract (*Aspergillus niger* and *Penicillium notatum*), yeast extract and chitosan enhance the synthesis of psoralen in *Psoralea corylifolia* suspension cultures.

The other side of the elicitation has also been used to elucidate the complex metabolic pathways (Moreno et al. 1996), to characterize the interaction between biotic and abiotic stress responses at a molecular level (Atkinson and Urwin 2012). There are also studies which have been published to mark the effects of elicitors on enzymes of secondary metabolism (Zenk 1991), important signal molecules that mediate plant resistance reactions (Gao et al. 2012; Bux et al. 2012), oxidative burst (Davis et al. 1993) phytoalexin signal transduction (Preisig 1994) and anion channels (Zimmermann et al. 1998). Beside that elicitors can be used to study the chemical nature and signaling pathways of natural substances discharged by microbes, herbivores, and plants during pathogenic contamination, herbivory, symbiosis and allelopathic interactions (El- Samra et al. 2011; Maffei et al. 2012).

Elicitors and Elicitation

Earlier the term elicitor was used for molecules, which induced the production of phytoalexins, but it is now simplified as a compound which promotes various kinds of plant defense system (Hahn 1996; Nürnberger 1999). Further an elicitor may also be defined as a material which, instigate or advances the biosynthesis of specific compounds when added in small amount to a living cell system. In this way elicitation can be explained as the stimulated or improved biosynthesis of compounds due to addition of trace amounts of elicitors (Radman et al 2003, Angelova et al. 2006).

On the basis of their origin, structure and type, elicitors can be classified as biotic and abiotic (Eilert 1987; Barz et al. 1988). There is one more benefit of the use of elicitors is that they also encourage release of the metabolites into the medium (Pitta-Alvarez et al. 2000).

Biotic Elicitors

These are chemically complex biological compounds with unknown composition like microbial cell-wall preparations and yeast extract. In some cases, and particularly in recent years, elicitors with known chemical structure have been selected which helped in more detailed investigation of the elicitation process. Carbohydrates and proteins are examples of such defined elicitors (Radman et al. 2003). Different workers further classified these substances on the basis of their similarity. Material from living organisms include different polysaccharides from plant cell walls (pectin or cellulose) and microbial extracts (chitin or glucans) and glycoproteins (Eilert 1987; Nishi 1994; Benhamou 1996; Shirsau et al. 1997); phytoalexins: which are low-molecular-mass antimicrobial secondary compounds synthesized by plants in reaction of fungal or bacterial attack and physical damage. Protein kinase: regulate growth and cellular development by phosphorylating a number of target proteins; *calmodulin*: intracellular Ca^{2+}-binding proteins consisting of at least two different peptides, with four Ca^{2+}binding sites; *calmodulin* has no enzyme activity of its own, but acts by binding to other proteins (Radman et al 2003, Angelova et al. 2006). Recently Siddiqui et al. (2010) described fungal elicitor as a potent approach for enhancing secondary metabolites in cultured cells.

Abiotic Elicitors

As compared to biotic elicitors the use of abiotic elicitors in plant cell cultures has received less interest (Radman et al 2003, Angelova et al. 2006). Abiotic elicitor or stress agents are non-biological substances which includes different kinds of inorganic salts and physical factors like UV radiation, heavy metal salts (Cu and Cd ions), Ca^{2+}, high pH and other chemical compounds with diverse mechanism of action (Eilert 1987; Radman et al 2003). Recently Zuccarini (2009) reviewed ozone as a fungal elicitor. Addition of $AgNO_3$ and $CdCl_2$ to the cultures of *Brugmansia candida* (angel's trumpet) enhanced significantly the accumulation and release of these alkaloids, but $CdCl_2$ inhibited the growth of the hairy roots (Pitta-Alvarez et al. 2000). Wu et al. (2001) investigated the synthesis of taxol in cell cultures of *Taxus* sp. under the influence of the rare-earth metal lanthanum and reported a considerable augmentation (280 %) of taxol. Although elicitation has been carried out in large number of medicinal plants but for this chapter we extensively studied it in *C. roseus*. There is a great importance of this plant as it is an important source of anticancer compounds with activity against various kinds of carcinoma (Svoboda and Blake 1975; Schmeller and Wink 1998) along with it also houses a large number of Terpenoid indole alkaloids (TIAs) (Svoboda and Blake 1975; Van der Heijden et al. 1989; 2004), which possess pharmacological activity against different other diseases. The aerial parts of the plant contain 0.2–1 % alkaloids (Bruneton 1993). About 130 alkaloids have been isolated from *C. roseus*, of particular interest is a group of 20 dimeric alkaloids (Evans 1996). However of this large number of chemical compounds, only few (~ 11) are regularly investigated and even lesser (~ 8) are commercially available (Hisiger and Jolicoeur 2007). VCR and VLB are two most important alkaloids which can be found in *C. roseus* and are known for their antineoplastic activity. Ajmalicine and serpentine are the antihypertensive agents (Shanks et al. 1998), used to combat heart arrhythmias and improves the blood circulation in the brain (Moreno et al. 1995; Schmeller and Wink 1998). Other alkaloidal fractions from the leaves including vindoline are antibacterial in nature. Pericalline, perivine, VLB, VCR, carosine etc. show antiviral activity *in vitro* against vaccinia and polio type III viruses with pericalline being the most effective. Vincolidine, lochrovicine, catharanthine and vindoline shows diuretic activity. In different Indian system of medicine it is used as an antidiabetic agent (Singh et al. 2001), catharanthine, leurosine, lochnerine, tetrahydroalstonine, vindolineproduce varying degrees of lowering of blood sugar in rats. An alkaloidal mixture obtained from flowers showed significant hypoglycaemic activity in rabbits (Wealth of India 1992). Velban® is a trade name for Vinblastine sulphate and is employed mainly for the treatment of Hodgkin's disease in addition lymphosarcoma, choricarcinoma and carcinoma of the lungs, breast and other organs in acute and chronic leukemia. Oncovin® sold as Vincristine sulphate, arrests mitosis in metaphase and is very successful for the treatment of acute leukemia in children and lymphocytic leukemia. It is also administered against Hodgkin's disease, Wilm's tumour, rhabdosarcoma, reticulum cell sarcoma and neuroblastoma. VCR is superior to VLB in the treatment of lymphosarcoma and has greater toxicity (Wealth of India 1992). In terms of production of VLB and VCR, India ranks third in the world and is exporting these alkaloids to European countries. High demand and low yield of these alkaloids in the plant has led to research for alternative means for their production.

Elicitation of *C. roseus*

The use of a variety of biotic and abiotic elicitors or signal molecules in cell cultures frequently increases the yield of certain secondary compounds, perhaps due to their role in defense (Zhao and Verpoorte 2007; Aijaz et al. 2011). The biosynthesis of secondary metabolites in *C. roseus* cell cultures was lucidly reviewed by Moreno et al. (1995) and important factors affecting the production of indole alkaloids were discussed. Further Moreno et al. (1996) studied the suspension cultures of *C. roseus* cell under the influence of fungal filtrate of *Phytium aphanidermatum* as an elicitor. Although there was no enhancement in alkaloid synthesis was observed but the authors studied the result of elicitation on different metabolic pathways and summarized that the biosynthesis of phenolics and the pathway to tryptamine are the two important fluxes of intermediates in *C. roseus* cell cultures. There are reports which suggest that indole alkaloids production is also affected by abiotic stresses of sorbitol and mannitol (osmotic stress) whereas NaCl and KCl were employed to create salt stress (Moreno et al. 1995; Zhao et al. 2000b). Zhao et al. (2000a) employed metal stress with the help of vanadyl sulphate, sodium orthovanadate and some rare earth elements whereas (Zhao et al. 2000c; 2001c) stimulated with various chemicals. Fungal elicitors and hormones were used by (Namdeo et al. 2000; Zhao et al. 2001d; El-Sayed and Verpoorte 2004). Thus elicitation of *C. roseus* cell cultures not only enhances indole alkaloid biosynthesis in short time, but it is also responsible for the excretion of the products into the medium (Zhao and Verpoorte 2007).

Rijhwani and Shanks (1998) reported the effects of pectinase and methyl jasmonate elicitor on growth and levels of several alkaloids in *C. roseus*. When pectinase (72 units) was added about 150% increase in tabersonine was observed whereas due to addition of jasmonic acid (JA) a progressive increase of 60, 80, 150 and 500% was observed respectively in serpentine, ajmalicine, lochnericine hörhammericine. The production of ajmalicine or catharanthine in cell suspension cultures of *C. roseus* was enhanced by cerium (CeO_2 and $CeCl_3$), yttrium (Y_2O_3) and neodymium ($NdCl_3$). The yield of ajmalicine in these treated-cultures reached 51 mg/l (CeO_2), 40 mg/l ($CeCl_3$), 41 mg/l (Y_2O_3) and 49 mg/l ($NdCl_3$) while catharanthine production reached to 36 mg/l (CeO_2) and 31 mg/l ($CeCl_3$). In these treatments a main part of improved alkaloids was released into medium (Zhao et al. 2000a). When 14–1 bioreactor was compared with shake flask culture of *C. roseus* cell line a 80% decrease in total alkaloid production was observed, but in the same culture when 1 mM trans-cinnamic acid was added the original alkaloid amounts was restored (Godoy-Hernandez et al. 2000). Zhao et al. (2001b) reported that indole alkaloid biosynthesis in *C. roseus* cell cultures due to the action of elicitor is associated to Ca^{2+} influx and the oxidative burst and up to some extent indole alkaloid accumulation was inhibited by calcium channel blockers which could be improved by re-addition of calcium chloride.

In *C. roseus* cell suspension culture Zhao et al. (2001a, b) employed biotic elicitors derived from 12 fungi in order to test their effect on indole alkaloid production. They reported that different indole alkaloids were stimulated by different fungal mycelium homogenates and an improvement (2–5-fold high than control) in alkaloid synthesis was observed. Ten Hoopen et al. (2002) marked the effect of temperature on growth and ajmalicine production and reported that 27.5^0 C was optimum temperature for biomass and secondary metabolite production. In shake flasks and bioreactors an improved catharanthine synthesis in *C. roseus* cell cultures was observed by combined elicitor treatment (Zhao et al. 2001a). A combination of fungal preparations and chemicals enhanced the alkaloid accumulation, a highest yield of ajmalicine with an enhanced catharanthine accumulation was observed in a combination of tetramethyl ammonium bromide and mycelial homogenate of *Aspergillum niger*. The combination of malate and sodium alginate proved to be beneficial for the highest yield of catharnthine with a high yield of ajmalicine production. Later the process was optimized and refined by the authors (Zhao et al. 2001b), after 10 days of *C. roseus* cells in shake flasks and in bioreactor they reported 25 mg/l, 32 mg/l and 22 mg/l catharanthine yields in 500 ml flasks, 1,000 ml flasks and in 20 l airlift bioreactor, respectively. The defense responses, such as lipid peroxidation was believed to be stimulated by the combination of malate and alginate treatment in all *C. roseus* culture processes which further mediate the catharanthine production via the jasmonate pathway. El-Sayed and Verpoorte (2002) tested 2, 4-D and abscisic acid (phytohormones), salicylic acid (SA) and MJ on growth and accumulation of secondary metabolites in *C. roseus* cell suspension culture upon feeding with the precursors loganin and tryptamine. Among the tested treatments only MJ enhanced the accumulation of alkaloids whereas due to addition of abscisic acid catabolism of strictosidine was delayed.

Zheng and Wu (2004) treated the *C. roseus* cell with different Cadmium (Cd) concentration (0.05 to 0.4mM) and ajmalicine yield was monitored. They reported that due to Cd treatment a higher yield of ajmalicine was recorded because of the increased level of tryptophan decarboxylase (Tdc) transcript, the cellular tryptamine concentration, and ajmalicine excretion. The effect of different elements namely Co, Zn, Ni, Mn, Cr, W, Cu, B, V, Fe, and Mo and various hormones including natural and synthetic auxins, cytokinins, and gibberellin on the production and accumulation of indole alkaloids in *C. roseus* was investigated (Lovkova et al. 2005) studied. These compounds modified different phases in the biosynthesis of catharanthine and vindoline and up to a certain extent a feasible mechanism of the effect of Zn and auxin on this process were simplified. In *C. roseus* cell suspension cultures indole alkaloid production through a protein kinase-dependent signal pathway was stimulated using nitric oxide (Xu and Dong 2005) whereas $CaCl_2$ enhanced MJ-induced ajmalicine production in *C. roseus* (Lee-parson and Ertürk 2005). About 160 % increase in ajmalicine production was noted in *C. roseus* cultures and the synthesis depends on intracellular Ca^{2+} concentration. When Ca^{2+} influx was increased after a certain level by the addition of extracellular Ca^{2+}, ajmalicine production was declined, similarly a decrease in the accumulation of alkaloids was noted down when Ca^{2+} influx was dropped off.

For the production of ajmalicine in *C. roseus* cultures different strategies of optimizing gas compositions were used (Lee-Parsons 2007). Guo et al. (2007) conducted an experiment to study the effect of various temperatures on variation of alkaloid metabolism in *C. roseus* seedlings. The authors observed that with relation to the treatment time, at high temperature biosynthesis of different alkaloids were elevated in *C. roseus* seedlings. In *C. roseus* cell suspension culture a low dose of UV-B irradiation was applied (Ramani and Jayabaskaran 2007), which stimulated the transcription of genes encoding tryptophan decarboxylase (*Tdc)* and strictosidine synthetase *(Str)* and induced enhanced amount of catharanthine. In another experiment Ramani and Chelliah (2008) evaluated the influence of UV-B treatment on cell suspension culture of *C. roseus* in different growth phase. The results suggested that in stationary phase cultures the response to UV-B irradiation was more than the late exponential phase. There was a 3 and 12-fold enhancement in catharanthine and vindoline respectively. An efficient and promising protocol for achievement and enhancement of anthocyanin production from calli cultures of *C. roseus* was developed by Taha et al. (2008). The highest values 78.73 μg/gm of total anthocyanin production were recorded with *Catharanthus* calli cultures, when the MS medium was amended with 3 and 0.5 μM of L-phenylalanine and $CaCl_2$ respectively.

Binder et al. (2009) recorded a considerable enhancement in the production of TIAs on *C. roseus* hairy roots by exposing UV-B light. Alkaloid concentrations were analyzed up to 168 h after UV-B exposure that shows a considerable increase in the accumulation of lochnericine and considerable decrease in the accumulation of hörhammericine over time. In *in vitro* cell suspension, rootless shoot cultures and hairy roots of *C. roseus* effects of different abiotic agents like SA, ethylene and MJ on alkaloid accumulation was described (Vázquez-Flota et al. 2009). Jasmonate and ethylene treatments promoted ajmalicine accumulation; catharanthine and ajmalicine were stimulated by jasmonate in hairy roots, catharanthine accumulation was only induced by ethylene. In shoot cultures a positive vindoline accumulation was noticed under the influence of jasmonate and ethylene whereas in any of the studied *in vitro* culture systems, SA did not spot any effect. Ruiz-May et al. (2009) noted enhanced accumulation of alkaloids (ajmalicine, serpentine, ajmaline and catharanthine) in *C. roseus* hairy roots with different concentrations of MJ elicitation. Senoussi et al. (2009) reported that the lack of oxygenation in the culture medium provoked a very strong inhibition in accumulation of alkaloids and the addition of BA to the culture medium restored accumulation by increasing the ajmalicine production and eliminated the inhibitory effect of hypoxia. In Egyptian *C. roseus*, suspension culture was induced from leaf explants and the influences of different amino acids (L-tryptophan L-glutamine; L-asparagine; L-cystine and L-arginine) at different concentrations were examined for enhanced production of indole alkaloids. With modified MS medium containing 300 mg/l of either L-glutamine or L-typtophan the highest biomass and indole alkaloids production were achieved (Taha et al. 2009). Poutrin et al. (2009) demonstrated that calcium regulated the auxin-dependent monoterpenoid indole alkaloid (MIA) biosynthesis in *C. roseus*. Mustafa et al. (2009) examined the metabolic profile of *C. roseus* suspension using NMR spectroscopy and multivariate data analysis under the effect of SA. The data revealed that in SA treated cells high level of sugars (glucose and sucrose) were accumulated after treatment and/thereafter an active variation in tryptamine, amino acids and phenylpropanoids were observed. VCR production in callus culture of *C. roseus* was affected by auxin and cytokinin (Kalidass et al. 2010). The accumulation of VLB, vindoline and catharanthine in *C. roseus* was monitored under the influence of several agents (Pan et al. 2010). A noteworthy increase in VLB, vindoline and catharanthine was recorded down under SA and ethylene treatments, while abscisic acid and gibberellic acid had shown a negative effect on the accumulation of the same alkaloids. MJ treatment was not effective on the synthesis of these alkaloids and chlormequat chloride lowered the concentration of vindoline and catharanthine but it improved the accumulation of VLB. In the presence of flavin mononucleotide and manganese ions coupling reaction between catharanthine and vindoline occurs non-enzymatically at NUV light irradiation *in vitro* (Asano et al. 2010). It was also noted that the catharanthine and vindoline synthesis reduced and those of dimeric indole alkaloids were enhanced under NUV light at 4^0 C in *C. roseus*. A homogenate of *Pythium aphanidermatum* and MJ was studied on *in vitro* cultures of *C. roseus* (cv. 'Dhawal') (Shukla et al. 2010). Due to elicitation transcriptional upregulation of strictosidine beta-D:-glucosidase (SGD) occur which in turn improved the synthesis of total alkaloids but did not produce vindoline. Recently in a pot experiment foliar application of SA was conducted, to find out the unfavorable effects of water stress on periwinkle and its amelioration by SA (Idrees et al. 2011). The result suggested that SA (10^{-5}M) foliar application minimized destructive effects of stress and enhanced growth parameters and simultaneously improved the anticancerous compound VCR and VLB in stressed plants. Aslam et al. (2011) reported effects of freezing and non-freezing temperature on somatic embryogenesis and vinblastine synthesis in *C. roseus*. At 15°C temperature maximum numbers of embryos were produced whereas the same were matured in maximum number at 4°C. They also reported that VLB synthesis was temperature dependent. By using cyclodextrins and methyljasmonate along with a short exposure of UV enhanced the ajmalicine accumulation in suspension culture of *C. roseus* (Almagro et al. 2011). In *C. roseus* (Guo et al. 2012) studied the physiological responses of different nitrogen forms including varying ratio of nitrate to ammonium (1:0, N_1; 1:1, N_2; 1:3 N_3). After long term incubation in N_2 nitrogen solution catharanthine and VLB synthesis was increased to two folds than in N_1 or N_3 nitrogen solution.

Mechanism of Elicitor Action

In order to find out the mechanism of elicitor action in plants a thorough research has been dedicated (Angelova et al. 2006; Siddiqui et al. 2010), however, the induction mechanisms conveyed by fungal elicitors and the plant signals, as a whole, is still undefined (Djonović et al. 2007). While understanding the mechanism a general outline for the biotic elicitation in plants may be abridged by different workers on the basis of elicitor-receptor interaction (Zhao et al. 2005; Namdeo 2007). Elicitors are of very large array of structure, and can be transmitted by the pathogen (exogenous elicitors) or formed by the plant as a result of the plant–pathogen interaction (endogenous elicitors); in both cases, they stimulate the defense reaction of the plant (Ebel and Cosio 1994). After pathogen or elicitor recognition, sequences of cytological variation and biochemical reactions have been identified in plant cells. The cytological variations include papilla formation, increased cytoplasmic streaming and nuclear migration, which are connected with depolymerization of microtubules and microfilaments (Kombrink and Schmelzer 2001). The biochemical reactions include transformation in the H^+, K^+, Cl^-, and Ca^{2+} fluxes across the plasma membrane, and the development of reactive oxygen species (ROS) that occur within 2–5 min after treatment of elicitor (Low and Merida 1996; Nürnberger 1999), oxidative cross linking of cell wall proteins, formation of phytoalexins, hydrolytic enzymes, incrustations of cell wall proteins with phenolics and hypersensitive death of plant cell (Namdeo 2007).

The first step during the contact between a pathogen and its host is species specific in which elicitor binds to plasma membrane receptor for the process of elicitation (Braun and Walker 1996; Hanania and Avni 1997), after that the genes of avirulence brought by the pathogen (the elicitors) faces resistance from the host genes of resistance. After proper recognition of the elicitors by the proper receptors of the plant, defenses mechanism comes into play and they activate its genes of resistance (Hahn 1996; Montesano et al. 2003) through a series of complex reaction which is still under investigation and might not be similar for all type of elicitors but it might give an insight about their mechanism of action (Radman et al. 2003, Namdeo 2007). After binding of the elicitor to plasma membrane a rapid ion fluxes across the membrane occur (Mathieu et al. 1991). As per other reports (Gelli et al. 1997; Pitta-Alvarez et al. 2000; Mithöfer et al. 2001) an influx of Ca^{2+} to the cytoplasm from the extracellular environment and intracellular Ca^{2+} reservoir and stimulation of K^+ and Cl^- efflux take place (Ivashikina et al. 2001). It is followed by speedy changes in protein phosphorylation and protein kinase activation pattern upon elicitor treatment (Yang et al. 1997; Roemis 2001); activation of protein kinases, in turn activate the mitogen-activated protein kinase (MAPK), MAP-kinases and calcium-dependant kinases catalyze mostly the Thr-Ser phosphorylation in the target proteins (Roemis 2001). The MAP kinase cascade involves MAP kinase kinase kinase (MAPKKK) proteins phosphorylating MAP kinase kinases that in turn phosphorylate MAP kinases. Upon activation, MAPKs are transported to the nucleus where they phosphorylate specific transcription factors. Activation of G-protein by some workers (Tyler 2002; Luan 1998; Roos et al. 1998) suggested their involvement in the early responses to elicitors (Legendre et al. 1992), followed by the production of secondary messengers Ins (1, 4, 5) *P*3 and diacylglycerol (DAG) (Mahady et al. 1998) mediating intracellular Ca^{2+} release, nitric oxide (Delledonne et al. 2002; Huang et al. 2002) and octadecanoid signalling pathway (Piel et al. 1997). Later on AOS and cytosol acidification is carried out by activation of NADPH oxidase (Leburun-Garcia et al. 1999) and reorganization of cytoskeleton take place (Kobayashi et al. 1995) and production of ROS like superoxide anion and H_2O_2 (Low and Merida 1996). It is followed by accumulation of pathogenesis-related proteins (PR proteins) (Mittler et al. 2004). Earlier Bol et al. (1990) and Bowles (1990) reported individually about the increase in PR proteins in response to elicitor. PR protein included chitinases, glucanases, endoploygalactouranases that contribute to the release of signaling pectic oligomers, hydroxyproline rich glycoproteins and protease inhibitors (Van Loon and Van Strien 1999). In hyper sensitive responses cell death occurs at the infection site (Luan 1998) and changes in the cell wall organization (lignification of the cell wall, callus deposition) take place (Kauss et al 1989). These all consequences are responsible for the transcriptional activation of the corresponding defence response genes (Memelnik et al. 2001; Huang et al. 2002) and accumulation of phytoalexins and tannins (Pedras et al. 2002) which stimulate the production of jasmonic and SA as secondary messengers (Memelink et al. 2001; Katz et al. 2002) and plant cell acquired systematic resistance (Lebrun-Garcia et al. 1999). It is known that the sequence of these events and arrangements is a very complex process and is still under investigation. It is further reported that this sequence of events is not followed by all elicitors, in some peptide elicitors plasma-membrane receptors are active (Nennstiel et al. 1998), certain peptides of bacterial origin act themselves as messenger of invasion signals and get transported to distal tissues. The systemic necrosis due to these proteins is not stimulated by rapid secondary signaling, but by transportation of the elicitor inside the tissue (Devergne et al. 1992). On the basis of available information, we summarize the following role/involvement of elicitor in the course of essential events in secondary metabolism.

- Binding of the elicitor to plasma membrane receptor.
- Changes in Ca^{2+} influx to the cytoplasm from extracellular and intracellular pools.
- Decrease of pH of the cytoplasm and activation of NADPH oxidases, protein phosphorylation patterns and protein kinase activation.
- Changes in cell wall structure (lignification) and in generating reactive oxygen species.
- Synthesis of JA and SA as secondary messengers.
- Activation of genes that produce defence-related proteins, plant defence molecules like phytoalexins and other secondary compounds including alkaloids.

In the same way elicitor-based signaling model has been proposed recently for enhanced activation of gene expression in *C. roseus* (Fig. 14.1) where alkalization of the medium by influxing proton molecule (particularly by calcium) was carried out by yeast elicitor. This in turn stimulates the octadecanoid pathway and synthesis o JA takes place. JA act as secondary messenger and influences the synthesis of nuclear proteins ORCA2 and ORCA3. These proteins interact with TDC- and 'JA-responsive STR' promoter of several biosynthetic genes and activate their gene expression (Memelink et al. 2001). Due to 'elicitor-receptor' mediated signaling, a large number of key gene's activities such as *Tdc*, *Str*, geraniol 10 dehydrogenase (*Gh*), anthranilate synthetase (*As*) increased (Van der Fits and Memelink 2000). We also believe that this architecture will evolve in future as refinement of cell biology and molecular biology approaches will allow further and full dissection of plant cell signaling pathways that effectively regulate elicitation events in *C. roseus*.

Conclusion and Future Perspectives

C. roseus is a very important plant due to large number of alkaloids which are used in treatment of various diseases including antineoplastic agents VLB and VCR. Present review provides an overview of external stress stimuli used for eliciting *C. roseus* cell to undergo a complex network of reactions which ultimately lead to the synthesis and accumulation of secondary metabolites. These secondary metabolites help out the plant to endure in stress challenge. They are also under high demands due to their pharmacological activities but due to the poor understanding of plant

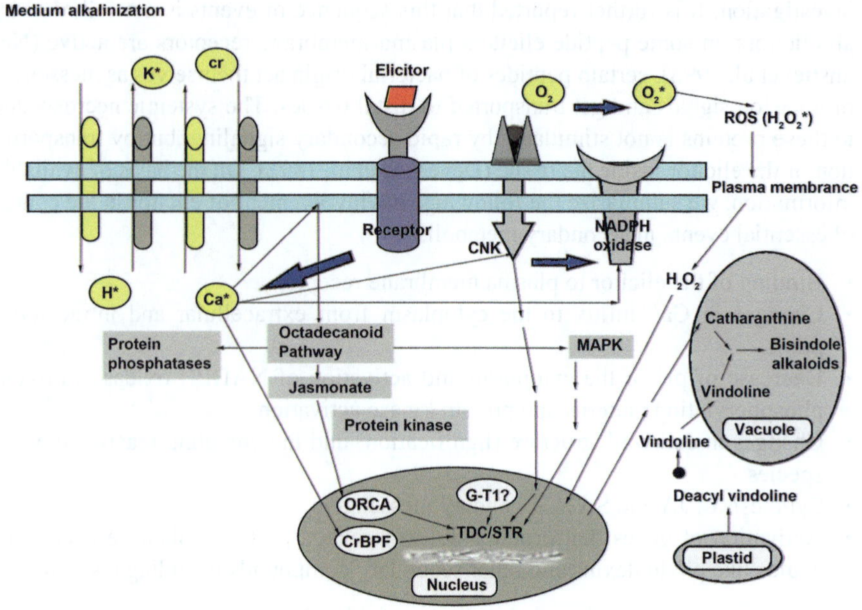

Fig. 14.1 Model for the elicitor mediated signal transduction leading to activation of genes in indole alkaloid synthesis pathways. (Modified from Memelink et al. 2001)

secondary metabolism, production is still far behind the target. By unraveling the signaling network would help in specific and competent engineering of the production of target secondary metabolites. The elicitation approaches not only enhances the synthesis of secondary metabolites in the plant system, it also helps us in better understanding and identifying the rate-limiting steps of complex biosynthetic pathways existing in secondary metabolite synthesis which in turn can contribute towards better productivity by utilizing metabolic engineering aspects.

Acknowledgements The first author is thankful to Council of Scientific and Industrial Research (CSIR) for receiving Senior Research Fellowship (SRF).

References

Abd El-Mawla, AMA (2012) Influence of certain abiotic elicitors on production of anthraquinones in cell cultures of *Rubia tinctorum*. Spatula DD 2:89–94

Ahmed SA, Baig MMV (2012) Elicitor enhanced production of psoralen in suspension cultures of *Psoralea corylifolia* L. www.apjtb.com/press/2012/B691.doc

Aijaz A, Jain S, Hariharan AG (2011) Effect of elicitation on the production of phyto-constituents through plant tissue culture technique—a review. Int. J Drug Disc Herb Res 1:84–90

Almagro L, López Perez AJ, Pedreño MA (2011). New method to enhance ajmalicine production in *Catharanthus roseus* cell cultures based on the use of cyclodextrins. Biotech Lett 33(2):381–385
Angelova Z, Georgiev S, Roos W (2006) Elicitation of plants. Biotech. Equip 20:72–83
Asano M, Harada K, Yoshikawa T, Bamba T, Hirata K (2010) Synthesis of anti-tumor dimeric indole alkaloids in *Catharanthus roseus* was promoted by irradiation with near-ultraviolet light at low temperature. Biosci Biotechnol Biochem 74:386–389
Aslam J, Mujib A, Sharma MP (2011) Influence of freezing and non-freezing temperature on somatic embryogenesis and vinblastine production in *Catharanthus roseus* (L.) G. Don. Acta Physiol Plant 33(2):473–480
Atkinson NJ, Urwin PE (2012) The interaction of plant biotic and abiotic stresses: from genes to the field. J Exp Bot. doi:10.1093/jxb/ers100
Balandrin MF, Klocke JA (1988) Medicinal, aromatic and industrial materials from plants. In: Bajaj YPS (ed) Biotechnology in agriculture and forestry, Vol. 40. Springer-Veralg, Berlin, pp 1–35
Barz W, Daniel S, Hinderer, W, Jaques U, Kessmann H, Koster J, Tiemann K (1988) Elicitation and metabolism of phytoalexins in plant cell cultures. In: Pais M, Mavituna F, Novais J (eds) "Plant cell biotechnology". "Biotechnology". NATO ASI series. Springer-Verlag, Berlin, pp 211–230
Benhamou N (1996) Elicitor induced plant defence pathways. Trends Plant Sci 1:233–240
Bhambhani S, Karwasara VS, Dixit VK, Banerjee S (2012) Enhanced production of vasicine in *Adhatoda vasica* (L.) Nees. cell culture by elicitation. Acta Physiol Plant 34:1571–1578
Binder BY, Peebles CA, Shanks JV, San, KY (2009) The effects of UV-B stress on the production of terpenoid indole alkaloids in *Catharanthus roseus* hairy roots. Biotechnol Prog 25(3):861–866
Bol JF, Linthorst HJM, Cornelissen BJC (1990) Plant pathogenesis-related proteins induced by virus infection. Annual Rev Phytopath 28:113–138
Bowles DJ (1990) Defense-related proteins in higher plants. Annual Rev Biochem 59:873–907
Braun DM, Walker JC (1996) Plant transmembrane receptors: new pieces in the signalling puzzle. Trends Biochem Sci 21:70–73
Bruneton J (1993) Pharmacognosy phytochemistry medicinal plants, 2nd edn. Lavoisier Publishers, Paris
Bux H, Ashraf M, Rasheed A, Poudyal DS, Kazi AG, Afzaal M (2012) Molecular basis of disease resistance in cereal crops: an overview. In: Ashraf M, Öztürk M, Ahmad MSA, Aksoy A (eds) Crop production for agricultural improvement. Springer, Netherlands, pp 477–489
Chong TM, Abdullah MA, Lai QM, Nor AFM, Lajis NH (2005) Effective elicitation factors in *Morinda elliptica* cell suspension culture. Process in Biochem 40:3397–3405
Coste A, Vlase L, Halmagyi A, Deliu C, Coldea G (2011) Effects of plant growth regulators and elicitors on production of secondary metabolites in shoot cultures of *Hypericum hirsutum* and *Hypericum maculatum*. Plant Cell Tiss and Org Cult 106:279–288
Davis D, Merida J, Legendre L, Low PS, Heinstein P (1993) Independent elicitation of the oxidative burst and phytoalexin formation in cultured plant cells. Phytochemistry 32:607–611
Delledonne M, Murgia I, Ederle D, Sbicego PF, Biondani A, Polverari A, Lamb C (2002) Reactive oxygen intermediates modulate nitric oxide signaling in the plant hypersensitive disease-resistance response. Plant Physiol Biochem 40:605–610
Devergne J-C, Bonnet P, Panabiers F, Blein JP, Ricci P (1992) Migration of the fungal protein cryptogein within tobacco plants. Plant Physiol 99:843–847
Djonović S, Walter A, Vargas, Michael V, Kolomiets, Michelle, Horndeski, Wiest A, Charles MK (2007) A proteinaceous elicitor Sm1 from the beneficial fungus *Trichoderma virens* is required for induced systemic resistance in Maize. Plant Physiol 145:875–889
Ebel J, Cosio EG (1994) Elicitors of plant defense responses. Int Rev Cytol 148:1–36
Eilert U (1987) Elicitation: methodology and aspects of application. In: Vasil IK, Constabel F (eds) Cell culture and somatic cell genetics of plants, Vol 4. Academic Press, San Diego, pp 153–196
El-Samra, I.A. Amer MA, Abd-El-Hamid MR, Kabeil SS, El-Alwany AM (2011) Chemical reaction in tomato plants in response to a biotic elicitors treatments. Nature and Science 9:169–185

El-Sayed M, Verpoorte R (2002) Effect of phytohormones on growth and alkaloid accumulation by a *Catharanthus roseus* cell suspension cultures fed with alkaloid precursors tryptamine and loganin. Plant Cell Tiss Org Cult 68:265–270

El-Sayed M, Verpoorte R (2004) Growth, metabolic profiling and enzymes activities of *Catharanthus roseus* seedlings treated with plant growth regulators. Plant Growth Regul 44:53–58

Evans WC (1996) Trease and evans, pharmacognosy, 14th ed. Harcourt Brace and Company, Asia Pvt Ltd

Gao F-k, Yong Y-h, Dai C-c (2011) Effects of endophytic fungal elicitor on two kinds of terpenoids production and physiological indexes in *Euphorbia pekinensis* suspension cells. J Med Plants Res 5(18):4418–4425

Gao F-K, Ren C-G, Dai C-C (2012) Signaling effects of nitric oxide, salicylic acid, and reactive oxygen species on isoeuphpekinensin accumulation in *Euphorbia pekinensis* suspension cells induced by an endophytic fungal elicitor. J Plant Growth Regul. doi:10.1007/s00344-012-9258-8

Gelli A, Higgins VJ, Blumwald E (1997) Activation of plasma membrane Ca^{2+}membrane channels by race specific fungal elicitors. Plant Physiol 113:269–279

Ghorpade R P, Chopra A, Nikam TD (2011) Influence of biotic and abiotic elicitors on four major isomers of boswellic acid in callus culture of *Boswellia serrata* Roxb. Plant Omics. 4:169–176

Godoy-Hernandez GC, Vazquez-Flota FA, Loyola-Vargas VM (2000) The exposure to trans-cinnamic acid of osmotically stressed *Catharanthus roseus* cells cultured in a 14–1 bioreactor increases alkaloid accumulation. Biotechnol Lett 22:921–925

Guo X-R, Yang L, Yu J-H, Tang Z-H, Zu Y-G (2007) Alkaloid variations in *C. roseus* seedlings treated by different temperatures in short term and long term. J Forestry Res 18:313–315

Guo X-R, Zu Y-G, Tang Z-H (2012) Physiological responses of *Catharanthus roseus* to different nitrogen forms. Acta Physiol Planta 34(2):589–598

Guoyin Kai G, Liao P, Xu H, Wang J, Zhou C, Zhou W, Qi Y, Xiao J, Wang Y, Zhang L (2012) Molecular mechanism of elicitor-induced tanshinone accumulation in *Salvia miltiorrhiza* hairy root cultures. Acta Physiol Plant 34(4):1421–1433

Gururaj HB, Giridhar P, Ravishankar GA (2012) Laminarin as a potential non-conventional elicitor for enhancement of capsaicinoid metabolites. Asian J Plant Sci Res 2:490–495

Hahn MG (1996) Microbial elicitors and their receptors in plants. Annual Rev Phytopath 34:387–412

Hamerski D, Beier RC, Kneusel RE, Matern U, Himmelspacht K (1990) Accumulation of coumarins in elicitor-treated cell suspension cultures of *Ammi majus*. Phytochemistry 29:1137–1142

Hanania U, Avni A (1997) High affinity binding site for the ethylene inducing xylanase elicitor on *Nicotiana tabacum* membrane. Plant J 12:113–120

Hisiger S, Jolicoeur M (2007) Analysis of *Catharanthus roseus* alkaloids by HPLC. Phytochem Rev 6:207–234

Huang X, Kiefer E, Von Rad, U, Ernst D, Foissner I, Durner J (2002) Nitric oxide burst and nitric oxide-dependent gene induction in plants. Plant Physiol Biochem 40:625–631

Hussain MS, Fareed S, Ansari S, Rahman MA, Ahmad IZ, Saeed M (2012) Current approaches toward production of secondary plant metabolites. J Pharm Bioallied Sci 4:10–20

Ivashikina N, Becker D, Ache P, Meyerhoff O, Felle HH, Hedrich R (2001) K^+ channel profile and electrical properties of *Arabidopsis* root hairs. FEBS Letters 508:463–469

Jha, S, Sahu, NP, Mahato, SB (1988) Production of the alkaloids emetine and cephaeline in callus cultures of *Cephaelis ipecacuanha*. Planta Medica 54:504–506

Kalidass C, Mohan VR, Arjunan Daniel A (2010) Effect of auxin and cytokinin on vincristine production by callus cultures of *Catharanthus roseus* L. (Apocynaceae). Trop Subtrop Agroeco 12:283–288

Karuppusamy S (2009) A review on trends in production of secondary metabolites from higher plants by *in vitro* tissue, organ and cell cultures. J Medi Plants Res 3:1222–1239

Katz V, Fuchs A, Conrath U (2002) Pretreatment with salicylic acid primes parsley cells for enhanced ion transport following elicitation. FEBS Letters 520:53–57

Kauss H, Jeblick W, Domard A (1989) The degrees of polymerization and N-acetylation of chitosan determine its ability to elicit callus formation in suspension cells and protoplasts of *Catharanthus roseus*. Planta 178:385–392

Kobayashi I, Murdoch LJ, Kunoch H, Hardham AR (1995) Cell biology of early events in the plant resistance response to infection by pathogenic fungi. Canadian J Botany 73:418–425

Kombrink E, Schmelzer E (2001) The hypersensitive response and its role in local and systemic disease resistance. European J Plant Path 107:69–78

Lebrun-Garcia A, Bourque S, Binet MN, Ouaked F, Wendehenne D, Chiltz A, Schaffner A, Pugin A (1999) Involvement of plasma membrane proteins in plant defense responses. Analysis of the cryptogein signal transduction in tobacco. Biochimie 81:663–668

Lee CWT, Shuler ML (1991) Different flask closures alter gas phase composition and ajmalicine production in *Catharanthus roseus* cell suspensions. Biotech Techn 5:173–178

Lee-Parsons CWT (2007) Gas composition strategies for the successful scale-up of *Catharanthus roseus* cell cultures for the production of Ajmalicine. Phytochem Rev 6:419–433

Lee-Parsons CWT, Ertürk S (2005) Ajmalicine production in methyl jasmonate-induced *Catharanthus roseus* cell cultures depends on Ca^{2+} level. Plant Cell Rep 24:677–682

Legendre L, Heinstein PF, Low PS (1992) Evidence for participation of GTP-binding proteins in elicitation of the rapid oxidative burst in cultured soybean cells. J Biol Chem 267:20140–20147

Lovkova MY, Buzuk GN, Sokolova SM, Kliment'eva NI (2001) Chemical features of medicinal plants (Review). App Biochem Micro 37:229–237

Lovkova M Y, Buzuk GN, Sokolova SM, Buzuk LN (2005) Role of elements and physiologically active compounds in the regulation of synthesis and accumulation of indole alkaloids in *Catharanthus roseus* L. App Biochem Micro 41:299–305

Low PS, Merida JR (1996) The oxidative burst in plant defense: function and signal transduction. Physiol Planta 96:533–542

Luan S (1998) Protein phosphatases and signaling cascades in higher plants. Trends Plant Sci 3:271–275

Maffei ME, Arimura G-I, Mithöfer A (2012) Natural elicitors, effectors and modulators of plant responses. Nat Prod Rep doi:10.1039/C2NP20053H

Mahady GB, Liu C, Beecher WW (1998) Involvement of protein kinase and G proteins in the signal transduction of benzophenanthridine alkaloid biosynthesis. Phytochemistry 48:93–102

Mathieu Y, Kurkdjian A, Xia H, Guern J, Koller, A, Spiro MD, Spiro M, O'Neill M, Albersheim P, Darvill A (1991) Membrane responses induced by oligogalacturonides in suspension-cultured tobacco cells. Plant J 1:333–343

Memelnik J, Verpoorte R, Kijne JW (2001) ORC Anization of jasmonate-responsive gene expression in alkaloid metabolism. Trends Plant Sci 6:212–219

Miao ZQ, Wei ZJ, Yuan, YJ (2000) Sheng Wu Gong Cheng Xue Bao Chinese. J Biotech 16:509–513

Mihai A, Cristina S, Helepciuc F, Brezeanu A, Stoian G (2011) Biotic and abiotic elicitors induce biosynthesis and accumulation of resveratrol with antitumoral activity in the *long—term Vitis vinifera* L. callus cultures. Romanian Biotech Lett 16:6683–6689

Mithöfer A, Fliegmann J, Daxberger A, Ebel C, Neuhaus-Url G, Bhagwat AA, Keister DL, Ebel J (2001) Induction of H_2O_2 synthesis by β-glucan elicitors in soybean is independent of cytosolic calcium transients. FEBS Letters 508:191–195

Mittler R, Vanderauwera S, Gollery M, Van BF (2004) Reactive oxygen gene network of plants. Trends Plant Sci 9:490–498

Mohd Idrees M, Naeem M, Aftab T, Khan MMA, Moinuddin (2011) Salicylic acid mitigates salinity stress by improving antioxidant defence system and enhances vincristine and vinblastine alkaloids production in periwinkle [*Catharanthus roseus* (L.) G. Don]. Acta Physiol Plant 33(3):987–999

Montesano M, Brader G, Palva ET (2003) Pathogen derived elicitors: searching for receptors in plants. Mol Plant Path 4:73–79

Moreno PRH, Poulsen C, Van Der Heijden R, Verpoorte R (1996) Effects of elicitation on different secondary metabolic pathways in *Catharanthus roseus* cell suspension cultures. Enzyme Microb Technol 18:99–107

Moreno PRH, Van Der Heijden R, Verpoorte R (1995) Cell and tissue cultures of *Catharanthus roseus*: a literature survey. 2. Updating from1988 to 1993. Plant Cell Tiss Org Cult 42:1–25
Mulabagal V, Tsay H-S (2004) Plant cell cultures—an alternative and efficient source for the production of biologically important secondary metabolites. Inter J App Sci Eng 2(1):29–48
Muranaka T, Ohakawa H, Yamada Y (1992) Scopalamine release into media by *Duboisia leichhardtti* hairy root clones. Appl Microbial Biotechnol 37:554–559
Mustafa NR, Kim HK, Choi YH, Verpoorte R (2009) Metabolic changes of salicylic acid-elicited *Catharanthus roseus* cell suspension cultures monitored by NMR-based metabolomics. Biotechnol Lett 31:1967–1974
Namdeo A, Patil S, Fulzele DP (2000) Influence of fungal elicitors on production of ajmalicine by cell cultures of *Catharanthus roseus*. Biotechnol Prog 18:159–162
Namdeo AG (2007) Plant cell elicitation for the production of secondary metabolites: A review. Pharmacog Rev 1:69–79
Nennstiel, D, Scheel D, Nürnberger T (1998) Characterization and partial purification of an oligopeptide elicitor receptor from parsley (*Petroselinum crispum*). FEBS Lett 431:405–410
Nigra HM, Caso OH, Guilietti AM (1987) Production of solasodine by calli form different parts of *Solanum eleaginifolium* Cav. Plants. Plant Cell Rep 6:135–137
Nishi A (1994) Effect of elicitors on the production of secondary metabolites. In: Ryu DDY, Furasaki S (eds) Advances in plant biotechnology. Amsterdam: Elsevier Science, pp 135–151
Nürnberger T (1999) Signal perception in plant pathogen defence. Cell Mol Life Sci 55:167–182
Pan Q, Chen Y, Wang Q, Yuan F, Xing S, Tian Y, Zhao J, Sun X, Tang K (2010) Effect of plant growth regulators on the biosynthesis of vinblastine, vindoline and catharanthine in *Catharanthus roseus*. Plant Growth Regul 60:133–141
Pande D, Malik S, Bora M, Srivastava PS (2002) Rapid protocol for *in-vitro* micropropagation of *Lepidium sativum* Linn and enhancement in the yield of lepidine. *In-vitro* Cell Dev Bio Plant 38:451–456
Pedras MSC, Nycholat CM, Montaut S, Xu Y, Khan AQ (2002) Chemical defenses of crucifers: elicitation and metabolism of phytoalexins and indole-3-acetonitrile in brown mustard and turnip. Phytochemistry 59:611–625
Piel J, Atzorn R, Gabler R, Kuhnemann F, Boland W (1997) Cellulysin from the plant parasitic fungus *Trichoderma viride* elicits volatile biosynthesis in higher plants via the octadecanoid signalling cascade. FEBS Letters 416:143–148
Pitta–Alvarez SI, Spollansky TC, Giulietti AM (2000) The influence of different biotic and abiotic elicitors on the production and profile of tropane alkaloids in hairy root cultures of *Brugmansia candida*. Enzyme Microb Technol 26:252–258
Poutrain P, Mazars C, Thiersault M, Rideau M, Pichon O (2009) Two distinct intracellular Ca^{2+}-release components act in opposite ways in the regulation of the auxin-dependent MIA biosynthesis in *Catharanthus roseus* cells. J Exp Bot 60:1387–1398
Pradel H, Dumkelehmann U, Diettrich B, Luckner M (1997) Hairy root cultures of *Digitalis lanata*. secondary metabolism and plant regeneration. J Plant Physiology 151:209–215
Prakash G, Srivastava AK (2008) Statistical elicitor optimization studies for the enhancement of azadirachtin production in bioreactor *Azadirachta indica* cell cultivation. Biochem Engi J 40:218–226
Preisig CL, Moreau RA (1994) Effects of potential signal transduction antagonists on phytoalexin accumulation in tobacco. Phytochemistry 36:857–863
Radman R, Saez T, Bucke C, Keshavarz T (2003) Elicitation of plants and microbial cell systems. Biotech Applied Biochem 37:91–102
Ramani S, Chelliah J (2008) Enhanced catharanthine and vindoline production in suspension cultures of *Catharanthus roseus* by ultraviolet-B light. J Mol Signal 3:9–14
Ramani S, Jayabaskaran C (2007) UV-B-induced signaling events leading to enhanced-production of catharanthine in *Catharanthus roseus* cell suspension cultures. BMC Plant Biol 7:61–78
Ramawat KG, Merillon JM (2007) Biotechnology; Secondary metabolites; Plants and microbes, 2nd edn. Science Publishers, Enfield

Rao RS, Ravishankar GA (2002) Plant cell cultures: chemical factories of secondary metabolites. Biotech Advan 20:101–153

Ravishankar GA, Grewal S (1991) Development of media for growth of *Dioscorea deltoidea* cells and *in-vitro* diosgenin production: Influence of media constituents and nutrient stress. Biotech Lett 13:125–130

Rijhwani SK, Shanks JV (1998) Effect of elicitor dosage and exposure time on biosynthesis of indole alkaloids by *Catharanthus roseus* hairy root culture. Biotechnol Prog 14:442–449

Robins RI (1994) Secondary products from cultured cells and organs: Molecular and cellular approaches. In: Dixon RA, Gonzales, RA (eds) Plant Cell Culture. IRI Press, NY

Roemis T (2001) Protein kinases in the plant defence response. Curr Opi Plant Biol 4:407–414

Roos W, Evers S, Hieke M, Tschöpe M, Schumann B (1998) Shifts of intracellular pH distribution as a part of signal mechanism leading to the elicitation of benzophenanthridine alkaloids. Plant Physiol 118:349–364

Ruiz-May E, Galaz-Ávalos RM, Loyola-Vargas VM (2009) Differential secretion and accumulation of terpene indole alkaloids in hairy roots of *Catharanthus roseus* treated with Methyl Jasmonate. Mol Biotechnol 41:278–285

Schmeller T, Wink M (1998) Utilization of alkaloids in modern medicine. In: Roberts MF, Wink M (eds) Alkaloids. Biochemistry, ecology, and medicinal applications. Plenum Press, New York, pp 435–459

Scragg AH (1992) Bioreactors for mass cultivation of plant cells. In: Fowler MW, Warren GS (eds) Plant biotechnology. Peragmon Press, England, pp 45–62

Senoussi MM, Bassa Nora B, Joêl C (2009) Impact of hypoxia on the growth and alkaloid accumulation in *Catharanthus roseus* cell suspension. Acta Physiol Plant 31:359–362

Shanks JV, Bhadra R, Morgan J, Rijhwani S, Vani S (1998) Quantification of metabolites in the indole alkaloid pathways of *Catharassnthus roseus*: implications for metabolic engineering. Biotechnol Bioeng 58:333–338

Sharma M, Sharma A, Kumar A, Basu SK (2011) Enhancement of secondary metabo;ites in cultured plant cells through stress stimulus. American J Plant Physiol 6:50–71

Shirsau K, Nakajima H, Krishnamachari RV, Dixon RA, Lamb C (1997) Salicylic acid potentiates an agonist-dependent gain control that amplifies pathogen signals in the activation of defense mechanisms. Plant Cell 9:261–270

Shukla AK, Shasany AK, Verma RK, Gupta MM, Mathur AK, Khanuja SP (2010) Influence of cellular differentiation and elicitation on intermediate and late steps of terpenoid indole alkaloid biosynthesis in *Catharanthus roseus*. Protoplasma 242:35–47

Siddiqui ZH, Mujib A, Ahmad MM, Ali A (2010) Fungal elicitors: A potent approach for enhancing secondary metabolites in cultured cells. In: Gupta VK, Tuohy M, Gaur RK (eds) Fungal biochemistry and biotechnology. Lap Lambert Academic Publishing AG & CO. KG, Germany, pp 88–104

Singh SN, Vats P, Suri S et al. (2001) Effect of an antidiabetic extract of *Catharanthus roseus* on enzymatic activities in streptozotocin induced diabetic rats. J Ethnopharmacol 76:269–277

Smetanska I (2008) Production of secondary metabolites using plant cell cultures. Advances in Biochem Engi Biotech 111:187–228

Smith DA, Banks SW (1986) Biosynthesis, elicitation and biological activity of isoflavonoid phytoalexins. Phytochemistry 25:979–995

Srividya N, Devi BPS (1998) Azadirachtin and nimbin content in in-vitrocultured shoots and roots of Azadirachta indica A. Juss. Indian J of Plant Physiol 3:129–134

Stafford A, Morris P, Fowler MW (1986) Plant cell biotechnology: a perspective. Enzyme Microb Technol 8:578–597

Svoboda GH, Blake DA,(1975) The phytochemistry and pharmacology of *Catharanthus roseus* (L.) G. Don. In: Taylor WI, Farnsworth NR (eds) The *Catharanthus* alkaloids. Marcel Dekker, Inc., New York, pp 45–83

Taha HS, Abd El-Rahman RA, Abd-El-Kareem FM, Aly UE (2008) Successful application for enhancement and production of anthocyantn pigment from calli cultures of some ornamental plants. AJBAS 2:1148–1156

Taha HS, El-Bahr MK, Seif-El-Nasr MM (2009) *In vitro* studies on Egyptian *Catharanthus roseus* (L.) G. Don. IV: Manipulation of some amino acids as precursors for enhanced indole alkaloids production in suspension cultures. AJBAS 3:3137–3144

Ten Hoopen HJG, Vinke JL, Moreno PRH, Verpoorte R, Heijnen JJ (2002) Influence of temperature on growth and ajmalicine production by *Catharanthus roseus* suspension culture. Enzyme Microb Technol 30:56–65

Threlfall DR, Whitehead IM (1988) Co-ordinated inhibition of squalene synthetase and induction of enzymes of sesquiterpenoid phytoalexin biosynthesis in cultures of *Nicotiana tabacum*. Phytochemistry 27:2567–2580

Toppel G, Witte L, Riebesehl B, Von Borstel K, Hartman T (1987) Alkaloid patterns and biosynthetic capacity of root cultures from some pyrrolizidine alkaloid producing *Senecio* spp. Plant Cell Rep 6:466–469

Toso RD (2010) Plant cell culture technology: a new ingredient source. Personal care 2:35–38

Tripathi L, Tripathi JN (2003) Role of biotechnology in medicinal plants. Trop J Pharma Res 2(2):243–253

Tyler BM (2002) Molecular basis of recognition between phytophthora pathogens and their hosts. Annual Rev Phytopath 40:137–167

Van Der Fits L, Memelink J (2000) ORCA3, a jasmonate-responsive transcriptional regulator of plant primary and secondary metabolism. Science 289:295–297

Van Der Heijden R, Jacobs DI, Snoeijer W, Hallared D, Verpoorte R (2004) The *Catharanthus* alkaloids: pharmacognosy and biotechnology. Curr Med Chem 11:607–628

Van Der Heijden R, Verpoorte R, Ten Hoopen HJG (1989) Cell and tissue cultures of *Catharanthus roseus* (L.) G. Don: a literature survey. Plant Cell Tiss Org Cult 18:231–280

Van Loon LC, Van Strien EA (1999) The families of pathogenesis-related proteins, their activities, and comparative analysis of PR-1 type proteins. Physiol Mol Plant Patho 55:85–97

Vázquez-Flota F, Hernández-Domínguez E, de Lourdes Miranda-Ham M, Monforte-González M (2009) A differential response to chemical elicitors in *Catharanthus roseus in vitro* cultures. Biotechnol Lett 31:591–595

Wealth of India (1992) A dictionary of Indian raw materials and industrial products. Volume-3 (Revised) Ca-Ci, Publication and information directorate, Council of scientific and industrial research, New Delhi

Wu J, Wang C, Mei X (2001) Stimulation of taxol production and excretion in *Taxus* spp. cell cultures by rare earth chemical lanthanum. J Biotechnol 85:67–73

Wyk BEV, Wink M (2004) Medicinal plants of the world. Pretoria, Briza.

Xu M, Dong J (2005) Nitric oxide stimulates indole alkaloid production in *Catharanthus roseus* cell suspension cultures through a protein kinase-dependent signal pathway. Enzyme Microb Technol 37:49–53

Yang J, Yu M, January YN, January LY (1997) Stabilization of ion selectivity alters by pore loop ion pairs in an inwardly rectifying potassium channel. Proc Nat Acad Sci USA 94:1568–1572

Zenk MH (1991) Chasing the enzymes of secondary metabolism: Plant cell cultures as a pot of gold. Phytochemistry 30:3861–3863

Zhang C, Jian-Yong W (2003) Ethylene inhibitors enhance elicitor-induced paclitaxel production in suspension cultures of *Taxus* spp. Cells. Enzyme Microb Technol 32:71–77

Zhao J, Verpoorte R (2007) Manipulating indole alkaloid production by *Catharanthus roseus* cell cultures in bioreactors: from biochemical processing to metabolic engineering. Phytochem Rev 6:435–457

Zhao J, Davis LC, Verpoorte R (2005) Elicitor signal transduction leading to production of plant secondary metabolites. Biotechnol Adv 23:283–333

Zhao J, Zhu WH, Hu Q (2000a) Promotion of indole alkaloid production in *Catharanthus roseus* cell cultures by rare earth elements. Biotech Lett 22:825–828

Zhao J, Zhu WH, Hu Q, Guo YQ (2000b) Improvement of indole alkaloid production in *Catharanthus roseus* cell cultures by osmotic shock. Biotechnol Lett 22:1227–1231

Zhao J, Zhu WH, Hu Q, He XW (2000c) Improved indole alkaloid production in *Catharanthus roseus* suspension cell cultures by various chemicals. Biotechnol Lett 22:1221–1226

Zhao J, Hu Q, Zhu WH (2001a) Enhanced catharanthine production in *Catharanthus roseus* cell cultures by combined elicitor treatment in shake flasks and bioreactors. Enzyme Microb Technol 28:673–681

Zhao J, Hu Q, Guo Y-Q, Zhu W-H (2001b) Elicitor-induced indole alkaloid biosynthesis in *Catharanthus roseus* cell cultures is related to Ca^{2+} influx and the oxidative burst. Plant Sci 161:423–431

Zhao J, Hu Q, Guo YQ, Zhu WH (2001c) Effects of stress factors, bioregulators and precursors on indole alkaloid production in the compact callus clusters culture of *Catharanthus roseus*. Appl Microbiol Biotechnol 55:693–698

Zhao J, Hu Q, Zhu WH (2001d) Selection of fungal elicitors to increase inadole alkaloid accumulation in *Catharanthus roseus* suspension cell cultures. Enzyme Microb Technol 28:666–672

Zhao J, Zhu WH, Hu Q, Guo YQ (2001e) Compact callus cluster suspension cultures of *Catharanthus roseus* with enhanced indole alkaloid biosynthesis. *In-vitro* Cell Dev Biol Plant 37:68–72

Zheng Z, Wu M (2004) Cadmium treatment enhances the production of alkaloid secondary metabolites in *Catharanthus roseus*. Plant Sci 166:507–514

Zimmermann S, Frachisse JM, Thomine S, Barbier-Brygoo H, Guern J (1998) Elicitor-induced chloride efflux and anion channels in tobacco cell suspensions. Plant Physiol Biochem 36:665–674

Zuccarini P (2009) Tropospheric ozone as a fungal elicitor. J Biosci 34:125–138

Chapter 15
Handling Soybean (*Glycine max* L.) Under Stress

Mohammad Miransari

Abstract Soybean is among the most important leguminous plants with the ability to establish symbiotic association with the N-fixing bacteria, *Bradyrhizobium japonicum*. With respect to the environmental and economical significance of N fixation, there has been extensive research work regarding the production of legumes including soybean under different conditions. Soils are usually subjected to some kind of stress including salinity, acidity and suboptimal root zone temperature. One of the most important processes, affecting the performance of soybean under stress is the inhibited exchange of the signal molecules, specifically genistein, between the host legume and *B. japonicum* during the initiation of symbiosis. Interestingly, inoculation of *B. japonicum* with the signal molecule genistein has partially or completely alleviated the stress. It is also of significance to determine the right combination of N-fertilization and rhizobium inoculums when planting leguminous including soybean. The use of breeding techniques may also be among the effective methods of improving soybean performance under stress. In this chapter some of the most important advances regarding the performance of soybean under different conditions including stress with respect to the molecular techniques are reviewed. Some future perspectives are also presented, the production of tolerant plants and microbes are among the most important ones.

Introduction

Soybean (*Glycine max* L.) is a crop plant belonging to the leguminous family and is widely cultivated in different parts of the world. It is a source of food, with high protein rate, containing useful nutrients. It is the major leguminous crop plant representing 50 % of the crop legume production and 68 % of the total crop production in the world. It is able to develop symbiotic association with the bacteria from the *rhizobium* family, *Bradyrhizobium japonicum*, fixing atmospheric N_2. The amount

M. Miransari (✉)
AbtinBerkeh Limited Co., Imam Blvd., Shariati Blvd., #107, 3973173831, Tehran, Iran
e-mail: miransari1@gmail.com

Mehrabad Rudehen, Imam Ali Blvd., Mahtab Alley, 55, 3978147395, Tehran, Iran

K. R. Hakeem et al. (eds.), *Crop Improvement*, DOI 10.1007/978-1-4614-7028-1_15,

of N fixed by soybean is annually equal to 16.4 Tg, accounting for 77% of the total N fixed by leguminous plants. With respect to the importance of soybean as a source of food for human and because N-fixation can contribute to a large part of necessary N for plant use, there has been extensive research work regarding the process of N-fixation in leguminous plants. Enhancing the efficiency of N-fixation can significantly increase crop yield and is of environmental and economical significance (Herridge et al. 2008).

The process of symbiotic N-fixation is between some soil bacteria, collectively called rhizobium, and the leguminous plants in which the atmospheric N_2 is fixed by the bacteria and reduced to ammonia. The ammonia is then assimilated by the host plant into the structure of amino acids and proteins. The process of N-fixation between rhizobium and leguminous plants is specific, indicating that only one strain of rhizobium is able to colonize the host plant roots and form nodules. Nodules are the place of rhizobium residence and hence N fixation (Hungria et al. 2005; Wang et al. 2011).

Between a hundred to a few hundred kg ha^{-1} atmospheric N_2 is fixed by rhizobium, providing a major part of necessary N for plant use. For example, in soybean, 50–60% of necessary N is supplied by biological N fixation (Salvagiotti et al. 2008). Although chemical fertilization can quickly supply the necessary nutrients for plant use, it has also some disadvantages including: (1) adversely affecting soil structure, and (2) being subjected to leaching and hence resulting in the pollution of water resources. This is why biological nitrogen fixation is important, as it can inhibit such un-favorable effects of chemical fertilization on the soil properties, and hence on the environment (Evans 1993; Salvagiotti et al. 2008).

Usually rhizobium can be found in the soil, especially under soil optimal conditions; however, its population may not be adequate to efficiently inoculate the host plant. For this reason, use of bacterial inoculum can be a useful method to inoculate the host plant with appropriate bacterial population. Inoculums have a carrier with a high bacterial population, used to inoculate seeds before planting or at the time of planting. The bacterial inoculum must have the ability to compete with the soil bacteria, adapted to the soil conditions (Miransari 2010, 2011).

Exploiting the potential genetic of plant and climate properties are among the most efficient methods, resulting in the enhancement of soybean yield (Salvagiotti et al. 2008). Although soybean is not a tolerant crop plant under stress, different methods have been tested to increase its tolerance under different stresses including salinity and drought, acidity, high amount of mineral N and sub-optimal root temperature. Among the most important hypotheses that have been successfully tested and approved by researchers is that soil stresses disrupt the process of molecular communications between rhizobium and the host plant. In the initial stages of N-fixation the two symbionts, rhizobium and host plant must exchange signal molecules to realize their presence and start the process of N-fixation. The disruption of such signaling exchange between the two symbionts can inhibit N-fixation by rhizobium and the host plant (Miransari et al. 2007, 2008, 2009).

Preincubation of *Bradyrhizobium japonicum* with the signal molecules, from the flavonoids biochemical group, have been shown to be a useful method to alleviate soil stresses on the process of N_2-fixation. In the case of soybean symbiotic bacteria, *Bradyrhizobium japonicum*, geneistin is among the dominant signal molecules, produced by the host plant root. Pretreatment of *Bradyrhizobium japonicum* with genistein can partially or completely alleviate the stress on the process of N-fixation. In this case, the bacterial genes become activated under the stress and proceed with the next stages of N-fixation process (Zhang and Smith 1995). Accordingly, some of the most recent findings regarding the methods used to alleviate soil stresses on soybean growth and yield production with some prospects for future research are presented.

Significance of Biological Nitrogen Fixation

Biological nitrogen fixation and specifically the symbiosis between legume plants and rhizobium, is one of the most important biological activities, globally. In the process of symbiotic association between legume plants and rhizobium, the atmospheric N is fixed by the bacteria and turned into available N (NH_3) by nitrogenase enzyme. Depending on the conditions, most part of the N necessary for legume growth and yield production is supplied by the process of N-fixation. However, it is likely to enhance the efficiency of N-fixation, especially under stress (Unkovich and Pate 2000; Yasmin et al. 2006).

With respect to the economical and environmental importance of biological nitrogen fixation, it can significantly contribute to the enhanced efficiency of ecosystem including crop production. Although N-chemical fertilization can rapidly provide the necessary N for plant growth, it is subjected to leaching and hence can adversely affect the water sources. N-fixation can also be economically considerable as there is a high annual rate of expenses, spent for the production and use of synthetic N (Miransari and Smith 2007, 2008, 2009; Miransari and Mackenzie 2010, 2011a, b).

There are different parameters affecting the process of N-fixation between legume and rhizobium including plant species, bacterial strains, N-chemical fertilization and soil and plant properties. There are some legume plant species and rhizobium strains, with specific genotypic properties, which can perform more effectively. Accordingly, some plant-bacterium combinations may be more efficient. Although N-fertilization at seeding can enhance the growth of legume seedlings, high amounts of N-fertilization can adversely affect the process of N-fixation. Under optimal soil conditions, the efficiency of N-fixation process can be at the highest, however, soil stresses can decrease the rate of N-fixation by negatively influencing both the host plant and the bacteria (Valliyodan and Nguyen 2008; Hamilton et al. 2011; Salah et al. 2011).

Climate properties including light, temperature, precipitation, and concentration of carbon dioxide can affect both plant and the symbiotic rhizobium. Light is necessary for the process of photosynthesis and at optimal rates can increase the photosynthetic process. Legume plants are adapted to mild temperatures and have the higher rate of N-fixation and yield production at the range of 20–30 °C. Higher or lower temperatures can reduce plant growth and yield production (Lynch and Smith 1993). Legume plants can grow well at the optimum soil moisture and hence precipitation. High or low level of soil moisture can decrease plant growth and yield production of legume plants (Sakthivelu et al. 2008; Sorensen et al. 2012).

The concentration of CO_2 can also affect plant performance by affecting the process of photosynthesis. Higher rates of CO_2 (500–1000 μmol mol^{-1}) decreased plant efficiency by decreasing the rate of protein. As a result of elevated CO_2 concentration, the concentration of Rubisco reduces. It is because the expression of photosynthetic genes, which are dependent on carbohydrate concentration, is affected. However, other mechanisms may also cause such alterations (Stitt and Krapp 1999; Taub et al. 2008). The reduction in the leaf protein concentration can decrease seed protein concentration, because usually the N content of senescing tissues is translocated to plants seeds (Fangmeier et al. 1999; Salon et al. 2001). However, compared with cereals, soybean grains indicated much smaller rates of protein reduction, which is mostly due to its symbiotic association with rhizobium. Root nodules are sinks for photosynthates, inhibiting plant leaf to increase the level of hexose, which can adversely affect Rubisco concentration. The fluctuations in ozone can also affect plant performance as higher rates of exposure can have negative effects on plant growth by adversely affecting the structure of leaf mesophyll, and hence decreased carbon assimilation and photosynthesis rate (Long and Naidu 2002; Garg and Bhandari 2012).

However, the adverse effects of ozone exposure on plant leaf can be inhibited by the elevated levels of CO_2 resulting in the enhanced protein seed concentration. It is because ozone can have negative effects on the process of N fixation, and hence decrease the translocation of photosynthates to the nodules (Pausch et al. 1996; Ti et al. 2012). Accordingly, elevated CO_2 levels may enhance the rate of N-fixation by increasing the level of C assimilation (De Graaff et al. 2006; Rogers et al. 2006).

Carbon cycling between soil and atmosphere is important affecting different biological processes such as N-fixation. The higher rate of carbon in soil can contribute to higher biological activities by soil microbes, improve soil structure and enhance soil fertility. However, for N-fixing rhizobium the atmospheric carbon may be of more importance as photosynthesis process assimilates it into carbohydrate. Rhizobium bacteria utilize hydrocarbons, supplied by plant as source of energy for their activities. Accordingly, higher rate of atmospheric C up to some level can increase the rate of photosynthesis and hence the process of N-fixation (Townsend et al. 2011; Finzi et al. 2011).

Hence, the process of N-fixation by legume plants and rhizobium is of high importance significantly contributing to the necessary N for plant growth and yield production while agriculturally and environmentally sustainable. This process is affected by different parameters and hence it is pertinent to find methods that can enhance its efficiency under different conditions including stress.

Soybean, Salinity and Drought

About one milliard hectares of agricultural soils are saline or subjected to some kind of salinity, worldwide (Flowers and Yeo 1995). Leguminous plants are classified among sensitive or moderately tolerant plants to salinity (Lauchli 1984). Legume's tolerance to salinity differs among different species (Lu et al. 2009) and usually under salinity they excrete ion salt from the leaf or localized the salt in different parts of the plant. Under high salinity rhizobium are not able to become dormant, and hence must have the ability to tolerate high salt levels. Parameters including soil fertility, N source, temperature, drought, relative humidity and physical properties can affect plant growth under saline conditions (Velagaleti et al. 1990; Munns 2002; Vercruysse et al. 2011; Meilhoc et al. 2011).

Higher concentration of salt in mature leaf, relative to the young leaf, results in the senescence of mature leaf. Plant ability to allocate salt to the cellular vacuoles is among the important parameters determining plant tolerance to salinity. The higher the plant ability to allocate salt to the vacuoles the higher its tolerance to salinity is. Plant hormones such as abscisic acid (ABA) can also regulate plant activities under stress by for example controlling the stomatal activities (Wolf et al. 1990; Yang et al. 2012). Salt adverse effects on plant growth under salinity also include cytoplasm malfunctioning, membrane leakage, and loss of turgor and water.

Drought and salinity adversely influence legume-host plant symbiosis by affecting the growth and survival of bacterium, delaying the infection process, suppressing nodule functionality, decreasing the photosynthesis rate, plant growth and N uptake in the host plant. There are usually interactions between drought/salinity and rhizobium as there are bacterial strains, which are more tolerant and hence efficient under stress. Physiological alterations under stress make the plants allocate more carbon to their roots (Miransari and Smith, 2007; Miransari et al. 2007; 2008).

Researchers hypothesized and proved that addition of genistein (4′,5,7-trihydroxyisoflavone), the plant to bacterium signal, under stress, enhanced the activation rate of bacterial Nod genes resulting in the increased production of nodulation (Nod) factors by *Bradyrhizobium japonicum* (Miransari et al. 2006; Wang et al. 2012) and hence faster formation of nodules (Zhang and Smith 1995; Pan and Smith 1998a; b). According to Miransari and Smith (2007, 2008, 2009) the effects of genistein became greater with time by more effectively influencing N-fixation and hence plant growth and yield in the second sampling compared with the first sampling. This also indicates that genistein persistence in soil is suitable.

Flavonoids are able to:

1. regulate the polar transport of auxin followed by the imbalance of auxin-cytokinin and initiation of nodule meristem formation (Schmidt et al. 1994),
2. enhance bacterial growth,
3. increase the production of Nod factors by bacteria as a result of higher Nod genes activation, resulting in the alteration of root morphological properties including root hair curling and bulging and eventual formation of root nodules by inducing cellular division at different sites, and

4. increase root exudates of isoflavonoids (Verma et al. 1992; Zhang and Smith 1995; Corradini et al. 2011).

Under stress root growth is less affected than nodulation (Abd-Alla et al. 1998; Miransari and Smith 2007), because under stress plant must spend more energy to alleviate the stress rather than developing a symbiotic association with the soil microbes (Miransari and Smith 2007; Miransari et al. 2007, 2008). Nodules are tissues with a high energy requirement and hence a part of energy produced by plant must be allocated to nodules for different activities such as respiration and development (Zhang and Smith 1995).

Under stress different concentrations of genistein ranging from 5 to 20 μM were tested and proved to be effective. In addition, the highest concentration of genistein (20 μM) did not adversely affect bacterial N fixation and hence plant growth. Results indicated that with increasing the level of stress genistein became more effective. The mathematical equations used to relate genistein concentration to nodulation and plant growth indicated that the most effective concentration of genistein ranged from 5 to 11 μM. Using multivariate equations, it is possible to predict the most optimum concentration of genistein under stress with respect to plant response to genistein affecting plant growth and yield production under field and greenhouse conditions (Miransari and Smith 2007).

During drought stress, water deficiency can adversely affect plant growth and yield production. Plant roots absorb water, which moves solutes to different plant parts for utilization and assimilation, and nutrients from the surrounding soil. Water is necessary for cellular expansion and development by producing the necessary turgor for cell growth. During the process of evapo-transpiration, water is evaporated from plant leaf creating the necessary potential for the uptake of water and nutrients by plant roots and their movement to different parts of the plant. Hence, with respect to the importance of water in plant its deficiency can significantly decrease plant growth and yield production (Asbjornsen et al. 2011).

Under drought plant utilizes different mechanisms to alleviate the stress. Such mechanisms result in the adjustment of plant growth and production of organic compounds as osmoprotectants regulating cellular water potential and the uptake of nutrients. The morphological and physiological alterations in plant growth under drought can alleviate the stress up to some extent. Yamaghuchi et al. (2010) indicated that different parts of soybean primary roots respond differently to drought stress. They attributed such a response to the regulation of phenypropanoid metabolism in different parts of the roots and hence the biosynthesis of isoflavonoids. However, contrary to this alteration, the production of caffeoyl-CoA *O*-methyltransferse, which is responsible for the production of lignin, highly increased resulting in the inhibition of root growth in the specific parts. In addition, different proteins were produced in the water stressed part of roots to alleviate the oxidative stress.

There are some responsive genes, being induced under salinity. For example, Li et al. (2008) found that a protein, which is homologous to oxysterol binding protein in soybean, was expressed under salinity stress resulting in cotyledon senescence. Under salinity, the isomer of a pathogenic related protein, as a responsive protein to high rate of salinity and drought was localized in the extracellular space of soybean roots. In addition, a lucine like protein was also identified in the mature tissues of soybean shoot under saline conditions. The production of enzyme, acid phosphatase, was related to plant response, under salinity by affecting the formation of reactive oxygen species as well as by affecting the transduction pathways, related to stress (Liao et al. 2003; Sobhanian et al. 2010). The salt responsive gene, *GmDREB2*, was expressed under salinity stress, resulting in the production of higher levels of proline, relative to the wild types, improving plant tolerance to salinity stress (Chen et al. 2007). It has been indicated that soybean genotypes, which are tolerant to salinity have the same gene (Lee et al. 2004).

Relative to the wild types, the salt tolerant of soybean genotypes was due to the inhibition of Cl^- transport from the soybean roots to the shoots. However, in the wild variety the tolerance was more related to the prevention of Na^+ movement from the plant roots to the aerial parts. This indicates that salt tolerance in the genetically modified varieties have been improved relative to wild types (Lee et al. 2009).

Under stresses like salinity and drought the level of proline increases in soybean nodules, as its synthesis is enhanced. High proline accumulation in the nodules results in the high ratio of NADP/NADPH and hence the activation of pentose phosphate pathway and eventual production of purine. The derivatives of purine can act as transporters of fixed N. Proline can be the transporter for the redox potential from plant cytoplasm to the bacteroid, verified by the high activity of pro dehydrogenase in the bacteroids of root nodules (Kohl 1988, 1990; Verbruggen and Hermans 2008; Sharma and Yadav 2012).

Although plant morphological and physiological properties are altered by salinity stress, the role of plant hormones is among the most important mechanisms by which plant can alleviate the stress of salinity (Velitcukova and Fedina 1998). Accordingly, the production of several proteins during salinity stress is induced by plant hormones such as jasmonates (Chao et al. 1999; Thaler 1999), salicylic acid (Hoyos and Zhang 2000) and abscisic acid (Jin et al. 2000; Wang et al. 2001; Kang et al. 2005; Miransari 2012a).

Yoon et al. (2009) found that salinity stress significantly decreased plant growth, gibberellins concentration, rate of photosynthesis and transpiration, and considerably increased ABA production as well as proline accumulation. However, application of methyl jasmonate (MeJA) significantly alleviated the stress of salinity on soybean growth, chlorophyll, photosynthesis and transpiration rate, and proline content, while enhancing the level of ABA and gibberellins (Miransari 2012a).

Calreticulin is a protein, which is able to bind calcium, and hence regulates calcium homeostasis and protein folding in the plant endoplasmic reticulum. Under salinity, osmotic stress resulted in the down regulation of calreticulin in rice, indicating the role of calcium under salinity stress as the main secondary messenger. Salinity also down regulated the activity of RuBisco activase, adversely affecting photosynthesis (Menegazzi et al. 1993; Rokka et al. 2001; Sobhanian et al. 2010).

The flow of cell cycling under stress must be maintained to alleviate the effects of stress on plant growth. Cellular cycling and integrity is important for cellular communication and signaling, particularly under stress. Using the proteomic analysis, Sobhanian et al. (2010) indicated the responsive proteins in soybean, which are expressed during salinity. They found that under salinity, the related proteins are mostly down regulated. The results of their study showed that NaCl down regulated the activity of Glyceraldehyde-3-phosphate dehydrogenase at both protein and m-RNA levels in soybean. In addition, Kinesin is a large family of proteins affecting microtubule activities and hence cell cycling (Liu et al. 1996) and its up regulation under saline conditions indicate its role in the alleviation of stress.

Using hydroponic growing medium, Martins et al. (2008) indicated that drought significantly decreased the mitotic activities of root cells as the expression of the related genes was altered. Drought stress can markedly decrease the rate of photosynthesis, followed by stomatal closure and increased temperature. As a result of cellular dehydration and increased leaf temperature, electron pathways are changed during respiration, significantly decreasing the rate of ATP production in mitochondria and hence the rate of photosynthesis (Flexas et al. 2004; Ribas-Carbo et al. 2005).

Plant initiates its response to stress at molecular level resulting in the alteration of the related genes. For the start of gene expression under stress some transcriptional elements like clone A2B3-2, which is a helix-loop-helix with a putative basic (bHLH) is necessary. Such bHLH proteins are transcription factors with so many genes (Dey and Harborne 1997; Lewin 2000; Chen et al. 2002). Accordingly, if the related transcription factors are modified, it may be likely to enhance plant tolerance to stress (Jaglo-Ottosen et al. 1998).

There are carrier proteins, which are able to move carrier monomers in the cellular bilayers (Cleves et al. 1991). Such kinds of proteins are able to influence the transduction pathways, the related signal molecules, and the movement of molecules across the cellular membrane (Kapranov et al. 2001). Such proteins can also regulate different cellular activities including perception, division and development as well as stomatal activities. However, drought may alter the structure and activities of such proteins and plant hormones (Martins et al. 2008).

Soybean and Acidity

Acidic and infertile soils including Oxisols and Ultisols are widely distributed in the humid areas covering about 1.6×10^9 ha of tropical soils worldwide (Sanchez and Salinas 1981). High rain leaches the alkaline cations in the soil and results in the high rate of weathering producing iron and aluminum oxides. Soils are subjected to pH fluctuations usually ranging from 4 to 10. There is a wide range of alkaline soils adversely affecting plant growth and the process of N-fixation. Under alkaline conditions low precipitation and the accumulation of anions such as carbonate and bicarbonate and cations such as calcium and magnesium increases soil pH, affecting different soil processes including microbial activities and nutrient availability (Joris et al. 2012).

In addition to the high H^+ concentration, the high level of Al can also adversely affect plant growth under acidic conditions. Under such conditions high Al decreases plant growth by reducing root growth, the photosynthetic ability and competing with nutrients such as N, Mg, P and Fe. In acid soils the reaction of P with Al and Fe, results in the precipitation of P compounds and hence significantly decreases P availability (Akaya and Takenaka 2001; Shamsi et al. 2008).

Al presence increased the activities of antioxidant enzymes including malondialdehyde (MDA), super oxidase dismutase (SOD) and peroxidase (POD). Shamsi et al. (2008) indicated that the differences in soybean genotypes under the high levels of elements such as Al are determined by the root ability to absorb such metals. Fluctuations in soil acidity can affect the physiological properties such as enzymatic structure and activities in plants and microbes. Accordingly, neutral pH is the most optimum acidity for the growth of crop plants and activity of soil microbes (Martinez et al. 2012; Bissoli et al. 2012).

The most suitable method to adjust high soil hydroxyl concentration is the use of elemental sulfur inoculated with the chemo-autotrophic bacteria, *Thiobacillus* spp. These bacteria are able to acquire the energy necessary for their activities by oxidizing sulfur, resulting in the production of hydrogen and sulfate ions, and hence decreasing soil pH. Use of tolerant plant species may also be another alternative to alleviate acidity stress on plant growth and yield production (Miransari and Smith 2007).

Soil acidity can adversely affect the process of N-fixation in leguminous plants by affecting both the host plant and rhizobium. In a 17-year experiment under acidic conditions Popescu (1998) found that pH's less than six decreased soybean yield. Usually the leguminous family is sensitive to high levels of soil acidity. High soil acidity decreased root nodulation in white clover (Wood et al. 1984) subclover (Whelan and Alexander 1986), pea (Lie 1969; Evans et al. 1980), cowpea (Keyser 1979), alfalfa (Munns 1968, 1970) and bean (Wolff et al. 1993; Vassileva et al. 1997) even in the presence of high rhizobial population. Soil acidity, lower than five, inhibits nodule formation (Appunu and Dhar 2006).

There are plants, which are able to accumulate high rate of alkaline cations in their tissues, while their growth and performance remains unaffected. Soybean plants grow the best at pH's around seven and under high or low acidity their growth may be adversely affected. Under acidic conditions plants produce lower amounts of hydrogen ions affecting the uptake of nutrients by plant roots (Marschner 1995). It can also affect the cytoplasmic pH and hence result in the reduction of shoot and root growth (Schubert et al. 1990; Yan et al. 1992).

Appunu and Dhar (2006) evaluated the survival of different strains of acid tolerant *Bradyrhizobium japonicum* under high acidity (pH = 4) and found that they can tolerate high acidity level. Interestingly, the strains showed higher tolerance to acidity in soil than in YEM broth medium. The slow growing *Bradyrhizobium* are more tolerant to acidity than the fast growing rhizobium (Cooper et al. 1985; Graham et al. 1994; Miransari and Smith 2007; Ferreira et al. 2012). All the tolerant species were able to noldulate the soybean roots with high differences in nodulation, nitrogenase activity, N uptake and plant growth (Zhang et al. 2002; Meghvanshi et al. 2005). Inoculation with the bacterium significantly enhanced the growth and N uptake of soybean plants.

The most sensitive stages of nodulation to acidity appear to be the early stages including attachment (Howieson et al. 1993), root morphological changes (Munns 1968; Evans 1980; Miransari et al. 2006) and the formation of infection tread (Evans 1980; Franco and Munns 1982). The followings may indicate the higher sensitivity of the early stages of N-fixation to low pH: decreased rhizobial number and growth, changes in the structure of plant roots adversely affecting bacterial recognition and higher uptake of some nutrients (Vassileva et al. 1997; Miransari et al. 2006; Miransari and Smith 2007; Ferreira et al. 2012).

The onset of N-fixation is with the exchange of signal molecules between the bacteria and the host plant. Plant roots produce different products with nutritional and non-nutritional values acting as secondary metabolites. Secondary metabolites, which are necessary with a threshold level of 10^{-12} M have different functioning in plant including the activation of microbial genes during the process of symbiosis (Boller 1995).

For the onset of N-fixation the specific host plant produces flavonoids, which are able to trigger the bacterial chemotaxis response and movement toward the plant roots. Subsequently, the nodulation genes (Nod genes) in bacteria are activated, resulting in the production of lipochitooligosacharides (LCO) by bacteria (Subramanian et al. 2004). LCO molecules are able to induce morphological changes in the roots of their host plant by curling or bulging the root hairs, inducing cell cycling, and stimulation of nodule formation. These stages are followed by the formation of infection thread, which results in the entrance of bacteria into the plant roots. Bacteria alter some of the root cellular, morphological and physiological activities, which can increase the division rate of root cortical cells resulting in the formation of root nodules (Aguilar et al. 1988; Long 2001; Miransari et al. 2006, Miransari and Smith 2007, 2008, 2009).

Miransari and Smith (2007) hypothesized that stress results in the disruption of the signal molecule exchange between the bacteria and the host plant at the onset of symbiosis under acidity. They also hypothesized and proved that preincubation of *Bradyrhizobium japonicum* inoculum with the signal molecule genistein may partially or completely inhibit the adverse effects of stress on the process of N-fixation under greenhouse and field conditions. By addition of salt or sulfur to the field soil they adjusted soil salinity and pH to the desired values, followed by the plantation of soybean seeds, and inoculation with the bacteria preincubated with genistein at μM levels. Different nodulation and soybean growth and yield parameters were measured both under salinity and acidity treatments. According to the results, genistein could alleviate the stress of salinity and acidity on soybean nodulation and hence N-fixation as well as soybean yield production. The enhancing effects of genistein on nodulation could be due to increasing the possibility of infection, resulting in the higher number of infections and nodulation (Zhang and Smith 1995). Under stress the production of signal molecules by soybean root as well as the sensitivity of *Badyrhizobium japonicum* to the signal molecule is decreased resulting in the inhibition of N-fixation process. (Miransari and Smith 2007, 2008, 2009).

Soybean and Suboptimal Root Zone Temperature

There are areas in the world, which are subjected to sub optimal root zone temperature, most of the year. Under such conditions plant growth as well as microbial activities is adversely affected. The symbiotic association between the host plant and the N-fixing bacteria is also influenced by suboptimal root temperature, resulting in the decline of N-fixation between the two symbionts. Similar to other stresses, suboptimal root temperature can also disrupt the initials stages of N-fixation, most importantly the exchange of the signal molecules between the two symbionts (Smith and Lynch 1993; Miransari 2012b).

The optimum temperature for soybean growth is between 25 and 30 °C (Lynch and Smith 1993). Falvonoids can act as Nod gene inducers and also enhance plant resistance to pathogens in soil. Higher rates of flavonoids are produced at higher temperature by plant roots. Accordingly, under lower temperature higher concentration of signal molecule may be necessary to induce the molecular changes in rhizobium as the effectiveness of the signal molecule is temperature dependent. The signal molecule by itself and its concentration can also affect the communications between the two symbionts (Miransari and Smith 2008; Miransari 2012b).

The activity of nitrogenase enzyme is oxygen dependent and under stress where plant growth is adversely affected the nodules permeability to oxygen may also change. This may influence nodule functionality by affecting the activity of nitrogenase enzyme (Wei and Layzell 2006). Temperature fluctuations can alter root respiratory demand and its permeability to oxygen. Hence, plant can use such a mechanism to regulate root permeability to oxygen under decreased temperature, which reduces root permeability to oxygen (Kuzma and Layzell 1994; Wang et al. 2012).

Using intact soil samples in cylinders, collected from the field soil, Miransari and Smith (2008) stimulated some of the field conditions under greenhouse conditions. They found that soil texture may also affect the signal communications between the host plant and rhizobium. This has been attributed mostly to the physical properties of the different soil textures. According to their results genistein addition was more effective under loamy and clay textures. Textures with finer particles have a higher rate of microporosity. Finer soil textures are higher in soil nutrients, because of their chemical structure and hence can provide more nutrients to the soil microbes as well as plant roots affecting their activities.

Different microbial activities are affected by soil texture including mineralization of organic matter, microbial biomass, respiration, nitrification and denitrification (Hassink 1992). Soils, with a more improved structure, result in the higher production of roots exudates influencing microbial activities more effectively. Van Gestel et al. (1996) indicated that relative to the bacterial population in sandy soils (7 %), significantly higher bacterial population was found in clay soils (More than 50 %).

There are some kind of interactions between genistein and some plant regulators such as the plant hormone, auxin. Falvonoids can adversely or positively affect the activity and transport of auxin (Brown et al. 2001). Geneistin may inhibit the transport of auxin in the plant. It has been indicated that the use of auxin transport inhibitors in soybean may result in the formation of nodule like organs, and expression of genes, which induce morphogenesis alteration in the roots resulting in the formation of root nodules (Fang and Hirsch 1998).

Soybean and N Fertilization

During the past 30 years, soybean yield has increased continuously, due to the improvement in the use of genetic techniques and agricultural practices with the yearly increase of 28 kg ha^{-1}, globally (Specht et al. 1999; Wilcox 2004). Biological N fixation and mineral N fertilization are the main sources of providing necessary N for soybean growth. However, biological N fixation by the residing bacteria in the root nodules and mineral N fertilization can behave antagonistically, especially when the other soil stresses are not present (Streeter 1988; Soares Novo et al. 1999; Purcell et al. 2004).

If there is not adequate amounts of N for plant use, plant will re-translocate N from the leaf to the grains, reducing plant photosynthetic potential and hence crop production. Van Kessel and Hartley (2000) indicated that rate of yield production can adjust nitrogenase activity, as higher amounts of yield results in higher nitrogenase activity. This is especially the case for when soybean plants yield at rates higher than 4.5 t. ha^{-1} (Mengel 1994).

In situations where the process of N-fixation is not able to supply the necessary N for plant growth and yield production, plant response to N-fertilization can be significant (Thies et al. 1995). However, usually there are antagonistic effects between N-fixation and N-fertilization. Nitrate present in the soil can decrease the process of N-fixation (Herridge and Rose 1994). Even little amounts of N-fertilization can suppress N-fixation during the first stages of plant growth; however at the same time N deficiency may delay the onset of nodule formation and hence, N-fixation by adversely affecting crop growth (Zhang and Smith 2002; Ikeda et al. 2011).

Salvagiotti et al. (2008) analysed a complete set of data regarding N-fixation and N-fertilization related to different parts of the world. They found that N-fixation is negatively and exponentially related to N-fertilization when it was applied to the soil surface at 20 cm. Their economical analyses, with respect to soybean and N-fertilization pricing, indicated that N-fertilization is advantageous when plant requirements are not met by the process of N-fixation. They accordingly suggested that to enhance the efficiency of N uptake by plant, future research must indicate the contribution of each N component including, N-fixation, N-fertilization and soil N. It is because, if N-fertilization is supplied adequately, while not affecting N-fixation, it can enhance the efficiency of plant N uptake (Ikeda et al. 2011).

It has been suggested and tested that plants be modified genetically for nitrate tolerant; however such a method has not been appropriate, because it resulted in the reduction of crop yield (Salvagiotti et al. 2008). Although the physiological alterations in plant metabolism is one of the main reasons for decreased N-fixation at the time of N-fertilization, however the other important reason is that the process of N-fixation is demanding and requires high rate of energy spent by the host plant; when N-fertilization provides mineral N for plan use, the plant will not be willing to develop a symbiotic association with the bacteria, adversely affecting the process of N fixation.

Conclusions and Future Perspectives

Handling crop plants under stress is among the most important research issues. Soybean is a leguminous crop plant most used by human. Although it is not considered a tolerant crop plant to stress, research has indicated that it is likely to alleviate the effects of different stresses such as acidity, salinity, sub optimal roots zone temperature and N-fertilization on soybean growth and yield using molecular and breeding methods. Researchers have tested and proved that use of the plant to bacteria signal, genistein, for the pre incubation of soybean symbiotic bacteria, *Bradyrhizobium japonicum* can be useful to alleviate the unfavorable effects of stress on soybean growth and yield. Identification of tolerant genes in different plant species and inserting them in soybean may also be effective to alleviate stress.

Although research work has indicated much detail related to the handling of soybean under stress, there are more, which must be elucidated. For example, using proteomic analysis, the production of different proteins under stress must be indicated and accordingly the related genes be recognized and inserted to produce tolerant varieties. The other important point is the production of tolerant rhizobium under stress, which may also be similarly recognized and produced.

References

Abd-Alla MH, Vuong TD, Harper JE (1998) Genotypic differences in dinitrogen fixation response to NaCl stress in intact and grafted soybean. Crop Sci 38:72–77

Aguilar JM, Ashby AM, Richards JM, Loake GJ, Watson MD, Shaw CH (1988) Chemotaxis of *Rhizobium leguminosarum* biovar phaseoli towards flavonoid inducers of the symbiotic nodulation genes. J Genetical Microbiol 134:2741–2746

Akaya M, Takenaka C (2001) Effects of aluminum stress on photosynthesis of *Quercus glauca* Thumb. Plant Soil 237:137–146

Appunu C, Dhar B (2006) Symbiotic effectiveness of acid-tolerant *Bradyrhizobium* strains with soybean in low pH soil. Afric J Biotechnol 5:842–845

Asbjornsen H, Alvardo-Barrientos MS, Rebel K, Van Osch FP, Rietkerk M, Chen J, Gotsch S, Tobon C, Geissert DR, Gomez-Tagle A, Kellie V, Dawson TE (2011) Ecohydrological advances and applications in plant–water relations research: a review. J Plant Ecol 4:3–22

Bissoli G, Niñoles R, Fresquet S, Palombieri S, Bueso E, Rubio L, Sánchez M, Fernández J, Mulet J, Serrano R (2012) Peptidyl-prolyl *cis-trans* isomerase ROF2 modulates intracellular pH homeostasis in Arabidopsis. Plant J 70:704–716

Boller T (1995) Chemoreception of microbial signals in plant cells. Ann Rev Plant Physiol Plant Mol Biol 46:189–214

Brown DE, Rashotte AM, Murphy A, Normanly J, Tague BW, Peer WA, Taiz L, Muday GK (2001) Flavonoids act as negative regulators of auxin transport in vivo in Arabidopsis. Plant Physiol 126:524–535

Chao WS, Gu YQ, Pautot V, Bray EA, Walling LL (1999) Leucine aminopeptidase RNAs, proteins, and activities increase in response to water deficit, salinity, and the wound signals systemin, methyl jasmonate, and abscisic acid. Plant Physiol 120:979–992

Chen W, Provart NJ, Glazebrook J, Katagiri F, Chang H-S, Eulgem T, Mauch F, Luan S, Zou G, Whitham SA, Budworth PR, Tao Y, Xie RT (2002) Expression profile matrix of *Arabidopsis* transcription factor genes suggest their putative functions in response to environmental stresses. Plant Cell 14:559–574

Chen M, Wang QY, Cheng XG, Xu ZS, Li LC, Ye XG, Xia LQ, Ma YZ (2007) *GmDREB2*, a soybean DRE-binding transcription factor, conferred drought and high-salt tolerance in transgenic plants. Biochem Biophys Res Commun 353:299–305

Cleves AE, Mcgee TP, Bankaitis VA (1991) Phospholipid transfer proteins: a biological debut. Trends Cell Biol 1:30–34

Corradini E, Foglia P, Giansanti P, Gubbiotti R, Samperi R, Lahana A (2011) Flavonoids: chemical properties and analytical methodologies of identification and quantitation in foods and plants. J Nat Prod 25:469–495

Cooper J, Wood M, Bjourson A (1985) Nodulation of *Lotus pedunculatus* in acid rooting solution by fast- and slow-growing rhizobia. Soil Biol Biochem 17:487–492

De Graaff M-A, Van Groenigen K-J, Six J, Hungate B, Van Kessel C (2006) Interactions between plant growth and soil nutrient cycling under elevated CO2: a meta-analysis. Glob Change Biol 12:2077–2091

Dey PM, Harborne JD (1997) Plant Biochemistry. Academic, San Diego, p 554

Evans L, Lewin K, Vella F (1980) Effect of nutrient medium pH on symbiotic nitrogen fixation by *Rhizobium leguminosarum* and *Pisum sativum*. Plant Soil 56:71–80

Evans L (1993) Crop evolution, adaptation, and yield. Cambridge University Press, Cambridge

Fang Y, Hirsch AM (1998) Studying early nodulin gene ENOD40 expression and induction by nodulation factor and cytokinin in transgenic alfalfa. Plant Physiol 116:53–68

Fangmeier A, De Temmerman L, Mortensen L, Kemp K, Burke J, Mitchell R, van Oijen M, Weigel HJ (1999) Effects of nutrients on grain quality in spring wheat crops grown under elevated CO2 concentrations and stress conditions in the European multiple-site experiment 'ESPACE-wheat'. Europ J Agron 10:215–229

Finzi AC, Austin AT, Cleland EC, Frey SD, Houlton BZ, Wallenstein MD (2011) Responses and feedbacks of coupled biogeochemical cycles to climate change: examples from terrestrial ecosystems. Front Ecol Environ 9:61–67

Flexas J, Bota J, Loreto F, Cornic G, Sharkey TD (2004) Diffusive and metabolic limitations to photosynthesis under drought and salinity in C3 plants. Plant Biol 6:269–279

Flowers TJ, Yeo AR (1995) Breeding for salinity resistance in crop plants—where next? Aust J Plant Physiol 22:875–884

Franco AA, Munns DN (1982) Acidity and aluminum restraints on nodulation, nitrogen fixation, and growth of *Phaseolus vulgaris* in solution culture. Soil Sci Soc Am J 46:296–301

Ferreira P, Bmfeti C, Soares B, Moreira F (2012) Efficient nitrogen-ixing *Rhizobium* strains isolated from amazonian soils are highly tolerant to acidity and aluminium. World J Microbiol Biotechnol 28:1947–1959

Garg N, Bhandari P (2012) Influence of cadmium stress and arbuscular mycorrhizal fungi on nodule senescence in *Cajanus Cajan* (L.) MILLSP. Inter J Phytorem 14:62–74

Graham PH, Draeger K, Ferrey ML, Conroy MJ, Hammer BE, Martinez-Romero E, Naarons SR, Quinto C (1994) Acid pH tolerance in strains of *Rhizobium* and *Bradyrhizobium*, and initial studies on the basis for acid tolerance of *Rhizobium tropici* UMR1899. Can J Microbiol 40:198–207

Hamilton TL, Lange RK, Boyd ES, Peters JW (2011) Biological nitrogen fixation in acidic high-temperature geothermal springs in Yellowstone National Park, Wyoming. Environ Microbiol 13:2204–2215

Hassink J (1992) Effects of soil texture and structure on carbon and nitrogen mineralization in grassland soils. Biol Fert Soils 14:126–134

Herridge DF, Rose IA (1994) Heritability and repeatability of enhanced N_2 fixation in early and late inbreeding generations of soybean. Crop Sci 34:360–367

Herridge DF, Peoples MB, Boddey RM (2008) Global inputs of biological nitrogen fixation in agricultural systems. Plant Soil 311:1–18

Howieson JG, Robson AD, Bottomley PJ (1993) External phosphate and calcium concentrations, and pH, but not the products of rhizobial nodulation genes affect the attachment of *Rhizobium meliloti* to roots of annual medics. Soil Biol Biochem 25:567–573

Hoyos M, Zhang SQ (2000) Calcium-independent activation of salicylic acid-induced protein kinase and a 40-kilodalton protein kinase by hyper osmotic stress. Plant Physiol 122:1355–1363

Hungria M, Franchini J, Campo R, Graham P (2005) The importance of nitrogen fixation to soybean cropping in South America. In: Werner D, Newton W (eds) Nitrogen fixation in agriculture, forestry, ecology, and the environment. Springer, Netherlands, pp 25–42

Jaglo-Ottosen K, Gilmour SJ, Zarka DG, Schabenberger O, Thomashow MF (1998) *Arabdopisis* CBF1 overexpression *induces COR* genes and enhances freezing tolerance. Science 280:104–106

Joris H, Caires E, Bini A, Scharr D, Haliski A (2012) Effects of soil acidity and water stress on corn and soybean performance under a no-till system. Plant Soil 365:409–424

Jin S, Chen C, Plant AL (2000) Regulation by ABA of osmotic-stress-induced changes in protein synthesis in tomato roots. Plant Cell Environ 23:51–60

Ikeda S, Anda M, Inaba S, Eda S, Sato S, Sasaki K, Tabata S, Mitsui H, Sato T, Shinano T, Minamisawa K (2011) Autoregulation of nodulation interferes with impacts of nitrogen fertilization levels on the leaf-associated bacterial community in soybeans. Appl Environ Microbiol 77:1973–1980

Kang DJ, Seo YJ, Lee JD, Ishii R, Kim KU, Shin DH, Park SK, Jang SW, Lee IJ (2005) Jasmonic acid differentially affects growth, ion uptake and abscisic acid concentration in salttolerant and salt-sensitive rice cultivars. J Agron Crop Sci 191:273–282

Kapranov P, Routt SM, Bankaitis VA, Bruijn FJ, Szczyglowski K (2001) Nodule-specific regulation of phosphatidylinositol transfer protein expression in *Lotus japonicus*. Plant Cell 13:1369–1382

Keyser H, Munns D, Hohenberg J (1979) Acid tolerance of rhizobia in culture and in symbiosis with cowpea. Soil Sci Soc Am J 43:719–722

Kohl DH, Schubert KR, Carter MB, Hagedorn CH, Shearer G (1988) Proline metabolism in N_2-fixing root nodules: energy transfer and regulation of purine synthesis. Proc Nat Acad Sci U S A 85:2036–2040

Kohl DH, Lin JJ, Shearer G, Schubert KR (1990) Activities of the pentose phosphate pathway and enzymes of proline metabolism in legume root nodules. Plant Physiol 94:1258–1264

Kuzma M, Layzell DB (1994) Acclimation of soybean nodules to changes in temperature. Plant Physiol 106:263–270

Lauchli A (1984) Salt exclusion: an adaptation of legumes for crops and pastures under saline conditions. In: Staples RC, Toeniessen GH (eds) Salinity tolerance in plants. Strategies for crop improvement. Wiley, New York, pp 171–187

Lee GJ, Boerma RH, Villagarcia RM, Zhu X, Carter TE Jr, Li Z, Gibbs MO (2004) A major OTL conditioning salt tolerance in S-100 and descendent cultivars. Theor Appl Genet 109:610–1619

Lee J, Shannon G, Vuong T, Nguyen H (2009) Inheritance of Salt Tolerance in Wild Soybean (Glycine soja Sieb. and Zucc.) accession PI483463. J Hered 100:798–801

Lewin B (2000) Genes VII. Oxford University Press, New York, p 990

Li DY, Inoue H, Takahashi M, Kojima T, Shiraiwa M, Takahara H (2008) Molecular characterization of a novel salt-inducible gene for an OSBP (oxysterol binding protein)-homologue from soybean. Gene 407:12–20

Liao H, Wong FL, Phang TH, Cheung MY, Li WYF, Shao G, Yan X, Lam HM (2003) *GmPAP3*, a novel purple acid phosphatase-like gene in soybean induced by NaCl stress but not phosphorus deficiency. Gene 318:103–111

Lie TA (1969) The effect of low pH on different phases of nodule formation in pea plants. Plant Soil 31:391–406

Liu B, Cyr RJ, Palevitza BA (1996) A Kinesin-like protein, KatAp, in the cells of *Arabidopsis* and other Plants. Plant Cell 8:119–132

Long S (2001) Genes and signals in the *Rhizobium*-legume symbiosis. Plant Physiol 125: 69–72

Long SP, Naidu SL (2002) Effects of oxidants at the biochemical, cell and physiological levels, with particular reference to ozone. In: Bell JNB, Treshow M (eds) Air Pollution and Plant Life. Wiley, UK, pp 69–88

Lu K, Cao B, Feng X, He Y, Jiang D (2009) Photosynthetic response of salt-tolerant and sensitive soybean varieties. Photosynthetica 47:381–387

Lynch DH, Smith DL (1993) Soybean (*Glycine max* (L.) Merr.) nodulation and N_2 fixation as affected by period of exposure to a low root zone temperature. Physiologia Plant 88:212–220

Marschner H (1995) Mineral Nutrition of Higher Plants. Academic, London

Martins P, Jordão B, Yamanaka N, Farias J, Beneventi M, Binneck E, Fuganti R, Stolf R, Nepomuceno A (2008). Differential gene expression and mitotic cell analysis of the drought tolerant soybean (*Glycine max* L. Merrill Fabales, Fabaceae) cultivar MG/BR46 (Conquista) under two water deficit induction systems. Genet Mol Biol 31:512–521

Martinez K, Kitko R, Mershon JP, Adcox H, Malek K, Berkmen M, Slonczewski J (2012) Cytoplasmic pH response to acid stress in individual cells of *Escherichia coli* and *Bacillus subtilis* observed by fluorescence ratio imaging microscopy. Appl Environ Microbiol 78:3706–3714

Meghvanshi MK, Prasad K, Mahna SK (2005) Identification of pH tolerant *Bradyrhizobium japonicum* strains and their symbiotic effectiveness in soybean (*Glycine max* (L.) Merr.) in low nutrient soil. Afric J Biotechnol 4:663

Meilhoc E, Boscari A, Bruand C, Puppo A, Brouquisse R (2011) Nitric oxide in legume-rhizobium symbiosis. Plant Sci 181:573–581

Mengel K (1994) Symbiotic dinitrogen fixation—its dependence on plant nutrition and its ecophysiological impact. Z Pflanzenernähr Bodenkd 157:233–241

Menegazzi P, Guzzo F, Baldan B, Mariani P, Treves S (1993) Purification of calreticulin-like protein(s) from spinach leaves. Biochem Biophys Res Commun 190:1130–1135

Miransari M, Balakrishnan P, Smith DL, Mackenzie AF, Bahrami HA, Malakouti MJ, Rejali F (2006) Overcoming the stressful effect of low pH on soybean root hair curling using lipochitooligosaccahrides. Commun Soil Sci Plant Anal 37:1103–1110

Miransari M, Smith DL (2007) Overcoming the stressful effects of salinity and acidity on soybean (*Glycine max* (L.) Merr.) nodulation and yields using signal molecule genistein under field conditions. J Plant Nutr 30:1967–1992

Miransari M, Bahrami HA, Rejali F, Malakouti MJ, Torabi H (2007) Using arbuscular mycorrhiza to reduce the stressful effects of soil compaction on corn (*Zea mays* L.) growth. Soil Biol Biochem 39:2014–2026

Miransari M, Smith DL (2008) Using signal molecule genistein to alleviate the stress of suboptimal root zone temperature on soybean-*Bradyrhizobium* symbiosis under different soil textures. J Plant Interact 3:287–295

Miransari M, Bahrami HA, Rejali F, Malakouti MJ (2008) Using arbuscular mycorrhiza to reduce the stressful effects of soil compaction on wheat (*Triticum aestivum* L.) growth. Soil Biol Biochem 40:1197–1206

Miransari M, Smith D (2009) Alleviating salt stress on soybean (*Glycine max* (L.) Merr.)—*Bradyrhizobium japonicum* symbiosis, using signal molecule genistein. Europ J Soil Biol 45:146–152

Miransari M (2010) Biological fertilization. In: Méndez-Vilas A (ed) Current research, technology and education topics in applied microbiology and microbial biotechnology. Microbiology book series—2010 Edition, Spain.

Miransari M, Mackenzie AF (2010) Wheat (*Triticum aestivum* L.) grain N uptake as affected by soil total and mineral N, for the determination of optimum N fertilizer rates for wheat production. Commun Soil Sci Plant Anal 41:1644–1653

Miransari M (2011) Soil microbes and plant fertilization. Review article. Appl Microbiol Biotechnol 92:875–885

Miransari M, Mackenzie AF (2011a) Development of a soil N-test for fertilizer requirements for corn (*Zea mays* L.) production in Quebec. Commun Soil Sci Plant Anal 42:50–65

Miransari M, Mackenzie AF (2011b) Development of a soil N test for fertilizer requirements for wheat. J Plant Nutr 34:762–777

Miransari M (2012a) Role of phytohormone signaling during stress. In: Ahmad P, Prasad MNV (eds) Environmental adaptations and stress tolerance of plants in the era of climate change, 1st edn. Springer, p 715, ISBN 978-1-4614-0814-7

Miransari M (2012b) Microbial Products and Soil Stresses. In: Maheshwari DK (ed) Bacteria in agrobiology: stress management. Springer, p 333, ISBN 978-3-642-23464-4

Munns DN (1968) Nodulation of *Medicago sativa* in solution culture. I. Acid sensitive steps. Plant Soil 28:129–146

Munns DN (1970) Nodulation of *Medicago sativa* in solution culture. V. Calcium and pH requirements during infection. Plant Soil 32:90–102

Munns R (2002) Comparative physiology of salt and water stress. Plant Cell Environ 25:239–250

Pan B, Smith DL (1998a) Genistein addition to the rhizosphere of soybean (*Glycine max* L. Merr.) at the onset of nitrogen fixation increases nodulation. J Plant Nutr 21:1631–1639

Pan B, Zhang F, Smith DL (1998b) Genistein addition to the soybean rooting medium increases nodulation. J Plant Nutr 21:1631–1639

Pausch RC, Mulchi CL, Lee EH, Meisinger JJ (1996) Use of 13C and 15N isotopes to investigate O_3 effects on C and N metabolism in soybeans. Part II. Nitrogen uptake, fixation, and partitioning. Agric Ecosys Environ 60:61–69

Popescu A (1998) Contributions and limitations to symbiotic nitrogen fixation in common bean (*Phaseolus vulgaris* L.) in Romania. Plant Soil 204:117–125

Purcell, L, Serraj R, Sinclair T, De A (2004) Soybean N_2 fixation estimates, ureide concentration, and yield responses to drought. Crop Sci 44:484–492

Ribas-Carbo M, Taylor NL, Giles L, Busquets S, Finnegan PM, Day DA, Lambers H, Medrano H, Berry JA, Flexas J (2005) Effects of water stress on respiration in soybean leaves. Plant Physiol 139:466–473

Rogers A, Gibon Y, Stitt M, Morgan PB, Bernacchi CJ, Ort DR, Long SP (2006) Increased C availability at elevated carbon dioxide concentration improves N assimilation in a legume. Plant Cell Environ 29:1651–1658

Rokka A, Zhang L, Aro EM (2001) Rubisco activase: an enzyme with a temperature dependent dual function? Plant J 25:463–471

Sakthivelu G, Devi M, Giridhar P, Rajasekaran T, Ravishankar G, Nedev T, Kosturkova G (2008) Drought-induced alterations in growth, osmotic potential and *in vitro* regeneration of soybean cultivars. Gen Appl Plant Physiol Special Issue 34:103–112

Salah IB, Slatni T, Gruber M, Messedi D, Gandour M, Benzarti M, Haouala R, Zribi K, Hamed KB, Perez-Alfocea F, Abdelly C (2011) Relationship between symbiotic nitrogen fixation, sucrose synthesis and anti-oxidant activities in source leaves of two *Medicago ciliaris* lines cultivated under salt stress. Environ Exp Bot 70:166–173

Salon C, Munier-Jolain NG, Duc G, Voisin AS, Grandgirard D, Larmure D, Emery RJN, Ney B (2001) Grain legume seed filling in relation to nitrogen acquisition: a review and prospects with particular reference to pea. Agronomie 21:539–552

Salvagiotti F, Cassman K, Specht J, Walters D, Weiss A, Dobermann A (2008) Nitrogen uptake, fixation and response to fertilizer N in soybeans: a review. Field Crops Res 108:1–13

Sanchez PA, Salinas JG (1981) Low-input technology for managing Oxisols and Ultisols in tropical America. Adv Agron 34:279–406
Schmidt P, Broughton W, Werner D (1994) Nod factors of *Bradyrhizobium japonicum* and *Rhizobium* sp. NGR34 induce flavonoid accumulation in soybean root exudate. Mol Plant-Microb Interact 7:384–390
Schubert S, Schubert E, Mengel K (1990) Effect of low pH of the root medium on proton release, growth, and nutrient uptake of field beans (*Vicia faba*). Plant Soil 124:239–244
Shamsi IH, Wei K, Zhang GP, Jilani GH, Hassan MJ (2008) Interactive effects of cadmium and aluminum on growth and antioxidative enzymes in soybean. Biologia Plant 52: 165–169
Sharma P, Yadav A (2012) Symbiotic characterization of mutants defective in proline dehydrogenase in *Rhizobium* sp. Cajanus under drought stress condition. Europ J Exp Biol 2:206–216
Sobhanian H, Razavizadeh R, Nanjo Y, Ehsanpour A, Rastgar Jazii F, Motamed N, Komatsu S (2010) Proteome analysis of soybean leaves, hypocotyls and roots under salt stress. Proteome Sci 8:1–15
Soares Novo MC, Tanaka RT, Mascarenhas HAA, Bortoletto N, Gallo PB, Alves Pereira JCVN, Teixeira Vargas AA (1999) Nitrogênio e potassio na fixação simbiotica de N_2 por soja cultivada no inverno. Scientia Agricola 56:143–156
Sorensen P, Lett S, Michelsen (2012) Moss-specific changes in nitrogen fixation following two decades of warming, shading, and fertilizer addition. Plant Ecol 213:695–706
Specht JE, Hume DJ, Kumudini SV (1999) Soybean yield potential—a genetic and physiological perspective. Crop Sci 39:1560–1570
Stitt M, Krapp A (1999) The interaction between elevated carbon dioxide and nitrogen nutrition: the physiological and molecular background. Plant Cell Environ 22:583–621
Streeter J (1988) Inhibition of legume nodule formation and N_2 fixation by nitrate. CRC Crit Rev Plant Sci 7:1–23
Subramanian S, Hu Z, Lu G, Odelland JT, Yu O (2004) The promoters of two isoflavone synthase genes respond differentially to nodulation and defense signals in transgenic soybean roots. Plant Mol Biol 54:623–639
Taub DR, Miller B, Allen H (2008) Effects of elevated CO_2 on the protein concentration of food crops: a meta-analysis. Glob Change Biol 14:565–575
Thaler JS (1999) Jasmonate-inducible plant defenses cause increased parasitism of herbivores. Nature 399:686–688
Thies JE, Singleton PW, Bohlool BB (1995) Phenology, growth, and yield of field-grown soybean and bush bean as a function of varying modes of N nutrition. Soil Biol Biochem 27:575–583
Ti C, Pan J, Xia Y, Yan X (2012) A nitrogen budget of mainland China with spatial and temporal variation. Biocheochem 108:381–394
Townsend AR, Cleveland CC, Houlton BZ, Alden CB, White JWC (2011) Multi-element regulation of the tropical forest carbon cycle. Front Ecol Environ 9:9–17
Unkovich MJ, Pate JS (2000) An appraisal of recent field measurements of symbiotic N_2 fixation by annual legumes. Field Crops Res 65:211–228
Valliyodan B, Nguyen H (2008) Genomics of Abiotic Stress in Soybean. In: Stacey G (ed) Genetics and genomics of soybean. Springer Science + Business Media, New York
Van Gestel M, Merckx R, Vlassak K (1996) Spatial distribution of microbial biomass in microaggregates of a silty-loam soil and the relation with the resistance of microorganisms to soil drying. Soil Biol Biochem 28:503–510
Van Kessel C, Hartley C (2000) Agricultural management of grain legumes: has it led to an increase in nitrogen fixation? Field Crops Res 65:165–181
Vassileva V, Milanov G, Ignatov G, Nikolov B (1997) Effect of nitrogen fixation of common bean grown at various calcium and nitrate levels. J Plant Nutr 20:279–294
Velagaleti RR, Marsh S, Kramer D, Fleischman D, Corbin J (1990) Genotypic differences and nitrogen fixation among soybean (*Glycine max* L.) cultivars grown under salt-stress. Tropic Agric 67:169–177
Velitcukova M, Fedina I (1998) Response of photosynthesis of *Pisum sativum* to salt stress as affected by methyl jasmonate. Photosynthetica 35:89–97

Verbruggen N, Hermans C (2008) Proline accumulation in plants: a review. Amino Acids 35:753–759

Vercruysse M, Fauvart M, Jans A, Beullens S, Braeken K, Cloots L, Engelen K, Marchal K, Michiels J (2011) Stress response regulators identified through genome-wide transcriptome analysis of the (p)ppGpp-dependent response in *Rhizobium etli*. Genome Biol 12:R17

Verma D (1992) Signals in root nodule organogenesis and endocytosis of *Rhizobium*. Plant Cell 4:373–382

Wang Y, Mopper S, Hasenstein KH (2001) Effects of salinity on endogenous ABA, IAA, JA and SA in *Iris hexagona*. J Chem Ecol 27:327–342

Wang X, Pan Q, Chen F, Yan X, Liao H (2011) Effects of co-inoculation with arbuscular mycorrhizal fungi and rhizobia on soybean growth as related to root architecture and availability of N and P. Mycorrhiza 21:173–181

Wang N, Khan W, Smith DL (2012) Changes in soybean global gene expression after application of lipo-chitooligosaccharide from *Bradyrhizobium japonicum* under sub-optimal temperature. PLoS ONE 7:e31571

Wei H, Layzell DB (2006) Adenylate-coupled ion movement. A mechanism for the control of nodule permeability to O_2 diffusion1[OA]. Plant Physiol 141:280–287

Whelan AM, Alexander M (1986) Effects of low pH and high Al, Mn and Fe levels on the survival of *Rhizobium trifolii* and nodulation of subterranean clover. Plant soil 92:363–371

Wilcox JR (2004) World distribution and trade of soybean. In: Boerma HR, Specht JE (eds) Soybeans: improvement, Production and Uses, ASA, CSSA, ASSA (2004), pp 1–13

Wolf O, Jeschke WD, Hartung W (1990) Long-distance transport of abscisic acid in NaCl-treated intact plants of *Lupinus albus*. J Exp Bot 41:593–600

Wolff AB, Singleton PW, Sidirelli M, Bohlool BB (1993) Influence of acid soil on nodulation and inter strain competitiveness in relation to tannin concentrations in seeds and roots of *Phaseolus vulgaris*. Soil Biol Biochem 25:715–721

Wood M, Cooper JE, Holding AJ (1984) Soil acidity factors and nodulation of *Trifolium repens*. Plant Soil 78:367–379

Yamaguchi M, Valliyodan B, Zhang J, Lenoble M, Yu O, Rogers E, Nguyen H, Sharp R (2010) Regulation of growth response to water stress in the soybean primary root. I. Proteomic analysis reveals region-specific regulation of phenylpropanoid metabolism and control of free iron in the elongation zone. Plant Cell Environ 33:223–243

Yan F, Schubert S, Mengel K (1992) Effect of low root medium pH on net proton release, root respiration, and root growth of corn (*Zea mays* L.) and broad bean (*Vicia faba* L.). Plant Physiol 99:415–421

Yang L, Ji W, Gao P, Li Y, Cai H, et al (2012) GsAPK, an ABA-Activated and Calcium-Independent SnRK2-Type Kinase from *G. soja*, mediates the regulation of plant tolerance to salinity and ABA stress. PLoS ONE 7:e33838

Yasmin K, Cadisch G, Baggs EM (2006) Comparing ^{15}N labelling techniques for enriching above- and below ground components of the plant–soil system. Soil Biol Biochem 38:397–400

Yoon Y, Hamayun M, Lee S, Lee I (2009) Methyl jasmonate alleviated salinity stress in soybean. J Crop Sci Biotechnol 12:63–68

Zhang F, Smith DL (1995) Preincubation of *Bradyrhizobium japonicum* with genistein accelerates nodule development of soybean (*Glycine max.* (L.) Merr.) at suboptimal root zone temperatures. Plant Physiol 108:961–968

Zhang F, Smith DL (2002) Interorganismal signaling in suboptimum environments: the legume-rhizobia symbiosis. Adv Agron 76:125–161

Zhang H, Daoust F, Charles TC, Driscoll BT, Prithiviraj B, Smith DL (2002) *Bradyrhizobium japonicum* mutants allowing improved nodulation and nitrogen fixation of field grown soybean in a short season area. J Agric Sci 138:293–300

Chapter 16
Environmental and Economical Opportunities for the Valorisation of the Genus *Atriplex*: New Insights

Maali Benzarti, Kilani Ben Rejeb, Ahmed Debez and Chedly Abdelly

Abstract *Atriplex* species are members of the Chenopodiaceae. There are more than 400 species growing naturally in arid and semi arid regions of the world, most of which are highly tolerant to drought and salt. *Atriplex* species contain high levels of protein and economically valuable compounds. These characteristics could make *Atriplex* a suitable food for livestock in saline or arid/ semi-arid area. Furthermore, *Atriplex* can take up salt ions from saline soil and sequester it into the salt glands at the leaf surface. This trait is of high significance since it allows them to be used for revegetation of saline or arid/semi-arid lands. *Atriplex* species have also been used for cloning some genes related to drought and salt tolerance. This review is a new contribution that updates knowledge on the ecological and socio-economical potential of some plant genus *Atriplex*.

Introduction

Many arid and semi-arid regions in the world have soils and water resources that are too saline for most of the common conventional crop systems (Pitman and Lauchli 2002). Halophytes are plants that have been naturally selected in saline environments and are distinguishable from glycophytes by their capacity to cope with ex-

C. Abdelly (✉) · M. Benzarti · K. B. Rejeb · A. Debez
Laboratoire des Plantes Extrêmophiles, Centre de Biotechnologie de Borj-Cedria (CBBC), BP 901, 2050 Hammam-Lif, Tunisia
e-mail: chedly.abdelly@cbbc.rnrt.tn

M. Benzarti
e-mail: maali23@yahoo.fr

A. Debez
e-mail: ahmed_debez@yahoo.com

K. B. Rejeb
Physiologie Cellulaire et Moléculaire des Plantes, UR5,
EAC 7180 CNRS, Université Pierre et Marie Curie (UPMC),
Case 156, 4 place Jussieu, 75252 Paris cedex 05, France
e-mail: kilanib@yahoo.com

K. R. Hakeem et al. (eds.), *Crop Improvement*, DOI 10.1007/978-1-4614-7028-1_16,

cessive levels of ions by various eco-physiological mechanisms. Some halophytes possess unique adaptations, such as salt glands or bladders that alleviate the deleterious effects of high ion concentrations. However, intrinsically cellular processes-must make the major contribution to the capacity of plants for salt adaptation. At the molecular level, the higher salt-adaptive plasticity of halophytes may be due to constitutive expression of genes that encode salt-tolerance determinants (Casas et al. 1992) or the better aptitude to regulate the expression of these genes in response to salt. This hypothesis makes halophytes a source of exclusive genes or new genetic mechanisms that could be applied in genetic manipulation of crops. Cultivation of salt-tolerant crops, or halophytes, on saline soil has significant social and economical potential that needs to be further explored and developed (Debez et al. 2011).

Among the halophytes extensively used in physiological and molecular biological investigations is the *Atriplex* genus. The genus *Atriplex* (Chenopodiaceae) contains various species distinguishable by different morphology, biological cycles and ecological adaptations (Le Houérou 1992). Tolerance to salinity, drought, heavy metals and temperature are important characteristics of species of *Atriplex*. However, the value of certain *Atriplex* species has been recognized by their incorporation in the rangelands improvement programs in many salt-affected regions throughout the world. In this contribution, we review the literature regarding the ecological and agronomic importance of the plant genus *Atriplex* in arid and semi arid regions.

Geographical Distribution of the Genus *Atriplex*

Atriplex species constitute the largest and most diversified genus of the family Chenopodiaceae (Kadereit et al. 2010). *Atriplex* species (saltbushes) are dominant in many arid and semi-arid regions of the world, particularly in habitats that combine relatively high soil salinity with aridity (Ortíz-Dorda et al. 2005). Over 400 species of *Atriplex* have been found to be geographically distributed on all continents. *Atriplex* species are mainly found in the deserts and semi-deserts in North America, South Australia, South Central Asia, West and South East America, and the Mediterranean basin. *A. nummularia*, and *A. halimus* are the most widely distributed species of the genus *Atriplex*. *A. halimus* is a perennial native shrub of the Mediterranean region (Ortíz-Dorda et al. 2005). This species has two subspecies: the subsp. *halimus*, which is present on the northern shores of the Mediterranean basin and the subsp. *schweinfurthii* (Boiss.) common on the southern shores of the Mediterranean basin, North Africa and Near East. *A. nummularia* occurs naturally in the semi arid and arid zone of southern and central Australia where it was divided in three subspecies (subsp. *nummularia*; subsp. *Omissa* and subsp. *Spathulata*). Molecular genetic and taxonomic evidence suggests that *Atriplex* was transported to Australia during the late Miocene (Kadereit et al. 2010). *A. nummularia* is proposed to have evolved from a common octoploid ancestor *A. paludosa* ssp. *moquiniana* (Moq.) Parr-Smith in the coastal semi-arid fringe of southwestern Australia. Sampson and Byrne

(2012) suggested that many species spread and diversified from this zone to exploit the arid and saline habitats that were increasingly becoming available as a result of changing climatic conditions through the Pliocene and Pleistocene in inland areas of Australia. There are two species that are considered to be closely related to *A. nummularia* that are found in the arid zones of Western and South Australia, respectively: *A. amnicola* Paul G. Wilson and *A. incrassate* F Muell (Sampson and Byrne 2012). *A. breweri* and *A. cansecens* are relatively close to *A. halimus*. Between 1920 and 1930, *A. nummularia, A. semibaccata* from Australia and *A. canescens* from USA were introduced to Tunisia and Morocco (Ben Salem et al. 2010). Only 13 species and subspecies are used for rangeland rehabilitation and fodder production: *A. halimus* subsp. *halimus, A. halimus* subsp. *schweinfurthii, A. mollis, A. glauca, A. leucoclada, A. nummularia, A. canescens* subsp. *canescens, A. canescens* subsp. *linearis, A. amnicola, A. undulata, A. repanda, A. semibaccata*, and *A. barclayana* (Ben Salem et al. 2010).

The Importance of *Atriplex* Species for Saline Soil Reclamation

Salt affected soils are widely spread in many arid and semi-arid regions of the world and increasingly threatening agricultural expansion and productivity. Yet, in many arid environments, high quality water in not available to support the establishment of plants for revegetation projects. The removal of sodium salts from saline soils by halophytes plants, as alternative for costly chemical amendments, has emerged as an efficient low cost technology (Gharaibeh et al. 2011). It is well known that *Atriplex* species actively accumulate soluble salts in leaves, especially sodium, in association with a drought tolerance mechanism. For this reason it is also considered as an excellent species for reducing soil salinity in drylands, if cut and collected (Ben Salem et al. 2005). It was found that the dehydrated *A. halimus* accumulated more Na^+ than the control plants even without the addition of NaCl to the stressed plants (Martinez et al. 2003). Glenn and Brown (1998) concluded that tolerance of *A. canescens* to water and salt stress was linked through a common mechanism of accumulating Na^+ for osmotic adjustment. By comparing salinity tolerance of three *Atriplex* species in well-watered and drying soil Glenn et al. (2012) found that *A. hortensis* was able to complete its life cycle on drying soil with a final salt content 85 g/l NaCl. *A. lentiformis* was able to survive on drying soils with salinities five times higher than seawater, whereas *A. canescens* had high survival on drying soils but was less salt tolerant than either *A. hortensis* or *A. lentiformis*. It has been demonstrated that there is a relationship between the habitat of the Mediterranean xero-halophyte species *A. halimus* and the strategy adopted for NaCl and osmotic stress resistance. The coastal (Monastir, salt-affected site) population is more tolerant of salinity than the inland (Sbikha, non saline semi-arid area) population and displays a higher ability to accumulate glycinebetaine (GB) in response to this constraint. In contrast, the inland population, exposed in its natural habitat to transient periods of drought, is more resistant to osmotic stress induced by 15% PEG, and mainly accumulates proline in response to this treatment (Ben Hassine et al. 2008). Some *Atriplex* species grown under rangeland conditions has leaf ash concentrations of 13–27% (Welch 1978; Hyder 1981; Berrett-Lennard 2002) and *Atriplex* species grown in saline soils can have leaf ash concentrations up to 39% (Malcolm et al. 1988). Khan et al. (2000) reported that *A. halimus* was more salt-tolerant than *A. calotheca* and *A. nitens*, when grown at 750 mM NaCl (−40, −67 and −80% of biomass production) but interestingly, all three species were able to survive at this salt concentration. Recently, Benzarti et al. (2012) found that *A. portulacoides* was able to grow in medium containing 1,000 mM NaCl without displaying salt-induced toxicity symptoms. The salinity resistance of some *Atriplex* species is often attributed to the presence of vesiculated trichomes covering the leaf surface and containing large amounts of salt (Smaoui et al. 2011). These trichomes play a significant role in removing salt from the the leaf tissues, thereby preventing the accumulation of toxic salts in the parenchyma and vascular tissues. For many *Atriplex* species, more than 50% of the salt transported to the shoots is excreted via these epidermic trichomes (Belkheiri and Mulas 2011). These attributes have led some workers to suggest that *Atriplex* species could be grown to remove salt from the soil (Barrett-Lennard 2002).

Atriplex species (*A. canescens*) has been especially recommended for arid zone restoration projects (Fitzsimmons et al. 1998). In a field experiment Chisci et al. (2001) demonstrated the use of *A. halimus* in improving physical characteristics of a clay soil in Italy and to provide environmental protection by controlling runoff and reducing soil erosion on slopes. *Atriplex* plant litter can modify the top soil salinity, along with other soil properties. Maganhotto de Souza Silva et al. (2008) found that soils cultivated with *A. nummularia* and irrigated with saline effluents, in semi-arid conditions in Brazil, improved their fertility (organic carbon, nitrogen and phosphorus contents) and microbiological properties (enzymes activity). Sameni and Soleimani (2007) studied the distribution of salinity and of some soil physico-chemical properties, observing significant changes in salinity and pH and found that *A. nummularia* may actually facilitate growth of plants under their canopy. The work by Zucca et al. (2011) in a study site in Morocco also confirmed that the significant relationship between soil properties and *A. nummularia* development can be mostly observed within the first 10 cm. *A. nummularia* is one of the most important species used for the revegetation of degraded land in low rainfall areas. Slavich et al. (1999) planted *A. nummularia* as a vegetative cover in a salt affected land in southeast Australia. Brown et al. (1999) showed that *A. barclayana* could be used as a biofilter to remove nutrients from saline aquaculture effluents. Gharaibeh et al. (2011) showed that amelioration of a calcareous saline sodic soil can be achieved efficiently by growing *A. halimus* without applying an amendment. Planting *A. halimus* reduced soil sodicity and electrical conductivity considerably to values comparable to that of gypsum treatments.

The *Atriplex* species may also be used for wildfire prevention purposes. The high salt concentration found in its leaves increases their moisture content, which makes this species behave as a fire retardant in the event of wildfire (Montgomery and Cheo 1969).

Table 16.1 Potential candidates for phytoremediation approach

Species	Metal contaminants	Reference
A. halimus	Cd/Zn/Pb	Lutts et al. 2004
A. portulacoides	Cd/Cu/Zn/Fe/Ti	Luque et al. 1999; Rebordea and Caçador 2007; Cambrollé et al. 2012a, b
A. canescens	Cd/Cr	Sawalha et al. 2006
A. hortensis spp purpurea	Pb/Zn	Kachout et al. 2012
A. hortensis spp rubra	Cu/Ni	Kachout et al. 2012
A. atacamensis Phil	Ar	Vromman et al. 2011
A. conodocarpa	Hg	Lomonte et al. 2010
A. palula	Se	Vickerman et al. 2002
A. spongiosa	Se	Vickerman et al. 2002

The Importance of *Atriplex* Species in Heavy Metal Phytoremediation

The contamination of soil by heavy metals is one of the most serious environmental problems and has significant implications for human health. The clean up of heavy metals contaminated soils is one of the most difficult tasks for environmental engineering. In most cases, traditional physiochemical methods are quite expensive and may lead to soil alterations (Gardea-Torresdey et al. 2005). Phytoremediation based on the use of plants to remove or degrade inorganic and organic pollutants, has been proposed as a promising, environmentally friendly and relatively cheap. The success of phytoremediation depends upon the identification of suitable plant species those hyperaccumulate heavy metals. The use of deep-rooting xero-halophyte species is may be of special interest in this context to remediate salty contaminated area, especially in arid and semi-arid regions. Several species of the genus *Atriplex*, which are naturally salt- and drought-tolerant have been also suggested as potential candidates for a phytoremediation approach (Table 16.1) because of their high biomass production associated with a deep root system. *A. portulacoides* was suggested as a suitable species for the phytoremediation owing to the high translocation rates of Cd and Cu towards the aboveground tissues (Reboreda and Caçador 2007). *A. portulacoides* can tolerate external Cu levels of up to 15 mmol/l (1,000 mg Cu/l) without suffering adverse physiological effects (Cambrollé et al. 2012a). Furthermore, despite the fact that Cu concentrations of between 20 and 100 mg Cu/kg DW in leaf tissue are generally considered excessive or toxic (Kabata-Pendias and Pendias 2001), growth parameters of this species are unaffected by leaf tissue concentrations as high as 80 mg Cu/kg DW. *A. portulacoides* is able to survive with external Cu levels of 35 mmol/l and can be found growing in sediments that contain 300–3,000 ppm Cu. Sousa et al. (2008) stated that compartmentalization and detoxification mechanisms are crucial to allow *A. portulacoides* to tolerate high levels of heavy metals, and found that this species is able to retain a considerable quantity of metals in the root cell wall. Cambrollé et al. (2012b) reported also that this salt-marsh shrub may represent a valuable tool in the restoration of Zn-polluted areas since the plant can tolerate high tissue concentrations of Zn without suffering adverse physiological effects, and can produce a significant amount of biomass while sequestrating high concentrations of this metal. Other species that belong to the *Atriplex* genus, such as *Atriplex halimus*, have already been studied for their level of resistance to heavy metals (Cd, Zn, Pb). These species have been recommended as a promising species for the phytoremediation of heavy-metal contaminated areas based on their high biomass production, deep root systems and ability to tolerate high concentrations of toxic elements (Lutts et al. 2004; Lefèvre et al. 2009; Manousaki and Kalogerakis 2009). Precipitation of heavy-metal with oxalate and/or its excretion into trichomes and increased synthesis of glycine betaine may contribute to the tolerance of *A. halimus*. Among the heavy metals frequently present in contaminated soils, mercury is arguably of the greatest environmental and public health concern. *A. conodocarpa* proved to be the most suitable candidates for mercury phytoextraction because of its ability to translocate mercury from roots to the above-ground tissues (Lomonte et al. 2010). Vickerman et al. (2002) evaluated 30 *Atriplex* lines for potential habitat improvement and phytoremediation of selenium contaminated sites. *A. patula* was found to be one of the top selenium accumulators and grew well in saline soil. *A. atacamensis Phil* has been proposed as possible candidates for phytoremediation of Ar (Vromman et al. 2011).

Atriplex Species Forage Production

The main limitations to animal production in the arid and semi-arid regions, is the scarcity of green forage. The use of native or introduced halophytes for livestock production is an important issue in many semi-arid and arid areas. Several halophytes plants have been used as fodder crops under saline conditions in order to produce green forage during the dry season (El Shaer 2010). These include planting perennial salt marsh plant species, mainly *Atriplex* species, in numerous regions. In several experiments, the Mediterranean *Atriplex* species has been the most successful shrub species in terms of establishment and productivity. Research conducted in north-east Morocco showed an average production of 920 Kg DM/ha (1,000 plants/ha density), with variations from 406 to 2,140 Kg DM/ha depending on the species studied. *A. vesicaria*, *A. semibaccata*, *A. nummularia* and *A. paludosa* scored the high levels of production. Other reports indicate that forage production from a 2-year old *Atriplex* plantation was 5 t under 150 and 200 mm rainfall.

The pastoral value of fodder shrubs depends not only upon their biomass but also upon their nutritive value and their palatability and digestibility (Salem et al. 2012). The low concentrations of metabolizable energy and high concentrations of soluble salt in herbage of *Atriplex* species as well as the presence of anti-nutritional compounds, including tannins, flavonoids, oxalate tends to reduce fodder palatability and feed intake of sheep and goats. Several methods have been devised to lessen adverse effects of phenols and to alleviate deleterious effects of sodium chloride found in tree and shrub fodder foliages. These include treatment with alkalis such as urea, ammonia and calcium hydroxide and oxidizing agents such as potassium dichromate. Addition of exogenous enzymes is also a method to improve the nutritive value of tree leaves. The study of Salem et al. (2012) showed that there are beneficial impacts of sun-drying and/or dietary exogenous enzyme addition for sheep fed *A. halimus*.

Despite the limited nutritional values, the use of *Atriplex* species as an important component of the diet should be considered in arid and semi-arid regions since this plant produces 2.5–20 t of dry matter per hectare per year and it is available most of the year. Given the low nutritional value of the *Atriplex* species, various authors have proposed supplementing them with other types of feed like barley grain, barly straw or spineless cactus so that the animals could obtain the energy from the hay and grain, and the protein and minerals from *Atriplex* species. Abu-Zanat (2005) reported that grazing a combination of salt-tolerant grasses, legumes and *Atriplex* species improved feeding value and maximize animal production (feed intake and growth rates) from saline land. Norman et al. (2008) reported that 170 g/day of grain was the minimum required to ensure sheep fed *A. nummularia* and *A. amnicola* maintained live weight. In the study of Ben Salem et al. (2005) lambs fed *A. nummularia* and supplemented with barley achieved a growth rate of 67 g/day over an 85-day period. Data presented by van der Baan et al. (2004) clearly demonstrate that supplementation with grains such as barley or maize significantly increases the digestibility of *A. nummularia* and this leads to an increase in growth rate of ruminant. Mixing alfalfa with *Atriplex* as green fodders to sheep may increase the palatability and consequently intake and utilization of *Atriplex* which lead to improvement of the performance of animal. Abu-Zanat (2005) reported that it is possible to replace up to 50 % of alfalfa hay by *A. nummularia* without negative effects on intake and digestibility of dry matter by Awassi lambs. Jacobs and Smith (1977) reported significant differences in chemical composition between (*Atriplex nummularia, A. Canescens, A. Brewerii* and *A. Lentiformis*) species and between seasons. Kandil and El-Shaer (1989) reported that *Atriplex nummularia* had higher nutritive value in spring and winter than in summer and autumn. Riasi et al. (2008) reported that *A. dimorphostegia* have more number of beneficial chemical nutritive components and digestible values than *Suaeda arcuata* as forage for ruminants. Farmers in some arid areas of the world have already begun to cultivate *Atriplex* as a salt-tolerant forage crop on lands where other crops are difficult to grow.

Characterization of Bioactive Compounds from Some *Atriplex* Species

Several salt marsh plants have traditionally been used for medical, nutritional, and even artisanal purposes. Currently, an increasing interest is granted to these species because of their high content in bioactive compounds (Ksouri et al. 2011). Various *Atriplex* species have medicinal values, e.g. *A. semibaccata* has been used as an antifungal agent and *A. vestita* in the traitement of bronchitis. *A. hortensis* has been

Table 16.2 Some *Atriplex* species and their isolated compounds

Species	Chemical content	Reference
A. portulacoides	Phenolic compounds	Boughalleb et al. 2009
A. inflata	Phenolic compounds	Boughalleb et al. 2009
A. halimus	Tanins, alkaloids, saponins, ascorbic acid	Benhammou et al. 2009
A. lentiformis	Quercetin 6,4′-dimethoxy-3-fructo rhamnoside, quercetin 4′-methoxy-3-fructo rhamnoside, kaempferol-4′-methoxy-rutinoside, kaempferol 7-rhamnoside, kaempferol 3,7-dirhamnoside, quercetin and kaempferol	Awaad et al. 2012
A. nummularia	Vit E, Vit A, saponins, polypodine, 20 hydroxyecdysone	Keckeis et al. 2000
A. hortensis	Kaempferol 3-*O*-sulphate-7-*O* arabinopyranoside quercetin 3-*O*-sulphate-7-*O*-arabinopyranoside	Bylka et al. 2001
A. littoralis	Patuletin 3-*O*-β-D-glucopyranoside, patuletin 3-*O*[5‴-*O*-feruloyl-β-D-apiofuranosyl (1‴→2″)-β-D-glucopyranoside	Bylka 2004
A. farinosa	Naringin, naringenin 7-O-glucoside, isorhamnetin 3-O-rhamnosyl (1-6) glucopyranoside and isorhamnetin 7-O-glucopyranoside	Al-Jaber et al. 1992
A. stocksii	Ursolic acid, oleanolic acid, β-amyrin, β-sitosterol, stigmasterol, atriplexinol	Siddiqui et al. 1994

regarded as a source of Vit A (Siddiqui et al. 1994). The arial parts of *A. hortensis* were used in flak medicine against diseases of respiratory tract, digestive and urinary systems, and due to their analgestic properties, in rheumatism (Bylka et al. 2001). *A. halimus* is effective in the treatment of type II diabetic patients (Bakhtiuary 2011) and it used in veterinary medicine to combat internal parasites. Extracts of *A.confertifolia* has significant bioactivity against human breast cancer cell lines; the bioactivity of *A. confertifolia* extract on these cells was compared to a FDA-approved cancer drug (Onxol®) and an industry-standard leukocyte control cell line. A dose-response curve of the extracts displayed significant cell death similar to Onxol® (Capua et al. 2010). Boughalleb et al. (2009) reported that many *Atriplex* species (*A. inflate* and *A. portulacoides*) may contain phytochemical compounds with fungicide properties.

Several studies attributed the anti-carcinogenic, anti-inflamatory, antifongique and antioxydants activities potential of plant extracts to their bioactive compounds compositions (Ksouri et al. 2011). Chemical investigation of the species of the genus *Atriplex* (Table 16.2) showed the presence of saponin glycosides, alkaloids, ascorbic acid and phytoecdysteroids (Keckeis et al. 2000). Benhammou et al. (2009) reported that *A. halimus* leaves and stems were characterized by the presence of the flavonoids, the tannins, the alkaloids and the sponins where the leaves exhibited the higher yields. Bylka et al. (2001) isolated from leaves of *A. hortensis* tow relatively rare sulphated flavonoids: kaempferol 3-*O*-sulphate-7-*O*-arabinopyranoside and quercetin 3-*O*-sulphate-7-*O*-arabinopyranoside (belong the group of compounds easily soluble in water) and a new acetylated flavonol glycoside from

Table 16.3 Genes isolated from some *Atriplex* species

Species	Gene	Gene Bank	Product	Reference
A. gmelini	*AgNHX1*	AB038492	Antiporter Na^+/H^+	Hamada et al. 2001
A. nummularia	*AnGAPDH*	U02886.1	Glyceraldehyde-3-phosphate dehydrogenase	Niu et al. 1994
	AnCMO	AB112481	Choline monooxygenase	Tabuchi et al. 2005
	AnPEAMT	AB196771	Phosphoethanolamine N-methyltransferase	Tabuchi et al. 2005
	AnSAMS	AB183565	S-adenosyl-L-methionine synthase 5	Tabuchi et al. 2005
	H⁺-ATPase	–	PM H^+-ATPase	Niu et al. 1993
A. hortensis	*AhCMO*	AF270651	Choline monooxygenase	Zhang et al. 2009; Shen et al. 2002
	AhBADH	DQ497233	Betaine aldehyde dehydrogenase	
A. prostrata	*ApCMO*	AY082068	Choline monooxygenase	Wang et al. 2004
A. centralasiatica	*AcBADH*	AY093684	Betaine aldehyde dehydrogenase	Yin et al. 2002
A. halimus	*AsDBRE*	JF451138	DRE binding transcription factor	Khedr et al. 2011

A. littoralis (Bylka 2004). Other earlier works suggested the presence the naringin, naringenin 7-O-glucoside, isorhamnetin 3-O-rhamnosyl (1–6) glucopyranoside and isorhamnetin 7-O-glucopyranoside in *A. farinose* (Al-Jaber et al. 1992). More recently, two new flavonoids, quercetin 6,4′-dimethoxy-3-fructo-rhamnoside and quercetin 4′-methoxy-3-fructo-rhamnoside in addition to another five known compounds were isolated from *A. lentiformis* (Torr.) S. Wats (Awaad et al. 2012). All of the extracts and the isolated compounds were tested for their antioxidant activity, the two new compounds were found to have the highest antioxidant activity with no side effect.

Isolation and Characterization of Genes from *Atriplex* Species

Atriplex species are among the most salt tolerant higher plants. And the elucidation of its salt tolerance mechanisms is of significance for generating salt tolerant crops via selective breeding or genetic engineering. Studying the regulation of stress inducible genes can lead to understanding of the mechanisms by which halophytes maintain growth and thrive under abiotic stress. Several stress related genes have been isolated from these halophytes (Table 16.3). Both glycophytes and halophytes cannot tolerate large amounts of salt in the cytoplasm. Plants maintain a low concentration of Na^+ in the cytosol by active exclusion of Na^+ ions into the apoplast and vacuole by using specific plasma membrane and tonoplast Na^+/H^+ antiporter

(NHX1) (Shi and Zhu 2002). In particular, the vacuolar Na^+/H^+ antiporter had been demonstrated to play a key role in salt tolerance of plants (Blumwald et al. 2000). The greater salt tolerance in *Atriplex* species is related to the transport to shoots of high quantities of Na^+ concomitant to an efficient vacuolar compartmentation of this ion, which prevents the ionic damage to the cytoplasm. The vacuolar Na^+/H^+ antiporter genes has been characterized and identified from *A. gmelili* (*AgNHX1*) (Hamada et al. 2001). The analysis and comparison of these genes showed that they were highly homologous with similar structural and conserved domains. Ohta et al. (2002) demonstrated that transgenic rice plants overexpressing *AgNHX1* gene could survive after short period of high concentration salt exposure (300 mM NaCl for 3 days). Evacuation of Na^+ from the cytoplasm is energy-dependent. A partial sequence of an isoform of the plasma membrane PM-H^+-ATPase was been isolated from *A. nummularia*. Increased H^+-ATPase mRNA abundance was reported in *A. nummularia* when NaCl adapted (342 mM NaCl) cells were re-exposed to NaCl after having been grown in media without additional NaCl (Niu et al. 1993). Which provide evidence that enhanced H^+-transport activity by NaCl in *A. nummularia* is mediated at least in part by transcriptional or post-transcriptional processes that result in higher mRNA accumulation.

Exposure to saline and drought stress results in the accumulation in the cytosol of low-molecular mass compounds, termed as compatible solutes, which do not interfere with normal biochemical reactions. It has been frequently reported that GB acts as the main stress-induced agent involved in the osmotic adjustment and protection of cellular structure in plant species belonging to the Chenopodiaceae (Rhodes and Hanson 1993). GB facilitates osmotic adjustment by lowering the internal osmotic potential that contributes to the water stress tolerance ability. In addition, it stabilizes both PSII complex and RuBisCO during photosynthesis under stress conditions (Sakamoto and Murata 2000). Yang et al. (2007) reported that genetically engineered tobacco with the ability to accumulate GB showed a higher content of ascorbate and reduced glutathione as well as an increase in the activity of superoxide dismutase (SOD). The positive effect of exogenous glycine betaine application in plant growing under salinity stress has been proven. Plant cell could be protected from the adverse effect of salinity induced oxidative stress by the exogenous application of glycine betaine (Demiral and Türkan 2004). In *A. nummularia* GB play a major role in cytosol osmotic adjustment in both leaves and roots, regardless of NaCl presence (Silveira et al. 2009). In higher plants, the first and second steps in the biosynthesis of GB are catalyzed by a rate-limiting enzyme choline monooxygenase (CMO) and betaine aldehyde dehydrogenase (BADH), respectively (Sakamoto and Murata 2000). *CMO* gene from *A. hortensis* (*AhCMO*) has been isolated and used for GB production in tobacco (Shen et al. 2002) and cotton plants (Zhang et al. 2009) to improve its abiotic stress tolerance. CMO homologs have been also identified in *A. prostrate* (Wang and Showalter 2004) and *A. nummularia* (Tabuchi et al. 2005). The gene encoding the second enzyme, *BADH*, has been cloned from *A. hortensis* (*AhBADH*) and introduced into rice, wheat, and turf grass (Xiao et al. 1995; Guo et al. 1997, 2000) and improvement of salt tolerance in transgenic plants was observed during growth. Similar result was also achieved in transgenic trifoliate orange (Fu et al. 2011). The enhanced salt tolerance was correlated, at least in part, with reduced lipid peroxidation, greater abundance in photosynthetic proteins, stimulation of K^+ uptake, and low Na^+/K^+ ratios. The *BADH* gene that originated from *A. hortensis* was also transformed into the most important forage crop alfalfa with *Agrobacterium*-mediated transformation method. The transgenic plants grew vigorous in salt stress condition, whereas the wild type plants was retarded and did not survive. The expression of *BADH* gene in alfalfa genome enhanced its salt tolerance through improved membrane protection as measured by relative electrical conductivity and malondialdehyde (MDA) content, scavenge of free radicals by increase of peroxidase (POD) and SOD activities, and the osmotic adjustment (Liu et al. 2011).

Transcription factor genes play important roles in stress survival by serving as master regulators of sets of downstream stress-responsive genes via binding to specific elements (*cis*-elements) in target genes. Functional analysis of the promoter regions of some of stress-inducible genes has led to identification of the *cis*-element DRE (Dehydration-responsive element), which is responsible for dehydration-inducible transcription (Yamaguchi-Shinozaki and Shinozaki 1994). Full-length DRE-binding transcription factor (*AhDREB1*) gene has been isolated from *A. hortensis* (Shen et al. 2003). In transgenic tobacco, *AhDREB1* led to the accumulation of its putative downstream genes and these transgenic lines showed an increased stress tolerance, suggesting that the AhDREB1 protein functions as a DRE-binding transcription factor and play roles in the stress tolerant response of *A. hortensis*. *DREB* in *A. halimus* (*AsDBRE*) is regulated by the osmotic component but not by the ionic one of salt stress (Khedr et al. 2011). It seemed that *DREB* was not involved in the regulation of sodium manipulating genes like *NHX1*, *SOS1* or *H^+-PPase*. Moreover, *DREB* could be involved directly of indirectly in *CMO* regulation because of timing of induction. Also, *DREB* was the most up-regulated gene under salt (fivefold) and drought (twofold) conditions, which reinforced the importance of this gene in *A. halimus* tolerance to stress. Moreover, its constitutive expression under normal conditions also indicated its involvement in other growth and developmental programs (Khedr et al. 2011).

Microsatellites are widely used in population genetic studies and may prove to be useful in studies of closely related species to infer relationships when sequence variation is very low or there are few or no genome resources available. Ortíz-Dorda et al. 2005 has evaluated the genetic structure of 51 populations of *A. halimus* from the Mediterranean Basin using RAPD (random amplified polymorphic DNA)-PCR technique. The authors found that there are a clear intrapopulational diversity of *A. halimus*. Such heterogeneity could be exploited to select clones or develop synthetic populations with a combination of good traits such as high palatability, high edible biomass production, and good adaptability to environmental limiting factors in semi-arid Mediterranean environments. 12 polymorphic loci were isolated in *A. nummularia* (Byrne et al. 2008) which will be useful to describe levels of genetic variability across the range of the species and in a breeding programme.

Xu et al. (2011) investigate the physiological responses and differential gene expression caused by salinity exposure in *A. centralasiatica* plants grown from two different seed morphs. *A. centralasiatica* widely distributed in China produce tow type of seeds, yellow and brown seeds. Seedlings derived from yellow seeds showed a greater salt tolerance than those derived from brown seeds. By using suppression subtractive hybridization (SSH) and subsequent microarray and RT-PCR analysis to isolate and compare genes that were differentially expressed, the authors suggest a major contribution of gene regulation to the salt resistant phenotype of seedlings derived from yellow seeds. These genes encoded proteins related to osmotic and ionic homeostasis, redox equilibrium and signal transduction. This study clearly links physiological responses with differential gene expression in seedlings derived from dimorphic seeds. Such dimorphism offers the advantage to halophytic plants to survive in highly variable environments.

Conclusion and Future Perspective

The global distribution of the genus *Atriplex* in arid and semi-arid areas and its abiotic stress tolerance combined with the utility of *Atriplex* species for restoration, remediation and forage for livestock have helped these plants to rank among the most widely studied native halophytes species. The major limitation of use *Atriplex* in livestock production is its high salt concentration. *Atriplex* species are best considered as a supplement rather forage. The plant used for these purposes are primarily wild type and there are little information available on the nutritive value of *Atriplex* species growing in greenhouse and irrigated with different concentration of NaCl. Strategies need to be devised to minimize the salt contents in the *Atriplex* leaves.

Although the biochemistry of *Atriplex* species tends to establish that they may be a source of novel compounds along with providing a new source for many already know biologically active compounds. Data of chemical composition of *Atriplex* species is still not completed.

Atriplex species are well adapted to both salt and drought stress and can serve as model species to understand the mechanisms of tolerance in plant (Flowers and Colmer 2008). Very little research has been carried out to identify the molecular mechanisms directly responsible for the specific tolerance of *Atriplex* species to abiotic stress. In this way, *Atriplex* may serve as a particularly useful model plant for studies of regulatory mechanisms related to the activation of the GB biosynthetic pathway in response to environmental stress.

With over 400 species of this genus, a significant opportunity then exists to explore the potential of other locally adapted *Atriplex* species. Research is required to select and breed potentially useful plants and identify the best species combining nutritional, agronomic and environmental potential.

Acknowledgments This work was supported by the Tunisian Ministry of Higher Education and Scientific Research (LR10CBBC02).

References

Abu-Zanat MMW (2005) Voluntary intake and digestibility of saltbush by sheep. Asian Aust J Anim Sci 18:214–220

Al-Jaber NAA, Mujahid TG, Al-Hazmi HMG (1992) Secondary metabolites of chenopodiaceace species. J Chem Scoc Pak 14:76–83

Awaad AS, Maitland DJ, Donia AERM, Alqasoumi SI, Soliman GA (2012) Novel flavonoids with antioxidant activity from a Chenopodiaceous plant. Pharmaceutical Biol 50:99–104

Bakhtiuary Z (2011) Herbal Medicines in Diabetes. IJDO 3:88–95

Barrett-Lennard EG (2002) Restoration of saline land through revegetation. Agric Water Manag 53:213–226

Belkheiri O, Mulas M (2011) The effects of salt stress on growth, water relations and ion accumulation in two halophyte *Atriplex* species. Environ Exp Bot (in press). http://dx.doi.org/10.1016/j

Ben Hassine A, Ghanem ME, Bouzid S, Lutts S (2008) An inland and a coastal population of the Mediterranean xero-halophyte species *Atriplex halimus* L. differ in their ability to accumulate proline and glycinebetaine in response to salinity and water stress. J Exp Bot 59:1315–1326

Ben Salem H, Nefzaoui A, El-Mastouri A, Ben Salem L (2005) Nutritive value, behavior, and growth of Barbarine lambs fed on oldman saltbush (*Atriplex nummularia L.*) and supplemented or not with barley grains or spineless cactus (*Opuntia ficus-indica F. inermis*) pads. Small Rumin Res 59:229–237

Ben Salem H, Norman HC, Nefzaoui A, Mayberry DE, Pearce KL, Revell DK (2010) Potential use of oldman saltbush (*Atriplex nummularia* Lindl.) in sheep and goat feeding. Small Rumin Res 91:13–28

Benhammou N, Fawzia AB, Tatjana KP (2009) Antioxidant activity of methanolic extracts and some bioactive compounds of *Atriplex halimus*. Comptes Rendus Chimie 12:1259–1266

Benzarti M, Ben Rejeb K, Debez A, Messedi D, Abdelly C (2012) Photosynthetic activity and leaf antioxidative responses of *Atriplex portulacoides* subjected to extreme salinity. Acta Physiol Plant. doi:1007/s11738-012-0963-5

Blumwald E, Aharon GS, Apse MP (2000) Sodium transport in plant cells. Biochim Biophys Acta 1465:140–151

Boughalleb N, Trabelsi L, Harzallah-Skhiri F (2009) Antifungal activity from polar and non-polar extracts of some Chenopodiaceae wild species growing in Tunisia. Nat Prod Res 23:988–997

Brown JJ, Glen EP, Fitzsimmons KM, Smith S (1999) Halophytes for the treatment of saline aquaculture effluent. Aquaculture 175:255–268

Bylka W (2004) A new acylated flavonol diglycosides from *Atriplex littoralis*. Acta Physiol Plant 24:393–398

Bylka W, Stobiecki M, Franski R (2001) Sulphated flavonoid glycosides from leaves of *Atriplex hortensis*. Acta Physiol Plant 23:285–290

Byrne M, Hankinson M, Sampson JF, Stankowski S (2008) Microsatellite markers isolated from a polyploidy saltbush, *Atriplex nummularia* Lindl. (Chenopodiaceae). Mol Ecol Resour 8:1426–1428

Cambrollé J, Mancilla-Leytón JM, Muñoz-Vallés S, Luque T, Figueroa ME (2012a) Tolerance and accumulation of copper in the salt-marsh shrub *Halimione portulacoides*. Mar Pollut Bull 64:721–728

Cambrollé J, Mancilla-Leytón JM, Muñoz-Vallés S, Luque T, Figueroa ME (2012b). Zinc tolerance and accumulation in the salt-marsh shrub *Halimione portulacoides*. Chemosphere 86:867–874

Capua CJ, Hopson NP, Malcolm MGS, Johnston GR, KimL GB, Lee CM, Booth GM (2010). Cytotoxicity of *Atriplex confertifolia*. J Toxicol 1–7. doi:10.1155/2010/976548

Casas AM, Nelson DE, Raghothama KG, D'Urzo MP, Singh NK, Bressan RA, Hasegawa PM (1992) Expression of osmotin-like genes in the halophyte *Atriplex nummularia* L. Plant Physiol 99:329–337

Chisci GC, Bazzoffi P, Pagliai M, Papini R, Pellegrini S, Vignozzi N (2001) Association of Sulla and *Atriplex* shrub for the physical improvement of clay soils and environmental protection in central Italy. Agric Ecosyst Environ 84:45–53

Debez A, Huchzermeyer B, Abdelly C, Koyro HW (2011) Current challenges and future opportunities for a sustainable utilization of halophytes. In: Öztürk M et al (ed) Sabkha ecosystems, tasks for vegetation science 34. Vol 46, pp 59–77

Demiral T, Türkan I (2004) Does exogenous glycinebetaine affect antioxidative system of rice seedlings under NaCl treatment. J Plant Physiol 6:1089–1100

El Shaer HM (2010) Halophytes and salt-tolerant plants as potential forage for ruminants in the Near East region. Small Rumin Res 91:3–12

Fitzsimmons K, Lovely C, Glenn E (1998) Growth differences among widely separated geographic accessions of fourwing saltbush (*Atriplex canescens*) in the Great Basin Desert, New Mexico, USA. Arid Soil Res Reah 12:87–94

Flowers TJ, Colmer TD (2008) Salinity tolerance in halophytes. New Phytol 179:945–963

Fu X-Z, Khan EU, Hu S-S, Fan Qi-J, Liu Ji-H (2011) Overexpression of the betaine aldehyde dehydrogenase gene from *Atriple hortensis* enhances salt tolerance in the transgenic trifoliate orange (Poncirus trifoliate L. Raf.). Environ Exp Bot 74:106–113

Gardea-Torresdey JL, Peralta-Videab JR, de la Rosa G, Parsons JG (2005) Phytoremediation of heavy metals and study of the metal coordination by X-ray absorption spectroscopy. Coord Chem Rev 249:1797–1810

Gharaibeh MA, Eltaif NI, Albalasmeh AA (2011) Reclamation of highly calcareous saline sodic soil using *Atriplex halimus* and by-product gypsum. Int J Phytoremediation 13:873–883

Glenn EP, Brown JJ (1998) Effects of soil salt levels on the growth and water use efficiency of *Atriplex canescens* (Chenopodiaceae) varieties in drying soil. Am J Bot 85:10–16

Glenn EP, Nelson SG, Ambrose B, Martinez R, Soliz D, Pabendinskas V, Hultine K (2012) Comparison of salinity tolerance of three Atriplex spp. in well-watered and drying soils. Environ Exp Bot 83:62–72

Guo BH, Zhang YM, Li HJ, Du LQ (2000) Transformation of wheat with a gene encoding for the betaine aldehyde dehydrogenase (BADH). Acta Bot Sin 42:279–283

Guo Y, Zhang L, Xiao G, Chen SY (1997) Expression of the BADH gene and salinity tolerance in rice transgenic plants. Sci China 27:151–155

Hamada A, Shono M, Xia T, Ohta M, Hayashi Y, Tanaka A, Hayakawa T (2001) Isolation and characterization of a Na^+/H^+antiporter gene from the halophyte *Atriplex gmelini*. Plant Mol Biol 46:35–42

Hyder SZ (1981) Preliminary observations on the performance of some exotic species of *Atriplex* in Saudi Arabia. J Range Manage 34:208–210

Jacobs GA, Smith CJ (1977) Utilization of four *Atriplex* species by sheep. Agroanimalia 9:37–43

Kabata-Pendias A, Pendias H (2001) Trace elements in soils and plants. CRC Press, Florida

Kachout SS, Ben Mansoura A, Mechergui R, Leclerc JC, Rejeb MN, Ouerghi Z (2012) Accumulation of Cu, Pb, Ni and Zn in the halophyte plant *Atriplex* grown on polluted soil. J Sci Food Agric 92:336–342

Kadereit G, Mavrodiev EV, Zacharias EH, Sukhorukov AP (2010) Molecular phylogeny of Atripliceae (Chenopodioideae, Chenopodiaceae): implications for systematics, biogeography, flower and fruit evolution, and the origin of C4 photosynthesis. Am J Bot 97:1664–1687

Kandil HM, El-Shaer HM (1989) The utilization of *Atriplex nummularia* by goats and sheep in Sinai. Proceedings of international symposium on the constraints and possibilities of ruminant production in the dry tropics, Cairo, Egypt, pp 71–73

Keckeis K, Sarker SD, Dinan LN (2000) Phytoecdysteroids from *Atriplex nummularia*. Fitoterapia 71:456–458

Khan MA, Ungar IA, Showalter M (2000) Effects of salinity on growth, water relation and ion accumulation of the subtropical perennial Halophyte, *Atriplex griffithii var. stocksii*. Ann Bot 85:225–232

Khedr AHA, Serag MS, Nemat-Alla MM, Abo-Elnaga AZ, Nada RM, Quick WP, Abogadallah GM (2011) A DREB gene from the xero-halophyte *Atriplex halimus* is induced by osmotic but not ionic stress and shows distinct differences from glycophytic homologues. Plant Cell Tiss Organ Cult 106:191–206

Ksouri R, Ksouri WM, Jallali I, Debez A, Magné C, Hiroko I, Abdelly C (2011) Medicinal halophytes: potent source of health promoting biomolecules with medical, nutraceutical and food applications. Crit Rev Biotechnol. doi:10.3109/07388551.2011.630647

Le Houérou HN (1992) The role of saltbushes (*Atriplex* spp.) in arid land rehabilitation in the Mediterranean Basin: a review. Agrofor Syst 18:107–148

Lefèvre I, Marchal G, Meerts P, Corréal E, Lutts S (2009) Chloride salinity reduces cadmium accumulation by the Mediterranean halophyte species *Atriplex halimus* L. Environ Exp Bot 65:142–152

Liu Z-H, Zhang H-M, Li G-L, Guo X-L, Chen S-Y, Liu G-B, Zhang Y-M (2011) Enhancement of salt tolerance in alfalfa transformed with the gene encoding for betaine aldehyde dehydrogenase. Euphytica 178:363–372

Lomonte C, Sgherri C, Baker AJM, Kolev SD, Navari-Izzo F (2010) Antioxidative response of *Atriplex codonocarpa* to mercury. Environ Exp Bot 69:9–16

Luque CJ, Castellanos EM, Castillo JM, Gonzalez M, Gonzalez-Vilches MC, Figueroa ME (1999) Metals in halophytes of a contaminated estuary (odiel saltmarshes, SW Spain). Mar Pollut Bull 38:49–51

Lutts S, Lefèvre I, Delpérée C, Kivits S, Dechamps C, Robledo A et al (2004) Heavy metal accumulation by the halophyte species Mediterranean saltbush. J Environ Qual 33:1271–1279

Maganhotto de Souza Silva CM, Vieira RF, Oliveira PR (2008) Salinity, sodicity and microbiological properties of an Ultisol cultivated with saltbush and irrigated with saline effluents. Pesqui Agropecu Bras 43:1389–1396

Malcolm CV, Clarke AJ, D'Antuono MF, Swaan TC (1988) Effects of plant spacing and soil conditions on the growth of five *Atriplex* species. Agric Ecosyst Environ 21:265–279

Manousaki E, Kalogerakis N (2009) Phytoextraction of Pb and Cd by the Mediterranean saltbush (*Atriplex halimus* L.) metal uptake in relation to salinity. Environ Sci Poll Res 16:844–854

Martinez JP, Ledent JF, Bajji M, Kinet JM, Lutts S (2003) Effect of water stress on growth, Na and K accumulation and water use efficiency in relation to osmotic adjustment in two populations of *Atriplex halimus* L. Plant Growth Regul 41:63–73

Montgomery KR, Cheo PC (1969) Moisture and salt effects on fire retardance in plants. Am J Bot 56:1028–1032

Niu X, Wang W, Bressan RA, Hasegawa PM (1994) Molecular cloning and expression of a glyceraldehydes-3-phosphate dehydrogenase gene in a desert halophyte, *Atriplex nummularia* L. Plant Physiol 104:1105–1106

Niu X, Zhu JK, Narasimhan ML, Bressan RA, Hasegawa PM (1993) Plasma- membrane H^+-ATPase gene expression is regulated by NaCl in cells of the halophyte *Atriplex nummularia* L. Planta 190:433–438

Norman HC, Masters DG, Wilmot MG, Rintoul AJ (2008) Effect of supplementation with grain, hay or straw on the performance of weaner Merino sheep grazing old man (*Atriplex nummularia*) or river (*Atriplex amnicola*) saltbush. Grass Forage Sci 63:179–192

Ohta M, Hayashi Y, Nakashima A, Hamada A, Tanaka A, Nakamura T, Hayakawa T (2002) Introduction of a Na^+/H^+antiporter gene from *Atriplex gmelini* confers salt tolerance to rice. FEBS Letters 532:279–282

Ortíz-Dorda J, Martínez-Mora C, Correal E, Simón B, Cenis JL (2005) Genetic Structure of *Atriplex halimus* Populations in the Mediterranean Bassin. Ann Bot 95:827–834

Pitman MG, Lauchli A (2002) Global impact of salinity and agricultural ecosystems. In: Lauchli A, Luttge V (eds) Salinity: environment plants molecules. Kluwer, The Netherlands, pp 3–20

Reboreda R, Caçador I (2007) Halophyte vegetation influences in salt marsh retention capacity for heavy metals. Environ Pollut 146:147–154

Rhodes D, Hanson AD (1993) Quaternary ammonium and tertiary sulfonium compounds in higher plants. Annu Rev Plant Physiol Plant Mol Biol 44:357–384

Riasi A, Mesgaran MD, Stern MD, Moreno MJR (2008) Chemical composition, *in situ* ruminal degradability and post-ruminal disappearance of dry matter and crude protein from the halophytic plants *Kochia scoparia*, *Atriplex dimorphostegia*, *Suaeda arcuata* and *Gamanthus gamacarpus*. Anim Feed Sci Technol 141:209–219

Sakamoto A, Murata N (2000) Genetic engineering of glycinebetaine synthesis in plants: current status and implications for enhancement of stress tolerance. J Exp Bot 51:81–88

Salem AZM, Hassan AA, Khalil MS, Gado HM, Alsersy H, Simbaya J (2012) Effects of sun-drying and exogenous enzymes on nutrients intake, digestibility and nitrogen utilization in sheep fed *Atriplex halimus* foliages. Anim Feed Sci Tech 171:128–135

Sameni AM, Soleimani R (2007) Effect of salinity and some chemical properties of the under- and intercanopy soils on range plants in a dry region of Southern Iran. Commun Soil Sci Plan 38:15–33

Sampson JF, Byrne M (2012) Genetic diversity and multiple origins of polyploidy *Atriplex nummularia* Lindl. (Chenopodiaceae). Biol J Linn Soc 105:218–230

Sawalha MF, Peralta-Videa JR, Romeor-Gonzalez J, Gardea–Torresdey JL (2006) Biosorption of Cd(II), Cr(II), and Cr(VI) by Saltbush (*Atriplex Carnescens*) biomass: thermodynamic and isotherm studies. Colloid Interf Sci 300:100–104

Shen YG, Du BX, Zhang WK, Zhang JS, Chen SY (2002) *AhCMO*, regulated by stresses in *Atriplex hortensis*, can improve drought tolerance in transgenic tobacco. Theor Appl Genet 105:815–821.

Shen YG, Zhang WK, Yan DQ, Du BX, Zhang JS, Liu Q, Chen SY (2003) Characterization of a DRE-binding transcription factor from a halophyte *Atriplex hortensis*. Theor Appl Genet 107:155–161

Shi H, Zhu JK (2002) Regulation of expression of the vacuolar Na^+/H^+antiporter gene AtNHX1 by salt stress and abscisic acid. Plant Mol Biol 50:543–550

Siddiqui BS, Ahmed S, Ghiasuddin M, Khan AU (1994) Triterpenoids of *Atriplex stocksii*. Phytochemistry 37:1123–1125

Silveira JAG, Araújo SAM, Lima GPMS, Viégas RA (2009) Roots and leaves display contrasting osmotic adjustment mechanisms in response to NaCl-salinity in Atriplex nummularia. Environ Exp Bot 66:1–8

Slavich PG, Smith KS, Tyerman SD, Walker GR (1999) Water use of grazed saltbush plantations with saline watertable. Agricu Water Manag 39:169–185

Smaoui A, Barhoumi Z, Rabhi M, Abdelly C (2011) Localization of potential ion transport pathways in vesicular trichome cells of Atriplex halimus L. Protoplasma 48:363–372

Sousa AI, Caçador I, Lillebø, AI, Pardal MA (2008) Heavy metal accumulation in *Halimione portulacoides*: intra- and extra-cellular metal binding sites. Chemosphere 70:850–857

Tabuchi T, Kawaguchi Y, Azuma T, Nanmori T, Yasuda T (2005) Similar regulation patterns of choline monooxygenase, phosphoethanolamine N-methyltransferase and S-adenosyl-L-methionine synthetase in leaves of the halophyte *Atriplex nummularia* L. Plant Cell Physiol 46:505–513

van der Baan A, van Niekerk WA, Rethman NFG, Coertze RJ (2004) The determination of digestibility of *Atriplex nummularia* cv. De Kock (Oldman's Saltbush) using different in invitro techniques. South Afr J Anim Sci 34:95–97

Vickerman DB, Shannon MC, Banuelos GS, Grieve CM, Trumble JT (2002) Evaluation of *Atriplex* lines for selenium accumulation, salt tolerance and suitability for a key agricultural insect pest. Environ Poll 120:463–473

Vromman D, Flores-Bavestrello A, Šlejkovec Z, Lapaille S, Teixeira-Cardoso C, Briceño M, Kumar M, Martínez J-P, Lutts S (2011) Arsenic accumulation and distribution in relation to young seedling growth in *Atriplex atacamensis* Phil. Sci Total Environ 412–413:286–295

Wang LW, Showalter AM (2004) Cloning and salt-induced, ABA-independent expression of choline mono-oxygenase in *Atriplex prostrata*. Physiol Plant 120:405–412

Welch BL (1978) Relationships of soil salinity, ash, and crude protein in *Atriplex canescens*. J Range Manage 31:132–133

Xiao G, Zhang GY, Liu FH, Chen SY (1995) The study of the BADH gene in *Atriplex hortensis*. Chin Sci Bull 40:741–745

Xu J, Yin H, Yang L, Xie Z, Liu X (2011) Differential salt tolerance in seedlings derived from dimorphic seeds of Atriplex centralasiatica: from physiology to molecular analysis. Planta 233:859–871

Yamaguchi-Shinozaki K, Shinozaki K (1994) A novel *cis*-acting element in an *Arabidopsis* gene is involved in responsiveness to drought, low-temperature, or high-salt stress. The Plant Cell 6:251–264

Yang X, Wen X, Gong H, Lu Q, Yang Z, Tang Y, Liang Z, Lu C (2007) Genetic engineering of the biosynthesis of glycinebetaine enhances thermotolerance of photosystem II in tobacco plants. Planta 225:719–733

Yin X, Zhao Y, Luo D, Zhang H (2002) Isolating the promoter of a stress-induced gene encoding betaine aldehyde dehydrogenase from the halophyte *Atriplex centralasiatica* Iljin. Biochim Biophys Acta 1577:452–456

Zhang H, Dong H, Li W, Sun Y, Chen S, Kong X (2009) Increased glycine betaine synthesis and salinity tolerance in AhCMO transgenic cotton lines. Mol Breed 23:289–298

Zucca C, Julitta F, Previtali F (2011) Land restoration by fodder shrubs in a semi-arid agro-pastoral area of Morocco. Effects on soils. Catena 87:306–312

Chapter 17
Dealing with Environmental Stresses: Role of Polyamines in Stress Responses

Rinukshi Wimalasekara and Günther F. E. Scherer

Abstract Extreme environmental conditions including drought, high- and low-temperatures, high salinity, mineral deficiency and heavy metal toxicity severely affect crop loss worldwide. Improvement of plants for enhanced resistance to adverse climatic conditions is a key issue in sustainable crop production, strengthening the global food safety. Understanding stress tolerance mechanisms of plants are a prime importance in crop improvement. Among the array of components involved, plant polyamines (PAs) are identified as one such group of components that play an important role in diverse environmental stress responses. PAs are small organic cations containing two or more amino groups. These are growth regulators present widely in all living organisms with varying quantities ranging from micromolar to milimolar. In plants, the most abundantly found PAs are di-amine putrescine, tri-amine spermidine and tetra-amine spermine. Accumulation of long chain and conjugated forms of PA occur under some environmental and growth conditions. Biosynthesis, transport, degradation and conjugation determine the level of PAs and vary throughout a plant life cycle. Catabolism of PAs by amine oxidases is trivial in the regulation of cellular levels of PAs. Apart from the essential functions in growth and development, PAs play a key role in environmental stress responses such as drought, chilling, salinity, mineral deficiencies such as potassium, nitrogen and magnesium deficiency, heavy metal toxicity, mechanical injuries and defence signalling against pathogens. Differential transcriptional regulation of several stress-related genes in PA-overexpressed transgenic plants suggests potential signalling function of PAs in stress responses. Genetic manipulation of crop plants for altered regulation of PA biosynthesis/catabolism may lead to improved stress tolerance potential. This article summarizes the recent findings on the involvement

R. Wimalasekara (✉) · G. F. E. Scherer
Institute of Floriculture and Wood Science, Section of Molecular Developmental Physiology, Leibniz University of Hannover, Herrenhäuser Str. 2, 30419, Hannover, Germany
e-mail: rinukshi@yahoo.co.in

G. F. E. Scherer
e-mail: scherer@zier.uni-hannover.de

K. R. Hakeem et al. (eds.), *Crop Improvement,* DOI 10.1007/978-1-4614-7028-1_17,

of PAs in abiotic stress responses in plants and possible means of manipulating PAs in the crop plants for enhanced stress tolerance.

Introduction

The impact of extreme climatic conditions such as atmospheric warming, desertification and soil salinisation on agriculture is growing, causing substantial crop loss worldwide. Climate change also modifies the risks of pest and pathogen outbreak, negatively affecting crop productivity. Devastating climate coupled with population growth exerts constant pressure on crop production, demanding more attention to find ways and means of crop adaptability to challenging environmental conditions. In order to secure the food requirements in changing climate scenario new technologies, efficient and sustainable farming practices have to be taken into consideration.

Plants are continuously being exposed to abiotic stress conditions such as drought, heat, chilling, freezing and high salinity and biotic stresses like pathogen invasion, insect predations and weeds. Understanding the mechanisms that trigger stress damage and adaptive machinery of plants to various stress situations is of prime importance in the progress of agriculture industry. Adaptation and acclimation to stresses is a result of a combination of events occurring at the anatomical and morphological levels to the cellular, biochemical and molecular levels. Signal transduction pathways that link the perception of stress signals with the appropriate cellular responses leading to stress tolerance are extensively studied (Taiz and Zeiger 2006) and profound knowledge in all these aspects is needed in production of plants with improved stress tolerance.

Among array of components involved, polyamines (PAs) are one of the components that play significant functions in plant stress responses. PAs are small organic cations containing two or more amino groups and known to be essential growth regulators present ubiquitously in both prokaryotic and eukaryotic cells (Evans and Malmberg 1989; Buchanan et al. 2000; Martin-Tanguy 2001). The most abundant free PAs in plants are di-amine putrescine (put), tri-amine spermidine (spd) and tetra-amine spermine (spm). PAs occur as free form or as conjugated forms attached to proteins, nucleic acids, hydroxycinnamic acid forming phenol amides, anionic components of phospholipids and cell wall components such as pectic polysaccharides (Buchanan et al. 2000; Kakkar et al. 2000; Martin-Tanguy 2001; Kakkar and Sawhney 2002; Takahashi and Kakehi 2010). The equilibrium between free and conjugated forms of PAs is crucial in development and their titer fluctuates in response to environmental conditions (Torrigiani et al. 1987; Galston and Sawhney 1990). In plants the PA content varies from micromolar to more than milimolar amounts.

The functions of PAs in growth and development (Kakkar et al. 2000; Martin-Tanguy 2001; Kakkar and Sawhney 2002; Kusano et al. 2008; Hussain et al. 2011) and in biotic and abiotic stress responses (Walter 2003; Alcázar et al. 2006b, 2010a; Groppa and Benavides 2008; Hussain et al. 2011; Marco et al. 2011; Nambeesan et al. 2012; Shelp et al. 2012) are extensively reviewed. PAs are crucial components in embryogenesis (Evans and Malmberg 1989; Galston et al. 1997; Kakkar et al. 2000; Bertoldi et al. 2004; Silveira et al. 2006; Pieruzzi et al. 2011), differentiation and morphogenesis (Kusano et al. 2007a, 2008; Hassannejad et al. 2012; Takano et al. 2012; Yoshimoto et al. 2012), floral initiation and development and fruit development (Galston et al. 1997; Bagni and Tassoni 2001; Kusano et al. 2007a, 2008) and senescence (Evans and Malmberg 1989; Takahashi and Kakehi 2010). Involvement of PAs in several types of abiotic stress responses such as salinity (Alcázar et al. 2006b; Liu et al. 2007; Alet et al. 2012; Hu et al. 2012), drought (Galston et al. 1997; Alcázar et al. 2006b; Groppa and Benavides 2008; Wang et al. 2011), extreme temperature tolerance (Evans and Malmberg 1989; Urano et al. 2003; Zhang et al. 2011; Cheng et al. 2012; Cvikrová et al. 2012; Lee et al. 2012;), and in mineral deficiencies and heavy metal toxicity (Martin-Tanguy 2001; Alcázar et al. 2010a; Shevyakova et al. 2011; Kumar et al. 2012) is extensively described.

PA catabolism is an important process in regulating PA levels in cells. In plants, copper-binding diamine oxidases (CuAO/DAO) and flavin adenine dinucleotide (FAD)-binding polyamine oxidases (PAO) have diversified roles in growth cycle and in environmental stress tolerance (reviewed in Cona et al. 2006; Moschou et al. 2008b). CuAO/DAO and PAO are involved in cell wall strengthening and rigidity during cell growth and development (Paschalidis and Roubelakis-Angelakis 2005; Cona et al. 2006; Delis et al. 2006; Tisi et al. 2011). In dormancy alleviation, differential regulation of genes involved in reactive oxygen species (ROS) production including amine oxidases is observed (Oracz et al. 2009). Further, association of PAO in developmental programmed cell death is reported (Paschalidis and Roubelakis-Angelakis 2005; Cona et al. 2006). Differential regulation of *PAO* genes and concomitant accumulation of proline accompanied with enhanced H_2O_2 production leading to salt tolerance is reported in some plant species (reviewed in Cona et al. 2006; Alcázar et al. 2010a; Campestre et al. 2011). Induction of PAO is described in water deficient conditions leading to stomatal regulation by ABA signalling (Moschou et al. 2008b) and osmotic stress tolerance (Aziz et al. 1997; Moschou et al. 2008b). Evidences indicate that H_2O_2 induced by DAO and PAO activity has a function in plant-pathogen/elicitor defence responses (reviewed in Cona et al. 2006; Moschou et al. 2008b; Wimalasekara et al. 2011a).

Biosynthesis of Polyamines

The intracellular levels of PAs are determined by their biosynthesis, degradation and conjugation. Spatial and temporal variations of intracellular PA levels are reported in many plant species. The initial step of the biosynthesis of PAs is the formation of di-amine put. Biosynthesis of put in plants occurs through two distinct pathways either directly from ornithine by ornithine decarboxylase (ODC) or from arginine by arginine decarboxylase (ADC) followed by two successive reactions catalysed by agmatine iminohydrolase (AIH) and N-carbamoylputrescine amidohydrolase

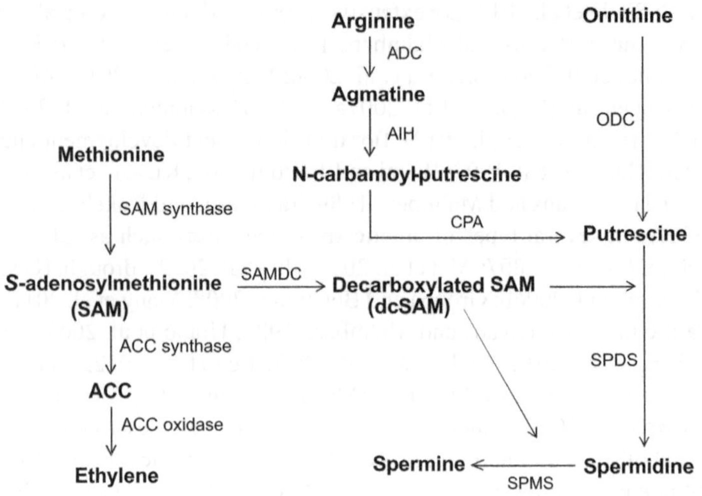

Fig. 17.1 Schematic representation of PA biosynthetic pathways in plants. *ADC* arginine decarboxylase, *AIH* agmatine iminohydrolase, *CPA* N-carbamoylputrescine amidohydrolase, *ODC* ornithine decarboxylase, *SPDS* spermidine synthase, *SPMS* spermine synthase, *SAM synthase* *S*-adenosylmethionine synthase, *SAMDC* *S*-adenosylmethionine decarboxylase, *ACC synthase* 1-amino-cyclopropane-1-carboxylic-acid synthase, *ACC oxidase* 1-amino-cyclopropane-1-carboxylic-acid oxidase. (Adapted from Wimalasekara et al. 2011a)

(CPA). The tri-amine spd and tetra-amine spm are synthesised by successive addition of aminopropyl groups to put and spd respectively in reactions catalyzed by spermidine synthase (SPDS) and spermine synthase (SPMS) (Slocum et al. 1984; Tiburcio et al. 1997). The aminopropyl groups are produced from decarboxylation of *S*-adenosylmethionine (SAM) catalyzed by SAM decarboxylase (SAMDC) (Slocum et al. 1984; Tiburcio et al. 1997) (Fig. 17.1). The presence of ODC pathway and characterization of genes coding for ODC is reported from many plant species (Michael et al. 1996). However, *Arabidopsis* lacks ODC pathway and biosynthesis of put occurs exclusively through the ADC pathway (Hanfrey et al. 2001). *ADC1* and *ADC2* are the two genes that encode ADC of *Arabidopsis* and they are shown to be expressed in a tissue specific manner (Soyka and Heyer 1999). *Arabidopsis* has single genes that code for AIH and CPA (Janowitz et al. 2003; Piotrowski et al. 2003), two genes namely *SPDS1* and *SPDS2* that code for spermidine synthase (Panicot et al. 2002), genes *SPMS* and *ACL5* coding, spermine synthase (Hanzawa et al. 2000) and at least four genes coding SAM decarboxylase (Urano et al. 2004). In some plant species, subcellular compartmentalization of ADC pathway is reported which probably lead to a concentration gradient of put in a cell (Borrell et al. 1995; Bortolotti et al. 2004). A number of studies indicate that genes coding for PA biosynthesis is highly regulated by an array of abiotic and biotic factors and PA metabolism interacts largely with other metabolic pathways (Bouchereau et al. 1999; Soyka and Heyer 1999; Urano et al. 2003; Alcázar et al. 2006b, 2010a).

Catabolism of Polyamines

PA catabolism is an important process that regulates the intra-cellular levels of PA. PAs are oxidatively catabolised by copper-binding diamine oxidases (CuAO)/ diamine oxidases (DAO) and FAD-binding polyamine oxidases (PAO) which are commonly found in all living organisms (Buchanan et al. 2000; Bagni and Tassoni 2001; Cona et al. 2006). Plant CuAO/DAO preferentially catalyses the oxidation of di-amine put, at the primary amine group producing 4-aminobutanal with concomitant production of NH_3 and H_2O_2. Resulting 4-aminobutanal is further converted to γ-aminobutyric acid (GABA) via Δ^1-pyrroline. PAOs preferentially catalyse the oxidation of spd and spm at the secondary amine group producing 4-aminobutanal and *N*-(3-aminopropyl)-4 aminobutanal, respectively, with concomitant production of 1,3-diaminopropane (DAP) and H_2O_2 (Buchanan et al. 2000; Cona et al. 2006; Moschou et al. 2008b, c) (Fig. 17.2). Experimental evidences indicate that CuAO/ DAO and PAO have important regulatory function in plant growth and in stress tolerance by modulating the levels of PAs and their reaction products. They are important regulators of ROS and GABA synthesis, which are key metabolites involved in various physiological processes. Both CuAO/DAO and PAO have species and tissue specific regulatory functions and sptio-temporal expression patterns (Cona et al. 2006). CuAO/DAO encoding genes have been isolated and characterized from some plants for example, from *Arabidopsis* (Moller and McPherson 1998), chick pea (Rea et al. 1998) and pea (Tipping and McPherson 1995). PAOs are identified from many plant species particularly in monocots (Sebela et al. 2001). Gene family of PAO from maize which consist of three members and proteins are bound to cell walls (Tavladoraki et al. 1998) and symplast localized barley PAO family consisting of two genes (Cervelli et al. 2001) are widely been studied. In *Arabidopsis,* twelve CuAO/DAO and five PAO coding genes have been identified and characterized. The proteins are localized in apoplast and in peroxisomes (Alcázar et al. 2006b; Cona et al. 2006; Tavladoraki et al. 2006). Studies on gain and loss of function mutants especially in *Arabidopsis* have provided new insights into molecular mechanisms of PA function and PA interaction with other signal transduction pathways. These findings with special emphasis on abiotic stress tolerance are discussed in detail in the following sections.

Abiotic Stresses

Experimental evidences indicate that differential regulation of plant PA biosynthesis, conjugation and catabolism are important in dealing with stress tolerance in response to a variety of environmental stresses ranging from drought, salinity, temperature extremes, mineral deficiency to wounding (reviewed in Cona et al. 2006; Groppa and Benavides 2008; Moschou et al. 2008b; Alcázar et al. 2010a; Angelini et al. 2010; Wimalasekara and Scherer 2010; Wimalasekara et al. 2011a). The balance between PA catabolism and anabolism is shown to play a significant role in

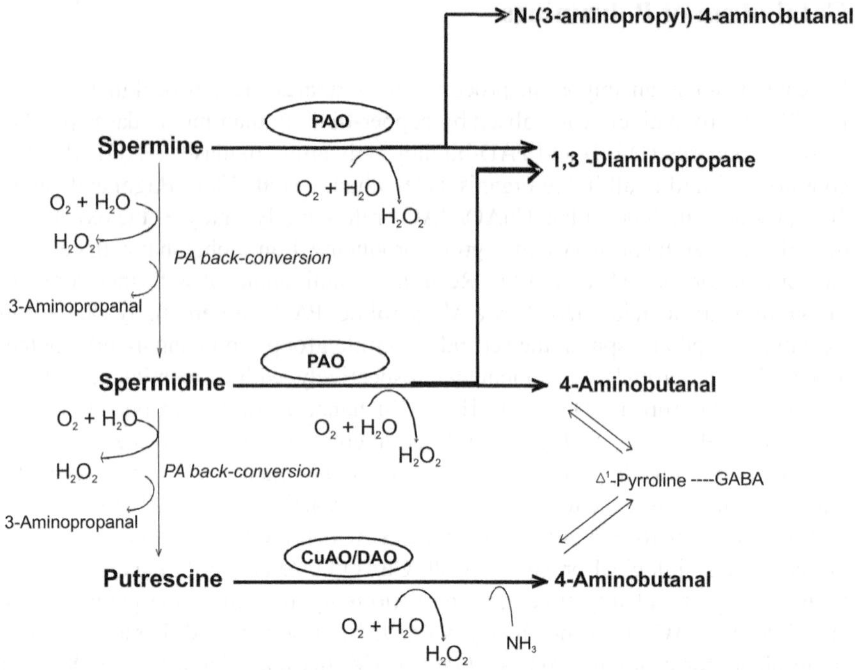

Fig. 17.2 Schematic representation of PA catabolic pathways in plants. Copper-binding diamine oxidases (CuAO)/diamine oxidases (DAO) catalyses the oxidation of di-amine putrescine, at the primary amine group producing hydrogen peroxide (H_2O_2), ammonia (NH_3) and 4-aminobutanal as reaction products. Polyamine oxidase (PAO) catalyses the oxidation of spermidine and spermine at the secondary amine group producing 4-aminobutanal and *N*-(3-aminopropyl)-4 aminobutanal respectively, in addition to 1,3-diaminopropane and H_2O_2. 4-aminobutanal produced through putrescine and spermidine oxidation is further converted to γ-aminobutyric acid (GABA) via Δ^1-pyrroline. Besides terminal catabolic pathway, back-conversion pathway takes place producing spermidine and putresine. H_2O_2 and 3-aminopropanal is produced as reaction products in the polyamine back conversion pathway. (Modified from Angelini et al. 2010)

PA-mediated stress tolerance. A number of recent studies indicate that cross talk of PA signalling with other signalling pathways in a variety of abiotic stress responses (Alcázar et al. 2010a).

Salt Stress

Plant growth is restricted when the soil contains excess amounts of minerals. Salinity affects a considerable amount of crop loss worldwide specifically in arid and semiarid regions. Salt injury is caused by decrease in water potential and by accumulation of ions to toxic levels. Tolerance levels to soil salinity differ dramatically among plant species. The mechanisms by which plants tolerate salinity are complex involving molecular synthesis, enzyme induction and membrane transport. Over the years, attempts were made to enhance the salt tolerance of several salt sensitive crop species using classical plant breeding and molecular biological approaches (Taiz and Zeiger 2006).

Significance of PAs, CuAO/DAO and PAO in salt stress tolerance is described in several plant species (reviewed in Alcázar et al. 2006b, 2010a; Cona et al. 2006; Liu et al. 2007). Accumulation of increased amounts of PAs in rice, tobacco and *Arabidopsis* is reported to enhance the tolerance to high salinity conditions. In salt-tolerant cultivars of rice, accumulated amount of spd and spm was higher than in the salt-sensitive rice cultivars while the latter showed higher amounts of put accumulation (Basu and Ghosh 1991). The quantification of PA levels in salt-sensitive rice cultivars showed that salt-sensitivity is related with the differences in PA accumulation in the shoot system under high salinity, specifically put to a higher level and spd and spm to lower levels (Krishnamurthy and Bhagwat 1989). In salt-tolerant cultivars of rice, root plasma membranes were rich in spd and spm while in salt-sensitive cultivars root plasma membranes were rich in put (Roy et al. 2005). In barley seedlings, salt injuries caused by high concentration of NaCl could be partially attenuate by exogenous application of put and spd (0.5 mM) (Zhao and Qin 2004). In roots and leaves of *Lupinus luteus* growing on high salinity conditions, accumulated increased amounts of PAs bound to microsomal membranes implying less injuries caused by salt stress (Legocka and Kluk 2005). There are more reports on adverse salinity-mediated increase in PA amounts in a number of plant species. For example, an increase in free, acid-soluble bound and total spm was observed in leaf tissues of sunflower exposed to increasing concentrations of NaCl (Mutlu and Bozcuk 2005), In spinach, lettuce, melon, pepper, broccoli, tomato and wheat high salt concentrations resulted in substantial accumulation of spd and spm (El-Shintinawy 2000; Zapata et al. 2004).

Experimental evidences indicate that salt stress induced PA-mediated responses mainly rely on the activation of arginine decarboxylase (*ADC2*) and amine oxidases. In *Arabidopsis*, strikingly increased expression level of *ADC2* and spermine synthase (*SPMS*) was observed under high salinity (Soyka and Heyer 1999). Further, mutants defect in PA biosynthesis displayed increased sensitivity to salt stress (Soyka and Heyer 1999; Yamaguchi et al. 2006). External supplementation of spm to spm-deficient mutants suppressed the salt sensitivity of these mutants (Yamaguchi et al. 2006). *Arabidopsis* mutants *spe1-1* and *spe2-1* with reduced *ADC* activity showed no accumulation of PAs in response to salt treatment, demonstrating the importance of *ADC* activity in salt tolerance (Kasinathan and Wingler 2004). Moreover, rice over-expressing oat *ADC* showed increased plant biomass under salinity indicating higher PA production by enhanced *ADC* activity (Roy and Wu 2001). In another study, salt stress resulted in an induction of *AtADC2* transcripts in *Arabidopsis* correlating with the accumulation of free put (Urano et al. 2004). Rice varieties exposed to high salinity showed an increase in transcript levels of *S*-adenosylmethionine decarboxylase (*SAMDC1*) and in the salt-tolerant variety, transcription of *SAMDC1* was higher than in the salt-sensitive variety (Li and Chen 2000). These experimental evidences indicate an obvious protective function of spd and spm in salt stress tolerance. Presence of a pool of put may be a prerequisite to spd and spm synthesis. spm and spd prevented the leakage of electrolytes and amino acids from roots and shoots of rice proposing a positive correlation between salt tolerance and increased accumulation of PAs (Chattopadhayay et al. 2002). The protective function of PAs in salt stress may be at least partly due to rigidifying microsomal membrane surfaces leading to stabilization against NaCl and osmotic stress (Legocka and Kluk 2005).

PA effect on salt-stress tolerance can also be attributed to the differential regulation of catabolism of PA by DAO and PAO (Moschou et al. 2008b; Angelini et al. 2010). The catabolic products of PA such as 4-aminobutanal, H_2O_2 and GABA are associated with various physiological processes involved in stress responses including salt stress tolerance. Enhanced DAO/CuAO and PAO activities was observed in the salt stressed oat seedlings and tomato leaf discs followed by accumulation of proline and proline accumulation was hindered when these plants were treated with DAO/CuAO inhibitors (Alcázar et al. 2010b).

In high salinity exposed soybean roots increase in CuAO and GABA activities was seen with concomitant decrease in put, cadaverine and spd levels (Xing et al. 2007). Decrease in CuAO activity, recovery of PA levels and a simultaneous reduction of GABA level was found in soybean during the recovery from NaCl stress proposing that GABA derived from PA catabolism is probably involved in salt-stress mediated defence reactions (Xing et al. 2007). Increased amounts of H_2O_2 produced by DAO/CuAO and PAO activities lead to PCD and expression of defence genes suggesting importance of amine oxidases in salt stress responses. Maize ZmPAO overexpressing tobacco plants showed high quantities of spd followed by increased activity of apoplastic PAO. As a result, an elevated amount of H_2O_2 was produced and PCD was induced (Moschou et al. 2008a). In contrast, tobacco plants down-regulating endogenous PAO accumulated lower amounts of H_2O_2 in response to salt stress exhibiting less PCD than wild type plants (Moschou et al. 2008a). Induction of several stress-responsive genes could be observed as a result of apoplastic PA catabolism by PAO (Moschou et al. 2008a). Moreover, downstream signalling substances such as mitogen activated protein kinases (MAPK) were induced by H_2O_2 produced through PAO activity. In maize, salt stress induced ROS generated by PAO was involved in signalling the adaptive responses to high salinity (Rodríguez et al. 2009). These studies indicate that ratio of PA catabolism and anabolism is crucial in salt stress tolerance.

Taken together, increased PA levels represent a salt-stress induced protective function in many different plant species. Genetic modification of PA biosynthesis and catabolism pathways is useful in enhancing salt-tolerance function of model plants as well as in crop plants.

Water Stress

Drought has a direct effect on agriculture leading to crop losses in different magnitudes in many parts of the world. Far beyond the crop loss, drought causes many socio-economic impacts and even famine in some less developed countries. It is being predicted that wide spread drought conditions may happen more regularly, severely and last for extended periods in coming decades as a result of global warming and climate change. Strategies for efficient management of droughts include strategic management of agriculture. Apart from efficient practices for water and soil management and crop planning, several attempts have been paid to improve the drought tolerant capacity of plants especially important crop plants by conventional as well as by modern genetic engineering techniques.

Drought resistance mechanisms vary with climate and soil conditions and the capacity of drought resistant vary greatly among plant species. Plants exhibit various responses to water stress such as inhibition of leaf expansion which occur as a result of decrease in turgor, limits in photosynthesis, leaf abscission, increased wax deposition on the leaf surface, altered energy dissipation from leaves, root extension into deeper wetter soil and stomatal closure. Sensing of water deficiency and signal transduction leads to the induction of genes responsible for acclimation and adaptation to stress. Several of these genes encode enzymes associated with osmotic adjustment (Taiz and Zeiger 2006). Considerable effort have been directed towards identifying traits associated with drought resistance of crop plants and with the availability of molecular techniques crops have been modified for improved drought resistance.

Among the many components, PAs are identified as one such component having regulatory function in water stress. In water stressed plants endogenous levels of PAs are considerably increased confirming stress-specific roles of PAs (Galston et al. 1997; Groppa and Benavides 2008). For example, in wheat seedlings exposed to drought, PAs especially spd accumulated to a higher level in comparison to the control plants (Kubis and Krzywanski 1989), chickpea plants responded to onset of drought by increasing the endogenous spd levels and total PA content in the roots (Nayyar and Chander 2004; Nayyar et al. 2005), and in water stressed *V. faba* leaves, free spd levels increased considerably (Liu et al. 2000). The *Arabidopsis* double mutant *acl5/spms* that produce lower spm is hypersensitive to drought stress (Kusano et al. 2007b). Manipulation of PA biosynthesis pathway caused altered drought resistant capacities in many plant species. In *Arabidopsis*, overexpression of *ADC2* increased put levels and drought tolerance was enhanced by stimulation of stomatal closure (Alcázar et al. 2010b). Rice plants overexpressing *ADC* gene of oat showed improved drought tolerance by increasing put levels and reducing chlorophyll loss (Capell et al. 1998, 2004). Up-regulation of *ADC* gene was observed in the osmotically-stressed oat leaves, indicating increased PA synthesis leading to stress tolerance (Galston et al. 1997). Overexpression of *S*-adenosyl methionine decarboxylase in *Arabidopsis* lead to increased spm levels and enhanced the expression of a key ABA biosynthesis gene *NCED3* (Alcázar et al. 2006a). During drought conditions, accumulation of put was impaired in the ABA-deficient and ABA-insensitive mutants (Alcázar et al. 2006a). During water stress, ABA is known to modulate PA metabolism by up-regulating the expression of *ADC2*, spermidine synthase (*SPDS1*) and spermine synthase (*SPMS*) genes (Alcázar et al. 2006a). A study by Lie et al. (2000) revealed that physiological function of elevated PA levels during water stress tolerance was attributed to PA-mediated regulation of inward K^+ channel in the plasma membrane of guard cells and modulation of stomatal aperture. Further, these findings indicate that PAs target a putative guard cell K^+ channel KAT1-like inward K^+ channels in guard cells and modulate stomatal movements, providing evidence for presence of link among stress conditions, polyamine levels, and stomatal regulation (Lie et al. 2000). When dehydrated *in vitro* grown *Citrus reticulata* plants were treated with exogenous spm, stomatal closure was promoted and exhibited a less wilted phenotype, reduction of water loss and electrolyte leakage compared to spm untreated plants (Shi et al. 2010), again indicating a function of PA in stomatal regulation.

The importance of PA catabolism in water stress responses is described in many plant species (reviewed in Cona et al. 2006; Moschou et al. 2008b; Angelini et al. 2010). Osmotic stress tolerance of rape leaf discs and whole rape seedlings subjected to drought is linked to increased amounts of put and DAP which is a product of PAO activity (Aziz et al. 1997). Experimental evidence of Toumi et al. (2010) indicated that drought-tolerant grapevine has increased PAO activity as compared to drought-sensitive variety. H_2O_2 produced by PA oxidation reaction was involved in signalling cascade of drought tolerance (Toumi et al. 2010). DAO/CuAO and PAO are considered as important controllers of ABA signalling pathway in stomatal regulation (Lie et al. 2000; An et al. 2008; Moschou et al. 2008a, b). In *Arabidopsis*, induction of *AtPAO2, AtPAO3* and *AtPAO4* by ABA suggests a role of *PAO* in ABA signalling (Moschou et al. 2008a). Gene expression analysis demonstrated that *PAO2* of *Arabidopsis* is upregulated during drought stress and shows similar expression kinetics as the ABA-inducible *RD29A* and *RD22* genes supporting a role of *PAO2* in drought resistance (Alcázar et al. 2011). An et al. (2008) showed that exogenous application of ABA to the *Vicia faba* leaf epidermis caused stimulation of apoplast CuAO activity followed by increased H_2O_2 production finally leading to induced stomatal closure. When CuAO inhibitors were applied these processes were impaired. These factors indicate that CuAO in *V. faba* guard cells is an essential enzymatic source for H_2O_2 production in ABA-induced stomatal closure. For the ABA-induced stomatal closure, second messengers such as Ca^{2+}, ROS, and nitric oxide (NO) are also important. CuAO/DAO and PAO may be important in regulating the other signalling substances probably via generation of H_2O_2 in stress situations like water stress. Involvement of DAO in the ABA induced H_2O_2 production in roots of rice seedling is reported (Lin and Kao 2001). Biosynthesis of H_2O_2 by ABA-stimulated CuAO activity resulted in root growth inhibition by cell wall stiffening by peroxidases (Lin and Kao 2001). Involvement of a copper amine oxidase gene, *COPPER AMINE OXIDASE1* (*CuAO1*) of *Arabidopsis* was tested for its role in ABA mediated stress responses using the knockouts *cuao1-1* and *cuao1-2* (Wimalasekara et al. 2011b). Compared to WT, the knockouts showed less sensitivity to ABA during germination, seedling establishment and root growth inhibition characterizing knockouts as ABA-insensitive. Further, PA-induced and ABA-induced NO production in *cuao1-1* and *cuao1-2* were impaired suggesting a function of *CuAO1* in PA and ABA-mediated NO production (Wimalasekara et al. 2011b).

Heat and Cold Stress

Global warming is becoming a worldwide concern and over the last 25 years an increasing rate of warming has taken place. Regional temperature anomalies, extreme high temperature with frequent heat waves and extreme cold weather conditions are also inevitable. Influence of unusually high and low temperatures on agriculture, forestry and fisheries are enormous finally leading to socio-economic impacts around the world. High temperature on crop production most of the times affect Asia and Africa. Prolonged winter and irregular low temperature conditions affect agriculture in many European and American countries. In terms of agriculture, apart from mitigation measures, adaptation mechanisms to such adverse impacts include development of cultivation techniques and development of resistant crop varieties.

Most plants are unable to survive extended exposure to temperatures above 45°C while succulent plants are adapted to high temperatures tolerating temperatures of 60–65°C. High temperatures most of the times lead to heat stress and heat shock and often the water and temperature stresses are interrelated. Heat stress inhibits photosynthesis and impairs membrane function and protein stability. Adaptive mechanisms to heat stress include responses that decrease light absorption by the leaves, heat shock protein synthesis which act as molecular chaperons in stabilizing and correcting protein folding and biochemical responses leading to pH and metabolic homeostasis (Taiz and Zeiger 2006).

Chilling and freezing stresses are triggered by low temperature. Typical chilling injuries such as slow growth, leaf lesions and wilting are caused primarily by loss of membrane properties ensuing from changes in membrane. Freezing injury is linked basically to damage caused by formation of ice crystals within cells. Mechanisms that confer freeze resistant include dehydration and supercooling of leaves, which limit the growth of ice crystals to extracellular spaces. Cold stress leads to osmotic stress and activate osmotic stress-related signalling pathways and accumulation of proteins involved in cold acclimation (Taiz and Zeiger 2006).

Several plant species including important crop species have been produced for better tolerance to high and low temperature regimes by modifying physiological and biochemical pathways concerned. There are number of reports showing differential accumulation of PAs in response to high and low temperatures (reviewed in Kakkar and Sawhney 2002; Alcázar et al. 2006b, 2010a; Cona et al. 2006; Groppa and Benavides 2008; Moschou et al. 2008b; Gill and Tuteja 2010; Wimalasekara et al. 2011a). Importance of put accumulation in freezing temperature tolerance in *Arabidopsis* and transcriptional upregulation of *ADC1* and *ADC2* upon cold treatment is described (Cuevas et al. 2008, 2009). Compared to wild type plants, mutants *adc1* and *adc2* displayed higher sensitivity to freezing and treating with put complemented the stress sensitivity (Cuevas et al. 2008, 2009). Moreover, this study revealed that detrimental consequences of put depletion during cold stress are due, at least in part, to alterations in the levels of ABA by modulating expression of ABA biosynthesis genes (Cuevas et al. 2008, 2009). Complementation analysis of *adc1* mutants with ABA and reciprocal complementation of *aba2-3* mutant with put revealed that diamine control the levels of ABA in response to cold by modulating ABA biosynthesis (Cuevas et al. 2008, 2009). In another study, PA levels specially put levels increased in chickpea subjected to chilling temperature (Nayyar 2005). Accumulation of put occurs as a rapid reaction to low temperature as seen in poplar seedlings grown at 4°C (Renaut et al. 2005). Supplementation of growth medium with spd prior to the cold treatment resulted higher cold tolerance in cucumber (He et al. 2002). In cold-tolerant cultivars of cucumber, markedly increased level of spd was observed during chilling as opposed to the cold sensitive cultivars (Shen et al. 2000). *OsSPDS2*, a novel *SPDS* gene from rice was involved in chilling responses in rice roots (Imai et al. 2004). Further, *Arabidopsis* plants overexpressing *Cucurbita ficifolia SPDS1* showed increased tolerance to chilling and freezing tolerance (Kasukabe et al. 2004). Microarray analysis revealed that increased transcription of several stress responsive genes in the transgenic plants under chilling stress suggesting an important role for spd as a signaling compound or as a regulator of stress signaling pathways, leading to developed stress tolerance mechanisms (Kasukabe et al. 2004).

Altered regulation of PA biosynthesis is observed as one of the mechanisms involved in high-temperature tolerance. Overexpression of *SAMDC* (from *Saccharomyces cerevisiae*) in tomato caused 1.7–2.4 fold higher levels of spd and spm production under high temperature stress, enhanced antioxidant enzyme activity and the protection of membrane lipid peroxidation (Cheng et al. 2009). The levels of free and conjugated PAs and ADC were higher in calli of heat-tolerant rice cultivars than in heat-sensitive cultivars under non-stressed conditions. Heat stress caused greater accumulation of free and conjugated polyamines in calli of the heat-tolerant cultivar (Roy and Ghosh 1996). PA catabolism was also associated with heat stress as shown by increased PAO activity in heat-tolerant rice cultivars (Roy and Ghosh 1996). In another study, tobacco plants over-producing proline showed a transient increase in the levels of free and conjugated Put and in the levels of free spd, norspermidine (N-Spd) and spm after a 2-h lag phase (Cvikrová et al. 2012). The findings indicate that proline and PA biosynthetic pathways act together in dealing with heat stress conditions.

All these experimental results indicate that transgenic approach, which increases PA biosynthesis could be a good strategy to improve the high and cold temperature tolerance.

Oxidative Stress

Oxidative stress is induced by various abiotic factors such as hyperoxia, light, drought, high salinity, cold, metal ions, pollutants, xenobiotics and toxins, biotic factors like pathogen infection and developmental transitions such as seed maturation and aging of plant organs. Oxidative stress produces ROS in plant cells (Grene 2002). Under physiological steady state, ROS are scavenged by antioxidative defence components that are often confined to particular cell compartments. In some instances, plants generate ROS as defence responses to various stresses. If the equilibrium between ROS production and scavenging is disturbed it is harmful to the plants (Apel and Hirt 2004). The most important ROS scavenging enzymes are superoxide dismutase (SOD), ascorbate peroxidase (APX), glutathione peroxidase (GPX), and catalase (CAT).

PAs play a role in oxidative stress tolerance by functioning as antioxidants under some environmental conditions (Groppa et al. 2001, Chattopadhayay et al. 2002; Kakkar and Sawhney 2002). Few examples are, leaf injury of the ozone-sensitive tobacco cultivar Bel W3 caused by ozone treatments was minimized when put, spd or spm was applied to the root system (Bors et al. 1989) and it was suggested that the antioxidative effect is due to a combination of their anionic and cationic-binding properties in radical scavenging function (Bors et al. 1989). Løvaas (1997) reported that PAs are involved in preventing photooxidative damage. But there are contrdicting evidences regarding lack of antioxidant activity of PAs.

A study using vesicles prepared with mixed soy bean phospholipids showed that PA mediated inhibition of lipid peroxidation by metal-catalysed the oxidative reactions (Tadolini et al. 1988). H_2O_2 produced as a result of catabolism of PA by CuAO/DAO and PAO act as a signalling molecule that promotes activation of defence responses but it can also act as a prooxidant agent (Groppa and Benavides 2008).

Mineral Deficiency and Heavy Metal Stress

Both, natural and agricultural ecosystems are most of the times known to have sub-optimal levels of mineral nutrients (Lynch and Clair 2004). Mineral stress conditions cause important, complex, and poorly understood interactions with global climate change. Factors such as changes in rainfall, temperature, solar radiation and atmospheric CO_2 concentration have significant impact on soil nutrient status and soil erosion rates. Inadequate supply of minerals to the plants results in nutritional disorders manifested by characteristic deficiency symptoms. Presence of excess minerals in the soil mostly associated with accumulation of heavy metals such as zinc, copper, cobalt, nickel, mercury, lead and cadmium causing severe toxicity in plants (Taiz and Zeiger 2006). In plants metal toxicity arise from the binding of metals to sulphydryl groups of proteins inhibiting enzymes activities or altering protein structure (Van Assche and Clijsters 1990), stimulation of ROS formation leading to oxidation of macromolecules and oxidative stress (Sandalio et al. 2001). The presence of excess minerals in soil also leads to saline conditions in which plant growth is restricted. Most of the times mineral deficiency and toxic stresses in plants are not clearly identifiable with other stress conditions making it even more complicated to predict and find solutions in improving crop yield. Genetic variation among plants has a significant influence in dealing with mineral stresses. Plants display variable morphological, physiological and biochemical adoptive responses in dealing with particular mineral deficient condition and heavy metal toxicity.

Considerable efforts have been made to improve the yield of plants subjected to these stresses by modifying the soil, selecting and breeding genotypes and by genetic modification.

Mineral deficient conditions influence differential regulation of hormones and growth promoting substances. Altered metabolism of PAs in response to K^+, Mg^+ and PO_4^+ deficiencies are reported (reviewed in Groppa and Benavides 2008). K^+ deficiency-induced PA metabolism is mostly studied. Young and Galston (1984) showed that accumulation of put under K-deficiency by long term growing of oat plants on a low-K nutrient medium. ADC and ODC activities in entire K-deficient shoots were 6-fold and 2-fold greater than in the K-sufficient grown plants respectively (Young and Galston 1984). *Arabidopsis thaliana* responded to K-deficiency by increasing ADC activity upto 10-fold over unstressed plants with a corresponding increase in put levels up to 20-fold (Watson and Malmberg 1996). Important function of put in maintaining cation-anion balance in plant tissues is well known (reviewed in Bouchereau et al. 1999). In birch leaves, increased amounts of put and DAP, a product of PAO activity was found in K-deficient situations (Sarjala and Kaunisto 2002). Put accumulation is considered as a good indicator of K-deficiency of forest trees (Kaunisto and Sarjala 1997; Sarjala and Kaunisto 2002). In tobacco plants a short-term boron deficiency caused an increase in free put in roots and conjugated put in leaves indicating a possible link between boron and PA levels (Camacho-Cristóbal et al. 2005). Source of Nitrogen dependent PA accumulation has been reported in wheat and pepper. Significant increase in put content was observed in ammonium nutrition and to a lesser extent in urea nutrition (Houdusse et al. 2005). However, put content was significant reduced with nitrate nutrition (Houdusse et al. 2005). Proline content of the plants also showed a similar pattern as put content but to a lesser degree (Houdusse et al. 2005). They hypothesized that put biosynthesis might be related to proline degradation by a specific pathway related to ammonium detoxification. It is also reported that Mg and N deficiencies induced changes in PA content in grapevine (Evans and Malmberg 1989).

Changes in PA metabolism occur by metal toxicity (Groppa et al. 2001, 2003, 2007; Lin and Kao 1999). Increased accumulation of total PAs was observed when tobacco BY-2 cells were exposed to 0.05 mM Cd^{2+} for 3 days (Kuthanová et al. 2004). Among all the PAs, put content was significantly higher and DAO activity was also stimulated (Kuthanová et al. 2004). Treating mungbean seedlings with $CdCl_2$ (0.1–1.5 mmol/L) resulted decreased DAO activities and subsequently restricted the accumulation of endogenous PAs (Choudhary and Singh 2000). Groppa et al. (2003) reported that PA metabolism is differentially regulated by Cd^{2+} and Cu^{2+} in sunflower and wheat leaves. Both metals increased put content in wheat leaves but reduced put content in sunflower leaves. Treating wheat leaves with Cd^{2+} raised the ADC and ODC activities while Cu^{2+} raised only ODC activity (Groppa et al. 2003). Groppa et al. (2007) showed that both ADC and ODC activities in shoots of sunflower plants were increased by 1 mM Cd, whereas 1 mM Cu enhanced only ADC activity. A protective role of PAs in Cd^{2+} and Cu^{2+}-induced oxidative stress in sunflower leaves is described (Groppa et al. 2001). Pre-treatment with 1 mM spd and spm prevented the Cd^{2+} and Cu^{2+}-mediated reduced activities of glutathione reductase (GR) and superoxide dismutase activities (Groppa et al. 2001). In accordance, Tang et al. 2005 showed increased tolerance of *Typha latifola* under Cd^{2+} stress by exogenous addition of spd, primarily by increasing GR activity and GSH level.

Taken together all these experimental evidences indicate a function of PA in mineral deficiency and heavy metal toxicity stress responses although the molecular mechanism is not well understood.

Mechanical Stress

Mechanical damage/wounding caused by abiotic factors such as rain and wind, and by biotic factors such as pathogen attack and insect bite is a continuous threat to plants and it leads to a reduction in crop yield. Mechanical stress initiates signalling pathways and known to generate signalling substances for example jasmonic acid, salicylic acid, NO and H_2O_2 for the induction of wound-responsive genes.

Recent studies indicate that PAs are also involved in mechanical stress responses. In mechanically injured *Arabidopsis* and oilseed rape a significant increase of put levels was observed (Perez-Amador et al. 2002; Cowley and Walters 2005). In response to mechanical wounding and methyl jasmonate treatment, *Arabidopsis* showed increased expression of *ADC2* and a transient increase in the level of free put (Perez-Amador et al. 2002). Moreover, after wounding a decrease in the level of free spm, coincident with the increase in put was observed (Perez-Amador et al. 2002). Cowley and Walters (2005) showed that mechanical wounding of the oilseed rape leaves led to significant, but transient, increases in *ADC* activity and decrease in DAO activity both locally and systemically (Cowley and Walters 2005). Several reports provide evidences regarding the wound-healing function of CuAO/DAO and PAO (reviewed in Angelini et al. 2010). In chickpea seedlings mechanical wounding induced a rapid increase in *CuAO* expression levels and treatment of plants with CuAO inhibitor decreased the accumulation of H_2O_2 and lignosuberization along the lesion (Rea et al. 2002). In *Zea mays,* involvement of PAO in increased H_2O_2 biosynthesis during wounding was observed and wound healing was accelerated by the deposition of lignin and suberin along the wound area (Angelini et al. 2008). Importance of PAO-mediated H_2O_2 in wound healing was observed and it was revealed that lignin and suberin deposition along the wound area was accelerated by increased PAO activity (Angelini et al. 2008). Furthermore, CuAO involvement in wound healing is demonstrated in pea seedlings and injury induced an increase of CuAO and peroxidase activities and increased levels of put, cadevarine, spd and GABA (Petrivalský et al. 2007). In general, the function of CuAO/DAO and PAO in wound healing is considered to be due to the intensified lignin and suberine deposition as a consequence of H_2O_2 release.

Other Stresses

Increased levels of UV radiation and ozone and herbicide are some of the factors that impose stress in plants and they have developed various physiological and biochemical adaptations to deal with these stress conditions. A number of components in plants involved in the stress tolerance reactions are identified. PAs, CuAO/DAO and PAO are among these components playing a role in these stress responses (reviewed in Groppa and Benavides 2008).

Increasing UV radiation in the atmosphere is a reality in Antarctic regions. A few of the examples of association of PA in UV stress are: in response to UV-B radiation total free PAs was decreased and conjugated PAs increased in *Phaseolus vulgaris* plants (Smith et al. 2001), in cucumber leaves UV-B radiation resulted in decreased total PA content concomitant with increased electrolyte leakage and weakening of plant growth (An et al. 2004), in tobacco cultivars subjected to UV-B radiation, PAs especially put was increased and after extended periods of UV-B exposure PA levels declined (Lutz et al. 2005). The influence of UV-B on PA metabolism is an indicator that plants indeed sense UV-B as a stress that also may affect crop yield.

Increasing ozone content in the biosphere is a great problem, which contributes to global crop losses and forest decline. For the year 2000 global crop yield loss due to ambient ozone was estimated to be worth \$14–26 billion, and 40% of this damage occurred in China and India (Van Dingenen et al. 2009). Exposure to ozone results in foliar injury, impaired photosynthesis, reduced growth and yield, and an accelerated onset of senescence in plants (Langebartels et al. 1991). Plants have evolved preventive mechanisms to minimize the damages caused by ozone. They can limit entry of ozone to interior through stomatal closure or they have tolerance mechanisms, which include the detoxification of ozone diffused into the leaf interior through chemical reactions with ascorbic acid or enzymatic conversion to H_2O_2 (Chen et al. 2003). Since the toxicity of ozone results mainly from oxidative stress, protecting plants through application of antioxidants is being investigated. PAs exert several functions, which counteract ozone effects (Langebartels et al. 1991). Accumulation of PAs in response to ozone exposure and protection against ozone-derived oxidative damage has been reported from different plant species for example from barley (Rowland-Bamford et al. 1989), wheat (Raab and Weinstein 1990) and tobacco (Langebartels et al. 1991). Increased ADC activity and accumulation of free and conjugated put was observed in tobacco cultivar Bel B (Langebartels et al. 1991). In accordance with this study, Van Buuren et al. (2002) showed that accumulation of free put in both tobacco cultivars, ozone-resistant and ozone-sensitive when exposed to ozone. In tissues undergoing cell death in ozone-sensitive cultivar, accumulation of conjugated put and transient increase of ADC and ODC activity was observed (Van Buuren et al. 2002). The protective function of PAs in ozone damage may be due to the control of the cellular redox state, though the precise mode of action remains unknown (Van Buuren et al. 2002).

A large number of herbicides are widely used in agriculture. Some experimental evidences exist on the relationship of PAs and herbicides such as paraquat, an intensively used oxidative stress inducer (Groppa and Benavides 2008). In *Arabidopsis,* treated with paraquat, there was an increase in put, but not spd and spm (Benavides et al. 2000). Exogenous addition of PAs was effective in protection against paraquat toxicity to various degrees in *Arabidopsis* (Benavides et al. 2000) and in maize leaves (Durmus and Kadioglu 2005). These results suggested that PAs protective function may be due to the antioxidative function.

Taken together, PAs are growth regulators present in plants implicated in vast number of physiological processes in growth and development and abiotic and biotic stress responses (reviewed in Bouchereau et al. 1999; Groppa and Benavides 2008; Alcázar et al. 2010a; Wimalasekara et al. 2011a). As discussed in this review, differential regulation of PA biosynthesis and catabolism play important roles in responding to several types of abiotic stresses leading to stress tolerance. Availability of information about key genes in biosynthetic and catabolic pathways provide useful information in manipulating the same for the production of gain and loss of function mutants which in turn provide the underlying molecular mechanism of PA functions. For example, overexpression of PA biosynthetic genes, *ADC1* and *ADC2* was successful in generating several plant species ranging from the model plant *Arabidopsis* to the important crop plant rice that exhibited enhanced tolerance to a variety of stresses. Further, transgenic plants (*Arabidopsis,* rice, pear) overexpressing spd and spm synthase genes *SPDS* and *SPMS* showed enhanced tolerance to a number of abiotic stresses (Groppa and Benavides 2008; Alcázar et al. 2010a). In most of the cases, these transgenic plants showed tolerance to a broad spectrum of abiotic stresses suggesting interaction in mechanisms of stress resistance common in different stress types.

Importance of PA catabolic enzymes CuAO/DAO and PAO in plant development and stress tolerance is extensively reviewed (Cona et al. 2006; Kusano et al. 2008; Moschou et al. 2008b; Angelini et al. 2010; Wimalasekara et al. 2011a). For example, overexpression of *CuAO* in tobacco showed enhanced tolerance to salt stress (Moschou et al. 2008b) and overexpression of ZmPAO in tobacco showed enhanced wound-healing response (Angelini et al. 2008). Most of the PA catabolic functions are associated with H_2O_2 that is produced by the activity of CuAO/DAO and PAO. In most of the instances, H_2O_2 produced in this manner is involved in reactions occurring during stress-induced cell wall modifications, in PCD, and as a second messenger in signalling stomatal regulation. Several experimental evidences exist regarding the roles of CuAO/DAO and POA in biotic stress tolerance especially in triggering the hypersensitive response and CPD (reviewed in Walter 2003; Angelini et al. 2010; Moschou et al. 2008a). Transfer of the knowledge obtained from rather limited plant species over to valuable crop species will be a future challenge in the agriculture industry despite the many constrains that exist.

Conclusion and Future Perspective

Plant growth is highly affected by the adverse environmental conditions such as drought, high salinity, extreme temperature regimes, mineral deficiencies and metal toxicity causing decreased crop yields. The dynamic climate changes have great impacts on global food security, emphasising vital solution for the crop improvement by enhancing the stress tolerance. Despite the improvement of conventional methods, considerable attention has been paid to the utilization of recent advancements such as transgenic approach, marker assisted screening methods and breeding in enhancing plant performance under stress conditions.

Apart from essential growth regulatory functions, plant PAs are known to play important role in stress tolerance by modulating the PA levels. Considerable evidences exist for the natural variations of PAs in different cultivars/accessions correlating with stress situations (Bouchereau et al. 1999). As described in this article, PA biosynthesis and catabolism are genetically manipulated in some plant species especially in model plants for enhanced environmental stress tolerance ranging from drought, salinity, temperature extremes, mineral deficiency to wounding. Further investigations are necessary in understanding the molecular mechanisms of PA action in response to multiple stress situations. Broader insight is also required on the interacting components and signalling pathways in the PA metabolism and catabolism to fully uncover the protective function of PAs for subsequent successful utilization in enhanced crop performance. Advanced techniques including microarray, proteomics and metabolomics will be helpful in gaining detailed understanding. Future challenge is to transfer knowledge obtained especially from model plants to a variety of important crop species for enhanced tolerance to adverse environmental stresses finally aiming at increased crop yields.

References

Alcázar R, Cuevas JC, Patrón M, Altabella T, Tiburcio AF (2006a) Abscisic acid modulates polyamine metabolism under water stress in *Arabidopsis thaliana*. Physiol Plant 128:448–455

Alcázar R, Marco F, Cuevas JC, Patrón M, Fernando P, Carraso A, Tiburcio AF, Altabella T (2006b) Involvement of polyamines in plant responses to abiotic stress. Biotechnol Lett 28:1867–1876

Alcázar R, Altabella T, Marco F, Bortolotti C, Reymond M, Koncz C, Carrasco P, Tiburcio AF (2010a) Polyamines: molecules with regulatory functions in plant abiotic stress tolerance. Planta 231:1237–1249

Alcázar R, Planas J, Saxena T, Zarza X, Bortolotti C, Cuevas JC, Bitrian M, Tiburcio AF, Altabella T (2010b) Putrescine accumulation confers drought tolerance in transgenic *Arabidopsis* plants overexpressing the homologous *Arginine decarboxylase 2* gene. Plant Physiol Biochem 48:547–552

Alcázar R, Bitrian M, Bartels D, Koncz C, Altabella T, Tiburcio AF (2011) Polyamine metabolic canalization in response to drought stress in *Arabidopsis* and the resurrection plant *Craterostigma plantagineum*. Plant Signal Behav 6:243–250

Alet AI, Sánchez DH, Cuevas JC, Marina M, Carrasco P, Altabella T, Tiburcio AF, Ruiz OA (2012) New insights into the role of spermine in Arabidopsis thaliana under long-term salt stress. Plant Sci 182:94–100

An LZ, Liu GX, Zhang MX, Chen T, Liu YH, Feng HY, Xu SJ, Qiang WY, Wang XL (2004) Effect of enhanced UV-B radiation on polyamine content and membrane permeability in cucumber leaves. Russ J Plant Physiol 51:658–662

An Z, Jing W, Zhang W (2008) Hydrogen peroxide generated by copper amine oxidase is involved in abscisic acid-induced stomatal closure in *Vicia faba*. J Exp Bot 59:815–825

Angelini R, Tisi A, Rea G, Chen MM, Botta M, Federico R, Cona A (2008) Involvement of polyamine oxidase in wound healing. Plant Physiol 146:162–177

Angelini R, Cona A, Federico R, Fincato P, Tavladoraki P, Tisi A (2010) Plant amine oxidases "on the move": an update. Plant Physiol Biochem 48:560–564

Apel K, Hirt H (2004) Reactive oxygen species: metabolism, oxidative stress, and signal transduction. Annu Rev Plant Biol 55:373–399

Aziz A, Martin-Tanguy J, Larher LF (1997) Plasticity of polyamine metabolism associated with high osmotic stress in rape leaf discs and with ethylene treatment. Plant Growth Regul 21:153–163

Bagni N, Tassoni A (2001) Biosynthesis, oxidation and conjugation of aliphatic polyamines in higher plants. Amino Acids 20:301–317

Basu R, Ghosh B (1991) Polyamines in various rice genotypes with respect to NaCl salinity. Physiol Plant 82:575–581

Benavides MP, Gallego SM, Comba ME, TomaroML (2000) Relationship between polyamines and paraquat toxicity in sunflower leaf discs. Plant Growth Regul 31:215–224

Bertoldi D, Tassoni A, Martinelli L, Bagni N (2004) Polyamines and somatic embryogenesis in two *Vitis vinifera* cultivars. Physiol Plant 120:657–666

Borrell A, Culiañez-Macià FA, Altabella T, Besford RT, Flores D, Tiburcio AF (1995) Arginine decarboxylase is localized in chloroplasts. Plant Physiol 109:771–776

Bors W, Langebartels C, Michel C, Sandermann JH (1989) Polyamine as radical scavengers and protectants against ozone damage. Phytochemistry 28:1589–1595

Bortolotti C, Cordeiro A, Alcazar R, Borrell A, Culiañez-Macià FA, Tiburcio AF, Altabella T (2004) Localization of arginine decarboxylase in tobacco plants. Physiol Plant 120:84–92

Bouchereau A, Aziz A, Larher F, Martin-Tanguy J (1999) Polyamines and environmental challenges: recent development. Plant Sci 140:103–125

Buchanan B, Gruissem W, Jones R (2000) Biochemistry and molecular biology in plants. American Society of Plant Physiologists, Maryland, pp 1–90

Camacho-Cristóbal JJ, Maldonado JM, Gonzláez-Fontes A (2005) Boron deficiency increases putrescine levels in tobacco plants. J Plant Physiol 162:921–928

Campestre MP, Bordenave CD, Origone AC, Menéndez AB, Ruiz OA, Rodríguez AA, Maiale SJ (2011) Polyamine catabolism is involved in response to salt stress in soybean hypocotyls. J Plant Physiol 168:1234–1240

Capell T, Escobar C, Liu H, Burtin D, Lepri O, Christou P (1998) Overexpression of the oat arginine decarboxylase cDNA in transgenic rice (*Oryza sativa* L.) affects normal development patterns *in vitro* and results in putrescine accumulation in transgenic plants. Theor Appl Gen 97:246–254

Capell T, Bassie L, Christou P (2004) Modulation of the polyamine biosynthetic pathway in transgenic rice confers tolerance to drought stress. Proc Natl Acad Sci USA 10:9909–9914

Cervelli M, Cona A, Angelini R, Polticelli F, Federico R, Mariottini P (2001) A barley polyamine oxidase isoform with distinct structural features and subcellular localization. Eur J Biochem 268:3816–3830

Chattopadhayay MK, Tiwari BS, Chattopadhayay G, Bose A, Sengupta DN, Ghosh B (2002) Protective role of exogenous polyamines on salinity-stressed rice (Oryza sativa) plants. Physiol Plant 116:192–199

Chen Z, Young TE, Ling J, Chang S-C, Gallie DR (2003) Increasing vitamin C content of plants through enhanced ascorbate recycling. Proc Natl Acad Sci USA 100:3525–3530

Cheng L, Zou YJ, Ding SL, Zhang JJ, Yu XL, Cao JS, Lu G (2009) Polyamine accumulation in transgenic tomato enhances the tolerance to high temperature stress. J Integr Plant Biol 51:489–499

Cheng L, Sun RR, Wang FY, Peng Z, Kong FL, Wu J, Cao JS, Lu G (2012) Spermidine affects the transcriptome responses to high temperature stress in ripening tomato fruit. J Zhejiang Univ Sci B 13:283–297

Choudhary A, Singh RP (2000) Cadmium-induced changes in diamine oxidase activity and polyamine levels in *Vigna radiata* Wilczek seedlings. J Plant Physiol 156:704–710

Cona A, Rea G, Angelini R, Federico R, Tavladoraki P (2006) Functions of amine oxidases in plant development and defense. Trends Plant Sci 11:80–88

Cowley T, Walters DR (2005) Local and systemic changes in arginine decarboxylase activity, putrescine levels and putrescine catabolism in wounded oilseed rape. New Phytol 165:807–811

Cuevas JC, Lopez-Cobollo R, Alcázar R, Zarza X, Koncz C, Altabella T, Salinas J, Tiburcio AF, Ferrando A (2008) Putrescine is involved in Arabidopsis freezing tolerance and cold acclimation by regulating abscisic acid levels in response to low temperature. Plant Physiol 148:1094–1105

Cuevas JC, Lopez-Cobollo R, Alcazar R, Zarza X, Koncz C, Altabella T, Salinas J, Tiburcio AF, Ferrando A (2009) Putrescine as a signal to modulate the indispensable ABA increase under cold stress. Plant Signal Behav 4:219–220

Cvikrová M, Gemperlová L, Dobrá J, Martincová O, Prásil IT, Gubis J, Vanková R (2012) Effect of heat stress on polyamine metabolism in proline-over-producing tobacco plants. Plant Sci 182:49–58

Delis D, Dimou M, Flemetakis IE, Aivalakis G, Katinakis P (2006) A root- and hypocotyl-specific gene coding for copper-containing amine oxidase is related to cell expansion in soybean seedlings. J Exp Bot 57:101–111

Durmus N, Kadioglu A (2005) Spermine and putrescine enhance oxidative stress tolerance in maize leaves. Acta Physiol Plant 27:515–522

El-Shintinawy F (2000) Photosynthesis in two wheat cultivars differing in salt susceptibility. Photosynthetica 38:615–620

Evans PT, Malmberg RL (1989) Do polyamines have roles in plant development? Annu Rev Plant Physiol Plant Mol Biol 40:235–269

Galston AW, Sawhney RK (1990) Polyamines in plant physiology. Plant Physiol 94:406–410

Galston AW, Kaur-Sawhney R, Altabella T, Tiburcio AF (1997) Plant polyamines in reproductive activity and response to abiotic stress. Bot Acta 110:197–207

Gill SS, Tuteja N (2010) Polyamines and abiotic stress tolerance in plants. Plant Signal Behav 5:26–33

Grene R (2002) Oxidative stress and acclimation mechanisms in plants. In Arabidopsis Book. American Society of Plant Biologists, Rockville, pp 1–20

Groppa MD, Benavides MP (2008) Polyamines and abiotic stress: recent advances. Amino Acids 34:35–45

Groppa MD, Tomaro ML, Benavides MP (2001) Polyamines as protectors against cadmium or copper-induced oxidative damage in sunflower leaf discs. Plant Sci 161:481–488

Groppa MD, Benavides MP, Tomaro ML (2003) Polyamine metabolism in sunflower and wheat leaf discs under cadmium or copper stress. Plant Sci 164:293–299

Groppa MD, Ianuzzo MP, Tomaro ML, Benavides MP (2007) Polyamine metabolism in sunflower plants under long-term cadmium or copper stress. Amino Acids 32:265–275

Hanfrey C, Sommer S, Mayer MJ, Burtin D, Michael AJ (2001) *Arabidopsis* polyamine biosynthesis: absence of ornithine decarboxylase and the mechanism of arginine decarboxylase activity. Plant J 27:551–560

Hanzawa Y, Takahashi T, Michael AJ, Burtin D, Long D, Pineiro M, Coupland G, Komeda Y (2000) ACAULIS5, an Arabidopsis gene required for stem elongation, encodes a spermine synthase. EMBO J 19:4248–4256

Hassannejad S, Bernard F, Mirzajani F, Gholami M (2012) SA improvement of hyperhydricity reversion in Thymus daenensis shoots culture may be associated with polyamines changes. Plant Physiol Biochem 51:40–46

He L, Nada K, Tachibana S (2002) Effects of spd pretreatment through the roots on growth and photosynthesis of chilled cucumber plants (*Cucumis sativus* L.). J Jpn Soc Hort Sci 71:490–498

Houdusse F, Zamarreno AM, Garnica M, Garcia-Mina J (2005) The importance of nitrate in ameliorating the effects of ammonium and urea nutrition on plant development: the relationships with free polyamines and plant proline contents. Func Plant Biol 32:1057–1067
Hu X, Zhang Y, Shi Y, Zhang Z, Zou Z, Zhang H, Zhao J (2012) Effect of exogenous spermidine on polyamine content and metabolism in tomato exposed to salinity-alkalinity mixed stress. Plant Physiol Biochem 57:200–209
Hussain SS, Ali M, Ahmad M, Siddique KH (2011) Polyamines: natural and engineered abiotic and biotic stress tolerance in plants. Biotechnol Adv 29:300–311
Imai R, Ali A, Pramanik MDHR, Nakaminami K, Sentoku N, Kato H (2004) A distinctive class of spermidine synthase is involved in chilling response in rice. J Plant Physiol 161:883–886
Janowitz T, Kneifel H, Piotrowski M (2003) Identification and characterization of plant agmatine iminohydrolase, the last missing link in polyamine biosynthesis of plants. FEBS Lett 544:258–261
Kakkar RK, Nagar PK, Ahuja PS, Rai VK (2000) Polyamines and plant morphogenesis. Biol Plant 43:1–11
Kakkar RK, Sawhney VK (2002) Polyamine research in plants-a changing perspective. Physiol Plant 116:281–292
Kasinathan V, Wingler A (2004) Effect of reduced arginine decarboxylase activity on salt tolerance and on polyamine formation during salt stress in *Arabidopsis thaliana*. Physiol Plant 121:101–107
Kasukabe Y, He L, Nada K, Misawa S, Ihara I , Tachibana S (2004) Overexpression of spermidine synthase enhances tolerance to multiple environmental stresses and up-regulates the expression of various stress-regulated genes in transgenic *Arabidopsis thaliana*. Plant Cell Physiol 45:712–722
Kaunisto S, Sarjala T (1997) Critical needle potassium concentrations indicated by diamine putrescine in Norway spruce growing on peat soils. Silva Fennica 31:383–390
Krishnamurthy R, Bhagwat KA (1989) Polyamines as modulators of salt tolerance in rice cultivars. Plant Physiol 91:500–504
Kubis J, Krzywanski Z (1989) The dynamics of polyamine accumulation in spring wheat leaves during increasing water stress. Acta Physiol Plant 11:157–164
Kumar M, Bijo AJ, Baghel RS, Reddy CR, Jha B (2012) Selenium and spermine alleviate cadmium induced toxicity in the red seaweed Gracilaria dura by regulating antioxidants and DNA methylation. Plant Physiol Biochem 51:129–138
Kusano T, Yamaguchi K, Berberich T, Takahashi Y (2007a) Advances in polyamine research. J Plant Res 120:345–350
Kusano T, Yamaguchi K, Berberich T, Takahashi Y (2007b) The polyamine spermine rescues *Arabidopsis* from salinity and drought stresses. Plant Signal Behav 2:251–252
Kusano T, Berberich T, Tateda C, Takahashi Y (2008) Polyamines: essential factors for growth and survival. Planta 228:367–381
Kuthanová A, Gemperlová L, Zelenková S, Eder J, Macháčková I, Opatrný Z, Cvikrová M (2004) Cytological changes and alterations in polyamine contents induced by cadmium in tobacco BY-2 cells. Plant Physiol Biochem 42:149–156
Langebartels C, Kerner K, Leonardi S, Schraudner M, Trost M, Heller W, Sandermann H Jr (1991) Biochemical plant responses to ozone. I. Differential induction of polyamine and ethylene biosynthesis in tobacco. Plant Physiol 95:882–889
Lee YP, Babakov A, de Boer B, Zuther E, Hincha DK (2012) Comparison of freezing tolerance, compatible solutes and polyamines in geographically diverse collections of Thellungiella sp. and Arabidopsis thaliana accessions. BMC Plant Biol 12:131
Legocka J, Kluk A (2005) Effect of salt and osmotic stress on changes in polyamine content and arginine decarboxylase activity in *Lupinus luteus* seedlings. J Plant Physiol 162:662–668
Li ZY, Chen SY (2000) Differential accumulation of the *S*-adenosylmethionine decarboxylase transcript in rice seedlings in response to salt and drought stresses. Theor Appl Genet 100:782–788
Lie K, Fu H, Bei Q, Luan S (2000) Inward potassium channel in guard cells as target for polyamine regulation of stomatal movements. Plant Physiol 124:1315–1326

Lin CH, Kao CH (1999) Excess copper induces an accumulation of putrescine in rice leaves. Bot Bull Acad Sin 40:213–218
Lin CC, Kao CH (2001) Abscisic acid induced changes in cell wall peroxidase activity and hydrogen peroxide level in roots of rice seedlings. Plant Sci 160:323–329
Liu K, Huihua F, Bei O, Luan S (2000) Inward potassium channel in guard cells as a target for polyamine regulation of stomatal movements. Plant Physiol 124:1315–1326
Liu J-H, Kitashiba H, Wang J, Ban Y, Moriguchi T (2007) Polyamines and their ability to provide environmental stress tolerance to plants. Plant Biotech 24:117–126
Løvaas E (1997) Antioxidative and metal-chelating effects of polyamines. Adv Pharmacol 38:119–149
Lutz C, Navakoudis E, Seidlitz HK, Kotzabasis K (2005) Simulated solar irradiation with enhanced UV-B adjust plastid- and thylakoid-associated polyamine changes for UV-B protection. Biochim Biophys Acta 1710:24–33
Lynch JP, Clair SB (2004) Mineral stress: the missing link in understanding how global climate change will affect plants in real world soils. Field Crops Research 90:101–115
Marco F, Alcázar R, Tiburcio AF, Carrasco P (2011) Interactions between polyamines and abiotic stress pathway responses unraveled by transcriptome analysis of polyamine overproducers. OMICS 15:775–781
Martin-Tanguy J (2001) Metabolism and function of polyamines in plants: recent development (new approaches). Plant Growth Reg 34:135–148
Michael AJ, Furze JM, Rhodes MJC, Burtin D (1996) Molecular cloning and functional identifcation of a plant ornithine decarboxylase cDNA. Biochem J 314:241–248
Moller SG, McPherson MJ (1998) Developmental expression and biochemical analysis of the *Arabidopsis atao1* gene encoding an H_2O_2-generating diamine oxidase. Plant J 13:781–791
Moschou PN, Paschalidis KA, Delis ID, Andriopoulou AH, Lagiotis GD, Yakoumakis DI, Roubelakis-Angelakis KA (2008a) Spermidine exodus and oxidation in the apoplast induced by abiotic stress is responsible for H2O2 signatures that direct tolerance responses in tobacco. Plant Cell 20:1708–1724
Moschou PN, Paschalidis KA, Roubelakis-Angelakis KA (2008b) Plant polyamine catabolism: The state of the art. Plant Signal Behav 3:1061–1066
Moschou PN, Sanmartin M, Andriopoulou AH, Rojo E, Sanchez-Serrano JJ, Roubelakis-Angelakis KA (2008c) Bridging the gap between plant and mammalian polyamine catabolism: a novel peroxisomal polyamineoxidase responsible for a fullback-conversion pathway in *Arabidopsis*. Plant Physiol 147:1845–1857
Mutlu F, Bozcuk S (2005) Effects of salinity on the contents of polyamines and some other compounds in sunflower plants differing in salt tolerance. Russ J Plant Physiol 52:29–34
Nambeesan S, AbuQamar S, Laluk K, Mattoo AK, Mickelbart MV, Ferruzzi MG, Mengiste T, Handa AK (2012) Polyamines attenuate ethylene-mediated defense responses to abrogate resistance to Botrytis cinerea in tomato. Plant Physiol 158:1034–1045
Nayyar H (2005) Putrescine increases floral retention, pod set and seed yield in cold stressed chickpea. J Agron Crop Sci 191:340–345
Nayyar H, Chander S (2004) Protective effects of polyamines against oxidative stress induced by water and cold stress in chickpea. J Agron Crop Sci 190:355–365
Nayyar H, Kaur S, Singh S, Kumar S, Singh KJ, Dhir KK (2005) Involvement of polyamines in the contrasting sensitivity of chickpea (*Cicer arietinum* L.) and soybean (*Glycine max* L. Merrill.) to water deficit stress. Bot Bull Acad Sin 46:333–338
Oracz K, Bouteau HM, Kranner I, Bogatek R, Corbineau F, Bailly C (2009) The mechanisms involved in seed dormancy alleviation by hydrogen cyanide unravel the role of reactive oxygen species as key factors of cellular signaling during germination. Plant Physiol 150:494–505
Panicot M, Minguet EG, Ferrando A, Alcázar R, Blázquez MA, Carbonell J, Altabella T, Koncz C, Tiburcio AF (2002) A polyamine metabolon involving aminopropyl transferases complexes in *Arabidopsis*. Plant Cell 14:2539–2551
Paschalidis KA, Roubelakis-Angelakis KA (2005) Sites and regulation of polyamine catabolism in the tobacco plant. Correlations with cell division/expansion, cell cycle progression and vascular development. Plant Physiol 138:2174–2184

Perez-Amador MA, León J, Green PJ, Carbonell J (2002) Induction of the arginine decarboxylase *ADC2* gene provides evidence for the involvement of polyamines in the wound responses in *Arabidopsis*. Plant Physiol 130:1454–1463

Petrivalský M, Brauner F, Luhová L, Gagneul D, Sebela M (2007) Aminoaldehyde dehydrogenase activity during wound healing of mechanically injured pea seedlings. J Plant Physiol 164:1410–1418

Pieruzzi FP, Dias LL, Balbuena TS, Santa-Catarina C, dos Santos AL, Floh EI (2011) Polyamines, IAA and ABA during germination in two recalcitrant seeds: Araucaria angustifolia (Gymnosperm) and Ocotea odorifera (Angiosperm). Ann Bot 108:337–345

Piotrowski M, Janowitz T, Kneifel H (2003) Plant C–N hydrolases and the identification of a plant N-carbamoylputrescine amidohydrolase involved in polyamine biosynthesis. J Biol Chem 278:1708–1712

Raab MM, Weinstein LH (1990) Polyamine and ethylene metabolism in *Triticum aestivum* (var. vona). In HE Flores, RN Arteca, JC Shannon (eds) Polyamines and ethylene: biochemistry, physiology, and interactions. American Society of Plant Physiologists, Rockville, pp 408–410

Rea R, Laurenzi M, Tranquili E, D'Ovidio R, Federico R, Angelini R (1998) Developmentally and wound-regulated expression of the gene encoding a cell wall copper amine oxidase in chickpea seedlings. FEBS Lett 23:177–182

Rea R, Metoui O, Infantino A, Federico R, Angelini R (2002) Copper amine oxidase expression in defence responses to wounding and *Ascochyta rabiei* invasion. Plant Physiol 128:865–875

Renaut J, Hoffmann L, Hausman JF (2005) Biochemical and physiological mechanisms related to cold acclimation and enhanced freezing tolerance in poplar plantlets. Physiol Plant 125:82–94

Rodríguez AA, Maiale SJ, Menéndez AB, Ruiz OA (2009) Polyamine oxidase activity contributes to sustain maize leaf elongation under saline stress. J Exp Bot 60:4249–4262

Rowland-Bamford AJ, Borland AM, Lea PJ, Mansfield TA (1989) The role of arginine decarboxylase in modulating the sensitivity of barley to ozone. Environ Pollut 61:95–106

Roy M, Ghosh, B (1996) Polyamines, both common and uncommon, under heat stress in rice (*Oryza sativa*) callus. Physiol Plant 98:196–200

Roy M, Wu R (2001) Arginine decarboxylase transgene expression and analysis of environmental stress tolerance in transgenic rice. Plant Sci 160:869–875

Roy P, Niyogi K, SenGupta DN, Ghosh B (2005) Spermidine treatment to rice seedlings recovers salinity stress-induced damage of plasma membrane and PM-bound H^+-ATPase in salt-tolerant and salt-sensitive rice cultivars. Plant Sci 168:583–591

Sandalio LM, Dalurzo HC, Gómez M, Romero-Puertas MC, del Río LA (2001) Cadmium-induced changes in the growth and oxidative metabolism of pea plants. J Exp Bot 52:2115–2126

Sarjala T, Kaunisto S (2002) Potassium nutrition and free polyamines of *Betula pendula* Roth and *Betula pubescens* Ehrh. Plant Soil 238:141–149

Sebela M, Radová A, Angellini R, Tavladoraki P, Frébort I, Pec P (2001) FAD-containing polyamine oxidases: a timely challenge for researchers in biochemistry and physiology of plants. Plant Sci 160:197–207

Shelp BJ, Bozzo GG, Trobacher CP, Zarei A, Deyman KL, Brikis CJ (2012) Hypothesis/review: contribution of putrescine to 4-aminobutyrate (GABA) production in response to abiotic stress. Plant Sci 193–194:130–135

Shen W, Nada K, Tachibana S (2000) Involvement of polyamines in the chilling tolerance of cucumber cultivars. Plant Physiol 124:431–440

Shevyakova NI, Il'ina EN, Stetsenko LA, Kuznetsov VV (2011) Nickel accumulation in rape shoots (Brassica napus L.) increased by putrescine. Int J Phytoremediation 13:345–356

Shi J, Fu X-Z, Peng T, Huang X-S, Fan Q-J, Liu J-H (2010) Spermine pretreatment confers dehydration tolerance of citrus *in vitro* plants via modulation of antioxidative capacity and stomatal response. Tree Physiol 30:914–922

Silveira V, Santa-Catarina C, Tun NN, Handro W, Scherer GFE, Handro W, Guerra M, Floh EIS (2006) Polyamine effects on the endogenous polyamine contents, nitric oxide release, growth and differentiation of embryonic suspension cultures of *Araucaria angustifolia* (Bert.) O. Ktze. Plant Sci 171:91–98

Slocum RD, Kaur-Sawhney R, Galston AW (1984) The physiology and biochemistry of polyamines in plants. Arch Biochem Biophys 235:283–303
Smith J, Burrit D, Bannister P (2001) Ultraviolet-B radiation leads to a reduction in free polyamines in Phaseolus vulgaris L. Plant Growth Regul 35:289–294
Soyka S, Heyer AG (1999) *Arabidopsis* knockout mutation of *ADC2* gene reveals inducibility by osmotic stress. FEBS Lett 458:219–223
Tadolini B (1988) Polyamine inhibition of lipid peroxidation. Biochem J 249:33–36
Taiz L, Zeiger E (2006) Plant physiology. Sinauer Association Inc. USA, pp 671–703
Takahashi T, Kakehi J (2010) Polyamines: ubiquitous polycations with unique roles in growth and stress responses. Ann Bot 105:1–6
Takano A, Kakehi J, Takahashi T (2012) Thermospermine is not a minor polyamine in the plant kingdom. Plant Cell Physiol 53:606–616
Tang CF, Liu YG, Zeng GM, Li X, Xu WH, Li CF, Yuan XZ (2005) Effects of exogenous spermidine on antioxidant system responses of *Typha latifolia* L. under Cd2+ stress. J Integ Plant Biol 47:428–434
Tavladoraki P, Schininà ME, Cecconi F, Di Agostino S, Manera F, Rea G, Mariottini P, Federico R, Angelini R (1998) Maize polyamine oxidase: primary structure from protein and cDNA sequencing. FEBS Lett 426:62–66
Tavladoraki P, Rossi MN, Saccuti G, Perez-Amador MA, Polticelli F, Angelini R, Federico R (2006) Heterologous expression and biochemical characterization of a polyamine oxidase from *Arabidopsis* involved in polyamine back conversion. Plant Physiol 141:1519–1532
Tiburcio AF, Altabella T, Borrel A, Masgrau C (1997) Polyamine metabolism and its regulation. Physiol Plant 100:664–674
Tipping AJ, McPherson MJ (1995) Cloning and molecular analysis of the pea seedling copper amine oxidase. J Biol Chem 270:16939–16946
Tisi A, Federico R, Moreno S, Lucretti S, Moschou PN, Roubelakis-Angelakis KA, Angelini R, Cona A (2011) Perturbation of polyamine catabolism can strongly affect root development and xylem differentiation. Plant Physiol 157:200–215
Torrigiani P, Altamura MM, Pasqua G, Monacelli B, Serafini-Fracassini D, Bagni N (1987) Free and conjugated polyamines during *de novo* floral and vegetative bud formation in thin cell layers of tobacco. Physiol Plant 70:453–460
Toumi I, Moschou PN, Paschalidis KA, Bouamama B, Salem-fnayou AB, Ghorbel AW, Mliki A, Roubelakis-Angelakis KA (2010) Abscisic acid signals reorientation of polyamine metabolism to orchestrate stress responses via the polyamine exodus pathway in grapevine. J Plant Physiol 167:519–525
Urano K, Yoshiba Y, Nanjo T, Igarashi Y, Seki M, Sekiguchi F, Yamaguchi-Shinozaki K, Shinozaki K (2003) Characterization *of Arabidopsis* genes involved in biosynthesis of polyamines in abiotic stress responses and developmental stages. Plant Cell Environ 26:1917–1926
Urano K, YoshibaY, Nanjo T, Ito T, Yamaguchi-Shinozaki K, Shinozaki K (2004) *Arabidopsis* stress-inducible gene for arginine decarboxylase *AtADC2* is required for accumulation of putrescine in salt tolerance. Biochem Biophys Res Commun 313:369–375
Van Assche F, Clijsters H (1990) Effects of metals on enzyme activity in plants. Plant Cell Environ 13:195–206
Van Buuren ML, Guidi L, Fornal S, Ghetti F, Franceschetti M, Soldatini GF, Bagni N (2002) Ozone-response mechanisms in tobacco: implications of polyamine metabolism. New Phytol 156:389–398
Van Dingenen R, Dentener FJ, Raes F, Krol MC, Emberson L, Cofala J (2009). The global impact of ozone on agricultural crop yields under current and future air quality legislation. Atmosp Environ 43:604–618
Walter DR (2003) Resistance to plant pathogens: possible roles for free polyamines and polyamine catabolism. New Phytol 159:109–115
Wang BQ, Zhang QF, Liu JH, Li GH (2011) Overexpression of PtADC confers enhanced dehydration and drought tolerance in transgenic tobacco and tomato: effect on ROS elimination. Biochem Biophys Res Commun 413:10–16

Watson MB, Malmberg RL (1996) Regulation of *Arabidopsis thaliana* (L.) Heynh arginine decarboxilase by potassium deficiency stress. Plant Physiol 111:1077–1083

Wimalasekara R, Scherer GFE (2010) Polyamines and cytokinin: is nitric oxide biosynthesis the key to overlapping functions? In: Hayat S, Mori M, Pichtel J, Ahmad A (eds) Nitric oxide in plant physiology. Wiley-Vch, Weinheim pp 65–76

Wimalasekara R, Tebartz F, Scherer GFE (2011a) Polyamines, polyamine oxidases, and nitric oxide (NO) in development, biotic and abiotic stresses. Plant Sci 181:593–603

Wimalasekara R, Villar C, Begum T, Scherer GFE (2011b) *COPPER AMINE OXIDASE1* (*CuAO1*) of *Arabidopsis thaliana* contributes to abscisic acid- and polyamine-induced nitric oxide biosynthesis and abscisic acid signal transduction. Mol Plant 4:663–678

Xing SG, Jun YB, Hau ZW, Liang LY (2007) Higher accumulation of gamma-aminobutyric acid induced by salt stress through stimulating the activity of diamine oxidases in *Glycine max* (L.) Merr.roots. Plant Physiol Biochem 45:560–566

Yamaguchi K, Takahashi T, Imai A, Miyazaki A, Takahashi T, Michael A, Kusano T (2006) The polyamine spermine protects against high salinity stress in *Arabidopsis*. FEBS Lett 580:67783–6788

Yoshimoto K, Noutoshi Y, Hayashi K, Shirasu K, Takahashi T, Motose H (2012) A chemical biology approach reveals an opposite action between thermospermine and auxin in xylem development in Arabidopsis thaliana. Plant Cell Physiol 53:635–645

Young ND, Galston AW (1984) Physiological control of arginine decarboxylase activity and potassium-deficient oat shoots. Plant Physiol 76:331–335

Zapata PJ, Serrano M, Pretel MT, Amorós A, Botella MA (2004) Polyamines and ethylene changes during germination of different plant species under salinity. Plant Sci 167:781–788

Zhang X, Shen L, Li F, Meng D, Sheng J (2011) Methyl salicylate-induced arginine catabolism is associated with up-regulation of polyamine and nitric oxide levels and improves chilling tolerance in cherry tomato fruit. J Agric Food Chem 59:9351–9357

Zhao FG, Qin P (2004) Protective effect of exogenous polyamines on root tonoplast function against salt stress in barley seedlings. Plant Growth Regul 42:97–103

Index

K. R. Hakeem et al. (eds.), *Crop Improvement*, DOI 10.1007/978-1-4614-7028-1,

9781461470274.